Crystalline Solids

Duncan McKie/Christine McKie
University of Cambridge

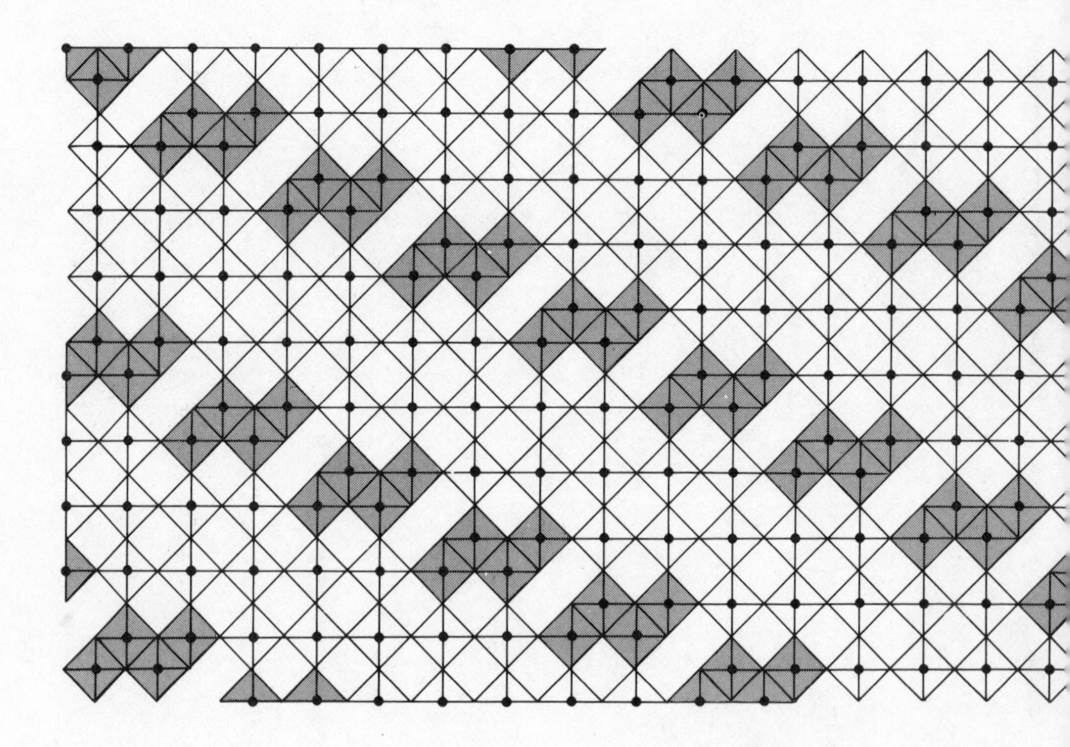

Crystalline Solids

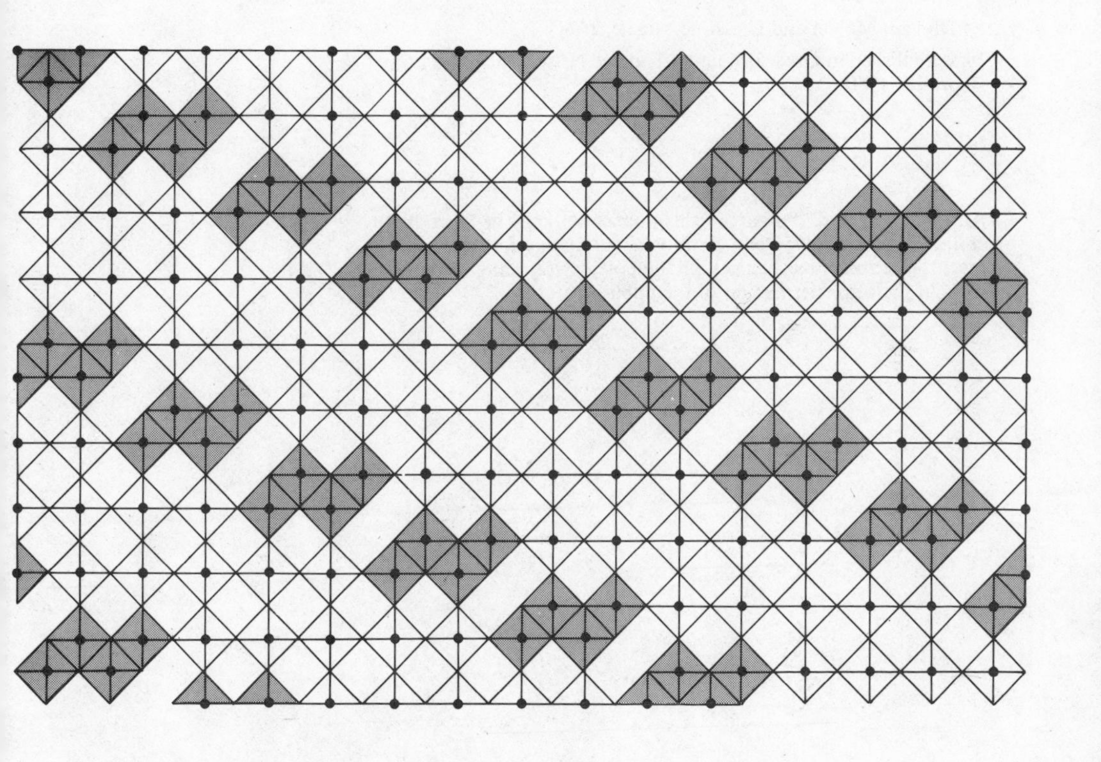

Nelson

Thomas Nelson and Sons Ltd,
Nelson House, Mayfield Road,
Walton-on-Thames, Surrey KT12 5PL
PO Box 18123, Nairobi, Kenya
Watson Estate, Block A, 13 Floor,
Watson Road, Causeway Bay, Hong Kong
116-D JTC Factory Building,
Lorong 3, Geylang Square, Singapore 14

Thomas Nelson Australia Pty Ltd,
19–39 Jeffcott Street, West Melbourne, Victoria 3003

Nelson Canada Ltd
81 Curlew Drive, Don Mills, Ontario M3A 2R1

Thomas Nelson (Nigeria) Ltd,
8 Ilupeju Bypass, PMB 21303, Ikeja, Lagos

First published in Great Britain by Thomas Nelson and Sons Ltd, 1974
Reprinted 1980

ISBN 0 17 761001 8
NCN 5630 42 2

Printed in Great Britain by A. Wheaton & Co. Ltd., Exeter

Contents

Preface

The study of crystalline solids enters at some stage and to some extent into the courses which almost every student of the physical sciences takes during his university career. The stage at which this topic is introduced and the depth in which it is explored varies with the student's primary subject of study and varies from university to university. In this book we have attempted to provide an introduction to the study of crystalline solids for students who may go on to develop their understanding and knowledge of the crystalline state in the fields of mineralogy, inorganic chemistry, metallurgy, or some other branch of physical science. Much of the book is suitable for those in their first year of study of crystalline solids, the more advanced sections being more suitable for the second year of study. Although the book is not in any sense intended to be a post-graduate text, it may be of use to those who have moved into this field from other disciplines.

Our treatment of the subject is based on our joint experience of teaching—in lectures and in supervisions—almost all the topics covered in the book over the past fifteen years. We do not attempt to provide a rigorous development either of crystallography or of the thermodynamics of crystalline solids, but prefer to concentrate on physical understanding of the subject. We may appear to the informed reader to labour some elementary matters; this is deliberate, because it is our experience that most students find conceptual difficulties in the early stages of the study of crystallography.

Our examples are drawn mainly from the fields of mineralogy and inorganic chemistry because these are the fields we know best. But we make no pretence of providing a systematic survey of either.

Part I (chapters 1–12) is essentially structural in its approach and is intended to provide a general introduction to the crystalline state by discussing the fundamentals of crystallography, diffraction by crystalline solids, crystal chemistry, and crystal physics. Part II (chapters 13–16) is essentially non-structural and is aimed specifically at the mineralogist and at those in other branches of physical science, such as the materials scientist, whose interests are generally similar.

We owe debts of gratitude to many. First, not only in sequence of time, to the late Sir Gavin de Beer, who persuaded us to write the book and unhappily did not live to see its completion. We have of necessity leant heavily on the authors of a number of lucid text-books, especially M. J. Buerger, the late E. A. Guggenheim, and F. C. Phillips; we have found their books invaluable in the preparation of ours and we wish to acknowledge our debt to them. We are particularly indebted to Dr Helen D. Megaw, whose approach to crystallography has been a constant source of stimulation.

The figures are the work of Mr K. O. Rickson, for whose skill in converting our rough sketches into intelligible diagrams we are deeply grateful; certainly without his knowledgeable and always cheerful help this book would never have come under starter's orders. Miss Vivien Gray typed almost all the text; for her ability to translate almost illegible manuscript accurately into typescript at high speed we are immensely grateful. We are particularly grateful also to Dr J. Gittins of the University of Toronto for reading the typescript of Part II and to Dr S. G. Fleet, our colleague in the

Department of Mineralogy and Petrology, for reading the whole typescript. Last, but not least, we wish to thank our publishers, especially for their forebearance and generally for their kind helpfulness.

D. McK

Cambridge, 3rd April 1973 C. H. McK

Part I

1
Crystal Lattices

A crystalline solid is essentially a solid whose atoms are disposed in regular three-dimensional array. The atoms in a solid are not static: each atom possesses thermal energy and vibrates about its mean position. It is the mean positions of the constituent atoms that are regularly arranged in space in a crystalline solid. Such regularity of mean atomic positions corresponds to a state of minimum free energy and is the fundamental characteristic of the crystalline state.

In its early development crystallography was confined to the study of *single crystals*, that is solid bodies bounded by natural plane faces within which the mean positions of all the constituent atoms are related to a single regular three-dimensional array of points. But there are in addition many other solid crystalline substances which can never, or only with difficulty, be obtained in single crystal form; such are the common metals, brass and steel, which are aggregates of interlocking randomly oriented crystals of varying shape and size. Such *polycrystalline* substances belong just as surely to the crystalline state as do the single crystals which exclusively formed the subject of the science of crystallography in its early days. Not all solids are crystalline however; glasses and other amorphous solids have, like liquids, only severely localized volumes of atomic order involving merely hundreds or thousands of atoms. Examples of solids with two-dimensional or one-dimensional atomic periodicity are known and are regarded as special cases within the crystalline state.

In this first chapter we develop the principles of geometrical crystallography by consideration of *perfect single crystals*. For a perfect single crystal, the regular arrangement of atoms in the crystal can be completely described by definition of a fundamental *repeat unit* coupled with a statement of the translations necessary to build the crystal from the repeat unit. For geometrical simplicity we exemplify this basic crystallographic concept first by consideration of a two-dimensional case.

The arrangement of atoms in a layer of graphite (the crystalline form of carbon stable at room temperature and atmospheric pressure) is shown in Fig 1.1. The carbon atoms, represented as small solid circles in the figure, are in a honeycomb pattern. The distance between the centres of adjacent hexagons of the 'honeycomb' is 2·46 Å so that a layer of area about 1 mm^2 will contain about $(4.10^6)^2 = 1{\cdot}6.10^{13}$ hexagons; the array of atoms in a layer of this size is thus effectively infinite. The repeat unit of the two-dimensional structure, containing two carbon atoms, is shown in the top left-hand corner of the figure enclosed in a parallelogram whose corners lie at the

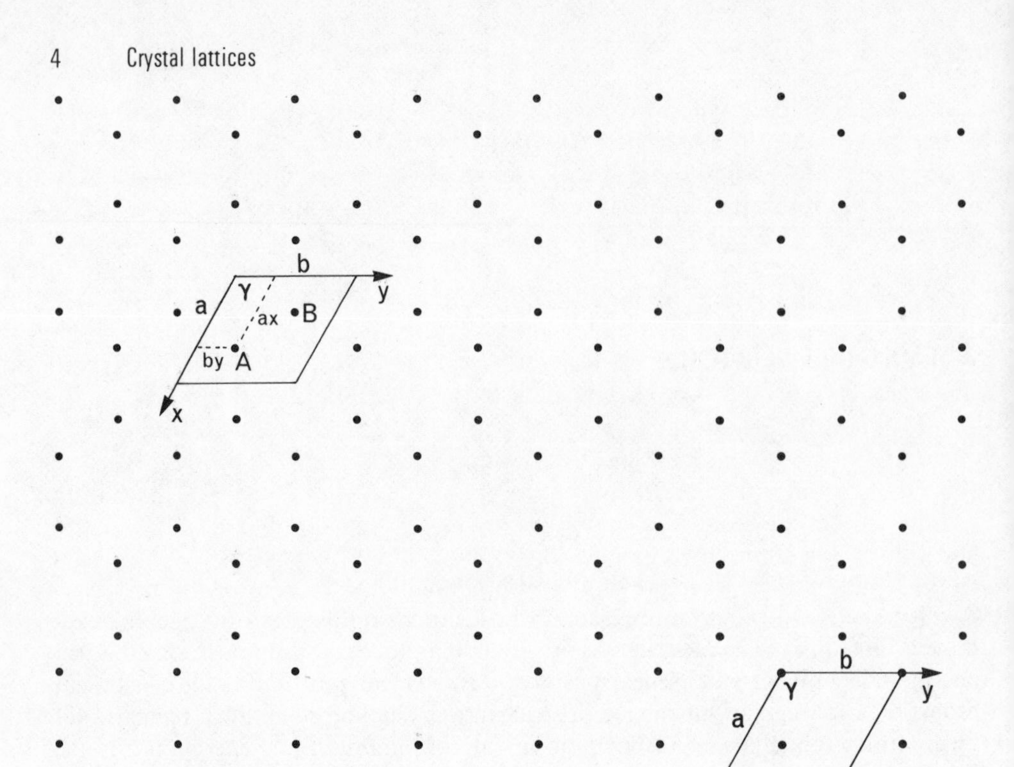

Fig 1.1 The arrangement of carbon atoms in one layer of the graphite structure. Each carbon atom is represented by a small solid circle. Two reasonable unit-meshes are outlined and labelled with the axial vectors **a**, **b**, and the inter-axial angle γ.

centres of four adjacent hexagons. The atomic pattern of the layer can be reconstructed by repeating this parallelogram in a regular manner so as to fill the plane of the atomic layer completely. The parallelogram, known as the *unit-mesh* of the layer, is completely specified by designating two of its sides as the reference axes x and y, stating the interaxial angle, and specifying the lengths of its edges. It is conventional to denote the lengths of the edges of the unit-mesh parallel to the x and y axes as a and b respectively and to denote the angle between the x and y axes as γ. In graphite the unit-mesh has $a = b = 2{\cdot}46$ Å, $\gamma = 120°$. A variety of parallelograms, all of the same area, could have been chosen as the unit-mesh of a graphite layer; but it is in general conventional and convenient to select a unit-mesh with a and b as short as possible and the angle $\geqslant 90°$. It is immaterial where the corners of the unit-mesh are placed in relation to the atoms of the graphite layer; the shape of the conventional unit-mesh is controlled by the atomic pattern to be constructed from it, a change of origin merely affecting the coordinates of the atoms within the unit-mesh. For the purpose of defining the positions of the atoms of the repeat unit within the unit-mesh we employ a coordinate system which has the edges of the unit-mesh as axes, the unit of length along each axis being taken as the length of the corresponding edge; atomic coordinates are thus given as fractions of the lengths of the edges of the unit-mesh referred conventionally to an origin at the top left-hand corner of the unit-mesh. The origin of each of the unit-meshes in Fig 1.1 is differently disposed with respect to the atomic array, but the reference axes are parallel and the area is the same in each case. The unit-mesh on the left of the figure contains an atom A with coordinates $\frac{2}{3}$, $\frac{1}{3}$ and an atom B with coordinates $\frac{1}{3}$, $\frac{2}{3}$. The periodic nature of the atomic arrangement naturally implies that an atom situated at a point with coordinates x, y, that is at a vector distance $x\mathbf{a}+y\mathbf{b}$ from the origin, will be repeated at vector distances $(m+x)\mathbf{a}+(n+y)\mathbf{b}$ from the origin, where m and n are integers; in this case the atom A at $\frac{2}{3}$, $\frac{1}{3}$ is repeated at$(m+\frac{2}{3})\mathbf{a}$, $(n+\frac{1}{3})\mathbf{b}$ and the atom B at $(m+\frac{1}{3})\mathbf{a}$, $(n+\frac{2}{3})\mathbf{b}$. The presence of an atom at the origin of the unit-mesh on the right of the figure implies the presence of other atoms of the same element at points with coordinates 1, 0; 0, 1; 1, 1; 2, 1; and so on: a statement of any one such pair of coordinates is sufficient for reconstruction of the structure. In the unit-mesh on the left of the figure the atom B has coordinates $\frac{1}{3}$, $\frac{2}{3}$ and there will be necessarily an equivalent atom with coordinates $-1+\frac{1}{3}$, $-1+\frac{2}{3}$, i.e. $-\frac{2}{3}$, $-\frac{1}{3}$, corresponding to the coordinates of the atom A in this unit-mesh with change of sign. The positions of the two carbon atoms in unit-mesh I can thus be neatly specified as $\pm(\frac{2}{3}, \frac{1}{3})$. In terms of this unit-mesh the structure of a layer of graphite can be completely specified by stating the dimensions of the unit-mesh, $a = b = 2{\cdot}46$ Å, $\gamma = 120°$, and the coordinates of the carbon atoms within it, $\pm(\frac{2}{3}, \frac{1}{3})$; the atomic layer can then be reconstructed by repetition of the unit-mesh in two non-parallel directions.

We now pass on to the next stage of complexity and consider in general terms a three-dimensional structure. Here the repeat unit can always be enclosed within a parallelepiped, known as the *unit-cell*, and the effectively infinite structure can be built up by repetition of the unit-cell in three non-coplanar directions which are conventionally taken as the reference axes x, y, and z. The lengths of the unit-cell edges parallel to the x, y, and z axes are respectively denoted a, b, and c. It is conventional also to take the positive directions of the reference axes so that the axial system is right-handed and the interaxial angles $\alpha = y \wedge z$, $\beta = z \wedge x$, $\gamma = x \wedge y$ are all three $\geqslant 90°$ as exemplified in Fig 1.2.[1] As in the two-dimensional example considered earlier the coordinates of atomic positions are conventionally stated as fractions of the unit-cell edges.

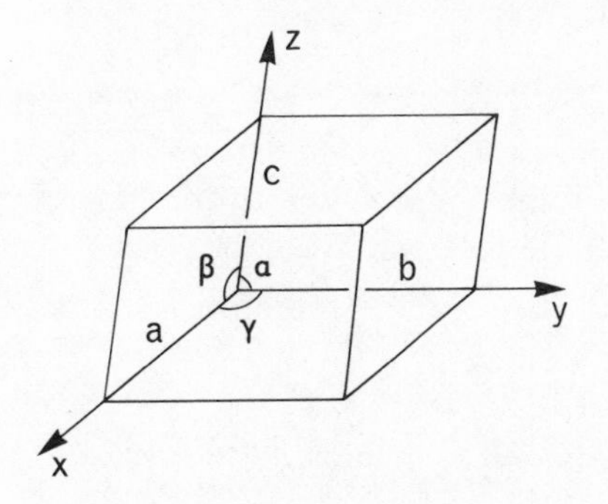

Fig 1.2 Unit-cell nomenclature. The reference axes x, y, z are right-handed, the length of the unit-cell edge parallel to each reference axis is respectively a, b, c, and the interaxial angles are denoted α, β, γ.

It is difficult to make easily intelligible perspective drawings of three-dimensional structures unless they are very simple and the task is virtually impossible for really complicated structures. It has consequently become common practice to use structural plans where the three-dimensional structure is projected down one of the reference

[1] Only very occasionally is it convenient to modify this simple convention.

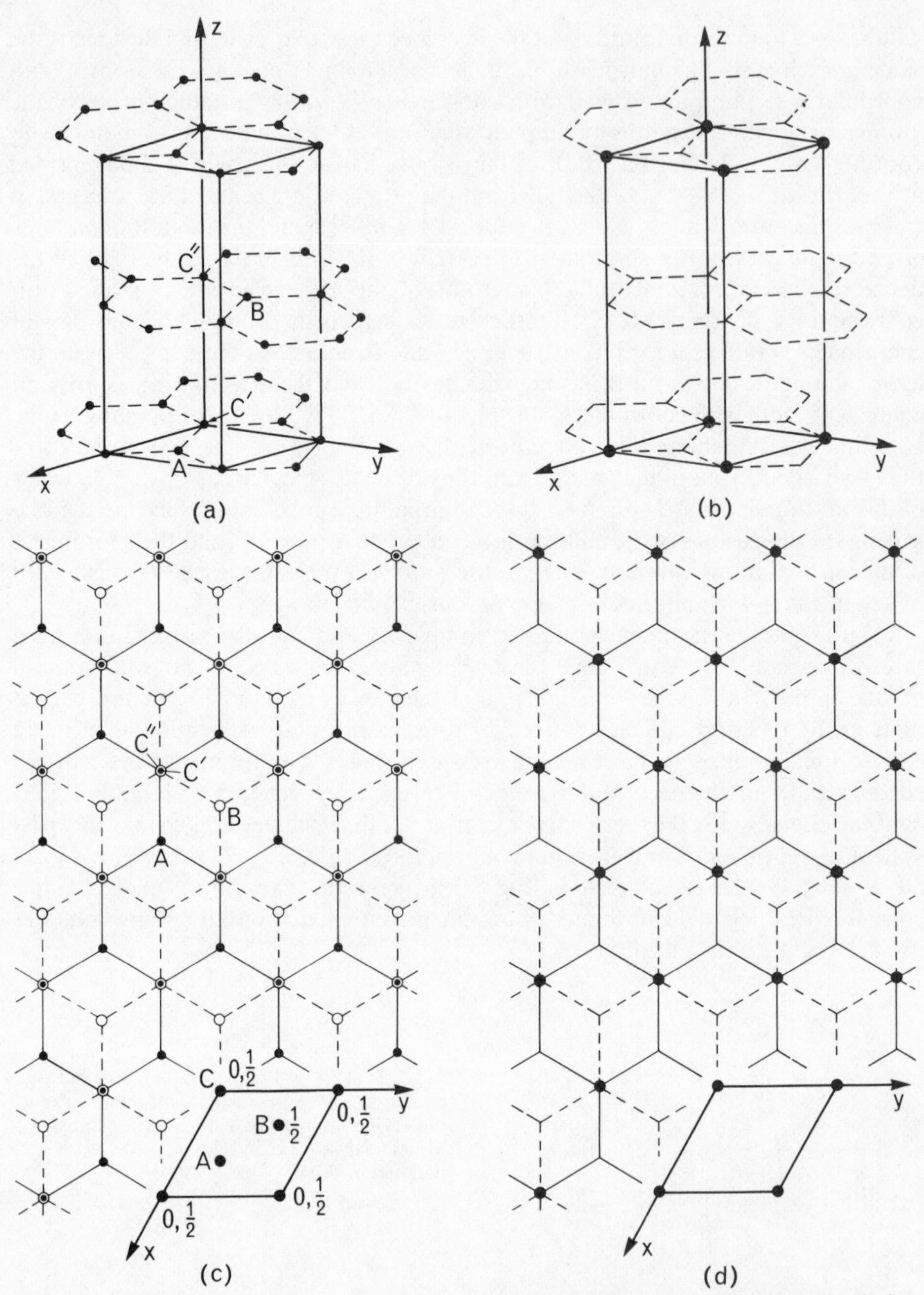

Fig 1.3 The structure of graphite. (a) and (b) are perspective drawings to show how identical two-dimensional layers are stacked to make the three-dimensional structure; in (a) the positions of carbon atoms are shown as small solid circles and C_6 rings are outlined; in (b) the C_6 rings are again outlined and lattice points are shown as large solid circles; in both (a) and (b) the unit-cell is outlined. The coordinates of the carbon atoms in the graphite unit-cell, are $0, 0, 0$; $0, 0, \frac{1}{2}$; $\frac{2}{3}, \frac{1}{3}, 0$; $\frac{1}{3}, \frac{2}{3}, \frac{1}{2}$. (c) and (d) are projections down the z-axis on to the xy plane; in (c) the carbon atoms with $z = 0$ are shown as small solid circles and the C_6 rings of this layer are outlined with solid lines while the carbon atoms of the superimposed layer at $z = \frac{1}{2}$ are shown as open circles and their linkage into C_6 rings is indicated by broken lines; in (d) the C_6 rings of the $z = 0$ and $z = \frac{1}{2}$ layers are similarly represented and lattice points are shown as large solid circles; in the lower right-hand corners of both (c) and (d) the unit-cell is shown in projection.

axes on to the plane containing the other two axes, which may or may not be perpendicular to the axis of projection. Atomic coordinates in the direction of the axis of projection are marked on the plan beside the symbol representing the atomic position. In Fig 1.3(a) and (c) a perspective drawing and a plan of the three-dimensional graphite structure are shown. The atom labelled A lies in the x, y plane and at distances $m\mathbf{c}$, where m is a positive or negative integer, above or below the plane. When the coordinate parallel to the axis of projection of an atom, such as A, is zero it is customary to omit the coordinate from the plan of the structure; an atom with no coordinate written beside it is to be taken as lying in the plane of projection. The atom labelled B lies at $\frac{1}{2}\mathbf{c}$ above the plane of projection and this is indicated by writing $\frac{1}{2}$ next to the symbol for the atom on the plan. At C, and related positions, two atoms, C′ and C″, are superimposed in projection, one with $z = 0$ and the other with $z = \frac{1}{2}$; in such a case it is customary to write both coordinates beside the symbol for the atom as $0, \frac{1}{2}$.

Lattices

Some crystal properties of interest and importance are dependent only on the shape of the unit-cell, that is to say they depend only on the way in which repeat units are related to one another. It is consequently useful to have a simple way of describing the periodicity of a crystal structure and for this purpose the concept of the *lattice* is introduced. The way in which the crystal structure is built up by repetition of the repeat unit can be completely, and very simply, described by replacing each repeat unit by a *lattice point* placed at an exactly equivalent point in each and every repeat unit. All such lattice points have the same environment in the same orientation and are indistinguishable from one another. We return to two dimensions to exemplify this matter in the first instance and again take as our example a layer of the graphite structure (Fig 1.4). Figure 1.4a shows a layer of the graphite structure with carbon atoms labelled A, B, C, . . . , a, b, c, . . . and a conventional unit-mesh outlined. The lattice of this structure can be constructed by placing a lattice point at the carbon atom A and at all equivalent points, that is at B, C, D, E, F, G, H, I, etc. The resultant two-dimensional lattice is shown in Fig 1.4(b). If a lattice point is placed at A, then it is not permissible to place a lattice point at a because, although A and a both represent carbon atoms they are not identically situated; both lie at the centroid of a triangle formed by their three nearest neighbours, but the triangles about A and a are disposed at 60° to each other so that although both atoms have identical environments, their environments are not similarly oriented. Either the carbon atoms at A, B, C, . . . or the atoms at a, b, c, . . . , but not both sets of atoms, may be taken as lattice points.

Figure 1.4(c) represents a layer of the structure of boron nitride, BN, boron atoms being represented by solid circles and nitrogen atoms by open circles. The two-dimensional lattices of graphite and BN are evidently identical except for the small difference in their unit-mesh dimensions: for graphite $a = b = 2{\cdot}46$ Å, while for BN $a = b = 2{\cdot}51$ Å. The repeat unit in graphite however consists of two carbon atoms, while in boron nitride it consists of one boron and one nitrogen atom.

In a lattice every repeat unit of the structure is represented by a lattice point. A graphite layer, for instance, can be built up by placing the repeat unit of two carbon atoms in the same orientation at each lattice point in such a manner that the corresponding point of every repeat unit is placed at a lattice point. It is of no consequence which point of the repeat unit is sited at the lattice point so long as it is the same point for every repeat unit. The lattice thus has, in two dimensions, the same unit-mesh as the structure to which it refers and is completely specified by a statement of the repeat lengths a and b parallel to its x and y axes and its interaxial angle γ.

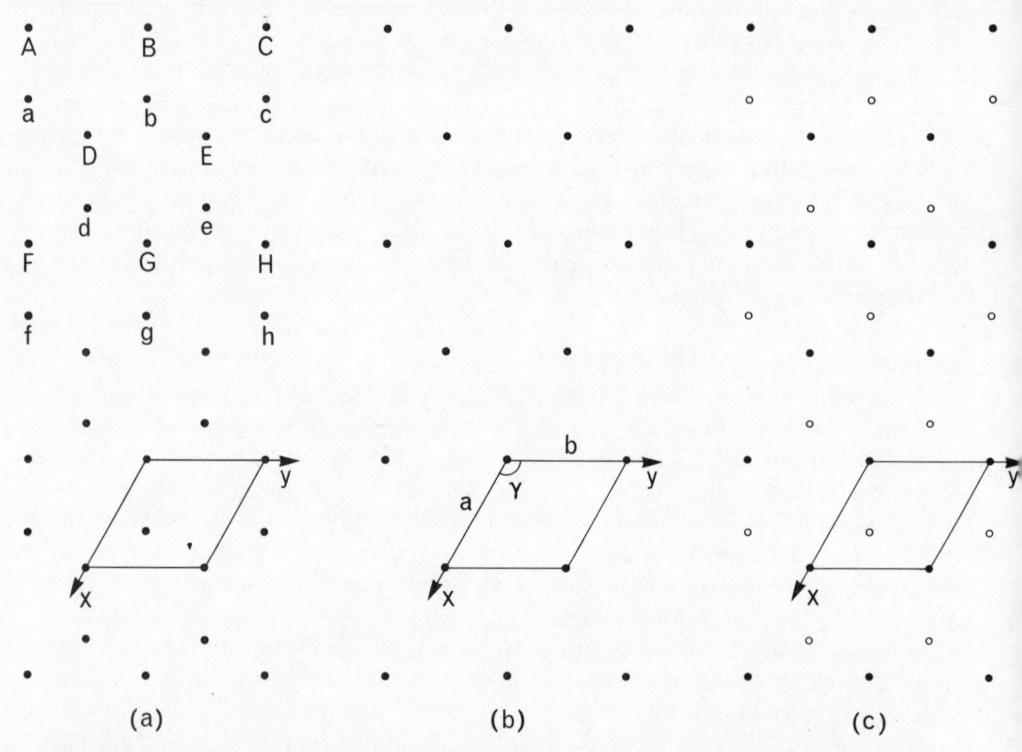

Fig 1.4 The two-dimensional lattice of a graphite layer. (a) shows the arrangement of carbon atoms (solid circles) in one layer of the graphite structure (Figs 1.1, 1.3); identically situated carbon atoms are labelled A, B, C, D, . . . ; the carbon atoms labelled a, b, c, d, . . . have a differently oriented environment but are each identically situated; the unit-mesh is outlined and it is apparent that the repeat unit consists of two carbon atoms, such as A and a. (b) shows the corresponding two-dimensional *lattice* with the dimensions $a = b$, γ of the unit-mesh indicated. (c) shows the structure of a layer of boron nitride, BN, which has the same lattice with $\gamma = 120°$ and $a = b$, but a is slightly different from a for graphite, the difference being too small to show on the diagram; boron and nitrogen atoms are represented respectively as solid and open circles.

A three-dimensional lattice can be derived in an exactly analogous manner. For instance, the three-dimensional structure of graphite has a repeat unit containing four carbon atoms (Fig 1.3(a) and (c)). The lattice of this structure may be simply obtained by placing lattice points at the site of the carbon atom C′ (Figs 1.3(a) and (c)) and at all equivalent points. Inspection of the figure shows that the atom B cannot be related to the atom C′ by a lattice translation; both atoms have identical environments in their own layer, but their environments in adjacent layers are different. The unit-cell of the graphite lattice has dimensions $a = b = 2{\cdot}46$ Å, $c = 6{\cdot}80$ Å, $\alpha = \beta = 90°$, $\gamma = 120°$. The lattice of the graphite structure is shown in perspective and in plan in Figs 1.3(b) and (d) respectively.

Having exemplified a crystal lattice, we are now ready to make a formal definition of a lattice as *an array of points in space such that each lattice point has exactly the same environment in the same orientation.* It follows immediately that any lattice point is related to any other by a simple lattice translation.

A plane passing through three non-colinear lattice points is known as a *lattice plane.* Since all lattice points are equivalent there will be equivalent parallel planes passing through all the other points of the lattice. Such a set of planes is known as a *set of lattice planes*; several sets are illustrated in Fig 1.5. A set of lattice planes divides each edge of the unit-cell into an integral number of equal parts; this property forms the basis of the very useful system of indexing of lattice planes developed by W. H. Miller, Professor of Mineralogy in the University of Cambridge from 1832 to 1880. If the lattice repeats along the x, y, z axes are respectively a, b, c and if the first plane out from the origin (at a lattice point) of a set of lattice planes makes intercepts a/h, b/k, c/l, where h, k, l are integers, on the x, y, z axes respectively, then the *Miller indices* of this set of lattice planes are (hkl), the three factors h, k, l being conventionally enclosed in round brackets. A set of lattice planes (hkl) thus divides a into $|h|$ parts, b into $|k|$ parts, and c into $|l|$ parts. The set of lattice planes labelled I in Fig 1.5 has Miller indices (122).

The equations to a set of lattice planes can be written in intercept form as $(hx/a)+(ky/b)+(lz/c) = n$, where n is an integer. If n is zero the lattice plane passes through the origin; if $n = 1$ the plane of the set makes intercepts a/h, b/k, c/l on the x, y, z axes respectively; if $n = 2$ the intercepts are $2a/h$, $2b/k$, $2c/l$; and if $n = -1$ the intercepts are $-a/h$, $-b/k$, $-c/l$. Thus the set of lattice planes (hkl) includes the plane with indices $(\bar{h}\bar{k}\bar{l})$, which makes intercepts $-a/h$, $-b/k$, $-c/l$ on the reference axes and is commonly spoken of as the 'bar h, bar k, bar l' plane.

Of course some sets of lattice planes will make intercepts that are not all positive or all negative: for instance the first plane out from the origin (taken as the front lower right-hand corner) of the set labelled II in Fig 1.5 makes intercepts $-a/2$, $-b$, $c/2$ on the x, y, z axes respectively so that the indices of this set are $(\bar{2}\bar{1}2)$. If a plane is parallel to one of the reference axes, its intercept on that axis is at infinity and the corresponding Miller index is zero; thus set III in Fig 1.5 being parallel to the z-axis has c/l infinite so that l must be zero and the Miller indices of the set are $(\bar{2}10)$. The set of planes labelled IV in Fig 1.5 is parallel to the x and z axes so that $h = l = 0$; since the intercept of the first plane out from the origin on the y-axis is b, the indices of the set are (010). In terms of Miller indices the unit-cell can be described as the parallelepiped bounded by adjacent lattice planes of the sets (100), (010), (001).

The line of intersection of any two non-parallel lattice planes is the row of lattice points common to both planes. The intersections of two sets of lattice planes will thus be a set of parallel rows of lattice points; for instance sets III and IV in Fig 1.5 intersect in lines parallel to the z-axis. It is convenient to index such rows by reference to the parallel row through the origin, which is itself the intersection of the lattice planes through the origin belonging to each of the two sets. The coordinates of the lattice points in such a row are 0, 0, 0 for the lattice point at the origin; Ua, Vb, Wc, where U, V, W are integers with no common factor other than unity, for the next lattice point out from the origin; and nUa, nVb, nWc, where n is an integer, for the other lattice points of the row. Such a row of lattice points is completely specified by the three integers U, V, W, which are conventionally enclosed in square brackets as $[UVW]$ in order to distinguish them from Miller indices for lattice planes, conventionally enclosed in round brackets as (hkl). The symbol $[UVW]$ represents not only the lattice point row passing through the origin and through the lattice point with coordinates Ua, Vb, Wc but all parallel lattice point rows, just as

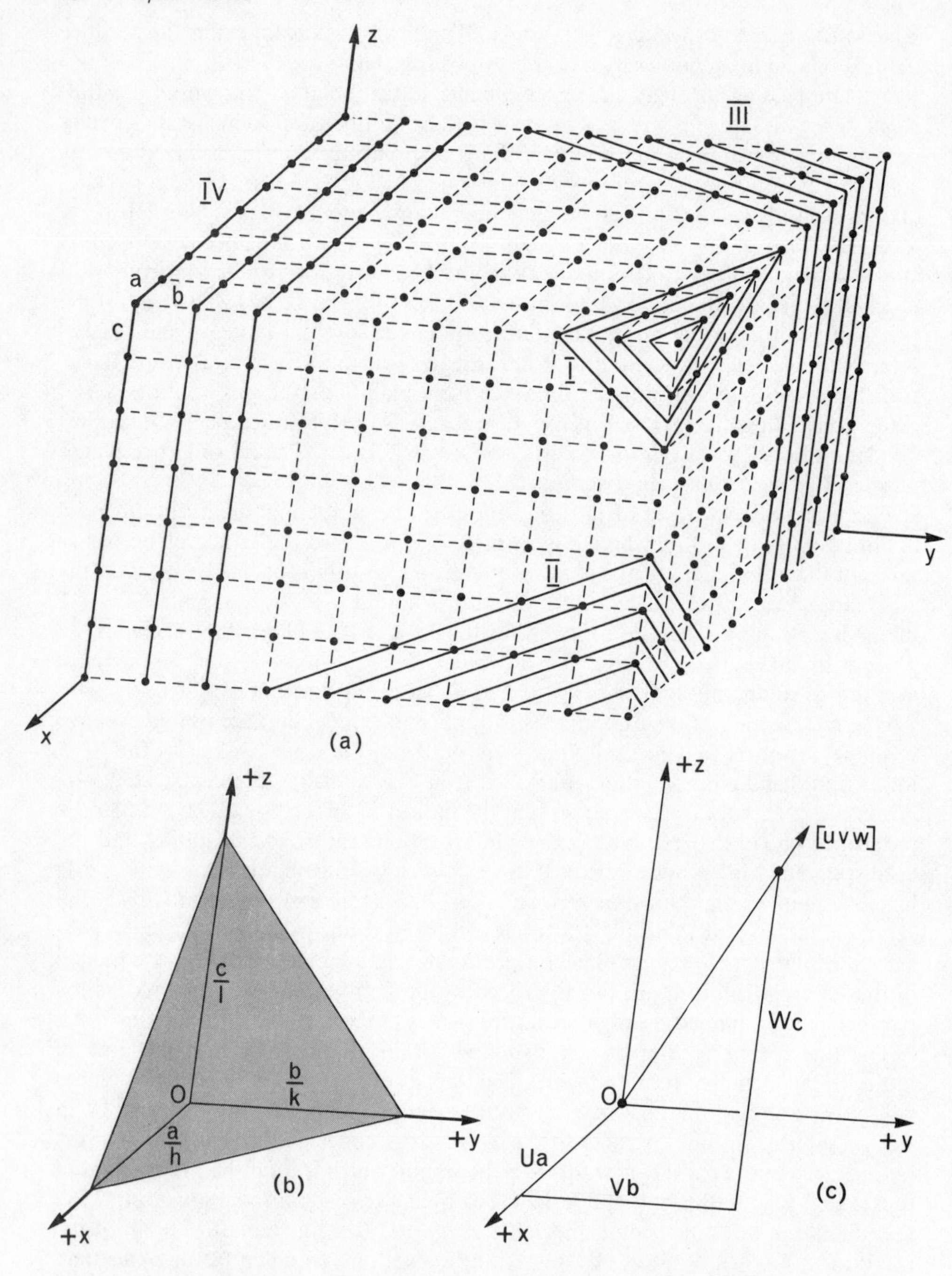

Fig 1.5 Lattice planes and zone axes. The array of lattice points exposed on the three visible faces of a parallelepiped whose edges are parallel to those of the unit-cell is displayed in (a) with solid circles to represent lattice points and thin broken lines parallel to axial directions. Four sets of lattice planes are indicated by thick solid lines representing the intersection of lattice planes with the visible faces of the parallelpiped; their Miller indices are I (122), II ($\bar{2}\bar{1}2$), III ($\bar{2}10$), IV (010). In (b) the definition of Miller indices is illustrated: the shaded plane (*hkl*) makes intercepts a/h, b/k, c/l on the x, y, z axes, a, b, c being the lattice repeat along each axis and h, k, l being integers. In (c) the definition of the zone axis symbol is illustrated: [*UVW*] is the direction parallel to the line through the origin and the point Ua, Vb, Wc.

the Miller indices (hkl) represent the lattice plane which makes intercepts a/h, b/k, c/l on the x, y, z axes respectively and all parallel lattice planes.

It is necessary for us to explore further the geometry and notation of lattice planes and rows because a thorough understanding of these matters is essential not only for the description of the external shape of single crystals but also, and more importantly, for the interpretation of the diffraction of X-radiation by crystals (chapter 6). In both these fields it is only the angular disposition of lattice planes and rows with respect to the reference axes of the unit-cell that is significant; the actual position in space of a given plane or row is of no consequence. For this reason the Miller indices (hkl) may be taken to represent a set of parallel lattice planes and the *zone axis symbol* $[UVW]$ to represent a set of parallel lattice point rows (the term 'zone' will be defined later). The symbol (hkl) thus denotes any plane of the set of lattice planes which satisfy the equation $(hx/a)+(ky/b)+(lz/c) = n$, where n is integral, with one qualification which enables a distinction to be made between planes of the set which lie on opposite sides of the origin. If it is desired to make this distinction, as is often the case, the symbol (hkl) is reserved for those planes of the set which make intercepts on the same side of the origin as the plane whose intercepts on the x, y, z axes are respectively a/h, b/k, c/l and the symbol $(\bar{h}\bar{k}\bar{l})$ is used to denote those planes of the set which make intercepts on the same side of the origin as the plane whose intercepts on the x, y, z axes are respectively $-a/h$, $-b/k$, $-c/l$. The plane $(\bar{h}\bar{k}\bar{l})$ is said to be the *opposite* of (hkl), the superscript 'bar' representing, as is usual in crystallography, a minus sign. In a precisely analogous way the zone axis symbol $[UVW]$ represents all directions parallel to the vector from the origin to the lattice point with coordinates Ua, Vb, Wc, with the proviso that opposite directions may be distinguished as $[UVW]$ and $[\bar{U}\bar{V}\bar{W}]$; $[UVW]$ is taken to be in the same sense as the vector from the origin to Ua, Vb, Wc and $[\bar{U}\bar{V}\bar{W}]$ in the same sense as the vector from the origin to the lattice point at $-Ua$, $-Vb$, $-Wc$. In general, of course, the indices in the symbols (hkl) or $[UVW]$ need not all be of the same sign.

The condition for a lattice point row $[UVW]$ to be parallel to a plane (hkl) amounts simply to the condition that the point row of the set $[UVW]$ through the origin should lie in the plane of the set (hkl) through the origin, that is that the point Ua, Vb, Wc should satisfy the equation $(hx/a)+(ky/b)+(lz/c) = 0$. This is so when $(hUa/a)+(kVb/b)+(lWc/c) = 0$, i.e. $hU+kV+lW = 0$ since a, b, c are necessarily non-zero. The equation $hU+kV+lW = 0$ is known as the *zone equation* for reasons which will be discussed in the next section.

The external shape of crystals

The regular nature of the spatial arrangement of the atoms within a crystal, whether simple or complicated, leads directly to the consequence that different directions in the crystal may not be equivalent. We take a very simple example, the structure of caesium chloride illustrated in Fig 1.6. The unit-cell of caesium chloride is a cube and thus has $a = b = c$, $\alpha = \beta = \gamma = 90°$; if a caesium atom (solid circle) is situated at 0, 0, 0, then a chlorine atom (open circle) lies at $\frac{1}{2}, \frac{1}{2}, \frac{1}{2}$. The direction parallel to the x-axis, [100], is evidently equivalent to the direction parallel to the y-axis [010], but neither of these in any way corresponds to the direction [110]. There is thus no apparent reason why any directional property of the crystal should have equal magnitude in the directions [100] and [110]. The reasons why some directional properties do and others do not have equal magnitudes in such crystallographically distinct directions will be developed in chapter 11; it suffices here to restrict the

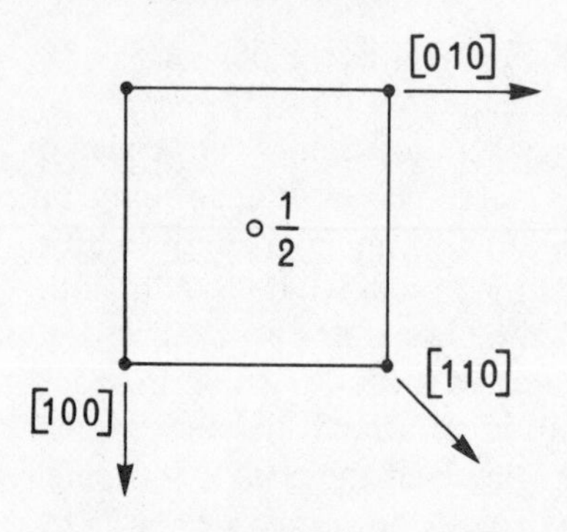

Fig 1.6 The structure of CsCl shown in plan on (001). The directions [100] and [010] are equivalent to each other but not to the direction [110]. Solid circle = Cs; open circle = Cl.

argument to the growth of crystals. There is no obvious physical reason why atoms should attach themselves to the growing crystal as readily in one direction as in a crystallographically distinct direction, and indeed one would intuitively expect that not to be so; if it were so, crystals would tend to grow towards a spherical shape and that is not found experimentally to happen. Crystals tend to grow with plane faces which are parallel to lattice planes, especially to lattice planes with a high density of lattice points per unit area. A high density of lattice points per unit area of a lattice plane implies a large interplanar spacing (Fig 1.7) and consequently large intercepts a/h, b/k, c/l on the reference axes; the indices of commonly well-developed faces on crystals thus tend to have small values of h, k, and l. Not only are crystal faces parallel to lattice planes, but in practice the Miller indices (hkl) of the faces on a natural crystal rarely involve an integer greater than six. Moreover it is customary to index crystal faces with reference to an origin within the crystal so that the faces (hkl) and ($\bar{h}\bar{k}\bar{l}$) are parallel faces on opposite sides of the crystal. The recognition of the comparative

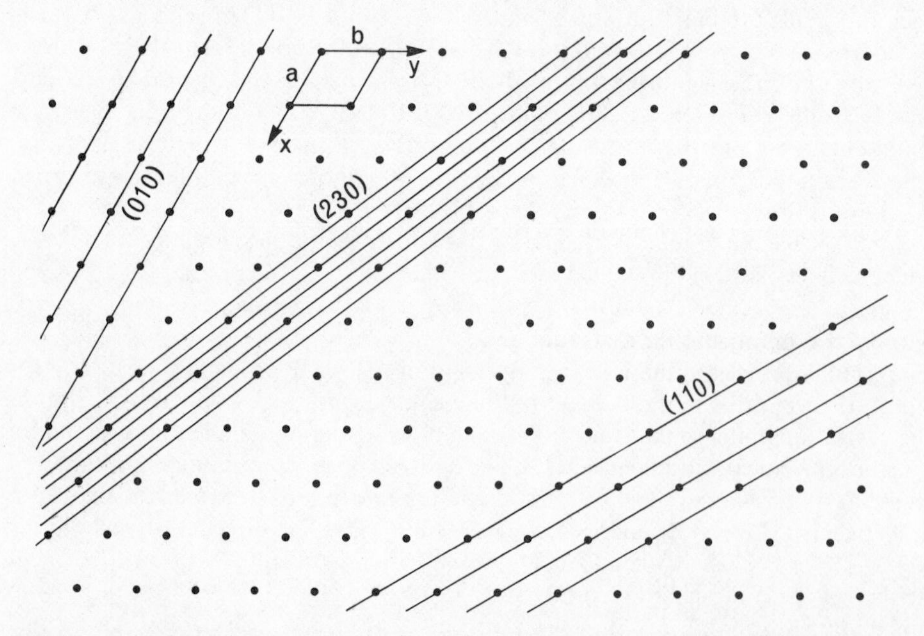

Fig 1.7 Projection of the lattice of graphite down the z-axis on to the xy plane to illustrate the decrease in the density of lattice points per unit area of lattice planes ($hk0$) as their indices h and k increase. The density of lattice points per unit length in projection decreases from (100) to (110) to (230).

geometrical simplicity of the angular relations of the faces of a crystal goes back to the writings of the Abbé René Just Haüy (1743–1822). Although Haüy's manner of explaining the simplicity of natural crystal forms has not withstood the test of time, it is apparent that his ideas were a helpful influence on those who came later with more powerful experimental tools to develop the modern science of structural crystallography.[2]

Two crystal faces intersect in an edge which, since the faces are parallel to lattice planes, must be parallel to a lattice point row $[UVW]$. Commonly a crystal displays several faces whose mutual edges of intersection are all parallel; such faces, which must all be parallel to a common lattice point row, are said to lie in a *zone* and the common direction of their edges of mutual intersection is known as a *zone axis.* Since the faces in a zone are all parallel to the zone axis their normals from any point, must be coplanar, a geometrical consequence that will be developed in chapter 2. The concept of zones has obvious significance for the study of the external shapes of crystals, that is *morphological crystallography*, and it is important too in a study of the diffraction of X-radiation by crystals (chapters 6–9). A zone is geometrically characterized by the symbol of its axis $[UVW]$, which may be used to indicate the group of faces whose edges of mutual intersection are parallel to the direction $[UVW]$ as well as the direction of the lattice point row $[UVW]$; this dual interpretation of the zone axis symbol $[UVW]$ will be developed in succeeding chapters.

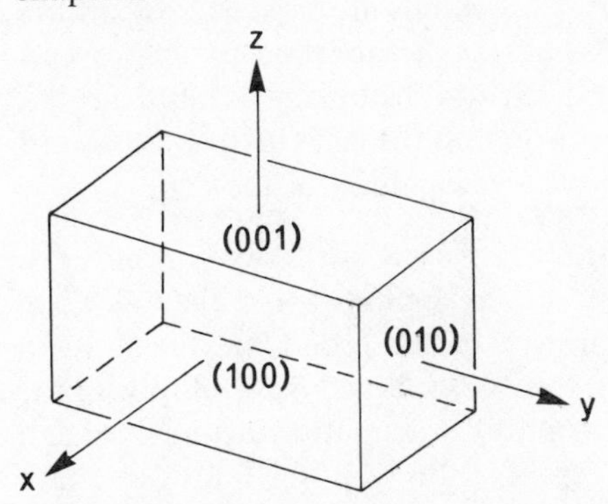

Fig 1.8 An hypothetical crystal with faces parallel to those of the chosen unit-cell. Face indices are shown for the front faces (100), (010), (001). Parallel faces on the back of the crystal, whose edges are shown by broken lines, have indices $(\bar{1}00)$, $(0\bar{1}0)$, $(00\bar{1})$.

We consider first a crystal of the utmost simplicity of form, a parallelepiped whose six faces are parallel to the faces of the chosen unit-cell of the lattice. The faces (Fig 1.8) of such a crystal have indices:

(100) parallel to the y and z axes and intersecting the positive x-axis;
$(\bar{1}00)$ parallel to the y and z axes and intersecting the negative x-axis;
(010) parallel to the x and z axes and intersecting the positive y-axis;
$(0\bar{1}0)$ parallel to the x and z axes and intersecting the negative y-axis;
(001) parallel to the x and y axes and intersecting the positive z-axis;
$(00\bar{1})$ parallel to the x and y axes and intersecting the negative z-axis.

The faces (100), (010), $(\bar{1}00)$, $(0\bar{1}0)$ are all parallel to the z-axis and thus lie in the zone which has the z-axis as its zone axis, that is the zone [001]. Likewise the faces (010), (001), $(0\bar{1}0)$, $(00\bar{1})$ are all parallel to the x-axis so that they belong to the [100] zone, while the faces (100), (001), $(\bar{1}00)$, $(00\bar{1})$, being parallel to the y-axis, lie in the [010] zone.

[2] A useful account of the relationship of Haüy's work to modern crystallography is to be found in Phillips (1971). For a critique of Haüy's work in its historical setting the reader is referred to Gillispie (1972).

The reference axes x, y, z are in this particular case and in general the [100], [010], [001] zone axes.

In general the condition for the crystal face (hkl) to lie in the zone characterized by the zone axis symbol $[UVW]$ is the condition for the plane through the origin of the set (hkl) to contain the lattice point with coordinates Ua, Vb, Wc. We have already shown that this condition is the *zone equation*

$$Uh+Vk+Wl=0.$$

In particular any face in the [001] zone must satisfy the zone equation

$$0.h+0.k+1.l=0$$

and so must be of the type ($hk0$). For instance the faces (110), ($2\bar{1}0$), ($1\bar{2}0$), (320) belong to the [001] zone.

The zone axis symbol $[UVW]$ for the zone containing the two generalized faces $(h_1k_1l_1)$ and $(h_2k_2l_2)$ is obtained by solving the simultaneous equations

$$h_1U+k_1V+l_1W=0$$

$$h_2U+k_2V+l_2W=0$$

for U, V, W. The solution is conveniently expressed in determinant form as

$$\frac{U}{\begin{vmatrix} k_1 & l_1 \\ k_2 & l_2 \end{vmatrix}} = \frac{V}{\begin{vmatrix} l_1 & h_1 \\ l_2 & h_2 \end{vmatrix}} = \frac{W}{\begin{vmatrix} h_1 & k_1 \\ h_2 & k_2 \end{vmatrix}}$$

i.e. $$[UVW]=[k_1l_2-k_2l_1,\ l_1h_2-l_2h_1,\ h_1k_2-h_2k_1].$$

U, V, W are then chosen so as to have no common factor other than unity. A simple way of evaluating such two-by-two determinants is by using the *cross-multiplication* format: the indices of the first face are written twice in the upper line, the indices of the second face are written directly below them twice in the second line, the first and last columns are ignored; the first determinant is evaluated by cross-multiplying the second and third columns, the second determinant by cross-multiplying the third and fourth columns and the third determinant by cross-multiplying the fourth and fifth columns, i.e.

$$\begin{array}{c|ccccc|c} h_1 & k_1 & \times & l_1 & \times & h_1 & \times & k_1 & l_1 \\ h_2 & k_2 & & l_2 & & h_2 & & k_2 & l_2 \\ \hline & \multicolumn{7}{c}{k_1l_2-k_2l_1,\ l_1h_2-l_2h_1,\ h_1k_2-h_2k_1} \end{array}$$

It is convenient to isolate the first and last columns by drawing strong vertical lines to separate them off. By way of example we take (Fig 1.9) the zone containing the faces (210) and (011). Cross-multiplication

$$\begin{array}{c|ccccccc|c} 2 & 1 & \times & 0 & \times & 2 & \times & 1 & 0 \\ 0 & 1 & & 1 & & 0 & & 1 & 1 \\ \hline & & 1 & & \bar{2} & & 2 & & \end{array}$$

yields $[1\bar{2}2]$ for the zone axis symbol of the zone containing the faces (210) and (011).

It is occasionally convenient to represent a zone by a statement of the Miller indices of two non-parallel faces lying in the zone; this is done by enclosing the face

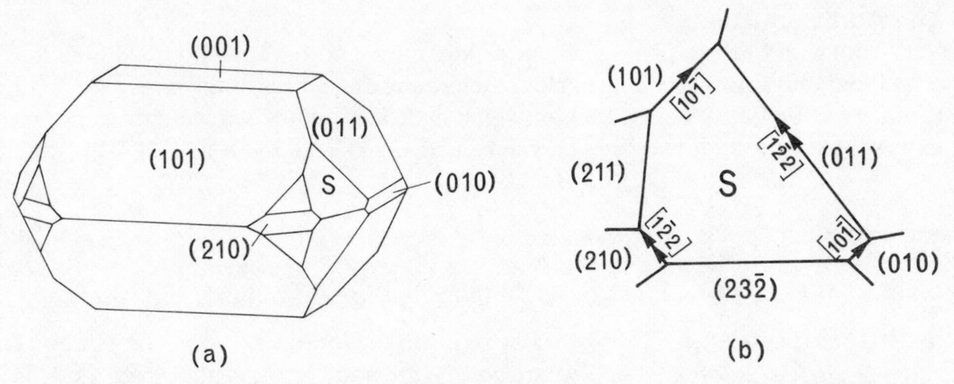

Fig 1.9 The *zone equation* and the *addition rule*. (a) is a perspective drawing of a crystal of the mineral *anglesite*, $PbSO_4$, to some faces of which Miller indices have been assigned and (b) is a plan of the face S parallel to its own plane with the neighbouring faces indexed. It is apparent from the plan that the edges of S against (210) and (011) are parallel and that the edges against (101) and (010) are also parallel so that S lies at the intersection of the zones $[10\bar{1}]$ and $[1\bar{2}2]$; the Miller indices of S are thus (232).

symbols within square brackets, thus $[(h_1k_1l_1), (h_2k_2l_2)]$. For example the zone $[1\bar{2}2]$ can be referred to alternatively as [(210), (011)].

Analogously the indices of the face (hkl) and its opposite $(\bar{h}\bar{k}\bar{l})$ which lie in the zones $[U_1V_1W_1]$ and $[U_2V_2W_2]$ are given by the solution of the simultaneous equations

$$hU_1 + kV_1 + lW_1 = 0$$
$$hU_2 + kV_2 + lW_2 = 0$$

i.e.
$$\frac{h}{\begin{vmatrix} V_1 & W_1 \\ V_2 & W_2 \end{vmatrix}} = \frac{k}{\begin{vmatrix} W_1 & U_1 \\ W_2 & U_2 \end{vmatrix}} = \frac{l}{\begin{vmatrix} U_1 & V_1 \\ U_2 & V_2 \end{vmatrix}}$$

i.e.
$$(hkl) \equiv (V_1W_2 - V_2W_1,\ W_1U_2 - W_2U_1,\ U_1V_2 - U_2V_1).$$

Here too h, k, l must have no common factor other than unity and the solution of the two-by-two determinants is most conveniently achieved by cross-multiplication ignoring the first and last columns:

$$\begin{array}{c|ccccccc|c} U_1 & V_1 & & W_1 & & U_1 & & V_1 & W_1 \\ U_2 & V_2 & \times & W_2 & \times & U_2 & \times & V_2 & W_2 \\ \hline & \multicolumn{7}{c}{V_1W_2 - V_2W_1,\ W_1U_2 - W_2U_1,\ U_1V_2 - U_2V_1} & \\ & \multicolumn{7}{c}{\| \qquad\qquad \| \qquad\qquad \|} & \\ & \multicolumn{7}{c}{n.h \qquad\quad n.k \qquad\quad n.l} & \end{array}$$

By way of example (Fig 1.9) we take the intersection of the zones $[10\bar{1}]$ and $[1\bar{2}2]$:

$$\begin{array}{c|ccccccc|c} 1 & 0 & & \bar{1} & & 1 & & 0 & \bar{1} \\ 1 & \bar{2} & \times & 2 & \times & 1 & \times & \bar{2} & 2 \\ \hline & & \bar{2} & & \bar{3} & & \bar{2} & & \end{array}$$

The faces common to the zones $[10\bar{1}]$ and $[1\bar{2}2]$ are $(\bar{2}\bar{3}\bar{2})$ and its opposite (232); the face labelled S in the figure is thus indexed as (232).

The Addition Rule

Very commonly in morphological crystallography, and occasionally in diffraction crystallography, one needs to be able to determine the indices of the faces that lie at the intersection of two zones, each of which is defined by the known indices of two faces belonging to it. If two faces $(h_1k_1l_1)$ and $(h_2k_2l_2)$ lie in the zone $[UVW]$,

$$h_1U+k_1V+l_1W=0$$

and $$h_2U+k_2V+l_2W=0$$

hence $$(ph_1+qh_2)U+(pk_1+qk_2)V+(pl_1+ql_2)W=0$$

so that the indices $(h_3k_3l_3)$ of any other face lying in the zone $[UVW]$ can be expressed as $(ph_1+qh_2, pk_1+qk_2, pl_1+ql_2)$, where p and q are positive or negative integers. This is known as the *addition rule*. For instance the zone [(101), (213)] includes the faces (314) for which $p=q=1$, (112) for which $p=-1$, $q=1$, (011) for which $p=-2$, $q=1$, $(\bar{1}10)$ for which $p=-3$, $q=1$, and so on. Thus it is quite a simple matter to determine the indices of a plane common to two zones, each defined by the Miller indices of two of its faces. For example in Fig 1.9 the face labelled S lies at the intersections of the zones [(210), (011)] and [(010), (101)]; S must have indices consistent with $(2p_1, p_1+q_1, q_1)$ and (q_2, p_2, q_2); therefore $2p_1=q_2=q_1$ and $p_2=p_1+q_1=\frac{3}{2}q_2$ so that $h{:}k{:}l=q_2{:}p_2{:}q_2=1{:}\frac{3}{2}{:}1$ so that (hkl) is identified as (232) or $(\bar{2}\bar{3}\bar{2})$ and the face S specifically as (232).

2
Representation in two dimensions: The stereographic projection

The interrelationships in space between lattice planes and lattice rows, crystal faces and zone axes are central to the study of crystal geometry. The point has already been made that angular relationships are of much greater importance than actual position in space in the limited range of topics discussed in chapter 1 and that is so generally. For graphical representation we have so far made use of perspective drawings and plans, but it is obvious that both these methods of representation will fail to give a clear picture of angular relationships in all but the simplest cases. Angular relationships are critical too in the study of crystal symmetry which we shall come to in chapter 3. Clearly then some means of representing three-dimensional angular relationships in two dimensions is needed. This need is met in crystallography by the *stereographic projection.*

We shall develop the stereographic projection in the context of crystal shape simply because a realistic perspective drawing of a crystal, showing faces and zone axes, can easily be made and compared with the angular morphological relations displayed in stereographic projection. But it must be emphasized that the stereographic projection is just as applicable and indeed more useful for the representation of angular relationships within a lattice; but for these a clearly intelligible perspective drawing cannot be made except in very simple cases. We shall therefore in this chapter, concerned with the technique of stereographic projection, confine ourselves to morphological examples; in later chapters we shall apply the stereographic projection in other and more significant fields.

The actual shape of a crystal of a given substance will depend on the conditions in which it grew. It is only in the rare circumstances of ideal conditions that a crystal will grow as a regular polyhedron or with equivalent faces equally developed: thus two crystals of the same substance may have quite different external appearances (Fig 2.1(a)). But crystal faces must be parallel to lattice planes, therefore the angle between any particular pair of faces will be the same in every crystal of the substance that exhibits those two faces. This is the *law of constancy of angle* discovered by Nicolaus Steno in 1669 and stated formally as: *In all crystals of the same substance the angles between corresponding faces have a constant value.* It is adequate therefore, if we are only concerned with angular relationships, to represent a crystal face by its normal drawn from an origin within the crystal. The normal to a given face may or may not intersect the face (Fig 2.1(b)), but this distinction is of no consequence. The whole crystal can thus be represented by a bundle of normals drawn from the origin, one to each face; angular relationships between the faces are preserved in the bundle of normals and the variation of aspect between different crystals of the same substance resulting from uneven development of faces is eradicated. It becomes convenient to quote the angle between the corresponding normals when the angle between two crystal faces is required; this is of course the supplement of the angle within the crystal which in single crystals is commonly a salient angle. Thus in Fig 2.1(b) the angle between (111) and $(1\bar{1}1)$, usually written as $(111){:}(1\bar{1}1)$, is quoted as 70° 32′; the angle between the face normals is 70° 32′ and the angle between the crystal faces is 180° − 70° 32′.[1]

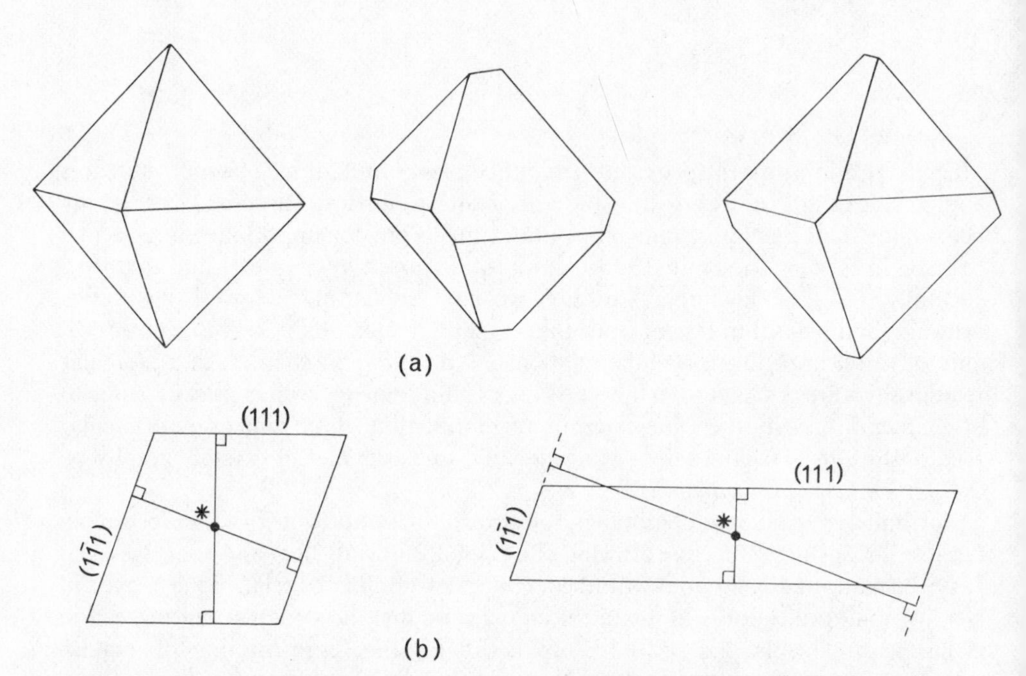

Fig 2.1 The law of constancy of angle. (a) shows a regular octahedron and two distorted octahedra: in the central drawing one pair of parallel faces is relatively overdeveloped and in the right-hand drawing the same pair is underdeveloped. (b) shows sections normal to (111) and $(1\bar{1}1)$ through a regular octahedron and through an octahedron with one pair of parallel faces overdeveloped; in each case the angle marked * is 70°32′.

The stereographic projection cannot be drawn immediately from the bundle of face normals; an intermediate stage of *spherical projection* is necessary. The spherical projection of a crystal is made on a sphere circumscribing the crystal and having its centre at the origin from which the face normals are drawn. On the surface of the sphere the point of intersection of every face normal is marked. These points of intersection are the *face poles* (usually abbreviated to *poles*) of the spherical projection. Fig 2.2(b) shows the spherical projection of the crystal drawn in perspective in Fig 2.2(a). In spherical projection crystal faces are represented by

[1] The angle between two planes is measured as the angle between their lines of intersection with a plane parallel to their normals.

points on the surface of the sphere of projection. It is of course not easy to draw an accurate picture of this three-dimensional projection in two dimensions and so the stereographic projection is introduced to project the spherical projection on to a plane.

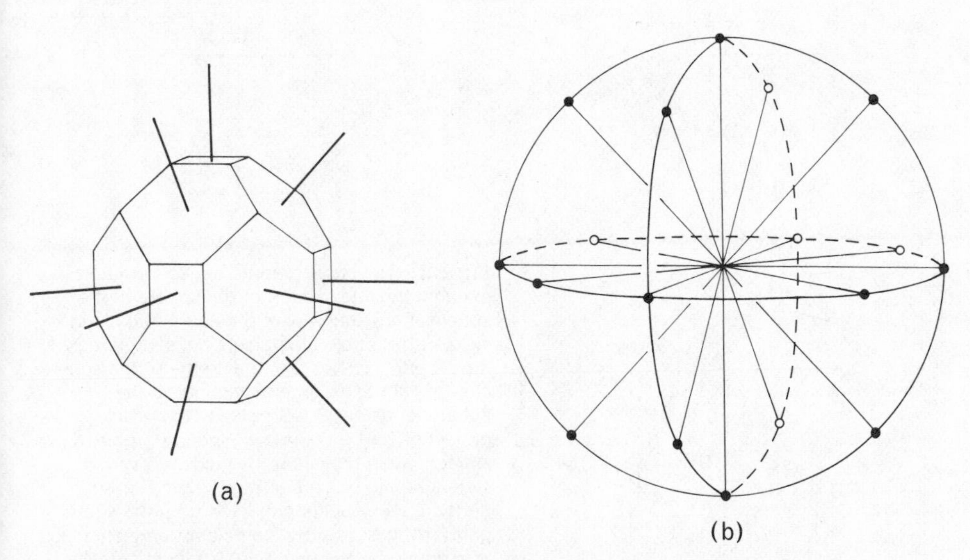

Fig 2.2 The spherical projection. (a) is a drawing of a crystal with face normals radiating from an origin within the crystal shown as bold lines. (b) shows the intersection of the resultant bundle of face normals with a sphere centred at the same origin: the array of intersections (shown as solid circles if on the front of the sphere, open circles if behind) is the spherical projection of the crystal drawn in (a).

The essentials of the stereographic projection are: a point of projection on the surface of the sphere and a plane of projection perpendicular to the diameter of the sphere passing through the point of projection. It is usual to select as the plane of projection the plane passing through the centre of the sphere and we shall always so place the plane of projection. By analogy with the earth the plane of projection is called the *equatorial plane* (Fig 2.3(a)), the projection point S is called the *south pole*, and the opposite end N of the diameter through S is called the *north pole.* The equatorial plane intersects the sphere in a circle known as the *primitive circle.* The line SP joining the pole P to the projection point S intersects the projection plane in p, which is the stereographic projection of the pole P. The stereographic projection of any pole lying above the equatorial plane, that is in the northern hemisphere, falls inside the primitive circle; the projection of a pole lying on the primitive circle is coincident with the pole itself (Fig 2.3(b)); the projection of a pole lying below the equatorial plane, that is in the southern hemisphere, falls outside the primitive circle (Fig 2.3(c)); and the projection of the south pole, the extreme case, is at infinity.

Poles lying in the southern hemisphere can more conveniently be projected within the primitive circle by taking the north pole as the projection point. Poles so projected from the north pole are distinguished on the stereogram from those projected from the south pole by representing the former as open circles and the latter as smaller solid circles.

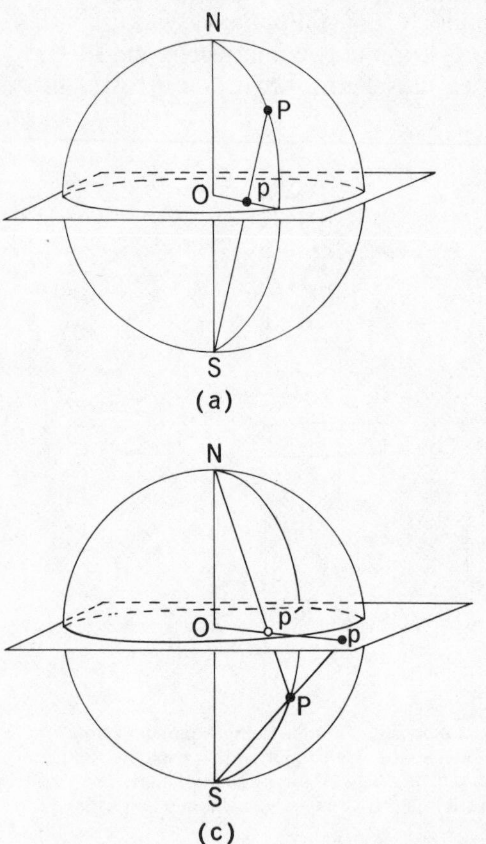

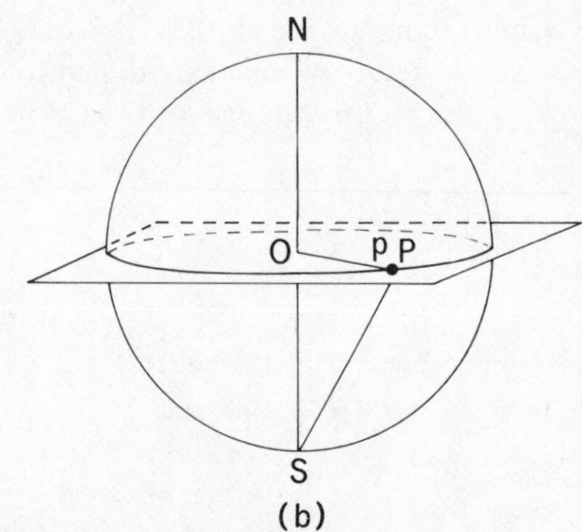

Fig 2.3 The stereographic projection. In each diagram the intersection of the sphere of the spherical projection with the equatorial plane is shown; this circle of intersection is known as the *primitive circle*; O is the centre of the sphere and the line SON is perpendicular to the equatorial plane; P is a pole in spherical projection and p is the corresponding pole in stereographic projection. (a) illustrates the stereographic projection p of a pole P in the northern hemisphere projected from the south pole. (b) illustrates the coincident projection p of a pole P on the primitive. (c) illustrates the projection of a pole P in the southern hemisphere projected from the south pole to p, lying outside the primitive, or projected from the north pole to p′ (represented by an open circle) within the primitive.

Figure 2.4(a) shows a central section through the sphere containing the projection point S and a pole P. The normal to the crystal face represented by the pole P is the radius OP. If this normal makes an angle ρ with the north–south diameter of the sphere, then $\widehat{NOP} = \rho$ and $\widehat{NSP} = \rho/2$. Therefore the distance of the projection p of the pole P from the centre of the primitive circle is $r \tan \rho/2$, where r is the radius of of the sphere and of the primitive circle. If the pole P′ lies in the southern hemisphere $180^\circ > \rho > 90^\circ$ and if the north pole is taken as the projection point, then $Op' = r \cot \rho/2$. If the crystal has two faces whose normals OP and OP′ are coplanar with and equally inclined to the north–south diameter NOS so that $\rho = 180^\circ - \rho'$, then if both normals lie on the same side of NOS their poles will project within the primitive circle at the same point; such pairs of poles are represented by an open circle concentric with a smaller solid circle in Figs 2.4(c) and (d). The normals to the pair of parallel faces (hkl) and $(\bar{h}\bar{k}\bar{l})$ are represented in spherical projection by the poles P and P_0 which lie at opposite ends of a diameter of the

Fig 2.4 Construction of the stereographic projection p of a pole P. In (a) the pole P lies in the upper hemisphere and its projection from S is p; in (b) the pole P′ lies in the lower hemisphere and its projection from N is p′. In (c) the projection (p and p′) of the poles P and P′, for which $\rho + \rho' = 180^\circ$, respectively from S and N is shown: the resultant stereographic projection with p represented by a solid circle and p′ by a concentric open circle is illustrated in (d). In (e) the projection (p and p_0) of the pole P and its opposite P_0 respectively from S and N is shown: the resultant stereographic projection with p represented by a solid circle and p_0 by an open circle with $p_0N = Np$ is illustrated in (f).

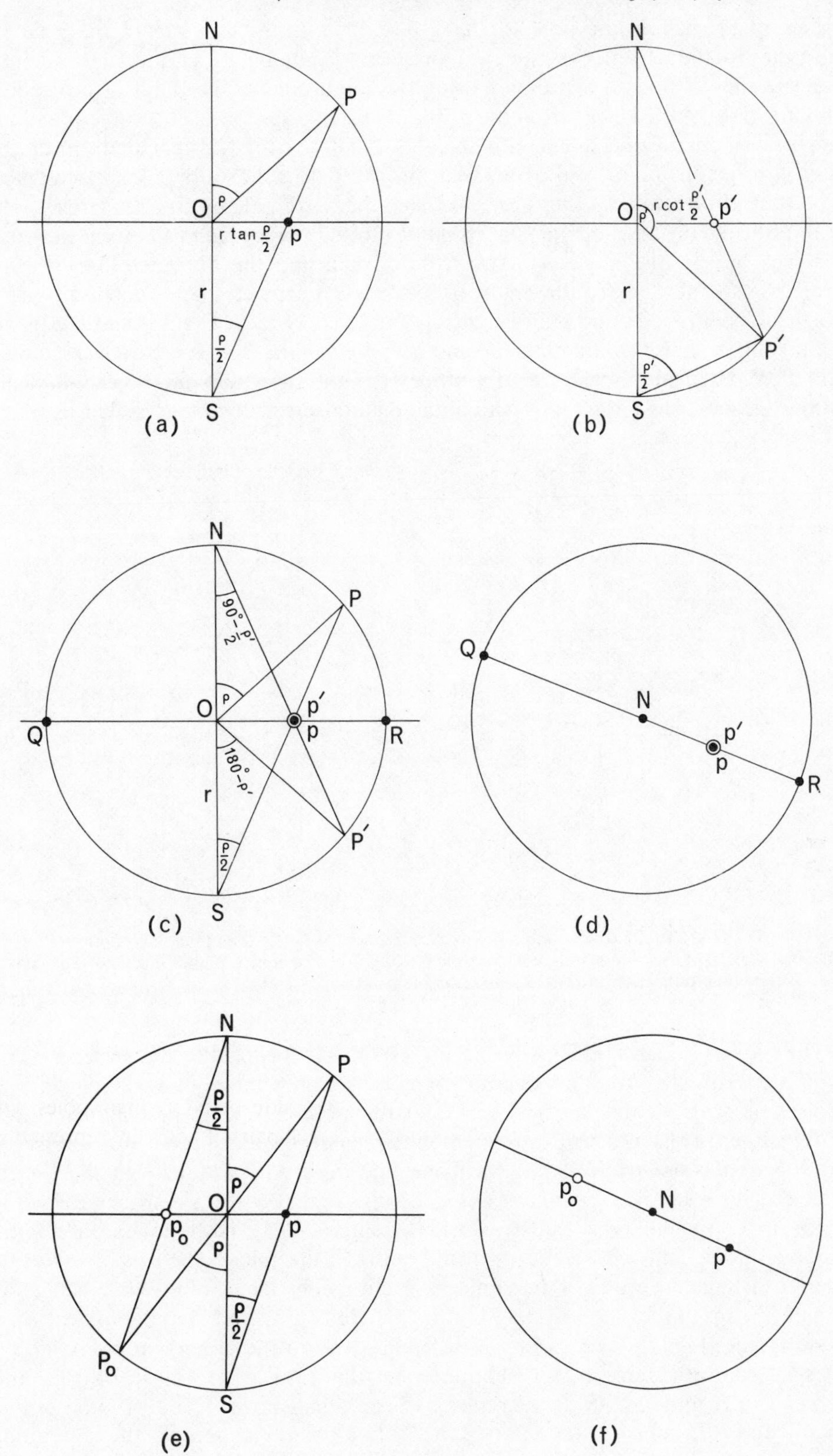
N
P
O
r tan ρ/2
p
r
ρ/2
S
(a)
N
O
r cot ρ′/2
p′
r
P′
ρ′/2
S
(b)
N
P
90°−ρ′/2
O
p′
p
Q
R
180°−ρ′
r
P′
ρ/2
S
(c)
Q
N
p′
p
R
(d)
N
P
ρ/2
ρ
O
p
ρ
ρ/2
P_o
S
(e)
p_o
N
p
(f)

sphere of projection, the pole P_0 being described as the *opposite* of the pole P. Inspection of the section of the sphere of projection containing P, S, and P_0 (Fig 2.4(e)) indicates that if P is projected as p using the south pole as the point of projection and if P_0 is projected as p_0 from the north pole, then $Op = Op_0$. The pole p and its opposite p_0 thus lie on a diameter pNp_0 of the primitive (Fig 2.4(f)), equidistant from the centre but on opposite sides; one being projected from the south pole is represented by a solid circle and the other projected from the north pole by an open circle.

In principle the faces of a crystal could be plotted stereographically by measuring for each face (i) the angle ρ between its normal and the diameter through the projection point, and (ii) the angle ϕ between a reference plane (defined by the north and south poles and the reference point R, i.e. NRS in Fig 2.5) and the plane containing the north–south diameter and the pole P of the face. It is however simpler and more practical to make use of a property of the stereographic projection, dealt with in the next few paragraphs, and of the relationships between crystal faces when plotting a stereogram of a crystal.

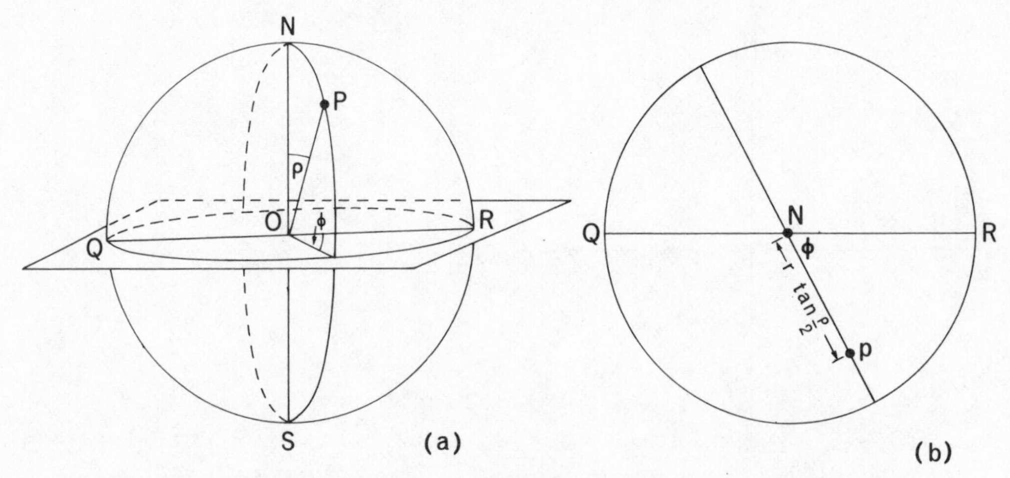

Fig 2.5 (a) serves to define the angle ρ between the normal OP to the plane (hkl) and the diameter NOS, and the angle ϕ between the plane NOP and a reference plane NRS. (b) represents the stereographic projection of P as p with coordinates $r \tan \rho/2$ and ϕ.

The property that is perhaps primarily responsible for making the stereographic projection attractive to crystallographers is that planes intersecting the sphere of projection project either as circles or as straight lines. Two types of planes are distinguished; a plane passing through the centre of the sphere of projection intersects the sphere in a *great circle*, while a plane that does not pass through the centre intersects the sphere in a *small circle* (Fig 2.6(a)). A great circle is a special case of a small circle. A small circle can be considered alternatively as the intersection with the sphere of a cone whose apex is at the centre of the sphere, the axis of the cone passing through the centre of the small circle and the semiangle of the cone being the angular radius of the small circle. The small circle is projected stereographically by drawing lines joining every point on the small circle to the projection point. These lines lie on the surface of an oblique cone (that is a cone whose base is not perpendicular to its axis) with a circular base, the small circle. The projection of the small circle (Fig 2.6(b)) is the intersection of this oblique cone with the equatorial

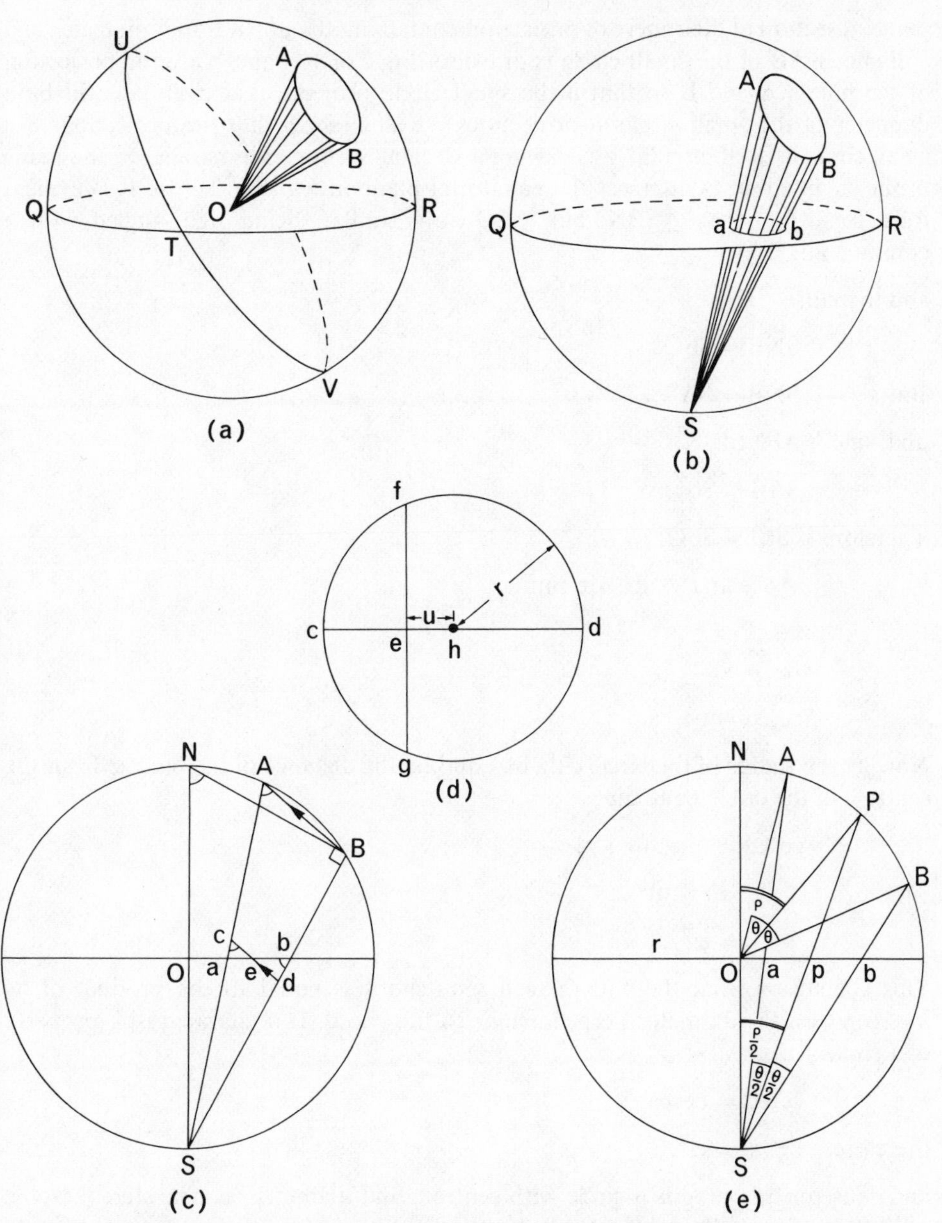

Fig 2.6 Stereographic projection of a small circle. In (a) QTR and UTV are examples of great circles and the cone AOB intersects the sphere in the small circle AB. In (b) the oblique cone ASB has a circular base, the small circle AB, and its apex is at the point of projection S; it intersects the equatorial plane QR in the figure ab. (c) is the section of the sphere of projection perpendicular to the equatorial plane and containing the diameter AB of the small circle; the stereographic projection of AB is ab. The circular section, shown in (d), of the oblique cone ASB is drawn parallel to the small circle AB and intersects the equatorial plane in feg. (e) is the same section of the sphere as (c) drawn to show the axis OP of the right cone AOB and the projection p of the centre of the small circle AB; in general ap ≠ pb.

plane. A section of the sphere of projection containing the north–south diameter and a diameter AB of the small circle is drawn in Fig 2.6(c): a and b are the projection of the points A and B so that if the small circle projects as a circle ab will be a diameter of the small circle in projection. We shall show that the projection of a small circle is itself circular by drawing a circular section cfdg parallel to the plane of the small circle to intersect the equatorial plane in feg (Fig 2.6(d)). It is evident from Fig 2.6(c) that $\triangle$SBN and $\triangle$SOb are similar, being right-angled with a common angle,

and therefore

$$\widehat{\mathrm{SNB}} = \widehat{\mathrm{SbO}}.$$

But $\widehat{\mathrm{SNB}} = \widehat{\mathrm{SAB}}$

and, since $\mathrm{AB} \parallel \mathrm{cd}$,

$$\widehat{\mathrm{SAB}} = \widehat{\mathrm{Scd}}$$

Therefore $\widehat{\mathrm{Scd}} = \widehat{\mathrm{SbO}}$,

$\triangle$ace and $\triangle$dbe are similar,

and $$\frac{\mathrm{ce}}{\mathrm{be}} = \frac{\mathrm{ea}}{\mathrm{ed}};$$

i.e. $$\mathrm{ce.ed} = \mathrm{be.ea}.$$

Now let the radius of the circle cfdg be r and let the distance of the point e from the centre h of the circle be u, then

$$\begin{aligned}\mathrm{ce.ed} &= (r-\mathrm{u})(r+\mathrm{u})\\ &= r^2-\mathrm{u}^2\\ &= \mathrm{ef}^2.\end{aligned}$$

This conclusion, that the square of a semi-chord is equal to the product of its intercepts on the diameter perpendicular to the chord, is a characteristic property of a circle.

But $\mathrm{ce.ed} = \mathrm{be.ea}$

therefore $\mathrm{be.ea} = \mathrm{ef}^2$,

and consequently afbg is a circle with centre h and diameter ba. The stereographic projection of a small circle is therefore a circle.

The centre P of a small circle (Fig 2.6(e)) will not in general project at the geometrical centre of the projected small circle, for if the centre of the small circle is such that $\widehat{\mathrm{NOP}} = \rho$ and if the angular radius of the small circle (that is, the semiangle of the right cone whose intersection with the sphere of projection is the small circle) is θ, then

$$\mathrm{ap} = \mathrm{Op}-\mathrm{Oa} = r\tan\frac{\rho}{2} - r\tan\frac{\rho-\theta}{2}$$

and $$\mathrm{pb} = \mathrm{Ob}-\mathrm{Op} = r\tan\frac{\rho+\theta}{2} - r\tan\frac{\rho}{2}.$$

In general therefore ap ≠ pb, but ap = pb when $\rho = 0$. The projection of the centre of a small circle coincides with the geometrical centre of the projected small circle only when the centre of the small circle lies at the point of projection or its opposite, i.e. at S or N.

It is obvious from Fig 2.7(a) that a small circle which passes through the south pole projects as a straight line because the lines joining every point on the circle to the point of projection are then coplanar with the small circle.

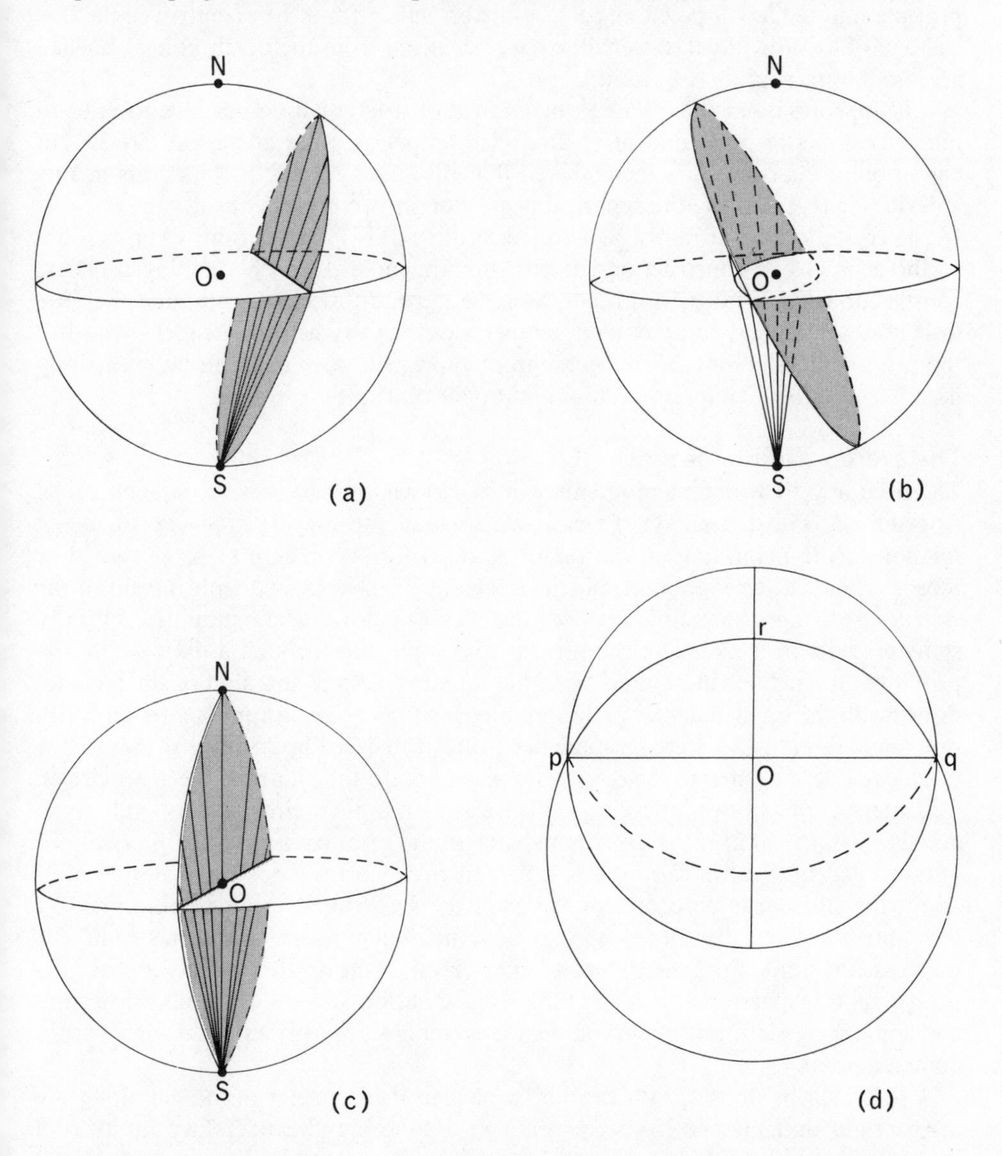

Fig 2.7 (a) shows that a small circle passing through the point of projection S projects as a straight line. (b) shows the projection of a great circle as an arc which intersects the primitive at the ends of a diameter of the primitive circle. (c) illustrates the special case of a great circle passing through the point of projection S; its projection is a diameter of the primitive. (d) is the stereographic projection of a great circle with S as the point of projection; it is customary to change the point of projection from S to N for that part of the great circle lying outside the primitive. Circles projected from S are shown conventionally as solid arcs and those projected from N as broken arcs.

Great circles, which are of course special cases of small circles, likewise project as circles; but since all great circles are coplanar with the centre of the sphere of projection they intersect the primitive circle at the opposite ends of a diameter of it (Fig 2.7(b)). If a great circle passes through the south pole (Fig 2.7(c)) it will project as a diameter of the primitive because the great circle is then coplanar with its projection. When a complete great circle is projected with the south pole as projection point, part of it must project outside the primitive circle. By changing the point of projection to the north pole this part can be brought within the primitive circle. It is customary to show great, or small, circles projected from the north pole as dashed arcs on the stereogram (Fig 2.7(d)).

It is appropriate at this point to mention that it is common practice to refer to the projections of poles and of great circles simply as poles and great circles. No confusion need arise if it is remembered that the stereographic projection is merely a device for representing the spherical projection in two dimensions.

The crystallographer rarely needs an accurately drawn stereogram; when he does so the geometrical constructions detailed in Appendix A can be employed. These constructions are dealt with at length in the appendix because, although they are little used in practice, study of their geometry can greatly help the reader towards a thorough understanding of the stereographic projection, which will be extensively used in subsequent chapters without additional explanation.

The stereographic net

The accuracy to which stereograms can be drawn by using the constructions of Appendix A will depend on the care with which the constructions are made in relation to the magnitude of the radius selected for the sphere of projection. For most purposes a stereogram of sufficient accuracy can be drawn with the aid of the *stereographic net*, sometimes called the Wulff net. A stereographic net is a stereographic projection of a set of great circles passing through a diameter of the primitive and inclined at 2° intervals to the equatorial plane and a set of small circles drawn with the same diameter of the primitive as their stereographic centre and with radii at 2° intervals. A stereographic net is illustrated in Fig 2.8. The planes of the small circles are normal to the equatorial plane and to the planes of the great circles. The intersections of two adjacent small circles with a great circle corresponds to an angular distance of 2° measured in the plane of the great circle and the intersections of two adjacent great circles on a given small circle corresponds to a rotation of 2° about the stereographic centre of the small circle. Printed stereographic nets are available commercially, either circular or semicircular and with various radii, the most useful radius for general work being about 2·5 in or 10 cm. Nets are usually printed on transparent paper so that they can be placed on a stereographic projection and the appropriate poles marked on the stereogram by pricking through with a compass point.

A pole may be plotted with the net by placing the diameter of the net along the appropriate diameter of the stereogram and counting the necessary number of intersections. The angle between two poles may be measured by rotating the net until both poles lie on the same great circle and then counting the number of intersections of small circles between them, each intersection corresponding to 2°. A rapid and convenient method of drawing the projection of a great circle which passes through two poles is to rotate the stereographic net until the two poles lie on the same great circle and to prick through the points at which the great circle

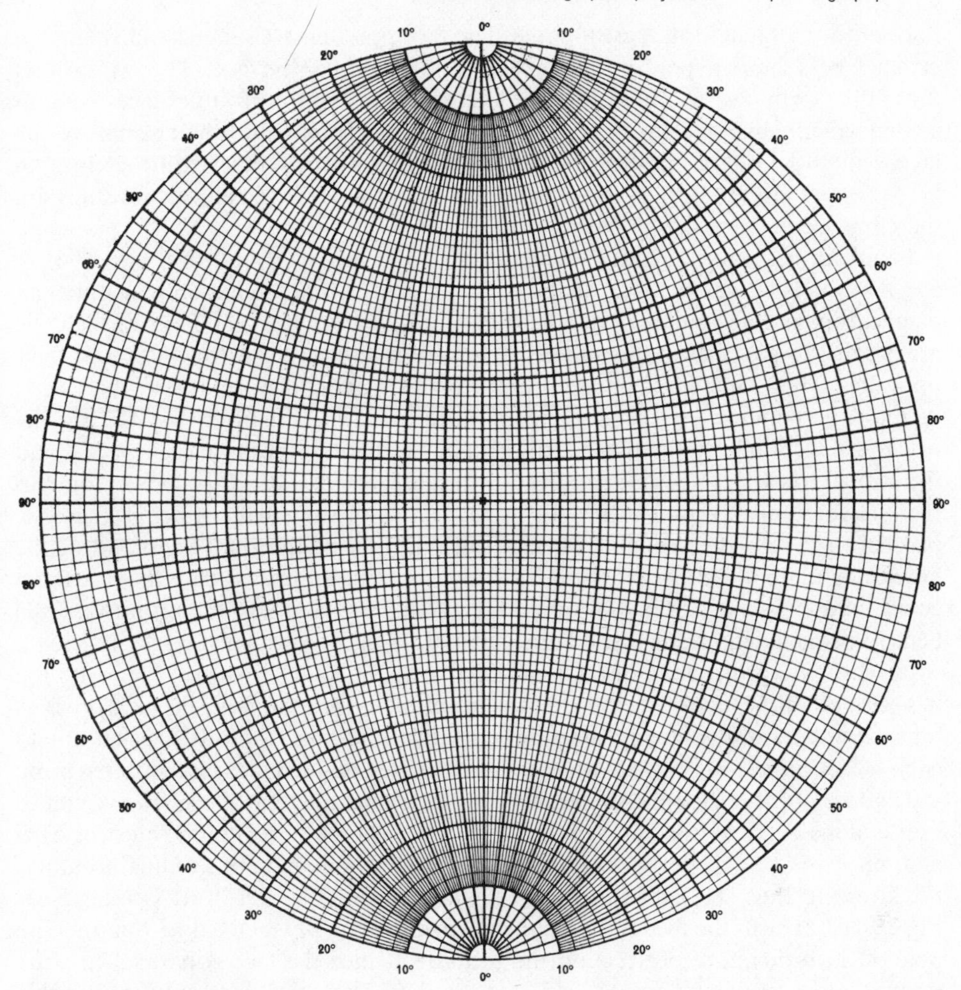

Fig 2.8 The stereographic net.

intersects the primitive. The required arc of a circle can then be drawn in using a device such as a *spring bow compass.* This is simply a thin strip of steel bent at each end so as to provide two finger holds, which will bend to give a reasonable approximation to an arc of a circle.

With full circle stereographic nets printed on cardboard stereograms can be drawn on tracing paper pinned through the centre of the net so that it can rotate freely relative to the net. This provides easily the most convenient means of drawing stereograms of moderate accuracy.

Use of the stereographic projection in crystallography

The stereographic projection is of value in the study of solids because it enables angular relationships between planes and directions to be represented. A plane is usually represented by its normal, although it is occasionally more convenient to represent it by the great circle to which it is parallel (chapter 12). It is necessary at this point to introduce the definition of the *pole of a great circle* as the direction normal to the plane of the great circle. The face pole representing the normal to a crystal face is then the pole of the great circle representing the face. (The two distinct uses of the term *pole* do not in practice lead to confusion.) A group of faces lying in a zone will all have one direction, the zone axis, in common and the normals to the faces will all lie in the plane normal to the zone axis. Their poles will therefore all lie on a great circle whose pole is the zone axis of the zone. Zonal relationships are therefore used extensively in plotting stereograms.

Figure 2.9(a) shows a crystal of copper, whose unit cell is a cube ($a = b = c = 3{\cdot}61$ Å, $\alpha = \beta = \gamma = 90°$). The external shape of the crystal is a combination of a cube, an octahedron, and a rhombic dodecahedron (Fig 2.9(b), (c), (d)). The faces of the cube are parallel to the faces of the unit cell and have indices (100), (010), (001), and their opposites. The face (100) is parallel to the y and z axes. The normal to the face is therefore perpendicular to y and z and, because γ and β are right-angles, it is parallel to the x axis. Similarly the normal to (010) is parallel to y because $\alpha = \gamma = 90°$ and the normal to (001) is likewise parallel to z because $\alpha = \beta = 90°$. It is conventional to plot stereograms of crystals with $+z$ in the centre of the northern hemisphere, i.e. parallel to the north pole, and the face normal (010) at the right-hand end of the horizontal diameter of the primitive as drawn. A stereogram showing the faces of a cube is drawn in Fig 2.9(e). (001) and (00$\bar{1}$) plot in the centre of the primitive, (001) being represented by a small solid circle and (00$\bar{1}$) by a larger open circle. In such a case, when the projections of (hkl) and ($hk\bar{l}$) are coincident it is usual to write the indices (hkl) of the face in the northern hemisphere beside the pole, it being obvious that the indices of the face in the southern hemisphere are ($hk\bar{l}$). It is usual not to enclose face indices in brackets when they are written beside poles on the stereogram.

The faces of the rhombic dodecahedron are all parallel to one of the reference axes and make equal intercepts on the other two. Therefore one index must be zero and, since $a = b = c$, the other two indices must be equal and are set equal to unity; the faces are thus (110), (101), (011), ($\bar{1}$10), and so on (Fig 2.9(d)). It is clear from Fig 2.9(f) that both the face (110) and its normal are equally inclined to x and y. The pole of (110) therefore projects on the primitive (since the face is parallel to z, its normal is perpendicular to z) at 45° to x and y. The other faces of the rhombic dodecahedron plot in similar positions (Fig 2.9(g)).

The faces of the octahedron make equal intercepts on the x, y, and z axes and therefore have indices such as (111), ($\bar{1}$11), and so on. It is clear from Fig 2.9(a) that the face (111) lies in the zone containing (001) and (110), since the edge of intersection of (001) and (111) is visibly parallel to the edge of intersection of (111) and (110). Therefore the normal to the face (111) lies in the plane containing the normals to (001) and (110). The face (111) also lies in the zones [(100), (011)] and [(010), (101)]; its position can be plotted on the stereogram by drawing the great circles representing these zones and marking their point of intersection. Only two such great circles are needed to plot (111) but in Fig 2.9(h) the three zones mentioned above have been drawn to emphasize the zonal relationships between the faces.

Figure 2.9(h) is a stereogram of the crystal of copper shown in Fig 2.9(a) in which the great circles necessary to show the zonal relationships between all the faces are drawn.

We turn now to another example to illustrate the use of the stereographic projection: the determination of the angle between (100) and (311) in a crystal of barium sulphate, the unit cell of which has dimensions $a = 8{\cdot}85$ Å, $b = 5{\cdot}44$ Å, $c = 7{\cdot}13$ Å, and $\alpha = \beta = \gamma = 90°$. The pole of (100) can be plotted directly (Fig 2.10(a));

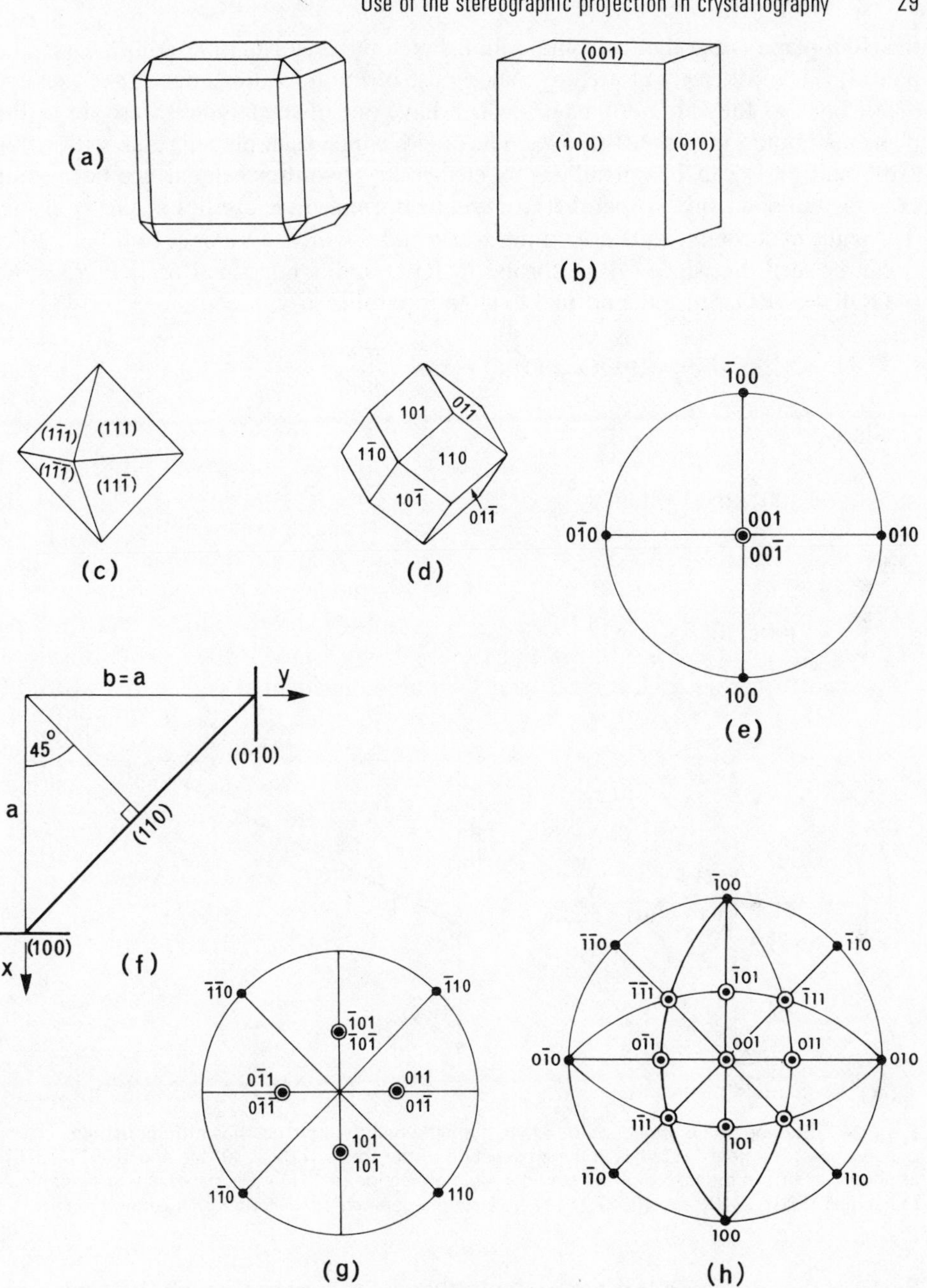

Fig 2.9 Stereographic projection of a cubic crystal. Copper forms crystals (a) which may be combinations of a cube (b), an octahedron (c), and a rhombic dodecahedron (d). A stereogram of the faces of a cube is shown in (e). The face (110) and its normal are equally inclined (f) to the *x* and *y* axes so that this and the other faces of the rhombic dodecahedron give rise to the stereographic projection (g). The projection of the octahedron faces is achieved by plotting intersecting zones with the aid of a stereographic net. The stereogram of the crystal drawn in (a) with all faces plotted but only those in the upper hemisphere indexed is shown as (h).

the (100) plane is parallel to y and z and therefore its normal is parallel to x. The pole (311) is most easily plotted by making use of zonal relationships. If the axes are orthogonal, as they are here, planes which have one of their indices zero lie in the plane containing the other two axes. The angles which such planes make with (100), (010), and (001) can be calculated by elementary two-dimensional geometry. For example the plane ($hk0$) is parallel to z and its normal therefore lies in the xy plane. The plane makes intercepts a/h on the x axis and b/k on the y axis. From Fig 2.10(b) it can be seen that since OP is normal to RPQ and x is normal to y, $\widehat{POQ} = 90° - \widehat{OQR} = \widehat{ORQ}$. Since the normal to (100) is parallel to x,

$$(100):(hk0) = \widehat{POQ} = \widehat{ORQ} = \tan^{-1}\frac{a/h}{b/k}. \tag{1}$$

Similarly,

$$(001):(h0l) = \tan^{-1}\frac{c/l}{a/h} \tag{2}$$

and

$$(001):(0kl) = \tan^{-1}\frac{c/l}{b/k}. \tag{3}$$

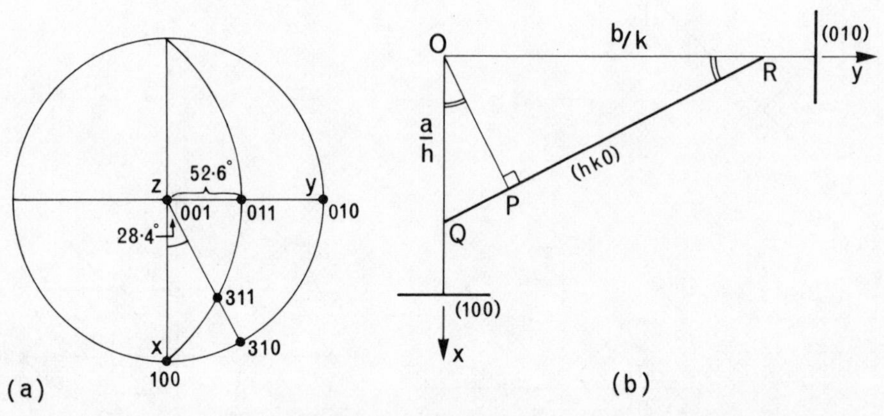

Fig 2.10 Stereographic projection of a face normal when the unit-cell has orthogonal but unequal axes. The pole (311) lies at the intersection of the zones [(100), (011)] and [(001), (310)] as shown in (a). In order to plot the stereographic projection of (311) it is necessary to determine the angles (100):(310) and (001):(011) by a simple geometrical calculation illustrated in (b).

By use of the addition rule it can be shown that in the zone containing (100) and (311) the plane with indices of the type ($0kl$) is (011). Therefore from equation (3) $(001):(011) = \tan^{-1}(7{\cdot}13)/(5{\cdot}44) = 52{\cdot}65°$. The zone [(100), (011)] can therefore be plotted by first plotting (011) and then drawing the great circle containing the two poles (100) and (011). The position of (311) can be found by drawing a second zone on which (311) must lie. A zone which also contains (001) projects as a diameter of the primitive and is therefore easy to construct. By use of the addition rule the face of the type ($hk0$) which lies in the zone [(001), (311)] is found to be (310). From equation (1) the angle $(100):(310) = \tan^{-1}[8{\cdot}85/(3 \times 5{\cdot}44)] = 28{\cdot}47°$. Therefore (310) can be plotted and the zone [(001), (310)] drawn on the stereogram. (311) lies at the intersection of the zones [(100), (011)] and [(001), (310)]. The angle (100):(311) can now be measured with the stereographic net or a more accurate value may be obtained by calculation from the stereogram (chapter 5).

Measurements occasionally have to be made of the angles between the faces actually developed on a crystal. Before the discovery of the diffraction of X-rays by crystals this was the principal method of studying crystals but its importance has declined over the past fifty years as the techniques for studying the internal structure of crystals have become increasingly available. However measurement of the angles between faces remains useful for determining the orientation of a crystal of known unit-cell dimensions prior to X-ray study. Two simple devices for the measurement of interfacial angles are described in Appendix B.

The gnomonic projection

The gnomonic projection (Fig 2.11), in which the centre of the projection sphere is used as projection point and the tangent plane at the north pole is the plane of projection, has certain limited uses in crystallography. Its advantage is that all zones project as straight lines because every great circle passes through the projection

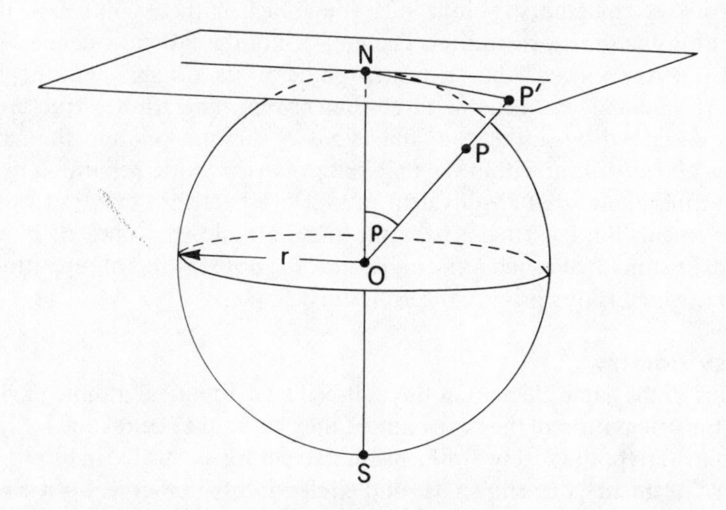

Fig 2.11 The gnomonic projection. The point of projection is the centre of the sphere and the plane of projection is tangential to the sphere.

point. Its main disadvantage is that a pole at an angle ρ from N projects at a distance $r \tan \rho$ from N and therefore only a small part of the surface of the projection sphere can be plotted within a manageable area. Poles lying on the equatorial plane project to infinity. Another disadvantage is that small circles do not project as circles.

3
Crystal symmetry

We have already explored in chapter 1 the description of the regular arrangement in space of the atoms in a crystalline solid in terms of the lattice concept. The lattice is a regular array of imaginary points such that each lattice point has the same environment in the same orientation. The unit-cell of the lattice is defined in such a manner that every unit-cell has a lattice point at its origin[1] and the lattice is produced by stacking unit-cells in three dimensions. The whole structure is then completely described by stating the dimensions of the unit-cell and the nature and coordinates of every atom within the unit-cell. Any two atoms separated by a lattice translation must therefore be equivalent in every respect: not only must they be of the same element, but each must have the same atomic environment in the same orientation. Distinct from such *lattice repetition* is another kind of repetition known as *symmetry*, which is our prime concern in this chapter.

Axes of symmetry

When atoms of the same element in the unit-cell have identical atomic environment except for the orientation of the environment they are said to be related by symmetry. For example the hypothetical two-dimensional structure illustrated in Fig 3.1 contains atoms of two elements, one shown as solid circles and the other as open circles. The lattice translations are such that all the atoms labelled A are equivalent. The atoms of the same element labelled B are likewise all equivalent to one another and have the same environment as those labelled A; the only difference is that the environment of an atom on a B site has to be rotated through 180° to bring it into the same orientation as the environment of an atom on an A site. Moreover if the whole structure is rotated through 180° about any point equivalent to O_1, then the rotated structure will be coincident with the structure as shown. There thus exists a *rotation axis of symmetry* perpendicular to the plane of the structure through O_1 and similar axes through all points related to O_1 by lattice translation, i.e. O_2, O_3, O_4, etc. The rotation axes in this example yield coincidence by rotation through 180° or $2\pi/n$ radians where $n = 2$; such axes are described as twofold or *diad* axes of symmetry.

Before going on to define the various kinds of symmetry axis it is convenient to distinguish between two types of symmetry, that of a finite body and that of an

[1] This statement requires amplification in the case of non-primitive lattices, which will be dealt with in chapter 4.

infinite body. In the crystallographic field the symmetry displayed by a finite body is apparent in the external shapes of crystals and in the physical properties of crystals (chapter 11) such as thermal and electrical conductivity and elastic properties. The shapes of crystals grown, either naturally or in the laboratory, under ideal conditions provide the clearest means of illustrating the symmetries possible in finite bodies; we shall therefore use crystal shape in the main to illustrate this chapter. It must always be borne in mind, however, that crystals commonly grow under non-ideal conditions even in nature so that symmetry related faces are unequally developed. In such cases the symmetry of the crystal will not be immediately apparent from its external shape (Fig 2.1); in this chapter we shall confine ourselves for purposes of illustration to perfectly developed crystals. In chapter 4 we shall be concerned with the symmetry of infinite bodies as exhibited by the arrangement of atoms in crystal structures.

An *n-fold rotation axis of symmetry* is defined as a line, rotation about which produces congruent positions (i.e. positions indistinguishable from the initial position) after rotation through $2\pi/n$. The crystal therefore comes into self-coincidence n times in a complete rotation through 2π. Rotation axes are described with reference to the value of n, which must of course be integral: onefold, twofold, threefold, fourfold, fivefold, sixfold, etc, rotation axes are known as monads, diads, triads, tetrads, pentads, hexads, etc. and are denoted by the symbols 1, 2, 3, 4, 5, 6, etc.

The monad, which brings the crystal into self-coincidence after rotation through 2π, is of course trivial. Crystals displaying diads, triads, tetrads, and hexads are shown in Fig 3.2, where the graphical symbol for each type of rotation axis is also displayed. The presence of an n-fold rotation axis of symmetry implies that a given plane or direction is repeated by being rotated through $2\pi/n$ radians about the axis, the operation being repeated until the initial position is reproduced. The operation of a rotation axis of symmetry on a single pole is conveniently displayed stereographically and this is done too in Fig 3.2.

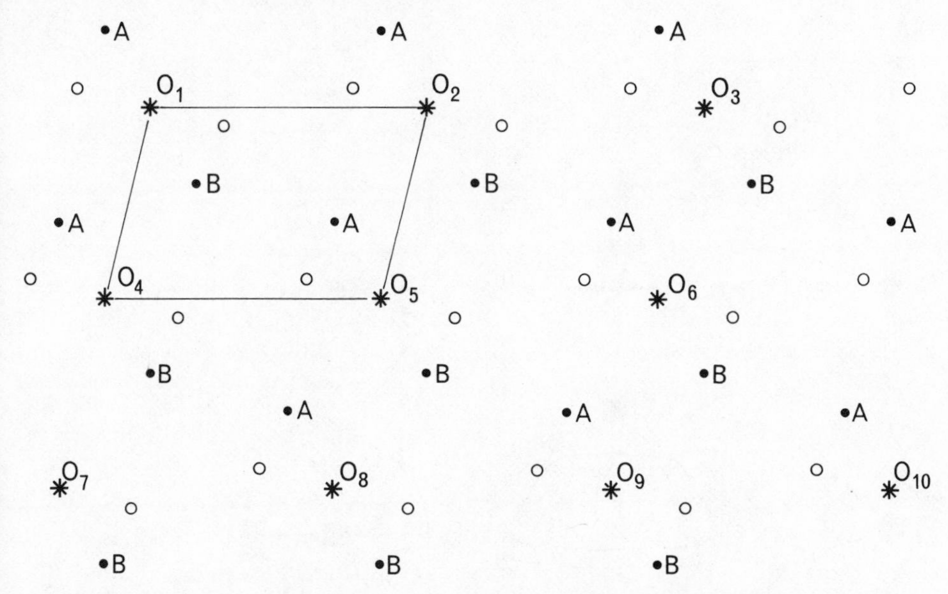

Fig 3.1 Hypothetical two-dimensional structure with identical atoms on sites A and B. Atoms of another element are represented by small open circles. The rotation diads, O_1, O_2, etc, are perpendicular to the plane of the diagram and related to one another by lattice translations. The unit-mesh is outlined.

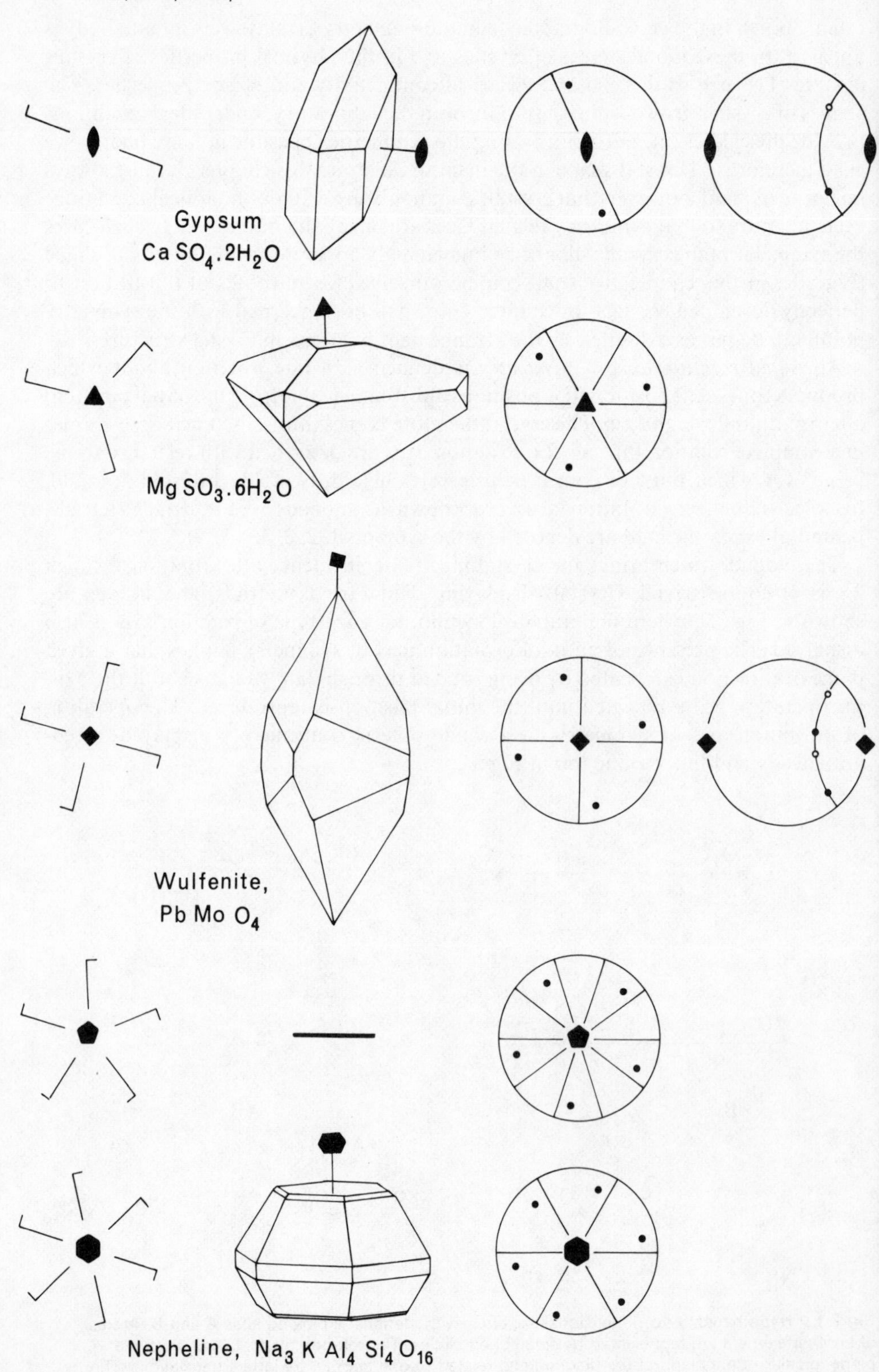
Gypsum
$CaSO_4.2H_2O$
$MgSO_3.6H_2O$
Wulfenite,
$PbMoO_4$
Nepheline, $Na_3KAl_4Si_4O_{16}$

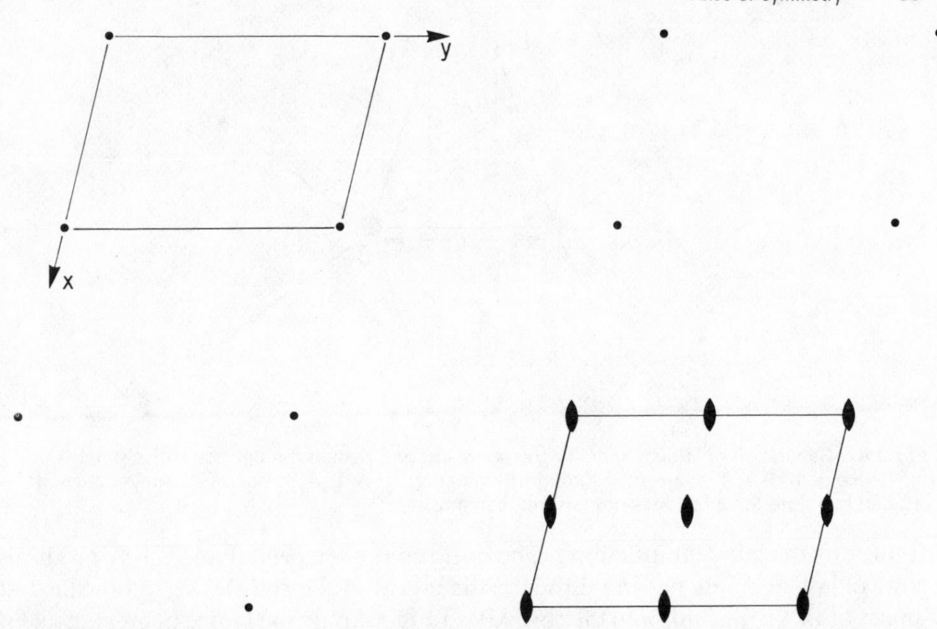

Fig 3.3 Lattice of the hypothetical two-dimensional structure shown in Fig 3.1. The unit-mesh, a general parallelogram is outlined upper left and the disposition of rotation diads on the unit-mesh is shown lower right.

That the only values of n possible for rotation axes of symmetry operative on crystalline solids are 2, 3, 4, and 6 (the monad being trivial) is a direct consequence of the regularity of atomic arrangement in a crystal. For a crystal to have an n-fold rotation axis implies that the atomic arrangement in the substance must likewise have n-fold symmetry. Consequently the shape of the unit-cell of the lattice of the structure must be consistent with the presence of an n-fold rotation axis of symmetry. In short the only rotation axes of symmetry that can operate on a crystal structure are those that can operate on a lattice. For example the hypothetical two-dimensional structure illustrated in Fig 3.1 has diad symmetry; the lattice of this structure likewise has diad symmetry (Fig 3.3) with diads through the lattice points and midway between adjacent lattice points. Comparison of Figs 3.1 and 3.3 shows that structure and lattice have identical arrangements of diads, which are separated by halved lattice translations. The example demonstrates that a diad can operate on a lattice and moreover that the operation of a diad imposes no geometrical restrictions on the lattice plane perpendicular to itself; the unit-mesh of the lattice plane perpendicular to a diad is thus usually a general parallelogram.

If a crystal structure is to have a fivefold rotation axis of symmetry, then the lattice on which the structure is based must also display pentad symmetry. If a pentad passes through one lattice point, such as A in Fig 3.4, then there must be a pentad through every other lattice point, all lattice points being by definition equivalent; in particular there has to be a pentad through the lattice point B which is separated

Fig 3.2 Rotation axes of symmetry. The left-hand column shows the operation of a diad, a triad, a tetrad, a pentad and a hexad on a point; the central column shows crystals displaying each type of axis, other than the pentad; and the right-hand column shows stereograms illustrating the operation of the rotation axes on a general pole.

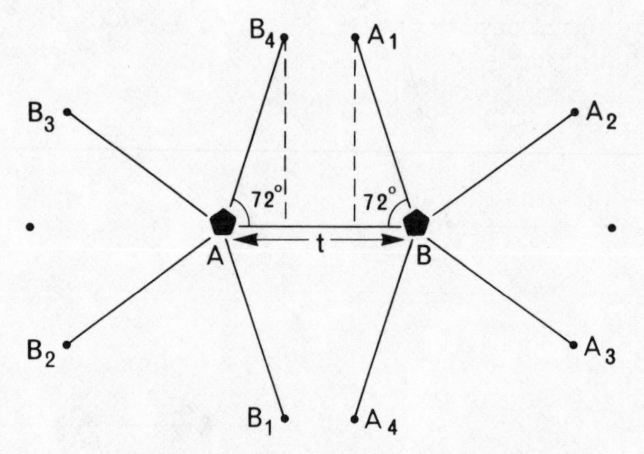

Fig 3.4 Generation of lattice points by the operation of a pentad through the lattice point A on the lattice point B and by a pentad through B operating on A. B_4A_1 is not an integral multiple of AB and therefore the array does not constitute a lattice.

from A by the lattice translation *t*. The operation of the pentad through A produces rows of lattice points passing through A identical to the row AB... and inclined at angles of $\frac{2}{5}\pi$, $\frac{4}{5}\pi$, $\frac{6}{5}\pi$, and $\frac{8}{5}\pi$ to the row AB. In Fig 3.4 only the lattice point equivalent to B in each of these rows, B_1, B_2, B_3, and B_4, is shown. Since there must likewise be a pentad through B, there must be rows of lattice points identical to the row BA passing through B and inclined to the row BA at angles of $\frac{2}{5}\pi$, $\frac{4}{5}\pi$, $\frac{6}{5}\pi$, and $\frac{8}{5}\pi$; again only the lattice points equivalent to A, that is A_1, A_2, A_3, and A_4, are shown in the figure. If the resultant array of points is to form a lattice, it must be a regular array; in particular the spacing of points on lines parallel to AB, such as B_4A_1, must be equal to *t* or some multiple thereof. It is evident from Fig 3.4 that

$$B_4A_1 = AB - AB_4 \cos 72° - BA_1 \cos 72°$$

and since

$$AB_4 = AB = BA_1$$

$$B_4A_1 = AB(1 - 2\cos 72°) = 0{\cdot}38\,AB$$

Therefore the array of points generated by pentads through adjacent lattice points is not regular and consequently not itself a lattice. In short pentads cannot be repeated on a lattice and never occur in crystals.

The argument that we have used in the special case of the pentad can simply be generalized to determine what values of *n* are permissible for an *n*-fold rotation axis

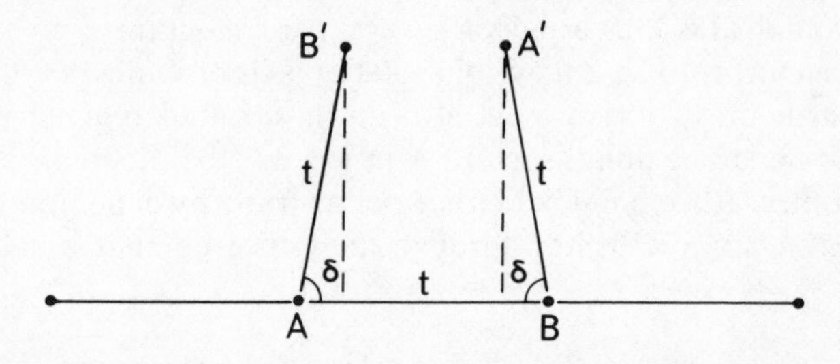

Fig 3.5 The point A′ is one of the points generated by an *n*-fold rotation axis through B operating on the point A; and the point B′ by a similar axis through A operating on B, the angle δ being equal to $2\pi/n$.

operating on a lattice. Consider a lattice with an n-fold rotation axis of symmetry. If the lattice point A is placed on an n-fold axis, then n-fold axes in the same orientation will pass through every lattice point and in particular through the adjacent lattice point B, selected so that the line AB is perpendicular to the axis of symmetry. The n-fold axis through A will repeat the lattice point B $n-1$ times in the plane perpendicular to the symmetry axis (Fig 3.5), the $(n-1)$th repetition being B′ where $AB' = AB = t$ and $\widehat{BAB'} = 2\pi/n = \delta$. Likewise the n-fold axis through B will repeat the lattice point A $n-1$ times in the same plane, the first repetition being A′ where $A'B = AB = t$ and $\widehat{ABA'} = \delta$. If the four points A, B, A′, B′ are to form a lattice, the spacing of points in the row containing A′ and B′ must be the same as that in the parallel row containing A and B; that is to say $B'A' = mt$ where m is an integer. Now

$$B'A' = AB - AB' \cos\delta - A'B \cos\delta$$

i.e.

$$B'A' = t(1-2\cos\delta)$$

Therefore the condition for formation of a lattice is

$$mt = t(1-2\cos\delta)$$

i.e.

$$\cos\delta = \frac{1-m}{2}$$

But $-1 \leqslant \cos\delta \leqslant 1$, so that the limits of the integer $1-m$ are $-2 \leqslant (1-m) \leqslant 2$, i.e. possible values of $1-m$ are -2, -1, 0, 1, 2. Possible values of n (Table 3.1) are therefore 2, 3, 4, 6, and the trivial monad with $n = 0$.

Table 3.1
Rotation axes of symmetry which can operate on a lattice

$1-m$	$\cos\delta$	δ	$n = 2\pi/\delta$	B′A′	Conventional unit mesh of lattice planes perpendicular to the axis
-2	-1	π	2	3AB	$a \neq b;\ \gamma \neq 90°$
-1	$-\frac{1}{2}$	$\pm\frac{2}{3}\pi$	3	2AB	$a = b;\ \gamma = 120°$
0	0	$\pm\frac{1}{2}\pi$	4	AB	$a = b;\ \gamma = 90°$
1	$\frac{1}{2}$	$\pm\frac{1}{3}\pi$	6	0	$a = b;\ \gamma = 120°$
2	1	0	–	–	

Note:
The symbol $\neq$ implies that equality is not required by symmetry.

Thus the rotation axes of symmetry that can operate on a lattice are restricted to diads, triads, tetrads, and hexads; these are the only rotation axes displayed by crystals. The operation of three, four, or sixfold rotation axes on a lattice plane imposes restrictions on the arrangement of lattice points in the plane. For example the presence of a triad, for which $\delta = \frac{2}{3}\pi$, restricts the arrangement of lattice points in the plane normal to the axis to the pattern shown in Fig 3.6(a). It is conventional to choose a rhombus as the unit-mesh so that $a = b$ and the interaxial angle between the x and y axes, $\gamma = 120°$. The same arrangement of lattice points is necessary in the plane perpendicular to hexads operating on the lattice. When a tetrad, with $\delta = \frac{1}{2}\pi$, is repeated on a lattice, the arrangement of lattice points in the plane normal to the axis must be as shown in Fig 3.6(b); the conventional unit-mesh is here a square with $a = b$, $\gamma = 90°$.

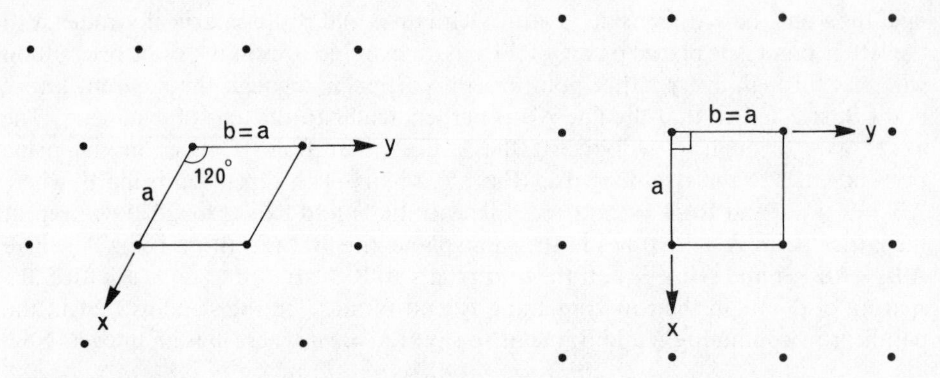

Fig 3.6 Two-dimensional lattices with (a) a triad or a hexad, (b) a tetrad perpendicular to the plane of the lattice.

Since the operation of a diad merely relates lattice points of the same row, the presence of diad symmetry places no restriction on the arrangement of lattice points in the plane perpendicular to itself; the unit-mesh is a general parallelogram with $a \neq b$, γ general. However, although symmetry places no restriction on the values of a, b, or γ, there may be fortuitous equality of a and b or γ may happen to have a special value. For example the hypothetical two-dimensional structure shown in Fig 3.7 has a square unit-mesh with $a = b$, $\gamma = 90°$, but there are no tetrads because the atomic arrangement shows only diad symmetry.

Combination of symmetry axes

A crystal may have more than one symmetry axis. When this is so the angular disposition of the axes relative to one another must be such that their operation is mutually consistent. We shall now proceed to establish all the combinations of rotation axes of symmetry that can operate on a lattice and can thus be displayed by a crystal.

It is a general principle that when two rotation axes are combined a third rotation axis is created. The principle is illustrated in Fig 3.8. The face I in Fig 3.8(a) is related to the face II and to four other faces that meet at *a* by a hexad. The face II is related to the face III by the diad which is perpendicular to the hexad and passes through the mid-point of the edge *bc*. The operation of the hexad on this diad will give rise to diads through the mid-points of the edges *de*, *fg*, *hi*, *jk*, and *lm* (*hi* and *jk* are at the back of the crystal and not shown in the figure) in the plane perpendicular to the hexad. Since the faces I and III are each equivalent to the face II they must be equivalent to each other; they are directly related by a diad perpendicular to the hexad and passing through the mid-point of the edge *mb*. This diad is inclined at 30° to the diad through the mid-point of the edge *bc*, both lying in the plane perpendicular to the hexad. In general then, the combination of a hexad and a diad perpendicular to it gives rise to five equivalent diads 60° apart in the plane perpendicular to the hexad and in addition a set of six diads in the same plane are inclined at 30° to those of the first set. Figure 3.8(b) shows a stereogram of the crystal drawn in Fig 3.8(a) and its rotation axes of symmetry.

The generation of additional rotation axes can be considered generally in terms of rotation axes A and B that respectively produce equivalence after rotation through $\delta_A = 2\pi/n_A$ and $\delta_B = 2\pi/n_B$ where n_A and n_B are integers. The operation of two such

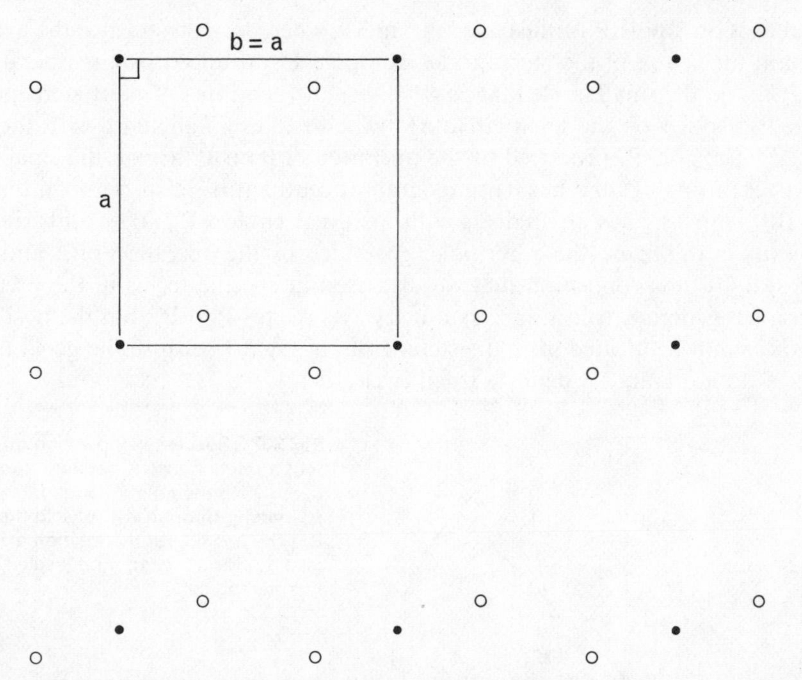

Fig 3.7 Hypothetical two-dimensional structure with two atomic types shown as solid and open circles respectively. The unit-mesh has $a = b$, $\gamma = 90°$ but the lattice has only diad symmetry.

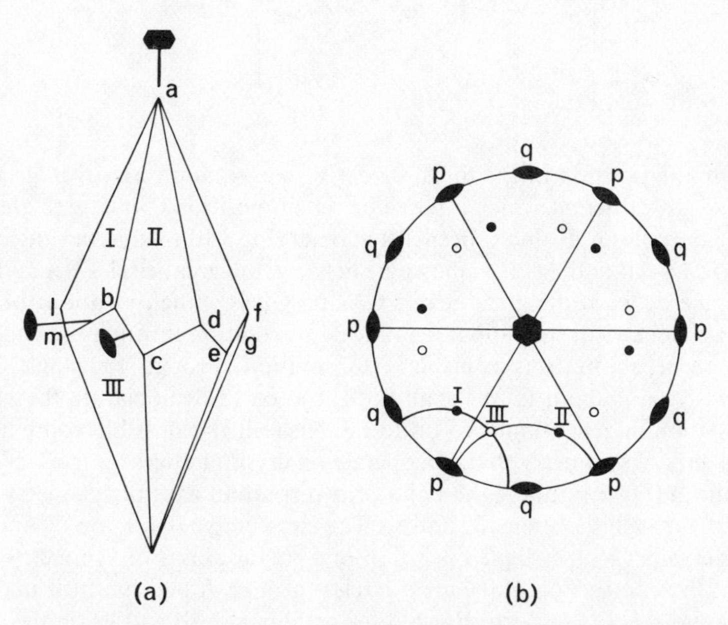

Fig 3.8 The combination of two rotation axes necessitates the presence of a third. (a) Faces I and II are related by the hexad; faces II and III are related by the diad through the edge *bc*; a diad through the edge *mb* therefore relates faces I and III. (b) A stereogram of the same crystal showing the disposition of the two sets (p and q) of six equivalent diads in the plane perpendicular to the hexad; the pole III is related to pole I by operation of a diad p and to pole II by operation of a diad q.

general axes on a pole P is illustrated in Fig 3.9, where for convenience the axis A is plotted at the centre of the stereogram. The pole P′ produced by the operation of A on P lies on the small circle that passes through P and has A as its stereographic centre; P′ also lies on the great circle AP′ which makes an angle δ_A with the great circle AP. The pole P″ produced by the operation of B on P′ lies on the small circle that passes through P′ and has its stereographic centre at B; P″ also lies on the great circle BP″ which makes an angle δ_B with the great circle BP′. (To avoid irrelevant complexity in the figure the other poles generated by the operation of A and B are not shown.) If the proposition that we discussed in specific terms in the preceding paragraph is generally true a third symmetry axis relates P to P″; but the position of this axis cannot be located from the stereogram of Fig 3.9 because the general poles P and P″ do not uniquely define a small circle.

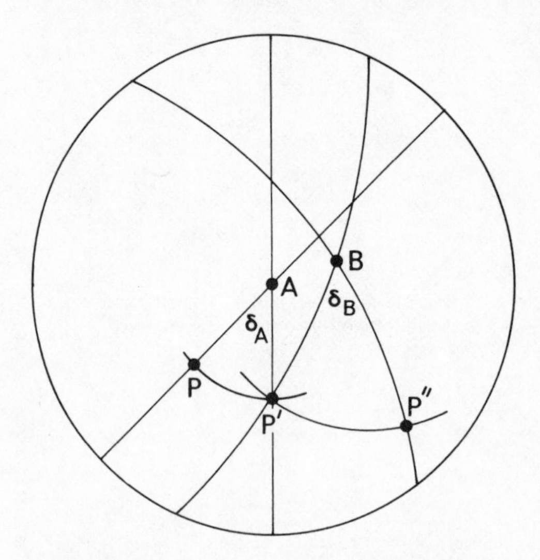

Fig 3.9 Successive operation on a pole P of a rotation axis A, rotating through δ_A, to yield the pole P′ and a rotation axis B, rotating through δ_B, to yield the pole P″. The position of the rotation axis relating P to P″ cannot be located from this stereogram.

To locate the position of the third, or derivative, rotation axis it is necessary to apply *Euler's construction*, which is based on the proposition that if three great circles intersect in the poles A, B, and C in such a manner (Fig 3.10(a)) that the angle between the great circles AB and AC is α, the angle between the great circles BA and BC is β, and the angle between the great circles CA and CB is γ, then rotation through the angle 2α in a clockwise sense about A followed by rotation through the angle 2β in a clockwise sense about B is equivalent to rotation through the angle 2γ in an *anti*clockwise sense about C. We shall now proceed to demonstrate the validity of this proposition in general and go on to establish all the possible combinations of rotational axes of symmetry that can operate on crystal lattices.

Let A and B (Fig 3.10(a)) be the poles of two rotation axes of symmetry, rotating respectively through the angles δ_A and δ_B. The consecutive operation of A and B will thus be equivalent to the single operation of a rotation axis of symmetry C whose pole lies at the intersection of the great circles through A and B which make angles of $\alpha = \frac{1}{2}\delta_A$ and $\beta = \frac{1}{2}\delta_B$ respectively with the great circle AB and lie on the same side of that great circle. If the rotation axis of symmetry whose pole is C produces equivalence by rotation through the angle δ_C, then the great circles AC and BC intersect at an angle $\gamma = \frac{1}{2}\delta_C$.

That this is so can be verified by operating A and B consecutively on the pole C

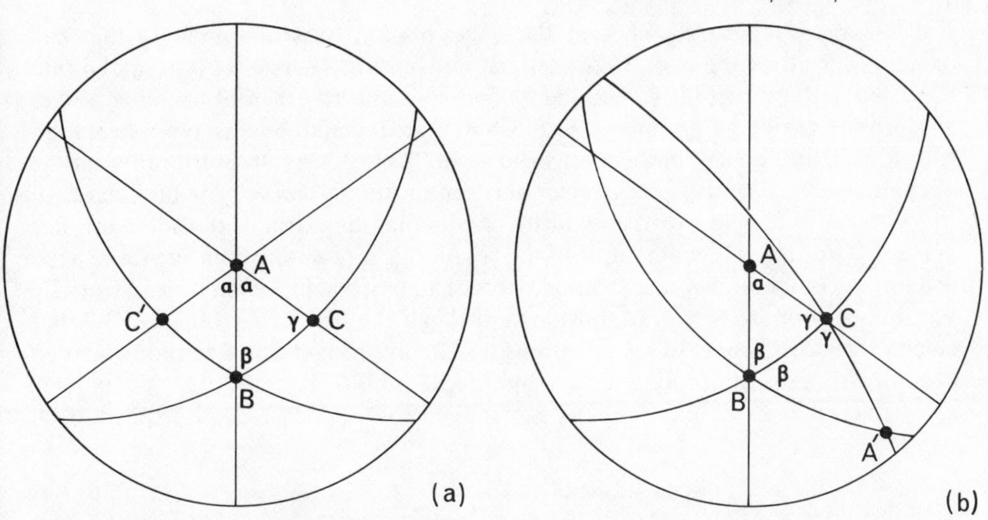

Fig 3.10 Euler's construction. The great circles AB, BC, CA are inclined to one another at angles $\alpha = \frac{1}{2}\delta_A$, $\beta = \frac{1}{2}\delta_B$, $\gamma = \frac{1}{2}\delta_C$. The poles A and B are rotation axes rotating respectively through 2α and 2β. (a) illustrates the successive clockwise operation of the axes A and B on the pole C. (b) illustrates the successive clockwise operation of the axes A and B on the pole A. C is shown to be a rotation axis rotating through 2γ.

(Fig 3.10(a)) and on the pole A (Fig 3.10(b)). When A operates on C in a clockwise sense the first equivalent pole produced will be C′ such that $\widehat{AC} = \widehat{AC'}$ and $\widehat{BAC} = \widehat{BAC'} = \frac{1}{2}\delta_A$. Since the side $\widehat{AB}$ is common to both, the spherical triangles[2] ABC, ABC′ are congruent and therefore $\widehat{ABC} = \widehat{ABC'} = \frac{1}{2}\delta_B$ and $\widehat{BC} = \widehat{BC'}$. Consequently operation of B in a clockwise sense will take the pole C′ back to C. Consecutive clockwise rotation of C through δ_A about A and through δ_B about B thus leaves C unmoved; C is therefore itself the pole of a rotation axis of symmetry whose operation is equivalent to the consecutive operations of A and B.

Now consider the consecutive operation of A and B on the pole A (Fig 3.10(b))). The operation of the rotation axis whose pole is A will leave A unmoved. The first equivalent pole produced by the operation of B on the pole A in a clockwise sense will be A′ such that $\widehat{BA} = \widehat{BA'}$ and $\widehat{ABC} = \widehat{A'BC} = \frac{1}{2}\delta_B$. Since the side $\widehat{BC}$ is common to both, the spherical triangles ABC, A′BC are congruent so that $\widehat{CA} = \widehat{CA'}$ and $\widehat{BCA} = \widehat{BCA'} = \frac{1}{2}\delta_C$. Therefore A′ is the first equivalent pole produced by anticlockwise operation of the symmetry axis whose pole is C and whose angle of rotation is δ_C. Consecutive clockwise rotation of A about itself and through δ_B about B is thus equivalent to anticlockwise rotation of A through δ_C about C. In general then one can say that if these symmetry axes have their poles at the vertices of a spherical triangle, each of whose angles is half the angle of rotation of the corresponding axis, then the consecutive clockwise operation of two axes is equivalent to the anticlockwise operation of the third axis.

A spherical triangle is uniquely determined by the three angles at its vertices; therefore if α, β, and γ are known, the angles between the three rotation axes can be evaluated from relations of the type (Appendix D)

$$\cos \widehat{BC} = \frac{\cos \alpha + \cos \beta \cos \gamma}{\sin \beta \sin \gamma}.$$

[2] A short account of spherical trigonometry is given in Appendix D.

It has already been shown that the only rotation axes of symmetry that can operate on a lattice are diads, triads, tetrads, and hexads. Therefore δ_A, δ_B, and δ_C are restricted to the values $\frac{2}{2}\pi$, $\frac{2}{3}\pi$, $\frac{2}{4}\pi$, and $\frac{2}{6}\pi$ and consequently the only possible values of α, β, and γ are $\frac{1}{2}\pi$, $\frac{1}{3}\pi$, $\frac{1}{4}\pi$, and $\frac{1}{6}\pi$. If A, B, and C are taken to be each type of rotation axis in turn, the twenty combinations shown in Table 3.2 result. Substitution of the magnitude of α, β, and γ for each combination in the expressions for the cosines of $\widehat{BC}$, $\widehat{CA}$, and $\widehat{AB}$ yields for the majority of combinations cosines outside the range $+1$ to -1, for a few combinations trivial solutions representing coincidence of axes, and for six combinations real solutions (rows 1 to 6 of Table 3.2 and Fig 3.11(a)–(f)). We do not propose to work laboriously through the derivation of Table 3.2, but merely to discuss the set of combinations which have a diad and a tetrad combined in turn with a diad, a triad, a tetrad, and a hexad. If we fix α and β at 90° and 45° respectively the expressions for the sides of the general spherical triangle (Appendix D) become

$$\cos\widehat{BC} = \frac{\cos 90^\circ + \cos 45^\circ \cos\gamma}{\sin 45^\circ \sin\gamma} = \cot\gamma$$

$$\cos\widehat{CA} = \frac{\cos 45^\circ + \cos\gamma \cos 90^\circ}{\sin\gamma \sin 90^\circ} = \frac{1}{\sqrt{2}\sin\gamma}$$

$$\cos\widehat{AB} = \frac{\cos\gamma + \cos 90^\circ \cos 45^\circ}{\sin 90^\circ \sin 45^\circ} = \sqrt{2}\cos\gamma$$

Table 3.2

A	B	C	$\widehat{BC}$	$\widehat{CA}$	$\widehat{AB}$	
2	2	2	90°	90°	90°	
2	2	3	90°	90°	60°	
2	2	4	90°	90°	45°	
2	2	6	90°	90°	30°	
2	3	3	70° 32′	54° 44′	54° 44′	
2	3	4	54° 44′	45°	35° 16′	
2	3	6	0	0	0	trivial
2	4	4	0	0	0	trivial
2	4	6	*	*	*	impossible
2	6	6	*	*	*	impossible
3	3	3	0	0	0	trivial
3	3	4	*	*	*	impossible
3	3	6	*	*	*	impossible
3	4	4	*	*	*	impossible
3	4	6	*	*	*	impossible
3	6	6	*	*	*	impossible
4	4	4	*	*	*	impossible
4	4	6	*	*	*	impossible
4	6	6	*	*	*	impossible
6	6	6	*	*	*	impossible

*indicates cosine > 1

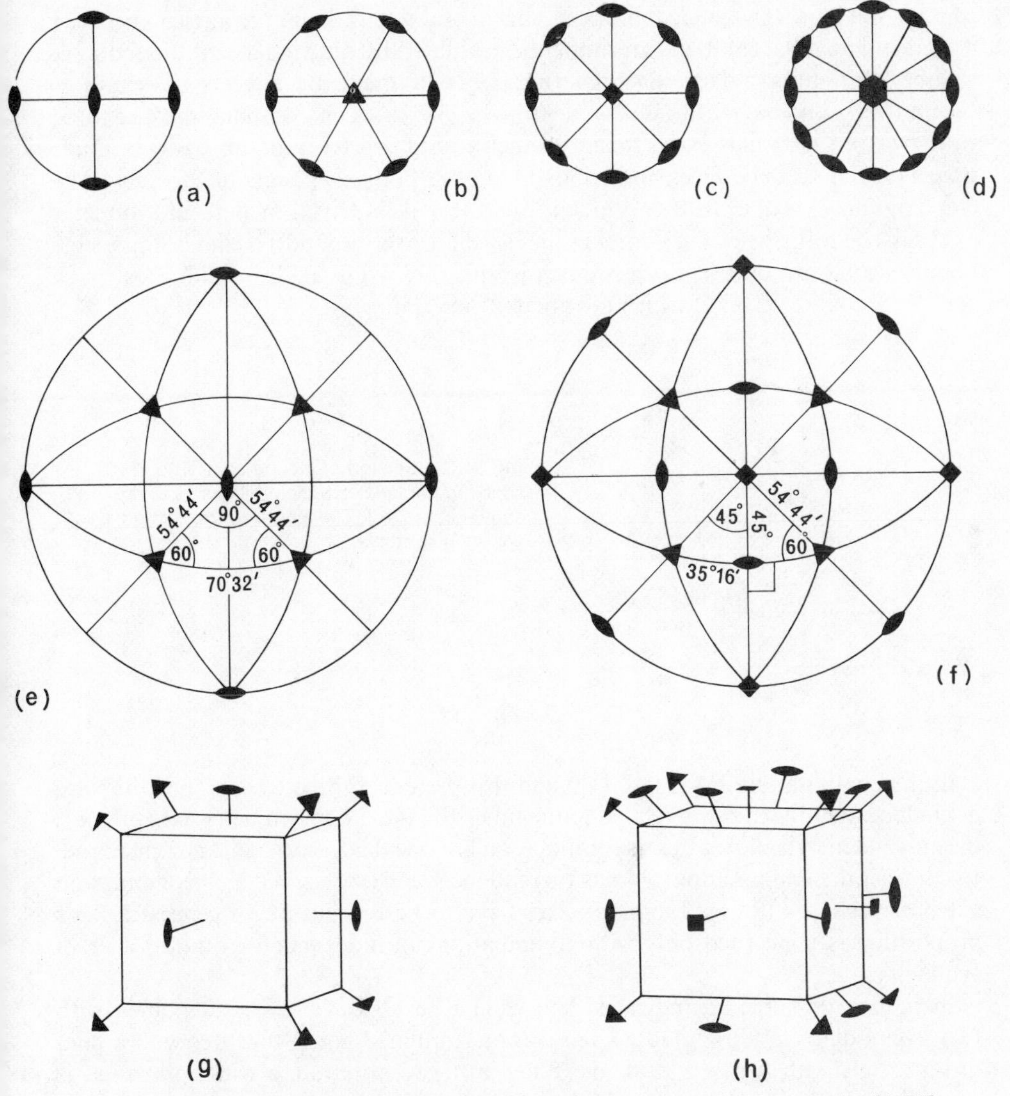

Fig 3.11 Combinations of rotation axes. (a) 222, three mutually perpendicular diads, (b) 223, a triad perpendicular to two diads inclined at 60° to each other, (c) 224, a tetrad perpendicular to two diads inclined at 45° to each other, (d) 226, a hexad perpendicular to two diads inclined at 30° to each other, (e, g) 233, a diad inclined at 54°44′ to two triads inclined at 70°32′ to each other, (f, h) 234, a diad inclined to a triad at 35°16′ and to a tetrad at 45°, the triad and tetrad being mutually inclined at 54°44′.

which on substitution of $\gamma = 90°, 60°, 45°, 30°$ in turn yield values of the interaxial angles in the combinations 242 (224), 243 (234), 244, 246 (the symbols shown in brackets are those shown in Table 3.2). The first of these combinations, a tetrad combined with two sets of diads, has its diads perpendicular to the tetrad and inclined at 45° to one another as illustrated in Fig 3.11(c). The combination of a diad, a tetrad, and a triad, illustrated stereographically in Fig 3.11(f), has a disposition of symmetry axes that is simply related to the geometry of the cube (Fig 3.11(h)): three mutually perpendicular tetrads are normal to the cube faces, four triads lie along the body diagonals of the cube, and six diads join the mid-points of opposite edges of the cube. The angle θ between a triad and a tetrad and the angle ω between a triad and an adjacent diad can simply be evaluated by drawing a central section of a cube containing a face diagonal (Fig 3.12): if the cube edge is of length a, $\theta = \tan^{-1}(\sqrt{2}a)/a = 54°44'$ and $\omega = 90° - \theta = 35°16'$. The combination of a diad with two sets of triads has similar geometry (Fig 3.11(e) and (g)) with its diads perpendicular to cube faces and triads along the body diagonals of the cube. The combination of a diad with two independent tetrads is trivial in that all symmetry axes must be coincident. The combination of a diad, a tetrad, and a hexad is impossible because solution of the expressions for the interaxial angles yields $\cos^{-1}\sqrt{3}$, $\cos^{-1}\sqrt{2}$, $\cos^{-1}\sqrt{\tfrac{3}{2}}$, all of which are greater than unity.

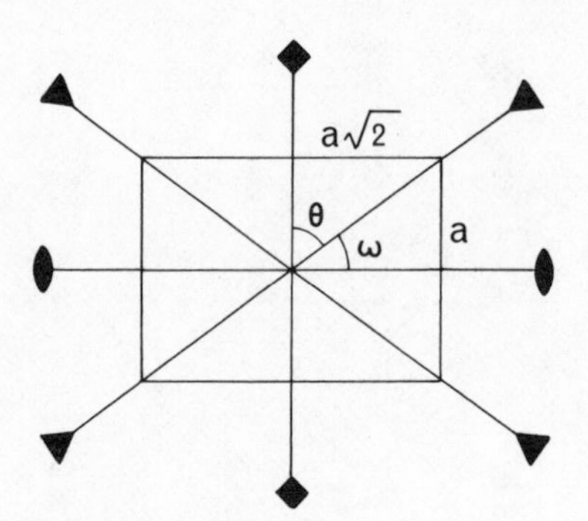

Fig 3.12 Section of a cube parallel to (110) showing coplanar symmetry elements in the combination 243. The length of the section is $a\sqrt{2}$ and its height is a.

In the combination 234 (Fig 3.11(f) and (h)) there are three tetrads, but they are not independent, each being related to the others by the other symmetry axes present; the only independent axes in this combination are one diad, one triad, and one tetrad. In contrast the combination 244 has two independent tetrads, but this combination is trivial because all three symmetry axes have to be coincident. In general Euler's proposition is concerned only with combinations of independent rotation axes of symmetry.

Inspection of Table 3.2 indicates that while a hexad can only be combined with two sets of diads, a tetrad and a triad can be combined either together with a diad or separately with two sets of diads. When a hexad, a tetrad, a triad, or a diad is combined with two independent diads, the diads lie in the plane normal to the axis of higher symmetry and are inclined to one another at an angle equal to half the rotation angle of that axis (Fig 3.11(a)–(d)).[3] When a diad and a triad are combined with either a triad or a tetrad the resultant disposition of symmetry axes is, as we have already indicated, related to the geometry of the cube (Fig 3.11(e)–(h)). These are the only six possible combinations of rotation axes of symmetry that can operate on crystal lattices.

Inversion axes of symmetry

A distinct type of symmetry axis is that which combines rotation about a line through $2\pi/n$ with inversion through a point. Such axes are known as *inversion axes of symmetry* and are designated as *inverse* n-fold axes, where $n = 1, 2, 3, 4, 6$. The operation of inversion can be considered at two levels. On inversion through the

[3] In the particular case of the combination of a triad (the only axis of odd order with which we are here concerned) with two diads (Fig 3.11(b)) the diads are not independent.

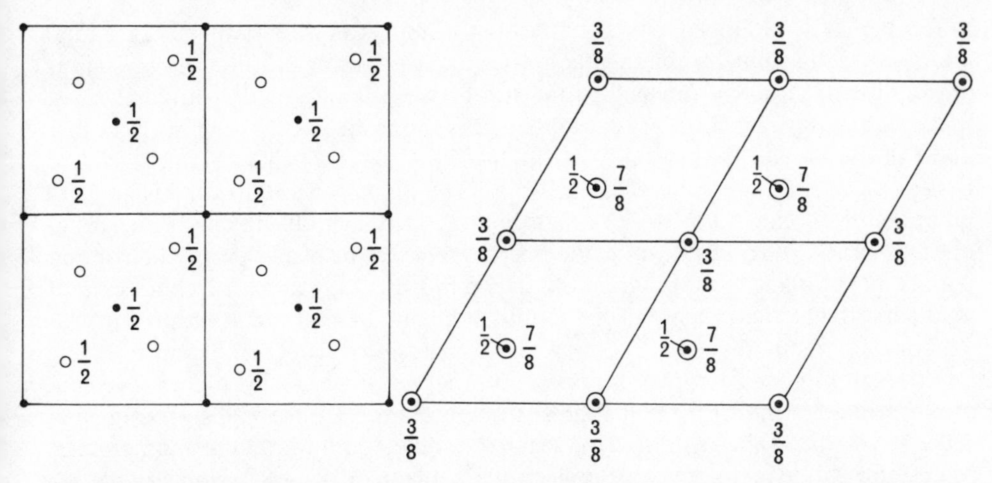

Fig 3.13 Centrosymmetric and non-centrosymmetric structures. (a) Rutile, TiO_2, with Ti atoms shown as solid circles and oxygens as open circles. (b) Wurtzite, ZnS, with Zn atoms shown as solid circles and sulphur atoms as open circles. Both structures are shown projected on the *xy* plane.

origin of coordinates every point with coordinates x, y, z becomes a point with coordinates $\bar{x}$, $\bar{y}$, $\bar{z}$. If in a crystal structure every atom with coordinates x, y, z, is duplicated by an atom of the same element with coordinates $\bar{x}$, $\bar{y}$, $\bar{z}$ the structure is said to possess a *centre of symmetry* at the origin.[4] The structure of rutile, TiO_2, which has a centre of symmetry at its origin is compared with that of wurtzite, which is non-centrosymmetric, in Fig 3.13. On the macroscopic scale a centre of symmetry causes the crystal faces (hkl) and $(\bar{h}\bar{k}\bar{l})$ to be equivalent so that in a perfectly developed crystal they will be equally developed (Fig 3.15(a)). The operation of a centre of symmetry on a general pole is illustrated in Fig 3.14, from which it is apparent that the operation of a centre of symmetry amounts to trivial rotation of the pole through $2\pi/1$ about any line through the centre followed by inversion through the centre. In conformity with the notation that we shall use for higher inversion axes of symmetry the centre of symmetry can be described as an inverse monad and assigned the symbol $\bar{1}$. In space group diagrams and occasionally in plans of crystal structures it is convenient to represent the positions of centres of symmetry; this is done by a small open circle.

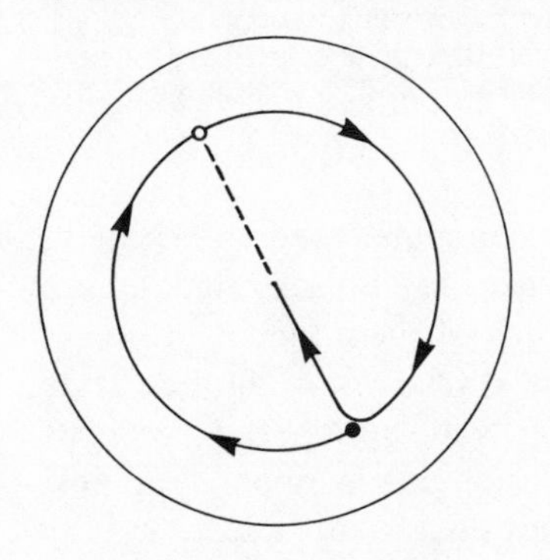

Fig 3.14 Stereogram to illustrate the operation of a centre of symmetry. The pole shown as a solid circle (in the upper hemisphere) is rotated through 360° and inverted through the centre to yield the pole shown as an open circle (in the lower hemisphere).

[4] It is worth noting that if the point x, y, z is equivalent to the point $\bar{x}$, $\bar{y}$, $\bar{z}$ then it must also be equivalent to the point $1-x$, $\bar{y}$, $\bar{z}$. Therefore if there is a centre of symmetry at 0, 0, 0 there must also be a centre of symmetry midway between x, y, z and $1-x$, $\bar{y}$, $\bar{z}$, i.e., at $\frac{1}{2}$, 0, 0, and likewise at 0, $\frac{1}{2}$, 0; 0, 0, $\frac{1}{2}$; and $\frac{1}{2}$, $\frac{1}{2}$, $\frac{1}{2}$; etc. Thus the spacing of centres of symmetry along any direction within the structure is half that of the lattice spacing in that direction.

The perfectly developed crystal of gypsum, $CaSO_4.2H_2O$, shown in Fig 3.15(a) has a centre of symmetry relating faces such as I and II. In addition the crystal, as drawn, displays a mirror image relationship between its left- and right-hand sides, that is to say the part of the crystal to the left of the plane abcd is a reflexion in that plane of the part of the crystal to the right of the plane. Such a crystal is said to possess a *plane of symmetry* or *mirror plane.* The mirror plane illustrated is an (010) plane and thus relates a face (hkl) to a face ($h\bar{k}l$), these faces being mirror images of one another. Stereograms showing the operation of mirror planes perpendicular and parallel to the plane of the diagram are shown in Fig 3.15(b) and (c). In both cases it is apparent that the operation of the mirror plane amounts to rotation through $2\pi/2 = 180°$ about the pole D followed by inversion through the centre of the stereogram. In conformity with the notation that we shall use for higher inversion axes of symmetry the mirror plane can be described as an inverse diad and designated $\bar{2}$, the inverse diad D being normal to the mirror plane; it is however common practice to describe this symmetry operator as a mirror plane or plane of symmetry and to assign to it the symbol m. Mirror planes are conventionally represented on stereograms as boldly drawn great circles.

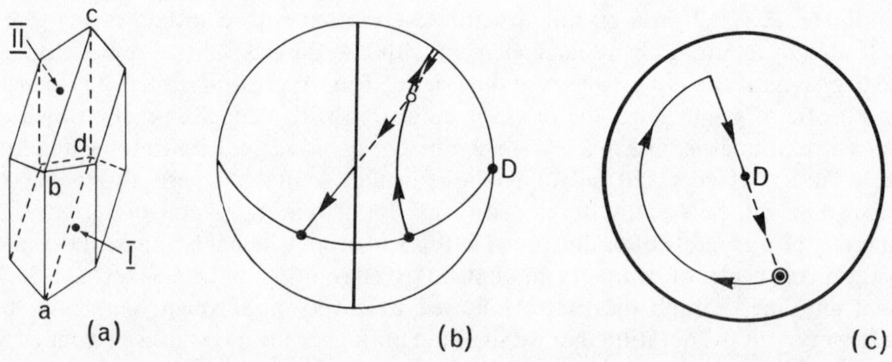

Fig 3.15 The operation of an inverse diad or mirror plane. (a) shows a crystal of gypsum; the faces I and II are related by a centre of symmetry; that part of the crystal to the left of the plane abcd is a reflexion in that plane of the part to the right. (b) shows stereographically the operation of an inverse diad D in the plane of the diagram; the inverse diad rotates the pole, shown in the upper hemisphere on the right-hand side of the diagram, on the small circle about D through 180° to the position shown by the open circle and this is followed by inversion through the centre. (c) shows stereographically the operation of an inverse diad D perpendicular to the plane of the diagram. In (b) and (c) the corresponding mirror planes are respectively a vertical great circle and the primitive.

The centre of symmetry and the mirror plane are commoner and rather more important in crystallography than the higher inversion axes, the inverse triad, tetrad, and hexad. Stereograms displaying the operation of each of these higher inversion axes on a general pole are shown in Fig 3.16(a)–(c) and crystals showing these axes are illustrated in Fig 3.16(d)–(f). The inverse triad, tetrad, and hexad are conventionally represented as $\bar{3}$, $\bar{4}$, and $\bar{6}$ respectively. Their conventional graphical symbols are shown in Fig 3.16; these are difficult to draw on a small scale, especially the $\bar{6}$ symbol, and are often shown as open triangles, squares, and hexagons. The operation of the inverse triad is equivalent to the operation of a triad combined with a centre of symmetry (i.e. $\bar{3} = 3+\bar{1}$); the operation of the inverse hexad is equivalent to that of a triad combined with a perpendicular mirror plane (i.e. $\bar{6} = 3+m$); but the operation of the inverse tetrad is not equivalent to any combination of other rotation or inversion axes although it includes a rotation diad.

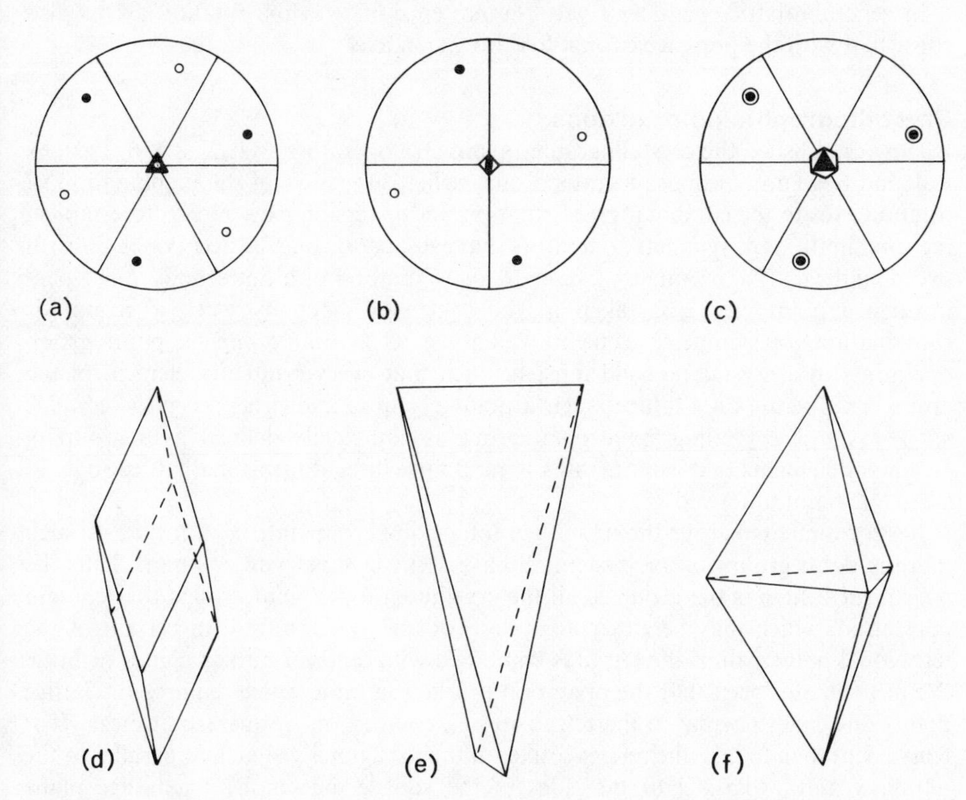

Fig 3.16 Inversion axes of symmetry. The stereograms show the operation of a $\bar{3}$, a $\bar{4}$, and a $\bar{6}$ axis on a general pole. The perspective drawings show crystals displaying inverse triad, tetrad, and hexad symmetry.

Since a lattice is a regular array of points in space every lattice point is associated with lattice points at vector distances $+t$ and $-t$ from it. That is to say all lattices are necessarily centrosymmetric with a centre of symmetry at every lattice point. A lattice consistent with fourfold symmetry thus has a centre of symmetry at every lattice point as well as a tetrad; the symmetry of such a lattice (Fig 3.17) can be described variously as a tetrad or an inverse tetrad combined with a perpendicular mirror plane or as a tetrad or an inverse tetrad combined with a centre of symmetry.

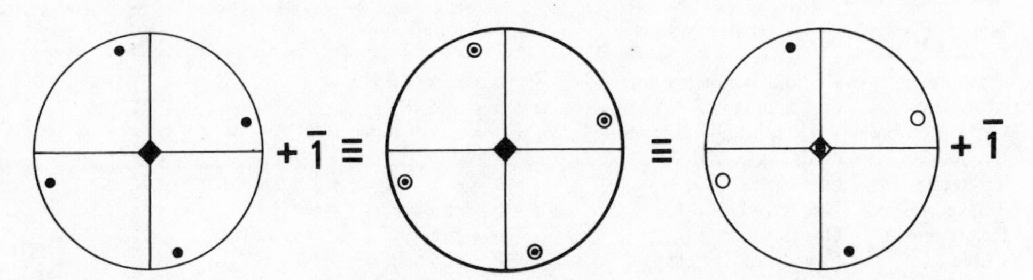

Fig 3.17 Stereograms to illustrate that the combination of a rotation tetrad with a centre of symmetry is equivalent to the combination of an inverse tetrad with a centre of symmetry and to the combination of a rotation tetrad with a perpendicular mirror plane.

The lattice is thus consistent with the presence not only of a rotation tetrad but also of an inverse tetrad.

In general a lattice consistent with the presence of an n-fold rotation axis is also consistent with the presence of an n-fold inversion axis.

Crystallographic point groups

Having established the crystallographic symmetry operators as 1, 2, 3, 4, 6, $\bar{1}$, $\bar{2}$ (m), $\bar{3}$, $\bar{4}$, and $\bar{6}$ we now propose to extend our earlier discussion of the combination of rotation axes to the combination of rotation and inversion axes and so to establish the combinations of symmetry operators that can operate on a lattice. We begin with two definitions. The symmetry elements (or operators) of a finite body must pass through a point, which is taken as the centre of the body; such a group (or combination) of symmetry elements is known as a *point group*. A point group operating on a crystalline solid must be such that every symmetry element of the group can operate on a lattice; such a point group is known as a *crystallographic point group*. A crystallographic point group is completely defined as a group of symmetry elements that can operate on an infinite three-dimensional lattice so as to leave one point unmoved.

It is convenient to group the crystallographic point groups into *crystal systems*, such that the point groups of one system all have some symmetry in common. Thus the *tetragonal* system is the group of all the crystallographic point groups that contain one tetrad, which may be either a rotation or an inverse tetrad; in certain of the tetragonal point groups the tetrad is combined with diads or mirror planes or both. We have already seen that the operation of a tetrad on a lattice requires the lattice points on planes normal to the tetrad to be arranged on a square unit-mesh. It is convenient then to take the reference axes of the tetragonal system as z parallel to the tetrad, x and y parallel to the sides of the square unit-mesh of a lattice plane perpendicular to the tetrad; the conventional unit-cell of the tetragonal system thus has $a = b \neq c$ and $\alpha = \beta = \gamma = 90°$.

In general it is convenient to select a unit-cell whose axes are, if possible, parallel or normal to symmetry axes for the good reason that this simplifies the description of symmetry relationships. The shape of the conventional unit-cell is then characteristic of the system to which it refers. In Table 3.3 the nomenclature of the crystal systems and the restrictions on the shape of the conventional unit-cell for each system are set out.

Table 3.3
The crystal systems

Name of system	Characteristic symmetry	Conventional unit-cell
Triclinic	Onefold symmetry only	$a \neq b \neq c;\ \alpha \neq \beta \neq \gamma$
Monoclinic	One diad ($\parallel y$)	$a \neq b \neq c;\ \alpha = \gamma = 90°,\ \beta > 90°$
Orthorhombic	Three mutually perpendicular diads ($\parallel x$, y and z)	$a \neq b \neq c;\ \alpha = \beta = \gamma = 90°$
Trigonal*	One triad ($\parallel [111]$)	$a = b = c;\ \alpha = \beta = \gamma < 120°,\ \neq 90°$
Tetragonal	One tetrad ($\parallel z$)	$a = b \neq c;\ \alpha = \beta = \gamma = 90°$
Hexagonal	One hexad ($\parallel z$)	$a = b \neq c;\ \alpha = \beta = 90°,\ \gamma = 120°$
Cubic	Four triads ($\parallel \langle 111 \rangle$)	$a = b = c;\ \alpha = \beta = \gamma = 90°$

The symbol $\neq$ implies that equality is not required by symmetry.
*The unit-cell of the hexagonal system is however commonly used for the trigonal system.

We turn now to consider what combinations of rotation and inversion axes are possible and so to derive all the distinct crystallographic point groups. We have already shown that a rotation axis of order 1, 2, 3, 4, or 6 can operate alone on a lattice and thus is itself a crystallographic point group. We have further shown that there are only six combinations of rotation axes capable of operating on a lattice and that each combination has a definite geometrical arrangement (Table 3.2); each of the combinations 222, 223, 224, 226, 233, and 234 is thus a distinct crystallographic point group.[5] We have also shown that the lattice consistent with the operation of an *n*-fold rotation axis is consistent with the operation of an inversion axis of the same order; each of the inversion axes, $\bar{1}$, $\bar{2}$ (= m), $\bar{3}$, $\bar{4}$, and $\bar{6}$, operating on its own thus constitutes a distinct crystallographic point group.

Since all lattices are necessarily centrosymmetric, a crystal may itself have a centre of symmetry. Further crystallographic point groups can thus be derived simply by adding a centre of symmetry to the point groups that we have already identified. Before doing this it may be helpful to the reader to be reminded of two points that have been made earlier in this chapter. Firstly, the operation of a rotation axis of odd order combined with a centre of symmetry is equivalent to an inversion axis of the same order (Figs 3.14 and 3.16); thus, trivially, $1+\bar{1}=\bar{1}$ and, more significantly, $3+\bar{1}=\bar{3}$. Secondly, the operation of a rotation axis of even order combined with a centre of symmetry is equivalent to the operation of an inversion axis of the same order combined with a centre of symmetry; we have shown this (Fig 3.17) for the tetrad and leave the reader to satisfy himself that it holds also for the diad and the hexad. Thus the combination of a centre of symmetry either with one of the rotation axes 2, 4, 6 or with one of the inversion axes, $\bar{2}$, $\bar{4}$, $\bar{6}$, yields a new crystallographic point group. Each of these point groups can alternatively be described as the combination of a 2, 4, or 6 rotation axis with a perpendicular mirror plane (Fig 3.17 illustrates one case); this alternative description is the basis of the conventional nomenclature for these three point groups $2/m$, $4/m$, $6/m$.

Before discussing the point groups obtained by substitution of inversion for rotation axes or by addition of a centre of symmetry in the point groups derived by means of Euler's proposition (Table 3.2) it is convenient to consider the question of *hand*. The operation of a rotation axis, being a simple rotation, is incapable of changing the hand of the object on which it operates. In contrast the operation of an inversion axis, because it involves inversion through the centre, must convert a right-handed object into a left-handed object. For example Fig 3.18(a) is a stereogram showing the operation of an inverse tetrad on a group of three poles. If the original group of poles is taken to be that in the upper right-hand quadrant a single operation of the inverse tetrad gives rise to the group in the lower left-hand quadrant; it is apparent from the figure that these two groups of poles cannot be superimposed, being mirror images of one another, the one right-handed and the other left-handed. Now in general terms an inversion axis A will produce a left-handed group P_L from a right-handed group P_R (Fig 3.18(b)) and a rotation axis B will produce a left-handed group P'_L from the left-handed group P_L. Therefore the axis consistent with A and B must be an axis C relating P_R to P'_L and, since a change of hand is produced, this must be an inversion axis. Thus in a combination of three axes either all three are rotation axes *or* one is a rotation axis and two are inversion axes; no other combination is possible.

[5] We shall postpone discussion of the conventional nomenclature for point groups and for the time being use a nomenclature related to that of Table 3.2.

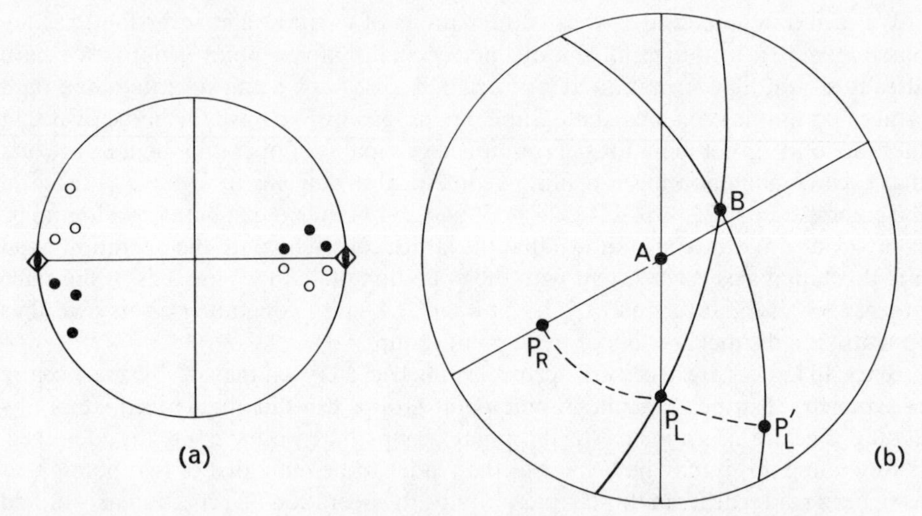

Fig 3.18 The significance of *hand* in the combination of axes. The stereogram (a) illustrates the change of hand resulting from the operation of an inverse tetrad. The stereogram (b) illustrates the general proposition that the successive operation of an inversion axis A, which produces the left-handed group P_L from the right-handed group P_R, and a rotation axis B, which produces the left-handed group P'_L from the left-handed group P_L, can only be consistent with the operation of an inversion axis producing the left-handed group P'_L from the right-handed group P_R.

We now proceed to apply the conclusion of the last paragraph to each of the point groups derived by means of Euler's proposition. From the point group 222 only one new crystallographic point group can be derived, $\bar{2}\bar{2}2$, which can alternatively be described as *mm*2.[6] This point group has two perpendicular mirror planes with a rotation diad along their line of intersection. From the point group 223 (conventionally described as 32) we derive $\bar{2}\bar{2}3$ (conventionally 3*m*) and $\bar{2}2\bar{3}$ (conventionally $\bar{3}m$), which are new point groups. From the point group 224 (conventionally 422) we derive $\bar{2}\bar{2}4$ (conventionally 4*mm*) and $\bar{2}2\bar{4}$ (conventionally $\bar{4}2m$), both of which are new.

Table 3.4
Derivation of the crystallographic point groups

Symmetry axes parallel to one direction only:

Rotation X	1	2	3	4	6
Inversion $\bar{X}$	$\bar{1}$	m	$\bar{3}$	$\bar{4}$	$\bar{6}$
$X+\bar{1}$	($\bar{1}$)	$2/m$	($\bar{3}$)	$4/m$	$6/m$

Symmetry axes in more than one direction:

XYZ	222	223 (32)	224 (422)	226 (622)	233 (23)	234 (432)
$\bar{X}\bar{Y}Z$	$mm2$	$mm3$ ($3m$)	$mm4$ ($4mm$)	$mm6$ ($6mm$)	$m\bar{3}\bar{3}$ ($m3$)	$m\bar{3}4$ ($m3m$)
$\bar{X}Y\bar{Z}$	($m2m$)	$m2\bar{3}$ ($\bar{3}m$)	$m2\bar{4}$ ($\bar{4}2m$)	$m2\bar{6}$ ($\bar{6}m2$)	($m3$)	$m3\bar{4}$ ($\bar{4}3m$)
$X\bar{Y}\bar{Z}$	($2mm$)	($\bar{3}m$)	($\bar{4}2m$)	($\bar{6}m2$)	($m3$)	($m3m$)
$XYZ+\bar{1}$	mmm	($\bar{3}m$)	$4/mmm$	$6/mmm$	($m3$)	($m3m$)

Note:
(1) Symbols in brackets standing alone refer to point groups higher up in the table.
(2) Symbols in brackets following symbols not in brackets are conventional point group symbols.

[6] To follow this and succeeding paragraphs the reader will find it helpful to refer to Table 3.4 and to look forward to Fig 3.20.

From the point group 226 (known conventionally as 622) we derive $2\bar{2}6$ (conventionally $6mm$) and $2\bar{2}\bar{6}$ (conventionally $\bar{6}m2$). From the point group 233 (conventionally known as 23) we derive $2\bar{3}3$ and $23\bar{3}$, which turn out to be identical and are conventionally described as $m3$. From the point group 234 (conventionally 432) we derive $2\bar{3}4$ and $23\bar{4}$, which are identical and known as $m3m$, and $\bar{2}3\bar{4}$ which is conventionally known as $\bar{4}3m$.

To complete our list of crystallographic point groups we have to add a centre of symmetry to each of the point groups derived from Euler's proposition and to each of those obtained in the preceding paragraph. This task can be simplified by bearing in mind that the combination of a centre of symmetry with a rotation triad is equivalent to an inverse triad and that the combination of a centre of symmetry with either a rotation or an inversion axis of even order is identical. Thus 222 and $mm2$ on combination with a centre of symmetry both yield the same new point group $\frac{2}{m}\frac{2}{m}\frac{2}{m}$, conventionally known as mmm. The addition of a centre of symmetry to 223 (32) and $\bar{2}\bar{2}3$ ($3m$) yields the same point group, which is identical with $\bar{2}\bar{2}\bar{3}$ ($\bar{3}m$) and already accounted for. The addition of a centre of symmetry to 224 (422), or $\bar{2}\bar{2}4$ ($4mm$), or $\bar{2}2\bar{4}$ ($\bar{4}2m$) yields the same new point group $\frac{2}{m}\frac{2}{m}\frac{4}{m}$, conventionally known as $4/mmm$. Likewise the addition of a centre of symmetry to 226 (622) or $\bar{2}\bar{2}6$ ($6mm$), or $\bar{2}2\bar{6}$ ($\bar{6}m2$) yields the same new point group $\frac{2}{m}\frac{2}{m}\frac{6}{m}$, conventionally known as $6/mmm$. Addition of a centre of symmetry to 233 (23) yields the point group $\frac{2}{m}\bar{3}\,\bar{3}$ which has already been accounted for and designated $m3$. Finally addition of a centre of symmetry to 234 (432) or $\bar{2}3\bar{4}$ ($\bar{4}3m$) yields the point group $\frac{2}{m}\bar{3}\frac{4}{m}$ which has already been accounted for and designated $m3m$.

We have derived in all thirty-two crystallographic point groups, five consisting of a single rotation axis, another five consisting of a single inversion axis, three by combination of a centre of symmetry with a rotation axis, six directly derivative from Euler's proposition and a further ten by mixing inversion and rotation axes in these geometrical combinations, and three more by combining a centre of symmetry with these combinations.

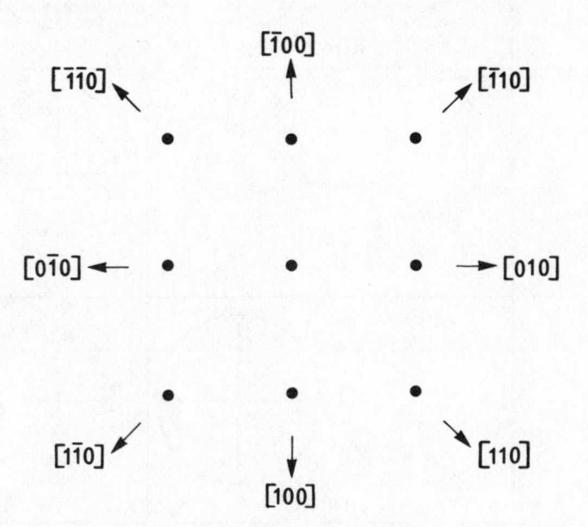

Fig 3.19 Plan of the (001) plane of the tetragonal lattice.

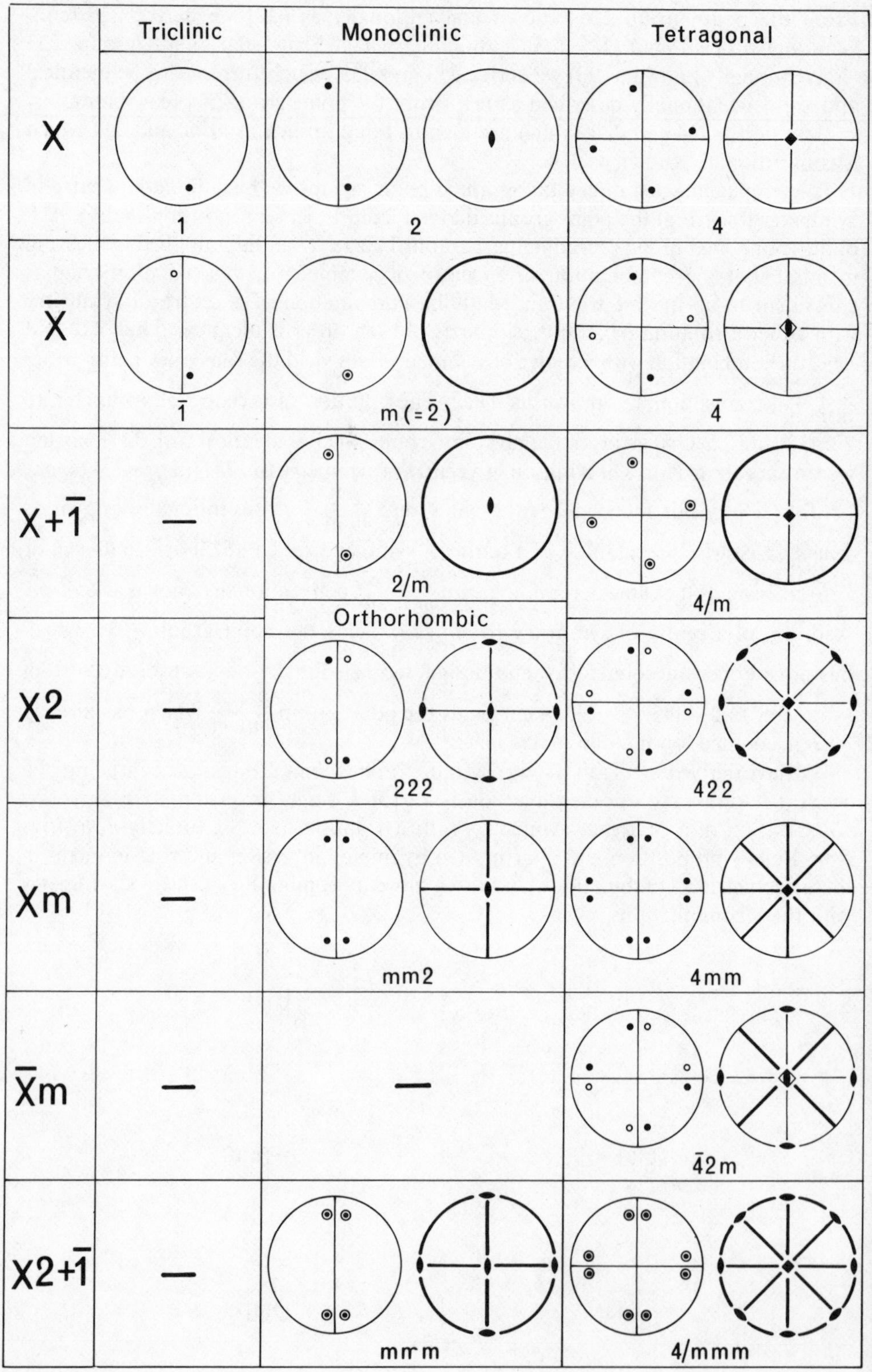

	Triclinic	Monoclinic	Tetragonal
X	1	2	4
$\bar{X}$	$\bar{1}$	m (= $\bar{2}$)	$\bar{4}$
X+$\bar{1}$	—	2/m	4/m
		Orthorhombic	
X2	—	222	422
Xm	—	mm2	4mm
$\bar{X}$m	—	—	$\bar{4}$2m
X2+$\bar{1}$	—	mmm	4/mmm

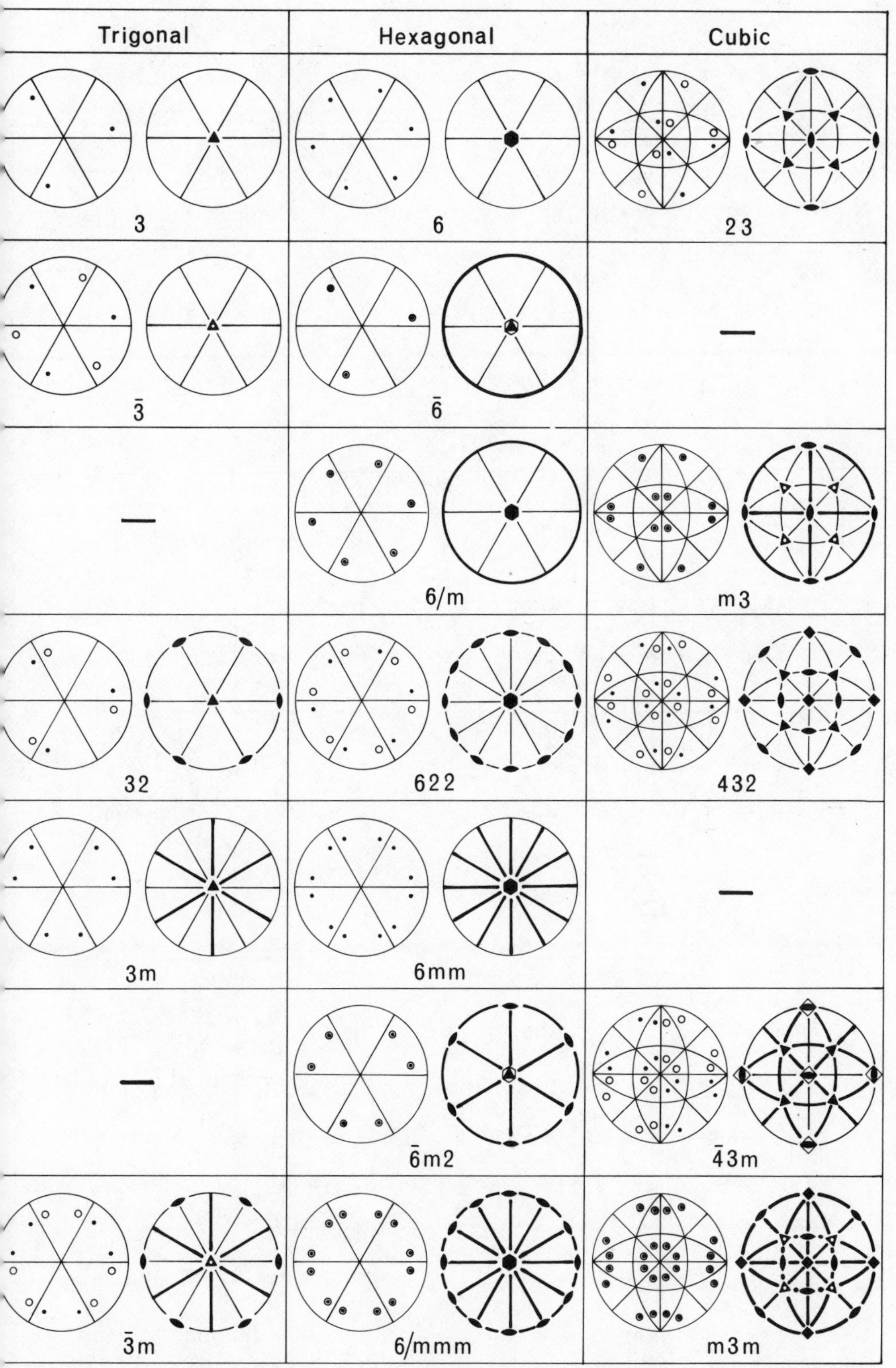

Fig 3.20 The thirty-two crystallographic point groups. Each pair of stereograms shows, on the left, the poles of a general form and, on the right, the symmetry elements of the point group. Planes of symmetry are indicated by bold lines.

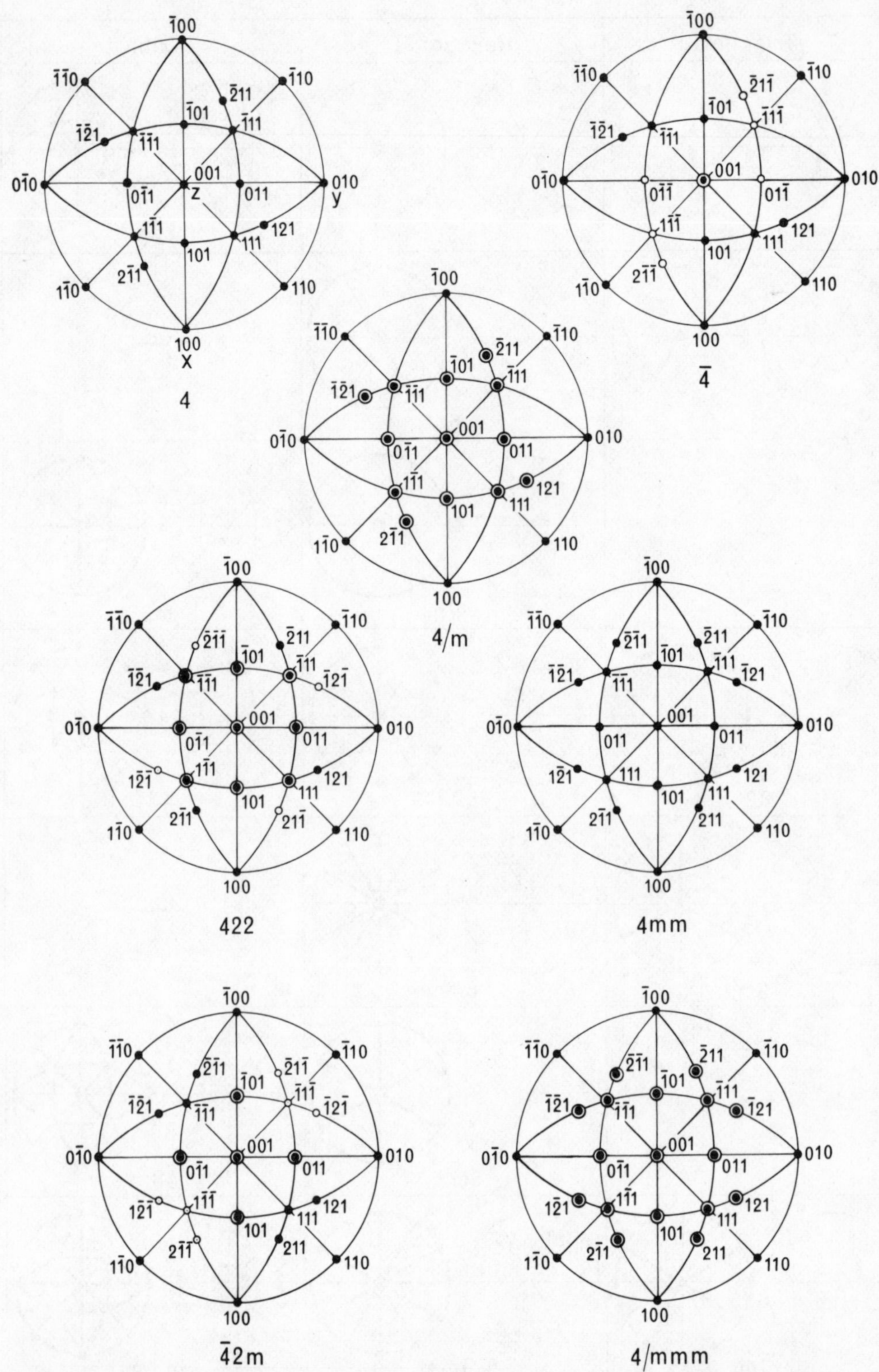

Fig 3.21 The tetragonal point groups. The stereograms show the poles of {100}, {001}, {110}, {101}, {111}, and {121} in each of the point groups. Where poles are superimposed only the pole in the upper hemisphere is indexed.

In the conventional nomenclature of point groups, which we have already used without explanation, emphasis is placed on the presence of mirror planes of symmetry. In point group symbols a mirror plane is associated with the direction of its normal whereas, naturally, a rotation or inversion axis is associated with its own direction in the crystal. Where all the symmetry elements of the point group are associated with a single direction the conventional point group symbol takes one of the forms X (an X-fold rotation axis only), $\bar{X}$ (an X-fold inversion axis only), or $\dfrac{X}{m}$ (an X-fold rotation axis combined with a perpendicular mirror plane). This last type of symbol is often printed for convenience of typesetting as X/m. Where the symmetry elements of the point group are associated with more than one direction those associated with each direction are stated in the conventional point group symbol in a standard order. By way of illustration we take the tetragonal system, where z is taken as the direction of the rotation or inversion tetrad and the x and y axes are chosen so as to be parallel to the shortest lattice repeats in the plane perpendicular to the tetrad; the unit-cell is thus a square prism with $a = b \neq c$. The geometry of the combinations of symmetry axes with which we are concerned in the tetragonal system is that shown for 224 in Table 3.2: a tetrad combined with two independent diads inclined at 45° to one another and both perpendicular to the tetrad. Inspection of the (001) plane of the tetragonal lattice (Fig 3.19) shows that the only possible diad directions[7] are $\langle 100 \rangle$ and $\langle 110 \rangle$. Where mirror planes occur in the point group in place of diads they are necessarily perpendicular to $\langle 100 \rangle$ or $\langle 110 \rangle$, that is parallel to $\{100\}$ or $\{110\}$. The conventional symbol for those tetragonal point groups which have symmetry elements associated with more than one direction consists of three terms: first, a statement of the symmetry elements associated with the [001] direction; second, a statement of the symmetry elements associated with the $\langle 100 \rangle$ directions; and third, a statement of the symmetry elements associated with the $\langle 110 \rangle$ directions.

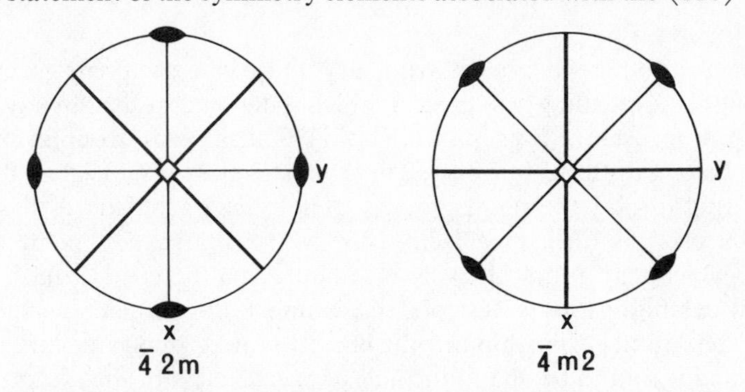

Fig 3.22 The point groups $\bar{4}2m$ and $\bar{4}m2$ which are related by rotation through 45° about [001].

Thus the symbol 422 refers to a point group (Figs 3.20 and 3.21) with a rotation tetrad parallel to [001] and diads parallel to $\langle 100 \rangle$ and $\langle 110 \rangle$. The symbol $4mm$ refers to the point group with a rotation tetrad parallel to [001] and mirror planes perpendicular to $\langle 100 \rangle$ and $\langle 110 \rangle$. In the symbol $\bar{4}2m$ however the $\langle 100 \rangle$ and $\langle 110 \rangle$ directions are distinguished: the inverse tetrad is parallel to [001], diads are parallel to $\langle 100 \rangle$ and mirror planes are perpendicular to $\langle 110 \rangle$ in this point group. The

[7] $\langle UVW \rangle$ and $\{hkl\}$ represent respectively all the zone axes and all the faces derived from $[UVW]$ and (hkl) by the operation of the point group.

symbol $\bar{4}m2$, which is illustrated in Fig 3.22, refers to the point group $\bar{4}2m$ in a different orientation; as far as point group symmetry is concerned the two are indistinguishable, being simply related by rotation through 45° about [001]. Only when it is known whether a diad or a mirror plane is associated with the shortest lattice repeat in the (001) plane is it realistic to distinguish between them; this takes us into the field of space groups, which will be discussed in chapter 4. Finally, the centrosymmetric point group $\frac{4}{m}\frac{2}{m}\frac{2}{m}$ has a rotation tetrad parallel and a mirror plane perpendicular to [001] combined with diads parallel and mirror planes perpendicular to both the sets of directions $\langle 100 \rangle$ and $\langle 110 \rangle$; this point group is commonly represented by its *short symbol* $4/mmm$.

Further discussion of point group nomenclature is postponed until the crystal systems have been explored and is then dealt with system by system. The basis of the nomenclature is summarized in Table 3.5.

Table 3.5
The conventions used for the symbols of the point groups

System	Directions associated with symmetry elements		
	First position	Second position	Third position
Triclinic		Centre only	
Monoclinic	[010]		
Orthorhombic	[100]	[010]	[001]
Trigonal	[0001]	$\langle 10\bar{1}0 \rangle$	$\langle 2\bar{1}\bar{1}0 \rangle$
Tetragonal	[001]	$\langle 100 \rangle$	$\langle 110 \rangle$
Hexagonal	[0001]	$\langle 10\bar{1}0 \rangle$	$\langle 2\bar{1}\bar{1}0 \rangle$
Cubic	$\langle 100 \rangle$	$\langle 111 \rangle$	$\langle 110 \rangle$

Crystal classes

A crystalline solid that exhibits the symmetry of a particular point group is said to belong to the corresponding *crystal class*, which is denoted by the same symbol as the point group. For instance gypsum, $CaSO_4.2H_2O$, has point group symmetry $2/m$; it is said to belong to the crystal class $2/m$. A crystal class is defined as the group of substances that display the point group symmetry characteristic of the class. The term 'crystal class' is often incorrectly used as a synonym for 'point group'; the distinction, which may at first sight seem pedantic, is in practice useful.

A crystal exhibiting the largest possible number of symmetry elements for its system may be described as exhibiting the highest point group symmetry possible for the system and is said to belong to the *holosymmetric* class of the system. The point group symmetry of each holosymmetric class is that of the corresponding lattice (the trigonal system is, as we shall show, an exception). The symmetry of every other point group of the system is lower because the repeat unit of the crystal has lower symmetry than the lattice. Figure 3.23, in which three two-dimensional patterns based on a square lattice are shown, illustrates this point: the lattice has a tetrad perpendicular to the plane of the mesh and lines of symmetry every 45°, so does pattern (b), but pattern (c) is entirely lacking in lines of symmetry, and pattern (d), for which the tetrad descends to a diad, has lines of symmetry only parallel to the sides of the unit-mesh. The pattern shown in Fig 3.23(d) has mutually perpendicular lattice repeats of equal length but unrelated by the symmetry elements of the pattern; formally the lattice on which this pattern is based is rectangular with a only *accidentally* equal to b.

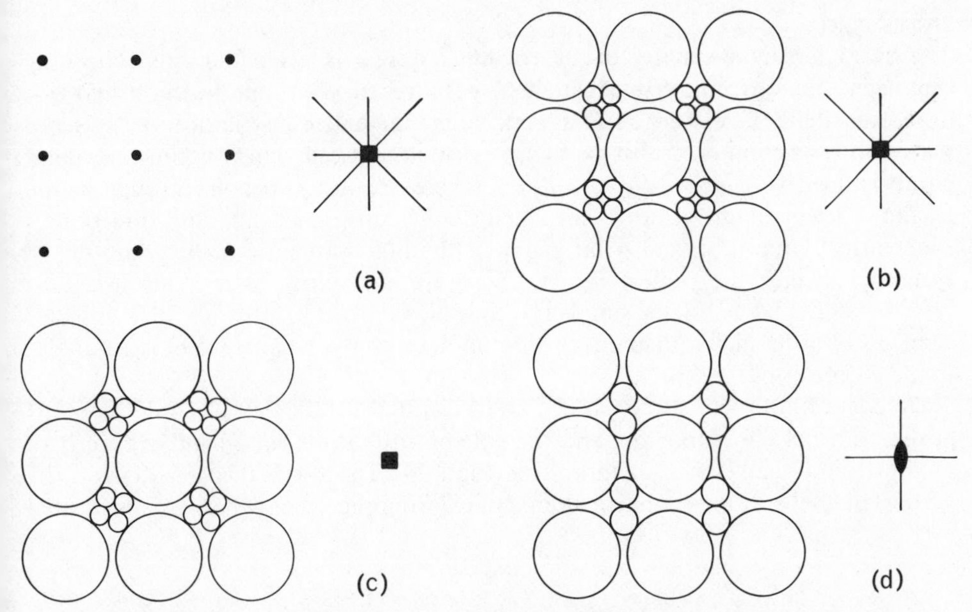

Fig 3.23 Two-dimensional patterns based on the square lattice (a). Pattern (b) has the full symmetry of the lattice; pattern (c) lacks lines of symmetry; and in pattern (d) the tetrad has become a diad and there are lines of symmetry parallel to the sides of the unit-mesh only.

Crystal systems

We have defined a crystal system as a group of point groups that have some symmetry in common and we have made use of the point groups of the tetragonal system to explain point group nomenclature. We now proceed to define each of the seven crystal systems and to specify and comment on all the point groups of each system.

Before examining each system in detail it may be helpful to name and make some general comments on the seven crystal systems. Three systems can be described as *orthogonal*, that is they can be referred to mutually perpendicular reference axes: these are the *cubic system*, for which the unit-cell is a regular cube, the *tetragonal system*, whose unit-cell is a square prism, and the *orthorhombic system*, whose unit-cell is a rectangular parallelepiped. Two systems can be referred to unit-cells that are 120° prisms: these are the *hexagonal* and *trigonal systems*. The remaining two systems are the least symmetrical, the *monoclinic system*, which has a unit-cell with $a \neq b \neq c$, $\alpha = \gamma = 90° \neq \beta$ conventionally, and the *triclinic system*, which has a unit-cell with $a \neq b \neq c, \alpha \neq \beta \neq \gamma$ in general.

In the course of exploring each system we shall deal with such nomenclatorial or representational specialities as may arise. But there are two general matters of nomenclature that are most conveniently dealt with now. The group of faces produced by the operation of all the symmetry elements of a point group acting on one face (hkl) is known as a *form* and represented as $\{hkl\}$. Similarly all the zone axes produced by the operation of all the symmetry elements of a point group on one zone axis $[UVW]$ is known as a *form of zone axes* and represented as $\langle UVW \rangle$. Thus in crystallography we distinguish between indices enclosed in four kinds of brackets: ( ) known simply as *brackets*, { } as *braces*, [ ] as *square brackets*, and $\langle\ \rangle$ as *carets*.

Triclinic system

The characteristic symmetry of the triclinic[8] system is a onefold axis. The only symmetry that a triclinic crystal can display is a centre of symmetry and it may not have even that. Since a centre of symmetry is inherent in any lattice, its presence places no restriction on the shape of the triclinic unit-cell, which is thus a general parallelepiped with $a \neq b \neq c$, $\alpha \neq \beta \neq \gamma$. There are just two point groups in the triclinic system: point group 1, in which neither any plane nor any direction is symmetrically repeated, and point group $\bar{1}$ (the holosymmetric point group of the system), in which the planes (hkl) and $(\bar{h}\bar{k}\bar{l})$ are equivalent, as are the directions $[UVW]$ and $[\bar{U}\bar{V}\bar{W}]$. The simplicity engendered by the absence of symmetry elements of order higher than one is offset in practice by the absence of right-angles in the geometry of the unit-cell.

Stereograms of triclinic crystals can conveniently be drawn by plotting the z-axis at the centre of the stereogram and the pole of (010) at the right-hand extremity of the horizontal diameter of the primitive (Fig 3.24). The x-axis then projects on the vertical diameter of the primitive at an angle β from the z-axis. The y-axis is located

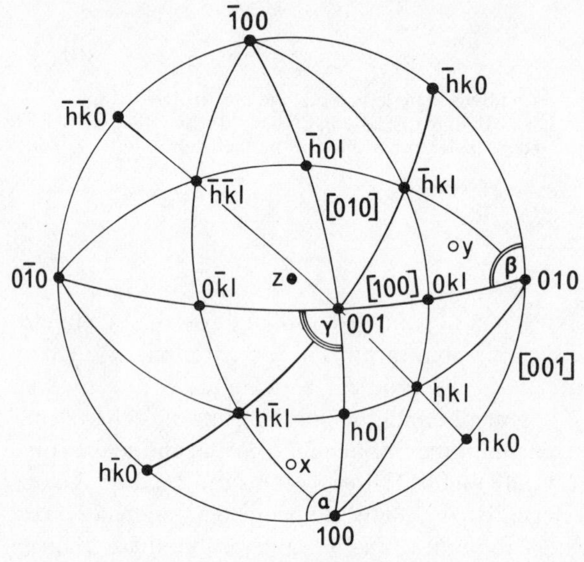

Fig 3.24 The triclinic system. The stereogram shows the relationship of various poles to the reference axes in a hypothetical crystal of class 1. In a crystal of class $\bar{1}$ symmetry would require the opposite of any pole shown to be shown also.

at the intersection of small circles of radius α and γ about the z and x axes respectively. The poles of (100) and (001) are respectively the poles of the great circles containing the y and z axes and the x and y axes.

Monoclinic system

The characteristic symmetry of the monoclinic system is a single axis of twofold symmetry, which is conventionally taken as the y-axis of the unit-cell. The axis of diad symmetry places no restriction on the shape of the unit-mesh of the lattice in the plane perpendicular to itself. It is reasonable therefore to take as the x and z axes directions in this plane so that a and c are as short as possible. The positive directions of the x and z axes are conventionally chosen so that the angle between them, β, is obtuse. The geometry of the monoclinic unit-cell is thus $a \neq b \neq c$, $\alpha = \gamma = 90°$, $\beta \geqslant 90°$ (Fig 3.25(a)).

[8] This system used to be known as the *anorthic system*. After many years of disuse the old name has recently been revived by some authors. Both triclinic and anorthic are currently in common use.

The plane (010) is by definition parallel to x and z; therefore, since $\alpha = \gamma = 90°$, the normal to the (010) plane is parallel to the y-axis. The plane (100) is necessarily parallel to the y and z axes, but, since $\beta \neq 90°$, its normal must be inclined to the x-axis (Fig 3.25(b)). Likewise the normal to the (001) plane must be inclined to the z-axis and perpendicular to the x and y axes. It is apparent from Fig 3.25(b) that $\widehat{ARC} = \widehat{AOC} = \beta$ and that $\widehat{ARC} + \widehat{POQ} = 180°$. Therefore $\widehat{POQ} = 180° - \beta$, that is to say the angle (100):(001) $= 180° - \beta$.

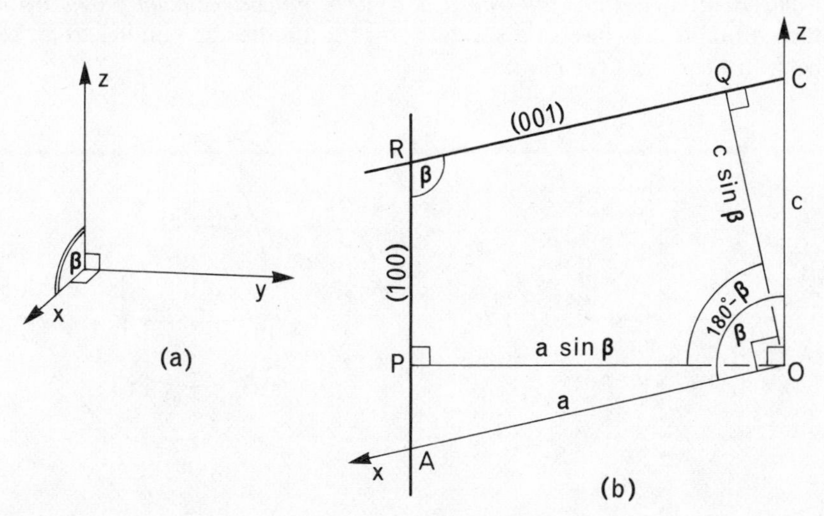

Fig 3.25 Monoclinic geometry: (a) shows the disposition of the reference axes and (b) is an (010) section through the origin. The angle between the normals to the planes (100) and (001) is $180° - \beta$, where β is the interaxial angle $x{:}z$.

The monoclinic system comprises three point groups. The point group 2 has a single symmetry element, a rotation diad parallel to the y-axis; the planes (hkl) and $(\bar{h}k\bar{l})$ are equivalent, as are the directions $[UVW]$ and $[\bar{U}V\bar{W}]$. The point group m has a mirror plane perpendicular to the y-axis, that is parallel to (010); the planes (hkl) and $(h\bar{k}l)$ are equivalent, as are the directions $[UVW]$ and $[U\bar{V}W]$. The point group $2/m$ is the holosymmetric point group of the system; its symmetry elements are a rotation diad parallel to the y-axis and a mirror plane parallel to (010), a combination that introduces a centre of symmetry. In point group $2/m$ the four planes (hkl), $(\bar{h}k\bar{l})$, $(h\bar{k}l)$, and $(\bar{h}\bar{k}\bar{l})$ are equivalent, as are the four directions $[UVW]$, $[\bar{U}V\bar{W}]$, $[U\bar{V}W]$, and $[\bar{U}\bar{V}\bar{W}]$. Stereograms showing selected forms in each monoclinic point group are presented in Fig 3.26.

It is immediately apparent from Fig 3.26 that certain forms in each point group comprise fewer faces than does the *general form* $\{hkl\}$. Thus in point group 2 the face-normal (010) is coincident with the diad so that the form {010} comprises only the face (010), whereas the general form $\{hkl\}$ comprises the two faces (hkl) and $(\bar{h}k\bar{l})$. The form $\{0\bar{1}0\}$ likewise consists of a single face $(0\bar{1}0)$. The general form in point group m likewise comprises two faces (hkl) and $(h\bar{k}l)$; but a face of the type[9] $(h0l)$, whose pole lies in the mirror plane, is not repeated by the mirror plane so that the form $\{h0l\}$ consists of a single face. Forms such as {010} and $\{0\bar{1}0\}$ in point group 2

[9] Now and subsequently we use the symbol $\{h0l\}$ to represent any form with $k = 0$, the indices h and l having any integral values including 0. The symbol $\{h0l\}$ thus includes {100} and {001}; but in point groups where these have fewer faces than $\{h0l\}$ they receive special mention.

and $\{h0l\}$ in point group m are said to be *special forms*. In point group $2/m$ the general form $\{hkl\}$ comprises four faces (hkl), $(\bar{h}k\bar{l})$, $(h\bar{k}l)$, and $(\bar{h}\bar{k}\bar{l})$; special forms are of two kinds, that not affected by the diad, i.e. $\{010\}$ and those not affected by the mirror plane, i.e. $\{h0l\}$. In this case both kinds of special form have the same number of faces, but we shall see in other systems that different kinds of special form may have different numbers of faces.

We take as our definition of a special form the statement: *a special form is any form comprising fewer faces than the general form in the same point group.* It follows that the normal to any face of a special form must either be parallel to an axis of symmetry or lie in a mirror plane.[10]

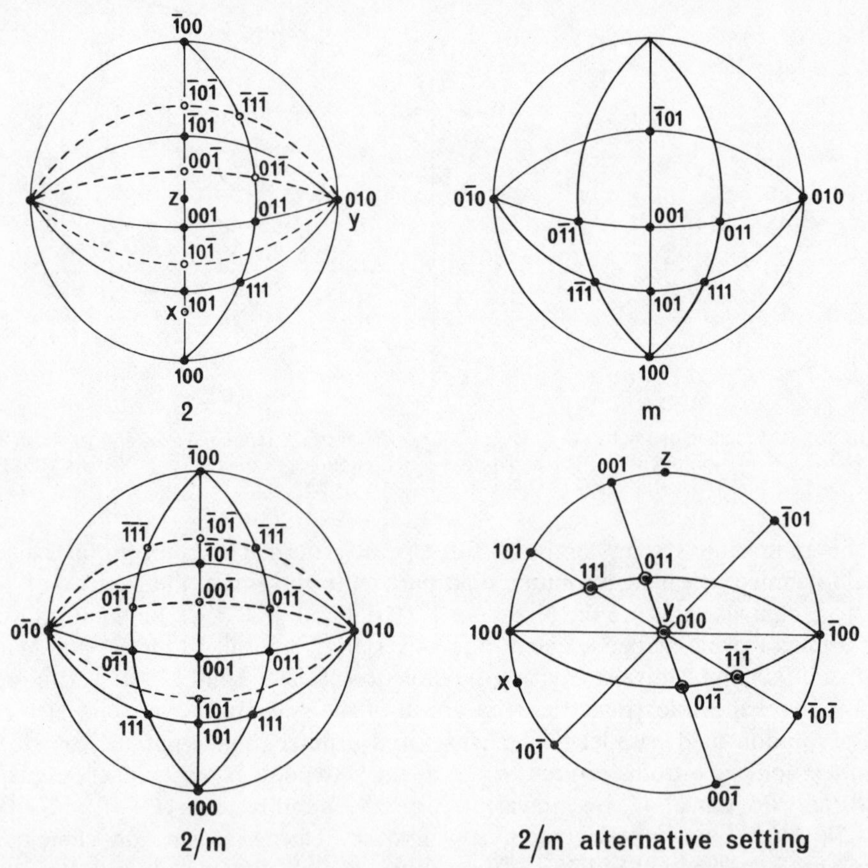

Fig 3.26 The monoclinic point groups. The stereograms show poles of the forms $\{100\}$, $\{010\}$, $\{001\}$, $\{101\}$, $\{\bar{1}01\}$, $\{011\}$, and $\{111\}$ in each of the point groups. The two lower stereograms illustrate alternative settings for $2/m$.

Stereograms of monoclinic crystals are usually plotted, as in Fig 3.26, with the z-axis at the centre of the stereogram and the y-axis at the right-hand end of the horizontal diameter of the primitive. The pole of the positive direction of the x-axis then projects on the vertical diameter of the primitive and, as β is obtuse, in the

[10] Other definitions of a special form have occasionally been employed by authors whose main concern was crystal morphology. The definition adopted here is to be preferred because it is analogous to the definition of a special equivalent position in a space group (chapter 4) and leads directly to the determination of multiplicity in powder photographs (chapter 7).

southern hemisphere. The pole of (010) is coincident with the pole of the *y*-axis. Since (100) is parallel to *y* and *z* its pole lies on the primitive at the lower extremity of the vertical diameter. Since (001) is parallel to *x* and *y* its pole lies at the intersection of the vertical diameter of the primitive with the great circle whose pole is *x*; the pole of (001) thus makes an angle $\beta - 90°$ with *z*.

An alternative setting for the monoclinic stereogram is occasionally employed. The *y*-axis is plotted in the centre of the stereogram so that the primitive represents the (010) plane, the mirror plane in point groups *m* and 2/*m*. The advantage of this setting in these two point groups is that the planes (*hkl*) and ($h\bar{k}l$), which are related by the mirror plane, project so as to be superimposed. Some simplification is thus achieved; in the first mentioned setting symmetry related poles cannot be superimposed. The stereogram of a crystal of point group 2/*m* plotted in the alternative setting is shown in the lower right-hand diagram of Fig 3.26.

Orthorhombic system

The characteristic symmetry of the orthorhombic system is the presence of three mutually perpendicular axes of twofold symmetry. It is obviously convenient to place the *x*, *y*, and *z* axes parallel to symmetry axes so that the unit-cell is obliged to be a rectangular parallelepiped. None of the axes is related to either of the other two so the unit-cell has, in general, edges unequal in length. The shape of the unit-cell is thus given by $a \neq b \neq c$, $\alpha = \beta = \gamma = 90°$.

The plane (100) is by definition parallel to the *y* and *z* axes and, since *x*, *y*, and *z* are mutually perpendicular, its pole is coincident with the *x*-axis. Similarly the normals to the planes (010) and (001) are respectively coincident with the *y* and *z* axes.

The orthorhombic system comprises three point groups. Each is represented by a

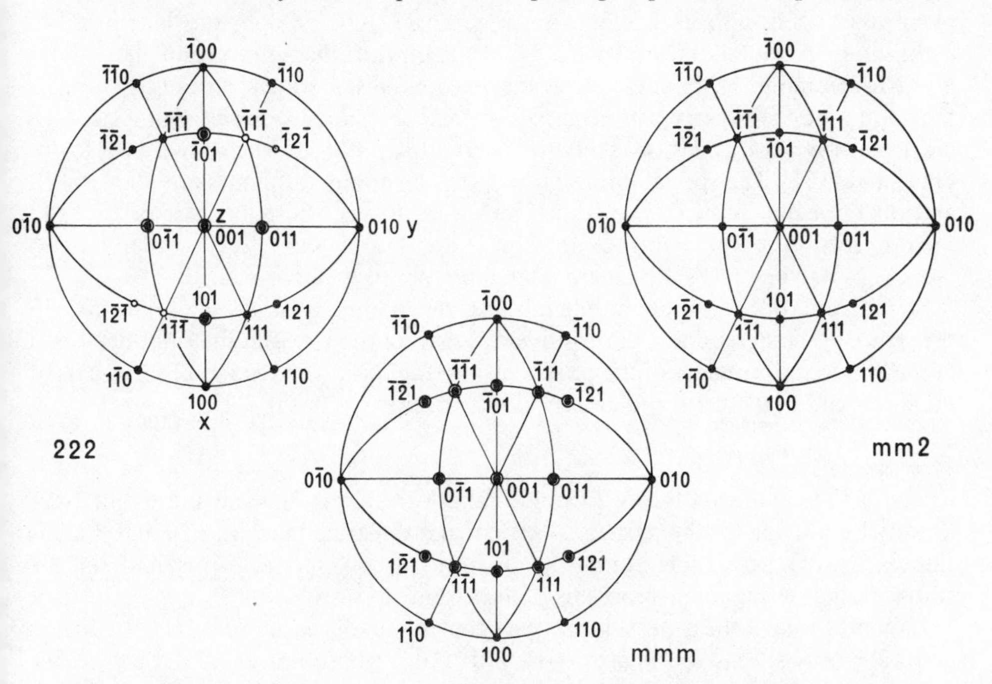

Fig 3.27 The orthorhombic point groups. The stereograms show poles of the forms {100}, {010}, {001}, {011}, {101}, {110}, {111}, and {121} in each of the point groups. Where poles are superimposed only the pole in the upper hemisphere is indexed.

symbol having three positions, which state in turn the symmetry elements associated with the *x*, *y*, and *z* axes.

The point group 222 has three mutually perpendicular rotation diads and is one of the point groups derived directly from Euler's proposition. The general form $\{hkl\}$ comprises four faces (hkl), $(h\bar{k}\bar{l})$, $(\bar{h}k\bar{l})$, and $(\bar{h}\bar{k}l)$; Fig 3.27 shows five such forms $\{011\}$, $\{101\}$, $\{110\}$, $\{111\}$ and $\{121\}$. The special forms in this point group are $\{100\}$, $\{010\}$, and $\{001\}$, each of which has its two faces perpendicular to a diad.

The point group *mm*2 has mirror planes perpendicular to two of the reference axes and a rotation diad parallel to the third. Conventionally the rotation diad is selected as the *z*-axis. The general form $\{hkl\}$ in this point group comprises four faces (hkl) and $(\bar{h}kl)$ related by the (100) mirror plane and $(h\bar{k}l)$ and $(\bar{h}\bar{k}l)$ related to the first two by the (010) mirror plane (Fig 3.27). The special forms $\{0kl\}$ and $\{h0l\}$ have poles lying in the (100) and (010) mirror planes respectively and each consists of two faces $(0kl)$ and $(0\bar{k}l)$ or $(h0l)$ and $(\bar{h}0l)$. The special forms, $\{001\}$ and $\{00\bar{1}\}$ have only a single face (001) and $(00\bar{1})$, the pole of which lies on the rotation diad at the intersection of the two mirror planes. Analogous to the special form $\{001\}$ is the special form of zone axes $\langle 001 \rangle$: the direction $[001]$ is not related by symmetry to the direction $[00\bar{1}]$ so that the positive and negative directions of the *z*-axis are unrelated by symmetry. In this point group therefore the *z*-axis is a *polar axis*, unlike the *x* and *y* axes whose positive and negative directions are related by mirror planes. A consequence of *z* being a polar axis is that poles in the northern and southern hemispheres of the stereogram are not symmetry related.

The holosymmetric point group of the orthorhombic system $\frac{2}{m}\frac{2}{m}\frac{2}{m}$ is derived by adding a centre of symmetry to 222 or to *mm*2, and has a rotation diad and a mirror plane associated with each of the reference axes, *x*, *y*, and *z*; the lattice of an orthorhombic crystal necessarily has the symmetry of this point group. This point group is commonly represented by its short symbol *mmm*, which sufficiently specifies the point group symmetry. The general form $\{hkl\}$ in point group *mmm* comprises the eight faces (hkl), $(\bar{h}kl)$, $(h\bar{k}l)$, $(\bar{h}\bar{k}l)$, $(hk\bar{l})$, $(\bar{h}k\bar{l})$, $(h\bar{k}\bar{l})$, $(\bar{h}\bar{k}\bar{l})$; Fig 3.27 shows two such forms $\{111\}$ and $\{121\}$. The special forms whose poles lie on mirror planes are $\{0kl\}$, $\{h0l\}$, and $\{hk0\}$; each comprises four faces. The special forms whose poles lie on diads at the intersection of mirror planes are $\{100\}$, $\{010\}$, and $\{001\}$; each comprises two faces, e.g. (100) and $(\bar{1}00)$, which are related by the (100) mirror plane.

Stereograms of orthorhombic crystals are conventionally drawn with the *z*-axis in the centre of the stereogram, the positive direction of the *y*-axis at the right-hand end of the horizontal diameter of the primitive, and the positive direction of the *x*-axis at the lower end of the vertical diameter of the primitive as in Fig 3.27.

Tetragonal system

Many of the comments that we would make for other systems at this point have already been made for the tetragonal system and illustrated in Figs 3.20 and 3.21. It suffices to add here some comments on general and special forms and on the conventional setting for stereograms of tetragonal crystals.

In point group 4 the general form $\{hkl\}$ consists of four faces (hkl), $(k\bar{h}l)$, $(\bar{h}\bar{k}l)$, and $(\bar{k}hl)$. The only special forms are $\{001\}$ and $\{00\bar{1}\}$, whose normal is parallel to the tetrad; each form consists of the single face. Point group $\bar{4}$ likewise has a general form $\{hkl\}$ consisting of four faces, but here they are (hkl), $(k\bar{h}\bar{l})$, $(\bar{h}\bar{k}l)$, and $(\bar{k}h\bar{l})$; the only special form is $\{001\}$, which here consists of the two parallel faces (001) and

$(00\bar{1})$. The rotation tetrad is thus a polar axis while the inverse tetrad is non-polar. In point group $4/m$ the general form $\{hkl\}$ consists of eight faces, the indices of which are simply derived from those of the general form in either 4 or $\bar{4}$ by adding a centre of symmetry, that is by including faces with indices opposite in sign to those already listed to give (hkl), $(\bar{h}\bar{k}\bar{l})$, $(k\bar{h}l)$, $(\bar{k}h\bar{l})$, $(\bar{h}\bar{k}l)$, $(hk\bar{l})$, $(\bar{k}hl)$, $(k\bar{h}\bar{l})$. There are two kinds of special form: $\{hk0\}$ consists of four faces whose poles lie in the mirror plane, $(hk0)$, $(k\bar{h}0)$, $(\bar{h}\bar{k}0)$, and $(\bar{k}h0)$, while $\{001\}$ consists of the two faces (001) and $(00\bar{1})$ so that in this point group the tetrad is not polar.

In point group 422 the general form $\{hkl\}$ consists of eight faces: operation of the [100] diad on (hkl) yields $(h\bar{k}\bar{l})$ and operation of the tetrad on these two faces yields $(\bar{k}hl)$ and $(\bar{k}\bar{h}\bar{l})$, $(\bar{h}\bar{k}l)$ and $(\bar{h}k\bar{l})$, and $(k\bar{h}l)$ and $(kh\bar{l})$. There are two kinds of special form, $\{100\}$ and $\{110\}$, each consisting of four faces whose poles are parallel to diads, and a third special form $\{001\}$ consisting of two opposite faces whose poles are parallel to the tetrad, which is thus non-polar.

The general form of point group $4mm$ consists again of eight faces, all of which lie in the same hemisphere (Fig 3.21). Special forms, the poles of whose faces lie in mirror planes, are $\{h0l\}$ and $\{hhl\}$; each comprises four faces. The special forms $\{001\}$ and $\{00\bar{1}\}$ consist each of a single face, the tetrad being polar.

Point group $\bar{4}2m$ likewise has a general form consisting of eight faces (Fig 3.21). The special form $\{100\}$ consists of four faces whose poles are parallel to the $\langle 100\rangle$ diads. The special form, which has $h = k$ and l unrestricted, that is $\{hhl\}$ likewise consists of four faces, the poles of its faces lying in mirror planes. The special form $\{001\}$ consists of two faces whose poles are parallel to the inverse tetrad.

The holosymmetric point group of the tetragonal system $4/mmm$ has a general form with more faces than that for any other point group of the system. Faces are superimposed on the northern and southern hemispheres of the stereogram by the (001) mirror plane so that the general form has the same faces as the general form in any of the point groups 422, $4mm$, or $\bar{4}2m$ duplicated by the (001) mirror plane to give sixteen faces (Fig 3.21). Special forms with eight faces are those whose poles lie in mirror planes, $\{h0l\}$, $\{hhl\}$, $\{hk0\}$; special forms with four faces are those whose poles are parallel to diads, $\{100\}$ and $\{110\}$; and $\{001\}$ is a special form consisting of two opposite faces whose poles are parallel to the tetrad.

This is a convenient point at which to make a brief digression into morphological crystallography. A well-developed crystal can be assigned to the correct system usually by inspection and always (with a few exceptions) by precise goniometric measurement. Whether it can be assigned unambiguously to a crystal class however depends on the sort of faces displayed. Reference to Fig 3.21 shows that the disposition of the faces of a general form such as $\{121\}$ enables the point group to be determined with certainty. But if the only faces present are those of the forms $\{001\}$ and $\{h0l\}$, unambiguous determination of class will not be possible unless the crystal belongs to class $\bar{4}$; the point groups 4 and $4mm$ will be indistinguishable as will the point groups $4/m$, 422, $\bar{4}2m$, and $4/mmm$. Now $\{001\}$ is a special form in all the tetragonal point groups but $\{h0l\}$ is a special form only in the point groups $4mm$ and $4/mmm$. Crystal morphologists have found it convenient to extend the definition of a special form to include forms whose faces are parallel to symmetry axes (i.e. poles of faces normal to axes). It then becomes possible to say that a crystal can only be assigned unambiguously to a class if it displays one or more general forms. On the extended definition $\{h0l\}$ is a special form in 422 and $\bar{4}2m$ so that the disposition of the faces of this form enables the three point groups, 4, $\bar{4}$, and $4/m$ in which $\{h0l\}$ is a general form to be distinguished. Since we are not primarily concerned with crystal morphology we shall make no further use of the extended definition of a special form.

Stereograms of tetragonal crystals are conventionally drawn, as in Figs 3.20 and 3.21, with the tetrad in the centre of the stereogram, the positive direction of the y-axis at the right-hand end of the horizontal diameter of the primitive, and the positive direction of the x-axis at the lower end of the vertical diameter of the primitive.

Cubic system

The cubic system is characterized by triads equally inclined to orthogonal reference axes so that x, y, and z are equivalent. The unit-cell must therefore be a cube; in consequence the angle between any pair of planes (hkl) and $(h'k'l')$ is independent of the magnitude of the unit-cell dimension a and the same for all cubic crystals, a point that we shall explore in detail later in this chapter.

The point groups of the cubic system (shown in Figs 3.20 and 3.28) are derived from the last two effective combinations of rotation axes listed in Table 3.2 and illustrated in Fig 3.11(e)–(h); triads are disposed parallel to the body diagonals of a cube in both, with diads parallel to cube edges in 233 and parallel to face diagonals of the cube in 234, which has tetrads parallel to cube edges. The cubic system, like the orthorhombic, lacks the sort of point group that merely has a single rotation or inversion axis with or without a centre of symmetry. The conventional symbol for a cubic point group consists of three terms: first a statement of the symmetry elements associated with the cube edges $\langle 100 \rangle$, second a statement of the symmetry elements associated with the body diagonals $\langle 111 \rangle$, and third a statement of the symmetry elements, if any, associated with face diagonals $\langle 110 \rangle$.

The cubic point group 23 is simply the Euler combination 233. Its symmetry elements are four triads parallel to $\langle 111 \rangle$ and three diads parallel to $\langle 100 \rangle$. The general form $\{hkl\}$ comprises twelve faces (Fig 3.28), whose indices are (hkl), (klh), (lhk), $(h\bar{k}\bar{l})$, $(k\bar{l}\bar{h})$, $(l\bar{h}\bar{k})$, $(\bar{h}\bar{k}l)$, $(\bar{k}\bar{l}h)$, $(\bar{l}\bar{h}k)$, $(\bar{h}k\bar{l})$, $(\bar{k}l\bar{h})$, $(\bar{l}h\bar{k})$. The special form $\{111\}$ consists of four faces whose poles are parallel to triads; it is a tetrahedron. A similarly shaped but distinct special form is $\{11\bar{1}\}$. The other special form $\{100\}$ consists of faces whose poles are parallel to diads and is a cube.

The point group[11] $m3$, which may be derived by addition of a centre of symmetry to 23, has four inverse triads parallel to $\langle 111 \rangle$ with three diads parallel to $\langle 100 \rangle$ and three mirror planes $\{100\}$. The general form (Fig 3.28) consists of twenty-four faces. The special form $\{111\}$ in this point group consists of eight faces and is an octahedron. The special form $\{100\}$ is again, as in all cubic point groups, the cube. The special form $\{hk0\}$ consists of twelve faces.

The remaining three cubic point groups are derived from the Euler combination 234, itself the point group 432. This point group has four triads parallel to $\langle 111 \rangle$, three tetrads parallel to $\langle 100 \rangle$, and six diads parallel to $\langle 110 \rangle$. The general form (Fig 3.28) consists of twenty-four faces. The special forms $\{111\}$ and $\{100\}$ are respectively the octahedron and the cube. A new kind of special form arises in this point group: $\{110\}$ which consists of twelve faces and is known as the rhombic dodecahedron (dodecahedron = a twelve-faced body; rhombic because in regular development each face is a rhombus). The rhombic dodecahedron can also occur in 23 and $m3$; in 23 it is the general form with $h = k$, $l = 0$ and in $m3$ it is the special form $\{hk0\}$ with $h = k$.

[11] Strictly the symbol for this point group is $m\bar{3}$, but $m3$ is always preferred. The same comment applies to the holosymmetric point group $m3m$.

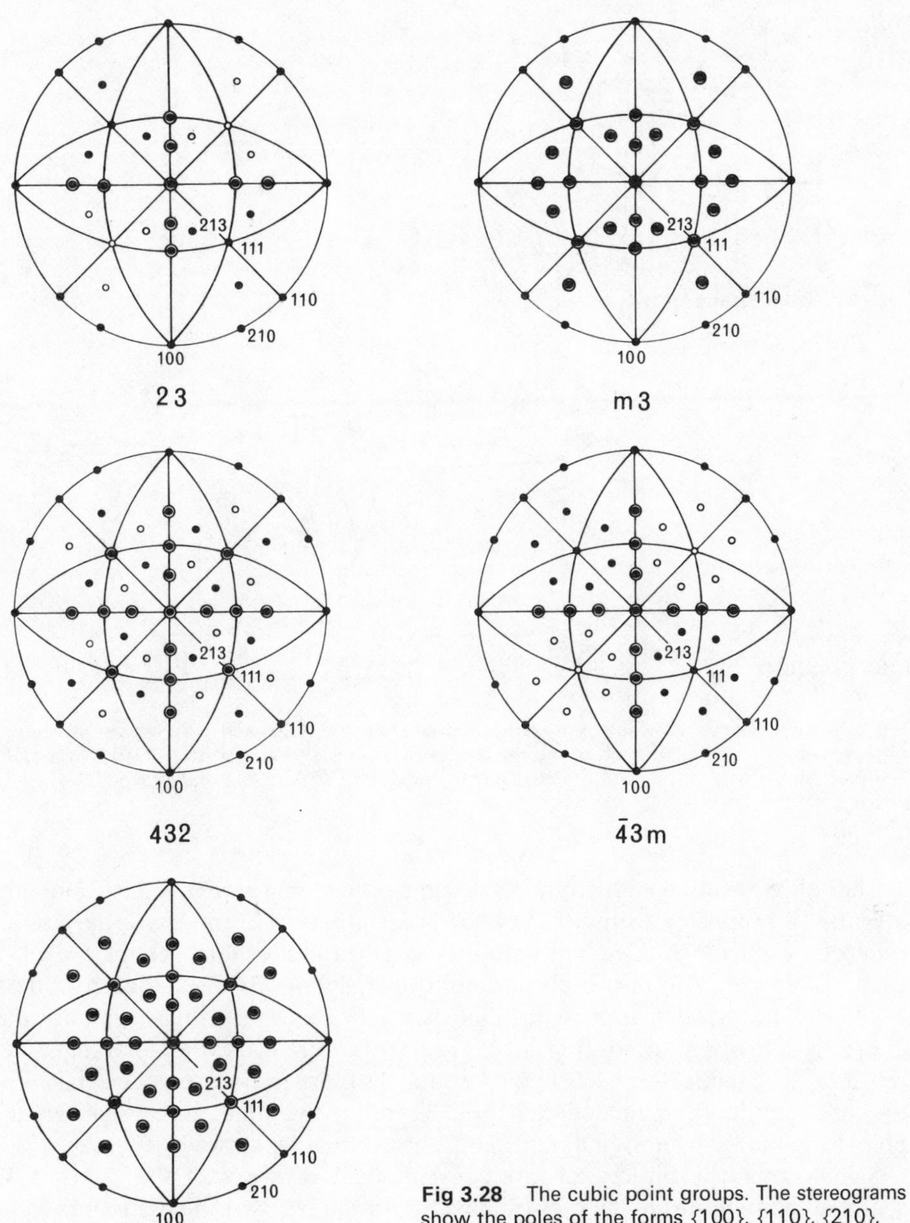

Fig 3.28 The cubic point groups. The stereograms show the poles of the forms {100}, {110}, {210}, {111}, and {213} in each of the point groups, only one face of each form being indexed.

The point group $\bar{4}3m$, obtained by allowing the tetrad and the diad in the Euler combination 234 to be inverse, has triads parallel to $\langle 111 \rangle$, inverse tetrads parallel to $\langle 100 \rangle$, and six mirror planes parallel to $\{110\}$. These $\{110\}$ mirror planes, which are known as diagonal mirror planes, are inclined at 45° to two of the reference axes and parallel to the third reference axis; each plane of the form is parallel to two of the $\langle 111 \rangle$ triads as shown in Fig 3.29 so that each triad is a line of intersection of three diagonal mirror planes which are mutually inclined at 60°. The general form (Fig 3.28) consists of twenty-four faces. The special forms are $\{111\}$ and $\{11\bar{1}\}$, tetrahedra as in 23; $\{100\}$, a cube; and $\{hhl\}$, which consists of twelve faces.

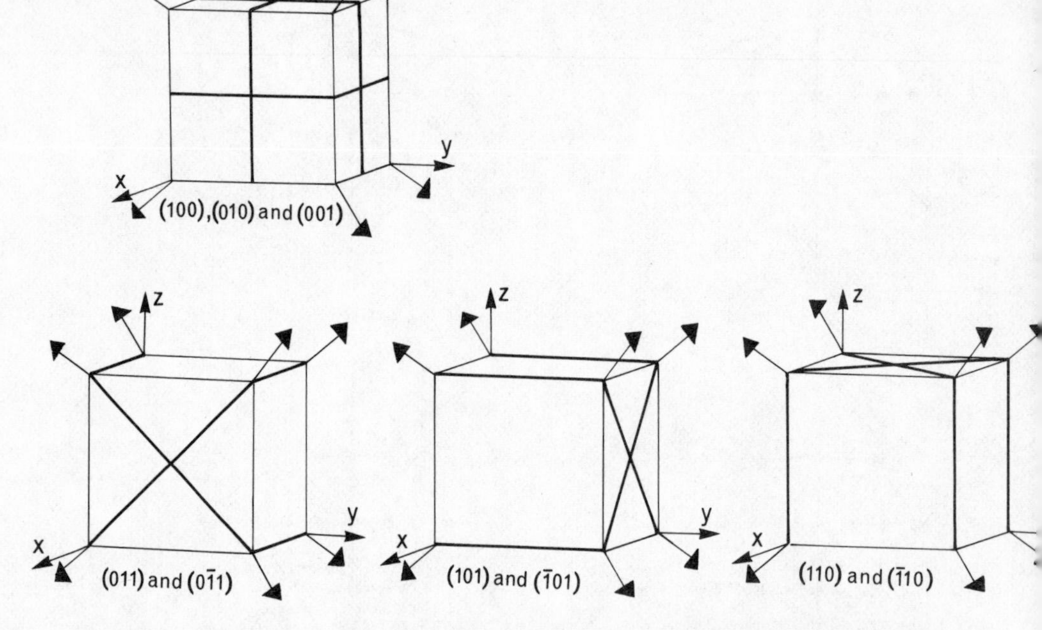

Fig 3.29 Perspective drawings of the cubic unit-cell to show {100} and {110} mirror planes (represented by bold lines). Each of the three lower drawings shows a pair of {110} mirror planes and their relationship to the ⟨111⟩ triads. Only the point groups $\bar{4}3m$ and $m3m$ have {110} mirror planes.

The holosymmetric point group of the cubic system, $m3m$, is derived by combining a centre of symmetry with either of the point groups 432 or $\bar{4}3m$. The essential symmetry elements of this point group, as specified in its short symbol, are ⟨111⟩ triads, {100} and {110} mirror planes; tetrads parallel to ⟨100⟩ and diads parallel to ⟨110⟩ may be regarded as consequential symmetry elements, as may the upgrading of the triads to inverse triads (Fig 3.20). The general form consists of forty-eight faces (Fig 3.28). The special forms {111}, {100}, and {110} are respectively the octahedron, the cube, and the rhombic dodecahedron. There are two other special forms whose poles lie in mirror planes, $\{h0l\}$ and $\{hhl\}$, each consisting of twenty-four faces.

Stereograms of cubic crystals are conventionally drawn with the z-axis in the centre of the stereogram, the positive direction of the y-axis at the right-hand end of the horizontal diameter of the primitive, and the positive direction of the x-axis at the lower end of the vertical diameter of the primitive. Occasionally it is convenient to draw a cubic stereogram with a triad axis perpendicular to the plane of the diagram, but this orientation does not provide a particularly clear statement of point group symmetry.

Hexagonal system

The hexagonal system is characterized by the presence of a rotation or inverse hexad, which is taken as the z-axis of the unit-cell. The unit-mesh of the lattice planes perpendicular to the hexad can conveniently be chosen as a rhombus with $a = b$, $\gamma = 120°$; x and y for the unit-cell are taken to be parallel to the sides of this rhombus.

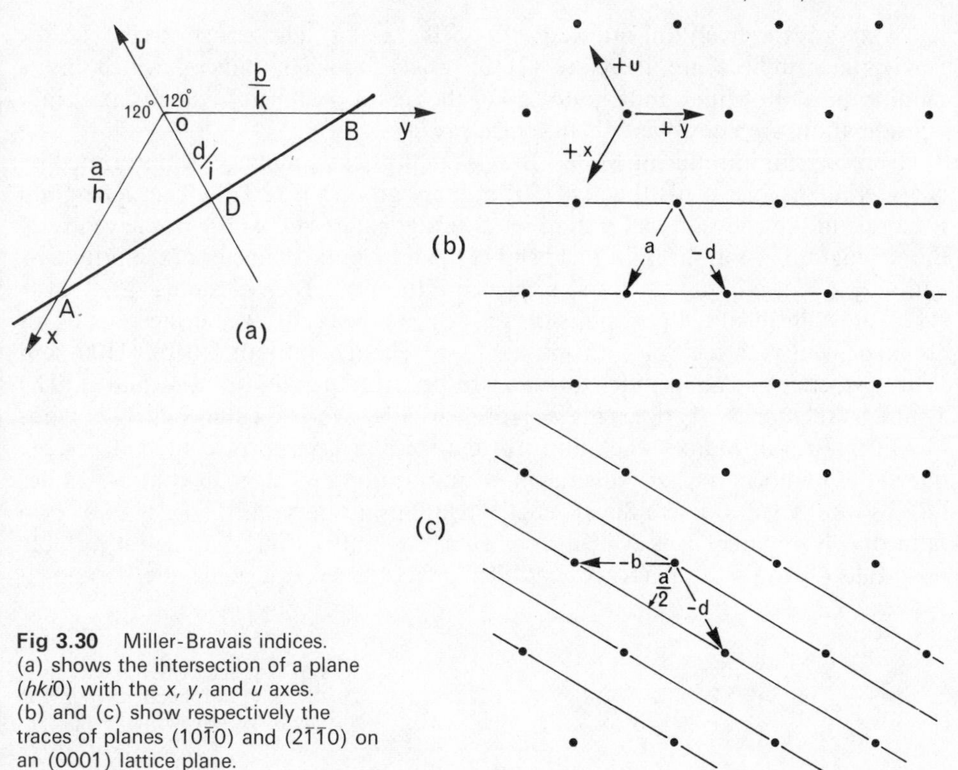

Fig 3.30 Miller-Bravais indices. (a) shows the intersection of a plane ($hki0$) with the x, y, and u axes. (b) and (c) show respectively the traces of planes ($10\bar{1}0$) and ($2\bar{1}\bar{1}0$) on an (0001) lattice plane.

The hexagonal unit-cell can thus be described as having $a = b \neq c$, $\alpha = \beta = 90°$, $\gamma = 120°$.

The hexagonal and trigonal systems differ from all other systems in that the operation of their principal symmetry axis generates a third axis equivalent to x and y, whose positive direction is inclined at 120° to $+x$ and to $+y$. This extra axis (Fig 3.30) is designated u, the lattice repeat along it being $d = a = b$. To take into account the generation of this extra axis a fourth index i is introduced into the symbol for a plane so that ($hkil$) represents a plane making intercepts a/h, b/k, d/i, c/l on the x, y, u, and z axes respectively. But three-dimensional geometry cannot be described in terms of four independent parameters so h, k, and i must be interrelated. Inspection of Fig 3.30(a) indicates that h, k, and i cannot all have the same sign; in the case illustrated h and k are positive while i is negative. The area of the triangle OAB is the sum of the areas of the triangles OAD and ODB, therefore

$$\frac{1}{2}.\frac{a}{h}.\frac{b}{k}\sin 120° = -\frac{1}{2}.\frac{a}{h}.\frac{d}{i}\sin 60° - \frac{1}{2}.\frac{b}{k}.\frac{d}{i}\sin 60°$$

since i is negative. But $a = b = d$,

therefore $$\frac{1}{hk} = -\frac{1}{hi} - \frac{1}{ki}$$

and $$h + k + i = 0.$$

Thus lattice planes parallel to y and z make equal intercepts on the $+x$ and $-u$ axes (Fig 3.30(b)); this set of planes is thus indexed as ($10\bar{1}0$). The set of planes whose normal is parallel to the x-axis (Fig 3.30(c)) makes intercepts $-b$ and $-d$ on the y

and u axes respectively, an intercept $a/2$ on the x-axis, and zero intercept on the z-axis; their indices are therefore $(2\bar{1}\bar{1}0)$. Such four-digit indices, which are a modification of Miller indices to take the extra symmetry-related axis into consideration, are known as *Miller–Bravais indices.*

The reason for introducing Miller–Bravais indices can simply be demonstrated by considering the forms $\{10\bar{1}0\}$ and $\{11\bar{2}0\}$ in point group 6 (Fig 3.31). The stereogram is plotted in the conventional setting for a hexagonal crystal with z in the centre of the stereogram, $+y$ at the right-hand end of the horizontal diameter of the primitive, and $+x$, $+y$, $+u$ in anticlockwise sequence 120° apart. The poles of $(2\bar{1}\bar{1}0)$, $(\bar{1}2\bar{1}0)$, $(\bar{1}\bar{1}20)$ are coincident with the poles of $+x$, $+y$, $+u$ respectively and their opposites are coincident with the poles of $-x$, $-y$, $-u$. The faces $(0\bar{1}10)$, $(10\bar{1}0)$, $(\bar{1}100)$ and their opposites are all parallel to z and respectively parallel to x, y, and u. The symmetry relationship between the faces of each of these forms is immediately obvious in Miller–Bravais indices: the indices of the faces of each form contain the same quartet of numbers regularly interchanged in position and sign. In contrast Miller indices wholly fail to make the symmetry relationship between the faces of a form immediately apparent as is evident from Table 3.6. In the sequel we shall invariably use Miller–Bravais in preference to Miller indices when discussing the hexagonal system.

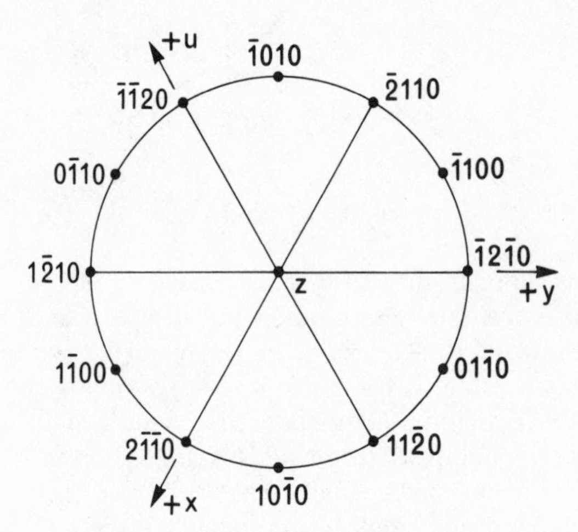

Fig 3.31 Stereogram of the forms $\{10\bar{1}0\}$ and $\{11\bar{2}0\}$ in point group 6 to illustrate Miller–Bravais indexing.

Table 3.6
Comparison of Miller–Bravais indices and Miller indices of the forms $\{10\bar{1}0\}$ and $\{11\bar{2}0\}$

Miller–Bravais	Miller	Miller–Bravais	Miller
$(10\bar{1}0)$	(100)	$(2\bar{1}\bar{1}0)$	$(2\bar{1}0)$
$(01\bar{1}0)$	(010)	$(11\bar{2}0)$	(110)
$(\bar{1}100)$	$(\bar{1}10)$	$(\bar{1}2\bar{1}0)$	$(\bar{1}20)$
$(\bar{1}010)$	$(\bar{1}00)$	$(\bar{2}110)$	$(\bar{2}10)$
$(0\bar{1}10)$	$(0\bar{1}0)$	$(\bar{1}\bar{1}20)$	$(\bar{1}\bar{1}0)$
$(1\bar{1}00)$	$(1\bar{1}0)$	$(1\bar{2}10)$	$(1\bar{2}0)$

When using Miller–Bravais indices the addition rule still holds for tautozonal faces: thus $(11\bar{2}0)$ lies in the same zone and between $(10\bar{1}0)$ and $(01\bar{1}0)$. But when using cross-multiplication to determine the indices of a zone axis it is necessary to get rid of the superfluous index i. For this purpose the Miller–Bravais index is written as $(hk.l)$ and the resultant zone axis symbol as $[UV\dagger W]$.

Analogous to Miller–Bravais face indices there is a system of four-digit zone axis symbols, known as *Weber symbols*; but these cannot be converted to Millerian three-digit zone axis symbols simply by omitting the superfluous index. The symbol $[UVW]$ specifies a line through the origin passing through a point with coordinates Ua, Vb, Wc; in three dimensions three indices are necessarily adequate. In contrast the four-digit symbol $[uvtw]$ contains an unnecessary index so that unless some condition is imposed to link u, v, and t a direction in three-dimensional space will not be uniquely represented by a four-axis symbol. For example if no such condition were imposed the x-axis could variously be described as $[1000]$, $[0\bar{1}\bar{1}0]$, $[2110]$, $[2\bar{1}\bar{1}0]$ as illustrated in Fig 3.32(a). The condition applied in the Weber nomenclature is $u+v+t=0$, analogous to $h+k+i=0$ for Miller–Bravais face indices; the positive direction of the x-axis is then represented by $[2\bar{1}\bar{1}0]$. The symbol $[UVW]$ represents the vector $U\mathbf{a}+V\mathbf{b}+W\mathbf{c}$ and the symbol $[uvtw]$ represents the vector $u\mathbf{a}+v\mathbf{b}+t\mathbf{d}+w\mathbf{c}$.

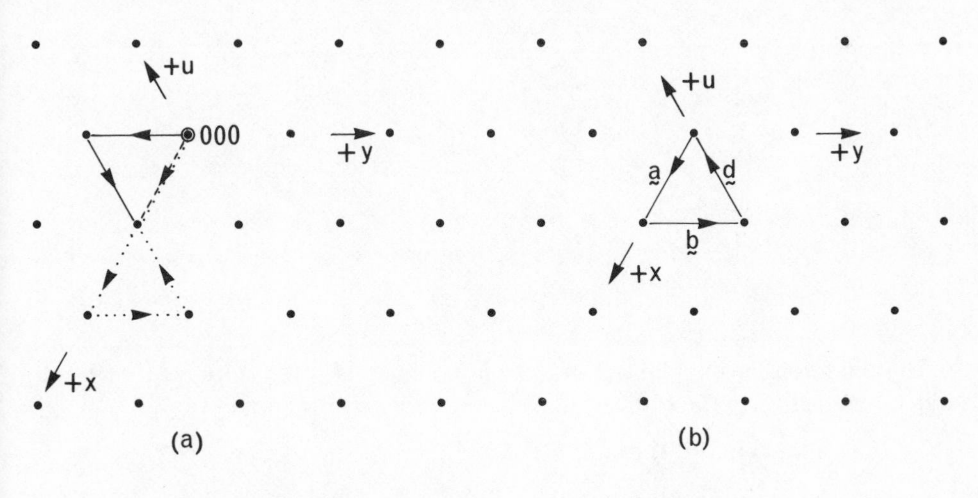

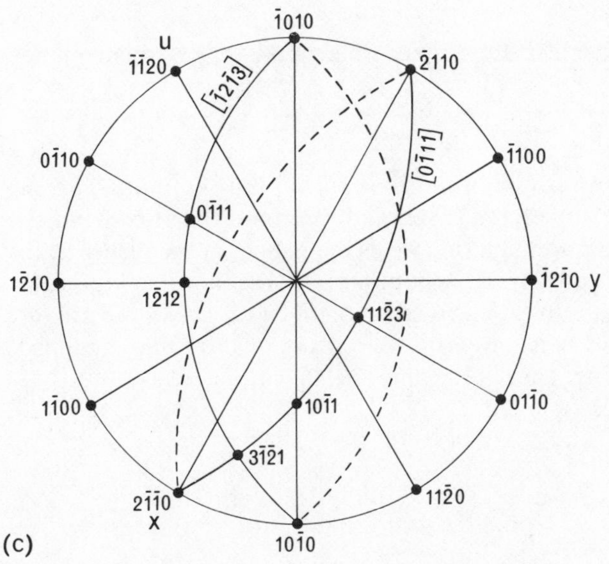

Fig 3.32 Weber symbols. (a) and (b) show an (0001) net of lattice points: (a) illustrates the various ways in which the $+x$ direction can be indexed in four-digit nomenclature and (b) the condition $\mathbf{a}+\mathbf{b}+\mathbf{d}=0$. The stereogram (c) illustrates the condition for the plane $(hkil)$ to lie in the zone $[uvtw]$, $hu+kv+it+lw=0$.

For these two vectors to be identical

$$U\mathbf{a}+V\mathbf{b}+W\mathbf{c} = u\mathbf{a}+v\mathbf{b}+t\mathbf{d}+w\mathbf{c}$$

Since the x, y, and u axes are inclined at 120° to one another and $a = b = d$,

$$\mathbf{a}+\mathbf{b}+\mathbf{d} = 0 \qquad \text{(Fig 3.32(b))}$$

Hence $\quad U\mathbf{a}+V\mathbf{b}+W\mathbf{c} = (u-t)\mathbf{a}+(v-t)\mathbf{b}+w\mathbf{c}$

and

$$U = u-t$$
$$V = v-t$$
$$W = w$$

These equations are adequate to convert Weber symbols to Millerian zone axis symbols, but to convert in the opposite direction it is necessary to apply the condition

$$u+v+t = 0$$

whence

$$U = 2u+v$$
$$V = u+2v$$
$$W = w$$

so that

$$u = \frac{2U-V}{3}$$
$$v = \frac{2V-U}{3}$$
$$t = -\frac{U+V}{3}$$
$$w = W.$$

The condition for the plane (hkl) to lie in the zone $[UVW]$ is $hU+kV+lW = 0$, which becomes when the equivalent Weber symbol $[uvtw]$ is used

$$h(u-t)+k(v-t)+lw = 0$$

i.e. $\quad hu+kv-(h+k)t+lw = 0.$

And if the Miller–Bravais index i is introduced, where $h+k+i = 0$, this equation becomes

$$hu+kv+it+lw = 0,$$

the condition for the plane $(hkil)$ to lie in the zone $[uvtw]$. For example the zone $[\bar{1}2\bar{1}3]$ contains the planes $(10\bar{1}0)$, $(0\bar{1}11)$, $(1\bar{2}12)$ and the zone $[0\bar{1}11]$ contains the planes $(2\bar{1}\bar{1}0)$, $(10\bar{1}1)$, $(11\bar{2}3)$. The addition rule yields the indices of the planes at the intersection of these two zones as $(3\bar{1}\bar{2}1)$ and its opposite $(\bar{3}12\bar{1})$ as shown in Fig 3.32(c).

To determine the indices of a plane at the intersection of two zones whose axes are represented by Weber symbols it is necessary to convert to three-digit zone axis symbols before cross-multiplying. Thus $[\bar{1}2\bar{1}3]$ and $[0\bar{1}11]$ become respectively $[011]$ and $[\bar{1}\bar{2}1]$. Cross-multiplication

$$\begin{array}{c|ccccc|c} 0 & 1 & & 1 & & 0 & & 1 & 1 \\ \bar{1} & \bar{2} & \times & 1 & \times & \bar{1} & \times & \bar{2} & 1 \\ \hline & & 3 & & \bar{1} & & 1 & & \end{array}$$

yields $(3\bar{1}.1)$ i.e. $(3\bar{1}\bar{2}1)$ and its opposite $(\bar{3}12\bar{1})$ for the faces common to both zones. Similarly the Weber symbol for the zone containing the faces $(2\bar{1}\bar{1}1)$ and $(01\bar{1}1)$ is determined by first cross-multiplying with the i index omitted

$$\begin{array}{c|cccccc|c} 2 & \bar{1} & & 1 & & 2 & & \bar{1} \quad 1 \\ & & \times & & \times & & \times & \\ 0 & 1 & & 1 & & 0 & & 1 \quad 1 \\ \hline & & \bar{2} & & \bar{2} & & 2 & \end{array}$$

to give the three-digit symbol $[\bar{2}\bar{2}2]$ which reduces to $[\bar{1}\bar{1}1]$, and then converting to the Weber symbol $[\frac{\bar{1}}{3}\frac{\bar{1}}{3}\frac{2}{3}1]$ which reduces to $[\bar{1}\bar{1}23]$.

The reason for introducing the Weber symbol is that it provides a clear expression of symmetry relationships between zone axes in the hexagonal system just as Miller–Bravais indices do for faces. But, as we have seen, conversion to normal three-digit symbols is a prerequisite to calculation. We shall therefore mostly use three-digit symbols $[UVW]$ modified in one respect. To avoid ambiguity about which two of the three symmetry-related axes x, y, and u are in use, it is convenient to indicate the position of the digit referable to the omitted axis by a dagger $[UV\dagger W]$. For example the symbol $[11\dagger0]$ represents the vector $\mathbf{a}+\mathbf{b}$, whereas the symbol $[\dagger110]$ represents the vector $\mathbf{b}+\mathbf{d}$. Symmetry relationships can satisfactorily be displayed, although this is not often done, by varying the omitted axis. For instance an inverse hexad makes the zone axes $[21\dagger0]$, $[\bar{1}1\dagger0]$, $[\bar{1}\bar{2}\dagger0]$ equivalent (Fig 3.33); the equivalence becomes apparent in the symbols on rewriting as $[1\dagger\bar{1}0]$, $[\bar{1}1\dagger0]$, $[\dagger\bar{1}10]$. One final comment has to be made on hexagonal zone axis symbols: it is conventional, convenient, and in no way misleading to represent the z-axis as $[0001]$ whatever symbolic notation is being used for other zone axes.

The point groups of the hexagonal system correspond exactly in nomenclature to those of the tetragonal system, which we have already explored. Those having symmetry elements associated only with the z-axis are denoted 6, $\bar{6}$, $6/m$. Those derived from the Euler combination 226 are denoted by symbols having three positions: first a statement of whether a rotation or inverse hexad is parallel to z, second a statement of whether a diad or a mirror plane is associated with the x, y, and u axes, and third a statement of whether a diad or a mirror plane is associated with $\langle 1\bar{1}\dagger0\rangle$ directions. The directions referred to in the third position are equally

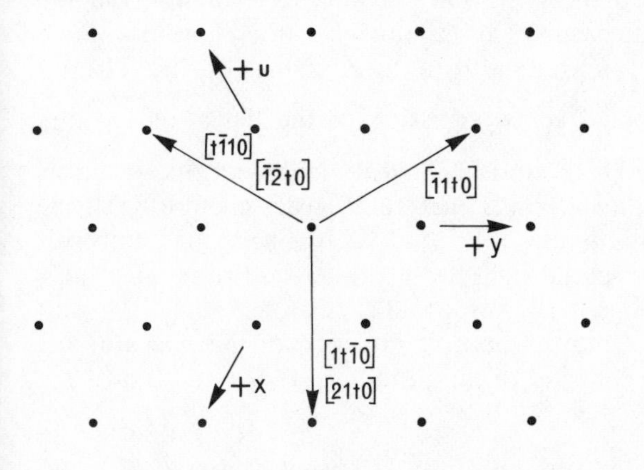

Fig 3.33 Symmetry relationships using three-digit zone axis symbols in the hexagonal system displayed by varying the axis omitted.

inclined in the (0001) plane to adjacent reference axes, thus $[1\bar{1}\bar{2}0]$ makes angles of 30° with $+x$ and $-y$.

In point group 6 the general form consists of six faces and the only special forms are $\{0001\}$ and $\{000\bar{1}\}$ which each consists of a single face; the hexad is thus polar. In point group $\bar{6}$ the general form again consists of six faces, but the special form $\{0001\}$ here has two faces; and, because the $\bar{6}$ axis includes a mirror plane (0001), $\{hk.0\}$ is a special form of three faces. Point group $6/m$ has a general form consisting of twelve faces and special forms $\{hk.0\}$, consisting of six faces, and $\{0001\}$ consisting of two faces. Stereograms showing selected forms in these three point groups are shown in Fig 3.34.

The Euler combination 226 becomes on rearrangement of its symbol into the conventional form the point group 622. This point group has a rotation hexad [0001] and two sets of rotation diads, one set parallel to x, y, and u, the other parallel to $\langle 1\bar{1}\bar{2}0\rangle$. The general form (Fig 3.34) consists of twelve faces; the special forms $\{2\bar{1}\bar{1}0\}$ and $\{10\bar{1}0\}$ each have six faces; and the special form $\{0001\}$ consists of the parallel faces (0001) and $(000\bar{1})$.

Point group $6mm$ has a rotation hexad [0001] and two sets of mirror planes $\{2\bar{1}\bar{1}0\}$ and $\{10\bar{1}0\}$. The general form once again comprises twelve faces, but here they all lie in the same hemisphere (Fig 3.34). The two special forms whose faces have poles lying in mirror planes are $\{2h\bar{h}\bar{h}l\}$ and $\{h0\bar{h}l\}$; each comprises six faces. The special forms $\{0001\}$ and $\{000\bar{1}\}$ in this point group each has a single face; the hexad is therefore polar.

Point group $\bar{6}m2$ has an inverse hexad [0001], mirror planes $\{2\bar{1}\bar{1}0\}$ and diads $\langle 1\bar{1}\bar{2}0\rangle$. The general form (Fig 3.34) comprises twelve faces, which appear on the stereogram in superimposed pairs because the $\bar{6}$ axis includes a mirror plane (0001) as is evident also from the stereogram for point group $\bar{6}$. Special forms related to mirror planes are thus $\{h0\bar{h}l\}$ and $\{hk.0\}$, each comprising six faces. There are two special forms associated with diads, $\{10\bar{1}0\}$ and its opposite $\{\bar{1}010\}$ each of which has three faces. The only other special form, of two faces, is $\{0001\}$. In this point group, as in $\bar{4}2m$, there are two orientations $\bar{6}m2$ and $\bar{6}2m$ depending on whether the x-axis is taken perpendicular to a mirror plane or parallel to a diad. The distinction here too can only be made on the basis of which gives x parallel to the shortest lattice repeat in the (0001) plane.

The holosymmetric point group of the hexagonal system has a rotation hexad and perpendicular mirror plane associated with [0001] combined with a diad and a mirror plane associated both with the x, y, u axes and with the directions $\langle 1\bar{1}\bar{2}0\rangle$. The full symbol $\frac{6}{m}\frac{2}{m}\frac{2}{m}$ is commonly abbreviated by omission of the diads and written $6/mmm$. The general form (Fig 3.34) here consists of twenty-four faces. Special forms associated with mirror planes are $\{2h\bar{h}\bar{h}l\}$, $\{h0\bar{h}l\}$, and $\{hk.0\}$, each comprising twelve faces. Special forms associated with diads, $\{2\bar{1}\bar{1}0\}$ and $\{10\bar{1}0\}$, have the number of their faces again halved to six. The special form $\{0001\}$ comprises two parallel faces.

It is conventional to draw stereograms of hexagonal crystals, as in Fig 3.34, with z in the centre of the stereogram and the positive direction of the y-axis at the right-hand end of the horizontal diameter of the primitive. The lower end of the vertical diameter of the primitive is then $[21\bar{2}0]$.

Fig 3.34 The hexagonal point groups. The stereograms show the poles of the forms $\{0001\}$, $\{10\bar{1}0\}$, $\{2\bar{1}\bar{1}0\}$, $\{10\bar{1}1\}$ $\{2\bar{1}\bar{1}1\}$, and $\{3\bar{1}\bar{2}1\}$. Only one face of each form is indexed; (0001) is not indexed.

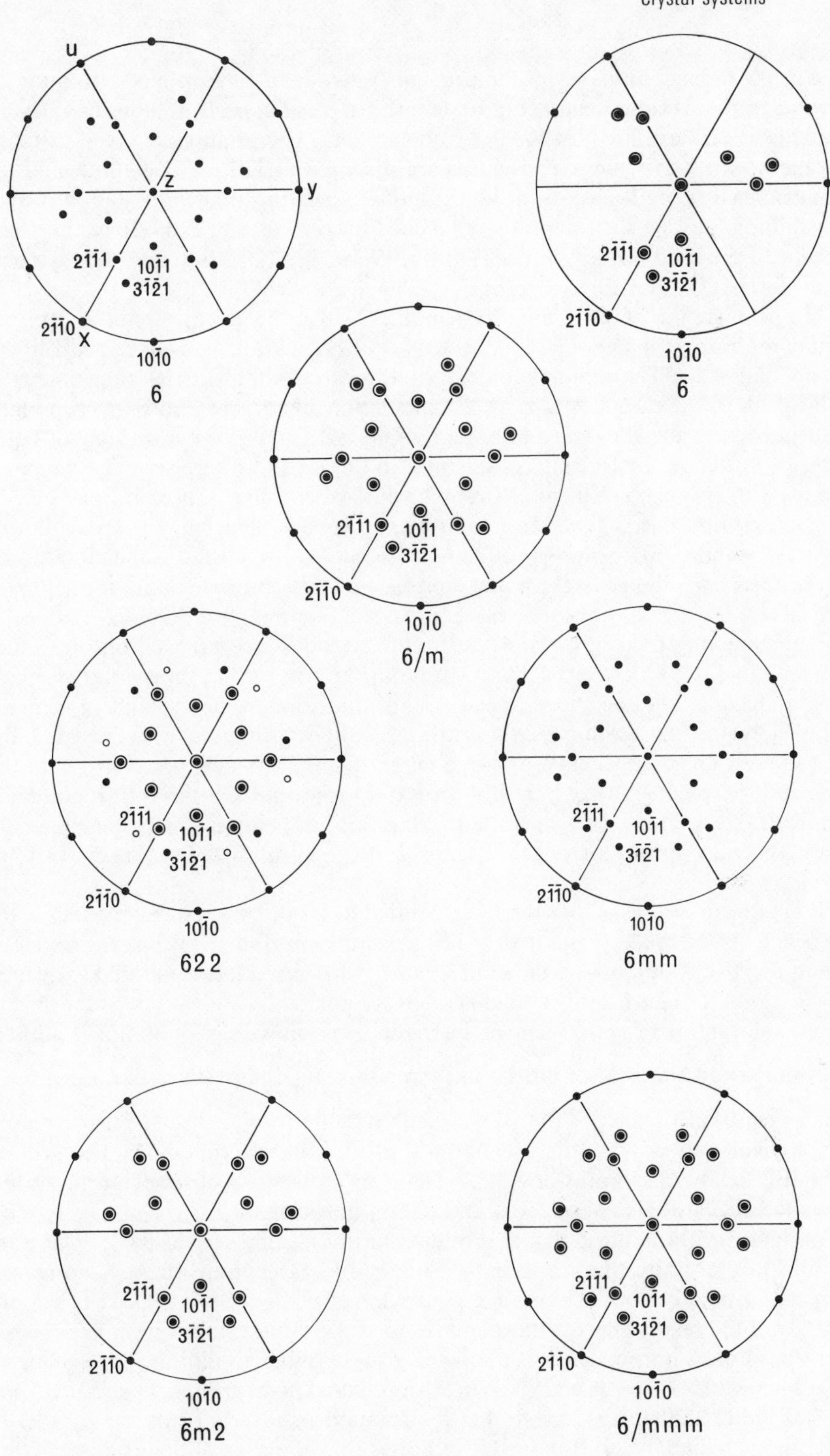
u
z
y
x
2$\bar{1}\bar{1}$1
10$\bar{1}$1
3$\bar{1}\bar{2}$1
2$\bar{1}\bar{1}$0
10$\bar{1}$0
6
$\bar{6}$
6/m
622
6mm
$\bar{6}$m2
6/mmm

Trigonal system

The characteristic symmetry of the trigonal system is the presence of a rotation or inverse triad. It is common practice to describe trigonal crystals in terms of the same unit-cell as we have used for the hexagonal system. The rotation or inverse triad is parallel to the z-axis; the x and y axes are disposed at 120° to one another in the plane normal to the z-axis. As in the hexagonal system it is customary to introduce an additional axis u such that $+u$ is inclined at 120° to $+x$ and $+y$ in the plane normal to the triad. Miller–Bravais indices, Weber symbols, and zone axis symbols of the form $[UV\dagger W]$ are used as in the hexagonal system.

The point groups of the trigonal system (Fig 3.20) follow a sequence different from that of the hexagonal and tetragonal systems because we are here concerned with an axis of odd order. The combination of a rotation triad with a centre of symmetry is equivalent to an inverse triad. The combination of a triad and a mirror plane associated with the same direction, that is $3/m$ or $\bar{3}/m$, has the symmetry of point group $\bar{6}$, higher symmetry than is admissible in this system. There are only two point groups with symmetry elements associated with a single direction, 3 and $\bar{3}$ (Fig 3.35). In point group 3 the general form comprises three faces and the only special forms are $\{0001\}$ and $\{000\bar{1}\}$, each of which has a single face; the triad is therefore polar. In point group $\bar{3}$ the general form comprises six faces and the special form $\{0001\}$ two faces.

The remaining point groups of the trigonal system are derived from the Euler combination 223, which is itself the point group 32. For each of these point groups a symbol having two positions specifies the combination of symmetry elements completely: the first position states whether $[0001]$ is a rotation or inverse triad, the second position states whether a diad, a mirror plane or both is associated with the x, y, and u axes. Point group 32 has a rotation triad parallel to $[0001]$ and diads parallel to x, y, and u. The general form (Fig 3.35) in point group 32 comprises six faces. There are special forms $\{2\bar{1}\bar{1}0\}$ and $\{\bar{2}110\}$, each of three faces, and $\{0001\}$ of two faces.

Point group $3m$ has a rotation triad parallel to $[0001]$ and mirror planes $\{2\bar{1}\bar{1}0\}$ perpendicular to the x, y, and u axes. The general form (Fig 3.35) contains six faces. There are two kinds of special form: $\{h0\bar{h}l\}$ with three faces and $\{000l\}$, where $l = \pm 1$, with a single face. The triad is therefore polar.

The holosymmetric point group of the trigonal system is derived by adding a centre of symmetry to either 32 or $3m$. Its full symbol is $\bar{3}\frac{2}{m}$ which states the presence of an inverse triad parallel to the z-axis, diads parallel to x, y, u and mirror planes perpendicular to x, y, u. It is usually known by its short symbol $\bar{3}m$. The general form (Fig 3.35) consists of twelve faces. There are three kinds of special form: $\{h0\bar{h}l\}$ of six faces, $\{2\bar{1}\bar{1}0\}$ also of six faces, and $\{0001\}$ of two faces.

We have tacitly assumed that the rotation or inverse diads in the (0001) plane are parallel to the shortest lattice repeats in that plane. Structurally, there is no reason why that should be so; the lattice has diad symmetry about $\langle 1\bar{1}\dagger 0\rangle$ as well as about $\langle 10\dagger 0\rangle$. If the relationship of the diad axes to the shortest lattice repeat is known a third position is introduced into the point group symbol to indicate the symmetry associated with $\langle 1\bar{1}\dagger 0\rangle$ directions as in the hexagonal point groups. Thus 321 or $3m1$ or $\bar{3}m1$ indicates that it is known that the rotation or inverse diads are parallel to x, y, and u, while 312 or $31m$ or $\bar{3}1m$ indicates that the axes of diad symmetry are parallel to $\langle 1\bar{1}\dagger 0\rangle$.

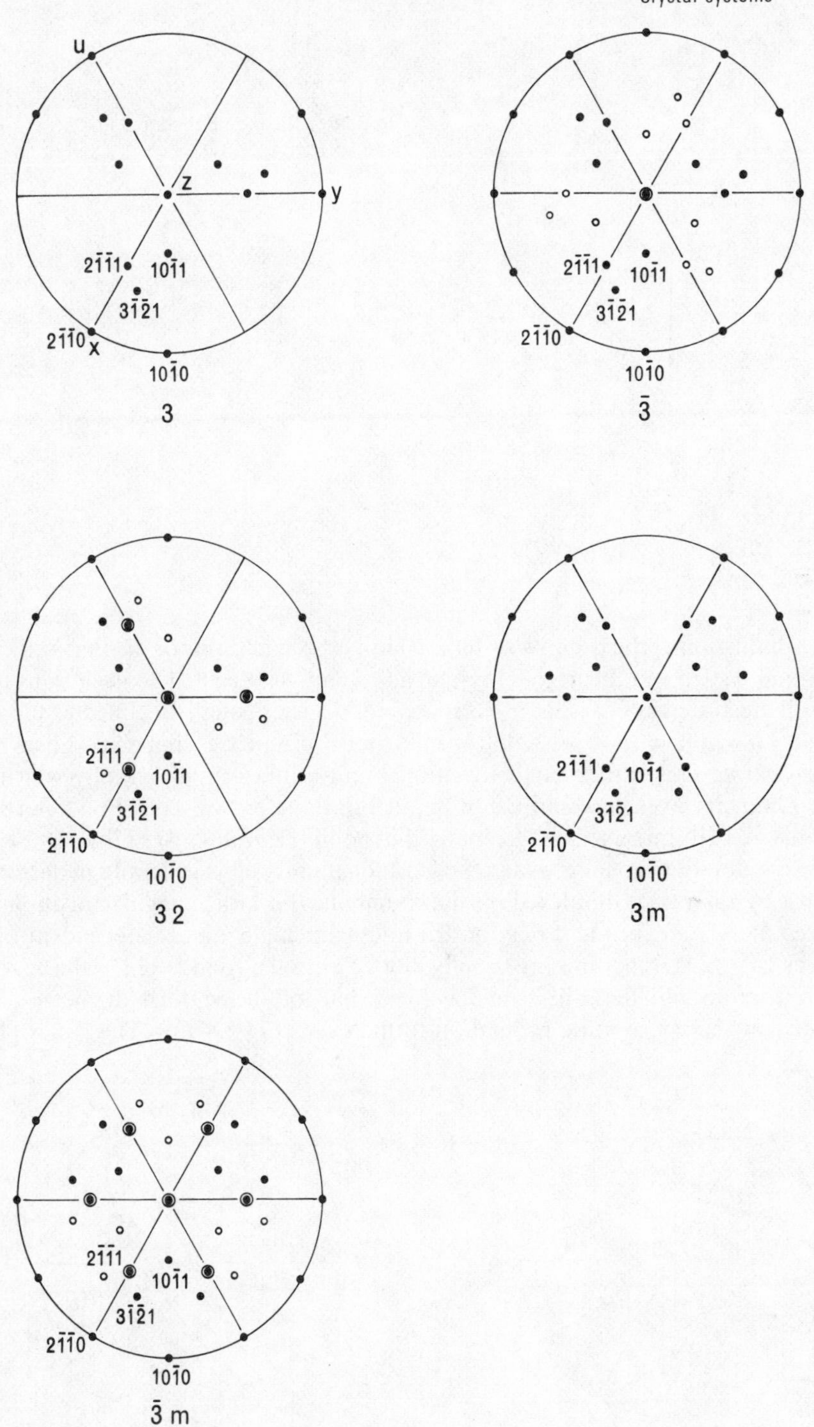

Fig 3.35 The trigonal point groups. The stereograms show the poles of the forms {0001}, {$10\bar{1}0$}, {$2\bar{1}\bar{1}0$}, {$10\bar{1}1$}, {$2\bar{1}\bar{1}1$}, and {$3\bar{1}\bar{2}1$}. Only one face of each form is indexed; (0001) is not indexed.

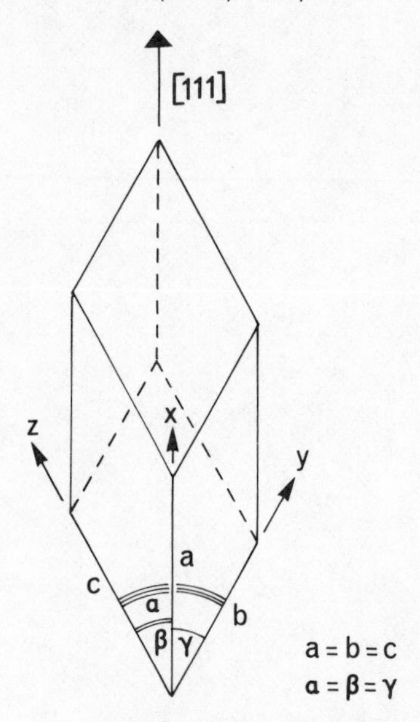

Fig 3.36 The rhombohedral unit-cell: $a = b = c$, $\alpha = \beta = \gamma \neq 90° < 120°$.

Our discussion of the trigonal system so far has been in terms of the unit-cell of the hexagonal system, but the trigonal system has a unit-cell peculiar to itself. This is the unit-cell illustrated in Fig 3.36; it has its x, y, and z axes equally inclined to the triad with $a = b = c$, $\alpha = \beta = \gamma < 120°$. A solid body of such a shape is known as a *rhombohedron.* This is the only conventional unit-cell in crystallography which has all its symmetry axes necessarily non-parallel to its reference axes. For this reason and because of its inconvenient geometry the rhombohedral unit-cell is little used for the description of trigonal crystals; the hexagonal unit-cell is generally preferred.

For a trigonal crystal indexed on the rhombohedral lattice angular relationships between faces are dependent only on the interaxial angle α and independent of the unit-cell edge a. Stereograms are usually plotted with the triad [111] in the centre of the stereogram and the x-axis on the lower half of the vertical diameter of the primitive at the appropriate inclination to the triad (Fig 3.37(a)). The zones [100],

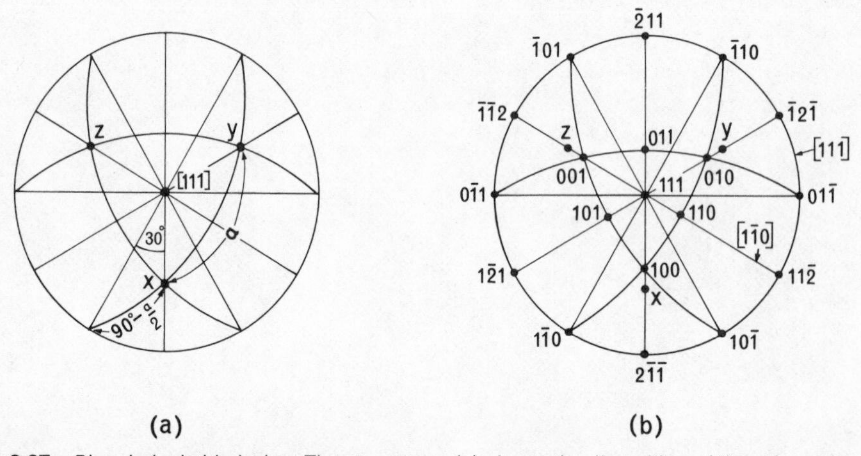

Fig 3.37 Rhombohedral indexing. The stereogram (a) shows the disposition of the reference axes about the triad, which is plotted centrally. The stereogram (b) shows how faces may be indexed on rhombohedral axes by use of intersecting zones.

[010], [001] can then be plotted by drawing the great circles whose poles are respectively *x*, *y*, and *z* (Fig 3.37(b)). It is evident from the symmetry of the stereogram that the normal to (111) is parallel to the triad [111] and faces (*hkl*) with $h+k+l=0$ lie on the primitive. The position of any pole can thereafter be found by intersection of zones: for instance (110) lies at the intersection of the zones [001] and [$1\bar{1}0$].

The plotting of trigonal crystals in terms of the hexagonal lattice needs no comment; it is exactly the same as for hexagonal crystals.

Interplanar and interzonal angles

To conclude this chapter we begin to consider in general terms how the angle between a pair of planes (*hkl*) and (*h'k'l'*) or between a pair of zones [*UVW*] and [*U'V'W'*] is related to the dimensions of the unit-cell in each of the crystal systems. At this stage it is possible only to set down certain basic equations and to indicate explicitly how certain simple calculations may be performed; it is only after the powerful methods of spherical trigonometry have been introduced in chapter 5 that the general problem can be solved in any system.

We begin with the least symmetrical system, the triclinic system, and progress to more symmetrical systems, obtaining more immediately useful results as the symmetry increases. Consider the triclinic stereogram shown in Fig 3.38(a) on which the poles of the general plane (*hkl*) and of the planes (100), (010), (001) and the zones containing these planes in pairs are plotted. The symbols of the axes of the zones shown in the figure are [$0\bar{l}k$], [$l0\bar{h}$], [$\bar{k}h0$], [100], [010], and [001]. The zone axes [$0\bar{l}k$], [010], and [001] are all parallel to the plane (100). The disposition of these three zone axes in the (100) plane is shown in Fig 3.38(b): the interaxial angle between the *y* and *z* axes is α (shown on the stereogram as the angle between the great circles whose poles are

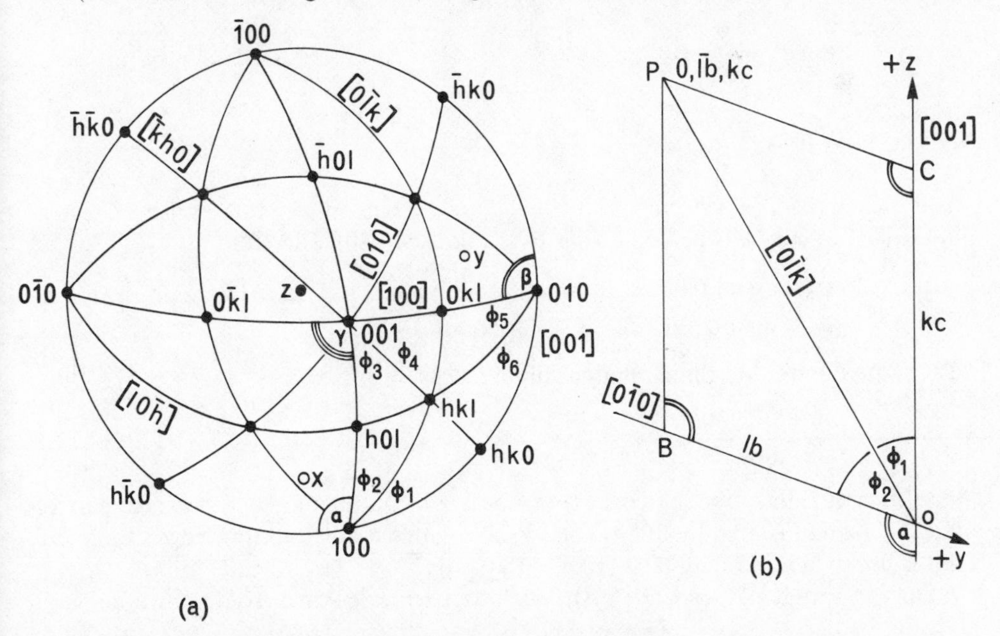

Fig 3.38 Triclinic geometry. The stereogram (a) shows the principal zones and the zones which serve to define the angles $\phi_1 \ldots \phi_6$. The disposition of zone axes in the (100) plane is shown in the lattice section (b), from which the relationship between ϕ_1, ϕ_2 and the axial ratio b/c can be demonstrated.

these axes) and $[0\bar{l}k]$ is drawn from the origin through a point with coordinates $0, -lb, kc$. Let the interzonal angles $[001]$:$[0\bar{l}k]$ and $[010]$:$[0\bar{l}k]$ be respectively ϕ_1 and ϕ_2. Then $\widehat{OPC} = \widehat{BOP} = \phi_2$ so that in the triangle OCP

$$\frac{\sin\phi_1}{\sin\phi_2} = \frac{lb}{kc} = \frac{b/k}{c/l}.$$

Now, returning to Fig 3.38(a), the angle between the great circle through the poles of (100) and (010) and the great circle through the poles of (100) and (*hkl*) is the angle ϕ_1; and likewise the angle between the great circle through the poles of (100) and (*hkl*) and the great circle through the poles of (100) and (001) is the angle ϕ_2. The analogous relationships

$$\frac{\sin\phi_3}{\sin\phi_4} = \frac{a/h}{b/k}$$

and
$$\frac{\sin\phi_5}{\sin\phi_6} = \frac{c/l}{a/h}$$

can be derived by identical arguments. All six angles ϕ_1 to ϕ_6 are marked on Fig 3.38(a). The reader will observe that the form of these three relationships is such that they are particularly easy to remember.

This is as far as relationships between interzonal angles and unit-cell dimensions can conveniently be taken in the triclinic system without recourse to spherical trigonometry (chapter 5).

In the *monoclinic system* some simplification is achieved because the interaxial angles α and γ are right-angles. Consequently (Fig 3.39) $\phi_1+\phi_2 = \phi_3+\phi_4 = 90°$ so that the first and second relationships for the triclinic system become

$$\tan\phi_1 = \frac{b/k}{c/l},$$

$$\tan\phi_3 = \frac{a/h}{b/k}.$$

And since the *y*-axis is normal to (010), the great circle on which the poles of planes in the $[010]$ zone lie is perpendicular to the pole of (010). Therefore

$$\phi_5 = (001):(h0l)$$
$$\phi_6 = (h0l):(100).$$

Therefore the third triclinic relationship becomes

$$\frac{\sin(001):(h0l)}{\sin(h0l):(100)} = \frac{c/l}{a/h}$$

for the monoclinic system. This expression enables the angle between any pair of faces $(h0l)$ and $(h'0l')$ in the $[010]$ zone to be evaluated from a knowledge of *a* and *c* if it is borne in mind that $(001):(100) = 180° - \beta$.

Once the angles ϕ_1, ϕ_2 and $(001):(h0l)$ have been calculated from known unit-cell

Fig 3.39 Monoclinic geometry. The stereogram (a) shows the principal zones and zones through the general pole (*hkl*); the angles $\phi_1+\phi_2 = \phi_3+\phi_4 = 90°$ and $\phi_5+\phi_6 = 180° - \beta$. The lattice sections (b)–(e) are referred to in detail in the text.

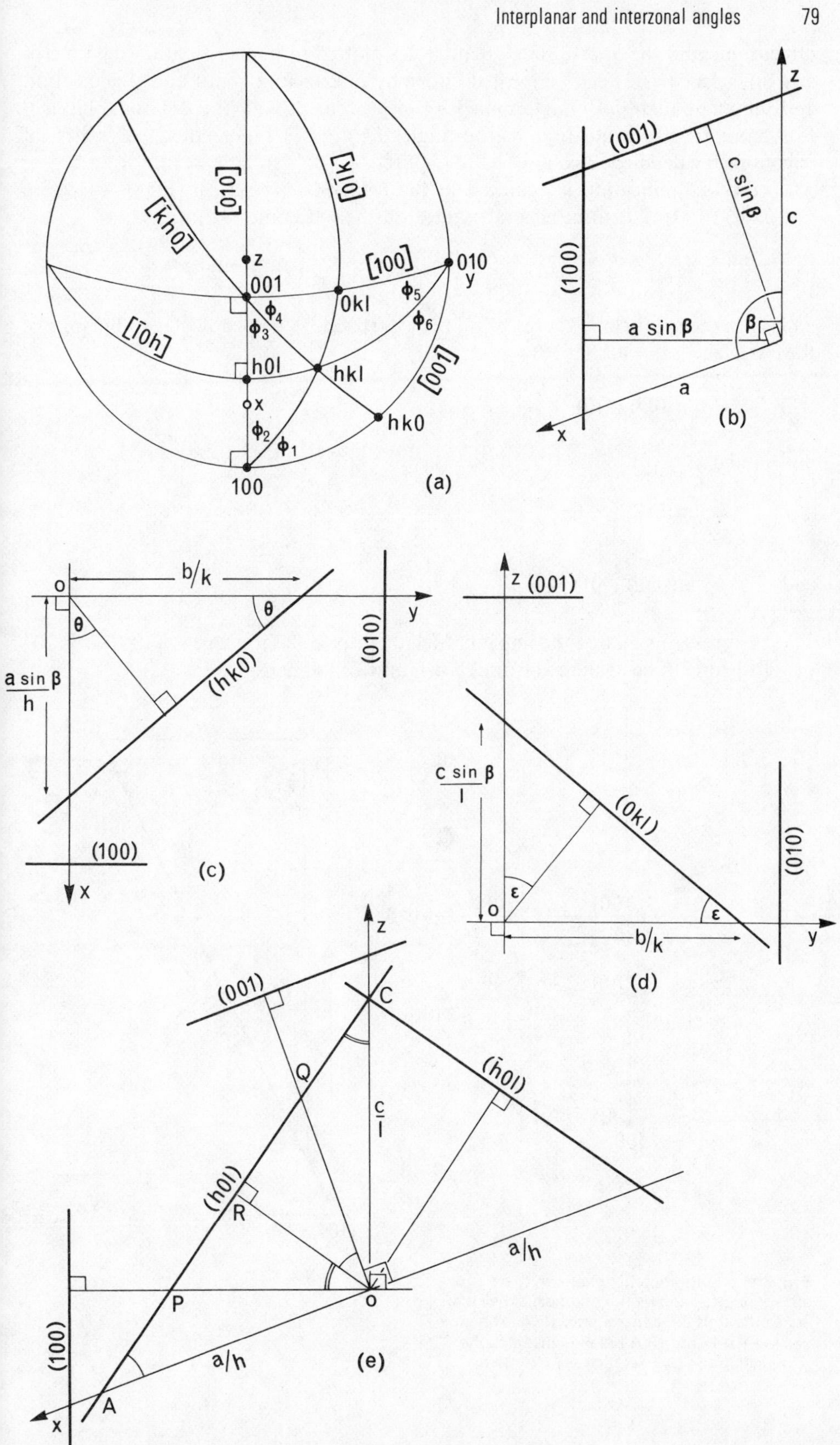
[010]
[0k̄k]
[k̄h0]
z
010
y
[100]
001
0kl
ϕ_5
ϕ_4
ϕ_6
ϕ_3
[1̄0h]
h0l
hkl
[001]
x
ϕ_2
ϕ_1
hk0
100
(a)
z
(001)
c sin β
c
(100)
β
a sin β
a
x
(b)
o
b/k
θ
y
(010)
a sin β / h
(hk0)
(100)
(c)
x
z
(001)
c sin β / l
(0kl)
(010)
ε
o
b/k
y
(d)
z
(001)
C
(h̄0l)
Q
c/l
(h0l)
R
a/h
(100)
P
o
a/h
(e)
A
x

dimensions and the appropriate great circles plotted on the stereogram the pole of any other face $(h'k'l')$ can be located either by intersecting zones or by calculation and the angle $(hkl){:}(h'k'l')$ determined by measurement with the stereographic net. For more precise evaluation of the angle $(hkl){:}(h'k'l')$ the methods of spherical trigonometry detailed in chapter 5 are required.

Further simplification is achieved in the *orthorhombic system* (Fig 3.40) where $\alpha = \beta = \gamma = 90°$. The three general expressions here become

$$\tan\phi_1 = \frac{b/k}{c/l}, \quad \tan\phi_3 = \frac{a/h}{b/k}, \quad \tan\phi_5 = \frac{c/l}{a/h}$$

and, since the great circle representing the [100] zone is perpendicular to the pole of the (100) face, $\phi_1 = (0kl){:}(010)$

therefore $$\tan(0kl){:}(010) = \frac{b/k}{c/l}$$

and similarly

$$\tan(hk0){:}(100) = \frac{a/h}{b/k}$$

and $$\tan(h0l){:}(001) = \frac{c/l}{a/h}.$$

These expressions enable the angle between any pair of faces in the [100] or [010] or [001] zone to be evaluated from known unit-cell dimensions.

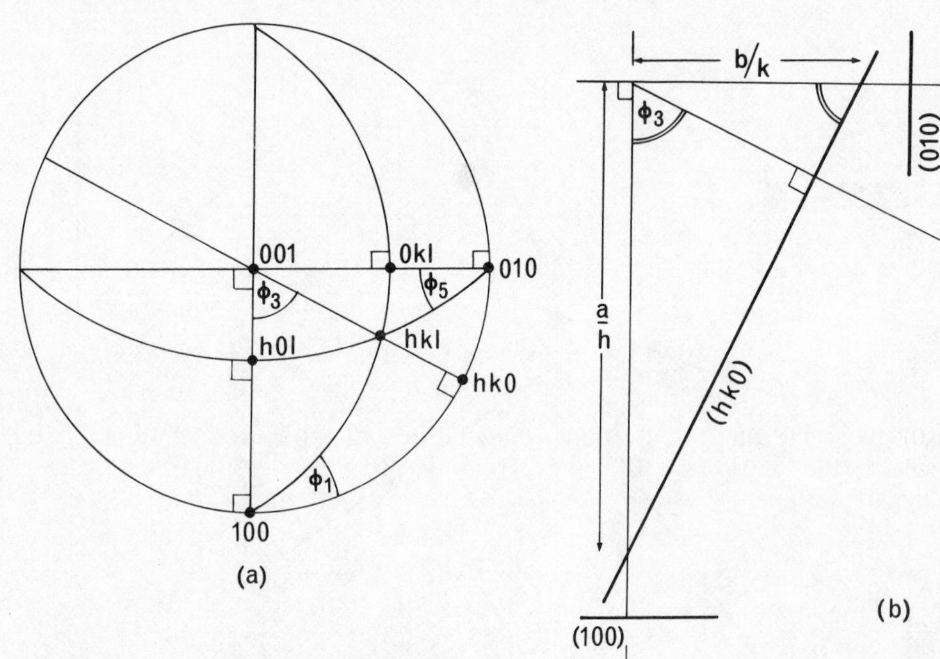

Fig 3.40 Orthorhombic geometry. The stereogram (a) shows the principal zones and zones through the general pole (hkl). The lattice section (b) in the (001) plane illustrates the relationship $\tan(hk0){:}(100) = (a/h)/(b/k)$.

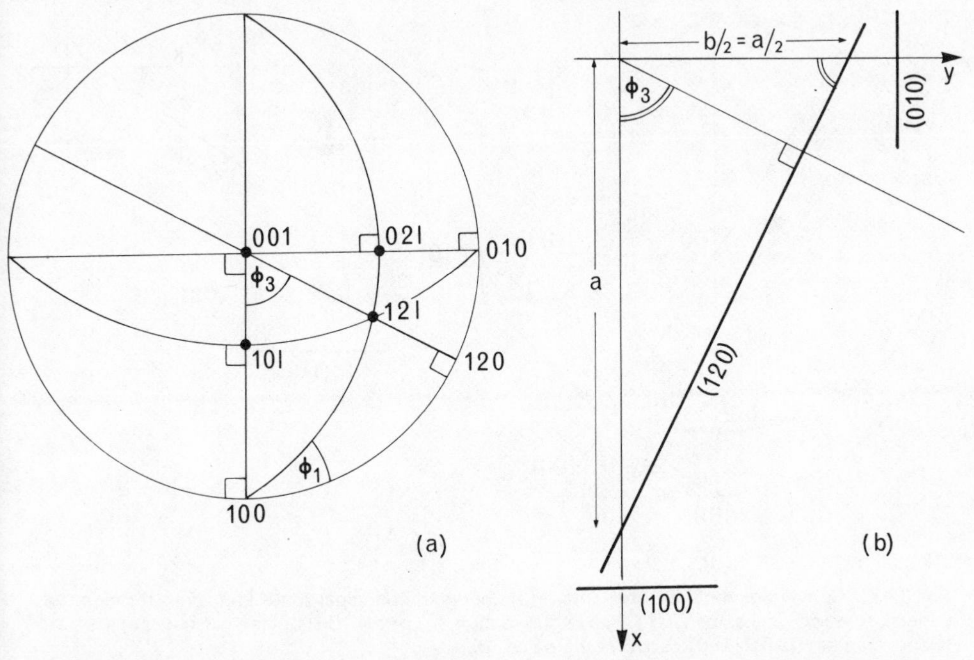

Fig 3.41 Tetragonal geometry. The stereogram (a) shows the principal zones and zones through the general pole (*hkl*). The lattice section (b) in the (001) plane illustrates the relationship tan (*hk*0) : (100) = *k*/*h* for the plane (120).

In the *tetragonal system* (Fig 3.41) the further simplification $a = b$ is introduced so that the three orthorhombic equations become

$$\tan(0kl):(010) = \frac{a/k}{c/l}$$

$$\tan(hk0):(100) = \frac{k}{h}$$

$$\tan(h0l):(001) = \frac{c/l}{a/h}$$

It is immediately apparent from the second of these that interplanar angles in the [001] zone are independent of the unit-cell dimensions, that is to say the angle $(hk0)$:(100) is the same for all tetragonal crystals and in particular (110):(100) = 45°. Angles in the [100] and [101] zones are symmetry related so that $(0kl)$:(001) = $(h0l)$:(001) when $h = k$.

In the *cubic system* (Fig 3.42) $a = b = c$ so that

$$\tan(0kl):(010) = \frac{l}{k}$$

$$\tan(hk0):(100) = \frac{k}{h}$$

$$\tan(h0l):(001) = \frac{h}{l}.$$

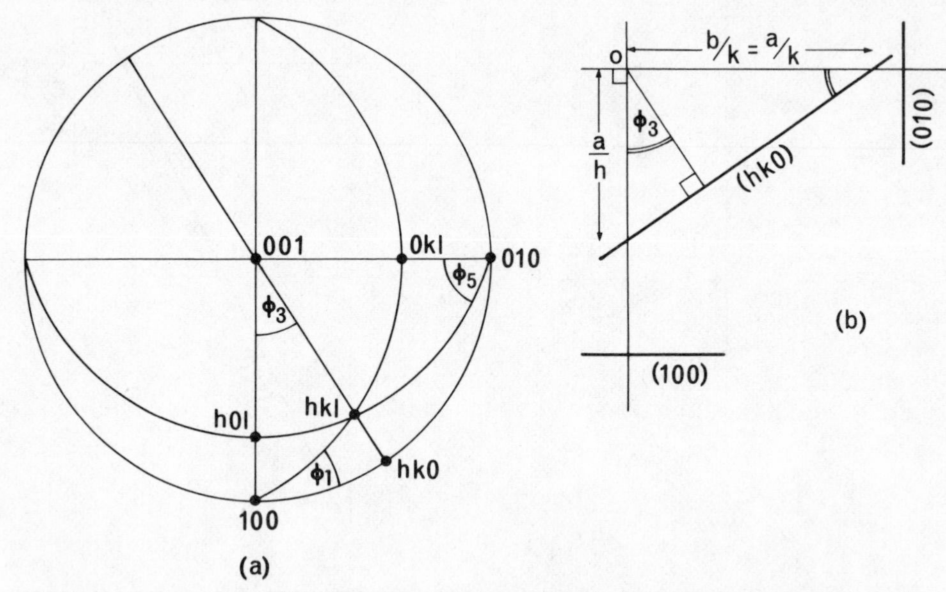

Fig 3.42 Cubic geometry. The stereogram (a) shows the principal zones and zones through the general pole (*hkl*), taken as (321). The lattice section (b) in the (001) plane illustrates the relationship tan (*hk*0) : (100) = *k*/*h* for the plane (320).

Interplanar angles in each of the zones [100], [010], [001] are thus independent of the magnitude of the unit-cell edge a; that the same is true for any zone $[UVW]$ in a cubic crystal should be self-evident, but will be shown explicitly in chapter 5.

We shall make no mention of the trigonal and hexagonal systems at this point except to say that this type of approach is not especially fruitful in these systems.

In the three systems that have orthogonal axes, certain interplanar angles can be calculated quite efficiently and straightforwardly by plane geometry. In the *orthorhombic system* the poles of planes in the [001] zone, that is planes with indices (*hk*0), lie in the (001) plane, which of course contains the x and y axes (Fig 3.40(a)). Therefore, since the (*hk*0) plane makes intercepts a/h and b/k on the x and y axes and since the normals to (100) and (010) are respectively the x and y axes (Fig 3.40(b)),

$$\tan(hk0){:}(100) = \frac{a/h}{b/k}.$$

The expressions for (0*kl*):(010) and (*h*0*l*):(001) can be obtained by analogous arguments.

The geometrical argument is precisely the same in the *tetragonal* and *cubic systems*, where it yields the simplified results previously noted.

In the *monoclinic system* (Fig 3.39(b)) the perpendicular distance of the (100) plane from the origin is $a\cos(\beta-90°) = a\sin\beta$. Therefore any plane (*hk*0) parallel to the z-axis will make an intercept $a/h.\sin\beta$ on the normal to (100). Now the planes (100), (*hk*0), (010) lie in the zone [001] so their normals are coplanar (Fig 3.39(c)). Moreover the normal to (010) is the y-axis, which is perpendicular to the normal to (100). Therefore

$$\tan(100){:}(hk0) = \frac{a}{h}\sin\beta \bigg/ \frac{b}{k}.$$

Similarly by consideration of the normals to the faces (001), ($0kl$), (010), which lie in the plane normal to [100] it can be shown (Fig 3.39(d)) that

$$\tan(001){:}(0kl) = \frac{c}{l}\sin\beta\bigg/\frac{b}{k}.$$

These two expressions are sometimes of greater practical utility than the corresponding expressions $\tan\phi_1 = (b/k)/(c/l)$ and $\tan\phi_3 = (a/h)/(b/k)$ derived earlier for the monoclinic system.

Angular relationships in the [010] zone in the monoclinic system are less conveniently established by plane geometry, but nevertheless useful results can be obtained. The x and z axes and the normals to the planes (100), ($h0l$), (001) lie in the (010) plane (Fig 3.39(e)). The ($h0l$) plane makes intercepts a/h and c/l on the x and z axes respectively. If OR is the normal to ($h0l$) and OQ the normal to (001),

$$\widehat{RQO}+\widehat{QOR} = \widehat{AQO}+\widehat{OAQ} = 90°$$

and therefore $\widehat{QOR} = \widehat{OAQ} = (001){:}(h0l)$

Similarly $\widehat{OPR}+\widehat{ROP} = \widehat{OPC}+\widehat{PCO} = 90°$

hence $\widehat{ROP} = \widehat{PCO} = (h0l){:}(100)$

In the triangle OAC

$$\frac{\sin\widehat{OAC}}{\sin\widehat{ACO}} = \frac{c/l}{a/h}$$

therefore

$$\frac{\sin(001){:}(h0l)}{\sin(h0l){:}(100)} = \frac{c/l}{a/h}.$$

If the plane ($\bar{h}0l$), also shown in Fig 3.39(e), is similarly considered the analogous relationship

$$\frac{\sin(001){:}(\bar{h}0l)}{\sin(\bar{h}0l){:}(\bar{1}00)} = \frac{c/l}{a/h}$$

is obtained; of course $(001){:}(\bar{h}0l) \neq (001){:}(h0l)$. These relationships are the same as those obtained earlier by consideration of interzonal angles.

We discuss finally *hexagonal* crystals and *trigonal* crystals indexed on the hexagonal unit-cell where the simple geometrical approach yields useful results. The normals to the faces ($2\bar{1}\bar{1}0$), ($2h, \bar{h}, \bar{h}, l$), (0001) are coplanar (Fig 3.43(a)) with the x and z axes. The ($2h, \bar{h}, \bar{h}, l$) plane makes intercepts $a/2h$ and c/l on the x and z axes respectively (Fig 3.43(b)). Therefore

$$\tan(2h, \bar{h}, \bar{h}, l){:}(0001) = \frac{c/l}{a/2h}$$

A face ($h0\bar{h}l$) makes intercepts a/h on the $+x$ and $-u$ axes (Fig 3.43(c)). It must therefore make an intercept $a/h.\cos 30°$ on the normal to ($10\bar{1}0$). Since the z-axis is coplanar with the normals to ($h0\bar{h}l$) and ($10\bar{1}0$), as illustrated in Fig 3.43(d).

$$\tan(h0\bar{h}l){:}(0001) = \frac{c}{l}\bigg/\frac{a}{h}\cos 30°.$$

A face $(hk.0)$ makes intercepts a/h and a/k on the $+x$ and $+y$ axes (Fig 3.43(e)) and its normal is coplanar with these axes. If OR is the normal to $(hk.0)$ and θ the angle $(2\bar{1}\bar{1}0)$:$(hk.0)$, then

$$\mathrm{OR} = \frac{a}{h}\cos\theta = \frac{a}{k}\cos(120° - \theta)$$

hence $$\frac{\cos\theta}{h} = -\frac{\cos\theta}{2k} + \frac{\sqrt{3}\sin\theta}{2k}$$

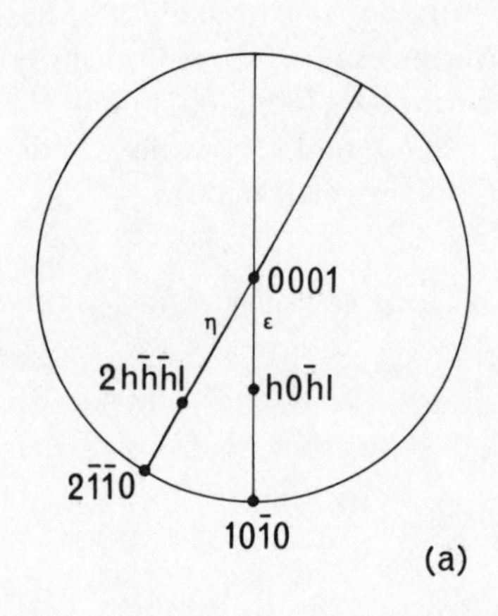

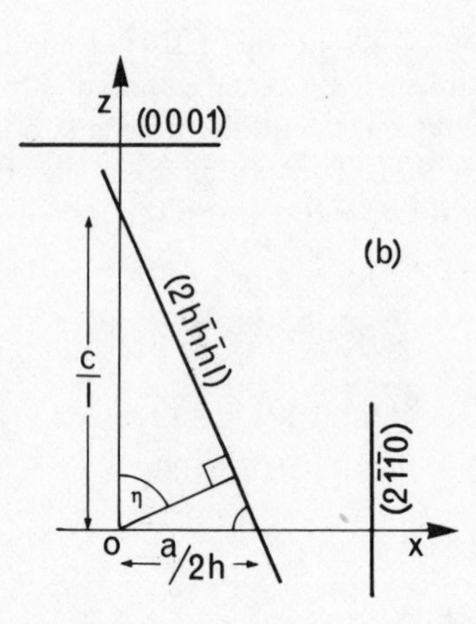

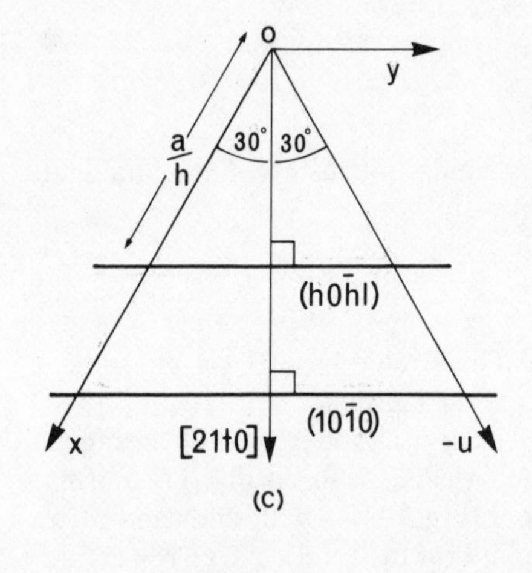

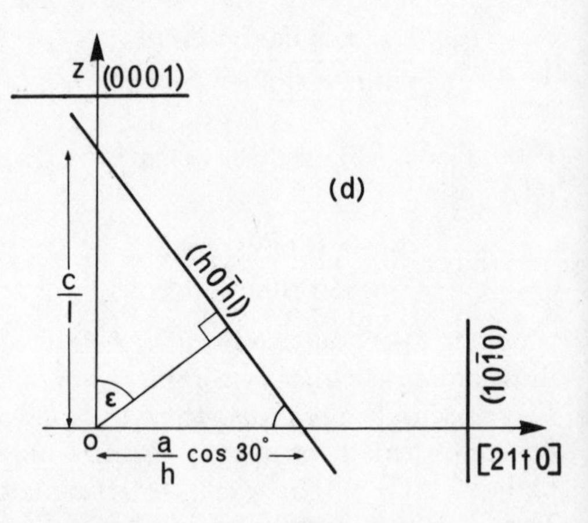

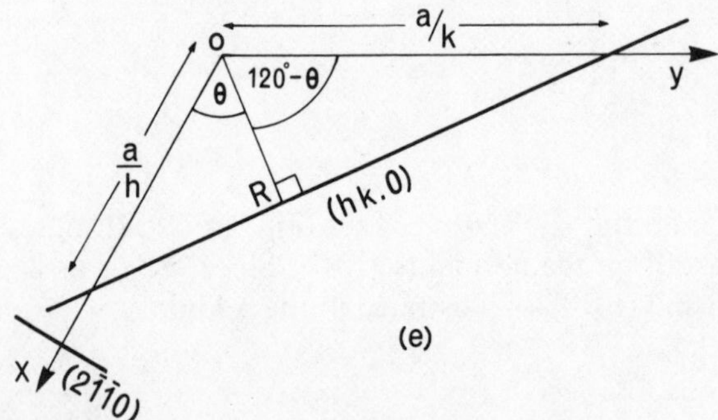

and $$\tan(2\bar{1}\bar{1}0):(hk.0) = \frac{h+2k}{h\sqrt{3}}.$$

Thus the angular relationships of faces in the [0001] zone are independent of the unit-cell dimensions of the hexagonal or trigonal lattice and are the same for all crystals referable to these systems.

In the next chapter we explore the symmetry of the internal structure of crystals and in chapter 5 resume consideration of the evaluation of interplanar and interzonal angles.

Two-dimensional point groups

By way of preparation for chapter 4 we now deal briefly with the two-dimensional point groups, that is the combinations of symmetry elements that can operate on a two-dimensional lattice. The symmetry operators for a two-dimensional lattice are the rotation axes 1, 2, 3, 4, and 6 and the *line of symmetry* (analogous to the plane of symmetry in three-dimensions and represented by the same symbol *m*). Rotation axes are necessarily perpendicular to the plane of the two-dimensional lattice and therefore cannot be combined with one another. Each type of rotation axis can however be combined with a line of symmetry in the plane of the lattice. There are thus ten combinations of symmetry elements that can operate on a two-dimensional lattice, the ten two-dimensional point groups; these comprise each of the five rotation axes on its own and combined with a line of symmetry. The ten two-dimensional point groups are listed in Table 3.7. When a rotation axis of even order (2, 4, or 6) is combined with a mirror line a set of mirror lines is produced equally inclined in the plane of the lattice to those of the primary set; these are noted in the third place of the conventional symbol for the point group, i.e. 4*mm* rather than just 4*m*. In the final column of Table 3.7 the shape of the conventional unit-mesh of the two-dimensional lattice is noted, oblique, rectangular, square, or hexagonal; this is a point that will be taken up in the next chapter.

Table 3.7
The two-dimensional point groups

Combination of symmetry elements	Conventional symbol	Angle between mirror lines of primary and secondary sets	Conventional unit-mesh of lattice	
1	1	—	$a \neq b, \gamma \neq 90°$	Oblique
1+*m*	1*m*	—	$a \neq b, \gamma = 90°$	Rectangular
2	2	—	$a \neq b, \gamma \neq 90°$	Oblique
2+*m*	2*mm*	90°	$a \neq b, \gamma = 90°$	Rectangular
3	3	—	$a = b, \gamma = 120°$	Hexagonal
3+*m*	3*m*	—	$a = b, \gamma = 120°$	Hexagonal
4	4	—	$a = b, \gamma = 90°$	Square
4+*m*	4*mm*	45°	$a = b, \gamma = 90°$	Square
6	6	—	$a = b, \gamma = 120°$	Hexagonal
6+*m*	6*mm*	30°	$a = b, \gamma = 120°$	Hexagonal

Note:
The symbol $\neq$ implies that equality is not required by symmetry

Fig 3.43 Hexagonal geometry. The stereogram (a) serves to define the angles η and ε used in the lattice sections (b)–(e); (b) is in the plane of the *x* and *z* axes, (c) in the (0001) plane, (d) in the plane of the *z* and [21†0] axes, and (e) in the (0001) plane.

Twinning

Some substances commonly crystallize as composite crystals of a sort known as *twinned crystals* or, colloquially, as *twins*. Well-known examples are copper, diamond, fluorite (CaF_2), and calcite ($CaCO_3$). A twinned crystal consists of two or more individual single crystals joined together in some definite mutual orientation; the lattice of one individual is related to that of the other individual or individuals in the composite crystal by some simple symmetry operation.

Twinned crystals may be produced in various ways. As a crystal grows from its initial nucleus some accident of growth may cause it to twin, such accidents being for a variety of reasons very much more probable in some structures than in others. Twinning may alternatively provide a means of relieving the strain induced by some applied stress. Twinning may also be produced as the result of polymorphic transformations when a structure of higher symmetry is converted to a structure of lower symmetry on cooling. These are the three principal types of twins and they are known respectively as *growth twins*, *deformation* (or glide) *twins*, and *transformation* (or inversion) *twins*. The anti-phase domains produced when a disordered alloy orders on cooling (chapter 10) are a special sort of transformation twinning. Here we shall concern ourselves with the geometry of twinning rather than with its physical origin and most of our examples will be growth twins.

The mutual relationship between the two components of a twinned crystal is described by a statement of the symmetry operation necessary to bring the lattice of one component into coincidence with that of the other component. The necessary operation is very commonly either a rotation through 180° about a direction known as the *twin axis* or reflexion in a plane known as the *twin plane*. A twin axis is always a zone axis or the normal to a lattice plane and a crystal twinned about such an axis is known as a *rotation twin*; a twin plane is always a lattice plane and a crystal twinned on such a plane is known as a *reflexion twin*. In a rotation twin where the twin axis is a zone axis rotation may be through 60°, 90°, 120°, or 180°, the first three cases being of very much less common occurrence than the last.

It is obvious that a twin axis cannot be parallel to a symmetry axis of even order in the point group of the crystal, nor can a twin plane be parallel to a mirror plane of the point group. Thus a diad, tetrad, or hexad cannot be a twin axis (at least not the common sort of twin axis rotating through 180°), but a twin axis may be parallel to a triad. In the case of a twin axis parallel to a triad, the twinning operation can variously be described as a 60°, a 180°, or a 240° rotation about the twin axis; it is conventional and convenient however always to consider such a twinning operation as a rotation of 180°.

Where the two components of a twinned crystal are joined in a plane, the crystal is called a *contact twin* and the plane of mutual contact, which is a lattice plane, is known as the *composition plane*. In general it is true to say that if the twin axis is a zone axis, then the composition plane is parallel to the twin axis, but if the twin axis is the normal to a lattice plane then the composition plane is normal to the twin axis; in reflexion twins the composition plane is parallel to the twin plane. Figure 3.44 illustrates contact twinning in a cubic crystal of point group $\bar{4}3m$ which exhibits the two complementary forms $\{111\}$ and $\{11\bar{1}\}$. The combination of these two tetrahedral forms in a truly single crystal is shown in Fig 3.44(a). A contact twin in which twinning is by rotation about the normal to $(11\bar{1})$, which is of course parallel to the zone axis $[11\bar{1}]$ in the cubic system, is shown in Fig 3.44(b). Another contact twin in which

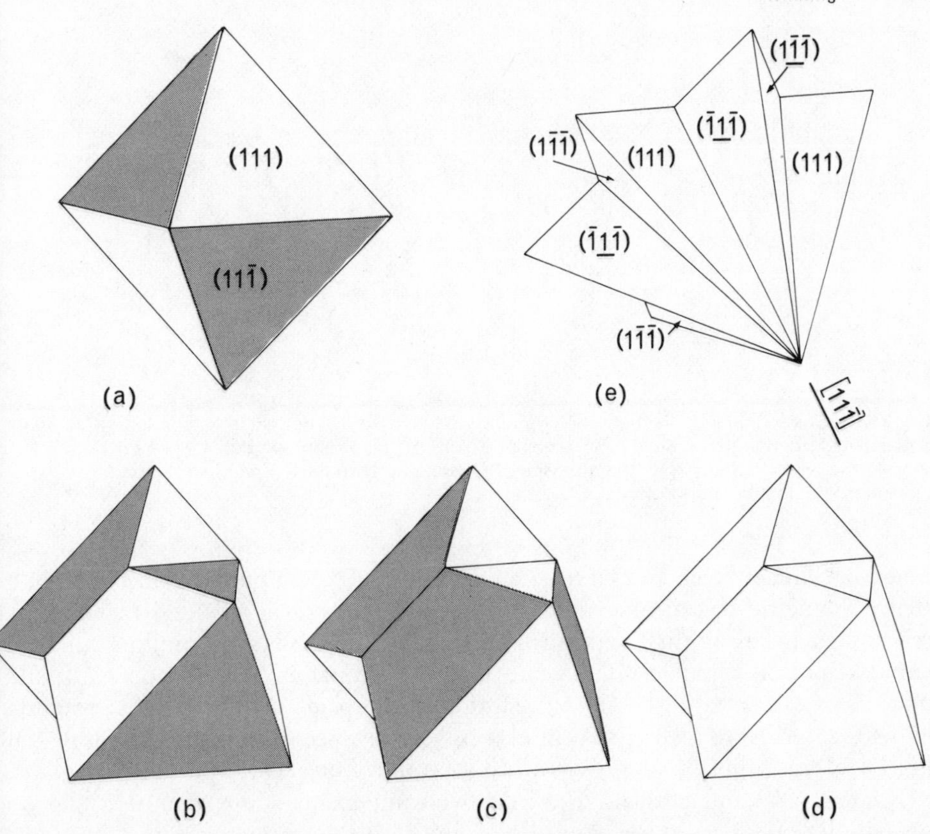

Fig 3.44 A crystal of a cubic substance of point group $\bar{4}3m$ exhibiting the forms $\{111\}$ and $\{11\bar{1}\}$ is shown in (a). Twinning of a similar crystal by rotation about the normal to $(11\bar{1})$ is shown in (b) and by reflexion in $(11\bar{1})$ in (c). In (a)–(c) faces of the form $\{111\}$ are shown clear and of the form $\{11\bar{1}\}$ shaded. A twin of a crystal of the centrosymmetric class $m3m$ which exhibits the form $\{111\}$ and is twinned on $(11\bar{1})$ is shown in (d). An interpenetrant rotation twin, twin axis $[11\bar{1}]$, of a crystal of class $\bar{4}3m$ exhibiting only the form $\{111\}$ is shown in (e): the face $(\bar{1}\bar{1}1)$, which lies at the back of the crystal as shown, is normal to the twin axis and is in consequence coplanar with $(\underline{\bar{1}\bar{1}1})$. Single crystals cannot display re-entrant angles; but twinned crystals, such as those shown here, frequently do so.

twinning is by reflexion in $(11\bar{1})$ is shown in Fig 3.44(c). Inspection of figures (b) and (c) shows that although the two twinned crystals are alike in shape, they differ in the disposition of symmetry-related faces; the difference is clearly displayed when the twinning operations are represented in stereographic projection. In Fig 3.45(a) the pole $\underline{\text{N}}$ is related to the pole N by the twin operation of rotation through 180° about the twin axis P. In Fig 3.45(b) the pole $\underline{\text{N}}'$ is related to the pole N by the twin operation of reflexion in the twin plane whose pole is P. Since N and $\underline{\text{N}}$ are related by 180° rotation about P, it follows that N, P, and $\underline{\text{N}}$ are coplanar, that is they lie on the same great circle, and $\text{N}{:}\text{P} = \text{P}{:}\underline{\text{N}}$. Since $\underline{\text{N}}'$ is the reflexion of N in the twin plane, whose normal is P, it follows that N, P, and $\underline{\text{N}}'$ lie in a plane normal to the twin plane and the angles which N and $\underline{\text{N}}'$ make with the twin plane are both equal to $90° - \text{N}{:}\text{P}$ so that $\underline{\text{N}}'{:}\text{N} = 180° - 2(\text{N}{:}\text{P})$. Therefore $\underline{\text{N}}'{:}\underline{\text{N}} = (\underline{\text{N}}'{:}\text{N}) + (\text{N}{:}\underline{\text{N}}) = (\underline{\text{N}}'{:}\text{N}) + 2(\text{N}{:}\text{P}) = 180°$, so that $\underline{\text{N}}'$ is the opposite of $\underline{\text{N}}$ and they are in general only equivalent if the crystal is centrosymmetric. The reader will recall that we have earlier shown that the operation of a mirror plane followed by the operation of a

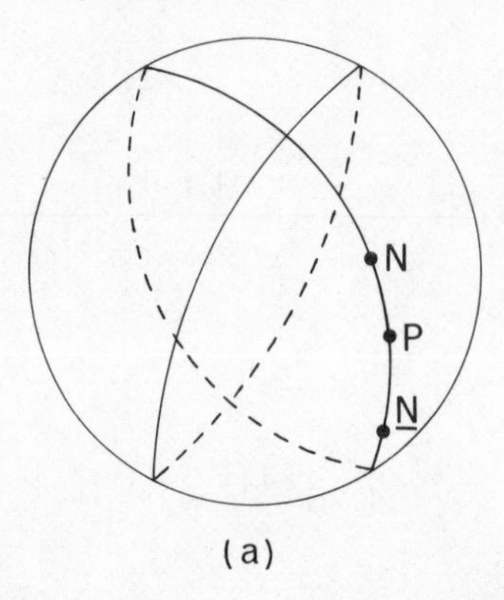

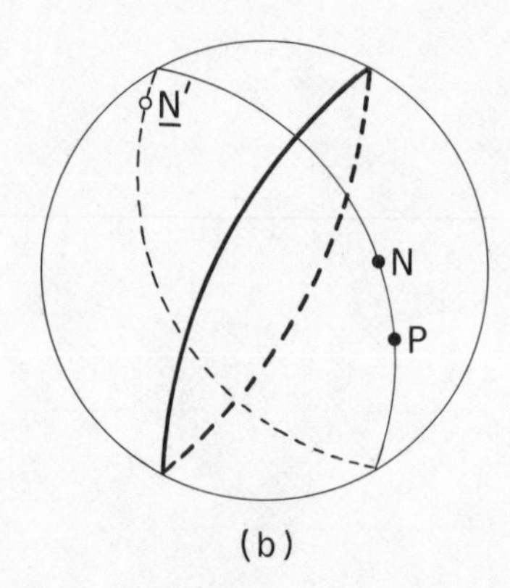

Fig 3.45 Stereograms to illustrate the twinning of a face whose normal is N by rotation (a) about a twin axis P to yield the face $\underline{\mathrm{N}}$ or by reflexion (b) in the plane (shown bold) whose normal is P to yield the face $\underline{\mathrm{N}}'$. Since $\underline{\mathrm{N}}'$ is the opposite of $\underline{\mathrm{N}}$ twinning by reflexion and by rotation are indistinguishable in a centrosymmetric crystal.

centre of symmetry generates a rotation diad normal to the mirror plane. Thus in a centrosymmetric crystal twinning by rotation about a twin axis and twinning by reflexion in a plane normal to the twin axis are indistinguishable operations; in these circumstances it is sufficient to state that the crystal is 'twinned on (hkl)' and unnecessary to specify whether by reflexion in the plane (hkl) or by diad rotation about the normal to (hkl). The twin of a cubic substance of class $m3m$ illustrated in Fig 3.44(d) may thus be simply described as 'twinned on $(11\bar{1})$'.

In contrast to contact twins, where the two components of the twinned crystal are joined only on an interface parallel to a lattice plane, in *interpenetrant twins* the interface between the twin components is irregular and the twin components in such twinned crystals are often intimately intergrown as illustrated in Fig 3.44(e).

The angular relationships between the faces of twinned crystals can, just as for truly single crystals, conveniently be displayed on a stereogram. One twin component is plotted in the standard orientation for its crystal system and the other in the orientation determined by the twinning operation (Fig 3.46). In plotting the poles of the faces of a twinned centrosymmetric crystal it is usually convenient to regard the second component as derived by rotation twinning from the first component because rotation about an inclined axis is very much more easily performed than reflexion in an inclined plane in the stereographic projection. The faces of each component of the twinned crystal are indexed separately in terms of the conventional orientation of the crystallographic reference axes within that component, the indices of the faces of one component being distinguished by underlining so that (hkl) refers to one component and $(\underline{hkl})$ to the other.

So far we have restricted our discussion to twinned crystals containing only two components; but *multiple twins* consisting of three or more components also occur. In some multiple twins the twinning operations which relate adjacent components are all identical; then the components tend to take the form of lamellae parallel to the composition plane so that such twins are known as *lamellar* or *polysynthetic twins* (Fig 3.47(a)). Polysynthetic twins may be on a macroscopic, microscopic, or submicroscopic scale.

Another sort of multiple twin is the *multiplet*, where several components are produced by the operation of symmetry-related twin planes or twin axes. For instance

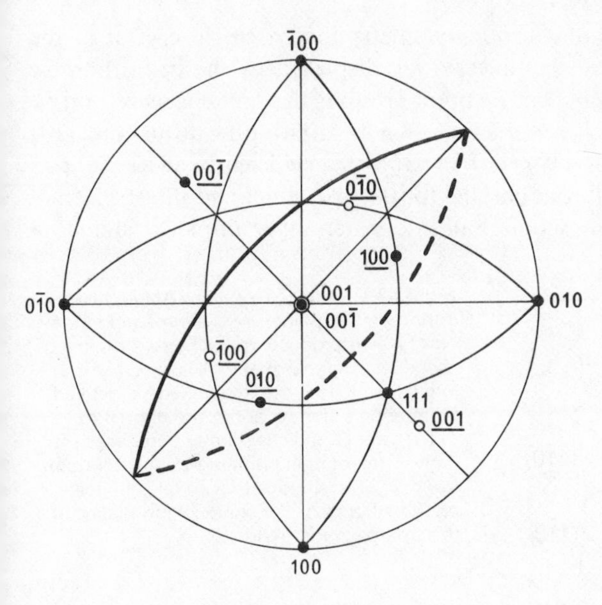

Fig 3.46 The stereogram shows the poles of the faces of a crystal exhibiting the form {100} in the point group $m3m$ and the poles of the faces {100} of its twin on (111). The faces of the twin have been indexed on the assumption that it is a rotation twin about the normal to (111); on the alternative assumption that the twinning is by reflexion in (111) the poles of the form {100} would be identically placed but the signs of their indices would be reversed. The plane (111) is shown as a bold great circle.

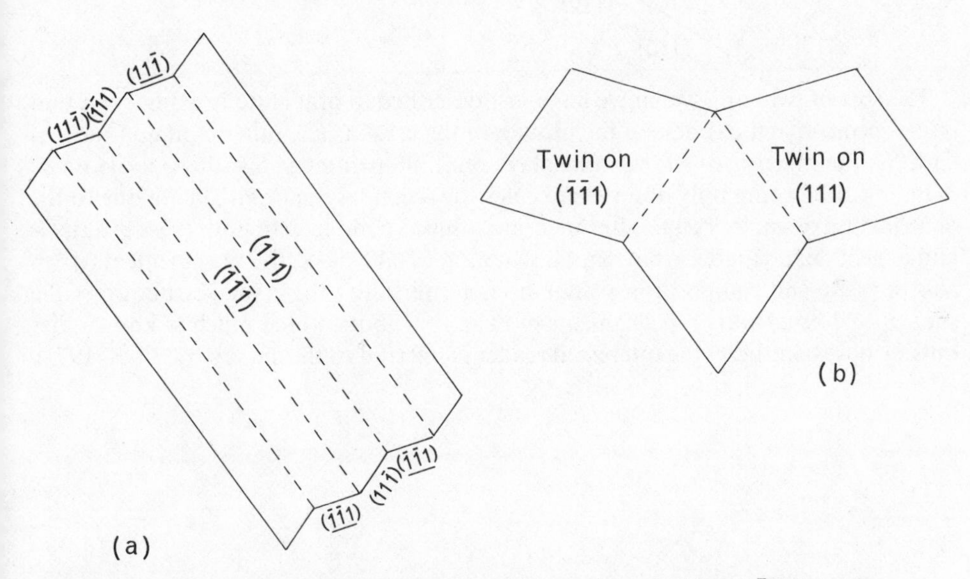

Fig 3.47 Polysynthetic and multiplet twinning. (a) is a section parallel to $(1\bar{1}0)$ through an octahedral crystal of point group $m3m$ twinned polysynthetically on (111). (b) is a section parallel to $(1\bar{1}0)$ through an octahedral crystal of point group $m3m$ twinned on the symmetry related planes (111) and $(\bar{1}\bar{1}1)$ so as to become a triplet. In each case the traces of composition planes are shown as broken lines.

if a cubic crystal of point group $m3m$ can twin on (111), it can also twin on other planes of the form {111}, such as $(\bar{1}11)$, $(1\bar{1}1)$, $(\bar{1}\bar{1}1)$, and may actually do so. In such circumstances some twinned crystals will consist of three or more components, each pair of components being related by a different twinning operation. Such multiplets may be distinguished as *triplets* when they contain three components, *quartets* when they contain four, and so on. A triplet produced by twinning on (111) and $(\bar{1}\bar{1}1)$ in a cubic substance of point group $m3m$ is illustrated in Fig 3.47(b).

Occasionally the geometry of the twinning operation may be such that the twinned crystal appears to have higher point group symmetry than a single crystal of the same substance. Such *mimetic twinning* is very well displayed by the orthorhombic form of $CaCO_3$, the mineral *aragonite*. The point group of aragonite is *mmm* and its unit-cell dimensions are $a = 4{\cdot}95$ Å, $b = 7{\cdot}95$ Å, $c = 5{\cdot}73$ Å. Twinning occurs on $\{110\}$ and, since the interfacial angle $(110){:}(1\bar{1}0) = 63°48'$ is sufficiently close to 60°, multiplets will appear to have a hexad parallel to [001]. Contact and interpenetrant multiplets are quite common in aragonite; an interpenetrant multiplet is shown in Fig 3.48.

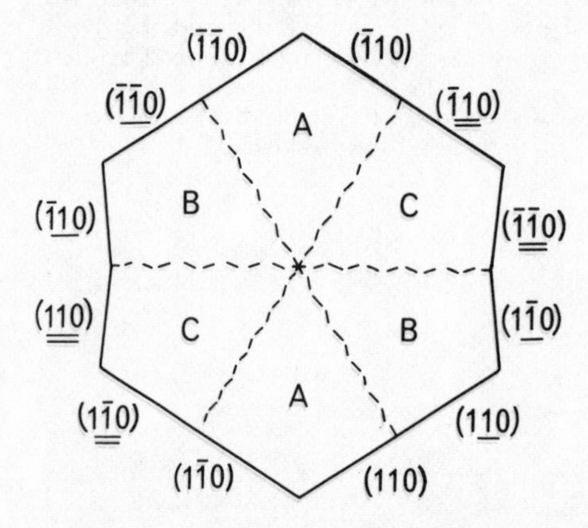

Fig 3.48 Mimetic twinning. Aragonite is orthorhombic, point group *mmm*, and twins on $\{110\}$ to produce interpenetrant triplets of pseudo-hexagonal shape. A is the parent individual, B (indices underlined) is derived by rotation twinning about the normal to (110), and C (indices doubly underlined) is derived by rotation twinning about the normal to $(1\bar{1}0)$. The sinuous broken lines indicate possible traces of the composition planes in this interpenetrant twin.

The sort of twinning which we have just described in aragonite is primary in that it has occurred in the course of the growth of the crystal. The other common form of $CaCO_3$, the mineral *calcite*, also displays multiple twinning; but the polysynthetic twinning quite commonly observed in calcite is secondary in origin, being due to the deformation of single crystals after their growth was complete. In both these examples and in general, no matter what caused twinning to take place, the twin elements (twin axis or plane and composition plane) are determined by the crystal structure of the substance. Discussion of mechanisms of twinning, about which much is known, lies outside our scope here; the interested reader is referred to the survey by Bloss (1971).

4 Internal structure of crystalline matter

In chapter 3 the shape of unit-cell appropriate to each crystal system was established and the crystallographic point groups were derived. We now proceed to relate these concepts in the two-dimensional or *plane lattices* as a preliminary to the development of the types of crystallographic three-dimensional or *Bravais lattices*. It is convenient that many of the significant features of the Bravais lattices are simply exemplified in the plane lattices. The reader is reminded of the restrictions imposed by symmetry on unit-cell shape, listed in Table 3.3.

Plane lattices

We now investigate systematically the operation of the two-dimensional point groups (listed in Table 3.7) on a lattice to establish the types of plane lattice. The most generalized unit-mesh ($a \neq b$; γ general) has, as has already been said, a diad perpendicular to the plane of the mesh through every lattice point and midway between adjacent lattice points; whether the hypothetical two-dimensional crystal has point group symmetry 1 or 2 depends on the atomic arrangement and not on the nature of the lattice. This lattice type is known as *oblique* and since the unit-mesh contains only one lattice point it is said to be a *primitive* lattice; the so-called *oblique* p-*lattice* is shown in Fig 4.1(a).

Next we examine the restrictions placed on lattice geometry by the presence of one set of parallel lines of symmetry. It is evident from Fig 4.2 that in a one-dimensional lattice, that is a lattice point row, lines of symmetry must either be of type I, passing through lattice points (Fig 4.2(a)), or of type II, passing midway between lattice points (Fig 4.2(b)). A two-dimensional lattice plane will be generated either by an array of rows all of type I, or all of type II, or an alternation of the two types (Fig 4.2(c), (d), and (e)); no other arrangement will preserve the essential lattice criterion that every lattice point must have the same environment in the same orientation. If all the rows are of the same type, whether I or II, identical lattices are generated, the unit-mesh being a rectangle with $a \neq b$, $\gamma = 90°$; this is the *rectangular* p-*lattice* (Figs 4.1(b) and 4.2(f)). The manner in which we were obliged to arrange the point rows has introduced additional symmetry elements so that the rectangular p-lattice has mutually perpendicular sets of lines of symmetry, one set perpendicular to x and the other perpendicular to y, each set being separated by half the appropriate

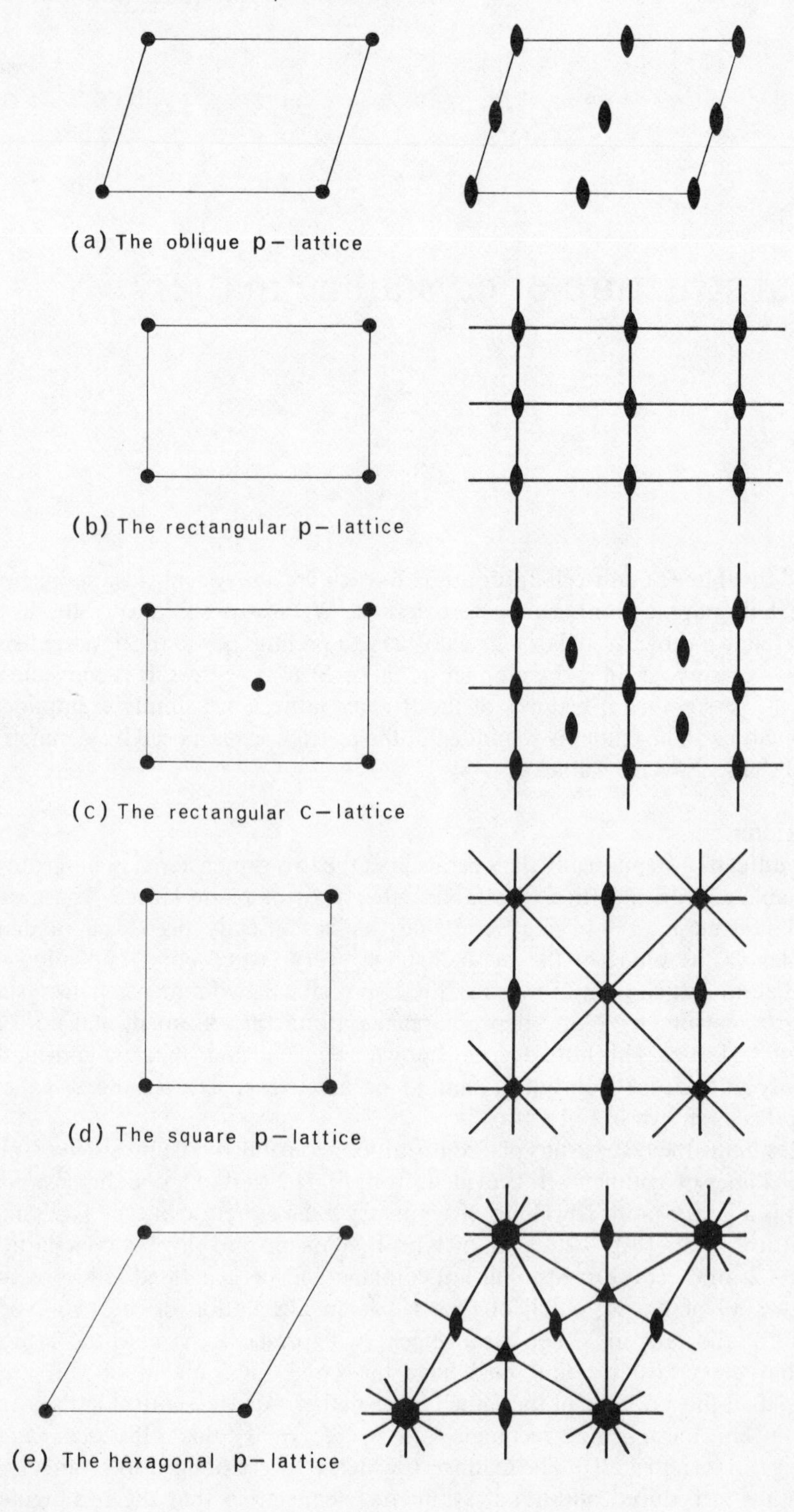

Fig 4.1 Unit-cells of the five plane lattice types.

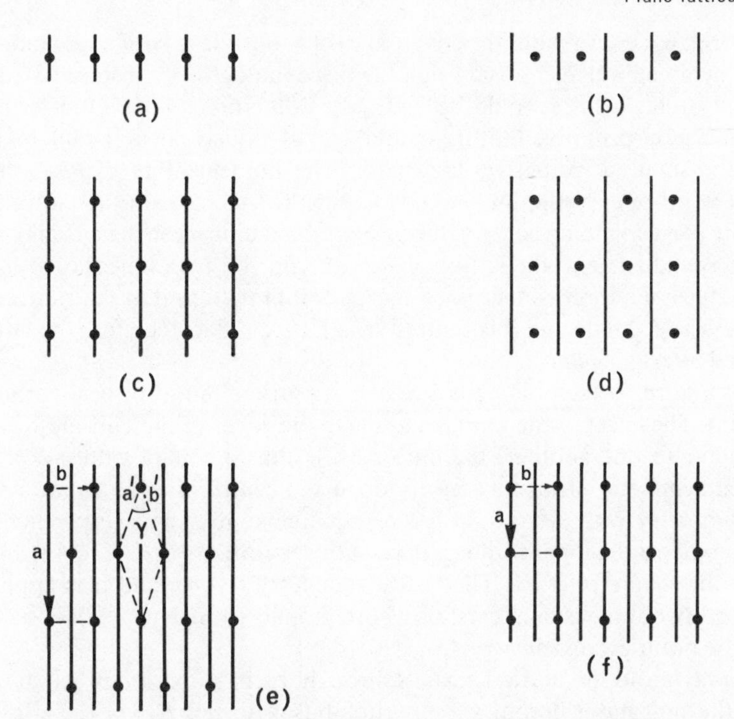

Fig 4.2 Restrictions on plane lattice geometry imposed by one set of parallel lines of symmetry. Lines of symmetry of two types operating on lattice point rows are shown in (a) and (b). The remaining diagrams show plane lattices generated by an array of rows all of type I (c), all of type II (d), or by alternation of types I and II (e). The plane lattice shown in (e) is rectangular-c and that shown in (f) is rectangular-p.

lattice spacing, $a/2$ and $b/2$ respectively; also of course a diad occurs perpendicular to the lattice plane at every intersection of lines of symmetry. The point group symmetry of this lattice is $2mm$; only the atomic arrangement in the hypothetical structure determines whether the crystal has point group symmetry $1m$ or $2mm$.

We now have to return to the lattice produced by alternation (Fig 4.2(e)). The smallest unit-mesh is a rhombus with $a' = b' = \sqrt{\{(a/2)^2 + (b/2)^2\}}$ and $\gamma = 2\tan^{-1} b/a$, which is not in general a special angle. But it must be noted that the symmetry of this lattice is closely related to that of the rectangular p-lattice: it has mutually perpendicular sets of lines of symmetry, one set $\perp x$ and $a/2$ apart, the other $\perp y$ and $b/2$ apart, a diad at every intersection of lines of symmetry, and in addition a diad at the mid-point of each side of the smallest unit-mesh (Fig 4.1(c)). But the relationship between lattice points and symmetry elements is different: whereas in the rectangular p-lattice alternate lines of symmetry of either set are devoid of lattice points, in this *rectangular c-lattice* every line of symmetry passes through lattice points. The lattice symmetry is again $2mm$ and hypothetical two-dimensional crystals of point group symmetry $1m$ or $2mm$ may have this lattice type. It is conventional, and convenient, in this lattice type to take as the unit-mesh not the smallest unit-mesh but a rectangular unit-mesh with edges perpendicular to each set of lines of symmetry and edges equal to the separation of lattice points in these directions. Such a unit-mesh has lattice points not only at its corners but also at its centre; it is described as *non-primitive* because the lattice it generates has more than one lattice point per unit-mesh. The larger area of a non-primitive unit-mesh is in some respects disadvantageous, but for the purposes of indexing lattice planes and specifying symmetry-related coordinates the advantages of a unit-mesh with edges more closely related to symmetry elements outweigh the disadvantages: for instance a certain set of symmetry-related lines would be indexed as (hk), $(\bar{h}k)$, $(\bar{h}\bar{k})$, $(h\bar{k})$ in terms of the conventional non-primitive unit-mesh and, less obviously, as (hk), (kh), $(\bar{h}\bar{k})$, $(\bar{k}\bar{h})$ in terms of the smallest (primitive) unit-mesh. The nomenclature of this lattice type requires a word of explanation: we have designated the axes in the lattice plane as x and y, the corresponding edges of the conventional unit-mesh being a and b, so that the axis perpendicular to the lattice plane is z and the two-dimensional unit-mesh is the (001) face of a three-dimensional lattice of infinite repeat in the z direction; the unit-mesh is a rectangle and it is centred by a lattice point; therefore the lattice type is described as *rectangular*-c.

We now return to the oblique lattice and suppose a tetrad to pass through each lattice point. The effect of the tetrad is to make the edges of the unit-mesh equal and perpendicular to one another; the unit-mesh is thus a square with $a = b$, $\gamma = 90°$. Introduced symmetry elements are tetrads in the centre of each square unit-mesh, lines of symmetry with separation $a/2$ perpendicular to x and perpendicular to y, lines of symmetry diagonal to the square with separation $\frac{1}{2}a\sqrt{2}$ (Fig 4.1(d)). This is known as the *square* p-*lattice*. The lattice symmetry is $4mm$ and it is applicable to hypothetical two-dimensional crystals whose atomic arrangement is consistent with either of the point groups $4mm$ or 4.

If either a hexad or a triad passes through each lattice point of the oblique p-lattice, the unit-mesh becomes a 60° rhombus with $a = b$, $\gamma = 120°$. Introduced symmetry elements are sets of lines of symmetry at 30° to one another such that one member of every set passes through each lattice point and a triad at the centroid of each of the two equilateral triangles that make up the unit-mesh (Fig 4.1(e)). The point group symmetry of the lattice, the *hexagonal* p-*lattice*, is $6mm$, but atomic arrangements of inadequate symmetry may produce degeneration of the symmetry of the structure to the point groups 6, $3m$, or 3 with retention of this lattice.

The relationship of the 5 plane lattices to the 10 two-dimensional point groups is summarized in Table 4.1.

Bravais lattices

There are fourteen three-dimensional or Bravais lattice types differentiated one from another by the symmetry of the arrangement of their lattice points, the actual dimensions of the lattice being of course unimportant. Before proceeding to the development of the Bravais lattices by considering the variety of ways in which plane lattices can be stacked so as to be consistent with three-dimensional point group

Table 4.1
The 5 plane lattices

Lattice type	Point group of lattice	Possible crystal point groups	Shape of conventional unit-mesh
Oblique p	2	1, 2	$a \neq b;\ \gamma$ general
Rectangular p, Rectangular c	$2mm$	$1m$, $2mm$	$a \neq b;\ \gamma = 90°$
Square p	$4mm$	4, $4mm$	$a = b;\ \gamma = 90°$
Hexagonal p	$6mm$	6, $6mm$; 3, $3m$	$a = b;\ \gamma = 120°$

symmetry it may be helpful to the reader if we first recapitulate some statements made in earlier chapters, in some cases developing their implications for this topic.

In chapter 3 we developed seven lattice types, each primitive, each consistent with the symmetry requirements of one crystal system, and each having unit-cell edges parallel to the crystallographic reference axes of the system. We now have to investigate whether other arrangements of lattice points may be consistent with crystal symmetry in any system. We shall, for example, show that there are two distinct lattice types consistent with tetragonal symmetry (Fig 4.7(b) and (c)). Taking orthogonal axes with z parallel to the tetrad, in conformity with the usual convention for the tetragonal system, the lattice shown in Fig 4.7(b) has a unit-cell with lattice points only at its corners, that is to say it is a P-lattice;[1] but the lattice shown in Fig 4.7(c) has a unit-cell with lattice points not only at its corners but in addition a lattice point at its centre. This latter type is known as the tetragonal I-lattice, I being the initial letter of the German word for body-centred, *innenzentrierte*. Of course the tetragonal I-lattice could alternatively be described in terms of a primitive unit-cell, but that would not have the characteristic tetragonal shape, a square prism, and consequently would not embody the characteristic symmetry of the lattice. The practical advantages to be gained by using such non-primitive unit-cells in general far outweigh the disadvantages that stem from their larger volume (the tetragonal I-cell contains two lattice points and so has twice the volume of the corresponding primitive unit-cell).

The task of discovering how many different arrangements of lattice points are possible is simplified by taking into account from the start a fundamental property of lattices: that they are centrosymmetric and that every lattice point lies at a centre of symmetry. Since a lattice is a regular array of points in space, it follows that if a lattice point A lies at a vector distance **t** from a lattice point B, then a lattice point C must lie at a vector distance $-\mathbf{t}$ from B; therefore the lattice point B is a centre of symmetry of the lattice and so is every other lattice point. Thus our discussion of possible lattice types can be limited to lattices consistent with the eleven centrosymmetric point groups, $\bar{1}$, $2/m$, mmm, $\bar{3}$, $\bar{3}m$, $4/m$, $4/mmm$, $6/m$, $6/mmm$, $m3$, and $m3m$.

Our task is further simplified by making use of the observation that a lattice plane normal to a threefold, a fourfold, or a sixfold symmetry axis must contain lines of symmetry, which become planes of symmetry parallel to the axis in the three-dimensional lattice. Thus a lattice consistent with point group symmetry $4/m$ must display the higher symmetry of point group $4/mmm$. The point groups $\bar{3}$, $4/m$, $6/m$, and $m3$ can thus be struck off our list and we are left with the seven point groups $\bar{1}$, $2/m$, mmm, $\bar{3}m$, $4/mmm$, $6/mmm$, and $m3m$, which are the holosymmetric point groups of the seven crystal systems. It follows that in deriving the Bravais lattices for the tetragonal system, for example, all we have to do is to discover the types of lattice consistent with the presence of a single tetrad; all such lattices will have point group symmetry $4/mmm$ and be characteristic of the tetragonal system.

We now consider the variety of ways in which plane lattices can be superimposed, or stacked, to produce three-dimensional lattices consistent with the characteristic symmetry of each crystal system. We start with the triclinic and proceed in sequence of increasing symmetry, leaving the trigonal and the hexagonal systems, which pose special problems, to the end. We shall consistently use **t** to represent the *stacking vector*

[1] We follow here the practice of *International Tables for X-ray Crystallography*, vol I (1969) in using lower case letters for plane lattice types, p and c, and capital letters for Bravais lattice types, P, C, I, F, etc.

between adjacent lattice planes and define **t** as the vector from a lattice point in one plane to a lattice point in the immediately superimposed plane.

Triclinic system

Suppose an oblique p-lattice plane is superimposed on another that is identical in such a manner that the diads perpendicular to the two planes do not coincide and that this mode of superposition is repeated indefinitely. The result is a primitive triclinic lattice; if the dimensions of the unit-mesh of the original lattice plane are $a \neq b$, γ general and the stacking vector has magnitude c and makes angles α and β with the directions of y and x respectively, then the dimensions of the general parallelpiped that is the unit-cell of this *triclinic* P-*lattice* will be $a \neq b \neq c$, $\alpha \neq \beta \neq \gamma$. Since the holosymmetric class of the triclinic system has only a centre of symmetry, which is the symmetry element common to all lattice types, there is no advantage to be gained by selecting non-primitive unit-cells in this system.

Monoclinic system

Again our starting point is the primitive oblique plane lattice, the axes of which we shall relabel as x and z, interaxial angle β, in order to produce the monoclinic Bravais lattices in conventional orientation. We consider the various ways in which two such lattice planes of identical unit mesh can be superimposed so that their diads are coincident. It is apparent from Fig 4.4(a) that there are four types of diad normal to the plane of any oblique p-lattice; these are labelled I–IV and are such that diads of type I pass through lattice points, diads of type II lie midway between lattice points that are a distance a apart, diads of type III lie midway between lattice points that are a distance c apart, and diads of type IV pass through the centre of each unit-mesh. There are four ways in which such plane lattices can be stacked so as to achieve coincidence of diads and these are illustrated in Fig 4.4: (i) the stacking vector **t** has no component in the xz plane so that each diad is superimposed on one of its own kind, i.e. I on I, II on II, III on III, IV on IV, (ii) the stacking vector **t** has a component $\frac{1}{2}\mathbf{a}$ in the xz plane so that diads are superimposed according to the scheme I on II, II on I, III on IV, IV on III, (iii) the stacking vector **t** has a component $\frac{1}{2}\mathbf{c}$ in the xz plane so that diads are superimposed I on III, II on IV, III on I, IV on II, and (iv) the stacking vector **t** has a component $\frac{1}{2}(\mathbf{a}+\mathbf{c})$ in the xz plane so that the stacking scheme is I on IV, II on III, III on II, IV on I.

The stacking sequence (i) with $\mathbf{t} = \mathbf{b}$ (Fig 4.3(a)) generates a primitive monoclinic lattice, the unit-cell of which has lattice points only at its corners and dimensions $a \neq b \neq c$, $\alpha = \gamma = 90°$, β obtuse. As in the oblique plane lattice (Fig 4.4(a)) there are four types of diads with coordinates (Fig 4.4(b)) $0, y, 0$; $\frac{1}{2}, y, 0$; $0, y, \frac{1}{2}$; $\frac{1}{2}, y, \frac{1}{2}$; where y is a variable. Mirror planes lie parallel to (010) and are of two types, the members of each type being separated from one another by **b**: one type comprises the (010) planes passing through points with coordinates $0, y, 0$, where y is integral and the planes of this set contain all the lattice points; the other type comprises the (010) planes passing through points $0, y, 0$, where $y = (2n+1)/2$. The point group symmetry of the lattice is $2/m$, but the point group symmetry of the crystal referred to it may be 2 or m or $2/m$. This lattice type is known as the *monoclinic* P-*lattice*.

The stacking sequence (ii) has a stacking vector of the form $\mathbf{t} = \frac{1}{2}\mathbf{a} + q\mathbf{b}$ where q is a simple fraction (Fig 4.3(b)). The stacking vector from the zeroth layer to the second layer will then be $2\mathbf{t} = \mathbf{a} + 2q\mathbf{b}$; this has an integral coefficient of **a** and alternate layers will then be directly superimposed when viewed down the y-axis. The resultant

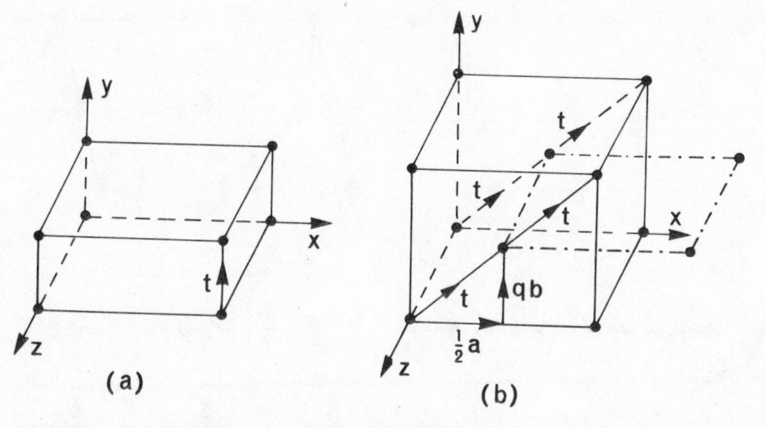

Fig 4.3 Stacking sequences for monoclinic P and C lattices.

non-primitive monoclinic lattice (Fig 4.4(c)) has a stacking vector $\mathbf{t} = \frac{1}{2}\mathbf{a} + \frac{1}{2}\mathbf{b}$ and two lattice points per unit-cell, one with coordinates 0, 0, 0 (i.e. $\frac{1}{8}$ of a lattice point at each of the 8 corners is associated with the chosen unit-cell) and the other with coordinates $\frac{1}{2}$, $\frac{1}{2}$, 0 (i.e. a lattice point lies at the centre of each of the two (001) faces and $\frac{1}{2}$ of each is associated with the chosen unit-cell). Since the (001) faces of the unit-cell are centred, this is known as the *monoclinic* C-*lattice*. Diads in this lattice type are of only two types: type I, with coordinates 0, y, 0 and $\frac{1}{2}$, y, 0, pass through lattice points and type II, with coordinates 0, y, $\frac{1}{2}$ and $\frac{1}{2}$, y, $\frac{1}{2}$, lie midway between lattice points that are **c** apart. Mirror planes are of one type only, parallel to (010) and $\frac{1}{2}\mathbf{b}$ apart. The point group symmetry of the monoclinic C-lattice is again $2/m$.

Stacking sequence (iii) with $\mathbf{t} = qb + \frac{1}{2}\mathbf{c}$ likewise generates a lattice with alternate layers directly above and below each other when viewed along [010] because $2\mathbf{t} = 2q\mathbf{b} + \mathbf{c}$ and, in this case, the coefficient of **c** is integral so that $q = \frac{1}{2}$. The resultant lattice type (Fig 4.4(d)) has lattice points at the corners (0, 0, 0) and at the centres of the (100) faces (0, $\frac{1}{2}$, $\frac{1}{2}$) of a conventionally shaped monoclinic unit-cell. This is the monoclinic A-lattice, but it is not a distinct lattice type because the x and z axes, which are not restricted to particular directions by symmetry, can be interchanged to convert it into a monoclinic C-lattice.

Stacking sequence (iv) with $\mathbf{t} = \frac{1}{2}\mathbf{a} + q\mathbf{b} + \frac{1}{2}\mathbf{c}$ generates a monoclinic lattice with lattice points at the corners (0, 0, 0) and at the body centre ($\frac{1}{2}$, $\frac{1}{2}$, $\frac{1}{2}$) of each unit-cell, q being equal to $\frac{1}{2}$. That this is not a distinct lattice type is evident from Fig 4.4(e), where it is shown that diagonal axes may be selected to define either an A-cell, with (100) faces centred, or a conventional C-cell, with (001) faces centred.

We have now exhausted all the stacking possibilities of the oblique plane lattice and seen that generalized stacking gives rise to loss of diads and the production of a triclinic lattice, while stacking with the restriction of coincidence of diads gives rise variously to monoclinic P- and C-lattices.

Orthorhombic system

Here we are concerned with stacking the two rectangular plane lattice types (Figs 4.5(a), 4.6(a)) which have symmetry $2mm$. The basic criterion that has to be satisfied is that lattice planes normal to x, y, and z in the three-dimensional lattice should have symmetry $2mm$. This can be achieved by superimposing either type of rectangular plane lattice so that diads, parallel to z, at the intersection of lines of symmetry are

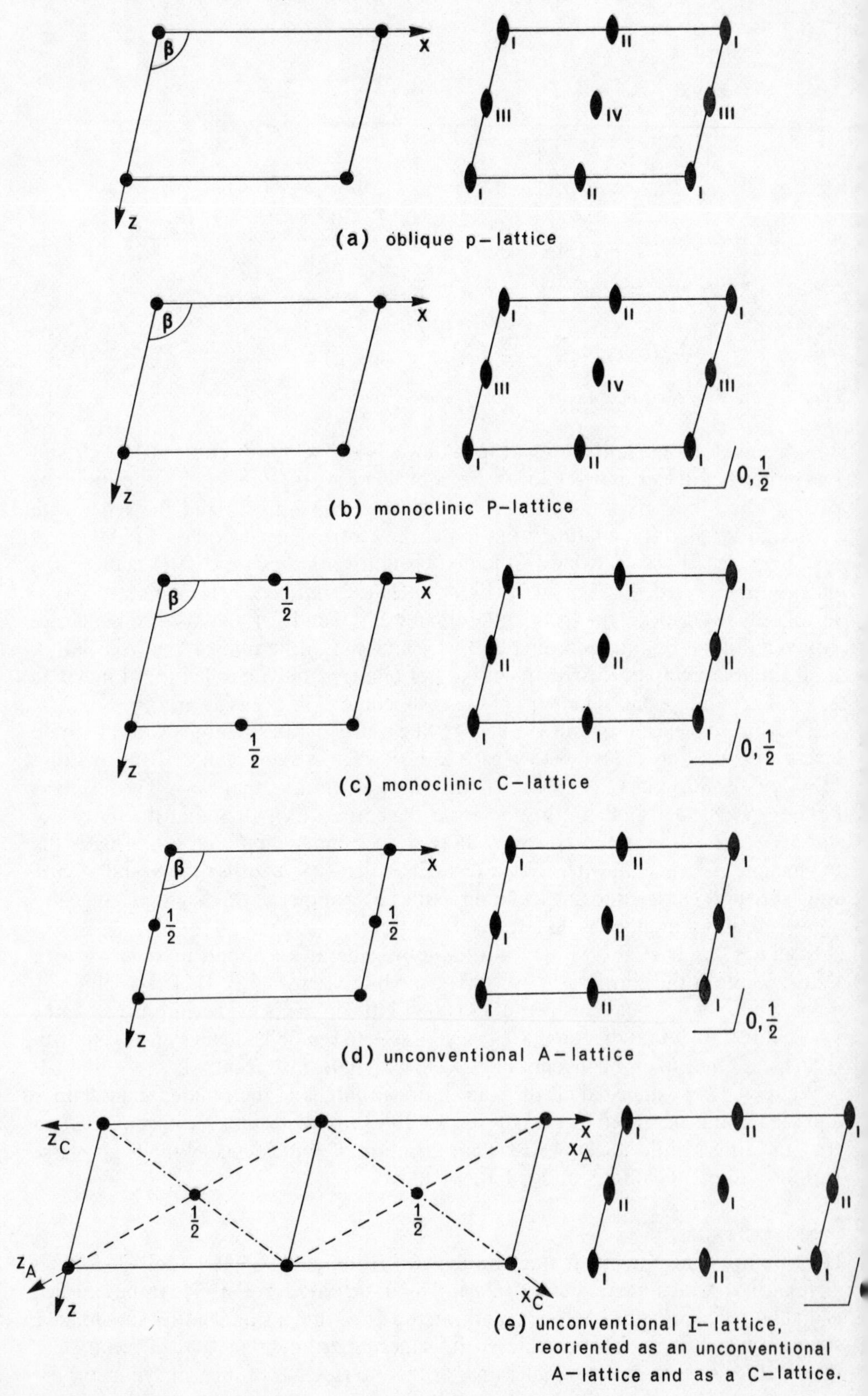

(a) oblique p-lattice

(b) monoclinic P-lattice

(c) monoclinic C-lattice

(d) unconventional A-lattice

(e) unconventional I-lattice, reoriented as an unconventional A-lattice and as a C-lattice.

coincident; the lines of symmetry parallel to x and y then generate (010) and (100) mirror planes and, because the lattice is necessarily centrosymmetric, [100] and [010] diads and (001) mirror planes are introduced; the resultant three-dimensional lattice types then have the required symmetry *mmm*.

The rectangular p-lattice has four types of diad, labelled I–IV on Fig 4.5(a), and four types of line of symmetry, labelled p, q, r, s. As for the oblique p-lattice there are four types (i)–(iv) of stacking sequence that maintain coincidence of diads, all of which lie at intersections of lines of symmetry. The stacking vector $\mathbf{t} = \mathbf{c}$ implies coincidence of diads of the same type in successive layers and gives rise to a three-dimensional lattice, the unit-cell of which is a parallelepiped $a \neq b \neq c$, $\alpha = \beta = \gamma = 90°$. This is a primitive unit-cell (Fig 4.5(b)) and has the shape characteristic of the orthorhombic system. The lattice, which is known as the *orthorhombic* P-*lattice*, has diads of four types parallel to [001] with coordinates $0, 0, z$; $\frac{1}{2}, 0, z$; $0, \frac{1}{2}, z$; $\frac{1}{2}, \frac{1}{2}, z$ respectively. The presence of lines of symmetry parallel to [010] in each (001) plane of lattice points generates mirror planes parallel to (100); these are of two types, p which pass through lattice points and constitute the faces of the unit-cell, and q, which pass centrally through each unit-cell. In precisely the same way mirror planes of two types, r and s, are generated parallel to (010). The lattice thus has mirror planes perpendicular to x and perpendicular to y and diads parallel to z and, since it is necessarily centrosymmetric, its point group symmetry is that of the holosymmetric class of the orthorhombic system, *mmm*. Detailed discussion of the full symmetry of this lattice type is postponed until after the introduction of space groups.

The other three ways of stacking rectangular p-lattices give rise to centred orthorhombic lattices in a manner generally corresponding to the monoclinic cases discussed in detail earlier. The stacking sequence (ii), with $\mathbf{t} = \frac{1}{2}\mathbf{a} + r\mathbf{c}$, where r is a fraction, superimposes diads of type I on those of type II and type III on type IV. Since lattice points on alternate lattice planes lie directly above or below each other along the z-axis the value of r is determined as $\frac{1}{2}$. The resultant three-dimensional lattice has lattice points at $0, 0, 0$ and $\frac{1}{2}, 0, \frac{1}{2}$, that is to say the (010) or B-face of the unit-cell (Fig 4.5(c)) is centred; this is the *orthorhombic* B-*lattice*. This lattice type has two types of diad parallel to z: type I pass through lattice points and type III[2] lie midway between lattice points a distance **b** apart. The (100) mirror planes are all of type p and pass through lattice points, whereas the (010) mirror planes are again of two types, r and s.

The stacking sequence (iii), with $\mathbf{t} = \frac{1}{2}\mathbf{b} + r\mathbf{c}$ superimposes diads of type I on those of type III and type II on those of type IV. The resultant three-dimensional lattice (Fig 4.5(d)) has lattice points at $0, 0, 0$ and $0, \frac{1}{2}, \frac{1}{2}$, that is to say the (100) or A-face of the unit-cell is centred; this is the *orthorhombic* A-*lattice*, which has the same symmetry elements as the B-lattice reoriented by interchange of x and y.

Stacking sequence (iv) with $\mathbf{t} = \frac{1}{2}\mathbf{a} + \frac{1}{2}\mathbf{b} + r\mathbf{c}$ superimposes diads of type I on those of type IV and those of type II on type III. The resultant orthorhombic unit-cell (Fig 4.5(e)) has lattice points at $0, 0, 0$ and $\frac{1}{2}, \frac{1}{2}, \frac{1}{2}$; it is a body-centred unit-cell and the lattice type is described as the *orthorhombic* I-*lattice*. Since in the orthorhombic system the reference axes are required to be parallel to the orthogonal diads, the

Fig 4.4 Unit-cells of the monoclinic Bravais lattices generated by stacking plane oblique p-lattices. In (e) the axes x_A, z_A refer to the A-cell and the axes x_C, z_C to the C-cell.

[2] We retain here the labels of Fig 4.5(a).

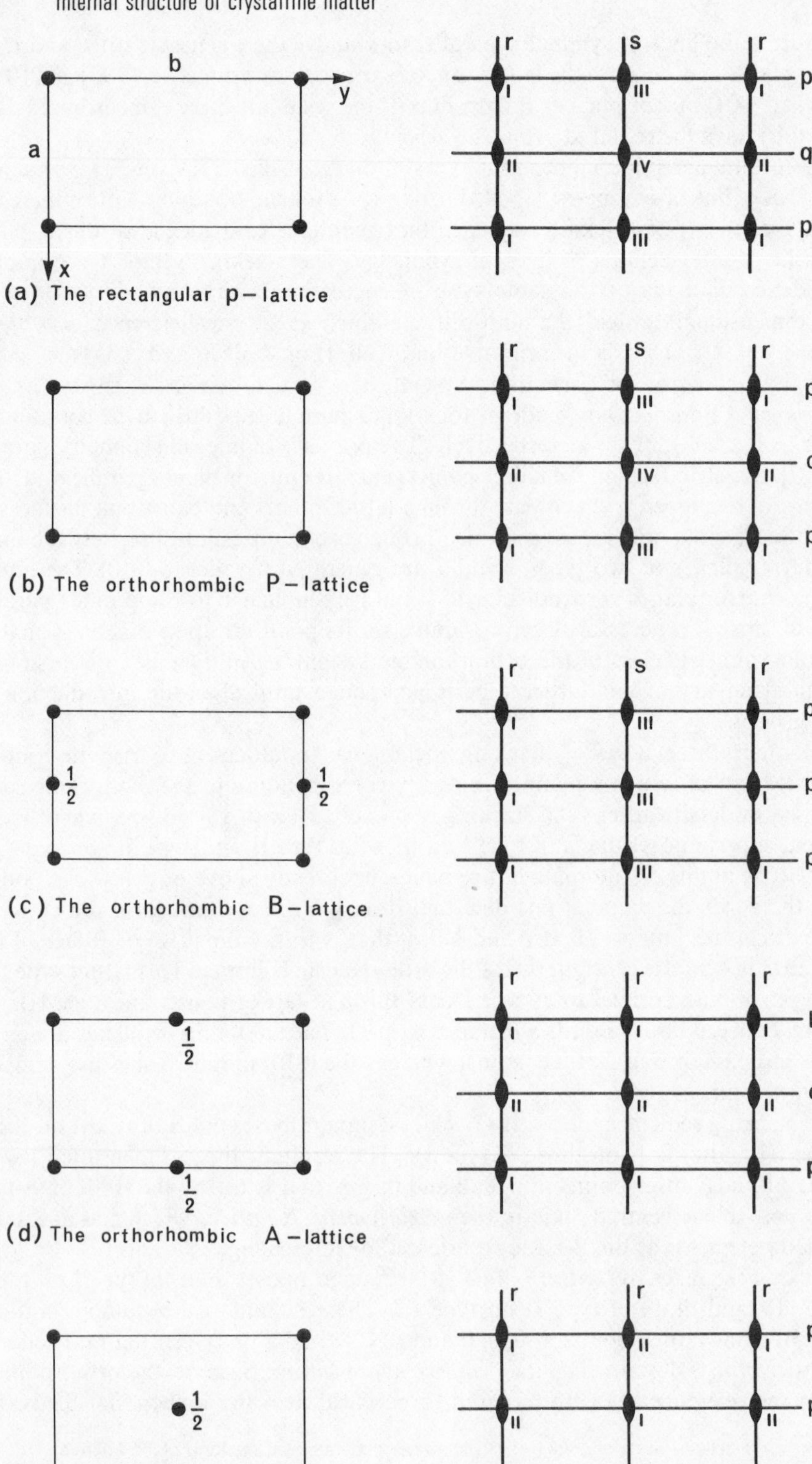

(a) The rectangular p-lattice

(b) The orthorhombic P-lattice

(c) The orthorhombic B-lattice

(d) The orthorhombic A-lattice

(e) The orthorhombic I-lattice

I-lattice is not equivalent either to the A- or the B-lattice; it will be recalled that, contrariwise, the monoclinic I-lattice was shown to be describable as an A- or a C-lattice. The body-centred orthorhombic lattice has two types of [001] diad: type I passes through lattice points while type II lies midway between lattice points on the x and y axes. All the (100) mirror planes pass through lattice points and are styled type p. Likewise all the (010) mirror planes pass through lattice points.

We now consider the variety of ways in which the rectangular c-lattice can be stacked with coincidence of diads. In this plane lattice type lines of symmetry are of two types, one perpendicular to x and designated type p, the other perpendicular to y and designated type r. This plane lattice type (Fig 4.6(a)) has two types of diad, I and II, at the intersection of lines of symmetry and two other types, III and IV, that do not lie on lines of symmetry. If diads of type I are superimposed on diads of type III, i.e. $\mathbf{t} = \frac{1}{4}\mathbf{a} + \frac{1}{4}\mathbf{b} + r\mathbf{c}$, planes of symmetry parallel to (100) and (010) will not be generated and the resultant three-dimensional lattice with $r = \frac{1}{2}$ would merely have diads parallel to z, that is to say it would be a monoclinic C-lattice in an unconventional orientation (Fig 4.6(b)). The dimensions of this monoclinic lattice would be such that two directions, x' and z' in the figure, would be of equal magnitude although unrelated by symmetry. A similar situation arises if diads of type I are superimposed on diads of type IV, i.e. $\mathbf{t} = \frac{3}{4}\mathbf{a} + \frac{1}{4}\mathbf{b} + r\mathbf{c}$.

Of the remaining stacking schemes possible for the rectangular c-lattice, one superimposes diads type for type, i.e. $\mathbf{t} = \mathbf{c}$, to yield an *orthorhombic C-lattice* (Fig 4.6(c)). This lattice type is identical with the orthorhombic A- and B-lattices; one can be converted to another merely by interchange of axial labels.

The fourth, and final, way in which rectangular c-lattices can be stacked is by superimposition of diads of type I on diads of type II; there are two possible stacking vectors $\mathbf{t} = \frac{1}{2}\mathbf{a} + r\mathbf{c}$ and $\mathbf{t} = \frac{1}{2}\mathbf{b} + r\mathbf{c}$, which, because the plane lattice is centred, are equivalent. Alternate layers are directly above and below one another in the z direction so that $r = \frac{1}{2}$. The resulting lattice has every face of its unit-cell centred, lattice points being sited at $0, 0, 0; 0, \frac{1}{2}, \frac{1}{2}; \frac{1}{2}, 0, \frac{1}{2}; \frac{1}{2}, \frac{1}{2}, 0$ (Fig 4.6(d)). This lattice type is all-face-centred and is known as the *orthorhombic F-lattice.* Here the [001] diads are restricted to two types (I and III); the (100) and (010) mirror planes are each of one type (p or r respectively).

By regular stacking of rectangular plane lattices we have derived six three-dimensional lattice types each of which has orthorhombic symmetry and can therefore be described in terms of the conventional orthorhombic unit-cell, $a \neq b \neq c$, $\alpha = \beta = \gamma = 90°$. The orthorhombic A-, B-, and C-lattices differ only in the labelling of their reference axes and together constitute a single Bravais lattice type; conventionally axes are chosen so that it is the (001) face that is centred and this lattice type is known as the orthorhombic C-lattice. The remaining three lattice types P, I, and F are clearly distinct. Of course any of the three non-primitive orthorhombic lattice types can be described in terms of primitive unit-cells which are dimensionally monoclinic or triclinic and do not reflect the lattice symmetry. Except for certain specialized computational purposes, there is no advantage to be gained by using a primitive unit-cell of lower symmetry and many disadvantages.

Fig 4.5 Unit-cells of the orthorhombic Bravais lattices generated by stacking plane rectangular p-lattices. In this and in the immediately following figures symmetry elements parallel to the plane of the figure are not shown.

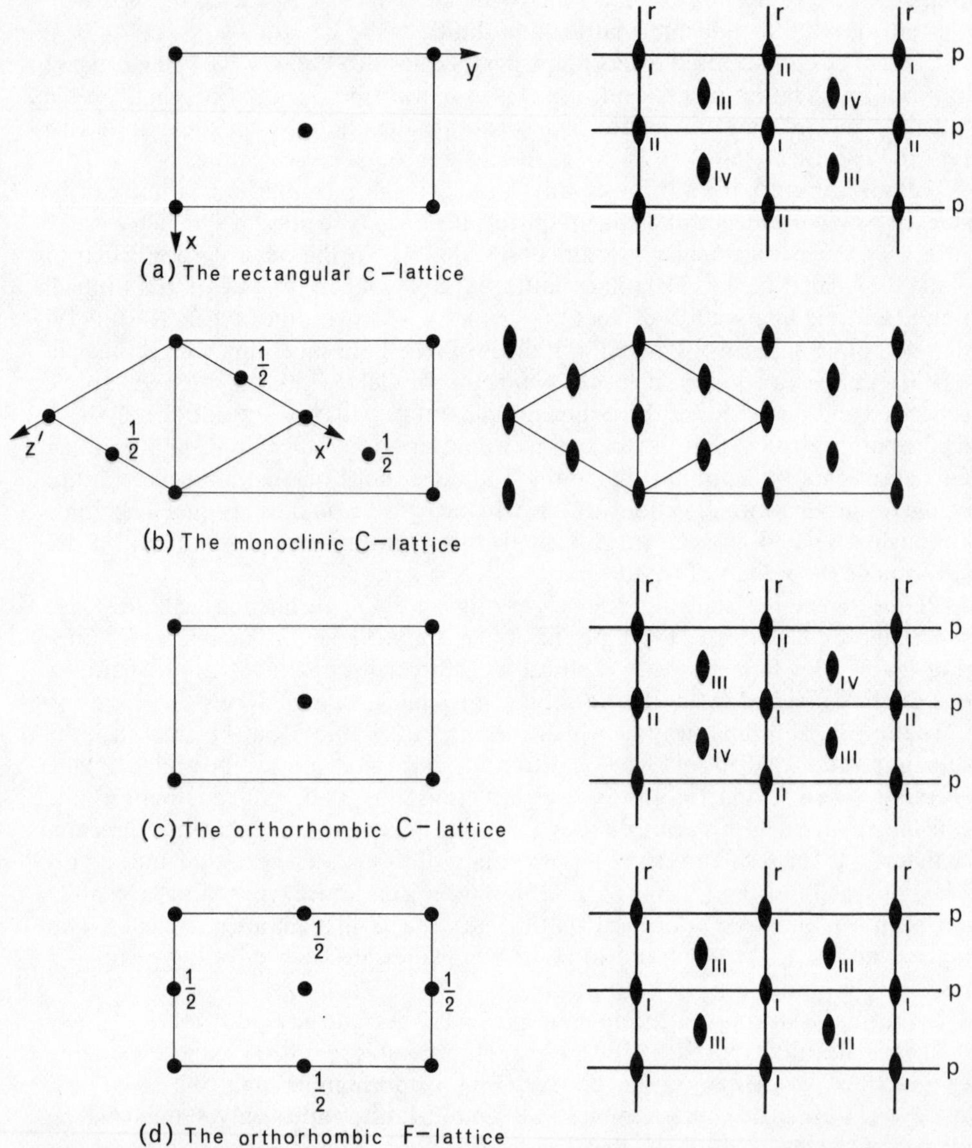

Fig 4.6 Unit-cells of the orthorhombic and monoclinic Bravais lattices generated by stacking plane rectangular c-lattices.

Tetragonal system

The next plane lattice to which we turn our attention is the square p-lattice (Fig 4.7(a)). This plane lattice type has tetrads of two kinds, type I which pass through lattice points, and type II, which pass through the centre of the unit-mesh. This is the only plane lattice type that has tetrad symmetry and in consequence lattice planes normal to the tetrad in three-dimensional tetragonal lattices must be of this type. In deriving the tetragonal lattice types it is adequate to consider stacking sequences that involve coincidence of tetrads and all other symmetry elements can safely be ignored; the resultant lattices will necessarily display the point group symmetry of the holosymmetric class of the tetragonal system, $4/mmm$. Since there are two tetrad types

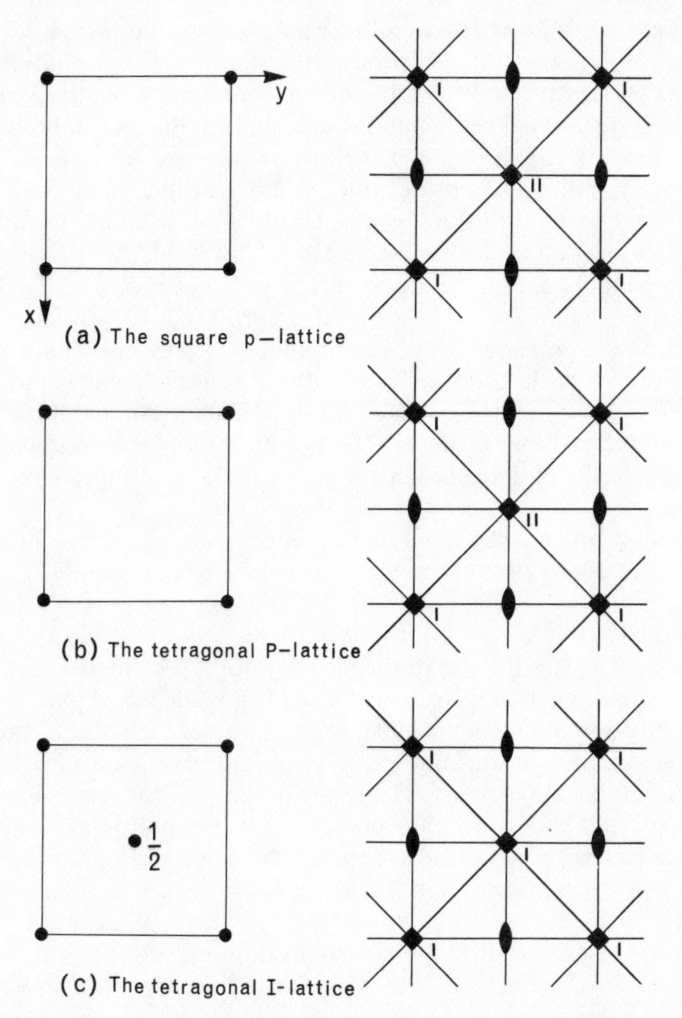

Fig 4.7 Unit-cells of the tetragonal Bravais lattices generated by stacking plane square p-lattices.

in the square p-lattice two stacking sequences are possible: (i) tetrads are superimposed type for type, (ii) tetrads of type I are superimposed on tetrads of type II. The first of these has a stacking vector $\mathbf{t} = \mathbf{c}$ and gives rise to a primitive unit-cell (Fig 4.7(b)) with dimensions $a = b \neq c$, $\alpha = \beta = \gamma = 90^\circ$ and lattice points at its corners. In this *tetragonal* P-*lattice* the tetrads are, as in the plane lattice, of two non-equivalent types. The other stacking sequence has $\mathbf{t} = \frac{1}{2}\mathbf{a} + \frac{1}{2}\mathbf{b} + r\mathbf{c}$ with $r = \frac{1}{2}$, as in the corresponding orthorhombic case, and gives rise to a body-centred lattice with lattice points at 0, 0, 0, and $\frac{1}{2}, \frac{1}{2}, \frac{1}{2}$. This *tetragonal* I-*lattice* (Fig 4.7(c)) has equivalent tetrads at 0, 0, z and $\frac{1}{2}, \frac{1}{2}, z$. These two lattices, P and I, are the only distinct tetragonal lattice types; both have $\{100\}$ and $\{110\}$ mirror planes and indeed all the symmetry of the holosymmetric class $4/mmm$.

Cubic system

Once again our starting point is the square p-lattice and our task is to discover all the stacking sequences consistent with cubic symmetry, the essential characteristic of which is the presence of four triads parallel to the $\langle 111 \rangle$ directions in a cubic unit-cell with $a = b = c$, $\alpha = \beta = \gamma = 90°$. Obviously the stacking vectors that generate the two tetragonal lattice types, P and I, will generate a *cubic* P-*lattice* and a *cubic* I-*lattice* when the additional restriction $a = c$ is applied (Fig 4.8). In both these cases the characteristic cubic triads lie in the mirror planes generated by the diagonal $\{11\}$ lines of symmetry of the square plane lattice. But if the cubic triads lie in the mirror planes generated by the $\{10\}$ lines of symmetry of the plane lattice, then the x and y axes of the cubic unit-cell have to be rotated through 45° relative to the x and y axes of the square plane lattice; the P-lattice becomes a C-lattice and the I-lattice an F-lattice and their unit-cells will be cubes if $c = a\sqrt{2}$ (the conventional tetragonal orientation of axes is preserved here). The resultant C-lattice has a unit-cell of cubic shape, but is not a cubic lattice type because the presence of the [111] triad requires that if the (001) face of the unit-cell is centred, then the (100) and (010) faces must likewise be centred and that is not so. The F-lattice is however consistent with the essential requirements of cubic symmetry and exists as a distinct lattice type, the *cubic* F-*lattice*.

All three cubic lattice types, P, I, and F, have holosymmetric cubic symmetry, $m3m$, and crystals of any class of the cubic system are referable to them.

Hexagonal system

Our concern now is with the last of the plane lattice types, the hexagonal p-lattice (Fig 4.1(e)). The unit mesh of this lattice contains only one hexad axis, which passes through the origin in a direction normal to the plane of the lattice. Thus only one stacking scheme, that with stacking vector $\mathbf{t} = \mathbf{c}$, can generate a three-dimensional lattice having hexagonal symmetry. The resultant unit-cell has dimensions $a = b \neq c$, $\alpha = \beta = 90°$, $\gamma = 120°$; the unit-cell is primitive with point group symmetry $6/mmm$ and the lattice type is known as the *hexagonal* P-*lattice*.

Trigonal system

We saw in the last chapter that the characteristic symmetry of this system is the presence of a single direction with threefold symmetry, that direction being conventionally taken as the z-axis to yield a unit-cell with $a = b \neq c$, $\alpha = \beta = 90°$, $\gamma = 120°$. Our starting point is again the hexagonal p-lattice, which is the only plane lattice type that can be consistent with trigonal symmetry. The hexagonal p-lattice (Fig 4.9(a)) has triads at $\frac{2}{3}, \frac{1}{3}$ and $\frac{1}{3}, \frac{2}{3}$, which are not lattice points, and hexads through the lattice points at the corners of the unit-mesh. Each hexad represents in essence the coincidence of the diad that necessarily passes through any lattice point of a plane lattice and a triad which is the basic symmetry element of this lattice type. There are thus two kinds of triad in the hexagonal p-lattice: type I passing through the lattice point at the origin and type II with coordinates $\frac{2}{3}, \frac{1}{3}$ and $\frac{1}{3}, \frac{2}{3}$, the two triads of type II being related by the diad through the centre of the unit-mesh and by the diad through the origin.

Two stacking sequences are possible; $\mathbf{t} = \mathbf{c}$ which superimposes triads type for type and $\mathbf{t} = \frac{2}{3}\mathbf{a} + \frac{1}{3}\mathbf{b} + r\mathbf{c}$ which superimposes triads of type I on the type II triad at $\frac{2}{3}, \frac{1}{3}$. We discuss the second of these first and illustrate the way in which a three-dimensional lattice is produced in Fig 4.9(b). Every third lattice plane is directly superimposed since $3\mathbf{t} = 2\mathbf{a} + \mathbf{b} + 3r\mathbf{c}$; therefore $r = \frac{1}{3}$. The point group symmetry of the resultant three-dimensional lattice is $\bar{3}m$ which is that of the holosymmetric class of the trigonal system. This lattice type is known variously as the *rhombohedral lattice*

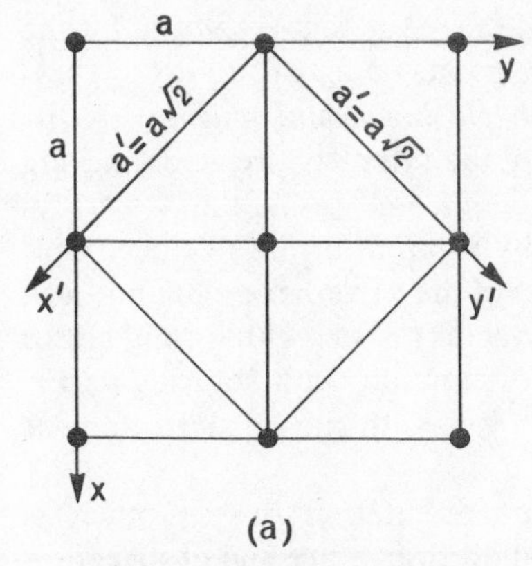

(a) The relationship between the tetragonal P-lattice ($a = b \neq c$) and the tetragonal C-lattice ($a' = b' \neq c$); the former becomes the cubic P-lattice when $a = b = c$.

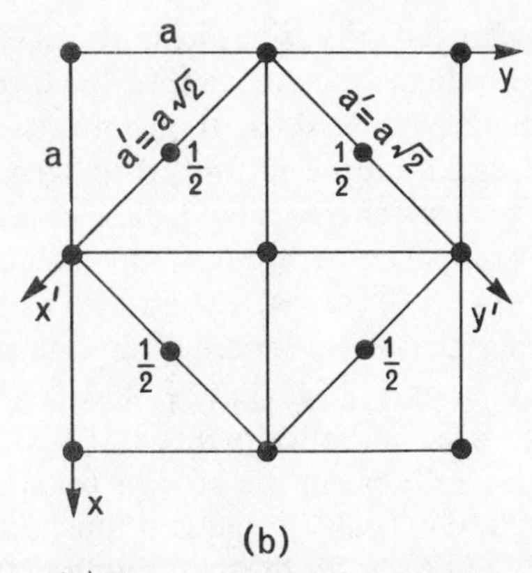

(b) The relationship between the tetragonal I-lattice ($a = b \neq c$) and the tetragonal F-lattice ($a' = b' \neq c$); the former becomes the cubic I-lattice when $a = b = c$ and the latter becomes the cubic F-lattice when $a' = b' = c$.

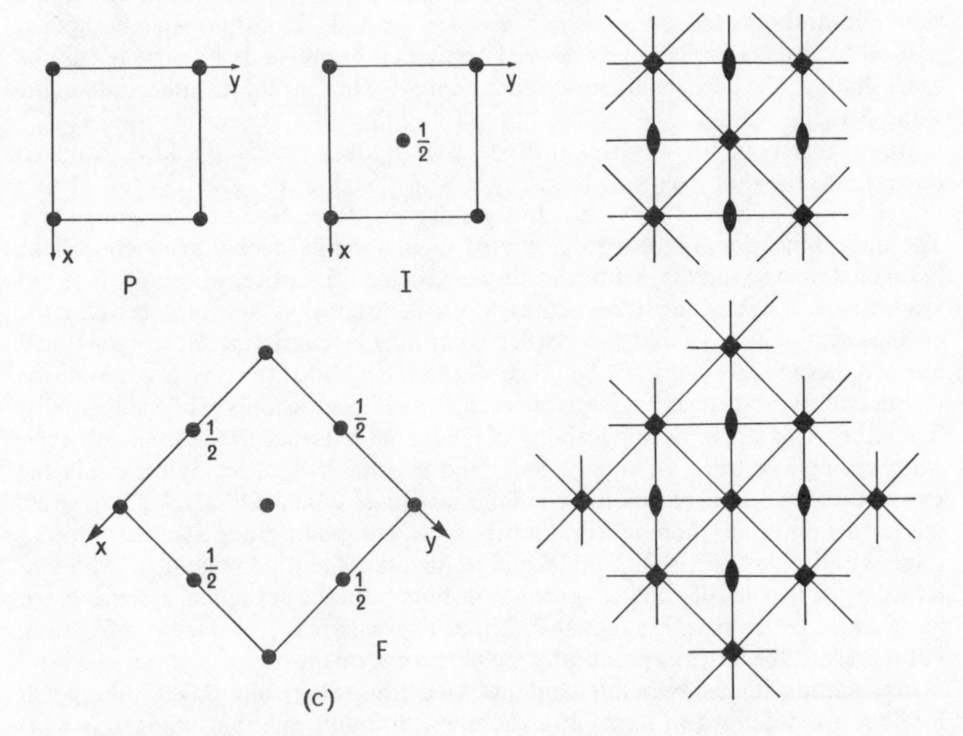

(c) The diagrams on the left show the arrangement of lattice points in the unit-cells of the cubic P, I, and F lattices; those on the right show symmetry elements parallel to [001] for the P and I lattices in the upper diagram, for the F-lattice in the lower diagram.

Fig 4.8 The cubic Bravais lattices generated by stacking plane square p-lattices in such ways as to satisfy the essential requirement for cubic symmetry, the presence of triads parallel to $\langle 111 \rangle$.

or the *trigonal* R-*lattice*; since its unit-cell is the same shape as that employed for the hexagonal system but contains (Fig 4.9) three lattice points (at 0, 0, 0; $\frac{2}{3}, \frac{1}{3}, \frac{1}{3}$; $\frac{1}{3}, \frac{2}{3}, \frac{2}{3}$) instead of one (at 0, 0, 0) it is known as the *triple hexagonal unit-cell* of the rhombohedral lattice. That the lattice generated by the symmetry related stacking sequence $\mathbf{t} = \frac{1}{3}\mathbf{a} + \frac{2}{3}\mathbf{b} + r\mathbf{c}$ is not a distinct type is evident from Fig 4.9; the two arrays of lattice points can be brought into coincidence by rotation of the reference axes x and y through 60° or 180°. These two orientations of the triple-hexagonal unit-cell are known as the *obverse* orientation, with stacking vector $\mathbf{t} = \frac{2}{3}\mathbf{a} + \frac{1}{3}\mathbf{b} + \frac{1}{3}\mathbf{c}$ and lattice points at 0, 0, 0; $\frac{2}{3}, \frac{1}{3}, \frac{1}{3}$; $\frac{1}{3}, \frac{2}{3}, \frac{2}{3}$, and the *reverse* orientation, with stacking vector $\mathbf{t} = \frac{1}{3}\mathbf{a} + \frac{2}{3}\mathbf{b} + \frac{1}{3}\mathbf{c}$ and lattice points at 0, 0, 0; $\frac{1}{3}, \frac{2}{3}, \frac{1}{3}$; $\frac{2}{3}, \frac{1}{3}, \frac{2}{3}$. In subsequent comment we shall generally prefer the obverse orientation.

Although the triple hexagonal unit-cell, in one orientation or the other, is commonly employed for the description of the rhombohedral lattice and of structures derivative therefrom, this lattice type does have a primitive unit-cell which embodies its characteristic symmetry and is in shape a rhombohedron. The reference axes for the *rhombohedral unit-cell* of the trigonal R-lattice are equally inclined to the triad axis and are parallel to the directions [21†1], [$\bar{1}$1†1], and [$\bar{1}\bar{2}$†1] of the triple hexagonal unit-cell in the obverse orientation (Fig 4.9); the dimensions of the rhombohedral unit-cell are $a = b = c$, $\alpha = \beta = \gamma < 120°$. That the triple hexagonal unit-cell is usually preferred, in spite of being non-primitive, is because it has the same shape as the hexagonal unit-cell and moreover has two of its interaxial angles, α and β, right-angles.

Up to this point in our treatment of the trigonal system we have confined ourselves to the rhombohedral lattice type with stacking sequence $\mathbf{t} = \frac{2}{3}\mathbf{a} + \frac{1}{3}\mathbf{b} + \frac{1}{3}\mathbf{c}$ or $\mathbf{t} = \frac{1}{3}\mathbf{a} + \frac{2}{3}\mathbf{b} + \frac{1}{3}\mathbf{c}$. It is now time to consider the other stacking sequence $\mathbf{t} = \mathbf{c}$. The three-dimensional lattice so generated retains the hexagonal symmetry of the hexagonal p-lattice and is identical with the hexagonal P-lattice, whose point group symmetry is $6/mmm$; but it has status too as a trigonal lattice type because the arrangement of atoms about each lattice point may be consistent with trigonal and not with hexagonal symmetry. Such use of the same lattice type by two systems is unique and gives rise to ambiguities of practice in different schools of crystallography. Some choose to stress the applicability of hexagonal reference axes to the unit-cells, whether single or triple, of all hexagonal and trigonal substances by regarding the trigonal system as a mere subdivision of the hexagonal system. Others prefer to stress the restriction of the rhombohedral lattice type, with point group symmetry $\bar{3}m$, to crystals of the classes 3, $\bar{3}$, 32, $3m$, $\bar{3}m$ and to maintain the trigonal system, with two lattice types, trigonal-R and hexagonal-P, distinct from the hexagonal system, which has a single lattice type, hexagonal-P. Either approach has its inherent difficulties, but we regard the latter as practically the more convenient.

In conclusion it must be pointed out that when trigonal crystals with the hexagonal P-lattice are described in terms of a rhombohedral unit-cell, that unit-cell is non-primitive and has lattice points at 0, 0, 0; $\frac{1}{3}, \frac{1}{3}, \frac{1}{3}$; $\frac{2}{3}, \frac{2}{3}, \frac{2}{3}$, i.e. the cell diagonal parallel to the triad is three times the lattice repeat in that direction. In such cases quite obviously the use of a primitive hexagonal unit-cell is preferable.[3] In the case of trigonal crystals with a rhombohedral lattice the use of the triple hexagonal unit-cell

[3] In chapter 3 (cf. Table 3.3) we took the rhombohedron as the conventional unit-cell for the trigonal system; that is correct in the morphological context of chapter 3 and moreover the reasons why the rhombohedral unit-cell has fallen into disuse could not be properly explained at that stage.

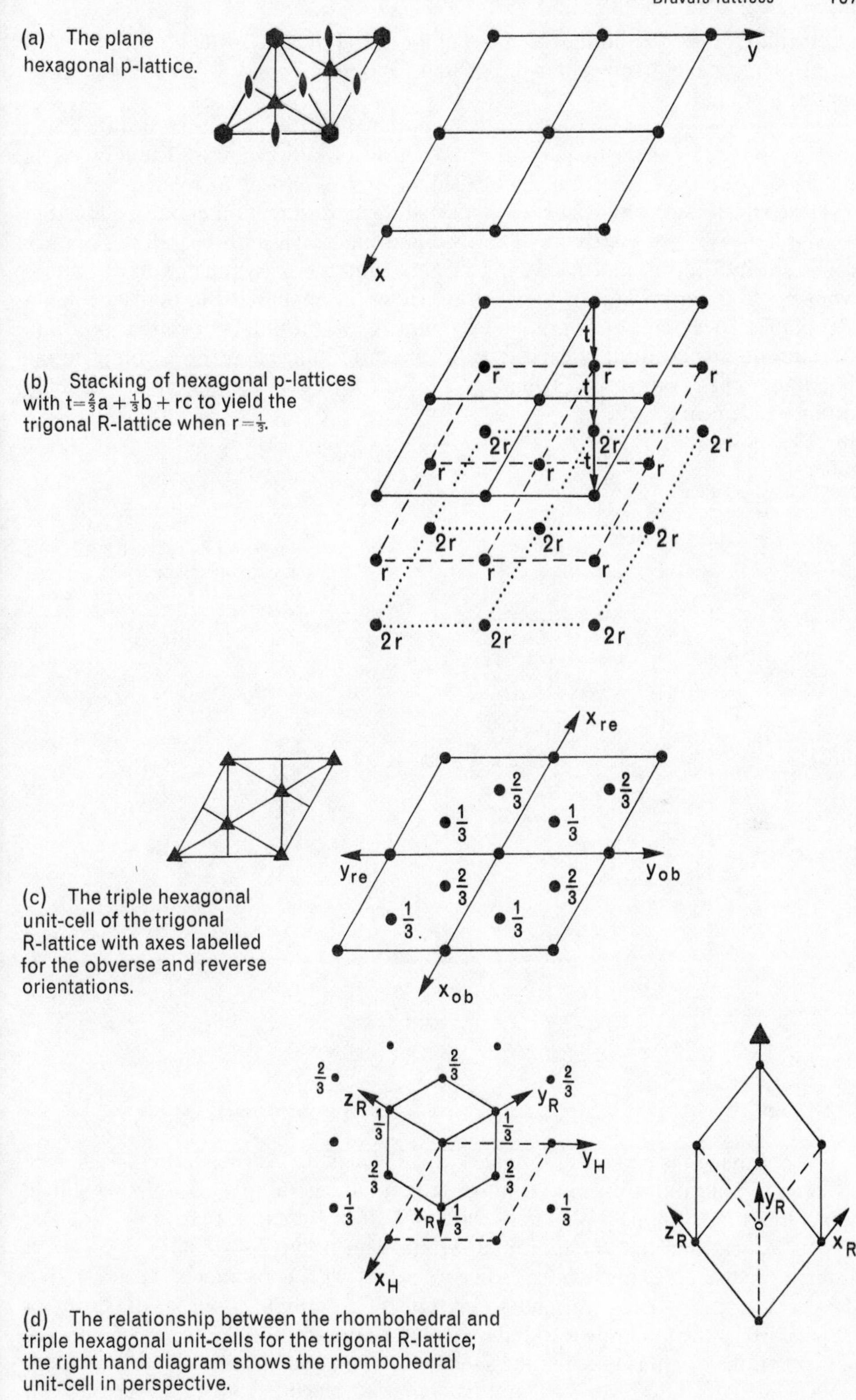

(a) The plane hexagonal p-lattice.

(b) Stacking of hexagonal p-lattices with $t = \frac{2}{3}a + \frac{1}{3}b + rc$ to yield the trigonal R-lattice when $r = \frac{1}{3}$.

(c) The triple hexagonal unit-cell of the trigonal R-lattice with axes labelled for the obverse and reverse orientations.

(d) The relationship between the rhombohedral and triple hexagonal unit-cells for the trigonal R-lattice; the right hand diagram shows the rhombohedral unit-cell in perspective.

Fig 4.9 The rhombohedral Bravais lattice.

is generally preferable and indeed the rhombohedral unit-cell, which was obviously useful in the days of morphological crystallography, is now of little more than historical interest.

We have now derived fourteen three-dimensional lattice types, each distinguished from the others by the symmetry of arrangement of lattice points. These fourteen, and there are no more, are commonly called the *Bravais lattices* after Auguste Bravais (1811–63) the French physicist who first listed them. Information about the fourteen Bravais lattices is summarized in Table 4.2 and Fig 4.10. That every Bravais lattice is necessarily centrosymmetric and has the point group symmetry of the holosymmetric class of the system to which it belongs is emphasized in the last column of the table. The immediately preceding column shows the number of lattice points in the conventional unit-cell of each type. At the end of this chapter we complete the description of Bravais lattice symmetry after the introduction of non-translational symmetry elements.

Table 4.2
The 14 Bravais lattices

System	Lattice symbol	Conventional unit-cell	Number of lattice points	Point group symmetry of lattice
Triclinic	P	$a \neq b \neq c, \alpha \neq \beta \neq \gamma$	1	$\bar{1}$
Monoclinic	P	$a \neq b \neq c, \alpha = \gamma = 90° = \beta$ ($\beta > 90°$)	1	$2/m$
	C(A)		2	
Orthorhombic	P	$a \neq b \neq c, \alpha = \beta = \gamma = 90°$	1	mmm
	C(A, B)		2	
	I		2	
	F		4	
Tetragonal	P	$a = b \neq c, \alpha = \beta = \gamma = 90°$	1	$4/mmm$
	I		2	
Cubic	P	$a = b = c, \alpha = \beta = \gamma = 90°$	1	$m3m$
	I		2	
	F		4	
Trigonal	R	$a = b \neq c, \alpha = \beta = 90°, \gamma = 120°$	3	$\bar{3}m$
		or $a = b = c, \alpha = \beta = \gamma < 120°$	1	
Hexagonal	P	$a = b \neq c, \alpha = \beta = 90°, \gamma = 120°$	1	$6/mmm$

Before leaving the subject of Bravais lattices it may be instructive to consider the structural implications of non-primitive lattices in general in terms of two simple cubic structures. Suppose the unit-cell has lattice points at 0, 0, 0; X_1, Y_1, Z_1; X_2, Y_2, Z_2, etc. and that an atom of a certain element has coordinates x, y, z, then atoms of the same element must occur at points with coordinates X_1+x, Y_1+y, Z_1+z; X_2+x, Y_2+y, Z_2+z; and so on. Our first example is one of the forms of metallic iron which is known to have a cubic I-lattice with two atoms of iron per unit-cell: if the origin is taken at the site of one iron atom, then the other iron atom must lie at $\frac{1}{2}, \frac{1}{2}, \frac{1}{2}$. Our second example is diamond, which has a cubic F-lattice with eight atoms of carbon per unit-cell; since an F-cell contains four lattice points, the

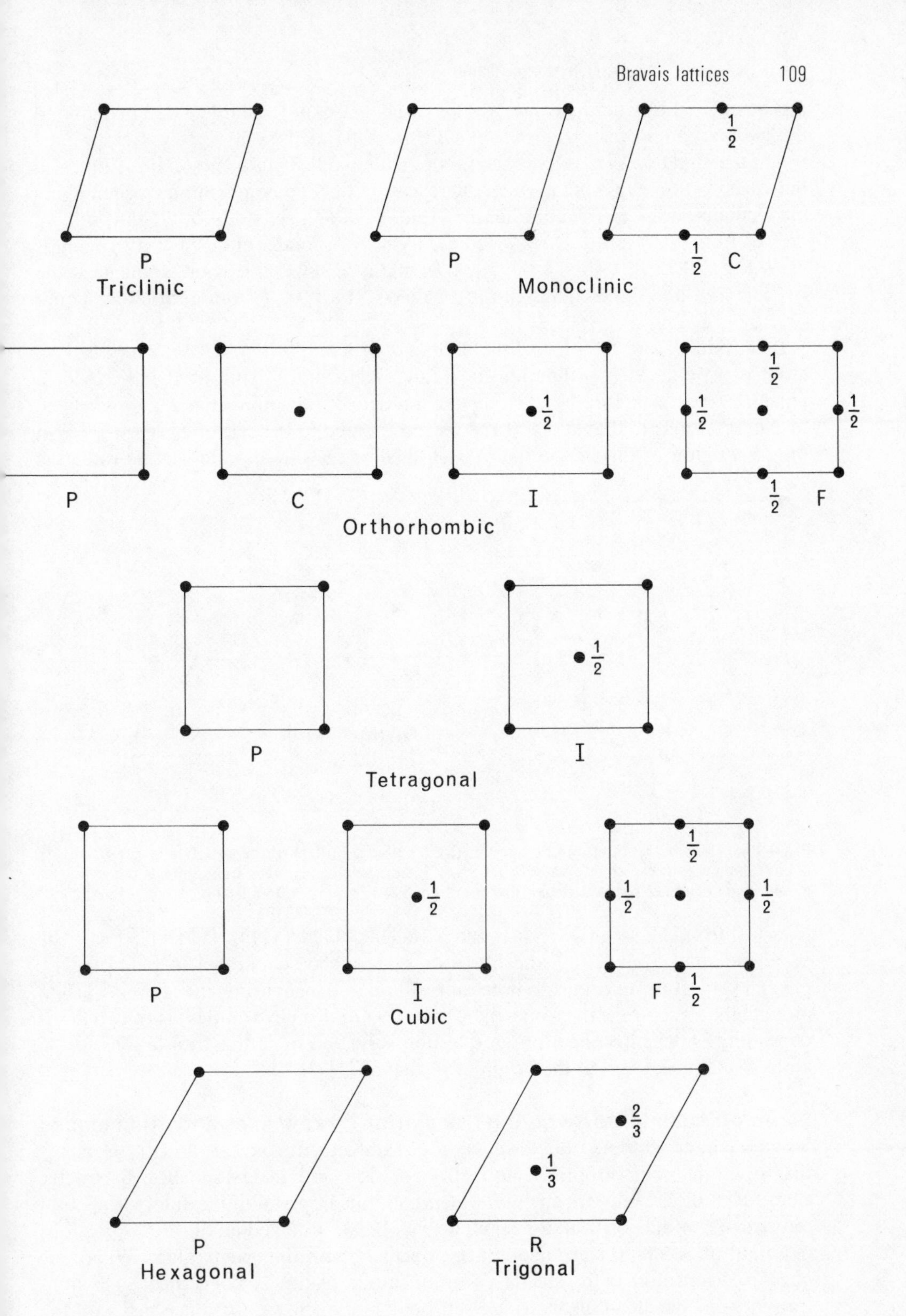

Fig 4.10 The unit-cells of the fourteen Bravais lattices.

repeat unit in this structure consists of two carbon atoms. It is known that in diamond the two carbon atoms of the repeat unit are separated by a vector of magnitude $\frac{\sqrt{3}}{4}a$ in the direction of a body diagonal of the unit-cell. Thus if the origin is taken at one carbon atom, the other atom of the repeat unit lies $\frac{\sqrt{3}}{4}a$ out from the origin along [111], that is at $\frac{1}{4}, \frac{1}{4}, \frac{1}{4}$. The coordinates of the remaining six atoms in the unit-cell are derived from 0, 0, 0 and $\frac{1}{4}, \frac{1}{4}, \frac{1}{4}$ by reference to the coordinates of the four lattice points of the F-cell, 0, 0, 0; 0, $\frac{1}{2}, \frac{1}{2}$; $\frac{1}{2}$, 0, $\frac{1}{2}$; $\frac{1}{2}, \frac{1}{2}$, 0; that is to say four carbon atoms lie at the lattice points and the remaining four are disposed at $\frac{1}{4}, \frac{1}{4}, \frac{1}{4}$ from each lattice point, i.e. $\frac{1}{4}, \frac{1}{4}, \frac{1}{4}$; $\frac{1}{4}, \frac{3}{4}, \frac{3}{4}$; $\frac{3}{4}, \frac{1}{4}, \frac{3}{4}$; $\frac{3}{4}, \frac{3}{4}, \frac{1}{4}$.

It is evident from the plan of the diamond structure, Fig 4.11, that every carbon atom is in fourfold coordination, its nearest neighbours lying at the apices of a regular tetrahedron. But there is a distinction between the four carbon atoms related by F-translations to that at 0, 0, 0 and the four related to the carbon atom at $\frac{1}{4}, \frac{1}{4}, \frac{1}{4}$ in the orientation of their coordination tetrahedra: the interatomic vectors radiating

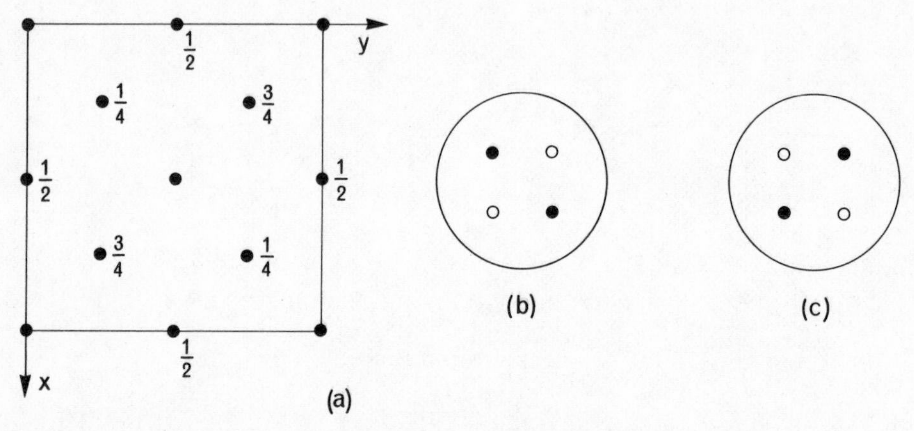

Fig 4.11 The crystal structure of diamond. (a) is a plan on (001) of the structure which has a cubic F-lattice with carbon atoms at 0, 0, 0; $\frac{1}{4}, \frac{1}{4}, \frac{1}{4}$ etc. (b) and (c) are stereograms of the interatomic vectors of nearest neighbours from the carbon atoms at 0, 0, 0 and $\frac{1}{4}, \frac{1}{4}, \frac{1}{4}$ respectively.

from 0, 0, 0 and like atoms are disposed parallel to [111], $[\bar{1}\bar{1}1]$, $[\bar{1}1\bar{1}]$, $[1\bar{1}\bar{1}]$ whereas those radiating from $\frac{1}{4}, \frac{1}{4}, \frac{1}{4}$ and like atoms are disposed parallel to $[11\bar{1}]$, $[\bar{1}\bar{1}\bar{1}]$, $[\bar{1}11]$, $[1\bar{1}1]$. It is important to note that, although the environment of every carbon atom in the structure is the same, the orientation of this environment is of two kinds depending on whether the atom in question is related by lattice translations to the atom of the repeat unit at the origin or to that at $\frac{1}{4}, \frac{1}{4}, \frac{1}{4}$.

Symmetry elements involving translation: screw axes and glide planes

So far we have restricted our discussion of symmetry to those symmetry elements that occur in point groups, that is the rotation and inversion axes. Symmetry elements of these kinds are such that their continued operation inevitably leads to a return to the initial position: for example two operations of a diad or three operations of a triad on a crystal face yield a face coincident with the original face. Where the angular disposition of crystal faces or of lattice planes is concerned such non-translational symmetry elements are sufficient, but when we are concerned with the symmetry of arrangement of atoms within a unit-cell an additional sort of symmetry element has to be introduced. This is the translational symmetry element which by its continued operation on a point cannot yield a point coincident with the original point but, after an appropriate number of operations, yields a point distant from the original point by an integral number of lattice translations. The simplest example of a translational symmetry element is the *screw diad*, whose operation is that of a diad (rotation through 180°) combined with translation through half the lattice repeat in the direction of the diad. Figure 4.12 contrasts the operation of a diad and a screw diad, each parallel to [001] through the origin; an atom at x, y, z is repeated by the diad at $\bar{x}, \bar{y}, z$ and by the screw diad at $\bar{x}, \bar{y}, \frac{1}{2}+z$. That the translation component is on the scale of a lattice repeat implies that the screw nature of the [001] diad is not discernible from observation of the macroscopic symmetry of the crystal; indeed a crystal that on the macroscopic scale has a diad parallel to [001] may on the lattice scale have diads or screw diads or both in that direction.

We now proceed to develop the variety of types of translational symmetry element, their nomenclature and conventional representation.

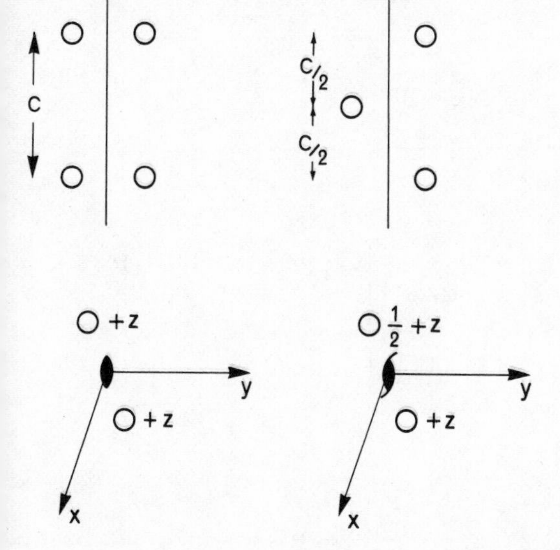

Fig 4.12 The contrasting effect of the operation of a diad (left-hand diagrams) and a screw diad (right-hand diagrams) parallel to [001] through the origin. In the upper diagrams the diad and screw diad (and the points on which they operate) lie parallel to the plane of the diagram; in the lower diagrams they are perpendicular to that plane.

Screw axes

An n-fold rotation axis is such that each operation of it is a rotation through $2\pi/n$, where $n = 1, 2, 3, 4, 6$. For an n-fold screw axis it is necessary to specify the value of n and in addition the magnitude of the pitch τ, which is restricted by the condition that the accumulated pitch after n operations, $n\tau$, must be an integral multiple m of the lattice repeat $\mathbf{t}$ in the direction of the axis, i.e. $n\tau = m\mathbf{t}$. When $m = 0$ there is no translation and the n-fold rotation axis appears as the special case of the n_m screw axis with $m = 0$. In general possible values of τ are given by $(m/n)\mathbf{t}$ where n may have the values given above and $0 < m < n$; the latter condition arises directly from the lattice concept which implies the presence of translations $\mathbf{t}+z\mathbf{t}$, $2\mathbf{t}+z\mathbf{t}$, etc. when the translation $z\mathbf{t}$ is specified. The symbol n_m can be used to give a complete description of a screw axis by adding to what has already been said the convention that the rotation through $2\pi/n$ is in the anticlockwise sense when the associated translation $(m/n)\mathbf{t}$ is in the positive direction along the screw axis (i.e. a right-handed screw).

All possible crystallographic screw axes are illustrated in Fig 4.13, from which we take as our examples for more detailed study the three screw tetrads, 4_1, 4_2, and 4_3. All the axes are taken to be parallel to z through the origin and symmetry related points (represented by open circles) are shown in projection on the xy plane. Beside each such *equivalent position* its z coordinate is written in terms of the code: $+$ = a position at a height $+z$ above the xy plane, $1/p+$ = a position at a height $(1/p)+z$ above the xy plane, where $1/p$ and z are, as in structure plans, fractions of the lattice repeat c in the direction of the z axis (Fig 4.14). For instance the screw tetrads shown in Fig 4.13 have equivalent positions indicated variously as $+$, $\frac{1}{4}+$,

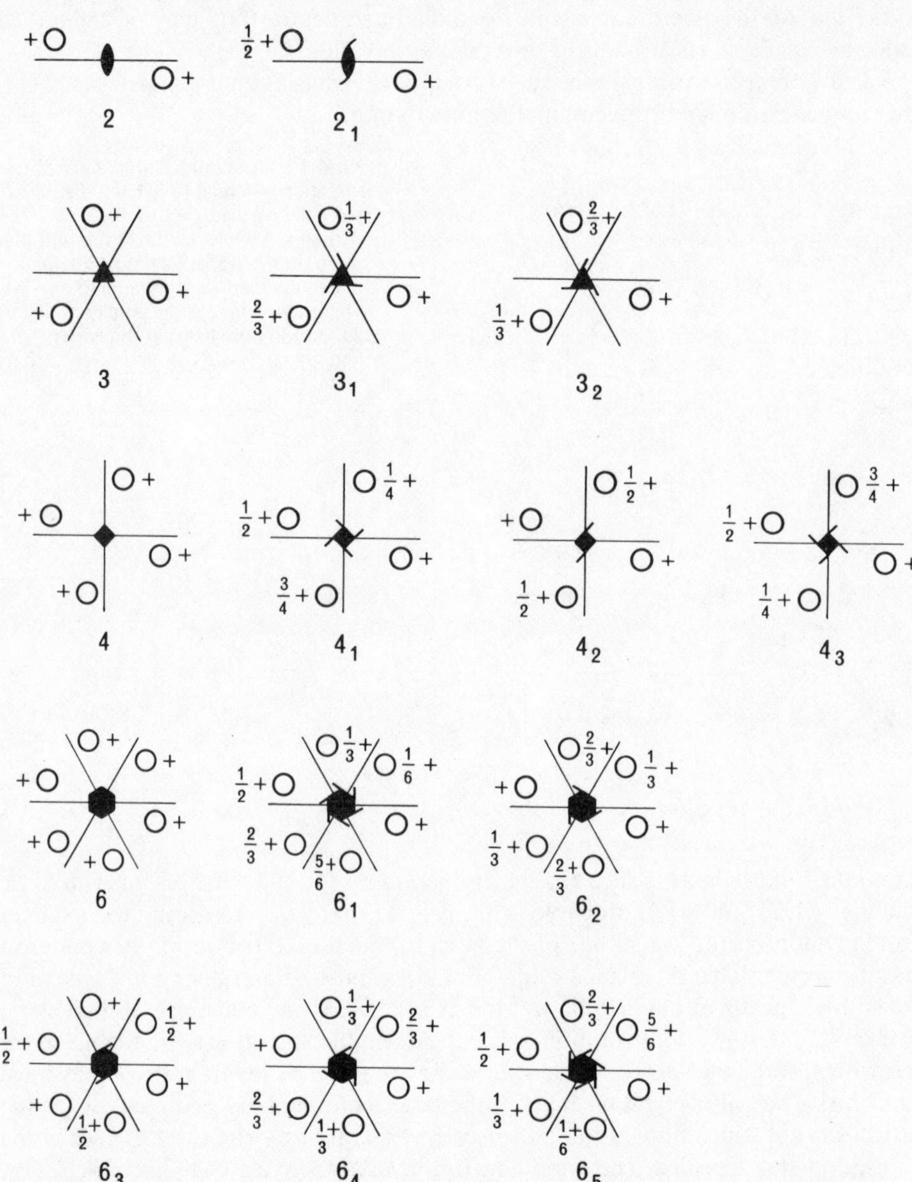

Fig 4.13 The operation of, and conventional symbols for, rotation and screw axes parallel to [001] shown in plans on (001).

$\frac{1}{2}+$, $\frac{3}{4}+$ to represent heights above the xy plane of z, $\frac{1}{4}+z$, $\frac{1}{2}+z$, $\frac{3}{4}+z$ respectively.

The rotation tetrad shown in Fig 4.13 operates on a position with coordinates x, y, z to yield equivalent positions with coordinates $\bar{y}, x, z$ after one operation, $\bar{x}, \bar{y}, z$ after the second operation, $y, \bar{x}, z$ after the third operation, and with the *fourth* operation returns to x, y, z. We have taken the rotation in the anticlockwise sense, but where there is no translation the sense does not affect the total pattern of equivalent positions. The 4_1 axis (Fig 4.14) rotates anticlockwise through $2\pi/4 = \frac{1}{2}\pi$ and simultaneously translates through $c/4$ so that the position xyz yields equivalent positions $\bar{y}, x, \frac{1}{4}+z$ by its first operation, $\bar{x}, \bar{y}, \frac{1}{2}+z$ by its second, $y, \bar{x}, \frac{3}{4}+z$ by its third, and by its fourth operation $x, y, 1+z$, which is necessarily equivalent to the initial position being one lattice repeat removed from it. The 4_2 axis likewise rotates anticlockwise through $\frac{1}{2}\pi$ and simultaneously translates through $2c/4 = \frac{1}{2}c$ so that the xyz position yields in succession equivalent positions at $\bar{y}, x, \frac{1}{2}+z$; $\bar{x}, \bar{y}, 1+z$; $y, \bar{x}, 1\frac{1}{2}+z$ and returns to superposition on the original position at $x, y, 2+z$; this introduces a new situation in that it now becomes necessary to reduce the equivalent positions produced by direct operation of the 4_2 screw axis by subtraction of integral lattice repeats parallel to z to bring them within one positive lattice repeat of the origin, i.e. $\bar{x}, \bar{y}, 1+z$ becomes $\bar{x}, \bar{y}, z$ and $y, \bar{x}, 1\frac{1}{2}+z$ becomes $y, \bar{x}, \frac{1}{2}+z$, these positions being themselves produced by direct operation of the 4_2 axis on the unit-cell immediately below the origin. The 4_3 axis rotates anticlockwise through $\frac{1}{2}\pi$ and simultaneously translates through $3c/4$ to yield from x, y, z successive equivalent positions $\bar{y}, x, \frac{3}{4}+z$; $\bar{x}, \bar{y}, 1\frac{1}{2}+z$; $y, \bar{x}, 2\frac{1}{4}+z$, which reduce to x, y, z; $\bar{y}, x, \frac{3}{4}+z$; $\bar{x}, \bar{y}, \frac{1}{2}+z$; $y, \bar{x}, \frac{1}{4}+z$ as shown in Fig 4.13.

Comparison of the diagrams in Fig 4.13 for the equivalent positions generated from x, y, z by 4_1 and 4_3 axes reveals that both can be regarded as screws of pitch $\frac{1}{4}c$, the former right-handed and the latter left-handed. It follows that a 4_1 and a 4_3 axis produce sets of equivalent positions that are mirror images of one another and cannot be superposed; such a pair of screw axes are said to be *enantiomorphous*. The 4_2 screw axis, like the rotation tetrad, is without hand, the same disposition of equivalent positions being obtained by an anticlockwise as by a clockwise rotation. In general one can say that the screw of smallest pitch that can be used to describe the disposition of equivalent positions related by a screw axis n_m is right-handed if $m < \frac{1}{2}n$, left-handed if $m > \frac{1}{2}n$, and without hand if $m = 0$ or $m = \frac{1}{2}n$.

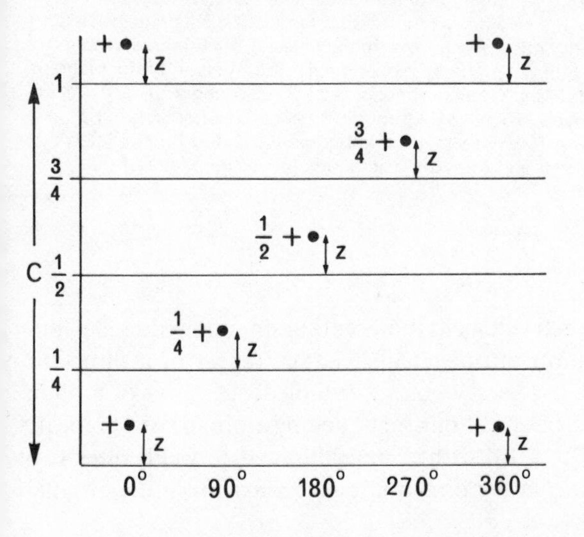

Fig 4.14 Equivalent positions generated by a 4_1 axis. The angle of rotation about a 4_1 axis parallel to [001] is represented as the horizontal coordinate and the translation in fractions of the lattice repeat c is plotted vertically.

It has already been stated that only screw axes involving 2, 3, 4, or 6-fold rotation elements can apply in crystals and it is appropriate at this point to justify that assertion. In general the operation of a screw axis on a group of atoms yields equivalent groups in identical environments, but differently orientated. For instance if a 4_1 axis is to be repeated on a lattice, a lattice point can be placed on that 4_1 axis and it follows that a 4_1 axis passes through every lattice point in the same direction. Consider two such lattice points P and Q, each of which represents a group of four atoms related by the 4_1 axis through the relevant lattice point (Fig 4.15): the 4_1 axis through P relates the atoms at A, B, C, and D and the 4_1 axis at Q relates the atoms at α, β, γ, and δ. But the 4_1 axis through P relates the atom at α to that at T, the atom at β to that at U, the atom at γ to that at V, and the atom at δ to that at W; in short a third lattice point arises at R related to that at Q by a *rotation tetrad* through P. This result may be generalized: when screw axes are repeated on a lattice, the lattice exhibits the symmetry of the rotation axis of the same order. This apparent anomaly arises from the fact that lattice points represent groups of atoms in the same environment in the same orientation—the basis of the lattice concept—and therefore if we are to be strictly rigorous a symmetry axis should be operated not on the lattice point itself but on the group of atoms that it represents; for rotation axes and mirror planes the distinction is immaterial, but for translational symmetry elements (screw axes and glide planes) the translational component is lost and the lattice exhibits only the symmetry of the corresponding rotation axis or mirror plane. It follows that only screw axes derived from diads, triads, tetrads, or hexads can operate on atomic groupings in crystal structures.

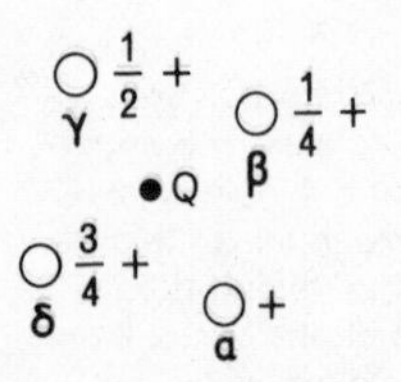

Fig 4.15 A lattice on which a screw tetrad 4_1 is repeated exhibits the symmetry of a rotation tetrad. The solid circles P, Q, R represent lattice points; the open circles, A, α, T, etc, represent atoms. Screw tetrads 4_1 perpendicular to the plane of the diagram through P and Q relate the atoms A, B, C, D and α, β, γ, δ respectively. The 4_1 axis through P relates the atomic groups $\alpha\beta\gamma\delta$ and TUVW so that the lattice points Q and R are related by a *rotation* tetrad through P perpendicular to the plane of the diagram.

Glide planes

A glide plane is a translational symmetry element representing simultaneous reflexion, as in a mirror plane, and translation through half a lattice repeat in a direction parallel to the plane. As in the case of screw axes, continued operation of a glide plane does not produce coincidence, but positions separated by a whole lattice repeat; for instance a glide plane parallel to (010) with a translation of $\frac{1}{2}c$ yields after two operations a position at a distance c in the direction of the z-axis from the original position. The magnitude of the translation associated with a glide plane is restricted, by the requirement of consistency with repetition on a lattice, to half a lattice spacing. The direction of the translation may be either parallel to a unit-cell edge (an *axial glide*) or parallel to a face-diagonal or body diagonal of the unit-cell (a *diagonal glide*). The translation of a diagonal glide plane is one half of the length of the relevant diagonal of the unit-cell except in the special case of the *diamond glide* where it is one quarter of the length of the diagonal.[4] The nomenclature of symmetry planes, mirror and glide, and the types of translation permitted in the latter are set out in Table 4.3.

Table 4.3
Symmetry planes

	Symbol	Translation
Mirror	m	none
Axial glide	a	$a/2$
	b	$b/2$
	c	$c/2$
Diagonal glide	n	$\frac{a+b}{2}, \frac{b+c}{2}, \frac{a+c}{2}$
		$\frac{a+b+c}{2}$ *
Diamond glide	d	$\frac{a\pm b}{4}, \frac{b\pm c}{4}, \frac{a\pm c}{4}$
		$\frac{a\pm b\pm c}{4}$ *

* cubic and tetragonal systems only

The graphical representation of glide planes on structural plans needs some explanation because a distinction has to be made between planes lying parallel and planes perpendicular to the plane of the diagram. A symmetry plane parallel to the plane of the diagram is conventionally represented (Fig 4.16) by the symbol ┐ beyond the top right-hand corner of the outline of the unit-cell and an arrow is incorporated in the symbol to indicate the direction of glide; the appropriate fraction is written beside the symbol to indicate height above the reference plane when the glide plane is not coincident with the reference plane. A symmetry plane perpendicular to the plane of the diagram is represented conventionally by a bold line (Fig 4.16) which is unbroken for a mirror plane, dashed for a glide plane with translation in the plane of the diagram, dotted for a glide plane with translation perpendicular to the plane of the diagram and alternate dashes and dots for a diagonal glide.

Equivalent positions are again represented by open circles, but here we have to distinguish between the hand of the atom group occupying each equivalent position

[4] Diamond glide planes are restricted to certain orientations in certain Bravais lattices such that a lattice point lies at the mid-point of the diagonal concerned so that the translation remains one half of the lattice repeat in the direction of the diagonal. These conditions are satisfied only in I and F lattices with the exception of the orthorhombic I lattice which cannot have planes of symmetry parallel to the body diagonals of the unit-cell, ⟨111⟩. A full account of the operation of diamond glides may be found in Buerger (1956).

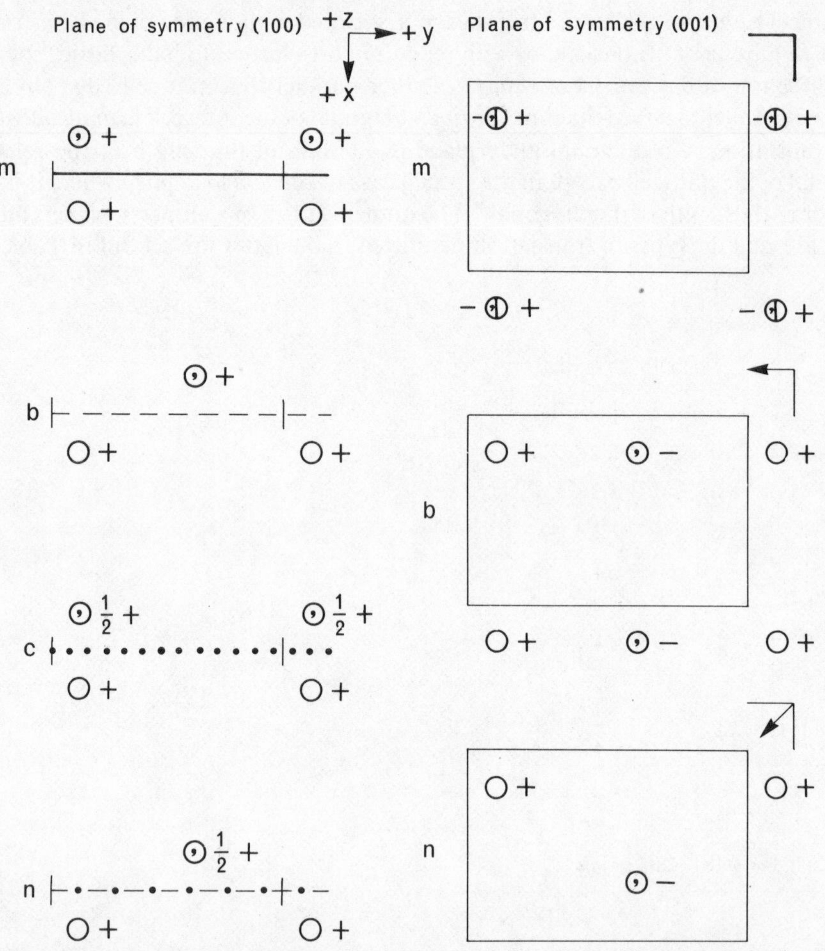

Fig 4.16 The representation and operation of mirror and glide planes. The orientation of axes is displayed centrally at the top of the figure. The left-hand and right-hand columns respectively show the operation of symmetry planes parallel to (100) and to (001), that is perpendicular and parallel to the plane of the diagram.

because the operation of a symmetry plane, unlike a rotation axis, produces a reversal of hand. The accepted convention is an open circle for the original position, at x, y, z, and an open circle enclosing a comma for positions related to it by an odd number of reflexions. Superimposition in projection of positions of different hand is represented by adjacent half circles. The heights of equivalent positions are denoted according to the convention explained in detail in the section on screw axes.

That the spacing of mirror and glide planes is necessarily half the lattice repeat in the direction perpendicular to the plane is illustrated in Fig 4.17, which shows two projections of a unit-cell with an (001) mirror plane passing through $x, y, \frac{1}{4}$. The position x, y, z on reflexion becomes $x, y, \frac{1}{2}-z$, which is itself related to the position $x, y, 1+z$ by a parallel mirror plane $x, y, \frac{3}{4}$.

Space groups

We have already seen that there are thirty-two groups of non-translational symmetry elements that can be repeated on a lattice, the so-called crystallographic *point groups.* When translational symmetry elements, screw axes and glide planes, are taken into

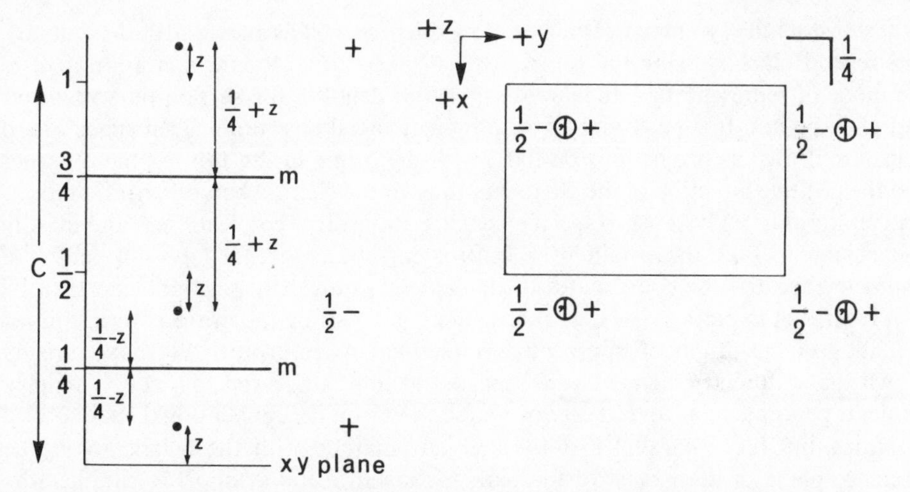

Fig 4.17 The generation of an (001) mirror plane at $z = \frac{3}{4}$ by an (001) mirror plane at $z = \frac{1}{4}$. The presence of an (001) mirror plane at $z = \frac{1}{4}$ requires an atom at a height z above the xy plane to be repeated at a height $\frac{1}{2} - z$ with the same x and y coordinates; the atom at $\frac{1}{2} - z$ is related to the lattice-repeated atom at $1 + z$ by an (001) mirror plane at $z = \frac{3}{4}$.

account the number of groups of symmetry elements that can be repeated on a lattice increases to two hundred and thirty, the so-called crystallographic *space groups.* The derivation and the complete listing of the 230 space groups are outside the scope of this book, and we attempt only an outline of nomenclature and graphical representation and, through examples, the use of space groups in the interpretation of crystal structures. For the reader who wishes to explore the fundamentals of this topic, excellent accounts of space group derivation are to be found in Buerger (1956) and Hilton (1963). An exhaustive tabulation of space groups with a most thorough explanatory introduction is provided by *International Tables for X-ray Crystallography,* vol I (1969). In what follows we adopt the conventions laid down in *International Tables* which are the current usage of most laboratories.

A space group symbol consists of two parts: first a letter to indicate lattice type and that is followed by a statement of the essential symmetry elements present. This second part of the space group symbol is of the same form as a point group symbol but may include reference to translational symmetry elements; if any translational elements referred to in the space group symbol are replaced by the corresponding non-translational elements, the point group, and thence the system, to which the space group belongs can immediately be read off. For instance the symbol C2/*c* represents a space group of the point group 2/*m*, which is one of the point groups of the monoclinic system; the lattice type is monoclinic C; there is a rotation diad parallel to *y* and a *c*-glide perpendicular to *y*. Taking another example, the space group P*mcn* belongs to point group *mmm* (orthorhombic), the lattice type is orthorhombic P, a mirror plane lies perpendicular to *x*, a *c*-glide perpendicular to *y*, and a diagonal glide perpendicular to *z*. In both examples other symmetry elements are present in addition, but those stated in the symbol are adequate for the complete description of the symmetry of the space group.

For the graphical representation of a space group two diagrams are employed: one shows the distribution of symmetry elements in the unit-cell as a plan and the other shows, on a plan in the same orientation, all the positions generated by the operation of the symmetry elements on a position with general coordinates *x*, *y*, *z*. The first plan shows all the symmetry elements, consequent as well as essential, in the unit-cell. The second plan, showing the *general equivalent positions* for the space group, uses the mode of representation that we discussed in detail in the section on screw axes and glide planes. It is customary not to indicate axial directions when space group plans are drawn as projections on (001) with the origin in the top left-hand corner and the positive direction of the y-axis pointing to the right; when other orientations are employed it is advisable to specify axial directions. It is conventional, and usually convenient, to take the origin of a centrosymmetric space group at a centre of symmetry because the coordinates of the general equivalent positions can then be simply written in pairs as $\pm(x, y, z)$ etc.: there is no straightforward convention for the siting of the origin of non-centrosymmetrical space groups. We have already shown that symmetry elements are repeated at half lattice spacings: all symmetry elements perpendicular to (001) are of course shown on the conventional space group diagrams, but those parallel to (001) are only marked with the height above the reference plane of the lowest of the pair, for instance the symbol ⌉$\frac{1}{4}$ implies the presence of (001) mirror planes at $z = \frac{3}{4}$ as well as at $z = \frac{1}{4}$.

Figure 4.18 shows the conventional diagrams for the space group P*mcn*, which will be taken to illustrate the points made in the previous paragraph. The (100) mirror planes are shown on the right-hand diagram intersecting the x-axis at $\frac{1}{4}a$ and $\frac{3}{4}a$; the (010) c-glide planes are shown intersecting the y-axis at $\frac{1}{4}b$ and $\frac{3}{4}b$; but the (001) diagonal glide planes, which intersect the z-axis at $\frac{1}{4}c$ and $\frac{3}{4}c$, are indicated briefly as $\frac{1}{4}$. Consequent symmetry elements are three non-intersecting sets of screw diads, one parallel to each reference axis, and eight centres of symmetry represented in the diagram by small open circles. Since the centres of symmetry, at $0, 0, 0$; $\frac{1}{2}, 0, 0$; $0, \frac{1}{2}, 0$; $0, 0, \frac{1}{2}$; $0, \frac{1}{2}, \frac{1}{2}$; $\frac{1}{2}, 0, \frac{1}{2}$; $\frac{1}{2}, \frac{1}{2}, 0$; $\frac{1}{2}, \frac{1}{2}, \frac{1}{2}$; have $z = 0, \frac{1}{2}$ the height above the reference plane is not indicated. The graphical representation of diads and screw diads when parallel to the reference plane requires explanation: diads are shown as full arrows and screw diads as one-armed arrows outside the unit-cell outline with the height, if not 0 or $\frac{1}{2}$, written alongside. It is worth noting at this point that although the space group symbol P*mcn* makes explicit reference only to planes of symmetry perpendicular to each reference axis the presence of diad axes, in this case all screw diads, parallel to each reference axis is clearly implied; the point group to which this space group belongs is *mmm* which has three mutually perpendicular diads as well as three mutually perpendicular mirror planes and therefore the space group P*mcn* must have three mutually perpendicular sets of rotation or screw diads as well as three mutually perpendicular sets of symmetry planes.

The left-hand diagram shows the disposition of the general equivalent positions for the space group. The number of general equivalent positions in a space group with a P lattice is equal to the number of planes in the general form $\{hkl\}$ of the corresponding point group; in this space group P*mcn* the general equivalent positions have eightfold *multiplicity* as does the general form $\{hkl\}$, e.g. $\{123\}$, in point group *mmm*. In the case of a C or I lattice the multiplicity of the general form of the point group has to be doubled and for an F lattice quadrupled, e.g. the general form in point group *m*3*m* has multiplicity 48 and the general equivalent positions in space group F*d*3*m* have multiplicity 192.[5] The coordinates of the general equivalent positions in space group P*mcn* are listed in Table 4.4. It is important to notice that the determination of the

[5] Correspondingly the number of sets of symmetry elements parallel to a given direction in a space group is equal to the number of lattice points in the unit-cell. This point is illustrated in the discussion of C2/*c* which follows.

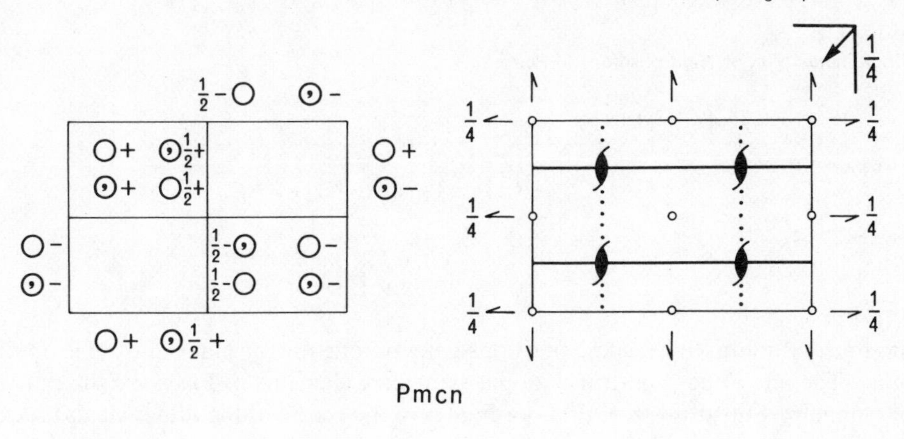

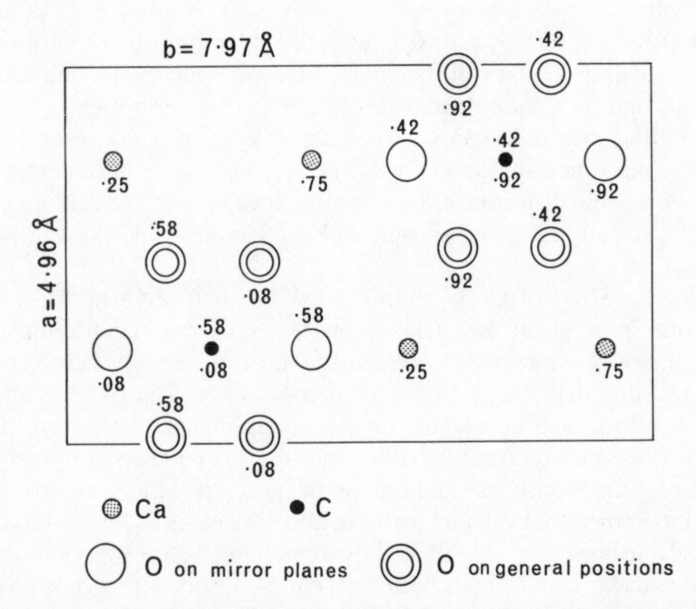

Fig 4.18 The crystal structure of aragonite, $CaCO_3$. The two upper diagrams are the conventional diagrams for the space group P*mcn* oriented so that $+z$ is upwards perpendicular to the plane of the figure, $+y$ is directed to the right and $+x$ downwards in the plane of the figure. The upper left-hand diagram shows the disposition of general equivalent positions and the upper right-hand diagram the disposition of symmetry elements in the space group. The lower diagram shows the structure of aragonite in plan on (001) in the same orientation: all the calcium and carbon atoms and four of the oxygen atoms lie on special equivalent positions on mirror planes with $x = \pm\frac{1}{4}$ and the remaining eight oxygen atoms lie on one set of general equivalent positions.

three parameters x, y, and z is sufficient to fix the coordinates of eight atoms of the same element occupying one set of general equivalent positions and that the whole set of eight atoms is represented by a single lattice point.

When the coordinates of a position are such that it lies on a non-translational symmetry element, two or more equivalent positions coalesce and the multiplicity of the set of equivalent positions is reduced. Such a set of *special equivalent positions* arises in P*mcn* when $x = \frac{1}{4}$, all the positions in this set then lie on mirror planes. Substitution of $x = \frac{1}{4}$ in the list of coordinates of the general equivalent positions yields immediately $\pm(\frac{1}{4}, y, z)$, $\pm(\frac{1}{4}, \frac{1}{2}-y, \frac{1}{2}+z)$; this set of special equivalent positions

Table 4.4
Coordinates of equivalent positions in P*mcn*

	Multiplicity	Point symmetry	
General	8	1	$\pm(x, y, z);\ \pm(\frac{1}{2}-x, y, z);$ $\pm(x, \frac{1}{2}-y, \frac{1}{2}+z); \pm(\frac{1}{2}-x, \frac{1}{2}-y, \frac{1}{2}+z)$
Special	4	m	$\pm(\frac{1}{4}, y, z);\ \pm(\frac{1}{4}, \frac{1}{2}-y, \frac{1}{2}+z)$
		$\bar{1}$	$0, 0, 0;\ \frac{1}{2}, 0, 0;\ 0, \frac{1}{2}, \frac{1}{2};\ \frac{1}{2}, \frac{1}{2}, \frac{1}{2}$
		$\bar{1}$	$0, \frac{1}{2}, 0;\ \frac{1}{2}, \frac{1}{2}, 0;\ 0, 0, \frac{1}{2};\ \frac{1}{2}, 0, \frac{1}{2}$

has fourfold multiplicity, some positions lying on one mirror plane and some on the other. The only other non-translational symmetry elements in P*mcn* are the centres of symmetry. Substitution of the coordinates of the centre at the origin yields $0, 0, 0;$ $\frac{1}{2}, 0, 0;\ 0, \frac{1}{2}, \frac{1}{2};\ \frac{1}{2}, \frac{1}{2}, \frac{1}{2}$; another set of special equivalent positions, again of multiplicity 4. The remaining four centres of symmetry likewise constitute a set of special equivalent positions, $0, \frac{1}{2}, 0;\ \frac{1}{2}, \frac{1}{2}, 0;\ 0, 0, \frac{1}{2};\ \frac{1}{2}, 0, \frac{1}{2}$. It is important to note that these two sets, each of four centres of symmetry, are not related to one another by symmetry; thus one set may be occupied by atoms of a certain element while the other set may either be occupied by atoms of a different element or be empty. The special equivalent positions associated with each type of non-translational symmetry element in space group P*mcn* have now been defined; all the remaining symmetry elements are translational in character and so cannot produce coalescence of equivalent positions.

We conclude our study of space group P*mcn* with some comments on a structure referable to this space group and take as our example the orthorhombic form of $CaCO_3$, the mineral *aragonite*. Measurements of the unit-cell dimensions and density of aragonite indicate that the unit-cell contains $4CaCO_3$. The calcium and carbon atoms must therefore each be situated on a set of special equivalent positions. The twelve oxygen atoms may be arranged either on three different sets of special positions or on one set of general positions and one set of special positions. Structure analysis of aragonite has shown that calcium, carbon, and four oxygens lie on mirror planes, y and z being determined for each set, and the remaining eight oxygens lie on general positions, for which x, y, and z have been determined. The resulting coordinates of all twenty atoms in the unit-cell are listed in Table 4.5 and the structure is shown in plan on Fig 4.18.

Table 4.5
Atomic coordinates in aragonite ($CaCO_3$)

Ca	4(m)	$\pm(\frac{1}{4}, 0{\cdot}42, 0{\cdot}75);$	$\pm(\frac{1}{4}, 0{\cdot}08, 0{\cdot}25)$
C	4(m)	$\pm(\frac{1}{4}, 0{\cdot}75, \overline{0{\cdot}08});$	$\pm(\frac{1}{4}, 0{\cdot}75, 0{\cdot}42)$
O(1)	4(m)	$\pm(\frac{1}{4}, \overline{0{\cdot}08}, \overline{0{\cdot}08});$	$\pm(\frac{1}{4}, 0{\cdot}58, 0{\cdot}42)$
O(2)	8(1)	$\pm(0{\cdot}48, 0{\cdot}67, \overline{0{\cdot}08});$	$\pm(0{\cdot}02, 0{\cdot}67, \overline{0{\cdot}08});$
		$\pm(0{\cdot}48, 0{\cdot}83, 0{\cdot}42);$	$\pm(0{\cdot}02, 0{\cdot}83, 0{\cdot}42)$

We take as our second example a non-primitive space group, C2/*c*, which belongs to the point group 2/*m* and has a monoclinic C lattice. Again the space group is centrosymmetrical and the origin is taken at a centre of symmetry (Fig 4.19). The (010) *c*-glide planes intersect the y-axis at the origin and at $y = \frac{1}{2}$. The [010] diads have $x = 0, \frac{1}{2}$ and $z = \frac{1}{4}, \frac{3}{4}$. The centring of the (001) face of the unit-cell introduces (010) diagonal glides at $y = \frac{1}{4}, \frac{3}{4}$ and [010] screw diads with $x = \frac{1}{4}, \frac{3}{4}$ and $z = \frac{1}{4}, \frac{3}{4}$;

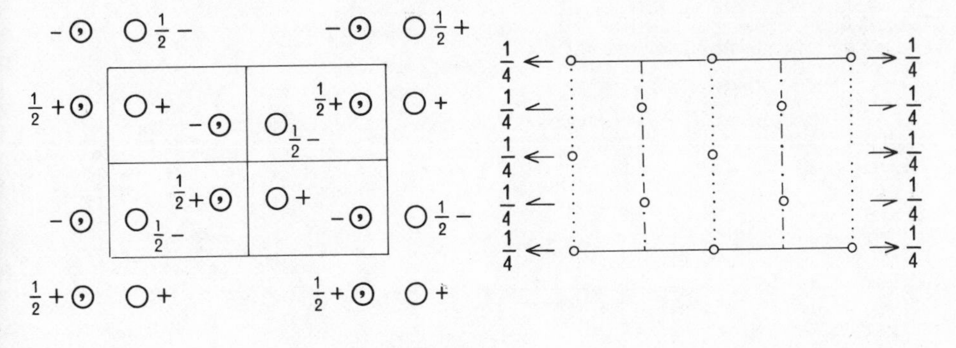

C 2/c

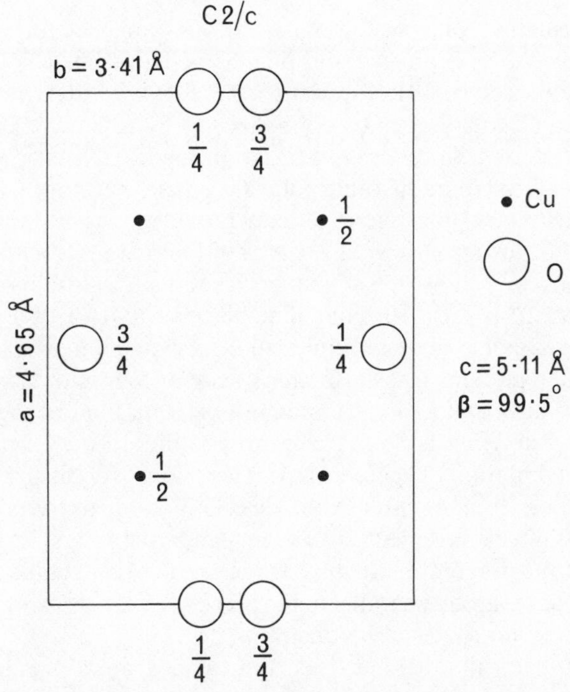

Fig 4.19 The crystal structure of tenorite, CuO. The two upper diagrams are the conventional diagrams for the monoclinic space group C2/c. The lower diagram is a plan of the structure of CuO on (001).

there are in all sixteen centres of symmetry. In a non-primitive space group a conveniently abbreviated way of listing the coordinates of equivalent positions is to state the coordinates of the several lattice points at the head of the list and then to give only the coordinates of the equivalent positions associated with one lattice point; this abbreviation is adopted in Table 4.6.

The only non-translational symmetry elements present are [010] diads and centres of symmetry. There is only one set of special equivalent positions on diads (with y as a variable parameter). The sixteen centres of symmetry are split into four groups, each of four symmetry related positions.

CuO, the mineral tenorite, has space group C2/c with only 4CuO in the unit-cell so that copper and oxygen must each occupy one set of special equivalent positions. The copper atoms are found to lie on centres of symmetry at $\frac{1}{4}, \frac{1}{4}, 0$ etc and the oxygen atoms lie on diads at $0, y, \frac{1}{4}$ etc, where $y = 0{\cdot}416$.

Table 4.6
Coordinates of equivalent positions in C2/*c*

	$(0, 0, 0; \frac{1}{2}, \frac{1}{2}, 0)+$ Multiplicity	Point symmetry	
General	8	1	$\pm(x, y, z); \pm(\bar{x}, y, \frac{1}{2}-z)$
Special	4	2	$\pm(0, y, \frac{1}{4})$
	4	$\bar{1}$	$0, 0, 0; 0, 0, \frac{1}{2}$
	4	$\bar{1}$	$0, \frac{1}{2}, 0; 0, \frac{1}{2}, \frac{1}{2}$
	4	$\bar{1}$	$\frac{1}{4}, \frac{1}{4}, 0; \frac{3}{4}, \frac{1}{4}, \frac{1}{2}$
	4	$\bar{1}$	$\frac{1}{4}, \frac{1}{4}, \frac{1}{2}; \frac{3}{4}, \frac{1}{4}, 0$

In the tabulation of space groups various nomenclatorial and orientational preferences have to be expressed; this point is well illustrated by the two space groups we have just considered. It is evident from Fig 4.19 that the space group there represented could equally well be symbolized as C2/*c*, $C2_1/c$, $C2_1/n$, or C2/*n* without reorientation of axes. Since a choice has to be made, C2/*c* is arbitrarily preferred. In the orthorhombic system the same point may arise, but not in P*mcn*, and moreover we are free to relabel the reference axes provided a right-handed axial system is maintained. For instance if the *x*, *y*, *z* axes of P*mcn* are relabelled x', y', z' according to the scheme $x \rightarrow y'$, $y \rightarrow z'$, $z \rightarrow x'$, the axes remain right-handed, the (100) mirror plane becomes (010), the (010) axial glide becomes (001) with a glide component $\frac{1}{2}a$, and the (001) diagonal glide becomes (100) so that the new space group symbol is P*nma*. Alternative axial transformations lead to four more space group symbols P*bnm*, P*nam*, P*mnb*, and P*cmn*; from these six orientations of the symmetry elements of the space group P*nma* is chosen arbitrarily as the *standard setting*. In compilations of data as in tabulation of space groups it is of course necessary to adhere rigidly to standard settings, but in discussion, especially comparative discussion, of actual structures it is often convenient to use non-standard settings; for instance in our brief discussion of the aragonite structure we chose the non-standard P*mcn* in order to have the carbonate groups parallel to the plane of a plan drawn with the conventional axial orientation for space group diagrams.

A note of the rules used in selecting standard settings of space groups in *International Tables for X-ray Crystallography* and of the conventions used in *Crystal Data*, the principal compilation of information about the unit-cells of real substances, is given in Appendix C.

Symmetry of the Bravais lattices

Earlier in this chapter it was pointed out that each Bravais lattice was distinguished from all others by the symmetry of its arrangement of lattice points. It becomes possible at this stage to argue the point more closely. We do so by considering first the two monoclinic Bravais lattices, P and C, whose unit-cells and symmetry elements are displayed in conventional space group orientation in Fig 4.20. It is to be noted that the C-cell has all the symmetry elements of the P-cell and in addition *a*-glides interleave the mirror planes, screw diads alternate with the [010] rotation diads, and another eight centres of symmetry appear. It is quite general in centred lattices that all the symmetry elements of the corresponding primitive lattice are present and in addition appropriate glide planes and screw axes relate the centring lattice points, whether C or I or F, to the lattice point at the origin.

The symmetry elements of the four orthorhombic Bravais lattices are displayed

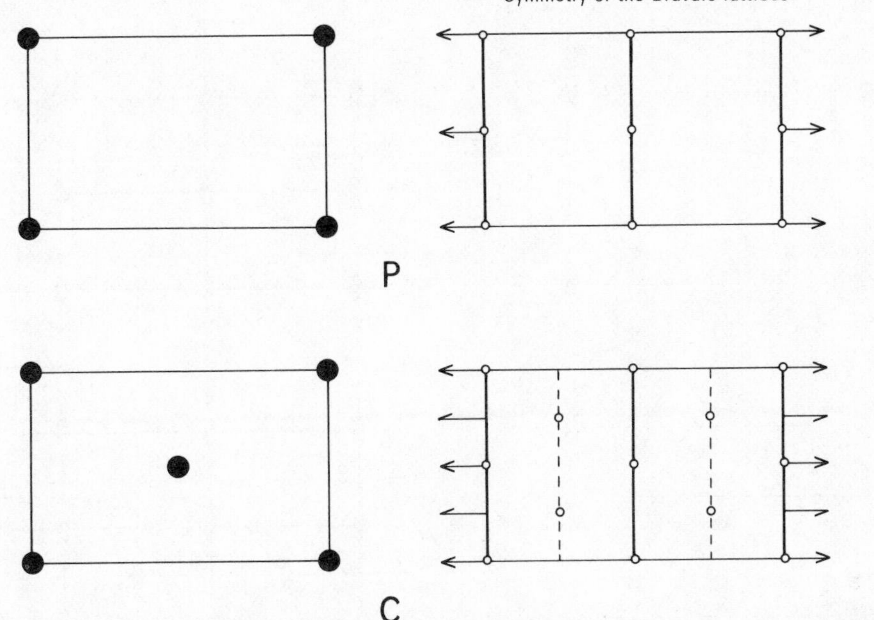

Fig 4.20 Unit-cells and symmetry elements of the two monoclinic Bravais lattices.

as Fig 4.21. The symmetry elements of the orthorhombic P-lattice are present in each of the non-primitive lattice types. The P-lattice has diads parallel to x, y, and z, mirror planes perpendicular to x, y, and z, and centres of symmetry at and midway between lattice points. In the centred lattices there are in addition translational symmetry elements and centres of symmetry that relate the lattice point at the origin to the centring lattice points. For instance in the I-lattice the lattice point at $\frac{1}{2}, \frac{1}{2}, \frac{1}{2}$ is related to the lattice point at the origin by a screw diad at $x, \frac{1}{4}, \frac{1}{4}$; this screw diad lies midway between the diads at $x, 0, 0$ and $x, \frac{1}{2}, \frac{1}{2}$, which are present also in the P-lattice; an n-glide interleaves the (100) mirror planes of the P-lattice and likewise relates the two lattice points; screw diads and n-glides correspondingly oriented with respect to the y and z axes similarly relate the two lattice points; finally, centres of symmetry at $\frac{1}{4}, \frac{1}{4}, \frac{1}{4}$, etc appear and likewise relate the lattice points at the origin and the body-centre. A significant distinction between the P and the I lattices is thus that, whereas in the former there is only one kind of diad parallel to and one kind of symmetry plane normal to each reference axis, there are in the latter two kinds of diad (2 and 2_1) parallel to and two kinds of symmetry plane (m and n) perpendicular to each reference axis; and moreover the number of centres of symmetry in the unit-cell increases from 8 to 16.

In the C-lattice an analogous situation exists. The centring of (001) faces gives rise to screw diads parallel to x and y, to additional rotation diads parallel to z, to b-glides and a-glides interleaving the (100) and (010) mirror planes respectively, and to eight additional centres of symmetry at $\frac{1}{4}, \frac{1}{4}, 0$, etc. The (001) n-glides that one would expect to be introduced are coincident with the mirror planes inherited from the P-lattice.

The F-lattice, like the P and the I lattices, has of course the same symmetry elements associated with each reference axis. Associated with the x-axis, for example, there are [100] rotation diads at $x, 0, 0$ and $x, \frac{1}{4}, \frac{1}{4}$, etc, and parallel screw diads at $x, 0, \frac{1}{4}$ and $x, \frac{1}{4}, 0$, etc, (100) mirror planes coincident with n-glides through $0, y, z$, etc, and b- and c-glides through $\frac{1}{4}, y, z$, etc; the additional translational symmetry elements relate the

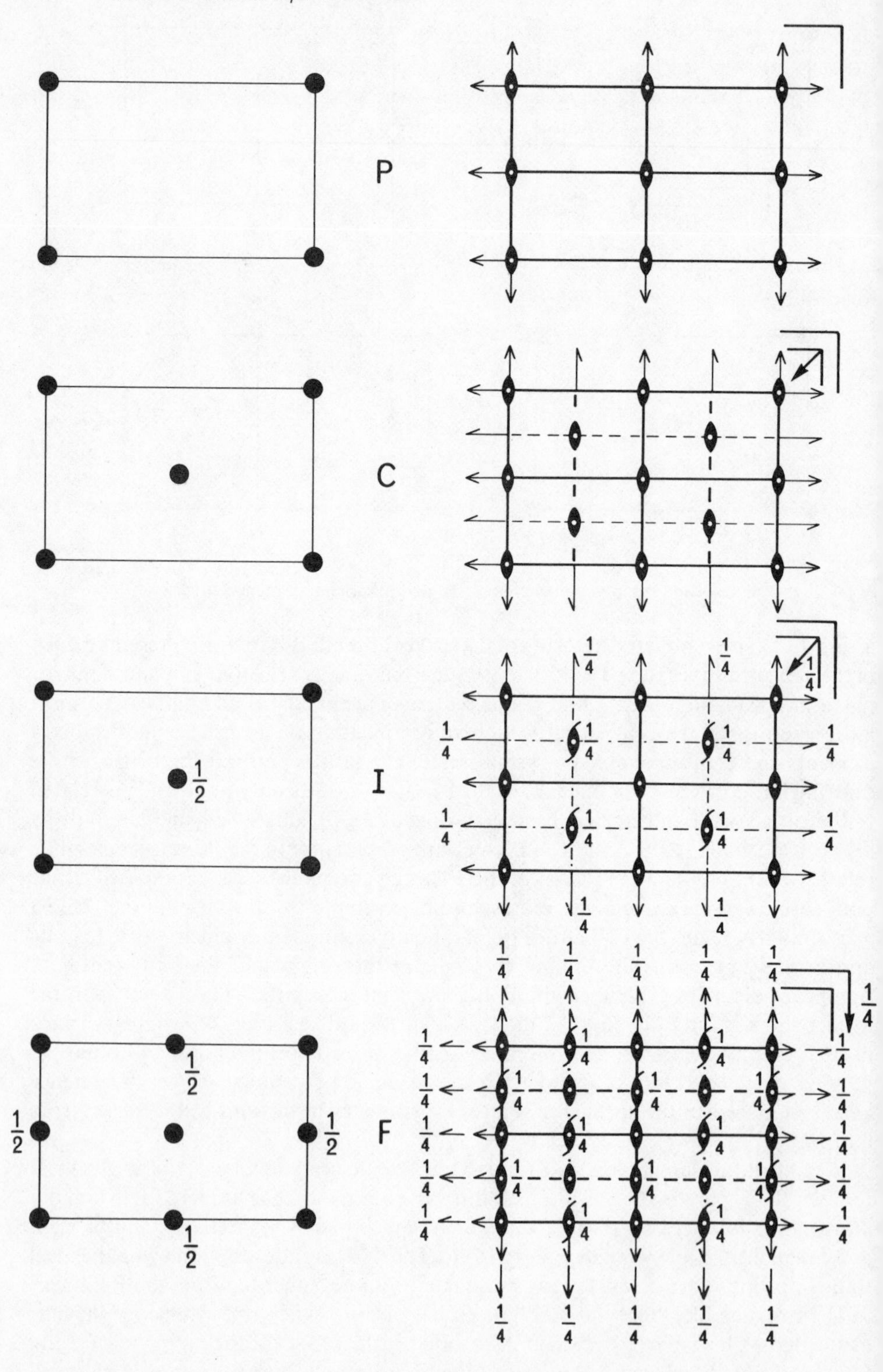

Fig 4.21 Unit-cells and symmetry elements of the four orthorhombic Bravais lattices.

face centring lattice points at $0, \frac{1}{2}, \frac{1}{2}$; $\frac{1}{2}, 0, \frac{1}{2}$; $\frac{1}{2}, \frac{1}{2}, 0$, to the lattice point at the origin and to one another. Centres of symmetry increase in number from 8 in the P-lattice to 32 in the F-lattice, where the additional centres have coordinates $0, \frac{1}{4}, \frac{1}{4}$; $\frac{1}{4}, 0, \frac{1}{4}$; $\frac{1}{4}, \frac{1}{4}, 0$; etc, midway between adjacent lattice points.

Each orthorhombic Bravais lattice thus has a characteristic set of symmetry elements as it has a characteristic disposition of lattice points. The same holds for the Bravais lattices of the remaining systems and, without exploring each in turn in tedious detail, we arrive at the general conclusion that a Bravais lattice is characterized not only by the disposition of its lattice points in space, but also by the nature and disposition of its symmetry elements.

5
Interplanar and interzonal angles: Some methods of calculation and transformation

In this chapter we deal with the mathematical techniques available for the calculation of interplanar and interzonal angles, beginning with an introduction to the elegant methods of spherical trigonometry and then dealing with the more cumbersome, but occasionally convenient, approach of three-dimensional coordinate geometry; the Miller formulae, which are of limited but very useful application, follow; and the chapter ends with a discussion of the transformation of axes, unit-cell coordinates, and face indices.

Spherical trigonometry

The calculation of interplanar and interzonal angles can usually be most easily programmed for computer calculation in the language of solid geometry, but for calculation 'by hand', that is with trigonometric tables and logarithms or a calculating machine, spherical trigonometry provides a means of attack that is at once elegant, rapid, and instructive.

A *spherical triangle* is defined as that portion of the surface of a sphere bounded by the intersection of the sphere with a three-sided pyramid whose apex is at the centre of the sphere. The sides of a spherical triangle are thus arcs of great circles. The *angles*, A, B, and C, of the spherical triangle ABC (Fig 5.1(a)) are the angles between the great circles whose planes are the faces of the pyramid OABC. The *sides*, a, b, and c, of the spherical triangle ABC are the angles between pairs of edges of the pyramid OABC and are therefore the angles subtended by the arcs, BC, CA, and AB, of the great circles at the centre, O, of the sphere. It follows that the sides, a, b, and c, are equal to the lengths of the arcs, BC, CA, and AB, when the sphere has unit radius (Fig 5.1(b)). Since the angles and sides of a spherical triangle are respectively the angles between faces and the angles between edges of a three-sided pyramid, every angle and side of a spherical triangle must be of magnitude less than π.

The general spherical triangle

In Appendix D we derive the eight fundamental relationships which relate the angles and sides of a general spherical triangle. There are three relationships between three sides and one angle,

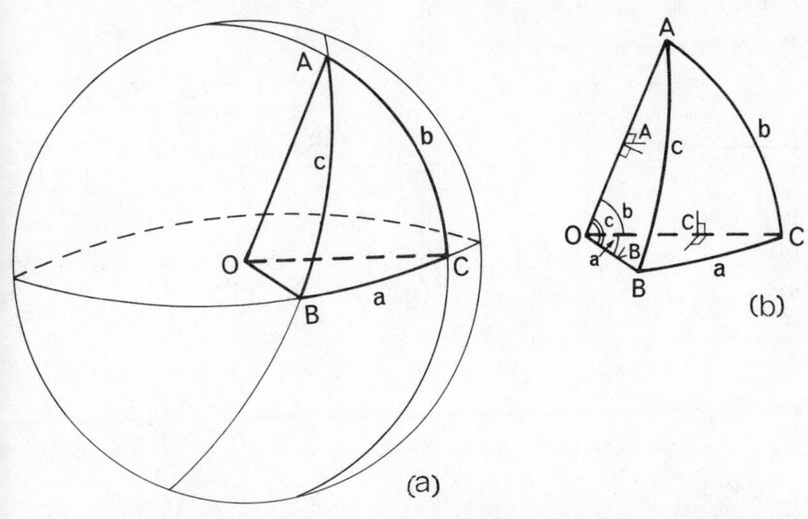

Fig 5.1 Spherical triangles. (a) shows the spherical triangle ABC as that portion of the surface of the sphere of centre O bounded by the great circles AB, BC, CA; the *angles* of the spherical triangle are denoted by A, B, C and its *sides* by *a*, *b*, *c*. (b) illustrates that the *angle* A is the angle between the normal to OA in the OAC plane and the normal to OA in the OAB plane, both normals being drawn from the same point on OA; and likewise for the *angles* B and C. The *sides*, *a*, *b*, *c*, of the spherical triangle ABC are equal to the lengths of the arcs BC, CA, AB respectively when the radius OA = OB = OC is unity.

$$\cos a = \cos b.\cos c + \sin b.\sin c.\cos \mathrm{A} \tag{1}$$

$$\cos b = \cos c.\cos a + \sin c.\sin a.\cos \mathrm{B} \tag{2}$$

$$\cos c = \cos a.\cos b + \sin a.\sin b.\cos \mathrm{C} \tag{3}$$

There are three relationships of similar aspect between three angles and one side,

$$\cos \mathrm{A} = -\cos \mathrm{B}.\cos \mathrm{C} + \sin \mathrm{B}.\sin \mathrm{C}.\cos a \tag{4}$$

$$\cos \mathrm{B} = -\cos \mathrm{C}.\cos \mathrm{A} + \sin \mathrm{C}.\sin \mathrm{A}.\cos b \tag{5}$$

$$\cos \mathrm{C} = -\cos \mathrm{A}.\cos \mathrm{B} + \sin \mathrm{A}.\sin \mathrm{B}.\cos c \tag{6}$$

And there is a set of relationships between angles and opposite sides,

$$\frac{\sin \mathrm{A}}{\sin a} = \frac{\sin \mathrm{B}}{\sin b} = \frac{\sin \mathrm{C}}{\sin c} \tag{7}$$

These relationships, which are adequate for the solution of any problem in spherical trigonometry, become greatly simplified if either one angle or one side of the spherical triangle is a right-angle and the simplified expressions that result are embodied in a set of easily remembered rules formulated by John Napier (1550–1617). The right-angled and right-sided cases are distinguished, the former being dealt with first.

Napier's Rules for right-angled triangles

Napier's Rules are stated in terms of the conventional diagram shown in Fig 5.2(a) where the angle at a point is divided into five parts by a horizontal radius running to the right, two vertical radii running upwards and downwards, and two inclined radii running to the left from the point. The horizontal line running to the right represents the right-angle and the remaining five elements of the spherical triangle are written on the diagram in cyclic order in the following manner: the compartments

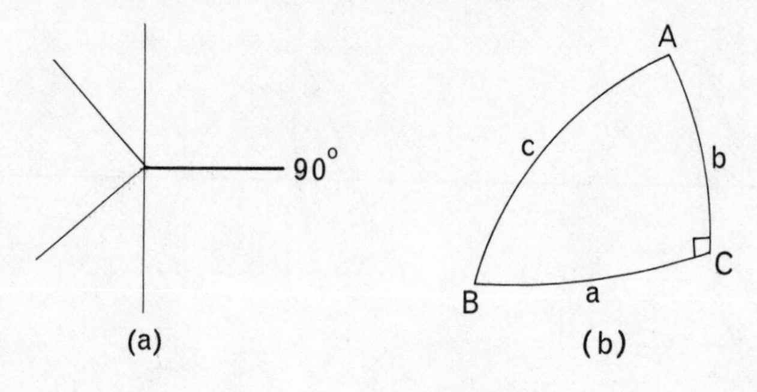

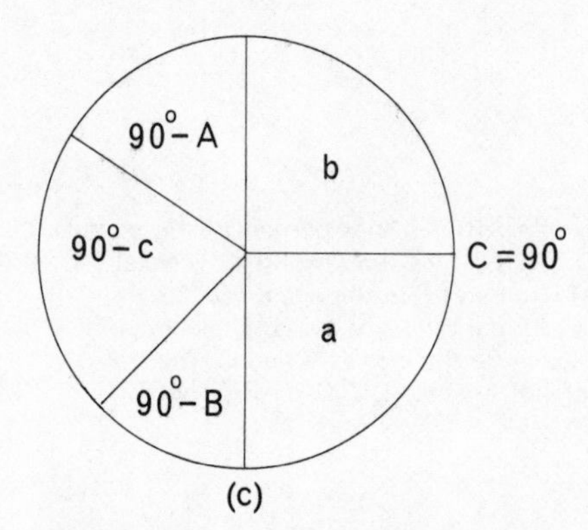

Fig 5.2 Napier's rules for right-angled spherical triangles. The thick line labelled 90° in (a) represents the right angle of the spherical triangle and the other sides and angles are written in cyclic sequence on this diagram. A spherical triangle with C = 90° is shown in (b) and the corresponding Naperian diagram is shown as (c).

to the right of the vertical line represent the sides adjacent to the right-angle and the three compartments to the left of the vertical line represent the *complements* of the remaining two angles and the side opposite the right-angle. Figure 5.2(b) shows a spherical triangle with $C = 90°$ and Fig 5.2(c) is the corresponding Napierian diagram. The magnitudes of any three elements of the spherical triangle are then related in terms of the conventional diagram by *Napier's Rules*:

the sine of a middle part = the product of the tangents of adjacent parts
= the product of the cosines of opposite parts[1]

Application of Napier's Rules to the spherical triangle of Fig 5.2(b), for which the conventional diagram[2] is Fig 5.2(c), yields the following ten equations:

$$\sin a = \tan b . \cot \mathrm{B} \tag{8}$$

$$\cos \mathrm{B} = \tan a . \cot c \tag{9}$$

$$\cos c = \cot \mathrm{B} . \cot \mathrm{A} \tag{10}$$

$$\cos \mathrm{A} = \cot c . \tan b \tag{11}$$

$$\sin b = \cot \mathrm{A} . \tan a \tag{12}$$

$$\sin a = \sin \mathrm{A} . \sin c \tag{13}$$

$$\cos \mathrm{B} = \cos b . \sin \mathrm{A} \tag{14}$$

[1] A mnemonic that may be useful is:

s**I**n m**I**ddle = t**A**n **A**dj. = c**O**s **O**pp.

[2] It is immaterial whether a clockwise sequence of elements on the triangle is represented by a clockwise or an anticlockwise sequence on the conventional diagram.

$$\cos c = \cos a . \cos b \tag{15}$$

$$\cos \mathrm{A} = \sin \mathrm{B} . \cos a \tag{16}$$

$$\sin b = \sin c . \sin \mathrm{B} \tag{17}$$

We now proceed to derive equations (8)–(17) from the relationships for a general spherical triangle by putting $\mathrm{C} = 90°$ in equations (1)–(7). This procedure constitutes a proof of Napier's Rules since equations (1)–(7) are symmetrical with respect to A, B, and C. For $\mathrm{C} = 90°$ equations (1) and (2) remain unchanged and (3)–(7) become

$$\cos c = \cos a . \cos b \qquad (3') \equiv (15)$$

$$\cos \mathrm{A} = \sin \mathrm{B} . \cos a \qquad (4') \equiv (16)$$

$$\cos \mathrm{B} = \sin \mathrm{A} . \cos b \qquad (5') \equiv (14)$$

$$-\cos \mathrm{A} . \cos \mathrm{B} + \sin \mathrm{A} . \sin \mathrm{B} . \cos c = 0$$

i.e.

$$\cos c = \cot \mathrm{A} . \cot \mathrm{B} \qquad (6') \equiv (10)$$

$$\frac{\sin \mathrm{A}}{\sin a} = \frac{\sin \mathrm{B}}{\sin b} = \frac{1}{\sin c}$$

i.e.

$$\sin a = \sin \mathrm{A} . \sin c \qquad (7') \equiv (13)$$

and

$$\sin b = \sin \mathrm{B} . \sin c \qquad (7'') \equiv (17)$$

Equations (10), (13), (14), (15), (16), and (17) are respectively identical to (6′), (7′), (5′), (3′), (4′), (7″) and it only remains to verify equations (8), (9), (11), and (12).

Elimination of $\cos a$ between equations (1) and (3′) yields

$$\cos \mathrm{A} = \frac{\cos c(1 - \cos^2 b)}{\cos b . \sin b . \sin c} = \cot c . \tan b$$

which is equation (11). And similarly elimination of $\cos b$ between (2) and (3′) yields equation (9). From equation (7)

$$\sin \mathrm{A} = \frac{\sin a . \sin \mathrm{B}}{\sin b}$$

which on substitution in (5′) yields

$$\cos \mathrm{B} = \sin a . \cot b . \sin \mathrm{B}$$

i.e.

$$\sin a = \tan b . \cot \mathrm{B}$$

which is equation (8). Similarly elimination of $\sin \mathrm{B}$ between (7) and (4′) yields equation (12), completing the verification of Napier's Rules for right-angled spherical triangles.

Napier's Rules provide a simple and straightforward means of calculating interplanar and interzonal angles in all systems other than the triclinic where recourse has to be made to the relationships for the general spherical triangle. The only problem facing the crystallographic user of Napier's Rules is the discovery of a right-angled spherical triangle containing two known angles, other than the right-angle, and the angle whose magnitude is required. In searching the stereogram for such a triangle it should always be remembered that a right-angled spherical triangle (or *Napierian triangle*) is formed by the intersection of mutually perpendicular zones. This can arise in two ways, either when the pole of one great circle representing a zone lies on another great circle which also represents a zone (three cases are illustrated in

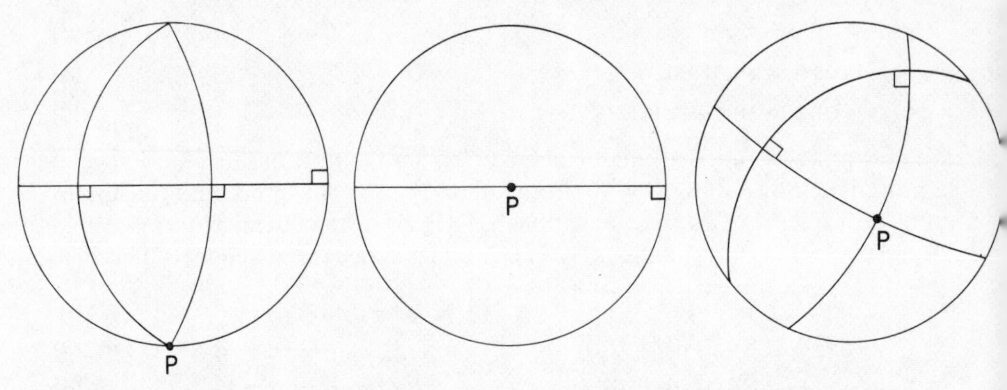

Fig 5.3 A right-angled spherical triangle arises when the pole of one great circle lies on another great circle. In the diagram at the left the pole P lies on the primitive and on two of the other great circles shown so that the horizontal diameter, of which P is the pole, intersects all three great circles orthogonally. In the middle diagram the primitive, whose pole is P, is perpendicular to all great circles passing through P. In the diagram at the right P lies at the intersection of two great circles which intersect the great circle whose pole is P orthogonally.

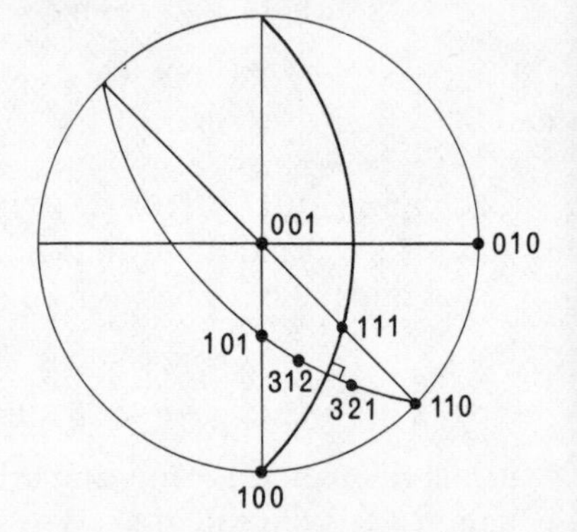

Fig 5.4 In the cubic system the lattice geometry is such that the positions of the poles (321) and (312) are related by the mirror plane through the poles (100) and (111) so that the great circle through (321) and (312) is perpendicular to the great circle through (100) and (111). Spherical triangles such as (100), (211), (101) and (211), (110), (111) are in consequence Naperian.

Fig 5.3), or when two poles are related by a mirror plane (the great circle through the two poles will then be perpendicular to the great circle representing the mirror plane, which is necessarily a zone, as illustrated in Fig 5.4). The latter case should always be borne in mind when dealing with the cubic system.

Example (i) To calculate the interfacial angle (100):(311) *in* $BaSO_4$ *which is orthorhombic with* $a = 8{\cdot}85$ Å, $b = 5{\cdot}44$ Å, $c = 7{\cdot}13$ Å

The first step is to draw a sketch stereogram showing all the zones relevant to the angle to be calculated (Fig 5.5(a)). Since the zones [(100), (010)] and [(310), (001)] are mutually perpendicular the triangle (100), (311), (310) is right-angled at (310) and Napier's Rules are applicable. This triangle is shown in isolation in Fig 5.5(b) and the conventional diagram is shown in Fig 5.5(c). Before Napier's Rules can be applied to determine the required angle, two angles must be evaluated from the given unit-cell dimensions. We have earlier shown that in general in the orthorhombic system

$$(100){:}(hk0) = \tan^{-1}\frac{a/h}{b/k}$$

Hence $$(100){:}(310) = \tan^{-1}\frac{8{\cdot}85}{3(5{\cdot}44)} = 28°28'.$$

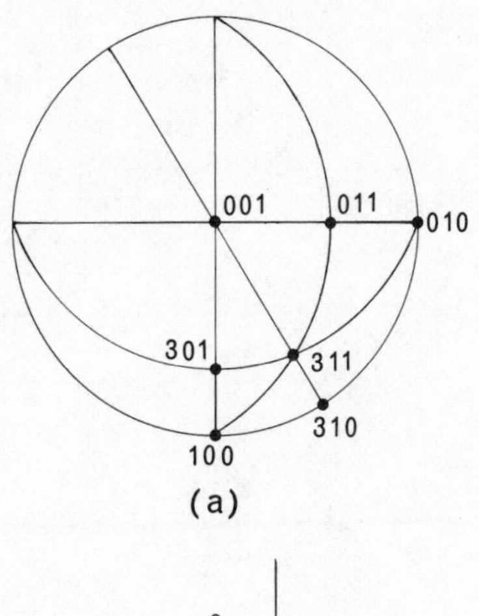

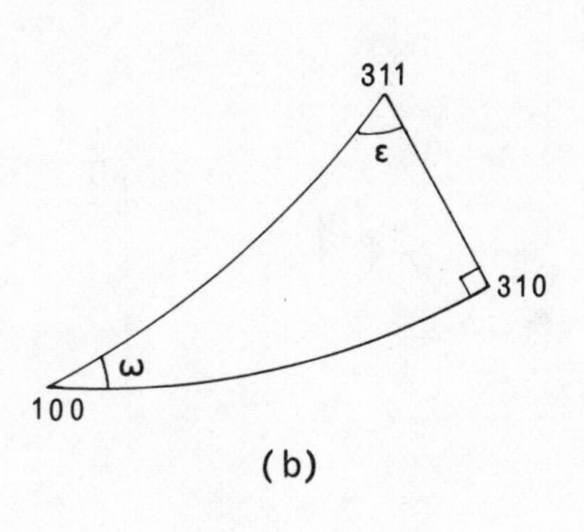

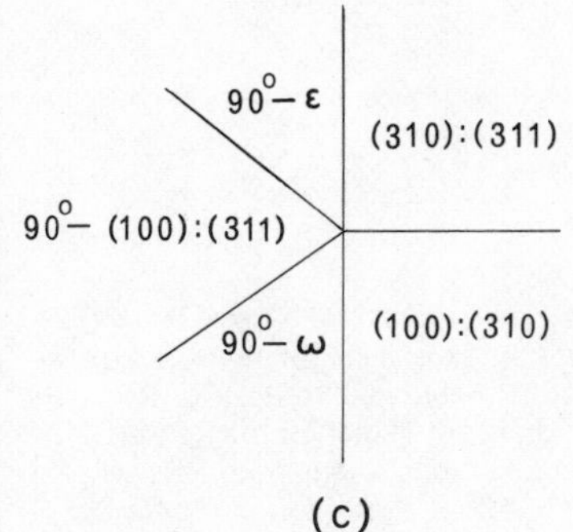

Fig 5.5 Calculation of the interfacial angle (100) : (311) in the orthorhombic system, the unit-cell dimensions being known. The stereogram (a) shows all the zones relevant to the calculation. The spherical triangle (100), (311), (310) is shown enlarged in (b) : tan (100) : (310) = $a/3b$ and $\tan \omega = b/c$. The Napier diagram for the right-angled spherical triangle (100), (311), (310) is shown in (c).

Since the normal to the plane (100) is parallel to the zone axis [100],

$$\omega = (011){:}(010)$$

Now in general in the orthorhombic system we have seen that

$$(010){:}(0kl) = \tan^{-1}\frac{b/k}{c/l}$$

Therefore $\omega = \tan^{-1}\dfrac{b}{c} = \tan^{-1}\dfrac{5{\cdot}44}{7{\cdot}13} = 37°21'$.

Application of Napier's Rule in the tangent form with $90° - \omega$ as the middle angle to which $90° - (100){:}(311)$ and $(100){:}(310)$ are adjacent (Fig 5.5(c)) yields

$$\cos\omega = \cot(100){:}(311) . \tan(100){:}(310).$$

Therefore
$$\tan(100){:}(311) = \frac{\tan(100){:}(310)}{\cos\omega}$$
$$= \frac{\tan 28°28'}{\cos 37°21'}.$$

Therefore $(100){:}(311) = 34°18'$

Alternatively the triangle (301), (311), (100), in which the angle between the zones [(100), (001)] and [(010), (301)] is 90°, may be employed, with intermediate evaluation of the angles $90 - \omega$ and $(100){:}(301)$.

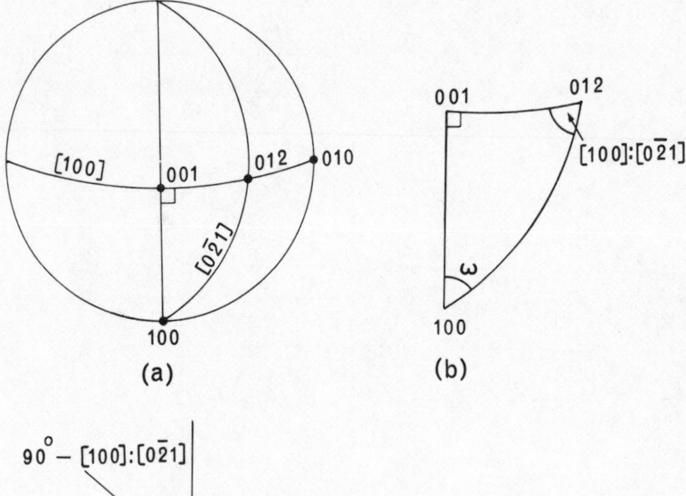

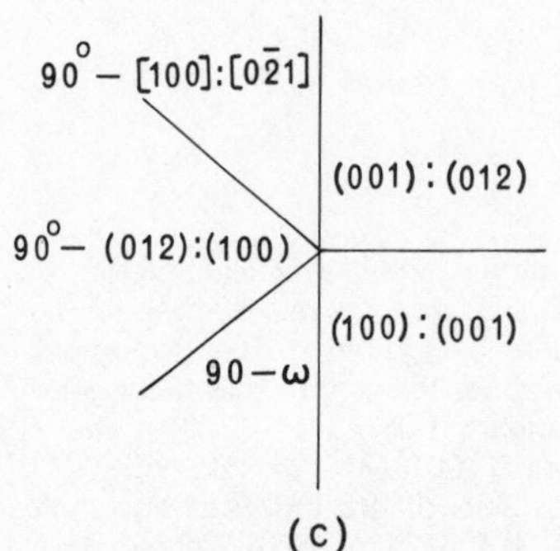

Fig 5.6 Calculation of the interzonal angle [100]:[0$\bar{2}$1] in the monoclinic system, the unit-cell dimensions being known. The stereogram (a) shows all the zones relevant to the calculation. The significant Naperian triangle (100), (001), (012) and its conventional diagram are shown in (b) and (c): (100):(001) = 180° – β and $\tan\omega = c/2b$.

Example (ii) To calculate the interzonal angle $[100]$:$[0\bar{2}1]$ *in gypsum,* $CaSO_4.2H_2O$, *which is monoclinic with* $a = 5{\cdot}68$ Å, $b = 15{\cdot}18$ Å, $c = 6{\cdot}29$ Å, $\beta = 113°50'$

The first step is to draw a sketch stereogram (Fig 5.6(a)) to display the principal zones and the zone $[0\bar{2}1]$. The indices of planes common to any two zones can be found by application of the *zone equation*: the zone $[100]$ contains all planes with indices of the type $(0kl)$ and in particular (010) and (001); the zone $[0\bar{2}1]$ contains all planes with indices such that $2k = l$, e.g. (100), (012); therefore the face common to both zones has $h = 0$, $2k = l$ and must be (012). The triangle (100), (001), (012) is right-angled at (001) since $[100] \perp [010]$ in the monoclinic system. This Napierian triangle and its conventional diagram are shown in Fig 5.6(b) and (c). It has already been shown that in the monoclinic system

$$(100){:}(001) = 180° - \beta$$

and
$$\tan\omega = \frac{c}{2b}$$

Application of Napier's Rule in the cosine form with $90° - [100]{:}[0\bar{2}1]$ as the middle angle to which $90° - \omega$ and (100):(001) are opposite (Fig 5.6(c)) yields

$$\cos[100]{:}[0\bar{2}1] = \sin\omega . \cos(100){:}(001)$$

Therefore
$$\cos[100]{:}[0\bar{2}1] = \sin\tan^{-1}\frac{c}{2b} . \cos(180° - \beta)$$

$$= \sin \tan^{-1} \frac{6{\cdot}29}{30{\cdot}36} . \cos 66°10'$$

$$= \sin 11°42' . \cos 66°10'$$

Hence $\qquad [100]{:}[0\bar{2}1] = 85°18'$

Example (iii) To calculate the angle between the polar edges of the cleavage rhombohedron $\{10\bar{1}4\}$ *in calcite*, $CaCO_3$, *which is trigonal* ($\bar{3}m$) *with* $a = 4{\cdot}990$ Å, $c = 17{\cdot}061$ Å

The *polar edges* of a rhombohedron are those edges that meet in the triad as shown for this case in the sketch Fig 5.7(a), where the required angle for the face $(10\bar{1}4)$ is marked α. The polar edges of the face $(10\bar{1}4)$ are the zone axes of the zone $[(10\bar{1}4), (0\bar{1}14)]$ and $[(10\bar{1}4), (\bar{1}104)]$ displayed on Fig 5.7(b). That the required angle α, being the angle between zone axes, is the supplement of the angle between the great circles representing the corresponding zones is

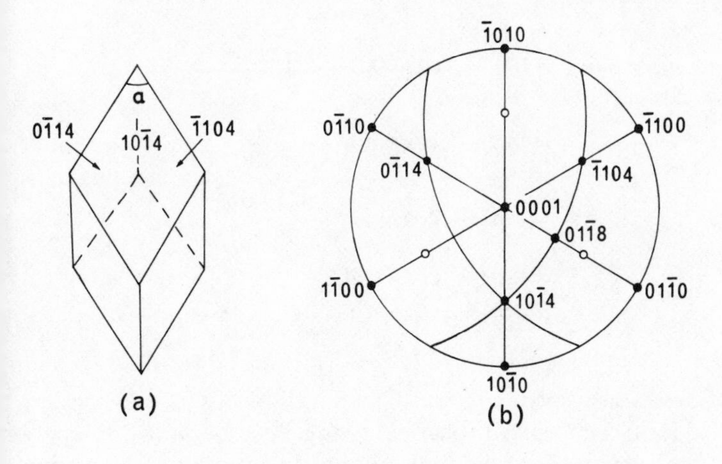

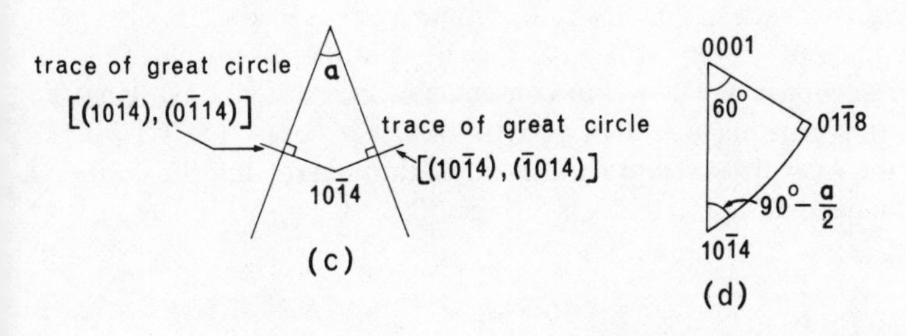

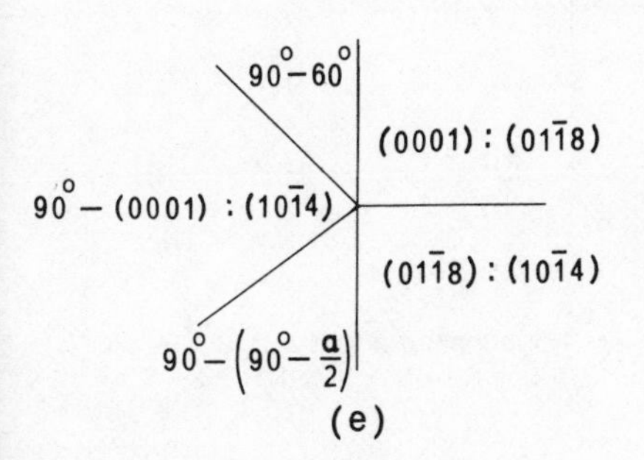

Fig 5.7 Calculation of the angle between the polar edges of the rhombohedron $\{10\bar{1}4\}$ in the trigonal system, the unit-cell dimensions being known. The rhombohedron is shown in perspective in (a) and in stereographic projection in (b). That the angle between the great circles $[(10\bar{1}4), (0\bar{1}14)]$ and $[(10\bar{1}4), (\bar{1}014)]$ is $180° - \alpha$, where α is the angle between the polar edges of the rhombohedron, is demonstrated in (c). The Naperian triangle (0001), $(10\bar{1}4)$, $(01\bar{1}8)$ and its conventional diagram are shown in (d) and (e): the angle at (0001) is 60° and $\tan (0001){:}(10\bar{1}4) = c/4a \cos 30°$.

evident from Fig 5.7(c). It is apparent from the stereogram that, by symmetry, the zone $[(0001), (01\bar{1}0)]$ is perpendicular to $[(10\bar{1}4), (\bar{1}104)]$ and by application of the *zone equation* the plane common to the two zones is seen to be $(01\bar{1}8)$. The triangle (0001), $(10\bar{1}4)$, $(01\bar{1}8)$ thus has a right-angle at $(01\bar{1}8)$, a 60° angle at (0001), and an angle $90°-(\alpha/2)$ at $(10\bar{1}4)$; this Napierian triangle and its conventional diagram are shown in Fig 5.7(d) and (e). It has already been shown that in the trigonal system

$$\tan(0001){:}(h0\bar{h}l) = \frac{c/l}{(a/h)\cos 30°}$$

$$\text{Therefore}\quad \tan(0001){:}(10\bar{1}4) = \frac{c}{4a\cos 30°}$$

Application of Napier's Rules yields

$$\cos(0001){:}(10\bar{1}4) = \cot 60° \tan\frac{\alpha}{2}$$

$$\text{Therefore}\quad \tan\frac{\alpha}{2} = \tan 60°.\cos(0001){:}(10\bar{1}4)$$

$$= \tan 60°.\cos\tan^{-1}\frac{c}{4a\cos 30°}$$

$$= \tan 60°.\cos\tan^{-1}\frac{17{\cdot}061}{19{\cdot}96\cos 30°}$$

$$= \tan 60°.\cos 44°38'$$

$$\text{Therefore}\quad \alpha = 101°54'$$

Napier's Rules for right-sided spherical triangles

Napier's Rules as stated for a right-angled spherical triangle can be shown to apply here too provided one change is made in the conventional diagram. For a Napierian triangle with $c = 90°$, as shown in Fig 5.8(a), the compartment in the middle of the left-hand side is occupied by $C-90°$ (Fig 5.8(b)) whereas in the case where $C = 90°$ this compartment is occupied by $90°-c$. In consequence expressions involving the sine or tangent of this part change sign $(\sin(C-90°) = -\cos C,\ \tan(C-90°) = -\cot C)$ relative to the case discussed previously while those involving the cosine of this part remain unchanged in sign $(\cos(C-90°) = \cos(90°-C) = \sin C)$. The validity

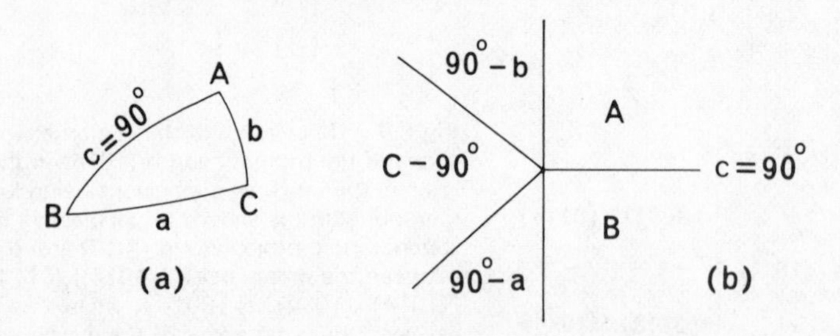

Fig 5.8 Napier's Rules for right-sided spherical triangles. A spherical triangle with $c = 90°$ is shown in (a); its conventional Napier diagram, shown in (b), has $C-90°$ instead of $90°-C$ in the middle compartment on the left-hand side.

of Napier's Rules for right-sided triangles can simply be demonstrated by substituting $c = 90°$ in the equations for a general spherical triangle, equations (1)–(7), and comparing the resultant expressions with those read off from Fig 5.8(b), the procedure we adopted to validate the right-angled rules. Usually in real crystallographic situations it is possible to discover an amenable spherical triangle with one of its angles equal to 90° and the conventional diagram for the right-sided case rarely has to be employed.

Analytical geometry

We restrict our treatment here to orthogonal axes, that is to the orthorhombic, tetragonal, and cubic systems. All the necessary relationships for extending this treatment to non-orthogonal systems are quoted without proof in Appendix E. For orthogonal axes the basic analytical equations (Appendix E) are the equation to a line whose direction cosines[3] are l, m, and n, where

$$l^2+m^2+n^2 = 1 \tag{18}$$

and the expression for the angle θ between two lines whose direction cosines are respectively $l_1m_1n_1$ and $l_2m_2n_2$

$$\cos\theta = l_1l_2+m_1m_2+n_1n_2 \tag{19}$$

Analytical expression for interfacial angles

The utmost generality that can be achieved with the restriction of orthogonality of axes is by considering an orthorhombic case. We take as mutually perpendicular reference axes x, y, and z springing from an origin at O (Fig 5.9). Let the plane (hkl) intercept the reference axes x, y, and z at A, B, and C so that $OA = a/h$, $OB = b/k$, and $OC = c/l$. If the normal from the origin to the plane (hkl) intersects the plane at N, then the direction cosines of ON will be $\cos\widehat{AON}$, $\cos\widehat{BON}$, $\cos\widehat{CON}$. It is evident from Fig 5.9 that $\cos\widehat{AON} = ON/(a/h)$, $\cos\widehat{BON} = ON/(b/k)$, $\cos\widehat{CON} = ON/(c/l)$, so that the direction cosines of the normal ON become $ON/(a/h)$, $ON/(b/k)$, $ON/(c/l)$. Therefore by equation (18)

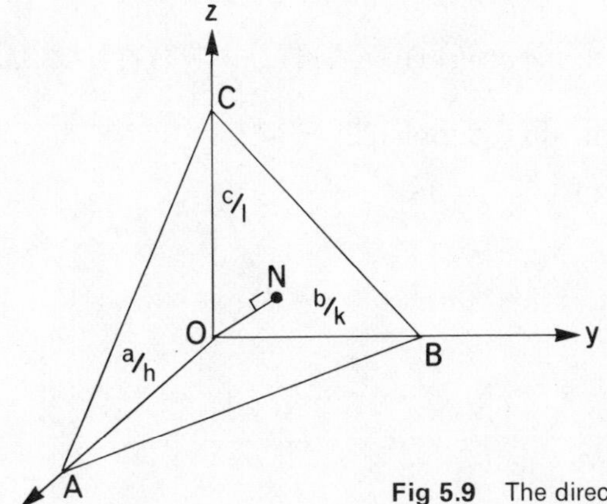

Fig 5.9 The direction cosines of the normal ON from the origin O to a plane (hkl) are $ON/(a/h)$, $ON/(b/k)$, $ON/(c/l)$.

[3] The use of l to represent one of the direction cosines is common to many well-known texts of analytical geometry but the possibility of confusion in crystallography with the third Miller index is obvious.

$$\frac{h^2}{a^2}+\frac{k^2}{b^2}+\frac{l^2}{c^2}=\frac{1}{\mathrm{ON}^2}$$

The direction cosines of ON can then be rewritten in terms of h/a, k/b, and l/c only as

$$\frac{h/a}{\sqrt{\left(\frac{h^2}{a^2}+\frac{k^2}{b^2}+\frac{l^2}{c^2}\right)}},\ \frac{k/b}{\sqrt{\left(\frac{h^2}{a^2}+\frac{k^2}{b^2}+\frac{l^2}{c^2}\right)}},\ \frac{l/c}{\sqrt{\left(\frac{h^2}{a^2}+\frac{k^2}{b^2}+\frac{l^2}{c^2}\right)}}$$

The angle between two planes $(h_1k_1l_1)$ and $(h_2k_2l_2)$ then follows from equation (19) as

$$\cos(h_1k_1l_1){:}(h_2k_2l_2)=\frac{\frac{h_1h_2}{a^2}+\frac{k_1k_2}{b^2}+\frac{l_1l_2}{c^2}}{\sqrt{\left(\frac{h_1^2}{a^2}+\frac{k_1^2}{b^2}+\frac{l_1^2}{c^2}\right)}\sqrt{\left(\frac{h_2^2}{a^2}+\frac{k_2^2}{b^2}+\frac{l_2^2}{c^2}\right)}} \tag{20}$$

Equation (20) is cumbersome and generally inconvenient, except when using a computer, relative to the spherical trigonometrical approach for the orthorhombic system. But in the tetragonal system, where $a = b$, some degree of simplification is achieved,

$$\cos(h_1k_1l_1){:}(h_2k_2l_2)=\frac{\{h_1h_2+k_1k_2+(a^2/c^2)l_1l_2\}}{\sqrt{(h_1^2+k_1^2+(a^2/c^2)l_1^2)}\sqrt{(h_2^2+k_2^2+(a^2/c^2)l_2^2)}} \tag{21}$$

which becomes particularly evident in the special case where $l_1 = l_2 = 0$,

$$\cos(h_1k_10){:}(h_2k_20)=\frac{h_1h_2+k_1k_2}{\sqrt{(h_1^2+k_1^2)}\sqrt{(h_2^2+k_2^2)}}$$

In the cubic system, where $a = b = c$, the expression becomes very much simpler and of considerable general utility,

$$\cos(h_1k_1l_1){:}(h_2k_2l_2)=\frac{h_1h_2+k_1k_2+l_1l_2}{\sqrt{(h_1^2+k_1^2+l_1^2)}\sqrt{(h_2^2+k_2^2+l_2^2)}} \tag{22}$$

Example To calculate the interfacial angles (100):(111) *and* (111):$(1\bar{1}1)$ *in the cubic system*

Substitute the indices of the planes in equation (22),

$$\cos(100){:}(111)=\frac{1\times1+0\times1+0\times1}{\sqrt{1}\ \ \sqrt{3}}$$

$$=\frac{1}{\sqrt{3}}$$

Therefore $(100){:}(111) = 54°44'$

$$\cos(111){:}(1\bar{1}1)=\frac{1\times1-1\times1+1\times1}{\sqrt{3}\ \ \sqrt{3}}$$

$$=\frac{1}{3}$$

Therefore $(111){:}(1\bar{1}1) = 70°32'$

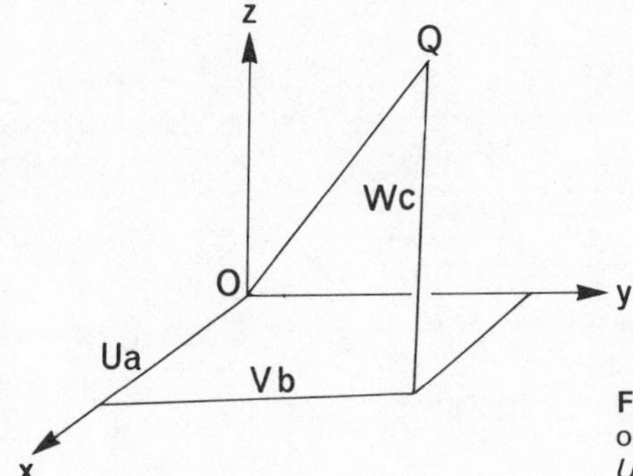

Fig 5.10 The zone axis [*UVW*] passes through the origin O and through a point Q with coordinates *Ua, Vb, Wc*.

Analytical expression for interzonal angles

The angle between two zones is the supplement of the angle between the corresponding zone axes (cf. Example (iii) on p. 133) and we shall confine our attention here to the angle between zone axes. Again we start with the orthorhombic system and consider a zone axis $[UVW]$ which passes by definition through the origin and a point Q with coordinates Ua, Vb, Wc (Fig 5.10). Therefore

$$\mathrm{OQ}^2 = U^2a^2 + V^2b^2 + W^2c^2$$

and the direction cosines of OQ, Ua/OQ, Vb/OQ, Wc/OQ can be expressed in terms of U, V, W, a, b, c only as

$$\frac{Ua}{\sqrt{(U^2a^2+V^2b^2+W^2c^2)}},\ \frac{Vb}{\sqrt{(U^2a^2+V^2b^2+W^2c^2)}},\ \frac{Wc}{\sqrt{(U^2a^2+V^2b^2+W^2c^2)}}$$

Application of equation (19) yields immediately the angle between the zone axes $[U_1V_1W_1]$ and $[U_2V_2W_2]$,

$$\cos[U_1V_1W_1]{:}[U_2V_2W_2] = \frac{U_1U_2a^2+V_1V_2b^2+W_1W_2c^2}{\sqrt{(U_1^2a^2+V_1^2b^2+W_1^2c^2)}\sqrt{(U_2^2a^2+V_2^2b^2+W_2^2c^2)}} \quad (23)$$

As in the case of the corresponding expression for interfacial angles in the orthorhombic system, equation (20), this expression is cumbersome and of little utility for calculation except by computer. In the tetragonal system it becomes adequately simplified for convenient use only in the special case where $W_1 = W_2 = 0$,

$$\cos[U_1V_10]{:}[U_2V_20] = \frac{U_1U_2+V_1V_2}{\sqrt{(U_1^2+V_1^2)}\sqrt{(U_2^2+V_2^2)}} \quad (24)$$

and in the cubic system the general expression

$$\cos[U_1V_1W_1]{:}[U_2V_2W_2] = \frac{U_1U_2+V_1V_2+W_1W_2}{\sqrt{(U_1^2+V_1^2+W_1^2)}\sqrt{(U_2^2+V_2^2+W_2^2)}} \quad (25)$$

is convenient to use.

Comparison of equations (22) and (25) makes it immediately apparent that in the cubic system the normal to the plane (pqr) is parallel to the zone axis $[pqr]$ and therefore the normals to planes in the zone $[pqr]$ are coplanar with (pqr). In the tetragonal system this is so only for $r = 0$ and in other systems only for more restrictive conditions. Figure 5.11 shows the poles of the (110) and (111) face normals

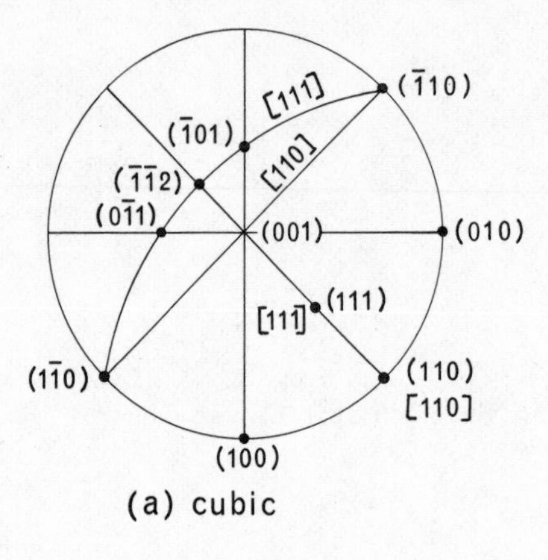

(a) cubic

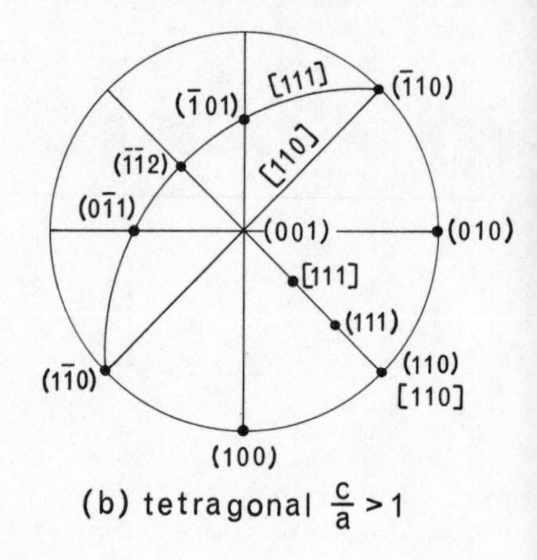

(b) tetragonal $\frac{c}{a} > 1$

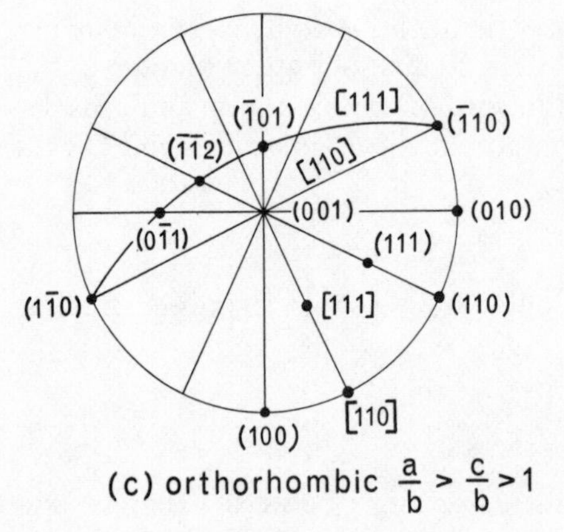

(c) orthorhombic $\frac{a}{b} > \frac{c}{b} > 1$

Fig 5.11 The normal to the plane (*pqr*) is parallel to the zone axis [*pqr*] generally in the cubic system (a), when $r = 0$ in the tetragonal system (b), and only for (100), (010), (001) and their opposites in the orthorhombic system (c).

together with the [110] and [111] zones and zone axes in cubic, tetragonal, and orthorhombic cases to emphasize this important point.

It is in practice in systems other than the cubic most convenient to calculate interzonal angles by applying Napier's Rules to appropriately chosen spherical triangles to evaluate the angle between the corresponding great circles.

Equations to the normal to a plane

Whether the reference axes x, y, z are orthogonal or not the plane (hkl) will make intercepts equal to a/h, b/k, and c/l on x, y, and z respectively. If the normal to the plane (hkl) through the origin O intersects the plane in N, then (Fig 5.9)

$$\mathrm{ON} = \frac{a}{h}\cos x\!:\!(hkl) = \frac{b}{k}\cos y\!:\!(hkl) = \frac{c}{l}\cos z\!:\!(hkl) \tag{26}$$

where $x\!:\!(hkl)$ is the angle between the x-axis and the normal to (hkl) and so on. Equation (26) is of limited application in crystallography but may provide a rapid means of evaluating two such angles when the other is known. When the reference axes are orthogonal equation (26) can be rewritten as

$$\frac{a}{h}\cos(100)\!:\!(hkl) = \frac{b}{k}\cos(010)\!:\!(hkl) = \frac{c}{l}\cos(001)\!:\!(hkl) \tag{27}$$

an expression that is occasionally quite useful.

Example To calculate the interfacial angles (010):(311) *and* (001):(311) *in* $BaSO_4$ *which is orthorhombic with* $a = 8{\cdot}85$ Å, $b = 5{\cdot}44$ Å, $c = 7{\cdot}13$ Å

The angle (100):(311) was evaluated by Napierian triangle on p. 130 and therefrom the angles (010):(311) and (001):(311) can quickly be evaluated by the equations to the normal.

$$\cos(010)\!:\!(hkl) = \frac{a}{b}\cdot\frac{k}{h}\cos(100)\!:\!(hkl)$$

Since $(100)\!:\!(311) = 34^\circ\,18'$

$$\cos(010)\!:\!(311) = \frac{8{\cdot}85}{5{\cdot}44}\cdot\frac{1}{3}\cdot\cos 34^\circ\,18'$$

Therefore $(010)\!:\!(311) = 63^\circ\,23'$

And $\cos(001)\!:\!(hkl) = \dfrac{a}{c}\cdot\dfrac{l}{h}\cos(100)\!:\!(hkl)$

Hence $\cos(001)\!:\!(311) = \dfrac{8{\cdot}85}{7{\cdot}13}\cdot\dfrac{1}{3}\cdot\cos 34^\circ\,18'$

and so $(001)\!:\!(311) = 70^\circ\,01'$

The sine ratio or Miller formulae[4]

The Miller formulae relate the angles between four faces in a zone (i.e., four *tautozonal* faces) and are of general application in all systems.

Let the tautozonal faces P_1, P_2, P_3, and P_4 (Fig 5.12) have indices $(h_1k_1l_1)$, $(h_2k_2l_2)$, $(h_3k_3l_3)$, and $(h_4k_4l_4)$ respectively and denote the interfacial angles $\theta_{12} = P_1\!:\!P_2$, $\theta_{13} = P_1\!:\!P_3$ and so on.

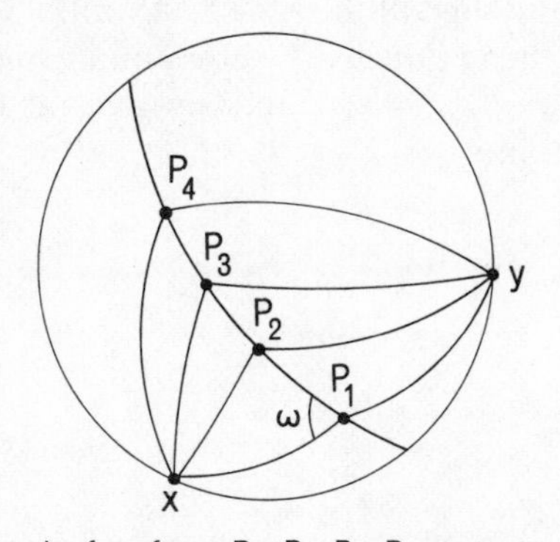

Fig 5.12 The Miller formulae: the four faces P_1, P_2, P_3, P_4 are tautozonal.

In the triangle xP_1P_2 application of equation (1) yields

$$\cos x\!:\!P_2 = \cos P_1\!:\!P_2 \,.\, \cos x\!:\!P_1 + \sin P_1\!:\!P_2 \,.\, \sin x\!:\!P_1 \cos\omega \qquad (28)$$

where ω is the angle at P_1. And correspondingly in triangle xP_1P_3

$$\cos x\!:\!P_3 = \cos P_1\!:\!P_3 \,.\, \cos x\!:\!P_1 + \sin P_1\!:\!P_3 \,.\, \sin x\!:\!P_1 \,.\, \cos\omega \qquad (29)$$

[4] Our treatment of this topic follows closely that of F. C. Phillips, *Introduction to Crystallography* (4th ed. 1971) which we have found to be most lucid.

Multiplication of (28) by $\sin P_1{:}P_3$ and of (29) by $\sin P_1{:}P_2$ followed by subtraction gives

$$\begin{aligned}\cos x{:}P_2 . \sin P_1{:}P_3 - \cos x{:}P_3 . \sin P_1{:}P_2 \\ = \cos x{:}P_1(\cos P_1{:}P_2 . \sin P_1{:}P_3 - \cos P_1{:}P_3 . \sin P_1{:}P_2) \\ = \cos x{:}P_1 . \sin P_2{:}P_3\end{aligned} \tag{30}$$

And similarly from triangles yP_1P_2 and yP_1P_3 it can be shown that

$$\cos y{:}P_2 . \sin P_1{:}P_3 - \cos y{:}P_3 . \sin P_1{:}P_2 = \cos y{:}P_1 . \sin P_2{:}P_3 \tag{31}$$

Elimination of $\sin P_2{:}P_3$ between (30) and (31) yields

$$\frac{\cos x{:}P_2 . \sin P_1{:}P_3 - \cos x{:}P_3 . \sin P_1{:}P_2}{\cos y{:}P_2 . \sin P_1{:}P_3 - \cos y{:}P_3 . \sin P_1{:}P_2} = \frac{\cos x{:}P_1}{\cos y{:}P_1}$$

i.e.

$$\begin{aligned}\sin P_1{:}P_2(\cos x{:}P_1 . \cos y{:}P_3 - \cos x{:}P_3 . \cos y{:}P_1) \\ = \sin P_1{:}P_3(\cos x{:}P_1 . \cos y{:}P_2 - \cos x{:}P_2 . \cos y{:}P_1)\end{aligned} \tag{32}$$

Now the equation to the normal for a face P is

$$\frac{a}{h}\cos x{:}P = \frac{b}{k}\cos y{:}P$$

which on substitution in (32) leads to

$$\sin P_1{:}P_2 . \frac{a}{b}\left(\frac{k_3}{h_3} - \frac{k_1}{h_1}\right) . \cos x{:}P_1 . \cos x{:}P_3 = \sin P_1{:}P_3 . \frac{a}{b}\left(\frac{k_2}{h_2} - \frac{k_1}{h_1}\right) . \cos x{:}P_1 . \cos x{:}P_2$$

i.e.

$$\frac{h_1k_3 - h_3k_1}{h_3} . \sin P_1{:}P_2 . \cos x{:}P_3 = \frac{h_1k_2 - h_2k_1}{h_2} . \sin P_1{:}P_3 . \cos x{:}P_2 \tag{33}$$

Now $h_1k_3 - h_3k_1$ is the third digit of the zone axis symbol $[UVW]$ of the zone with which we are concerned when evaluated by cross-multiplication from the indices of P_1 and P_3. Putting $h_1k_3 - h_3k_1 = W_{13}$ and so on, and $\theta_{12} = P_1{:}P_2$ etc for the interfacial angles, (33) becomes

$$\frac{\sin\theta_{12}}{\sin\theta_{13}} = \frac{W_{12}}{W_{13}} . \frac{\cos x{:}P_2}{\cos x{:}P_3} . \frac{h_3}{h_2} \tag{34}$$

And from faces P_4, P_2, P_3 correspondingly

$$\frac{\sin\theta_{42}}{\sin\theta_{43}} = \frac{W_{42}}{W_{43}} . \frac{\cos x{:}P_2}{\cos x{:}P_3} . \frac{h_3}{h_2} \tag{35}$$

Division of (34) by (35) yields

$$\frac{\dfrac{\sin\theta_{12}}{\sin\theta_{13}}}{\dfrac{\sin\theta_{42}}{\sin\theta_{43}}} = \frac{\dfrac{W_{12}}{W_{13}}}{\dfrac{W_{42}}{W_{43}}}$$

And by similar argument this may be extended to the complete *Miller formulae*

$$\frac{\sin\theta_{12}/\sin\theta_{13}}{\sin\theta_{42}/\sin\theta_{43}} = \frac{U_{12}/U_{13}}{U_{42}/U_{43}} = \frac{V_{12}/V_{13}}{V_{42}/V_{43}} = \frac{W_{12}/W_{13}}{W_{42}/W_{43}} \tag{36}$$

The Miller formulae, equation (36), fail only when two of the tautozonal faces are parallel; if any of the angles $\theta_{12}, \theta_{13}, \theta_{42}, \theta_{43}$, is 180° the division which produces (36) becomes indeterminate and if $\theta_{14} = 180°$ equations (34) and (35) become identical. If two or one of the digits, U, V, W, is zero two or one of the equations labelled (36) become indeterminate, but the remaining equation (or equations) holds.

Since U, V, and W are of necessity integers, $(\sin\theta_{12}/\sin\theta_{13})/(\sin\theta_{42}/\sin\theta_{43})$ must be rational. This is a property that can be used, as will be seen in the examples that follow, to determine the indices of one of the four tautozonal faces when those of the other three and all four interfacial angles are known.

The Miller formulae can be expressed in a variety of forms, some leading to greater ease of computation than others; the form, equation (36), in which they are set down here has in our experience the advantage of being the most symmetrical and consequently the easiest to memorize.

The cotangent formula

This is the most useful of the rearrangements of equations (36) and leads to a special case of particular utility. Since U, V, and W are necessarily integers,

$$\frac{U_{12}/U_{13}}{U_{42}/U_{43}} = \frac{p}{q}$$

where p and q are integers.

$$\text{Therefore} \quad \frac{\sin\theta_{12}/\sin\theta_{13}}{\sin\theta_{42}/\sin\theta_{43}} = \frac{p}{q}$$

$$\text{Therefore} \quad \sin\theta_{12}.\sin(\theta_{14}-\theta_{13}) = \frac{p}{q}\sin\theta_{13}.\sin(\theta_{14}-\theta_{12})$$

$$q\sin\theta_{12}(\sin\theta_{14}\cos\theta_{13}-\cos\theta_{14}\sin\theta_{13}) = p\sin\theta_{13}(\sin\theta_{14}\cos\theta_{12}-\cos\theta_{14}\sin\theta_{12})$$

$$\text{and} \quad q\sin\theta_{12}\sin\theta_{13}\sin\theta_{14}(\cot\theta_{13}-\cot\theta_{14}) = p\sin\theta_{12}\sin\theta_{13}\sin\theta_{14}(\cot\theta_{12}-\cot\theta_{14})$$

$$\text{Hence} \quad p\cot\theta_{12}-q\cot\theta_{13} = (p-q)\cot\theta_{14} \tag{37}$$

Equation (37) is appropriate for the determination of one interfacial angle when the indices of all four tautozonal faces and two interfacial angles are known.

An important special case of (37) arises when $\theta_{14} = 90°$, then

$$\tan\theta_{12} = \frac{p}{q}\tan\theta_{13} \tag{38}$$

Equation (38) is of considerable utility in the solution of problems involving interfacial angles.

The additional restriction that one face should have two indices zero and another face have the remaining index zero, i.e. that the zone is $[(100), (0kl)]$, leads to further simplification. Suppose that P_1 is (100), P_2 $(h_2k_2l_2)$, P_3 $(h_3k_3l_3)$, and P_4 $(0k_4l_4)$, then

$$\frac{p}{q} = \frac{W_{12}/W_{13}}{W_{42}/W_{43}} = \frac{k_2/k_3}{-h_2k_4/-h_3k_4} = \frac{k_2/k_3}{h_2/h_3}$$

whence, simultaneously,

$$\left.\begin{aligned}\tan(100){:}(h_2k_2l_2) &= \frac{k_2/k_3}{h_2/h_3}\tan(100){:}(h_3k_3l_3)\\ &= \frac{l_2/l_3}{h_2/h_3}\tan(100){:}(h_3k_3l_3)\end{aligned}\right\} \quad (39)$$

For the zone [(010):(*h*0*l*)] the corresponding expression is

$$\left.\begin{aligned}\tan(010){:}(h_2k_2l_2) &= \frac{h_2/h_3}{k_2/k_3}\tan(010){:}(h_3k_3l_3)\\ &= \frac{l_2/l_3}{k_2/k_3}\tan(010){:}(h_3k_3l_3)\end{aligned}\right\} \quad (40)$$

and for the zone [(001):(*hk*0)],

$$\left.\begin{aligned}\tan(001){:}(h_2k_2l_2) &= \frac{h_2/h_3}{l_2/l_3}\tan(001){:}(h_3k_3l_3)\\ &= \frac{k_2/k_3}{l_2/l_3}\tan(001){:}(h_3k_3l_3)\end{aligned}\right\} \quad (41)$$

The symmetry of equations (39)–(41) is apparent: the denominator is the ratio of unique indices, the numerator the ratio of either of the other two indices, in each case for the faces (here P_2 and P_3) with all indices non-zero.

Example (i) To evaluate (hkl) given that P_1 $(1\bar{1}0)$, P_2 (112), P_3 (*hkl*), P_4 (011) *are tautozonal and that* $\theta_{12} = 66°17'$, $\theta_{24} = 42°32'$, $\theta_{23} = 13°33'$ (Fig 5.13)

$$\frac{\sin\theta_{12}/\sin\theta_{13}}{\sin\theta_{42}/\sin\theta_{43}} = \frac{\sin 66°17'/\sin 79°50'}{\sin 42°32'/\sin 28°59'} = \frac{2}{3}$$

Cross-multiplications:

$$\begin{array}{c}P_1,\ P_2\\ \begin{array}{c|ccccc|c} 1 & \bar{1} & & 0 & & 1 & \bar{1} \; 0\\ & & \times & & \times & \times & \\ 1 & 1 & & 2 & & 1 & 1 \; 2\end{array}\\ \hline [\bar{2} \quad \bar{2} \quad 2]\end{array} \qquad \begin{array}{c}P_1,\ P_3\\ \begin{array}{c|ccccc|c} 1 & \bar{1} & & 0 & & 1 & \bar{1} \; 0\\ & & \times & & \times & \times & \\ h & k & & l & & h & k \; l\end{array}\\ \hline [\bar{l} \quad \bar{l} \quad k+h]\end{array}$$

$$\begin{array}{c}P_4,\ P_2\\ \begin{array}{c|ccccc|c} 0 & 1 & & 1 & & 0 & 1 \; 1\\ & & \times & & \times & \times & \\ 1 & 1 & & 2 & & 1 & 1 \; 2\end{array}\\ \hline [1 \quad 1 \quad \bar{1}]\end{array} \qquad \begin{array}{c}P_4,\ P_3\\ \begin{array}{c|ccccc|c} 0 & 1 & & 1 & & 0 & 1 \; 1\\ & & \times & & \times & \times & \\ h & k & & l & & h & k \; l\end{array}\\ \hline [l-k \quad h \quad \bar{h}]\end{array}$$

Therefore $\dfrac{U_{12}/U_{13}}{U_{42}/U_{43}} = \dfrac{-2/-l}{1/(l-k)} = \dfrac{2}{3}$

i.e. $\dfrac{2(l-k)}{l} = \dfrac{2}{3}$

i.e. $3(l-k) = l$

i.e. $2l = 3k$

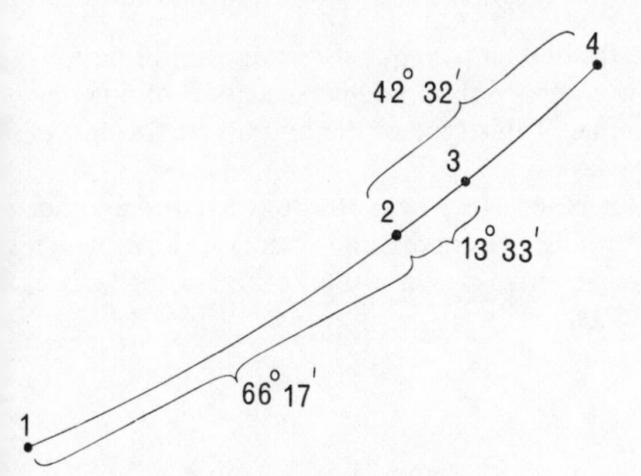

Fig 5.13 Determination of the indices of $P_3(hkl)$ by the Miller formulae, given that P_1 is $(1\bar{1}0)$, P_2 is (112), P_4 is (011) and that the angles θ_{12}, θ_{23}, and θ_{24} are known.

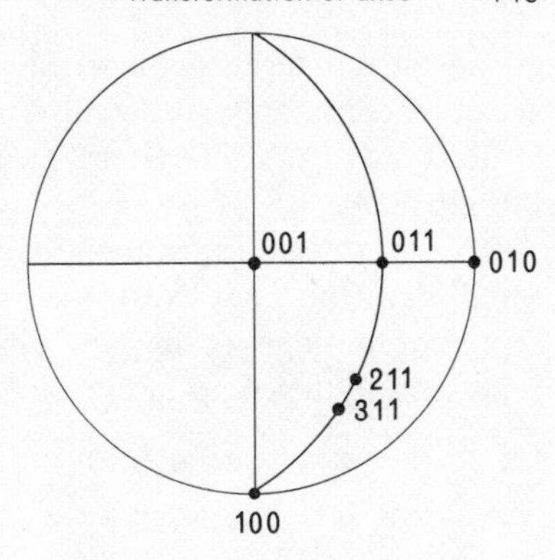

Fig 5.14 Calculation of (100):(211) for an orthorhombic crystal, given (100):(311), by the Miller formulae.

And
$$\frac{V_{12}/V_{13}}{V_{42}/V_{43}} = \frac{-2/-l}{1/h} = \frac{2}{3}$$

i.e.
$$\frac{2h}{l} = \frac{2}{3}$$

i.e.
$$3h = l$$

Therefore (hkl) is identified as (123).

Check:
$$\frac{W_{12}/W_{13}}{W_{42}/W_{43}} = \frac{2/(k+h)}{(-1)/(-h)} = \frac{2/3}{(-1)/(-1)} = \frac{2}{3}.$$

Example (ii) To calculate (100):(211) *in* $BaSO_4$ *(barite) which is orthorhombic and has* (100):(311) = 34° 18′ (cf. example (i), p. 130)

The faces (100), (311), (211), (011) are tautozonal (Fig 5.14) with (100):(011) = 90°. Application of equation (39)

$$\tan(100):(h_2k_2l_2) = \frac{k_2/k_3}{h_2/h_3}\tan(100):(h_3k_3l_3)$$

yields for P_2 (311) and P_3 (211)

$$\tan(100):(311) = \frac{1}{3/2}\tan(100):(211)$$

Therefore $(100):(211) = \tan^{-1}(\tfrac{3}{2}\tan 34°\,18')$
$$= 45°\,39'$$

Transformation of axes

In the course of investigation of a crystalline solid it may be necessary, for one reason or another, to change the orientation of the reference axes. For instance the crystal morphology of a substance of class 422 may have been referred to reference axes x, and y, which are shown by subsequent X-ray study to be parallel to the diagonal diads of the smallest unit mesh in the (001) plane; a procedure for reindexing the crystal faces in terms of the true unit-cell will then be required. Axial transformations are quite commonly needed also in the comparison of related structures of different lattice type or crystal system; an example of this type of use appears at the end of this chapter.

In order to change from one set of reference axes x, y, z, (the 'old' axes) to another set x', y', z' (the 'new' axes), it is necessary first to express the vectors $\mathbf{a}'$, $\mathbf{b}'$, $\mathbf{c}'$, which represent the edges of the 'new' unit-cell, in terms of vector sums of the vectors $\mathbf{a}$, $\mathbf{b}$, $\mathbf{c}$, representing the edges of the 'old' unit-cell,

i.e.
$$\mathbf{a}' = p_1\mathbf{a}+q_1\mathbf{b}+r_1\mathbf{c}$$
$$\mathbf{b}' = p_2\mathbf{a}+q_2\mathbf{b}+r_2\mathbf{c}$$
$$\mathbf{c}' = p_3\mathbf{a}+q_3\mathbf{b}+r_3\mathbf{c}$$

where $p_1, q_1, r_1, p_2 \ldots r_3$ are integral or simple fractional coefficients. Such coefficients completely specify the relationship of the 'new' to the 'old' unit-cell vectors and can conveniently be represented in matrix form:

$$\begin{matrix} p_1 & q_1 & r_1 \\ p_2 & q_2 & r_2 \\ p_3 & q_3 & r_3 \end{matrix} \qquad \text{or} \qquad p_1q_1r_1/p_2q_2r_2/p_3q_3r_3$$

the latter representation being most commonly used in print for obvious typographical reasons. Such a *transformation matrix* simply states the coefficients in the three equations that give $\mathbf{a}'$, $\mathbf{b}'$, $\mathbf{c}'$ in terms of $\mathbf{a}$, $\mathbf{b}$, $\mathbf{c}$.

In an exactly similar manner the 'old' unit-cell vectors can be expressed in terms of the 'new' unit-cell vectors, the transformation matrix being

$$\begin{matrix} P_1 & Q_1 & R_1 \\ P_2 & Q_2 & R_2 \\ P_3 & Q_3 & R_3 \end{matrix}$$

so that $\mathbf{a} = P_1\mathbf{a}'+Q_1\mathbf{b}'+R_1\mathbf{c}'$ and so on.

Transformation of coordinates

Let us suppose that the coordinates of an atom in the 'old' unit-cell with $\mathbf{a}$, $\mathbf{b}$, $\mathbf{c}$, are known and that it is desired to find the coordinates of that atom in the 'new' unit-cell with $\mathbf{a}'$, $\mathbf{b}'$, $\mathbf{c}'$. If the coordinates of the atom in the 'old' unit-cell are x, y, z then the vector from the origin to x, y, z, is

$$\begin{aligned} & x\mathbf{a}+y\mathbf{b}+z\mathbf{c} \\ & \quad = x(P_1\mathbf{a}'+Q_1\mathbf{b}'+R_1\mathbf{c}')+y(P_2\mathbf{a}'+Q_2\mathbf{b}'+R_2\mathbf{c}')+z(P_3\mathbf{a}'+Q_3\mathbf{b}'+R_3\mathbf{c}') \\ & \quad = (xP_1+yP_2+zP_3)\mathbf{a}'+(xQ_1+yQ_2+zQ_3)\mathbf{b}'+(xR_1+yR_2+zR_3)\mathbf{c}' \end{aligned}$$

and the vector from the origin to x', y', z' is

$$x'\mathbf{a}'+y'\mathbf{b}'+z'\mathbf{c}'$$

Therefore

$$x' = P_1x+P_2y+P_3z$$
$$y' = Q_1x+Q_2y+Q_3z$$
$$z' = R_1x+R_2y+R_3z$$

The matrix for the transformation from 'old' to 'new' coordinates is thus

$$\begin{matrix} P_1 & P_2 & P_3 \\ Q_1 & Q_2 & Q_3 \\ R_1 & R_2 & R_3 \end{matrix}$$

which is related by interchange of rows and columns to the matrix for the *reverse* transformation (i.e. from 'new' to 'old') for axes

$$\begin{matrix} P_1 & Q_1 & R_1 \\ P_2 & Q_2 & R_2 \\ P_3 & Q_3 & R_3 \end{matrix}$$

derived previously.

Since the zone axis symbol $[UVW]$ is a statement of the coordinates of a point on the zone axis, zone axes transform as coordinates. The matrix for the transformation from $[UVW]$ in terms of **a**, **b**, **c** to $[U'V'W']$ in terms of **a**′, **b**′, **c**′ is thus

$$\begin{matrix} P_1 & P_2 & P_3 \\ Q_1 & Q_2 & Q_3 \\ R_1 & R_2 & R_3 \end{matrix}$$

Transformation of Miller indices

Since the Miller symbol (*hkl*) represents the reciprocal ratio of the intercepts made by a plane on the three reference axes a different matrix is required for the transformation of face indices. A set of planes (*hkl*) can be regarded as dividing the vectors **a** into h parts, **b** into k parts, and **c** into l parts; the general vector $p_1\mathbf{a}+q_1\mathbf{b}+r_1\mathbf{c}$ is therefore divided into $p_1h+q_1k+r_1l$ parts. That this is so is apparent from the two-dimensional case illustrated in Fig 5.15 where a set of lattice lines divides **a** into two parts and **b** into three parts (i.e. $h = 2$, $k = 3$ and the lines have indices (23)); the line $\mathbf{OM} = 2\mathbf{a}+\mathbf{b}$ (i.e. $p_1 = 2$, $q_1 = 1$) is seen to be divided into $2\times2+1\times3 = 7$ parts and the line $\mathbf{ON} = -\mathbf{a}+\mathbf{b}$ (i.e. $p_1 = -1$, $q_1 = 1$) into $-2+3 = 1$ part. Thus if new axes x', y' are taken respectively parallel to **OM** and **ON** the lattice lines, (23) on the old axes, become (71) when referred to these new axes.

It is evident also from Fig 5.15 that whatever path is taken in passing from the origin O to a lattice point M the same number of lattice lines will be crossed, if all lines that are crossed twice in the course of the path are discounted. This provides the basis for the statement of the transformation matrix for the Miller indices ($h'k'l'$) referred to the 'new' unit-cell (defined by **a**′, **b**′, **c**′) from the indices (*hkl*) of the same set of planes referred to the 'old' unit-cell (defined by **a**, **b**, **c**). We have already shown that the transformation matrix for axes from **a**, **b**, **c**, to **a**′, **b**′, **c**′, is

$$\begin{matrix} p_1 & q_1 & r_1 \\ p_2 & q_2 & r_2 \\ p_3 & q_3 & r_3 \end{matrix}$$

whence $\mathbf{a}' = p_1\mathbf{a}+q_1\mathbf{b}+r_1\mathbf{c}$.

The planes ($h'k'l'$) divide the vector **a**′ into h' parts and the same set of planes (*hkl*), referred now to the 'old' unit-cell, divide the vectors **a**, **b**, and **c**, into h, k, and l parts respectively. Since the number of crossings is independent of the path, it follows that

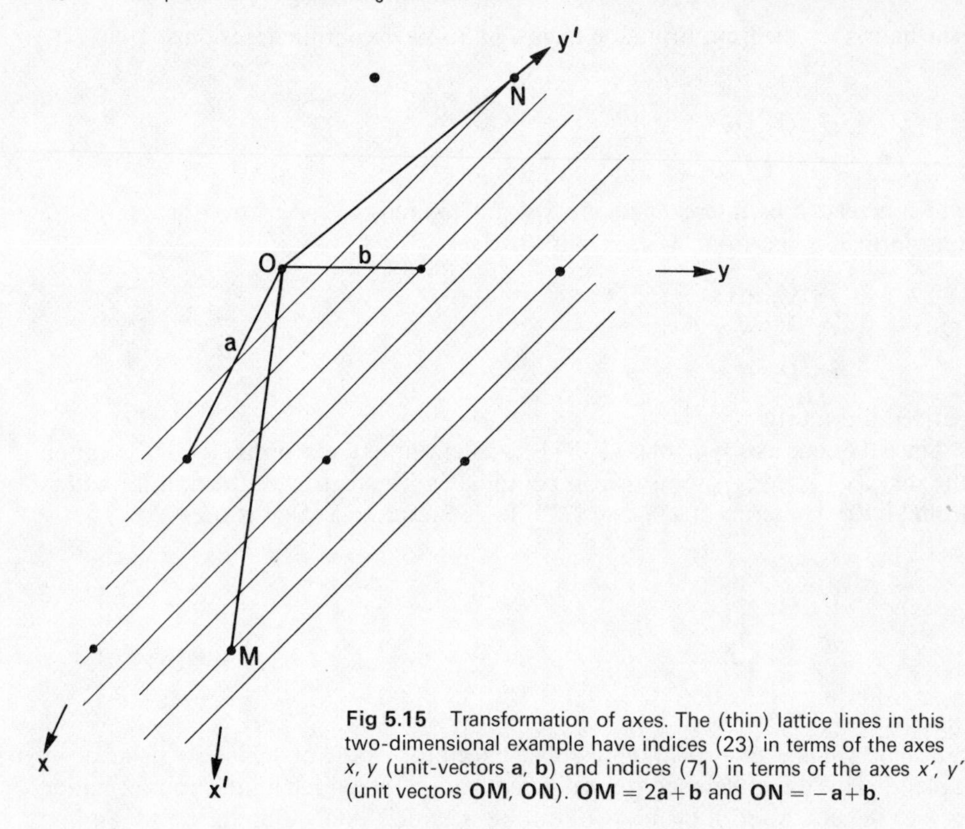

Fig 5.15 Transformation of axes. The (thin) lattice lines in this two-dimensional example have indices (23) in terms of the axes x, y (unit-vectors **a**, **b**) and indices (71) in terms of the axes x', y' (unit vectors **OM**, **ON**). **OM** = 2**a**+**b** and **ON** = −**a**+**b**.

$$h' = p_1 h + q_1 k + r_1 l.$$

And similarly by consideration of **b**′ and **c**′,

$$k' = p_2 h + q_2 k + r_2 l$$
$$l' = p_3 h + q_3 k + r_3 l.$$

The transformation matrix for Miller indices is thus

$$\begin{matrix} p_1 & q_1 & r_1 \\ p_2 & q_2 & r_2 \\ p_3 & q_3 & r_3 \end{matrix}$$

identical with the matrix for the transformation of unit-cell vectors.

We shall not attempt to explore the relationship between the matrix

$$\begin{matrix} p_1 & q_1 & r_1 \\ p_2 & q_2 & r_2 \\ p_3 & q_3 & r_3 \end{matrix}$$

for the transformation of unit-cell vectors from **a**, **b**, **c**, to **a**′, **b**′, **c**′ and the matrix for the reverse transformation from **a**′, **b**′, **c**′ to **a**, **b**, **c**. This matter is treated thoroughly by Buerger (1942, p. 22).

The matrices applicable to the various types of transformation are shown in Table 5.1. We conclude with a word of advice: after setting up a transformation matrix

Table 5.1
Relationships between transformation matrices

$p_1q_1r_1$ $p_2q_2r_2$ $p_3q_3r_3$ from old to new axes from old to new Miller indices	$P_1Q_1R_1$ $P_2Q_2R_2$ $P_3Q_3R_3$ from new to old axes from new to old Miller indices
$P_1P_2P_3$ $Q_1Q_2Q_3$ $R_1R_2R_3$ from old to new coordinates from old to new zone axes	$p_1p_2p_3$ $q_1q_2q_3$ $r_1r_2r_3$ from new to old coordinates from new to old zone axes

and applying it to a given set of Miller indices, set up the matrix for the reverse transformation and transform back to check.

Example Relationships between three forms of $BaTiO_3$
$BaTiO_3$ exists in four structural modifications: below −80 °C it is trigonal, from −80° to 5 °C orthorhombic, from 5° to 120 °C tetragonal, and above 120 °C cubic. We first consider the relationship between the cubic and orthorhombic forms. The cubic modification has space group P*m*3*m*, unit-cell $a \sim 4$ Å, and one formula unit per unit-cell; its structure is identical with ideal *perovskite* and is illustrated in Fig 5.16(a). The orthorhombic modification has space group B*mm*2, unit-cell dimensions, $a = 5{\cdot}656$ Å, $b = 3{\cdot}986$ Å, $c = 5{\cdot}675$ Å, and two formula units per unit-cell. That there is a simple approximate dimensional relationship between the cubic and orthorhombic unit-cells is evident from Fig 5.16(b); this relationship is expressed by the axial transformation matrices:

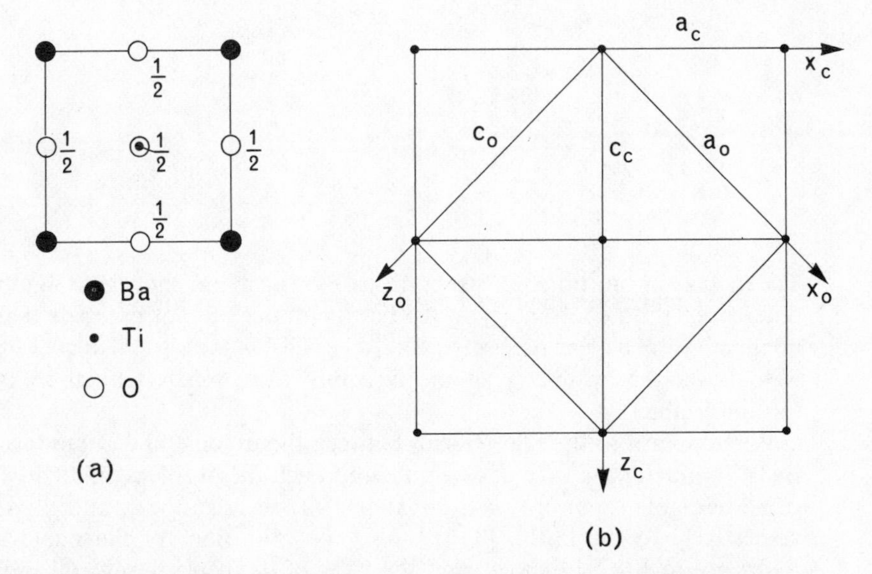

Fig 5.16 The $BaTiO_3$ structures. The plan (a) on (001) shows the *cubic* form of $BaTiO_3$, which has the ideal *perovskite* structure. The plan (b) shows the dimensional relationship between the unit-cells of the orthorhombic (a_o, b_o, c_o) and cubic (a_c, b_c, c_c) forms of $BaTiO_3$.

(I) from cubic to orthorhombic

$$\begin{matrix} 1 & 0 & 1 \\ 0 & 1 & 0 \\ \bar{1} & 0 & 1 \end{matrix}$$

(II) from orthorhombic to cubic

$$\begin{matrix} \frac{1}{2} & 0 & \frac{\bar{1}}{2} \\ 0 & 1 & 0 \\ \frac{1}{2} & 0 & \frac{1}{2} \end{matrix}$$

The appearance of fractions in matrix II is because it refers to a transformation from a non-primitive to a primitive unit-cell. To determine the coordinates of the atoms in the orthorhombic B-cell our first step is to operate the appropriate matrix for transformation of coordinates on the coordinates of the atoms in one cubic unit-cell. Reference to Table 5.1 shows that the matrix for this purpose is matrix II above taken in columns, i.e. from cubic to orthorhombic coordinates

$$\begin{matrix} \frac{1}{2} & 0 & \frac{1}{2} \\ 0 & 1 & 0 \\ \frac{\bar{1}}{2} & 0 & \frac{1}{2} \end{matrix}$$

Our second step is to apply the B-centring translation $\frac{1}{2}$, 0, $\frac{1}{2}$, to obtain the coordinates of the other half of the atoms in the orthorhombic unit-cell. Thus Ti, which lies at $\frac{1}{2}, \frac{1}{2}, \frac{1}{2}$ in the cubic unit-cell, has coordinates $(\frac{1}{2}.\frac{1}{2}+0.\frac{1}{2}+\frac{1}{2}.\frac{1}{2})$, $(0.\frac{1}{2}+1.\frac{1}{2}+0.\frac{1}{2})$, $(-\frac{1}{2}.\frac{1}{2}+0.\frac{1}{2}+\frac{1}{2}.\frac{1}{2})$, i.e. $\frac{1}{2}, \frac{1}{2}, 0$ and $(\frac{1}{2}+\frac{1}{2})$, $(\frac{1}{2}+0)$, $(0+\frac{1}{2})$, i.e. $0, \frac{1}{2}, \frac{1}{2}$ in the orthorhombic unit-cell. Table 5.2 lists the coordinates of all the atoms in the orthorhombic unit-cell derived in this way. These coordinates obviously refer to the cubic structure described on the orthorhombic unit-cell.

Table 5.2
Derivation of atomic coordinates in orthorhombic $BaTiO_3$ from those in cubic $BaTiO_3$

	Cubic	Orthorhombic	
		x, y, z	$\frac{1}{2}+x, y, \frac{1}{2}+z$
Ba	0, 0, 0	0, 0, 0	$\frac{1}{2}, 0, \frac{1}{2}$
Ti	$\frac{1}{2}, \frac{1}{2}, \frac{1}{2}$	$\frac{1}{2}, \frac{1}{2}, 0$	$0, \frac{1}{2}, \frac{1}{2}$
O_1	$\frac{1}{2}, \frac{1}{2}, 0$	$\frac{1}{4}, \frac{1}{2}, \frac{3}{4}$	$\frac{3}{4}, \frac{1}{2}, \frac{1}{4}$
O_2	$\frac{1}{2}, 0, \frac{1}{2}$	$\frac{1}{2}, 0, 0$	$0, 0, \frac{1}{2}$
O_3	$0, \frac{1}{2}, \frac{1}{2}$	$\frac{1}{4}, \frac{1}{2}, \frac{1}{4}$	$\frac{3}{4}, \frac{1}{2}, \frac{3}{4}$

The actual orthorhombic structure is very similar, but has significant differences: the z coordinates of the barium atoms are not precisely 0 and $\frac{1}{2}$, nor are those of oxygen precisely $\frac{1}{4}$ and $\frac{3}{4}$; the effect of these small displacements is to lower the symmetry of the structure from P*m*3*m* (cubic) to B*mm*2 (orthorhombic).

We now turn to the relationship between the trigonal and cubic forms of $BaTiO_3$ and, for the sake of diversity, approach the problem differently. The triple hexagonal unit-cell of trigonal $BaTiO_3$ has axes x, y, and z parallel respectively to the $[\bar{1}01]$, $[1\bar{1}0]$, and $[111]$ directions of the cubic form. Reference to Fig 5.17 shows that the faces of the triple hexagonal unit-cell will be parallel to the cubic planes $(\bar{1}\bar{1}2)$, $(1\bar{2}1)$, (111), and their opposites. Reference to Table 5.1 indicates that these cubic planes can be re-indexed in

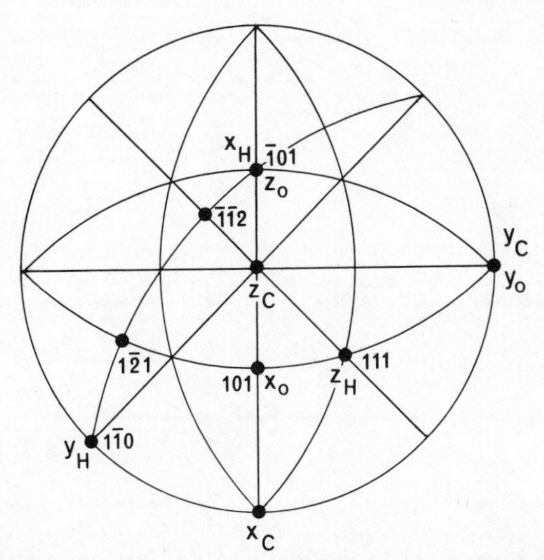

Fig 5.17 The stereogram shows the angular relationship between the axes of the triple hexagonal (x_H, y_H, z_H), orthorhombic (x_o, y_o, z_o) and cubic (x_c, y_c, z_c) unit-cells of three $BaTiO_3$ polymorphs. Indices shown refer to the cubic unit-cell.

terms of the orthorhombic unit-cell by applying the cubic→orthorhombic axial transformation matrix (I) directly, thus $(\bar{1}\bar{1}2)$ on cubic axes becomes $((-1.1-0.1+1.2), (-0.1-1.1+0.2), (+1.1-0.1+1.2))$, i.e. $(1\bar{1}3)$ on orthorhombic axes; $(1\bar{2}1)_c$ becomes $(2\bar{2}0)_o$ and $(111)_c$ becomes $(210)_o$. That we obtain indices that are not prime to one another in the case of $(2\bar{2}0)_o$ arises simply because the first plane out from the origin that makes intercepts a, $-\frac{1}{2}b$, c on the cubic axes makes intercepts $\frac{1}{2}a$ and $\frac{1}{2}b$ on the x and y axes of the larger orthorhombic unit-cell (Fig 5.16). Conversely the $(1\bar{1}0)$ plane of the orthorhombic lattice transforms, by direct application of the corresponding axial transformation matrix (II), to the $(\frac{1}{2}\bar{1}\frac{1}{2})$ plane of the cubic lattice. We leave it to the reader to show that the axes x_h, y_h, z_h of the triple hexagonal unit-cell correspond to the directions [001], $[\frac{1}{2}\bar{1}\frac{\bar{1}}{2}]$, [110] on the orthorhombic lattice by application of matrix II taken in columns to the cubic zone axes $[\bar{1}01]$, $[1\bar{1}0]$, [111]. Once again we get fractional coordinates, for the equivalent of y_h, because the orthorhombic unit-cell is non-primitive; it would conventionally be correct to write for the orthorhombic equivalent of y_h $[1\bar{2}\bar{1}]_o$. The reader may care to confirm the transformation of the axes of the triple hexagonal unit-cell to orthorhombic zone axes by transforming the indices of the faces of the triple hexagonal unit-cell from cubic to orthorhombic axes and then indexing the zone axes of pairs of faces.

6
Diffraction of X-rays by crystals

X-radiation is the name given to that part of the electromagnetic spectrum in the wavelength range 0·1 to 500 Å, sandwiched between γ-radiation at shorter wavelengths and ultraviolet radiation on the high wavelength side. We are not concerned immediately either with the general theory of electromagnetic radiation, which crops up in a different context and another wavelength range in chapter 12 where references to the general theory are given, or with details of the generation of X-rays (chapter 7). We assume that a source of X-rays, which for crystallographic purposes is usually restricted to the wavelength range 0·5–2·5 Å, is available and proceed to consider the nature of their interaction with matter in general and with crystalline solids in particular.

Interaction of X-rays with matter

When an X-ray beam passes through a material medium its intensity is reduced by the operation of a variety of effects which may be grouped under two general headings, *absorption* and *scattering*.

The *absorption* of X-ray photons by an atom leads among other effects to the ejection of electrons from the inner shells (K, L, or M) of the atom and consequent 'falling in' of electrons from lower energy levels to fill the vacancies so created. Such electronic transitions are accompanied by emission of X-rays of definite wavelength which is determined by the difference in energy between the initial and final state of the electron filling the vacancy in an inner shell. Such *fluorescent* X-rays may be reabsorbed by another atom to produce ejection of electrons from shells of lower energy, followed by 'falling in' of electrons from shells of even lower energy and emission of fluorescent X-rays of lower energy and therefore of longer wavelength. Absorption phenomena are utilized in various ways, including X-ray fluorescence analysis to which reference is made in chapter 15.

The scattering of X-rays by an atom may occur in either of two ways, both of which again involve interaction between X-radiation and extranuclear electrons. An X-ray photon passing close to an electron belonging to one of the constituent atoms of the material medium will be deflected by the electromagnetic field of the electron and will impart some of its energy to the electron as kinetic energy. The energy of the deflected X-ray photon will be correspondingly decreased and its wavelength increased. This is *incoherent* scattering which is not our immediate concern.

The second way in which X-rays are scattered by atoms is best considered by treating the incident X-ray beam as a plane wave-front. As the plane wave-front passes through an extranuclear electron belonging to an atom of the material medium it causes the electron to vibrate. The vibrating electron radiates X-rays of the same frequency as the incident beam. Such vibrating electrons act as secondary sources of X-radiation of fixed wavelength and give rise to interference effects. The interference phenomena associated with such *coherent* scattering are the basis of X-ray crystallography and will concern us in this and succeeding chapters.

That X-rays in the wavelength range 0·5–2·5 Å incident on crystalline solids give rise to observable diffraction patterns is because the distances between adjacent atoms in crystalline solids are on the same scale and because the X-rays are scattered coherently from the extranuclear electrons of the constituent atoms of the solid substance. Simultaneously incoherent scattering, that is scattering with change of wavelength, occurs and that contributes to the background of the diffraction pattern. The study of the diffraction patterns produced by X-rays incident on crystalline materials, in particular single crystals, has made possible the determination of the size and shape of the relevant unit-cell and the coordinates of the atoms within the unit-cell. X-ray diffraction studies have proved to be the most powerful tool for the study of the internal structure of crystalline solids. The remainder of this chapter will be devoted to an elementary treatment of the diffraction of X-rays by single crystals.

Simplifying assumptions

The refractive index of most substances is less than unity by a very small amount, of the order of 10^{-6}, so that refraction of incident and scattered X-radiation at air/crystal interfaces can be neglected except when extremely precise measurement of unit-cell dimensions is required.

Absorption by the crystal of incident and scattered radiation affects the intensities and the directions of the scattered beams. Thermal vibration of the constituent atoms of the crystal will also modify the intensity of the scattered X-radiation. Corrections for both effects are necessary in structure determination and can quite simply be applied. Correction for absorption is necessary when the angular disposition of the scattered beams is to be measured very accurately. In the elementary treatment given in this chapter, it will be assumed that absorption is negligible and that all the constituent atoms of the crystal are at rest.

A scattered X-ray beam may, in the course of its travel through the crystal, be scattered a second time, but only in certain special circumstances is the resultant beam of appreciable intensity.

In general scattered X-ray beams have a phase difference π relative to the incident beam that generated them. Only when the wavelength of the X-radiation is close to an absorption edge (chapter 11) of one of the constituent elements of the crystalline substance is this not so. It will be assumed in this elementary treatment that scattering always introduces a phase change π and, since the phase change is the same for all scattered beams it can be ignored. The origin of the phase change on scattering is discussed later in this chapter.

Finally, it may be noticed that the effects of coherent scattering of X-rays by a crystal are going to be observed at distances from the crystal that are very large compared with X-ray wavelengths; that is to say we are dealing with an example of Fraunhofer, as distinct from Fresnel, diffraction.

Combination of X-rays

Consider a narrow pencil of monochromatic X-rays of wavelength λ, angular frequency ω, and amplitude a travelling in the x direction. A disturbance will be produced at a point x such that[1]

$$\psi = a\cos(\omega t + \phi) \tag{1}$$

where ψ is the displacement in the plane normal to the direction of propagation at time t, and ϕ is the phase of the wave when $t = 0$. It is evident from Fig 6.1 that a wave with phase ϕ is in *advance* of a wave with $\phi = 0$.

Suppose now that such a beam of monochromatic X-rays is divided into two beams, one with zero phase (this merely requires a suitable choice of the zero of the time scale) and the other with phase ϕ. It follows that the second beam must have travelled a shorter distance than the first by an amount equal to $\lambda\phi/2\pi$.

When a number, N, of such waves arrives simultaneously at a given point, the resultant wave is given by the *principle of superposition* as the sum of the individual waves. Thus the resultant ψ_R of the waves $\psi_1, \psi_2, \ldots \psi_{N-1}, \psi_N$ arriving simultaneously at the point is given, if the waves all have the same frequency, ω, by

$$\begin{aligned}
\psi_R &= \psi_1 + \psi_2 + \ldots + \psi_{N-1} + \psi_N \\
&= \sum_1^N \psi_n \\
&= \sum_1^N a_n \cos(\omega t + \phi_n) \\
&= \sum_1^N \{a_n \cos\omega t \cos\phi_n - a_n \sin\omega t \sin\phi_n\} \\
&= \cos\omega t \sum_1^N a_n \cos\phi_n - \sin\omega t \sum_1^N a_n \sin\phi_n
\end{aligned} \tag{2}$$

But the resultant wave motion is described by

$$\begin{aligned}
\psi_R &= a_R \cos(\omega t + \phi_R) \\
&= a_R \cos\omega t \cos\phi_R - a_R \sin\omega t \sin\phi_R
\end{aligned} \tag{3}$$

Therefore $\quad a_R \cos\phi_R = \sum_1^N a_n \cos\phi_n$

and $\quad a_R \sin\phi_R = \sum_1^N a_n \sin\phi_n.$

But, since $\cos^2\phi_R + \sin^2\phi_R = 1$

$$a_R^2 = \left(\sum_1^N a_n \cos\phi_n\right)^2 + \left(\sum_1^N a_n \sin\phi_n\right)^2 \tag{4}$$

and

$$\tan\phi_R = \frac{\sum_1^N a_n \sin\phi_n}{\sum_1^N a_n \cos\phi_n}. \tag{5}$$

[1] The equation to a simple harmonic wave-motion can be written in various ways: $\psi = a\cos(\omega t + \phi)$ is preferred in X-ray crystallography, $y = A\sin(\omega t + \alpha)$ is one of several forms of the equation in common use in optics. It will be seen later in this chapter that $y = a\cos(\omega t + \phi)$ leads to the most convenient form of the expression for the structure factor.

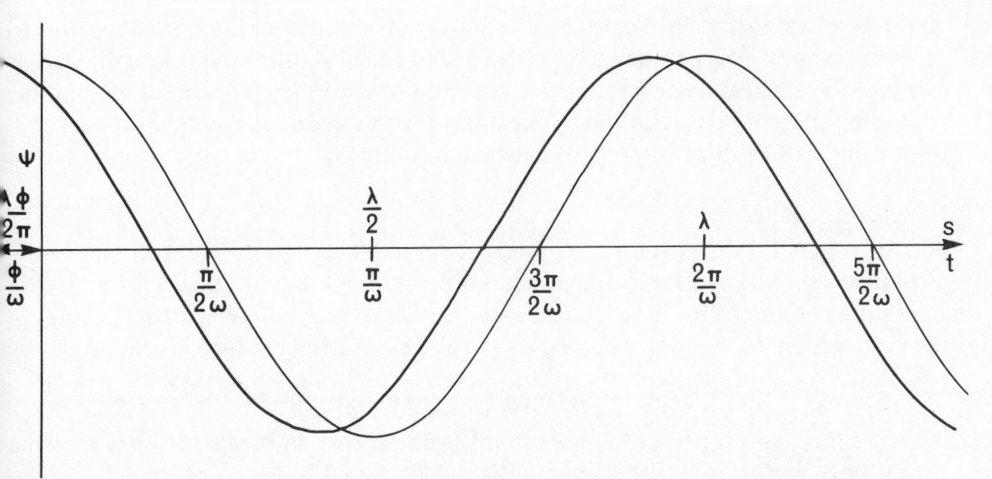

Fig 6.1 Two sinusoidal wave motions with the same frequency and velocity. Displacement ψ in the plane normal to the direction of propagation is plotted vertically while distance traversed s and time taken t are plotted horizontally. The wave motion represented by the thick line is in advance of that represented by the thin line by the phase difference ϕ, which corresponds to a time difference of ϕ/ω and a difference in distance traversed of $\lambda\phi/2\pi$.

Now, since there is no lens capable of refracting X-rays it is impossible to form an image of the crystal structure with the X-rays scattered by it and all that can be done in practice is to make measurements of the intensities and angular disposition of the scattered X-ray beams. Equation (4) provides an expression[2] for the resultant intensity I_R of N waves:

$$I_R = a_R^2 = \left(\sum_1^N a_n \cos \phi_n\right)^2 + \left(\sum_1^N a_n \sin \phi_n\right)^2 \tag{6}$$

To calculate the intensity of the resultant X-ray beam we thus need to know the amplitude and phase of the X-rays scattered by every electron in the crystal in the given direction.

Our ultimate goal is the calculation of the diffraction pattern produced by the interaction of an incident monochromatic X-ray beam with a single crystal and this is achieved by summing over all directions the waves scattered by every electron in the crystal. The summation is taken in steps, the first step being to consider the interaction of the X-ray beam with a single electron. The second step is to consider the scattering produced by all the electrons associated with each atomic species present in the crystal. The third step is to consider the scattering produced by all the atoms in a single unit-cell, having regard to their differing nature and to their spatial distribution within the unit-cell. The fourth and final step is to sum the scattering effect of one unit-cell over all the unit-cells in the crystal. In the course of this stepwise argument it is necessary to assume as a first approximation that each electron in the crystal can be assigned to a particular atom. This is not strictly true because valency electrons may be involved in bond formation and in consequence be shared between atoms; but the majority of electrons in the crystal will be core electrons, each belonging unambiguously to a particular atom, so that the assumption is valid as a first approximation at least.

The procedure outlined above is strictly logical but the steps are not progressive

[2] This form for the intensity expression is preferred at this stage because it gives emphasis to the way in which the amplitude of the scattered beam may be calculated. For the more elegant form of the expression in complex number rotation see later in this chapter.

in order of difficulty. In particular the final step is relatively easy, fundamental to crystallography, and immediately productive of useful results even before the earlier steps in the total summation have been taken. We shall therefore choose to take the final step first and consider the implications for the diffraction pattern of a crystal of a regular three-dimensional arrangement of unit-cells.

Laue Equations

Since a crystal is a regular three-dimensional arrangement of unit-cells it can be regarded as acting as a three-dimensional diffraction grating for X-rays. The effect of a grating is to limit the directions in which an observable diffracted beam occurs; in the diffraction of X-rays by crystals it will be shown that the directions of the diffracted X-ray beams depend on the dimensions of the unit-cell and their intensities on the nature and disposition of atoms within the unit-cell.

It is a familiar property of diffraction by gratings that diffracted intensity maxima occur only in directions for which the waves scattered by corresponding points in each grating element are in phase; in all other directions the scattered waves interfere destructively more or less. In other words the path difference in directions of intensity maxima between the waves scattered by corresponding points in different elements of the grating is, for all elements, an integral number of wavelengths (Fig 6.2).[3] By analogy with the corresponding points in the grating elements we can take corresponding points, one in each unit-cell of the crystal and so obtain an array of

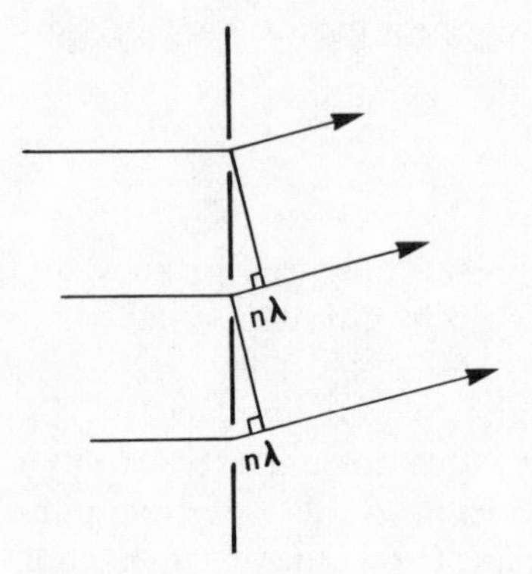

Fig 6.2 The path difference between waves scattered by corresponding points in each element of a diffraction grating is an integral number of wavelengths.

lattice points. Therefore if we are interested only in the angular disposition of intensity maxima of the X-radiation diffracted by a crystal, it is valid to make the assumption that the scattering is produced by the array of lattice points. This is a highly restrictive model that can give no information about the relative intensities of the local intensity maxima but it is adequate for our immediate purpose. Diffracted beams will occur in directions for which X-rays scattered by all lattice points are in phase and, because the lattice is a regular three-dimensional array of lattice points, this condition is satisfied if the X-rays scattered by pairs of adjacent lattice points lying on three non-coplanar rows are in phase.

We suppose a parallel beam of X-rays of wavelength λ to be incident on a row of lattice points of spacing t at an angle of incidence i (Fig 6.3) and consider a direction

[3] For thorough accounts of optical diffraction gratings see Jenkins and White (1957) and Longhurst (1957).

of scattering at an angle δ to the lattice row.[4] The path difference for X-rays scattered by adjacent lattice points is then, referring to Fig 6.3,

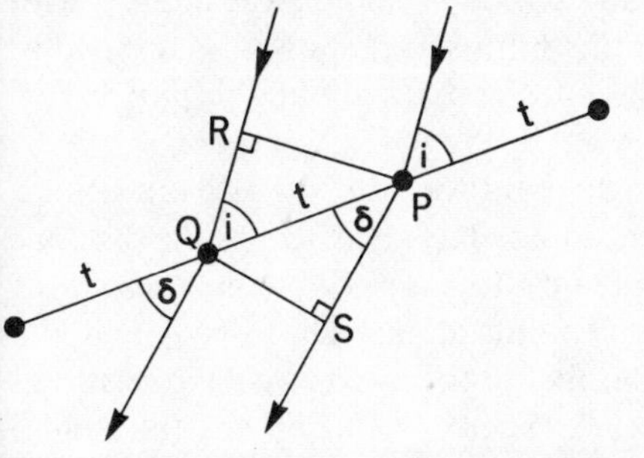

Fig 6.3 A parallel beam of X-rays of wavelength λ, represented by the wave front PR, is incident at an angle of incidence i on a row of lattice points, P, Q, . . . , of separation t. The condition for the X-rays scattered by adjacent lattice points to be in phase is $PS-RQ=n\lambda$, where n is an integer.

$$\begin{aligned} PS-RQ &= PQ(\cos\delta-\cos i) \\ &= t(\cos\delta-\cos i). \end{aligned}$$

For a diffracted beam to occur the X-rays scattered by adjacent lattice points have to be in phase, that is to say their path difference must be an integral number of wavelengths. Therefore the condition for diffraction by a lattice point row is

$$t(\cos\delta-\cos i) = n\lambda,$$

where n is an integer. Figure 6.4 illustrates the result for four small values of n.

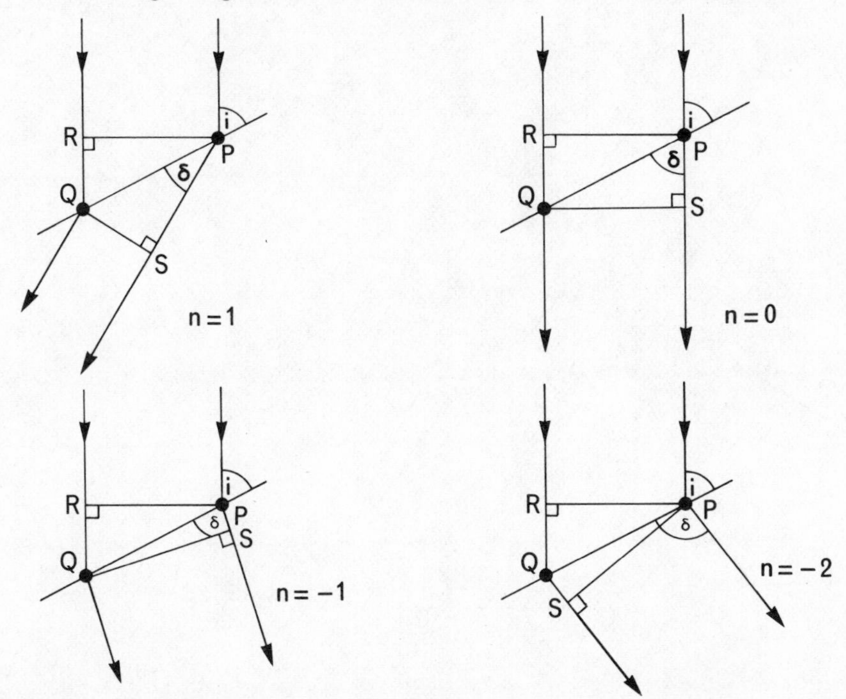

Fig 6.4 The condition for an intensity maximum in the X-radiation scattered by a lattice row of which P and Q are two adjacent lattice points. The condition $PS-RQ=n\lambda$, i.e., $t(\cos\delta-\cos i)=n\lambda$, is illustrated for $n=1, 0, -1, -2$.

[4] It is usual to measure i and δ with respect to opposite directions of the lattice row. We shall adopt the convention of taking i as the angle between the incident beam and the *negative* direction of the row and δ as the angle between the diffracted beam and the *positive* direction of the row.

Permissible directions of the diffracted beam are of course not confined to the plane defined by the incident beam and the lattice point row. The condition for diffraction, $t(\cos\delta - \cos i) = n\lambda$, is satisfied by any direction making the angle δ with the lattice row. Therefore for a given value of n the diffracted radiation is confined to the surface of a cone of semiangle δ; a set of cones coaxial about the lattice row represents solutions of the diffraction condition for $n = 0, \pm 1, \pm 2$, etc (Fig 6.5).

So far we have considered diffraction by a single point row, but a lattice is a regular three-dimensional array of points and as such is completely specified by the distance apart of adjacent lattice points in three non-coplanar directions. We label these three axes x, y, and z and take the separation of lattice points along each to be a, b, and c respectively. For a point row of separation a parallel to the x-axis we have seen that the diffraction condition is $a(\cos\delta_a - \cos i_a) = h\lambda$, where i_a and δ_a are the angles between the x-axis and the incident and diffracted beams respectively and h is an integer. If such a set of point rows is repeated successively with translation b parallel to the y-axis, each point row becomes a grating element and the diffraction condition

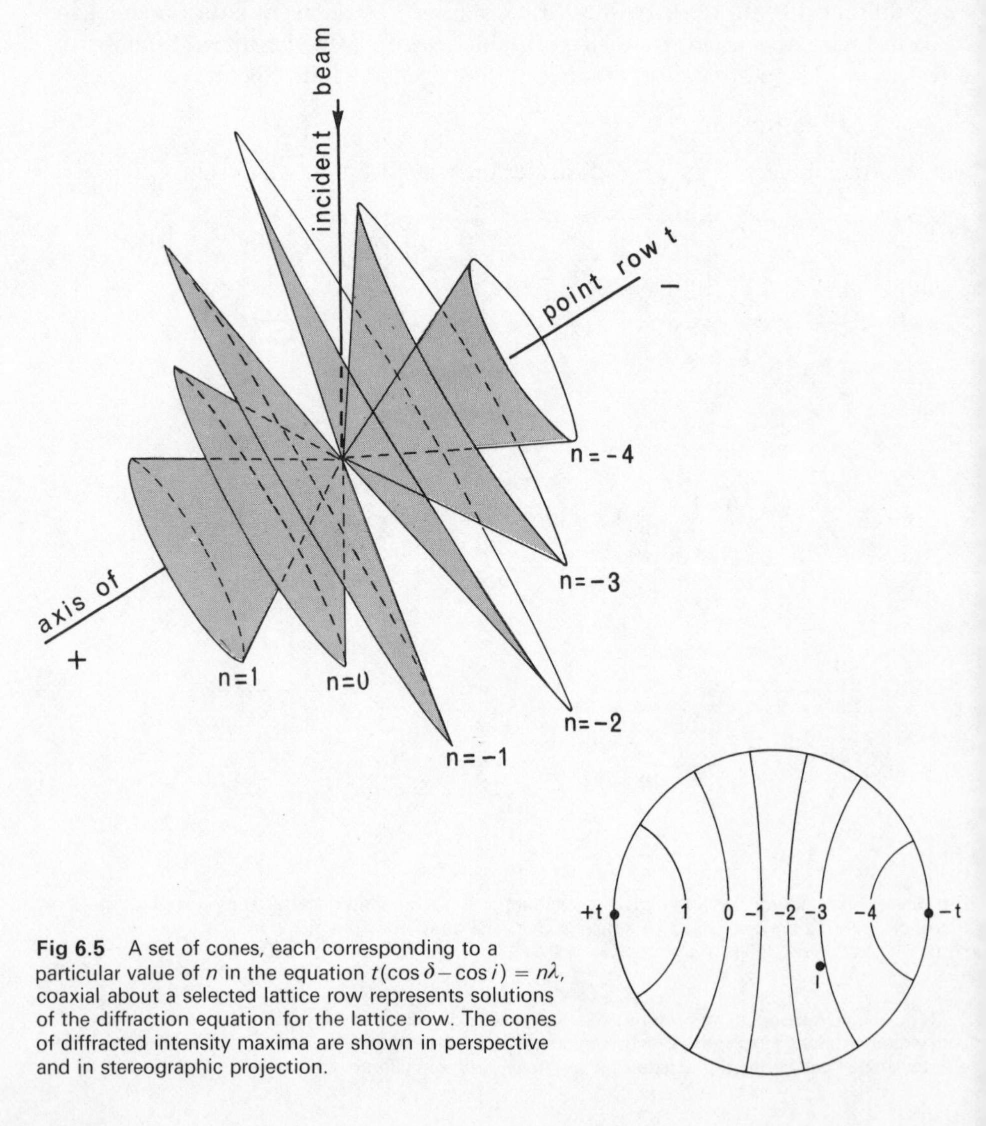

Fig 6.5 A set of cones, each corresponding to a particular value of n in the equation $t(\cos\delta - \cos i) = n\lambda$, coaxial about a selected lattice row represents solutions of the diffraction equation for the lattice row. The cones of diffracted intensity maxima are shown in perspective and in stereographic projection.

for this grating is $b(\cos\delta_b - \cos i_b) = k\lambda$, where k is an integer. The condition for the production of an observable diffracted beam by all the lattice points in the xy plane is then $a(\cos\delta_a - \cos i_a) = h\lambda$ and $b(\cos\delta_b - \cos i_b) = k\lambda$ simultaneously. If such a set of lattice planes is repeated successively with translation c parallel to the z-axis, each plane constitutes an element of a third linear diffraction grating for which the diffraction condition is $c(\cos\delta_c - \cos i_c) = l\lambda$, where l is an integer. The simultaneous operation of all three conditions is necessary for the production of an observable diffracted beam by all the lattice points of the three-dimensional lattice. This statement implies that if X-rays scattered by any pair of adjacent lattice points in a three-dimensional lattice are to be in phase, then the X-rays scattered by adjacent lattice points along each of the reference axes must be in phase; this constitutes a necessary and sufficient condition for diffraction by a three-dimensional array of lattice points.

The diffraction condition for a three-dimensional lattice can thus be written as:

$$\left.\begin{aligned} a(\cos\delta_a - \cos i_a) &= h\lambda \\ b(\cos\delta_b - \cos i_b) &= k\lambda \\ c(\cos\delta_c - \cos i_c) &= l\lambda \end{aligned}\right\} \qquad (7)$$

where the incident beam is inclined at angles i_a, i_b, i_c and the diffracted beam at angles δ_a, δ_b, δ_c with the x, y, z axes respectively, λ is the X-ray wavelength, and h, k, l are integers. These three equations (7) are known as the *Laue Equations* after Max von Laue, who in 1912, suggested that a crystal should act as a diffraction grating for X-rays.

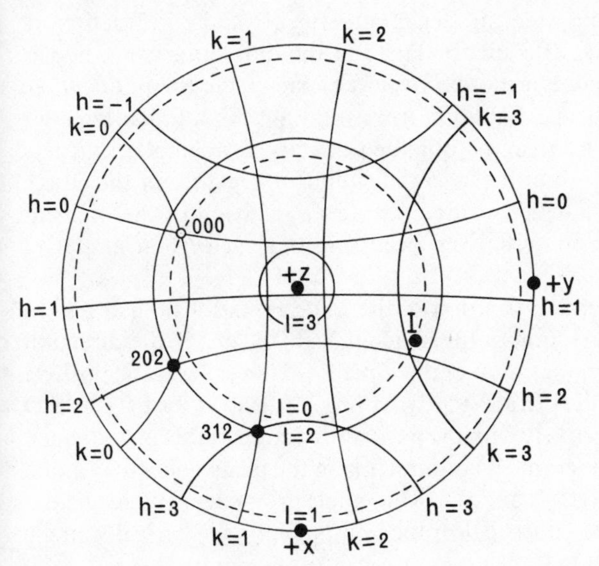

Fig 6.6 Solution of the Laue equations for a three-dimensional orthorhombic lattice for $h = k = l = 0$ (trivial), for $h = 2$, $k = 0$, $l = 2$ and for $h = 3$, $k = 1$, $l = 2$. The direction of the incident X-ray beam is represented in the stereographic projection by the pole I.

The first Laue Equation, $a(\cos\delta_a - \cos i_a) = h\lambda$, restricts the directions of observable diffracted beams to the surfaces of a set of cones coaxial about the x-axis and having semiangles δ_a consistent with the equation. This statement is illustrated in Fig 6.6 where, for a given direction I of the incident beam, the restriction placed by the first Laue Equation on the directions of diffracted beams is represented by a set of small circles of radius δ_a centred on the x-axis. The second Laue Equation, $b(\cos\delta_b - \cos i_b) = k\lambda$, further constrains the directions of observable diffracted beams to a set of small circles of radius δ_b centred on the y-axis. The simultaneous operation of these two Laue Equations thus restricts observable diffracted beams produced by X-rays incident in a particular direction to the intersections of pairs of small circles, the attitude of one small circle being dependent on h and of the other on k. Each pair of h, k values leads in general to two common directions. The third Laue Equation, $c(\cos \delta_c - \cos i_c) = l\lambda$, provides an additional restriction; for an observable diffracted beam to occur, a small circle of the set of radius δ_c centred on the z-axis must pass through the intersection of small circles of the x and y sets. An observable diffracted beam thus lies in a direction common to three cones, each coaxial with one of the reference axes and of semiangle consistent with the appropriate Laue Equation; such a direction is completely specified for a given lattice and for X-rays of given λ by the integers h, k, l. The orientation of the X-ray beam shown in Fig 6.6 leads to intersection of three small circles only for $h = k = l = 0$, for $h = 2$, $k = 0$, $l = 2$, and for $h = 3$, $k = 1$, $l = 2$. The first of these solutions of the Laue Equations is trivial, representing merely the forward direction of the incident beam.

It is worth noticing at this point that a three-dimensional grating, such as a crystal, differs from a one- or two-dimensional grating in the small number of diffracted beams produced by any particular orientation of the incident beam. Other diffracted intensity maxima can be observed only by changing the orientation of the incident beam relative to the crystallographic reference axes; in practical X-ray crystallography this is most conveniently done by rotating the crystal and keeping the attitude of the incident beam fixed.

Bragg Equation

Although the Laue Equations provide an elegant treatment of the diffraction of X-rays by crystals, they are difficult to manipulate and the diffraction condition is provided in what is, for most purposes, a more convenient form by the *Bragg Equation.* We now proceed to derive the Bragg Equation from the Laue Equations; later we shall show how the Bragg Equation can be obtained directly in a simpler, but less rigorous manner. It will be shown that the essential simplifying feature of the Bragg Equation is that a particular diffracted beam appears as a reflexion of the incident beam by a particular lattice plane, 'reflexion' occurring only at certain angles of incidence given by the equation.

We begin by considering[5] a general solution of the Laue Equations illustrated in Fig 6.7, where I represents the direction of the incident X-ray beam, D the direction of the diffracted X-ray beam for one solution of the Laue Equations, and z the positive direction of the z-axis of the lattice. Consistently with our previous usage I and D make angles $180° - i_c$ and δ_c respectively with the positive direction of the z-axis, which is placed in the centre of the stereogram. N is a direction in the plane defined by I and D and bisects the angle between I and D. The angle θ is defined such that $\mathrm{I{:}D} = 180° - 2\theta$ and $\mathrm{I{:}N} = \mathrm{N{:}D} = 90° - \theta$. For the non-Napierian spherical triangle INz equation (1) of chapter 5 yields

$$\cos(180° - i_c) = \cos z{:}\mathrm{N}.\cos(90° - \theta) + \sin z{:}\mathrm{N}.\sin(90° - \theta).\cos\omega$$

i.e.

$$-\cos i_c = \cos z{:}\mathrm{N}.\sin\theta + \sin z{:}\mathrm{N}.\cos\theta.\cos\omega. \qquad (8)$$

Similarly for the spherical triangle DNz,

$$\cos\delta_c = \cos z{:}\mathrm{N}.\cos(90° - \theta) + \sin z{:}\mathrm{N}.\sin(90° - \theta).\cos(180° - \omega)$$

i.e.

$$\cos\delta_c = \cos z{:}\mathrm{N}.\sin\theta - \sin z{:}\mathrm{N}.\cos\theta.\cos\omega. \qquad (9)$$

[5] We are indebted to Dr Helen D. Megaw for this simple ingenious argument.

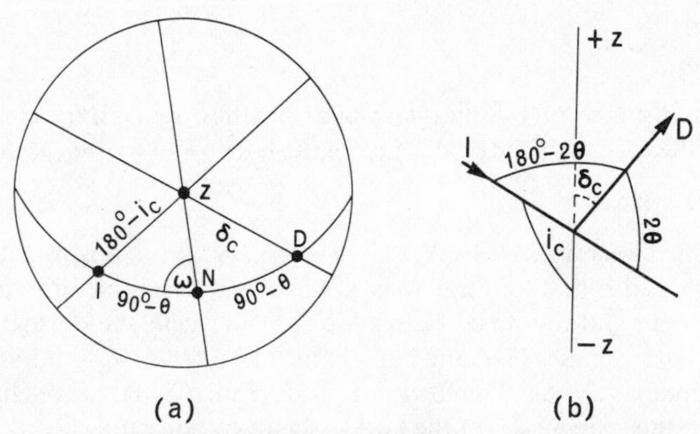

Fig 6.7 Derivation of the Bragg Equation from the Laue Equations. In the stereogram (a) and the perspective drawing (b) I represents the direction of the incident X-ray beam, D the diffracted beam and the pole *z* the positive direction of the *z*-axis; N is coplanar with I and D and bisects the angle ID.

Addition of equations (8) and (9) yields

$$\cos \delta_c - \cos i_c = 2 \cos z\,\dot{:}\,\mathrm{N}\,.\,\sin \theta. \tag{10}$$

But for an observable diffracted beam to occur the Laue Equations impose the restriction

$$c(\cos \delta_c - \cos i_c) = l\lambda, \tag{7}$$

which on substitution in equation (10) becomes

$$2 \cos z\,\dot{:}\,\mathrm{N}\,.\,\sin \theta = \frac{l\lambda}{c}$$

i.e.
$$\frac{c}{l}.\cos z\,\dot{:}\,\mathrm{N} = \frac{\lambda}{2 \sin \theta}.$$

Similarly the other two Laue Equations can be rewritten in terms of the angle θ and the angles between the direction N and the x and y axes as

$$\frac{b}{k}.\cos y\,\dot{:}\,\mathrm{N} = \frac{\lambda}{2 \sin \theta}$$

and
$$\frac{a}{h}.\cos x\,\dot{:}\,\mathrm{N} = \frac{\lambda}{2 \sin \theta}$$

so that
$$\frac{\lambda}{2 \sin \theta} = \frac{a}{h}.\cos x\,\dot{:}\,\mathrm{N} = \frac{b}{k}.\cos y\,\dot{:}\,\mathrm{N} = \frac{c}{l}.\cos z\,\dot{:}\,\mathrm{N}. \tag{11}$$

Now, the length of the normal ON from the origin O to the plane (*hkl*) at N is given by equation (26) of chapter 5 as

$$\mathrm{ON} = \frac{a}{h}.\cos x\,\dot{:}(hkl) = \frac{b}{k}.\cos y\,\dot{:}(hkl) = \frac{c}{l}.\cos z\,\dot{:}(hkl).$$

where $x\,\dot{:}(hkl)$ is the angle between ON and the x-axis and so on. Comparison with equation (11) indicates that the direction N shown on Fig 6.7 is the normal to the plane (*hkl*), where *h*, *k*, and *l* are the integral factors in the three Laue Equations, and

$$\frac{\lambda}{2 \sin \theta} = \text{ON}.$$

The perpendicular distance, ON, of the (*hkl*) plane from the origin is more conveniently described now as the spacing of the (*hkl*) planes of the lattice and written as d_{hkl}, so that

$$\lambda = 2d_{hkl} \sin \theta. \tag{12}$$

Equation (12), which was derived by W. L. Bragg in 1912, is known as the *Bragg Equation.* It simply states that for directions in which an observable diffracted beam can occur the incident and diffracted beams are coplanar with the normal from the origin to a set of lattice planes (*hkl*) and equally inclined at $90° - \theta$ to it and, further, that the angle θ (commonly called the *Bragg angle*) is related to the wavelength of the radiation λ and to the spacing d_{hkl} of the lattice planes by equation (12).

It is apparent from Fig 6.8 that the Bragg Equation provides an alternative way of looking at the diffraction of X-rays by a lattice: the diffracted beam can be regarded as a reflexion of the incident beam by the set of lattice planes (*hkl*), reflexion occurring only when the angle of inclination of the incident beam to the lattice planes, i.e. the Bragg angle θ, the spacing of the planes and the wavelength of the X-rays, satisfy the equation $\lambda = 2d_{hkl} \sin \theta$.

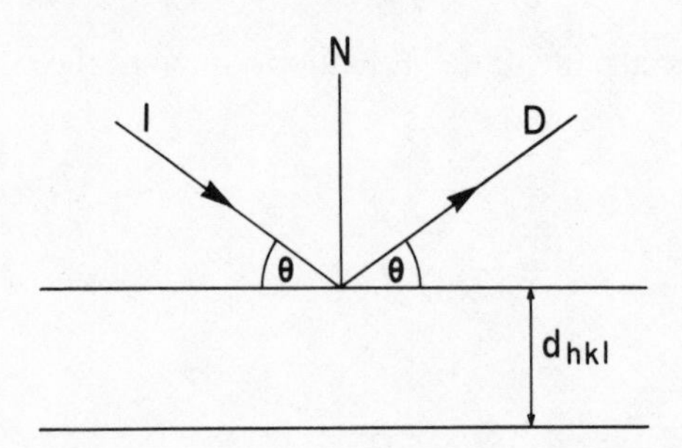

Fig 6.8 The Bragg Equation. The diffracted beam D can be regarded as a reflexion of the incident beam I by a set of lattice planes (*hkl*) when the angle of incidence θ satisfies the equation $\lambda = 2d_{hkl} \sin \theta$.

In his derivation of the Bragg Equation W. L. Bragg considered first how the X-rays scattered by all the lattice points in a plane (*hkl*) might be in phase and then established the condition for the X-rays scattered by the lattice points in all lattice planes parallel to (*hkl*) to be in phase. As the incident-plane wave-front passes over a set of points (which need not be lattice points or even regularly spaced) in a lattice plane, secondary wavelets build up a reflected wave-front according to the Huygens construction familiar in optics (see Fig 12.52). A small part of the energy of the incident wave-motion is transferred to the reflected wave-motion, but most of it passes on. The condition for optical reflexion, that the angle of incidence is equal to the angle of reflexion, ensures that the waves scattered by all points in the lattice plane are in phase with one another (Fig 6.9(a)). In general the waves reflected from successive lattice planes will not be in phase. A plane wave-front incident on an adjacent pair of lattice planes is shown in Fig 6.9(b), where the lattice planes are shown as horizontal lines. The waves reflected from the upper plane have an optical path that is shorter than that for those reflected from the lower plane by $\text{PB} + \text{BQ} = 2d_{hkl} \sin \theta$. For reinforcement the path difference must be a whole number of wavelengths; therefore the condition for an observable diffracted beam is

$$n\lambda = 2d_{hkl} \sin \theta. \tag{13}$$

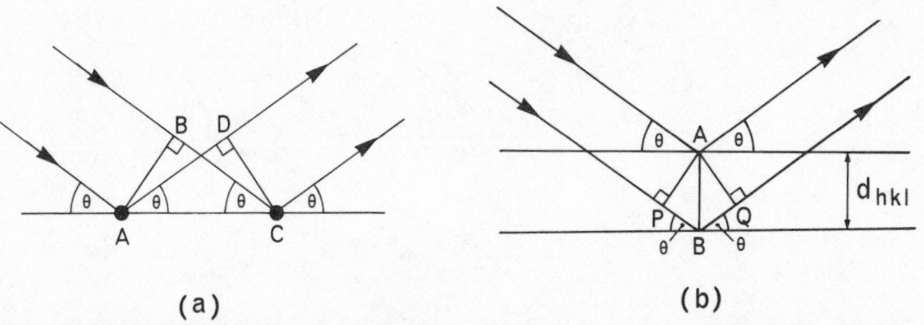

Fig 6.9 The Bragg Equation: the optical analogy. (a) illustrates the condition for optical reflexion, that the angle of incidence $\widehat{ACB} = \theta$ is equal to the angle of reflexion $\widehat{CAD}$ so that the waves scattered from all points in the plane AC are in phase with one another. (b) illustrates Bragg diffraction from successive lattice planes (*hkl*), AP and AQ being respectively incident and diffracted plane wave fronts; in general the waves reflected from successive lattice planes will not be in phase.

Equation (13) is a form of the *Bragg Equation* in common use in spectroscopy. In crystallographic usage it is rewritten as

$$\lambda = \frac{2d_{hkl}}{n}\sin\theta = 2d_{nh,nk,nl}\sin\theta$$

and the indices are divided through by the common factor n to give the form derived directly as equation (12), $\lambda = 2d_{hkl}\sin\theta$. In the derivation of equation (12) from the Laue Equations the only restriction placed on the values of h, k, and l was that they should be integral; they were not forbidden to have a common factor. However when the indices of planes (hkl) have a common factor n, the planes are not strictly lattice planes, but a set of planes parallel to the lattice planes (h/n, k/n, l/n) with a spacing of one nth of that of the lattice planes. This point is illustrated in Fig 6.10, where (210) lattice planes, (420) and (630) planes are shown in relation to an orthorhombic lattice. Every (210) lattice plane passes through lattice points and all planes are equivalent. However, only alternate planes of the (420) set pass through lattice points and in consequence not all planes of the set are equivalent. Likewise the planes (630) are not all equivalent, only one in every three passing through lattice points. In Fig 6.10 incident and diffracted beams that satisfy the Bragg Equation are shown for each set of planes and it is convenient at this point to comment on the statement, which has already been implied, that diffracted X-ray beams are commonly known as *X-ray reflexions*, it being understood that 'reflexion' implies a solution of the Bragg Equation and therefore a specific angle of incidence. The diffracted beams shown in Fig 6.10 would be described as the 210, 420, and 630 reflexions, the indices for reflexions being distinguished from those for planes by omission of brackets ( ). Reflexions are indexed so that the path difference for X-rays scattered by adjacent planes is λ, a point to which we shall return later. The statement that a particular diffracted beam is the *hkl* reflexion for X-rays of wavelength λ for a lattice of given dimensions completely specifies the angular relationship between incident and diffracted beam: the incident beam is required to lie on a cone of semi-axis $90° - \theta$ about the normal to the (hkl) plane and the diffracted beam lies on the same cone so as to be coplanar with the normal to the plane and the incident beam.

The Bragg Equation thus provides a simple and convenient statement of the geometry of the diffraction of X-radiation by crystals. It is the fundamental equation of X-ray crystallography. Its application in a variety of situations will be explored in the later parts of this chapter and in the two following chapters.

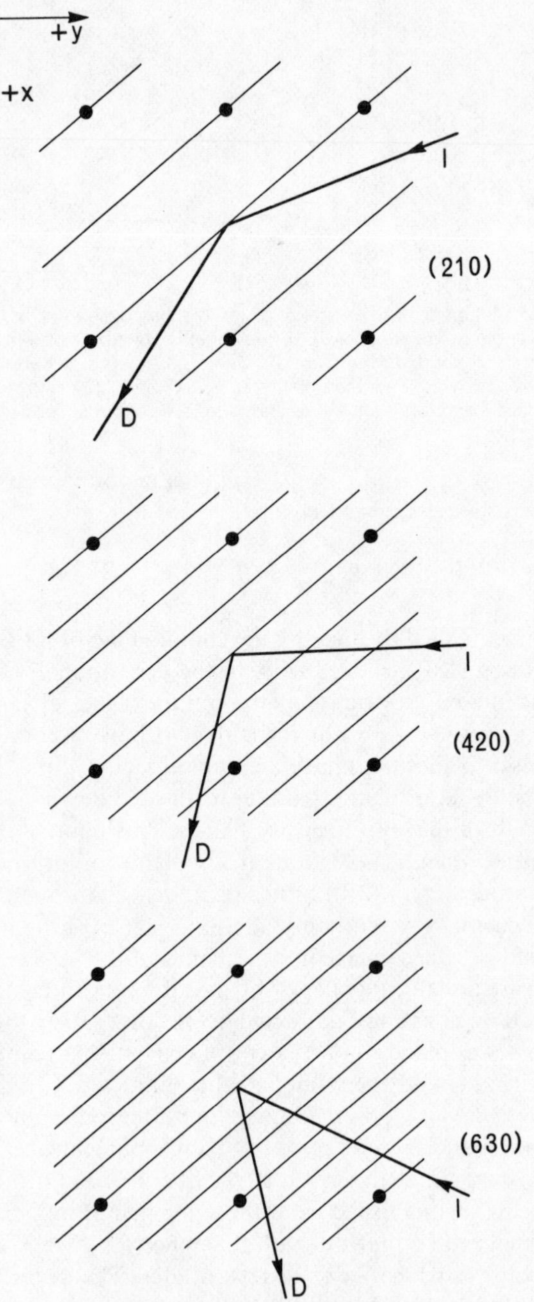

Fig 6.10 The Bragg Equation, $\lambda = 2d_{hkl} \sin\theta$. The top diagram shows an (001) projection of an orthorhombic P-lattice with lattice planes (210), and incident (I) and diffracted (D) X-ray beams. The central and lowermost diagrams show planes (420) and (630) for the same lattice and the directions of incident and diffracted X-ray beams. Since $d_{210} = 2d_{420} = 3d_{630}$, $\sin\theta_{210} = \frac{1}{2}\sin\theta_{420} = \frac{1}{3}\sin\theta_{630}$.

Intensities of X-ray reflexions

The regular arrangement of atoms in a crystal restricts diffracted intensity maxima to certain directions, which can be described either as directions that satisfy the Laue Equations or, alternatively, as directions in which reflexions from lattice planes satisfy the Bragg Equation. We need concern ourselves therefore only with the intensity of radiation scattered in the directions of X-ray reflexions, a much simpler problem than the investigation of the intensity of radiation scattered in any general direction.

Since the scattering produced by a crystal is almost entirely due to interaction between X-rays and electrons, the density of scattering matter at a point x, y, z in the unit-cell can be equated with the electron density[6] at that point. Strictly, electron density must be regarded as varying continuously throughout the unit-cell, but some simplifying assumptions can be made: since the electron density associated with an atom falls off rapidly with distance from the centre of the atom, it will be assumed that the electron cloud associated with any atom does not overlap that belonging to any other atom and is moreover spherically symmetric. All the atoms of a given element in the same state of ionization are thus taken to have identical distribution of electron density and to scatter X-rays similarly. We have deliberately ignored, at this first stage of approximation, valency electrons, which are involved in bond formation, because they are relatively few in number compared with the core electrons, each of which is unambiguously associated with one atomic nucleus; in short we are assuming that the core electrons are predominantly responsible for the scattering of X-rays and consequently that the radiation scattered by an atom will be independent of its environment. Thus the scattering produced by an atom of a given element in a given state of ionization will always be the same wherever it lies in the unit-cell and in all substances in which the element occurs.

In the foregoing we have referred qualitatively to the amount of radiation scattered by an atom; clearly we need to make this quantitative and a convenient unit is provided by the classical electrodynamic treatment of the scattering of X-radiation by a single free electric charge. When a wave falls on an electric charge it causes the charge to vibrate and to act as a secondary source of waves of the same frequency so that the charge can be regarded as scattering a small fraction of the radiation incident upon it. If unpolarized radiation of amplitude A and wavelength λ is incident on a free classical electron of charge e and mass m, it can be shown that the amplitude of the scattered radiation at a distance R from the electron, where $R \gg \lambda$, is

$$\frac{A}{R} \cdot \frac{e^2}{mc^2} \left\{ \frac{1+\cos^2 2\theta}{2} \right\}^{\frac{1}{2}},$$

where c is the velocity of light and 2θ is the angle between the scattered beam and the forward direction of the incident beam.[7] The factor $\{(1+\cos^2 2\theta)/2\}^{\frac{1}{2}}$ arises from the partial polarization of the scattered beam, which is out of phase with the incident beam by the amount π.

We now have to define the *atomic scattering factor*, f, of an atom as the ratio of the amplitude scattered in a particular direction by that atom to the amplitude scattered by a free classical electron in the same direction. The amplitude of the

[6] If ψ is the electronic wave function at a point, then the electron density at that point will be $|\psi|^2$.

[7] This result is proved in standard textbooks of electrodynamics and in James (1967) p. 29.

X-radiation scattered by an atom is thus

$$\frac{A}{R}\cdot\frac{e^2}{mc^2}f\left\{\frac{1+\cos^2 2\theta}{2}\right\}^{\frac{1}{2}}.$$

In the idealized, but impossible, case of a 'point atom', where all the electrons associated with the atomic nucleus are situated at a point, the atomic scattering factor of the atom would be equal to its atomic number,[8] Z. In reality the electrons associated with an atomic nucleus occupy a finite volume, the dimensions of which are of the same order of magnitude as the wavelength of X-rays. Each small element of volume ΔV within the electron cloud of the atom will give rise to scattered X-radiation of amplitude proportional to $\rho(r).\Delta V$, where $\rho(r)$ is the average electron density over the volume ΔV situated at a distance r from the centre of the atom. There will in general be a path difference between the X-rays scattered by any pair of volume elements within the electron cloud and this path difference will vary with scattering angle (Fig 6.11); the path difference will be zero for $2\theta = 0$ and will increase smoothly with increasing 2θ. We now suppose the radiation scattered by all volume elements of the electron cloud of the atom to be summed for a direction of scattering angle 2θ:

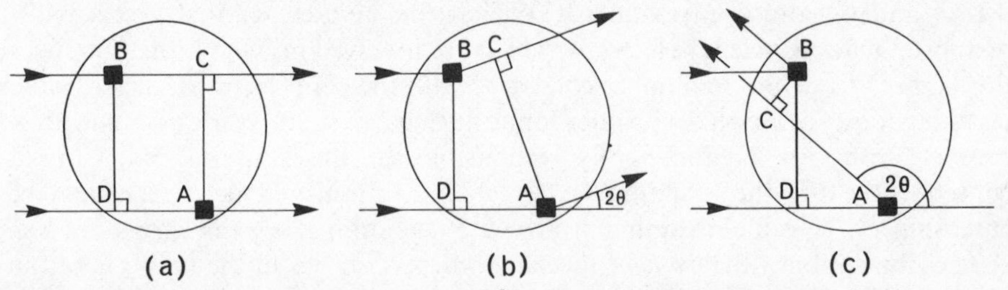

Fig 6.11 The dependence on scattering angle 2θ of the path difference between the X-rays scattered by two volume elements A and B within the electron cloud of the atom. The boundary surface of the electron cloud of the atom is represented by the large circle. (a) $2\theta = 0$; path difference $\delta = BC - DA = 0$. (b) 2θ small; path difference $\delta = DA - BC$. (c) 2θ large; path difference $\delta = DA + AC$.

for $2\theta = 0$ there can be no destructive interference and the atomic scattering factor, f, for the atom will be equal to its atomic number, Z; for small scattering angles all path differences will be small and there can be little destructive interference so that f will be only just less than Z; as the scattering angle increases the path difference for any pair of volume elements will increase and in general there will be a greater likelihood of destructive interference so that f will steadily decrease with increasing θ and at high θ, $f \ll Z$. The more tightly the electrons are bound to the nucleus, the greater will be the concentration of electron density towards the centre of the atom so that for given θ less destructive interference will be possible; for cations, especially for cations of high formal charge, the value of f will remain close to Z to higher θ values than for anions with a similar number of extranuclear electrons. In exploring the variation in magnitude of f we must also bear in mind that the phase difference consequent on a given path difference is dependent on the wavelength λ of the incident radiation: at given θ more pairs of volume elements will be able to produce destructive interference if λ is small. It is found that in general the atomic scattering factors of all elements and ions vary similarly with $\sin\theta/\lambda$, f decreasing from its value Z at $\theta = 0$ more slowly for cations with tightly bound electrons than for anions

[8] The atomic number is the charge on the nucleus which for a neutral atom is equal to the number of extranuclear electrons. It is the number of extranuclear electrons that concerns us here and we define this number as Z. For an ion Z is equal to the charge on the nucleus less the formal charge of the ion.

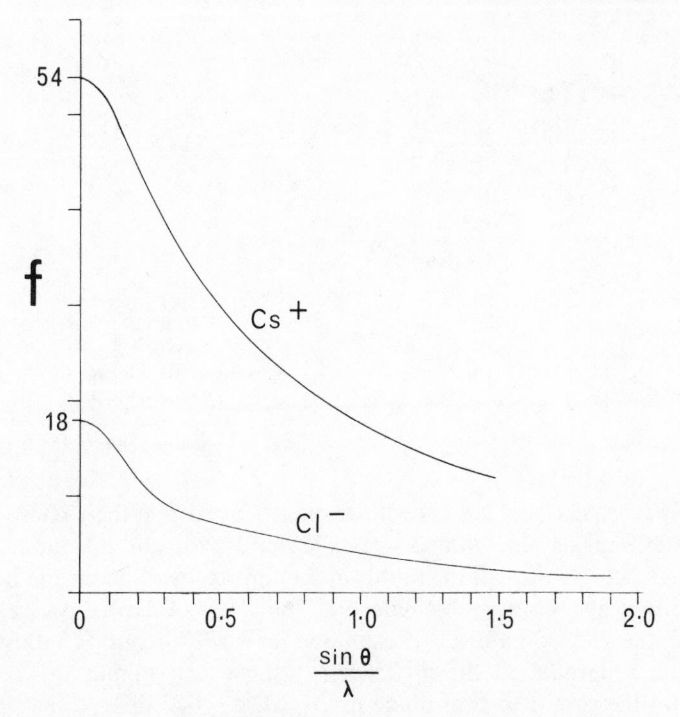

Fig 6.12 The dependence of atomic scattering factor f on $\sin\theta/\lambda$ for a cation Cs^+ and an anion Cl^-. The rate of fall-off for cations is generally less rapid than for anions.

with more diffuse electron clouds. The calculation of the electron density distribution within an atom is a problem in quantum mechanics; atomic scattering factors for X-radiation calculated by such methods are tabulated in *International Tables for X-ray Crystallography*, vol. III as functions of $\sin\theta/\lambda$. The dependence of f on $\sin\theta/\lambda$ is shown in Fig 6.12 for two examples, the cation Cs^+, in which the electrons are rather tightly bound, and the anion Cl^-, in which the electrons are relatively loosely bound.

Now although the amplitude of the X-rays scattered by each atom in a unit-cell is proportional to its atomic scattering factor, we cannot state the amplitude of the X-radiation scattered by the whole unit-cell until we know the phase of the X-rays scattered by each atom and that will depend on position in the unit-cell. We take the origin of the unit-cell as our reference point for phase and suppose that the X-rays scattered by an atom at the origin have phase $\phi = 0$. The phase of the X-rays scattered by an atom with coordinates x, y, z can then be deduced from the Bragg Equation by the following argument. The contributions to the *hkl* reflexion of X-rays scattered by atoms lying on the (*hkl*) plane through the origin will be in phase with one another and with the X-rays scattered by the atom at the origin, for which $\phi = 0$, because the path difference, Δ, between X-rays scattered by any point in such a plane is zero. Since the condition for an *hkl* reflexion is that the X-rays scattered from adjacent (*hkl*) lattice planes should have a path difference, $\Delta = \lambda$, which in phase terms amounts to saying a phase difference, $\phi = 2\pi$, the X-rays scattered from the first (*hkl*) plane out from the origin will have a path length less by λ than those scattered from the *hkl* plane through the origin and their phase will consequently be $\phi = 2\pi$ (Fig 6.13). Likewise the path length for X-rays scattered from the second plane out will be shorter again by λ and their phase will be 4π. In general for the nth plane out from the origin, the

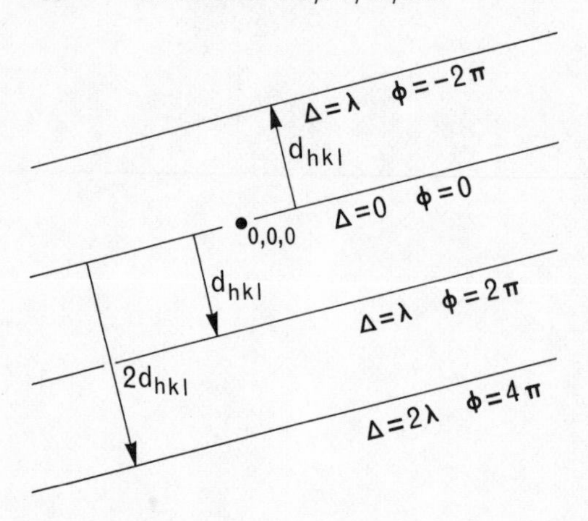

Fig 6.13 The phase difference ϕ between X-rays of wavelength λ scattered into the *hkl* reflexion by successive (*hkl*) lattice planes. The lattice planes (*hkl*) are perpendicular to the plane of the diagram.

path length of the scattered X-rays will be shorter by $n\lambda$ and their phase will be $2\pi n$. But the scattering is done by atoms, not by lattice planes, and it is unlikely except in very simple structures that all the atoms in the unit-cell will lie on the lattice planes corresponding to every observable reflexion. The phase of the X-rays scattered by an atom with general coordinates x, y, z can however easily be found by drawing a plane through x, y, z parallel to the (*hkl*) lattice planes and taking the perpendicular distance from the origin to that plane as D_{hkl} (Fig 6.14); then, if the separation of (*hkl*) lattice planes is d_{hkl}, the path difference between the X-rays scattered by the atom at x, y, z and an atom at the origin will be $(D_{hkl}/d_{hkl}).\lambda$ and the phase of the X-rays scattered by the atom at x, y, z will correspondingly be $2\pi(D_{hkl}/d_{hkl})$.

At this point it may be instructive to consider in some detail an example of phase calculation for a simple structure, caesium chloride (which is cubic with only two atoms in its unit-cell, Cs at 0, 0, 0 and Cl at $\frac{1}{2}, \frac{1}{2}, \frac{1}{2}$ (Fig 6.15)). We consider first the 100 reflexion. The Cs atom, being at the origin, lies on the (100) lattice plane through the origin so that the phase of the X-rays scattered by it is zero. The amplitude of the X-rays scattered by the caesium atom will therefore simply be proportional to f_{Cs}, the atomic scattering factor of caesium; moreover this will be so for all X-ray reflexions because the Cs atom lies at the origin. The chlorine atom however does not lie on the (100) lattice plane through the origin, but at a perpendicular distance $\frac{1}{2}d_{100} = \frac{1}{2}a$ from that plane so that the X-rays scattered by the Cl atom have a path difference $\frac{1}{2}\lambda$ and a phase difference π relative to the X-rays scattered by the Cs atom. The amplitude of the X-rays scattered by the chlorine atom will of course be proportional to f_{Cl}, the atomic scattering factor of chlorine. The principle of superposition gives

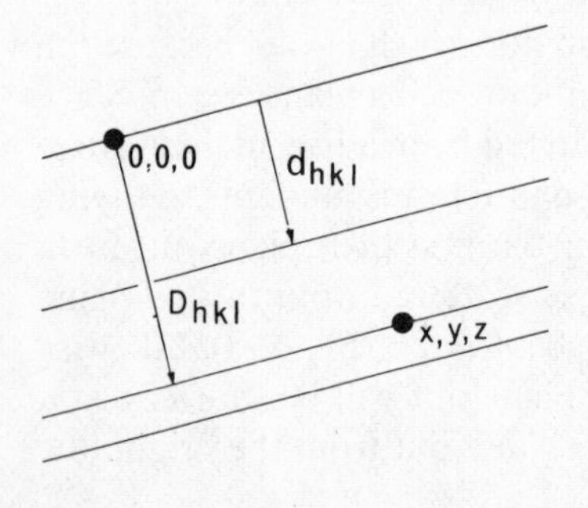

Fig 6.14 The phase of the X-rays scattered into the *hkl* reflexion by an atom at *xyz* is given by $2\pi(D_{hkl}/d_{hkl})$, where D_{hkl} is the perpendicular distance from the origin to the plane parallel to (*hkl*) passing through *xyz*. The plane of the diagram is perpendicular to (*hkl*) and does not necessarily contain the point *xyz*.

the intensity[9] of the X-rays scattered by one unit-cell in general as

$$I(hkl) = \left(\sum_{1}^{N} f_n \cos \phi_n\right)^2 + \left(\sum_{1}^{N} f_n \sin \phi_n\right)^2 \tag{6}$$

and for the 100 reflexion of CsCl as

$$\begin{aligned} I(100) &= (f_{\text{Cs}} \cos 0 + f_{\text{Cl}} \cos \pi)^2 + (f_{\text{Cs}} \sin 0 + f_{\text{Cl}} \sin \pi)^2 \\ &= (f_{\text{Cs}} - f_{\text{Cl}})^2. \end{aligned}$$

It should be noticed that we have taken our summation only up to $n = 2$, because there are only two atoms in the unit-cell, one Cs and one Cl; the other Cs atoms shown in Fig 6.15 belong to other unit-cells if the Cs atom at the origin is assigned wholly to the reference unit-cell.

We now turn to the 200 reflexion. The (200) lattice planes have an interplanar spacing $d_{200} = \frac{1}{2}d_{100}$ so that the Cl atom scatters X-rays of phase $2\pi(\frac{1}{2}a/\frac{1}{2}a) = 2\pi$ into the 200 reflexion. The phase of the X-rays scattered by Cs remains zero. The expression for the intensity of the 200 reflexion is thus

$$\begin{aligned} I(200) &= (f_{\text{Cs}} \cos 0 + f_{\text{Cl}} \cos 2\pi)^2 + (f_{\text{Cs}} \sin 0 + f_{\text{Cl}} \sin 2\pi)^2 \\ &= (f_{\text{Cs}} + f_{\text{Cl}})^2. \end{aligned}$$

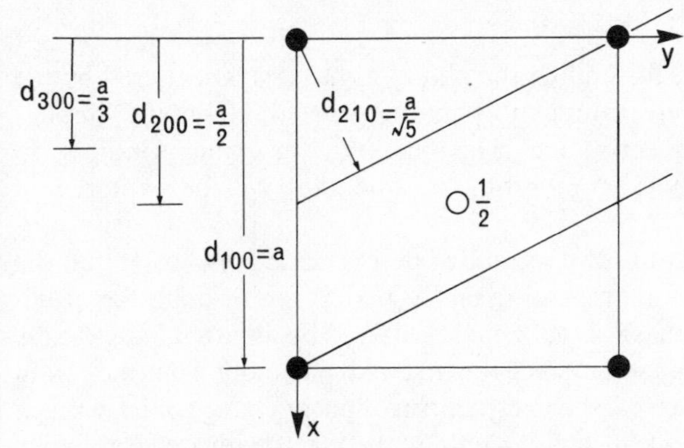

Fig 6.15 Projection of the structure of CsCl on (001) with Cs atoms represented as solid circles and Cl atoms as open circles. The spacings and orientation of (100), (200), (300) and (210) planes are indicated.

The Cl atom lies midway between (210) lattice planes with $D_{210} = \frac{3}{2}d_{210}$ so that the phase of the X-rays scattered by the Cl atom at $\frac{1}{2}, \frac{1}{2}, \frac{1}{2}$ into the 210 reflexion is $2\pi . \frac{3}{2} = 3\pi$. The intensity of the 210 reflexion is thus

$$\begin{aligned} I(210) &= (f_{\text{Cs}} \cos 0 + f_{\text{Cl}} \cos 3\pi)^2 + (f_{\text{Cs}} \sin 0 + f_{\text{Cl}} \sin 3\pi)^2 \\ &= (f_{\text{Cs}} - f_{\text{Cl}})^2. \end{aligned}$$

We have now expressed the intensities of the three reflexions 100, 200, and 210 of CsCl in terms of the atomic scattering factors of caesium and chlorine, f_{Cs} and f_{Cl}. But it must be borne in mind that atomic scattering factors are dependent on $\sin\theta/\lambda$ for the relevant expression and, by the Bragg Equation, $\sin\theta/\lambda = \frac{1}{2}d$. In Table 6.1 we give the cell edge, a, for CsCl; $\frac{1}{2}d$ for a selection of reflexions including 100, 200, and 210; f_{Cs} and f_{Cl} derived from the atomic scattering factor curves of Fig 6.12 for each reflexion; and the calculated intensity $I(hkl)$ of each reflexion.

[9] For practical purposes $I(hkl)$ is known as the 'intensity' of the *hkl* reflexion as will be explained later in this chapter.

Table 6.1
Intensities of some reflexions of CsCl

$a = 4{\cdot}123$ Å		Cs: 0, 0, 0			Cl: $\frac{1}{2}, \frac{1}{2}, \frac{1}{2}$
hkl	d_{hkl}	$\frac{\sin\theta}{\lambda} = \frac{1}{2d_{hkl}}$	f_{Cs}	f_{Cl}	$I(hkl)$
100	a	$\frac{1}{2a} = 0{\cdot}12$	50	15	$(f_{Cs} - f_{Cl})^2 = (35)^2 = 1225$
200	$\frac{a}{2}$	$\frac{1}{a} = 0{\cdot}24$	42	11	$(f_{Cs} + f_{Cl})^2 = (53)^2 = 2809$
300	$\frac{a}{3}$	$\frac{3}{2a} = 0{\cdot}36$	36	8	$(f_{Cs} - f_{Cl})^2 = (28)^2 = 784$
210	$\frac{a}{\sqrt{5}}$	$\frac{\sqrt{5}}{2a} = 0{\cdot}27$	41	10	$(f_{Cs} - f_{Cl})^2 = (31)^2 = 961$
420	$\frac{a}{2\sqrt{5}}$	$\frac{\sqrt{5}}{a} = 0{\cdot}54$	28	7	$(f_{Cs} + f_{Cl})^2 = (35)^2 = 1225$

The amplitude of the X-rays scattered by a free electron is taken as unity. Atomic scattering factors for the ions Cs^+ and Cl^- are used rather than those for the neutral atoms because the structure is known to be ionic.

The procedure outlined above for finding the phase of the X-rays scattered by the various atoms of a crystal structure is impossibly cumbersome for all but the simplest structures and is then only convenient for reflexions with very simple indices. In general an analytical approach is more profitable and that will be explored in succeeding paragraphs.

We consider a general hkl reflexion and recall two points made previously: that the phase of the X-rays scattered by an atom lying on the (hkl) plane through the origin is taken to be zero and that the phase of the X-rays scattered by an atom lying on the first (hkl) plane out from the origin is 2π. Since the first (hkl) plane out from the origin makes an intercept a/h on the x-axis and corresponds to a phase change of 2π, we can, by simple proportions, associate a translation ax along the x-axis with a phase change

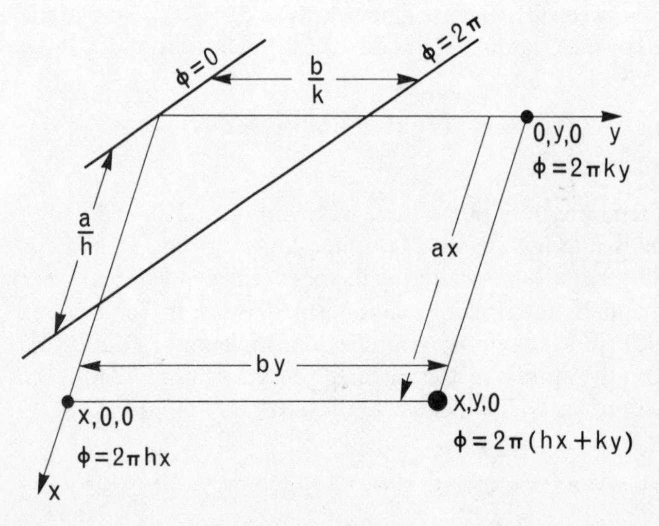

Fig 6.16 Atoms with coordinates $x00$, $0y0$, $xy0$ respectively scatter X-rays of phase $2\pi hx$, $2\pi ky$, $2\pi(hx+ky)$ into the hkl reflexion. The (hkl) plane through the origin, with phase difference $\phi = 0$, and the first (hkl) plane out from the origin, with $\phi = 2\pi$, are indicated by bold lines.

$(2\pi ax)/(a/h) = 2\pi hx$; thus an atom with coordinates x, 0, 0 will scatter X-rays of phase $2\pi hx$ into the hkl reflexion.

By analogy we can say that an atom with coordinates 0, y, 0 will scatter X-rays of phase $2\pi ky$ into the hkl reflexion. We can extend this idea to an atom with coordinates x, y, 0 (Fig 6.16): such an atom is associated with translations ax along the x-axis and by along the y-axis so that the phase of the X-rays scattered by it will be $[2\pi xa/(a/h)]+[2\pi yb/(b/k)] = 2\pi(hx+ky)$. Extension of the argument to three dimensions gives the phase of the X-rays scattered by an atom with coordinates x, y, z as $2\pi(hx+ky+lz)$, the amplitude of the scattered wave motion being proportional to the atomic scattering factor, f, of the atom at x, y, z. We can now apply the principle of superposition to a unit-cell containing N atoms of elements with atomic scattering factors $f_1, f_2, \ldots f_n$ and coordinates x_1, y_1, z_1; x_2, y_2, z_2; $\ldots x_N, y_N, z_N$. The intensity of the hkl reflexion will be given by

$$I(hkl) = \{f_1 \cos 2\pi(hx_1+ky_1+lz_1)+\ldots+f_N \cos 2\pi(hx_N+ky_N+lz_N)\}^2$$
$$+\{f_1 \sin 2\pi(hx_1+ky_1+lz_1)+\ldots+f_N \sin 2\pi(hx_N+ky_N+lz_N)\}^2$$

Therefore
$$I(hkl) = \left\{\sum_1^N f_n \cos 2\pi(hx_n+ky_n+lz_n)\right\}^2 + \left\{\sum_1^N f_n \sin 2\pi(hx_n+ky_n+lz_n)\right\}^2 \tag{14}$$

where the summation is taken over all the N atoms of the unit-cell.

From equation (14) the intensity of any X-ray reflexion can be calculated provided the coordinates of all atoms in the unit-cell and the relevant atomic scattering factors are known. Again taking CsCl as our example, we have two atoms in the unit-cell, Cs at 0, 0, 0 and Cl at $\frac{1}{2}, \frac{1}{2}, \frac{1}{2}$. The expression for the intensity of a general reflexion becomes

$$I(hkl) = \{f_{Cs} \cos 2\pi(h.0+k.0+l.0)+f_{Cl} \cos 2\pi(h.\tfrac{1}{2}+k.\tfrac{1}{2}+l.\tfrac{1}{2})\}^2$$
$$+\{f_{Cs} \sin 2\pi(h.0+k.0+l.0)+f_{Cl} \sin 2\pi(h.\tfrac{1}{2}+k.\tfrac{1}{2}+l.\tfrac{1}{2})\}^2$$
$$= \left\{f_{Cs}+f_{Cl} \cos 2\pi.\frac{h+k+l}{2}\right\}^2$$

The intensity of, for instance, the 420 reflexion is then given immediately by substitution as

$$I(420) = \{f_{Cs}+f_{Cl} \cos 2\pi.\tfrac{6}{2}\}^2$$
$$= \{f_{Cs}+f_{Cl}\}^2.$$

This expression is evaluated in Table 6.1.

It will have been noticed that in the case of CsCl the sine terms in the expression for $I(hkl)$ vanish because $\sin p\pi = 0$ for integral values of p. This is a rather special case of the general proposition that when the origin of the unit-cell of a centro-symmetric structure is taken at a centre of symmetry, the sine terms in the expression for $I(hkl)$ vanish. When the origin lies at a centre of symmetry the atoms in the unit-cell are related in pairs so that if an atom of a particular element has coordinates x_n, y_n, z_n an atom of the same element will lie at $\bar{x}_n, \bar{y}_n, \bar{z}_n$. Therefore for every term $f_n \sin 2\pi(hx_n+ky_n+lz_n)$ there will also be a term

$$f_n \sin 2\pi(-hx_n-ky_n-lz_n) = -f_n \sin 2\pi(kx_n+ky_n+lz_n)$$

and $$\sum_{1}^{N} f_n \sin 2\pi(hx_n+ky_n+lz_n) = 0$$

The intensity expression then becomes,

$$I(hkl) = \left\{\sum_{1}^{N} f_n \cos 2\pi(hx_n+ky_n+lz_n)\right\}^2 \tag{15}$$

In the case of CsCl both atoms lie on centres of symmetry so that there is no explicit pairing but the sine terms vanish immediately.

The trigonometric form of the intensity expression, equation (14), is obviously cumbersome; a more elegant and convenient formulation can be achieved by use of complex number notation. The expression

$$\psi = a\cos(\omega t+\phi) \tag{1}$$

for the disturbance produced at a given point by an incident wave-motion is the real part of the expression

$$\begin{aligned}\psi &= |a|\exp\{i(\omega t+\phi)\}\\ &= |a|\exp(i\phi)\exp(i\omega t).\end{aligned}$$

Since we are concerned in diffraction with radiation of constant frequency, $\exp(i\omega t)$ is a common factor in all our expressions and need not be considered further. The complex amplitude $a = |a|\exp(i\phi)$ expresses both the amplitude $|a|$ and the phase ϕ of the disturbance; it is represented in the complex plane (Fig 6.17) by a line of length $|a|$ inclined at an angle ϕ to the real axis. Defining x as the real and y as the imaginary axis, then

$$\begin{aligned}a &= x+iy\\ &= |a|\cos\phi+i|a|\sin\phi.\end{aligned}$$

The modulus $|a|$ of a thus represents the amplitude of the wave-motion,

$$|a| = \sqrt{(x^2+y^2)},$$

and its intensity is given by aa^*, where $a^* = x-iy$ is the complex conjugate of a,

i.e. $$aa^* = (x+iy)(x-iy) = x^2+y^2 = |a|^2.$$

The phase ϕ of the wave-motion is given by

$$\tan\phi = \frac{y}{x}.$$

Now the principle of superposition gives the resultant disturbance ψ_R at a point as the sum of the N disturbances, $\psi_1, \psi_2, \dots \psi_N$, arriving simultaneously at the point. Therefore

$$\begin{aligned}a_R &= \sum_{1}^{N} a_n\\ &= \sum_{1}^{N} |a_n|\exp(i\phi_n)\\ &= \sum_{1}^{N} |a_n|\cos\phi_n + i\sum_{1}^{N} |a_n|\sin\phi_n.\end{aligned}$$

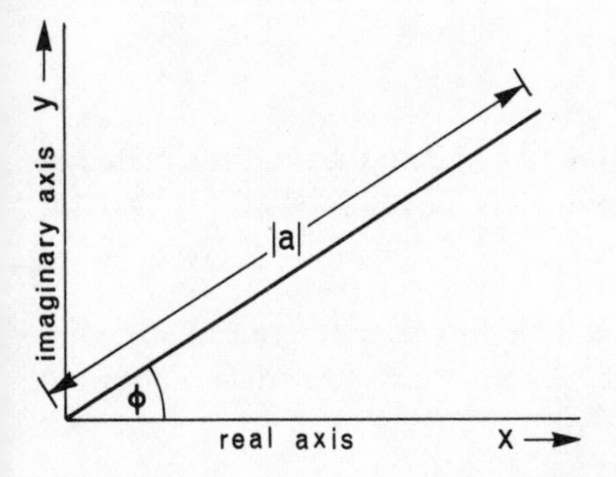

Fig 6.17 Representation of $a = |a|e^{\phi}$ in the complex plane by a line of length $|a|$ inclined at the angle ϕ to the real axis.

We have already seen that the wave-motion scattered by an atom with coordinates x_n, y_n, z_n has an amplitude proportional to the atomic scattering factor f_n of the element concerned and phase $2\pi(hx_n+ky_n+lz_n)$, so that the wave-motion can be expressed as $f_n \exp 2\pi i(hx_n+ky_n+lz_n)$. The wave-motion scattered by all the atoms in one unit-cell into the hkl reflexion can then be represented as

$$\left.\begin{aligned} F(hkl) &= \sum_1^N f_n \exp 2\pi i(hx_n+ky_n+lz_n) \\ &= \sum_1^N f_n \cos 2\pi(hx_n+ky_n+lz_n) + i\sum_1^N f_n \sin 2\pi(hx_n+ky_n+lz_n) \end{aligned}\right\} \quad (16)$$

$F(hkl)$ being known as the *structure factor*. The amplitude of the resultant wave-motion is proportional to $|F(hkl)|$, which is known as the *structure amplitude*. Since the atomic scattering factor of an element is the ratio of the amplitude scattered by one atom of that element into the hkl reflexion to the amplitude scattered by a free classical electron, the structure amplitude $|F(hkl)|$ is therefore the ratio of the amplitude scattered into the hkl reflexion by the contents of one unit-cell to the amplitude scattered by a free classical electron in the same direction. Both quantities, f_n and $|F(hkl)|$, are thus pure numbers which represent the number of electrons that would have to be situated at a point, if that were possible, to produce a scattered wave of the same amplitude as that scattered by, in one case, an atom and, in the other, the whole contents of a unit-cell.

The intensity of the X-rays scattered by one unit-cell into the hkl reflexion is then

$$\begin{aligned} I(hkl) &= |F(hkl)|^2 \\ &= \left\{\sum_1^N f_n \cos 2\pi(hx_n+ky_n+lz_n)\right\}^2 + \left\{\sum_1^N f_n \sin 2\pi(hx_n+ky_n+lz_n)\right\}^2 \end{aligned}$$

The reader will observe that this expression is identical with equation (14) which was derived without recourse to complex numbers.

We have also seen that when the origin of the unit-cell of a centrosymmetric structure is taken at a centre of symmetry there will be atoms of the same element at x_n, y_n, z_n and $\bar{x}_n$, $\bar{y}_n$, $\bar{z}_n$; the expression for the structure factor then becomes

$$F(hkl) = \sum_1^{\frac{1}{2}N} f_n\{\exp 2\pi i(hx_n+ky_n+lz_n) + \exp 2\pi i(-hx_n-ky_n-lz_n)\}$$

i.e.

$$F(hkl) = 2\sum_1^{\frac{1}{2}N} f_n \cos 2\pi(hx_n+ky_n+lz_n)$$

i.e. $$F(hkl) = \sum_{1}^{N} f_n \cos 2\pi(hx_n + ky_n + lz_n) \tag{17}$$

The last form is preferred because it eliminates the risk of counting atoms situated at centres of symmetry twice. An important consequence of equation (17) is that for centrosymmetric structures the structure factor is always real and the phase of the X-rays scattered into the *hkl* reflexion can only take the values 0 or π.

In general the intensities of X-ray reflexions will be discussed in terms of structure factors $F(hkl)$ in preference to intensities $I(hkl)$, except when numerical values are required.

Systematic absences in non-primitive lattice types

In our discussion of the intensities of X-ray reflexions we have hitherto tacitly assumed that we have been dealing with a primitive unit-cell, that is a unit-cell that contains only one repeat unit. We have shown that diffracted intensity maxima occur only in directions for which the X-rays scattered by corresponding points in repeat units have path differences of integral numbers of wavelengths; such corresponding points constitute a lattice and our simple assumption that lattice points may be regarded as scatterers is justified. We have shown that the direction of a diffracted intensity maximum is a reflexion of the incident X-ray beam in a plane which must be parallel to a lattice plane but may belong to a set of planes with an interplanar spacing that is a simple sub-multiple of the interplanar spacing of the parallel set of lattice planes, reflexion being restricted to angles that satisfy the Bragg Equation, $\lambda = 2d_{hkl} \sin\theta$. The spacing d_{hkl} of the planes involved is independent of the choice of unit-cell, but the indices (hkl) of the set of planes does depend on the choice of unit-cell. The diffraction pattern produced by a crystal is obviously unaffected by any arbitrary choice of unit-cell; it must therefore always be referable to a primitive, if unconventional, unit-cell and we can say that all diffracted beams are reflexions from planes indexed on a primitive unit-cell.

Suppose that a C-cell, with axes x_c, y_c, z_c, is related to a primitive unit-cell, with axes x_p, y_p, z_p, by the axial transformation matrix

$$\begin{matrix} 2 & 1 & 0 \\ 0 & 1 & 0 \\ 0 & 0 & 1 \end{matrix}$$

The transformation matrix from the C-cell to the P-cell is

$$\begin{matrix} \frac{1}{2} & -\frac{1}{2} & 0 \\ 0 & 1 & 0 \\ 0 & 0 & 1 \end{matrix}$$

The lattice, with both unit-cells outlined, is shown in Fig 6.18. It follows from the axial transformation matrix that the indices $(h_p k_p l_p)$ of a set of planes referred to the P-cell are related to indices referred to the C-cell by the equations

$$h_p = \tfrac{1}{2}(h_c - k_c)$$
$$k_p = k_c$$
$$l_p = l_c$$

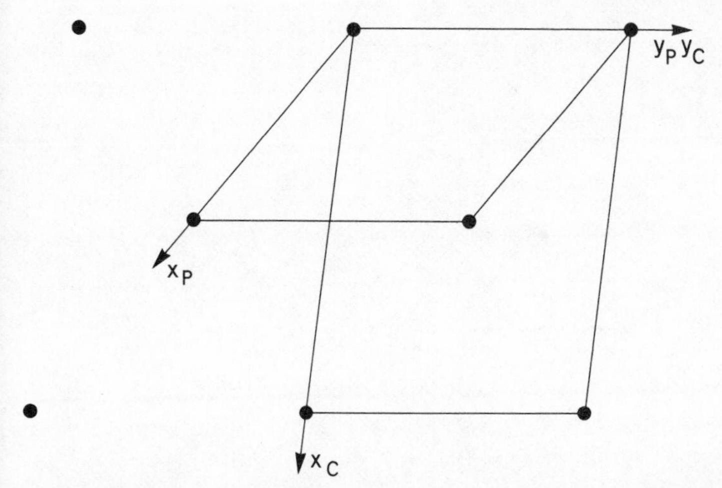

Fig 6.18 Relationship between a C-cell (x_c, y_c, z) and an unconventional P-cell (x_p, y_p, z) shown in projection on (001). Lattice points are represented by solid circles.

Since any diffracted beam produced by the crystal structure must be describable as a reflexion from a set of planes whose indices referred to the P-cell are integral, all X-ray reflexions must be describable as h_p, k_p, l_p reflexions. If h_p is to be integral, $h_c - k_c$ must be even; therefore $h_c + k_c$ must be even. When reflexions are indexed in terms of the C-cell, no reflexion can be observed for which $h_c + k_c$ is odd, i.e. the reflexions 100, 010, 120, etc will be systematically absent. A C-lattice is thus said to display a *systematic absence* or *extinction* for reflexions with $h + k = 2n + 1$, where n is an integer.

We can approach this sort of systematic absence, due to lattice type, in an alternative way. Let us assume that each lattice point scatters a wave equivalent to the resultant wave scattered by the atoms which it represents and consider the arrangement of lattice points in relation to lattice planes. Figure 6.19 shows the (001) projection of a C-lattice with the traces of (100), (200), and (120) planes outlined. Lattice points at the corners of the C-cell lie on (100) planes, but those at the centre of the (001) face of the unit-cell lie midway between (100) planes. The condition for occurrence of a 100 reflexion is that X-rays reflected from adjacent (100) lattice planes should have a phase difference of 2π; but the X-rays scattered by the (001)-face centring lattice points will then have a phase of π relative to the X-rays scattered by the lattice points at the corners of the unit-cell. Destructive interference will occur and, since in an effectively infinite lattice there will be equal numbers of both kinds of lattice point, the intensity of the 100 reflexion will be zero. From the traces of the (200) set of planes shown on the same lattice projection it is apparent that all lattice points lie on (200) planes and in consequence the X-rays scattered by all lattice points into the 200 reflexion are in phase. A 200 reflexion will therefore be observed unless, fortuitously, the resultant wave scattered in this direction by the atoms associated with a lattice point happens to be of zero amplitude. Figure 6.19 also shows the traces of the (120) set of planes on the same lattice projection. In this case the (001)-face centring lattice points lie midway between the planes of the set and scatter X-rays into the 120 reflexion with a phase $2\pi\frac{3}{2}$ relative to the X-rays scattered by the corner lattice points. Destructive interference takes place and the 120 reflexion has zero intensity. But, as in the case of the (200) planes, all lattice points lie on (240) planes, for which $d_{240} = \frac{1}{2}d_{120}$, and the 240 reflexion will, in general, have finite intensity.

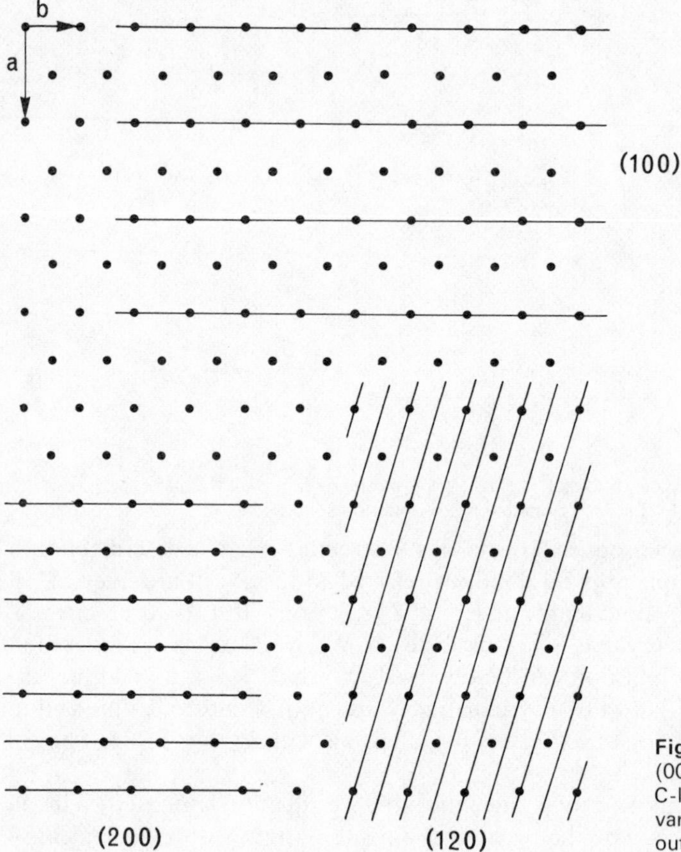

Fig 6.19 Projection on (001) of an orthorhombic C-lattice with the traces of various sets of lattice planes outlined.

Such arguments can be extended to general *hkl* reflexions: a reflexion *hkl* will be of non-zero intensity only if all lattice points lie on (*hkl*) planes. The equation (chapter 1) to a set of planes (*hkl*) is, in intercept form,

$$\frac{h}{a}x+\frac{k}{b}y+\frac{l}{c}z = n,$$

where n is an integer (positive, zero, or negative). The condition for a lattice point with coordinates x_n, y_n, z_n to lie on a plane of the (*hkl*) set is then

$$hx_n+ky_n+lz_n = n$$

If this equation is satisfied by all lattice points, then the *hkl* reflexion will have non-zero intensity; if the equation is not satisfied, a systematic absence will occur. In the case of a C-lattice there are lattice points with coordinates 0, 0, 0 and $\frac{1}{2}$, $\frac{1}{2}$, 0. The lattice point at the origin necessarily lies on all sets of planes; that at $\frac{1}{2}$, $\frac{1}{2}$, 0 lies only on planes for which $\frac{1}{2}h+\frac{1}{2}k$ is equal to an integer, that is on planes for which $h+k$ is even. Thus a systematic absence occurs for reflexions which have $h+k$ odd. A body-centred lattice has lattice points at 0, 0, 0 and $\frac{1}{2}$, $\frac{1}{2}$, $\frac{1}{2}$; the condition for the body-centring lattice points to lie on (*hkl*) planes is $\frac{1}{2}h+\frac{1}{2}k+\frac{1}{2}l = n$, where n is an integer, i.e. $h+k+l$ must be even for the *hkl* reflexion to be of non-zero intensity. An I-lattice thus gives rise to a systematic absence when $h+k+l$ is odd. All non-primitive

Table 6.2
Systematic absences displayed by conventional lattice types

	Coordinates of lattie points	Systematic absence
P	0, 0, 0	None
A	0, 0, 0; 0, $\frac{1}{2}$, $\frac{1}{2}$	$k+l = 2n+1$
B	0, 0, 0; $\frac{1}{2}$, 0, $\frac{1}{2}$	$h+l = 2n+1$
C	0, 0, 0; $\frac{1}{2}$, $\frac{1}{2}$, 0	$h+k = 2n+1$
F	0, 0, 0; 0, $\frac{1}{2}$, $\frac{1}{2}$; $\frac{1}{2}$, 0, $\frac{1}{2}$; $\frac{1}{2}$, $\frac{1}{2}$, 0	h, k, l *neither* all odd *nor* all even
I	0, 0, 0; $\frac{1}{2}$, $\frac{1}{2}$, $\frac{1}{2}$	$h+k+l = 2n+1$
R (hexagonal axes)	0, 0, 0; $\frac{2}{3}$, $\frac{1}{3}$, $\frac{1}{3}$; $\frac{1}{3}$, $\frac{2}{3}$, $\frac{2}{3}$	$-h+k+l = 3n \pm 1$
R (rhombohedral axes)	0, 0, 0	None

unit-cells exhibit such systematic absences; those arising from conventional non-primitive unit-cells are listed in Table 6.2.

The conditions that have to be fulfilled if a reflexion is to be of non-zero intensity when referred to a non-primitive unit-cell can be derived formally from the expression for the structure factor. If the lattice is referred to a C-cell, the N atoms in the unit-cell fall into two equivalent groups, each containing $M = \frac{1}{2}N$ atoms, related to each other by a translation $\frac{1}{2}a + \frac{1}{2}b$. For every atom with coordinates x_n, y_n, z_n there is an atom of the same element with coordinates $\frac{1}{2}+x_n$, $\frac{1}{2}+y_n$, z_n. The expression for the structure factor then becomes:

$$F(hkl) = \sum_{1}^{N} f_n \exp 2\pi i(hx_n + ky_n + lz_n)$$

$$= \sum_{1}^{M} f_n[\exp 2\pi i\{hx_n + ky_n + lz_n\} + \exp 2\pi i\{h(x_n + \tfrac{1}{2}) + k(y_n + \tfrac{1}{2}) + lz_n\}]$$

$$= \sum_{1}^{M} f_n \exp 2\pi i\{hx_n + ky_n + lz_n\} . \left\{1 + \exp 2\pi i \frac{h+k}{2}\right\}$$

Now
$$\exp 2\pi i \frac{h+k}{2} = \cos \pi(h+k) + i \sin \pi(h+k)$$
$$= (-1)^{h+k}$$

since h and k are integers. Therefore

$$F(hkl) = \sum_{1}^{M} f_n \exp 2\pi i\{hx_n + ky_n + lz_n\} . \{1 + (-1)^{h+k}\}$$

Therefore for $h+k = 2n$,

$$F(hkl) = 2 \sum_{1}^{M} f_n \exp 2\pi i\{hx_n + ky_n + lz_n\}$$

and for $h+k = 2n+1$,

$$F(hkl) = 0$$

Thus there is a systematic absence when $h+k$ is odd and when $h+k$ is even the amplitude of the wave scattered by one unit-cell is twice the resultant amplitude of the wave scattered by the atoms associated with any lattice point.

An F-lattice has four lattice points so that an atom at x_n, y_n, z_n is accompanied by atoms of the same element at $x_n, \frac{1}{2}+y_n, \frac{1}{2}+z_n$; $\frac{1}{2}+x_n, y_n, \frac{1}{2}+z_n$; and $\frac{1}{2}+x_n, \frac{1}{2}+y_n, z_n$. If the number of atoms associated with each lattice point is $M = \frac{1}{4}N$, the structure factor is

$$\begin{aligned} F(hkl) &= \sum_1^M f_n[\exp 2\pi i\{hx_n+ky_n+lz_n\} \\ &\quad +\exp 2\pi i\{hx_n+k(\tfrac{1}{2}+y_n)+l(\tfrac{1}{2}+z_n)\} \\ &\quad +\exp 2\pi i\{h(\tfrac{1}{2}+x_n)+ky_n+l(\tfrac{1}{2}+z_n)\} \\ &\quad +\exp 2\pi i\{h(\tfrac{1}{2}+x_n)+k(\tfrac{1}{2}+y_n)+lz_n\}] \\ &= \sum_1^M f_n \exp 2\pi i\{hx_n+ky_n+lz_n\} \\ &\quad \times\{1+\exp \pi i(k+l)+\exp \pi i(h+l)+\exp \pi i(h+k)\} \\ &= \sum_1^M f_n \exp 2\pi i\{hx_n+ky_n+lz_n\}.\{1+(-1)^{k+l}+(-1)^{h+l}+(-1)^{h+k}\} \end{aligned}$$

If h, k, and l are all odd or all even, $k+l$, $h+l$, and $h+k$ will all be even and

$$F(hkl) = 4\sum_1^M f_n \exp 2\pi i\{hx_n+ky_n+lz_n\}$$

But if h, k, and l are not all odd or all even, two of the sums $k+l$, $h+l$, and $h+k$ will be odd and the third even so that

$$F(hkl) = 0$$

Thus there is a systematic absence when h, k, and l are neither all odd nor all even; but when h, k, and l are all odd or all even the structure factor is four times that for the group of atoms associated with one lattice point.

Evidently then the labour of calculating structure amplitudes for non-primitive unit-cells will be appreciably lessened if all atoms that are related by lattice translations are first grouped together. By way of example we consider the cubic form of ZnS, the mineral *blende*, which has four formula units in the unit-cell and an F-lattice (Fig 6.20). One formula unit is necessarily associated with each lattice point and we choose to take as the repeat unit a zinc atom at the origin and a sulphur atom at $\frac{1}{4}, \frac{1}{4}, \frac{1}{4}$. Since we are dealing with an F-lattice reflexions of non-zero

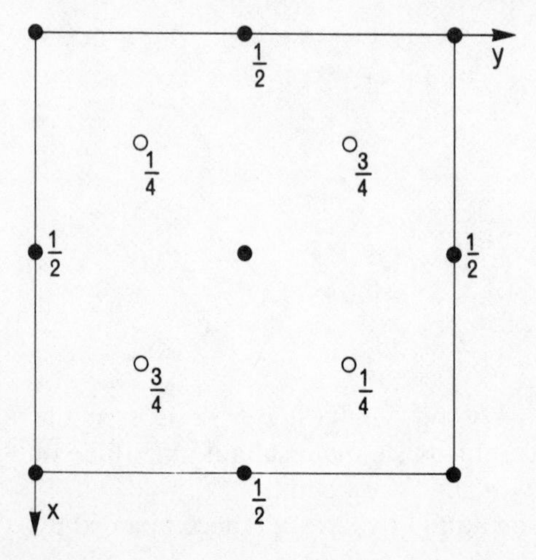

Fig 6.20 Projection of the structure of blende (ZnS) on (001). Solid circles represent Zn and open circles S atoms.

intensity will occur only when h, k, and l are all odd or all even, when the structure factor will be

$$F(hkl) = 4\sum_{1}^{M} f_n \exp 2\pi i(hx_n + ky_n + lz_n)$$

Here $M = 2$, therefore

$$F(hkl) = 4\left\{f_{\mathrm{Zn}} \exp 2\pi i.0 + f_{\mathrm{S}} \exp 2\pi i \frac{h+k+l}{4}\right\}$$

Therefore $$|F(hkl)|^2 = 16\left\{\left(f_{\mathrm{Zn}} \cos 0 + f_{\mathrm{S}} \cos 2\pi \frac{h+k+l}{4}\right)^2 + \left(f_{\mathrm{Zn}} \sin 0 + f_{\mathrm{S}} \sin 2\pi \frac{h+k+l}{4}\right)^2\right\}$$

$$= 16\left\{\left(f_{\mathrm{Zn}} + f_{\mathrm{S}} \cos 2\pi \frac{h+k+l}{4}\right)^2 + \left(f_{\mathrm{S}} \sin 2\pi \frac{h+k+l}{4}\right)^2\right\}$$

Only three possibilities arise:

if $h+k+l = 4n$ $\quad |F(hkl)|^2 = 16(f_{\mathrm{Zn}} + f_{\mathrm{S}})^2$

if $h+k+l = 4n+2$ $\quad |F(hkl)|^2 = 16(f_{\mathrm{Zn}} - f_{\mathrm{S}})^2$

if $h+k+l = 2n+1$ $\quad |F(hkl)|^2 = 16(f_{\mathrm{Zn}}^2 + f_{\mathrm{S}}^2)$

In the foregoing pages we have explored the systematic absences due to lattice type by examining the structure factors for the various lattice types. The resulting list (Table 6.2) of systematic absences provides the means of determining the lattice type of a crystalline substance by inspection of its diffraction pattern for systematically absent reflexions. Later we shall return to systematic absences to consider those due to the presence of translational symmetry elements.

Symmetry of X-ray diffraction patterns

Friedel's Law

We have seen that the intensities of X-ray reflexions are dependent on the positions of the atoms in the unit-cell and this would lead us to expect that the diffraction pattern produced by a crystal would exhibit a symmetry related to that of the disposition of the atoms in the unit-cell. Such a relationship would provide a means of determining the symmetry of the arrangement of atoms in the crystal by examination of its diffraction pattern. There is however a difficulty: there is no known means of determining the relative phases of X-ray reflexions. All that can be determined about an X-ray reflexion is its intensity and its angular position; the information about the symmetry of the atomic arrangement in a crystal that can be deduced from the symmetry of its diffraction pattern is in consequence limited.

We have already shown that the intensity of an X-ray reflexion is the product of the relevant structure factor and its complex conjugate,

i.e. $$F(hkl) = \sum_{1}^{N} f_n \cos 2\pi(hx_n + ky_n + lz_n) + i\sum_{1}^{N} f_n \sin 2\pi(hx_n + ky_n + lz_n)$$

and $$F(hkl)^* = \sum_{1}^{N} f_n \cos 2\pi(hx_n + ky_n + lz_n) - i\sum_{1}^{N} f_n \sin 2\pi(hx_n + ky_n + lz_n)$$

But $\cos(-\theta) = \cos\theta$ and $\sin(-\theta) = -\sin\theta$

Therefore
$$F(hkl)^* = \sum_1^N f_n \cos 2\pi(-hx_n - ky_n - lz_n) + i\sum_1^N f_n \sin 2\pi(-hx_n - ky_n - lz_n)$$
$$= F(\bar{h}\bar{k}\bar{l})$$

and similarly
$$F(hkl) = F(\bar{h}\bar{k}\bar{l})^*$$

Therefore
$$I(hkl) = F(hkl)F(hkl)^* = F(\bar{h}\bar{k}\bar{l})^* F(\bar{h}\bar{k}\bar{l})$$
$$= I(\bar{h}\bar{k}\bar{l}).$$

The intensities of the *hkl* and $\bar{h}\bar{k}\bar{l}$ reflexions are thus necessarily equal and in consequence an X-ray diffraction pattern always displays a centre of symmetry[10] or, in other words, the reflexions from either side of a set of (*hkl*) planes are invariably equal in intensity (Fig 6.21); this result is often called *Friedel's Law*.

If the crystal structure does not possess a centre of symmetry, although the intensities of the *hkl* and $\bar{h}\bar{k}\bar{l}$ reflexions must be equal their phases will be unequal. Consider the 111 and $\bar{1}\bar{1}\bar{1}$ reflexions of blende illustrated in Fig 6.21(a). The structure of this form of ZnS (Fig 6.20) is non-centrosymmetric, having a repeat unit consisting of Zn at 0, 0, 0 and S at $\frac{1}{4}, \frac{1}{4}, \frac{1}{4}$. The wave scattered by zinc into all reflexions thus has zero phase and amplitude f_{Zn}, while that scattered by sulphur has phase $2\pi(h+k+l)/4$ and amplitude f_S. Since the (111) and ($\bar{1}\bar{1}\bar{1}$) planes are of the same set, the atomic scattering factors of Zn and S will be the same for both. It will be apparent from Fig 6.21(b) and (c) that the amplitude and consequently the intensity of the 111 and $\bar{1}\bar{1}\bar{1}$ reflexions will be equal and that their phases will be equal in magnitude but opposite in sign.

Friedel's Law holds only so long as the assumption that every atom in the structure scatters X-rays with a phase change of π remains valid. The assumption breaks down only when the X-ray wavelength is close to the wavelength of an absorption edge of one of the constituent atomic species of the crystal. Discussion of such *anomalous scattering* is outside the scope of this book; the reader is referred to James (1967), for a detailed treatment. It is important however to bear in mind that, in favourable circumstances, anomalous scattering may provide a means of establishing the absence of a centre of symmetry in a crystal structure (chapter 11).

Laue symmetry

If two or more planes are related by symmetry then the X-ray reflexions to which they give rise will have equal Bragg angles (but will in general require different orientations of the incident X-ray beam) and equal intensities. Therefore examination of the symmetry of the total diffraction pattern produced by a crystal should yield information about the point group of the crystal; but because diffraction patterns are, by Friedel's Law, necessarily centrosymmetric a unique determination of the point group is not possible. The point group symmetry of the diffraction pattern will be the same as that of the crystal if the crystal is centrosymmetric; but, if the crystal

[10] The reader should be aware that this statement refers to the whole diffraction pattern. Since a crystal is effectively a three-dimensional grating the whole diffraction pattern cannot be recorded with a single orientation of the crystal relative to the incident X-ray beam and consequently the symmetry of the whole diffraction pattern may not be evident in a single X-ray photograph. This point will be amplified in chapter 8.

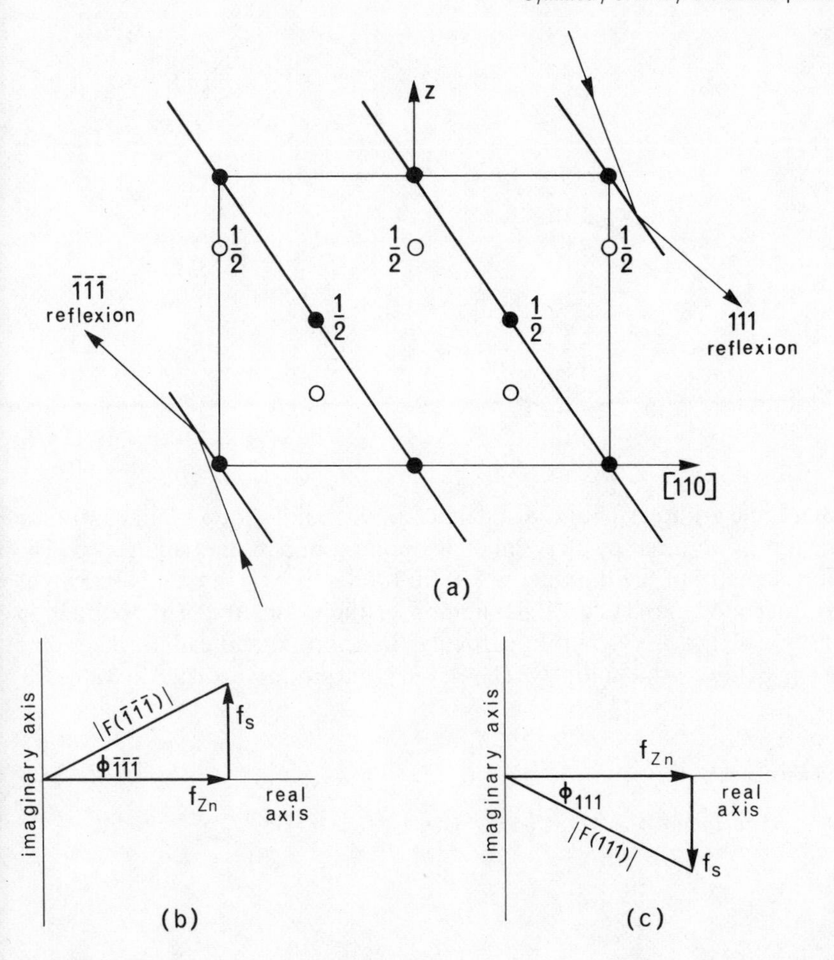

Fig 6.21 Friedel's Law. (a) is a projection of the structure of blende (Zn solid circles, S open circles) on the $(1\bar{1}0)$ plane; atomic heights above that plane are shown as fractions of the lattice repeat $(a/\sqrt{2})$ in the $[1\bar{1}0]$ direction. (b) and (c) are vector diagrams representing the amplitude and phase of the $\bar{1}\bar{1}\bar{1}$ and 111 reflexions.

is non-centrosymmetric, the point group of the diffraction pattern will be the point group obtained by adding a centre of symmetry to the point group of the crystal. For instance the symmetry of the diffraction pattern produced by a crystal of class 4 will be that of the point group $4/m$ (Fig 6.22). Conversely, if the diffraction symmetry of a crystal is identified as $4/m$, then the point group of the crystal may be $4/m$ or any point group that becomes $4/m$ when a centre of symmetry is added to it; reference to Fig 3.17 enables the non-centrosymmetric point groups to be identified as 4 and $\bar{4}$. Such a crystal is said to belong to the *Laue group* $4/m$, a statement which implies that the point group of the crystal is either 4 or $\bar{4}$ or $4/m$.

There are eleven centrosymmetric crystallographic point groups and it is evident from Fig 3.20 that addition of a centre of symmetry to the symmetry elements of any of the twenty-one non-centrosymmetric point groups yields one of these eleven. A centrosymmetric point group and all those non-centrosymmetric point groups which on addition of a centre of symmetry become identical with it constitute a *Laue group*. Each such group of point groups is assigned the symbol of its centrosymmetric point

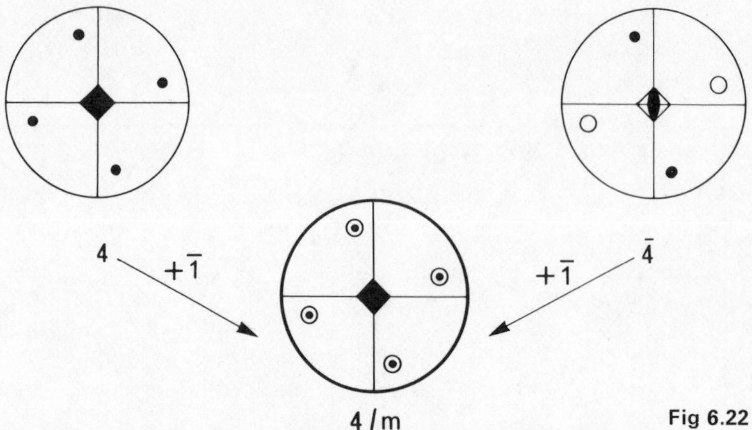

Fig 6.22 Laue group $4/m$.

group; such a Laue group symbol is a statement of the point group symmetry of the diffraction pattern produced by a crystal of any point group in the Laue group. The point group symmetry of the diffraction pattern produced by a crystal is known as the *Laue symmetry* of the crystal. Examination of the symmetry of its diffraction pattern can only assign a crystal to a particular Laue group and cannot determine the point group of the crystal uniquely. The eleven Laue groups are listed in Table 6.3.

Table 6.3
The eleven Laue Groups

System	Laue group	Constituent point groups of the Laue group			
Triclinic	$\bar{1}$	1	$\bar{1}$		
Monoclinic	$2/m$	2	m	$2/m$	
Orthorhombic	mmm	222	$mm2$	mmm	
Trigonal	$\bar{3}$	3	$\bar{3}$		
	$\bar{3}m$	32	$3m$	$\bar{3}m$	
Tetragonal	$4/m$	4	$\bar{4}$	$4/m$	
	$4/mmm$	422	$4mm$	$\bar{4}2m$	$4/mmm$
Hexagonal	$6/m$	6	$\bar{6}$	$6/m$	
	$6/mmm$	622	$6mm$	$\bar{6}m2$	$6/mmm$
Cubic	$m3$	23	$m3$		
	$m3m$	432	$\bar{4}3m$	$m3m$	

Systematic absences due to translational elements

We have already seen that selection of a non-primitive unit-cell gives rise to systematically absent reflexions and that the rules governing such absences apply to all reflexions from the crystal. But systematic absences also arise from the presence of glide planes or screw axes in the space group of a crystal; such absences are restricted to one zone of planes, in the case of a glide plane, or to one set of planes, in the case of a screw axis.

The presence of an a-glide through the origin parallel to (001) causes an atom at x, y, z to be duplicated at $\frac{1}{2}+x, y, \bar{z}$. The expression for the structure factor then becomes

$$F(hkl) = \sum_{1}^{\frac{1}{2}N} f_n\{\exp 2\pi i(hx_n + ky_n + lz_n) + \exp 2\pi i(\tfrac{1}{2}h + hx_n + ky_n - lz_n)\}$$

$$= \sum_{1}^{\frac{1}{2}N} f_n \exp 2\pi i(hx_n + ky_n)\{\exp 2\pi i\, lz_n + \exp 2\pi i(\tfrac{1}{2}h - lz_n)\}$$

which becomes when $l = 0$,

$$F(hk0) = \sum_{1}^{\frac{1}{2}N} f_n \exp 2\pi i(hx_n + ky_n)\{1 + \exp \pi i\, h\}$$

$$= \sum_{1}^{\frac{1}{2}N} f_n \exp 2\pi i(hx_n + ky_n)\{1 + (-1)^h\}$$

$$= 0 \quad \text{for} \quad h = 2n+1.$$

Thus an a-glide through the origin parallel to (001) produces a systematic absence in the $hk0$ reflexions when h is odd. The restriction that the glide plane should pass through the origin is not significant because it is evident that the phase of the X-rays scattered into the $hk0$ reflexions must be independent of the z coordinates of the atoms in the unit-cell, the phase difference between the X-rays scattered into any $hk0$ reflexion by two atoms related by an a-glide parallel to (001) being

$$2\pi(\tfrac{1}{2}h + hx_n + ky_n) - 2\pi(hx_n + ky_n) = \pi h;$$

wherever the glide plane intersects the z-axis the two atoms will scatter exactly out of

Table 6.4
Systematic absences produced by glide planes parallel to (001)

Type of glide	Translation	Systematic absences in $hk0$ reflexions
a	$\frac{a}{2}$	$h = 2n+1$
b	$\frac{b}{2}$	$k = 2n+1$
n	$\frac{a+b}{2}$	$h+k = 2n+1$
d	$\frac{a \pm b}{4}$	$h+k = 4n+2$ with $h = 2n$ and $k = 2n$

phase when h is odd to give a systematic absence and exactly in phase when h is even. Such a glide plane in effect halves the lattice spacing parallel to the x-axis where reflexions in the [001] zone are concerned.

All glide planes give rise to systematic absences in the reflexions of the zone whose axis is normal to the glide plane. The systematic absences produced by all possible types of (001) glide plane are listed in Table 6.4. Analogous systematic absences arise from glide planes parallel to (100) and (010). A complete list of the conventional glide planes and their associated systematic absences is to be found in *International Tables for X-ray Crystallography*, vol. I (1969), p. 54.

The presence of a screw diad through the origin parallel to the z-axis causes an atom at x, y, z to be duplicated at $\bar{x}, \bar{y}, \frac{1}{2}+z$. The expression for the structure factor then becomes

Table 6.5
Systematic absences produced by screw axes parallel to [001]

Screw axis	Translation	Systematic absences in $00l$ reflexions
2_1	$c/2$	$l = 2n+1$
4_1 and 4_3	$\pm c/4$	$l \neq 4n$
4_2	$c/2$	$l = 2n+1$
3_1 and 3_2	$\pm c/3$	$l \neq 3n$
6_1 and 6_5	$\pm c/6$	$l \neq 6n$
6_2 and 6_4	$\pm c/3$	$l \neq 3n$
6_3	$c/2$	$l = 2n+1$

$$F(hkl) = \sum_{1}^{\frac{1}{2}N} f_n\{\exp 2\pi i(hx_n + ky_n + lz_n) + \exp 2\pi i(-hx_n - ky_n + \tfrac{1}{2}l + lz_n)\}$$
$$= \sum_{1}^{\frac{1}{2}N} f_n \exp 2\pi i\, lz_n\{\exp 2\pi i(hx_n + ky_n) + \exp 2\pi i(-hx_n - ky_n + \tfrac{1}{2}l)\}$$

which becomes when $h = k = 0$,

$$F(00l) = \sum_{1}^{\frac{1}{2}N} f_n \exp 2\pi i\, lz_n\{1 + \exp \pi i\, l\}$$
$$= \sum_{1}^{\frac{1}{2}N} f_n \exp 2\pi i\, lz_n\{1 + (-1)^l\}$$
$$= 0 \quad \text{for} \quad l = 2n+1$$

Thus a screw diad parallel to z produces a systematic absence in the $00l$ reflexions when l is odd. The restriction that the 2_1 axis should pass through the origin is not significant because the phase of the X-rays scattered into the $00l$ reflexions must be independent of the x and y coordinates of the atoms in the unit-cell, the phase difference between the X-rays scattered into any $00l$ reflexion by two atoms related by a 2_1 axis parallel to z being $2\pi(\frac{1}{2}l + lz_n) - 2\pi lz_n = \pi l$; wherever the screw axis intersects the xy plane the two atoms will scatter exactly out of phase when l is odd and exactly in phase when l is even. Such a screw diad effectively halves the lattice spacing parallel to the z-axis for the $00l$ reflexions.

Screw axes parallel to other axes and with different translations give rise to analogous systematic absences. A list of systematic absences produced by all possible types of screw axis parallel to [001] is given in Table 6.5.

Diffraction symbols

We have already seen that observation of the symmetry of the diffraction pattern produced by a crystal enables the crystal to be assigned to a particular Laue group. Observation of systematic absences in the diffraction pattern enables the lattice type to be determined uniquely and the presence of translational symmetry elements to be detected. All this information can conveniently be expressed as the *diffraction symbol* of the crystal. Since non-translational symmetry elements do not give rise to systematic absences the diffraction symbol will represent a group of space groups; occasionally this will be a group of one and then the space group is uniquely determined.

A diffraction symbol consists of three parts: first, a statement of the Laue group; second, a statement of the lattice type; third, three spaces in which the symbols of any screw axes or glide-planes that have been detected are written in the same order as

Table 6.6
Diffraction symbols derived from *mmm*P.*cn* by reorientation of axes

The original axes are denoted $x\ y\ z$ and the new axes $x'\ y'\ z'$.

x'	y'	z'	Diffraction symbol
x	y	z	*mmm* P.*cn*
z	x	y	*mmm* P*n*.*a*
y	z	x	*mmm* P*bn*.
x	$\bar{z}$	y	*mmm* P.*nb*
y	x	$\bar{z}$	*mmm* P*c*.*n*
$\bar{z}$	y	x	*mmm* P*na*.

the corresponding non-translational symmetry element symbols appear in the relevant conventional point group symbol, absence of information about any of these three directions being indicated by a full stop. By way of example we take the diffraction symbol *mmm* P*bcn*; this means that the crystal belongs to Laue group *mmm*, has a primitive lattice, and has a *b*-glide parallel to (100), a *c*-glide parallel to (010), and an *n*-glide parallel to (001). In this case the space group is uniquely determined as P*bcn*. Suppose now that another crystal yields a diffraction pattern of symmetry *mmm* with systematic absences in $h0l$ for $l = 2n+1$, $hk0$ for $h+k = 2n+1$, $h00$ for $h = 2n+1$, $0k0$ for $k = 2n+1$, $00l$ for $l = 2n+1$. The diffraction symbol of this crystal will be *mmm* P.*cn* and this implies that its point group must be either *mmm* or 2*mm*; the systematic absences observed in the $h00$, $0k0$, and $00l$ reflexions are merely consequent on the more general conditions that apply to the $h0l$ and $hk0$ reflexions and do not necessarily imply the presence of screw diads parallel to x, y, and z. In this case however it can readily be seen by drawing out the space group diagrams for P.*cn* that this pair of glide-planes generates screw diads parallel to x. We know therefore that there is no (100) glide-plane and that there must be a [100] screw diad; there may or may not be a (100) mirror-plane. The determination of space group is not unique; there are two possibilities, $P2_1cn$ (which belongs to point group 2*mm*) and P*mcn* (which belongs to point group *mmm* and is illustrated in Fig 4.18).

When the diffraction symbol of a crystal has been obtained it should be compared with the list of diffraction symbols of the space groups in *International Tables for X-ray Crystallography*, vol. I (1969), pp. 349–352; but a word of caution is necessary. The list has been drawn up in terms of certain arbitrary conventions so that our observed diffraction symbol may not appear in the list but be represented by the corresponding diffraction symbol for a different axial orientation; this difficulty, which is most likely to occur in the orthorhombic system, is simply resolved by transforming the observed diffraction symbol for all possible axial orientations and selecting the setting consistent with the *International Tables* convention. Alternative settings for *mmm* P.*cn* are listed in Table 6.6.

In the case of monoclinic diffraction symbols the first term, the Laue group symbol, is written in such a way as to indicate unambiguously the choice of unique axis: as $12/m1$ if y is the unique axis and as $112/m$ if z is the unique axis. Also in this system three spaces are provided in the last term; but only one can be used, the second if y is unique and the third if z is unique. For instance consider a crystal with Laue symmetry $2/m$, the diad being parallel to y, and systematic absences for hkl when $h+k = 2n+1$, for $h0l$ when $h = 2n+1$, or $l = 2n+1$, for $0k0$ when $k = 2n+1$. Its diffraction symbol is $I2/m1C.c.$; the significant absences are in *hkl* for $h+k$ odd and in *h*0*l* for *l* odd, the other observed absences being consequent on these. In the monoclinic system it is important to bear in mind that symmetry elements are associated with one axial direction only; if that is chosen as *y*, then systematic absences in the 0*kl* and *hk*0 reflexions can only be due to systematic absences in the general *hkl* reflexions and systematic absences in *h*00 and 00*l* reflexions may be derivative from systematic absences in either *hkl* or *h*0*l* reflexions. Returning to our example, $I2/m1C.c.$, two space groups are possible, C*c* or C2/*c*. The former belongs to point group *m*. The latter belongs to point group 2/*m*, has diads and screw diads parallel to *y*, and has an *n*-glide as well as a *c*-glide parallel to (010) (Fig 4.19); the diad does not give rise to any systematic absences, the systematic absences due to the screw diad, 0*k*0 for $k = 2n+1$, are masked by those due to the C-lattice, *hkl* for $h+k = 2n+1$, and the systematic absences due to the *n*-glide, *h*0*l* for $h+l = 2n+1$, are masked by the conditions for reflexion by a C-lattice and a *c*-glide, which require *h* and *l* to be even for a reflexion to be observed.

The reciprocal lattice[11]

The interpretation of X-ray diffraction patterns by direct application of the Laue Equations or the Bragg Equation can be highly tedious. The reciprocal lattice concept provides however a means of interpretation that is at once elegant, powerful and—once one has overcome the initial conceptual difficulties—easy to use.

A reciprocal lattice is constructed by representing each set of (*hkl*) lattice planes by a *reciprocal lattice point* at a distance from the origin d^*_{hkl} inversely proportional to the interplanar spacing d_{hkl} in the direction of the normal from the origin to the planes of the set. We can thus put $d^*_{hkl} = K/d_{hkl}$. K is usually taken as unity in theoretical work and equal to the wavelength λ of the incident X-radiation in the practical task of interpretation of diffraction patterns.

Figure 6.23 shows a section through a crystal in the plane defined by the *x*-axis and the normal OP from the origin O to a set of planes (*hkl*). If the trace AP of the first (*hkl*) plane out from the origin intersects the *x*-axis in A, then $OA = a/h$, $OP = d_{hkl}$, and ω, the angle between the normal to the planes and the *x*-axis, is given by $\cos\omega = d_{hkl}/(a/h)$. The reciprocal lattice point representing the set of (*hkl*) planes is P* lying on OP produced at a distance from the origin $OP^* = K/d_{hkl}$. If a line perpendicular to the *x*-axis is drawn through P* and intersects the *x*-axis in A*, then $OA^* = OP^*\cos\omega = (K/d_{hkl}).d_{hkl}/(a/h) = hK/a$. All reciprocal lattice points with a given value of the index *h* thus lie in a plane normal to the *x*-axis, the perpendicular distance of the plane from the origin being hK/a. It follows that any reciprocal lattice point must lie on a plane of the parallel set of planes normal to the *x*-axis with interplanar spacing K/a, each plane of the set being characterized by a particular value of *h* (Fig 6.23(b)). Similarly it can be shown that reciprocal lattice points are confined to a set of planes normal to the *y*-axis with interplanar spacing K/b, and to a set of planes normal to the *z*-axis with interplanar spacing K/c. Reciprocal lattice points thus lie at the points of mutual intersection of planes of three parallel sets, normal to the *x*, *y*, and *z* axes and with interplanar spacings K/a, K/b, and K/c respectively. The array of reciprocal lattice points is thus itself a lattice, the *reciprocal lattice*.

[11] This section, which extends to the end of the chapter, can be omitted at a first reading. Understanding of subsequent chapters is not dependent on knowledge of this section, except for certain advanced sections that are similarly indicated.

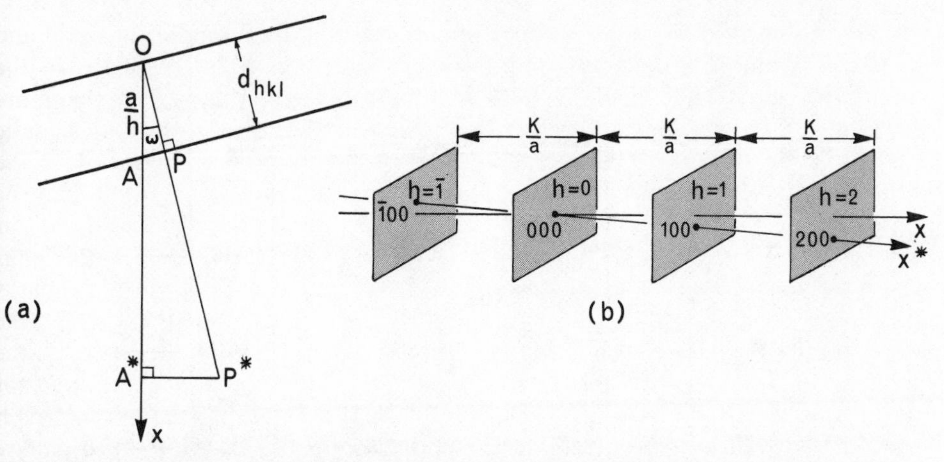

Fig 6.23 The reciprocal lattice. The (hkl) planes through the origin O and the first one out from the origin AP are shown as bold lines in (a). The reciprocal lattice point P* representing the (hkl) planes lies at a distance K/d_{hkl} from the origin on OP produced. A* lies on OA produced, that is the x-axis of the direct lattice, such that $O\widehat{A^*}P^* = 90°$; since $OA^* = hK/a$ all reciprocal lattice points with the same value of h will lie in a plane normal to the x-axis of the direct lattice. Reciprocal lattice points thus lie on planes perpendicular to the x-axis, each plane being characterized by a particular value of h and the interplanar separation being K/a, as shown in (b). Analogous planes with separation K/b, K/c occur perpendicular to the y and z axes respectively.

A reciprocal lattice point is identified by the indices of the set of (hkl) planes that it represents; conventionally the indices of a reciprocal lattice point are, like those for X-ray reflexions, written without parentheses, i.e. the reciprocal lattice point hkl represents the (hkl) set of planes in reciprocal space. The reference axes of a reciprocal lattice are denoted x^*, y^*, z^* and taken respectively in the directions of the normals to the (100), (010), and (001) planes. The reciprocal lattice points 100, 200, 300, ... thus lie on the x^*-axis, the repeat distance along which is designated a^*; $a^* = K/d_{100}$. The angles between the reciprocal axes are denoted α^*, β^*, and γ^*; that they are respectively equal to the angles (010):(001), (001):(100), and (100):(010) is evident from Fig 6.24. Reciprocal lattice constants are thus dependent on those of the corresponding *direct lattice*, which we have hitherto referred to as the crystal lattice.

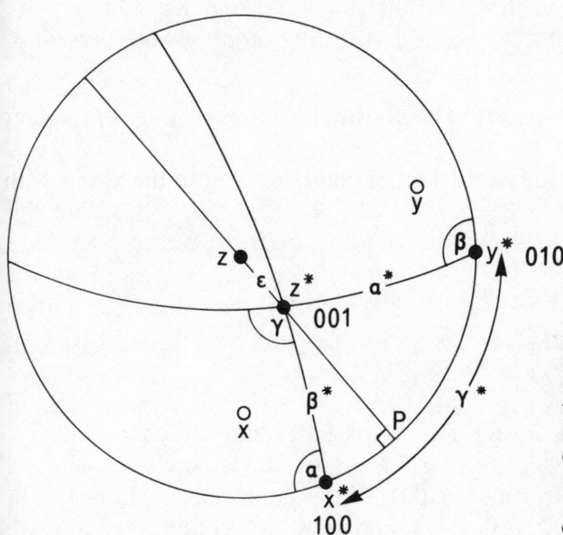

Fig 6.24 Stereogram to show the relationship between the direct and reciprocal axes, x, y, z and x^*, y^*, z^* in the general triclinic case. The reciprocal lattice angles α^*, β^*, γ^* are respectively equal to the angles between the planes (010) : (001), (001) : (100), (100) : (010). P is the intersection of the orthogonal great circles zz^* and x^*y^*.

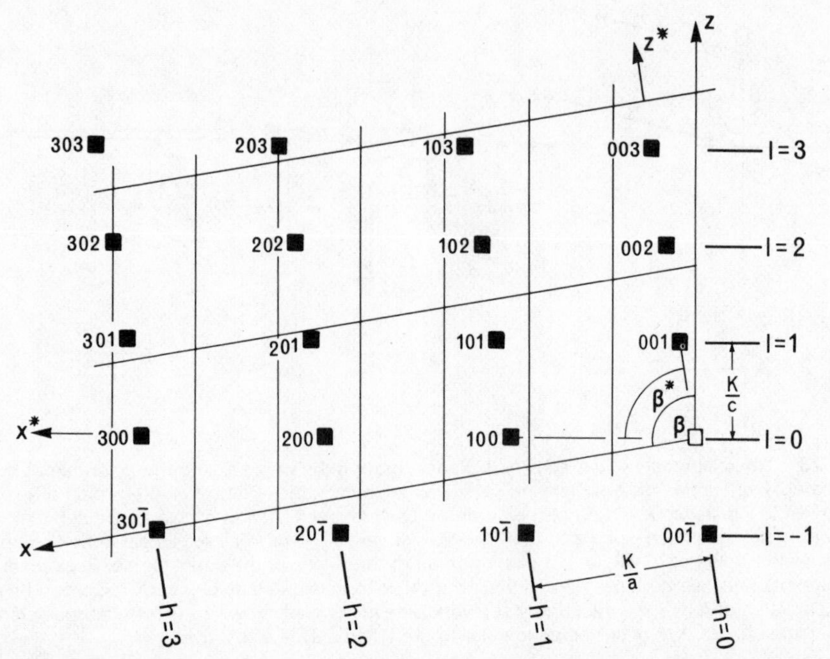

Fig 6.25 Relationship between reciprocal lattice points (solid squares except 000 which is an open square) and the direct lattice for the *xz* plane of a primitive monoclinic lattice. The thin lines represent the (100) and (001) planes of the direct lattice. The bold lines on the edges of the diagram indicate reciprocal lattice point rows of constant *h* or constant *l* in the *h*0*l* reciprocal lattice plane. Reciprocal and direct axial directions are indicated.

The relationship between reciprocal lattice points and direct lattice planes is illustrated for the xz plane of a typical primitive monoclinic lattice in Fig 6.25. In the next paragraph we develop the interrelationships between reciprocal and direct lattice constants for the general triclinic case; in systems of higher symmetry these relationships are simplified and those specific to each crystal system are shown in Table 6.7.

The external angles at the apices of the spherical triangle $x^*y^*z^*$ (Fig 6.24) are of course the interaxial angles of the direct lattice (Fig 3.38) and its sides are by definition the interaxial angles of the reciprocal lattice. Therefore by equation (6) of chapter 5,

$$\cos(180° - \gamma) = -\cos(180° - \alpha)\cos(180° - \beta) + \sin(180° - \alpha)\sin(180° - \beta)\cos\gamma^*$$

i.e.
$$\cos\gamma^* = \frac{\cos\alpha\cos\beta - \cos\gamma}{\sin\alpha\sin\beta}$$

And from equation (3) of chapter 5,

$$\cos\gamma^* = \cos\alpha^*\cos\beta^* + \sin\alpha^*\sin\beta^*\cos(180° - \gamma)$$

i.e.
$$\cos\gamma = \frac{\cos\alpha^*\cos\beta^* - \cos\gamma^*}{\sin\alpha^*\sin\beta^*}$$

Similar expressions can be written down for $\cos\alpha^*$, $\cos\beta^*$, $\cos\alpha$, and $\cos\beta$.

Now $c^* = K/d_{001} = K/c\cos\varepsilon$, where ε (Fig 6.27) is the angle between z and z^*, that is, the angle between z and the normal to (001). If P is defined as the intersection of the great circle through z and z^* with the primitive, which contains x^* and y^*,

Table 6.7
Relationships between direct and reciprocal lattice constants

Triclinic	$a^* = \dfrac{K}{a \sin\beta \sin\gamma^*} = \dfrac{K}{a \sin\beta^* \sin\gamma}$	
	$b^* = \dfrac{K}{b \sin\gamma \sin\alpha^*} = \dfrac{K}{b \sin\gamma^* \sin\alpha}$	
	$c^* = \dfrac{K}{c \sin\alpha \sin\beta^*} = \dfrac{K}{c \sin\alpha^* \sin\beta}$	
	$\cos\alpha^* = \dfrac{\cos\beta \cos\gamma - \cos\alpha}{\sin\beta \sin\gamma}$	$\cos\alpha = \dfrac{\cos\beta^* \cos\gamma^* - \cos\alpha^*}{\sin\beta^* \sin\gamma^*}$
	$\cos\beta^* = \dfrac{\cos\gamma \cos\alpha - \cos\beta}{\sin\gamma \sin\alpha}$	$\cos\beta = \dfrac{\cos\gamma^* \cos\alpha^* - \cos\beta^*}{\sin\gamma^* \sin\alpha^*}$
	$\cos\gamma^* = \dfrac{\cos\alpha \cos\beta - \cos\gamma}{\sin\alpha \sin\beta}$	$\cos\gamma = \dfrac{\cos\alpha^* \cos\beta^* - \cos\gamma^*}{\sin\alpha^* \sin\beta^*}$
Monoclinic	$a^* = \dfrac{K}{a \sin\beta}$	$\alpha^* = \alpha = 90°$
	$b^* = \dfrac{K}{b}$	$\beta^* = 180° - \beta$
	$c^* = \dfrac{K}{c \sin\beta}$	$\gamma^* = \gamma = 90°$
Orthorhombic ($a \neq b \neq c$)	$a^* = \dfrac{K}{a}$	$\alpha^* = \alpha = 90°$
Tetragonal ($a = b \neq c$)	$b^* = \dfrac{K}{b}$	$\beta^* = \beta = 90°$
Cubic ($a = b = c$)	$c^* = \dfrac{K}{c}$	$\gamma^* = \gamma = 90°$
Hexagonal, Trigonal ($a = b \neq c$, $\gamma = 120°$)	$a^* = \dfrac{K}{a \sin 60°}$	$\alpha^* = \alpha = 90°$
	$b^* = \dfrac{K}{b \sin 60°}$	$\beta^* = \beta = 90°$
	$c^* = \dfrac{K}{c}$	$\gamma^* = (180° - \gamma) = 60°$

the spherical triangle $z^*\mathrm{P}y^*$ is right-angled at P (Fig 6.24) and application of Napier's Rules to this triangle (Fig 6.26) gives

$$\cos\varepsilon = \sin\alpha^* \sin\beta$$

Therefore $$c^* = \frac{K}{c \sin\alpha^* \sin\beta}$$

This expression, which involves one interaxial angle of the reciprocal lattice and one of the direct lattice, is simple to use. Analogous expressions for a^* and b^* may be derived in an identical manner and are listed in Table 6.7. A more complicated expression, which relates c^* to direct lattice constants only can be derived in the

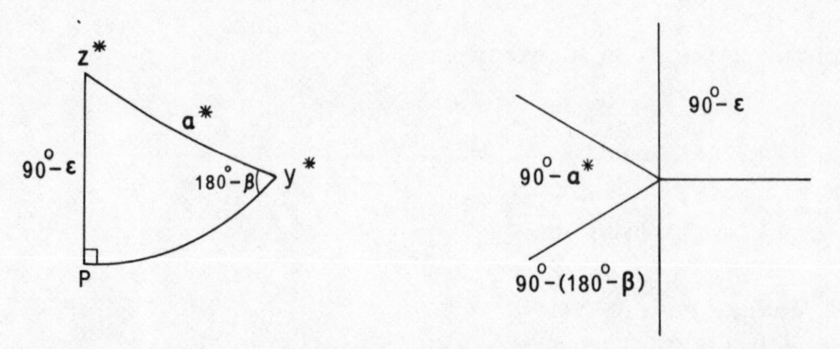

Fig 6.26 The spherical triangle z^*Py^* of Fig 6.24 and its Naperian diagram.

following way. The volume V of a triclinic unit-cell is given by the product of the area of its base $ab\sin\gamma$ and its height normal to the xy plane d_{001} (Fig 6.27).

Therefore
$$\begin{aligned} V &= ab\sin\gamma . d_{001} \\ &= abc\sin\gamma\cos\varepsilon \\ &= abc\sin\alpha^*\sin\beta\sin\gamma \end{aligned}$$

But
$$\cos\alpha^* = \frac{\cos\beta\cos\gamma - \cos\alpha}{\sin\beta\sin\gamma}$$

Therefore
$$\begin{aligned} V &= abc\sin\beta\sin\gamma\sqrt{\left(1 - \frac{(\cos\beta\cos\gamma-\cos\alpha)^2}{\sin^2\beta\sin^2\gamma}\right)} \\ &= abc\sqrt{\{(1-\cos^2\beta)(1-\cos^2\gamma)} \\ &\qquad -(\cos^2\beta\cos^2\gamma - 2\cos\alpha\cos\beta\cos\gamma + \cos^2\alpha)\} \\ &= abc\sqrt{(1-\cos^2\alpha-\cos^2\beta-\cos^2\gamma+2\cos\alpha\cos\beta\cos\gamma)} \end{aligned}$$

Now since $c^* = \dfrac{K}{c\cos\varepsilon}$ and $V = abc\sin\gamma\cos\varepsilon$,

we can write
$$c^* = \frac{K\,ab\sin\gamma}{V}$$

Therefore
$$c^* = \frac{K\sin\gamma}{c\sqrt{(1-\cos^2\alpha-\cos^2\beta-\cos^2\gamma+2\cos\alpha\cos\beta\cos\gamma)}}$$

Analogous relationships exist between a^* and a, b^* and b.

In those systems where the axes of the direct lattice are orthogonal, the

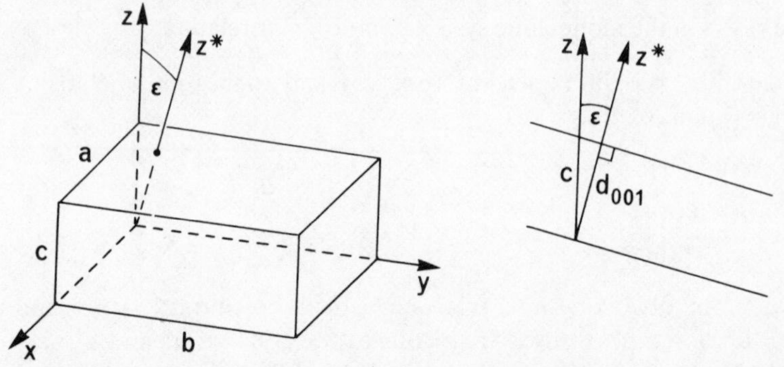

Fig 6.27 (a) is a clinographic projection of a triclinic unit-cell showing the inclination ε of z^* to z. (b) is a section of the same unit-cell in the zz^* plane to show that $d_{001} = c\cos\varepsilon$.

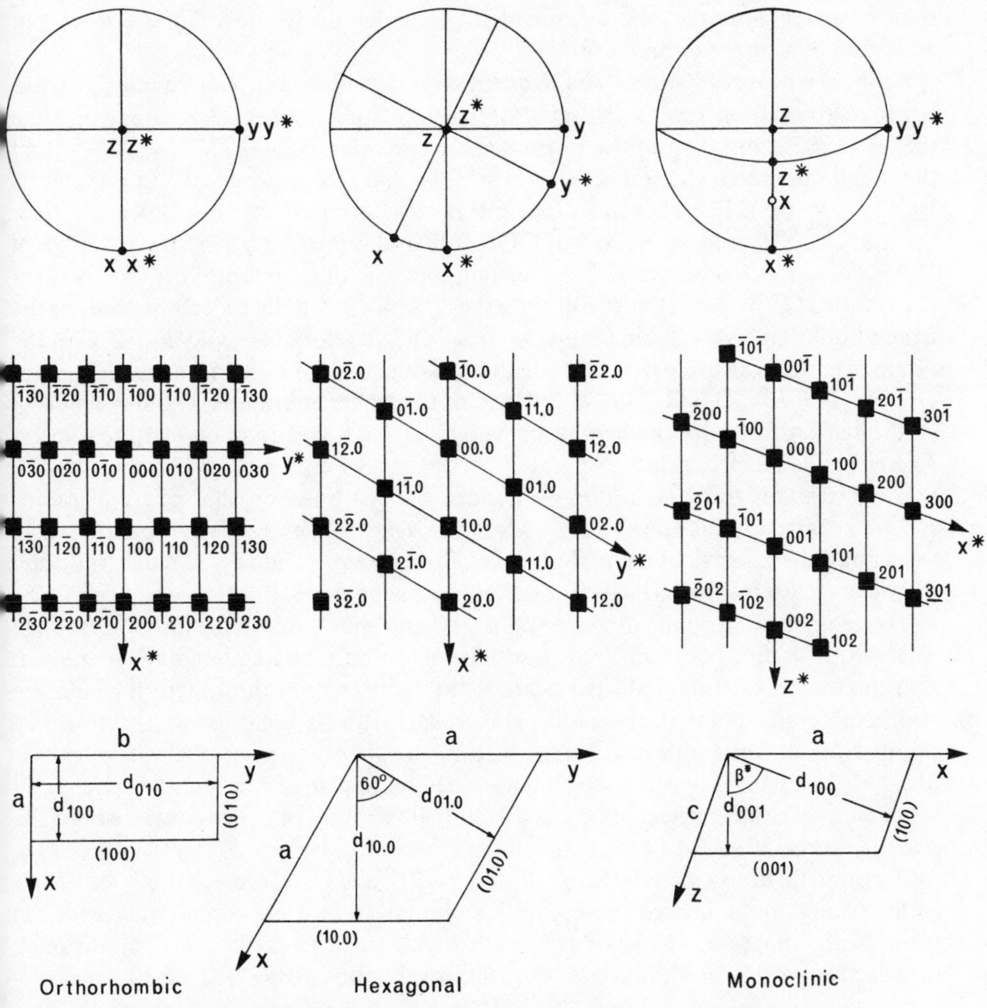

Fig 6.28 The angular relationships between reciprocal and direct axes in the orthorhombic, hexagonal and monoclinic systems are shown in the top row (a). Selected reciprocal lattice sections for these three systems are shown in (b), with diagrams to illustrate the relationship between direct and reciprocal lattice axial lengths in the bottom row.

orthorhombic, tetragonal, and cubic systems, the reciprocal lattice axes are parallel to the direct lattice axes. This is so also for the z-axis in the hexagonal and trigonal systems and for the y-axis in the monoclinic system. The angular relationship between reciprocal and direct axes in the orthorhombic, hexagonal, and monoclinic systems is illustrated in Fig 6.28(a) and selected reciprocal lattice sections for these systems are displayed in Fig 6.28(b).

The reflecting sphere

Having developed the concept of the reciprocal lattice we now go on to show how it provides a simple but powerful means of solving the Laue Equations or the Bragg Equation for a given incident X-ray direction. It is appropriate at this point to mention that the most elegant treatment of the reciprocal lattice and the reflecting sphere is in terms of vector algebra (see, for instance, Lipson and Taylor, 1958); we shall however continue our treatment in mathematically more cumbersome but conceptually simpler terms.

Suppose a beam of X-rays to be incident on a crystal in such a direction as to give rise to reflexion from a particular set of (hkl) planes. Then in Fig 6.29 the angle between the incident beam QO and the normal OP to an (hkl) plane is $\widehat{QOP} = 90° - \theta$ and the angle between the reflected beam OR and the normal to the plane is $\widehat{POR} = 90° - \theta$. If P is chosen so that QP is parallel to OR, then $\widehat{PQO} = 2\theta$. Now the triangle PQO is isosceles so that QO = QP and OP = 2QO.$\sin\theta$. If the length of QO is chosen as K/λ where K is a constant and λ is the wavelength of the incident X-rays, then OP = $2(K/\lambda)\sin\theta$. But since the plane (hkl) is in the reflecting position the Bragg Equation $\lambda = 2d_{hkl}\sin\theta$ must be satisfied; therefore OP = K/d_{hkl}. If O is the origin of reciprocal space then P is identified as the reciprocal lattice point hkl. Since QO = QP = K/λ, O and P lie on the surface of a sphere of radius K/λ and centre Q on the line parallel to the incident beam passing through the origin of reciprocal space. Therefore the reciprocal lattice point P corresponding to a plane hkl that gives rise to a reflexion for this orientation of the incident beam lies on the sphere of radius K/λ and centre Q; this sphere is known as the *reflecting sphere*. This property of the reflecting sphere provides a simple means of interpreting and in particular indexing diffraction patterns; its use will be exemplified in chapter 8.

The geometrical argument of the last paragraph shows that reflexion occurs when a reciprocal lattice point lies on the surface of the reflecting sphere and, moreover, that the direction of the reflected beam is the radius QP of the sphere through the reciprocal lattice point P. If now the reciprocal lattice is regarded as composed of points lying on planes perpendicular to the z-axis so that reciprocal lattice points $hk0$, $hk1$, $hk2$, etc, lie on successive planes, then these planes intersect the sphere in a set of parallel circles; reflected beams therefore lie on a set of cones coaxial about the z-axis. Figure 6.30 shows a central section of the reflecting sphere containing the z-axis. QO represents the incident beam (cf. Fig 6.29); RO and ST represent the intersection of the planes containing reciprocal lattice points $hk0$ and hkl respectively with the plane of the diagram; QRS is parallel to z and, since RS represents the distance between the zeroth and lth layer of the reciprocal lattice, RS = lK/c. But,

$$\begin{aligned} RS &= QS - QR \\ &= QT\cos\widehat{SQT} - QO\cos\widehat{SQO} \end{aligned}$$

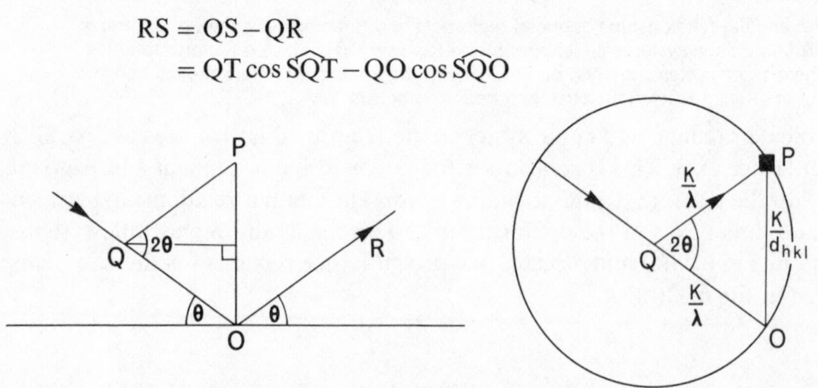

Fig 6.29 The Bragg Equation and the reflecting sphere. The left-hand diagram is a section containing the incident X-ray beam QO and the normal OP to a lattice plane (hkl); OR represents the diffracted beam and Q is chosen so that QP || OR. The right-hand diagram is the corresponding central section of the *reflecting sphere* with centre Q; O is the origin of reciprocal space and P is identified as the reciprocal lattice point hkl. Bragg reflexion occurs when a reciprocal lattice point lies on the surface of the reflecting sphere and the direction of the reflected beam is given by the radius through it.

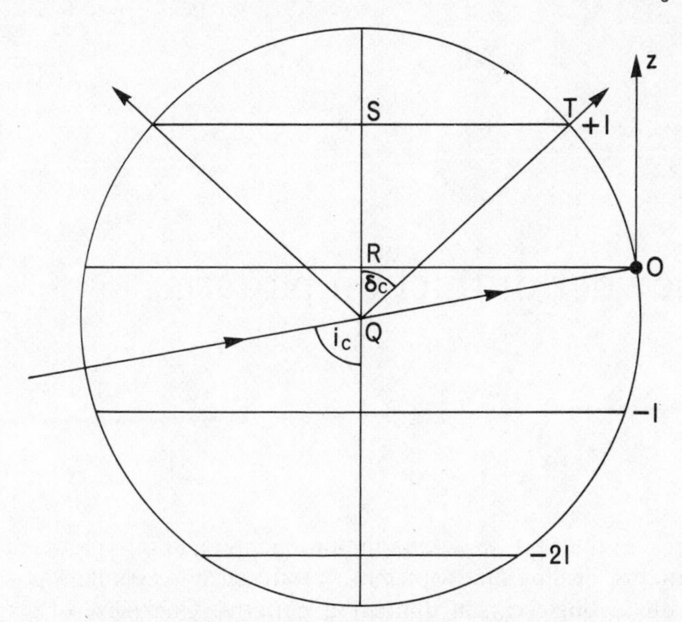

Fig 6.30 The Laue Equations and the reflecting sphere. Central section of the reflecting sphere containing the z-axis and the incident beam QO; planes of reciprocal lattice points hkl, $hk0$, $hk\bar{l}$, $hk\overline{2l}$ are indicated. A reciprocal lattice point T on the lth layer lies on the surface of the reflecting sphere, whence $c(\cos\delta_c - \cos i_c) = l\lambda$, one of the Laue Equations.

Now QT and QO are radii of the reflecting sphere,

Therefore $\mathrm{RS} = \dfrac{K}{\lambda}(\cos\delta_c - \cos i_c)$

Therefore $\dfrac{lK}{c} = \dfrac{K}{\lambda}(\cos\delta_c - \cos i_c)$

Therefore $c(\cos\delta_c - \cos i_c) = l\lambda.$

This is one of the Laue Equations (7); the other two Laue Equations can be derived by analogous arguments.

Our development of the reciprocal lattice and the reflecting sphere was in terms of the Bragg Equation; by deriving the Laue Equations in the manner of the last paragraph we have sought to emphasize the essential unity and the interconvertibility of these three approaches to the geometry of X-ray diffraction by crystals.

7
X-ray powder diffraction patterns

In this chapter and the next we discuss the production and interpretation of X-ray diffraction patterns, dealing first with those produced by crystalline powders and then moving on to single crystal diffraction patterns. One piece of groundwork remains to be done by way of preamble, that is the description of the spectrum emitted by an X-ray tube.

Emission spectrum of an X-ray tube

Our purpose in providing this account of the X-ray emission from an X-ray tube is to enable the reader to understand the general principles of the various types of X-ray goniometer as well as those features of diffraction patterns that are dependent on the wavelength distribution in the incident X-ray beam. Our treatment of this topic will be in outline only; for a detailed account the reader is referred to Klug and Alexander (1954).

A modern X-ray tube designed for crystallographic use (Fig 7.1) is in essence a permanently evacuated glass envelope into which is sealed a tungsten *filament* separated by about one centimetre from a *target* composed of a metallic element. The tungsten filament is heated by the passage of an electric current and emits electrons. A potential difference of the order of 50 kV applied between the filament, acting as cathode, and the target, as anode, accelerates the electrons emitted by the hot filament towards the anode so that a stream of high-energy electrons impinges on the anode. Most of the energy of the electron stream, about 98 per cent, is converted into heat so it is essential that the anode should be made of a material of high thermal conductivity and cooled from behind by a fast flowing stream of water. Anodes are usually made of copper; if a less well conducting element is required as target, either it is electroplated on to the copper anode or a small disc of it is soldered on. The filament is surrounded by a *focusing hood* which has a slot in its front face parallel to the length of the filament. The effect of the hood is to cause the electron stream to form a line focus ~ 1 cm $\times$ $\sim 0{\cdot}01$ cm on the target; the dissipation of heat from such a line focus is relatively efficient so that the tube can be run at a higher electron current and so produce X-radiation of higher intensity. X-rays are emitted from the target in all possible directions, but only a narrow beam (in the angular range of highest intensity) making an angle from 3° to 6° with the face of the target is utilized by being allowed to pass out of the evacuated envelope through a *window* made of a

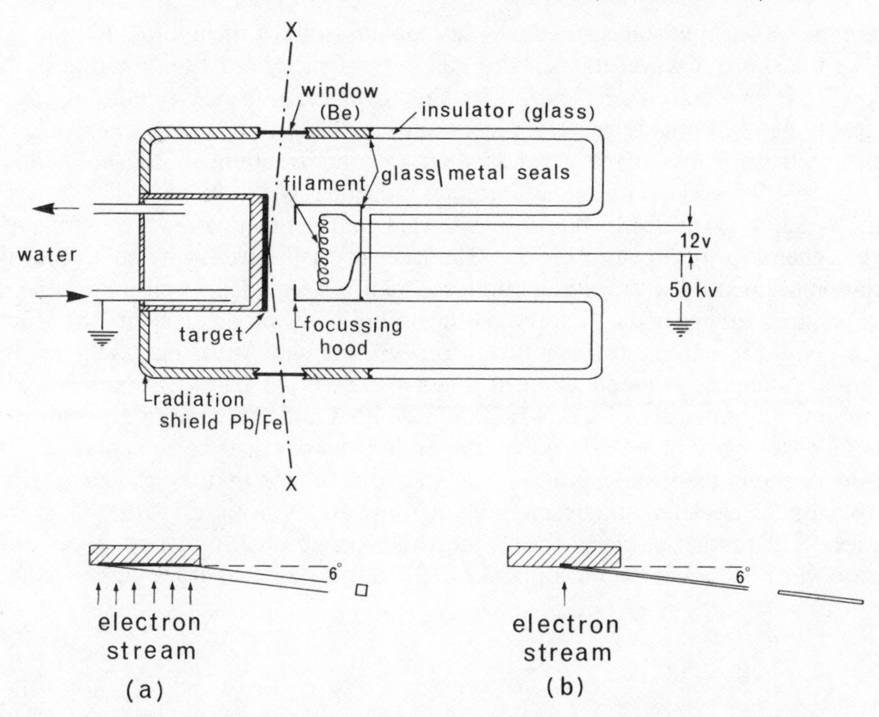

Fig 7.1 A crystallographic X-ray tube. The two lower diagrams are mutually perpendicular sections of a target, the thick black line representing the area irradiated by electrons, (a) in the plane containing the length of the filament and (b) in the plane perpendicular to the length of the filament. To the right of each diagram the cross-section of the emitted X-ray beam is shown; in (a) it is approximately square, a *spot focus*, and in (b) it is strongly elongated, a *line focus*.

substance with a very low absorption coefficient for X-rays; windows are usually made of beryllium. X-ray tubes are equipped with four windows, one situated on either side of the tube in line with the length of the line focus and two others at right-angles to these. Through the former the line focus appears as a nearly equidimensional spot and the X-ray beam that passes through the window is suitable for use with pin-hole collimators; but through the windows situated normal to the length of the line focus, the elongation of the focus is retained so that the X-ray beam transmitted has dimensions of about 1 cm × 0·01 cm and is suitable for use with a slit system of collimation.

The X-radiation emitted by the target is never monochromatic, but covers a considerable spectral range. In the X-ray spectrum it is convenient to distinguish between *white* (or *continuous*) radiation and *characteristic* radiation. We deal with the generation of white radiation first.

When an electron strikes the target it loses energy and part of the energy lost is converted into X-radiation of wavelength λ according to the equation $\lambda = hc/\Delta E$, where ΔE is the amount of energy lost by the electron, h is Planck's constant, and c is the velocity of light *in vacuo*. If the electron loses all the energy it has acquired by dropping through a potential V, then $\Delta E = eV$, where e is the charge on an electron, and the wavelength of the X-rays emitted will be

$$\lambda = \frac{hc}{eV} = \frac{12{\cdot}398}{V}, \qquad (1)$$

where the wavelength is measured in Ångstrom units and V is in kilovolts; this will will be the shortest wavelength in the spectrum of the X-radiation emitted by the target. That an electron will lose all its kinetic energy in a single collision with a target atom is improbable; most will lose their kinetic energy in a series of collisions, each involving a loss of energy less than eV and resulting in the emission of X-radiation of wavelengths longer than that indicated by equation (1). The intensity of the X-radiation emitted by the target will vary continuously with wavelength and is of the general form shown in Fig 7.2. The intensity at a given wavelength and the variation of intensity with wavelength in such a *white radiation* spectrum depend on the operating voltage of the tube and on the nature of the target element. The kinetic energy of the electrons striking the target will increase with increasing applied potential difference between filament and target, so that there will be an overall increase in the intensity of the X-radiation emitted and moreover a movement to shorter wavelengths of both the maximum in the intensity distribution curve and the cut-off (or minimum wavelength). In general it can be said that the efficiency of the conversion of electron kinetic energy to X-radiation increases with the atomic number of the target element; thus a molybdenum ($Z = 42$) target produces more intense white radiation than a copper ($Z = 29$) target operated at the same voltage.

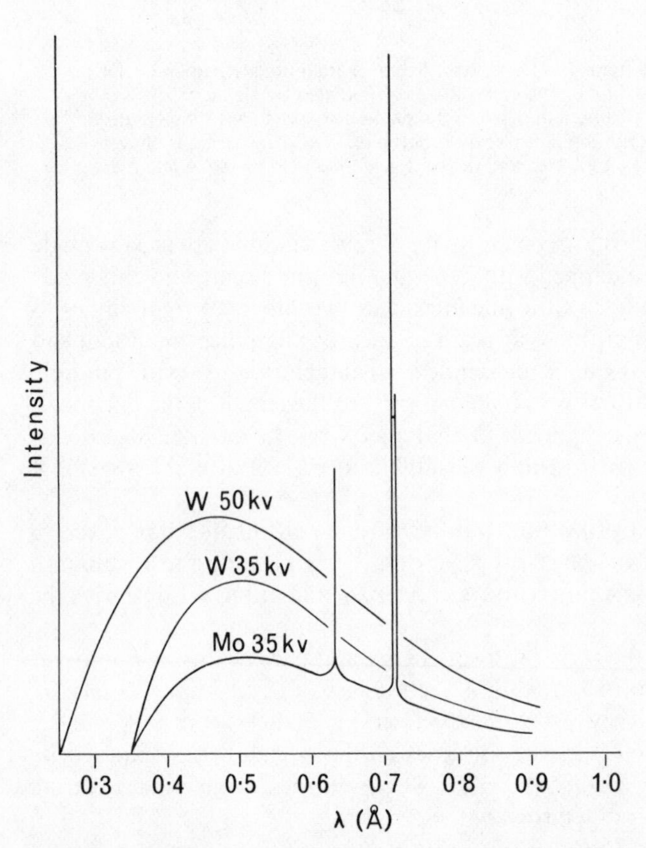

Fig 7.2 Emission from an X-ray tube showing the white radiation 'hump' and the characteristic α_1, α_2, and β lines for a molybdenum target operating at 35 kv; the characteristic lines of tungsten are at too high a wavelength to be shown.

We turn now to the generation of characteristic radiation. If an electron of sufficiently high energy strikes the target it may eject an electron from the K-shell of one of the atoms of the target element. Ejection of a K electron will be followed by the transfer of an electron from an electronic shell of higher energy to fill the vacant energy level; such a transfer will be accompanied by the emission of an X-ray photon whose energy is equal to the difference in energy between the two energy levels of the target atom. The X-rays emitted as a result of such a process will thus have a fixed wavelength characteristic of the target element and will constitute a line spectrum commonly described as the *characteristic* radiation of the target element. When the lower of the two energy levels concerned is in the K-shell of the target atom, the resulting spectral line is described as a K line.

K lines are classified in terms of the other energy level concerned. An electronic transition from the L-shell to the K-shell is said to give rise to a Kα line, while a transition from the M-shell to the K-shell gives rise to a Kβ line. Since the L-shell is of lower energy than the M-shell, the Kα line in the X-ray spectrum of a given element has a longer wavelength than the Kβ line of the same element. Moreover Kα lines are generally of higher intensity than the corresponding Kβ lines; in practice therefore Kα lines are invariably selected for isolation when monochromatic X-radiation is required. All elements give rise to two Kα lines, denoted Kα_1 and Kα_2, which have a very small difference in wavelength[1] and are resolved in diffraction only at high Bragg angle. The Kα_1 line has the shorter wavelength and is of about twice the intensity of the Kα_2 line; when the two lines are not resolved their wavelengths can be weighted to a fair approximation in the ratio 2:1 to give the wavelength of the unresolved doublet, usually written as Kα*, as $\lambda_{K\alpha^*} = \frac{2}{3}\lambda_{K\alpha_1} + \frac{1}{3}\lambda_{K\alpha_2}$. Of course the emission spectrum of an X-ray source includes characteristic spectral lines due to electronic transitions to the L-shell, Lα, Lβ lines, etc; such characteristic radiations are not commonly utilized crystallographically and need not be further discussed here.

The wavelengths of some commonly-used characteristic radiations are shown in Table 7.1. While the wavelength of characteristic radiation is dependent only on the nature of the target, its intensity is dependent on the magnitude of the voltage applied across the tube; in particular if the applied voltage is below a certain threshold value, none of the electrons incident on the target will have sufficient energy to eject a K-electron from a target atom and no K lines will be excited. The threshold voltage for excitation of K lines in any element is such as will impart to the electrons incident on the target kinetic energy equal to the photon wavelength of the K absorption edge of the target element. The operating voltage of an X-ray tube must therefore be greater than this threshold value and is chosen to give an optimum ratio of characteristic intensity to white radiation intensity.

When monochromatic radiation is required it is sufficient for most purposes merely to remove the Kβ line, the ratio of characteristic to white intensity being such that

[1] A Kα line is produced by an electronic transition from a $2p$ to a vacant $1s$ orbital. The reader familiar with X-ray spectroscopy will be aware that the $2p$ orbitals of an atom comprise two shells, denoted L*II* and L*III*, of slightly different energy because the total orbital angular momentum and the spin angular momentum of the $2p$ electrons can be combined in two different ways. Electrons in the L*II* shell have slightly lower energy than those in the L*III* shell and so electronic transitions L*II* → K give rise to the longer Kα_2 wavelength compared with electronic transitions L*III* → K which give rise to the slightly shorter Kα_1 wavelength of X-radiation. The L*II* shell ($J = \frac{1}{2}$) contains two electrons whereas the L*III* shell ($J = \frac{3}{2}$) contains four electrons so that the Kα_2 line is only half as strong as the Kα_1 line. Formally the Kα_2 line is produced by an electronic transition from a $1^2S_{\frac{1}{2}}$ to a $2^2P_{\frac{1}{2}}$ state whereas the Kα_1 line is produced by a transition from a $1^2S_{\frac{1}{2}}$ to a $2^2P_{\frac{1}{2}}$ state.

Table 7.1 Data for some common targets and filters

Element	Atomic Number	Line	Wavelength (Å)	Filter: Element	Atomic Number	Wavelength of *K* absorption edge (Å)
Mo	42	$K\alpha_1$ $K\alpha_2$ $K\beta$	0·70926 0·71354 0·63225	Zr	40	0·6888
Cu	29	$K\alpha_1$ $K\alpha_2$ $K\beta$	1·54050 1·54434 1·39217	Ni	28	1·4869
Co	27	$K\alpha_1$ $K\alpha_2$ $K\beta$	1·78890 1·79279 1·62073	Fe	26	1·7429
Fe	26	$K\alpha_1$ $K\alpha_2$ $K\beta$	1·93597 1·93991 1·75654	Mn	25	1·8954
Cr	24	$K\alpha_1$ $K\alpha_2$ $K\beta$	2·28962 2·29352 2·08479	V	23	2·2676

the latter can be ignored. The Kβ line can simply be removed from the X-ray emission of a target by placing immediately outside the window a thin foil of an element with an absorption edge of wavelength just less than that of the required Kα line. When the beam emitted from the X-ray tube passes through such a *filter* the intensity of the Kα line is reduced by a small factor and the Kβ line is reduced to negligible intensity. The appropriate filter for a target of an element of atomic number Z is an element of atomic number $Z-1$ usually, $Z-2$ in some cases, e.g. for CuKα the appropriate filter is nickel ($Z-1$) whereas for MoKα zirconium ($Z-2$) is employed. The dependence of mass absorption coefficient on wavelength is shown for one element in Fig 7.3. The optimum thickness of a filter varies from element to element; it should be adequate to place the intensity ratio $I(\mathrm{K}\alpha)/I(\mathrm{K}\beta)$ between 150 and 350, but no thicker. In the case of emission from a copper target, the former value of the $I(\mathrm{K}\alpha)/I(\mathrm{K}\beta)$ ratio corresponds of a reduction in $I(\mathrm{K}\alpha)$ of 45 per cent and the latter of 60 per cent. Although the primary purpose of such a filter is to cut out the Kβ line, white radiation of wavelength less than the absorption edge of the filter will also be drastically reduced in intensity, but white radiation with λ from just less than $\lambda_{\mathrm{K}\alpha}$ to high wavelengths will pass the filter.

When strictly monochromatic radiation is required the X-ray beam emitted from the tube is reflected from a crystal face set at the appropriate Bragg angle for the Kα line. Crystals with faces parallel to planes that yield very strong X-ray reflexions are employed, but even so the Kα beam is much reduced in intensity. The reduction in intensity can be made less severe by bending the crystal plate to give a focusing effect; this was the usual way of achieving strictly monochromatic X-radiation until recent developments in the synthetic growth of perfect graphite crystals enabled flat crystals to be produced which give a much smaller reduction in intensity. Such *monochromators* are employed only when strictly monochromatic radiation is essential for the purpose in hand.

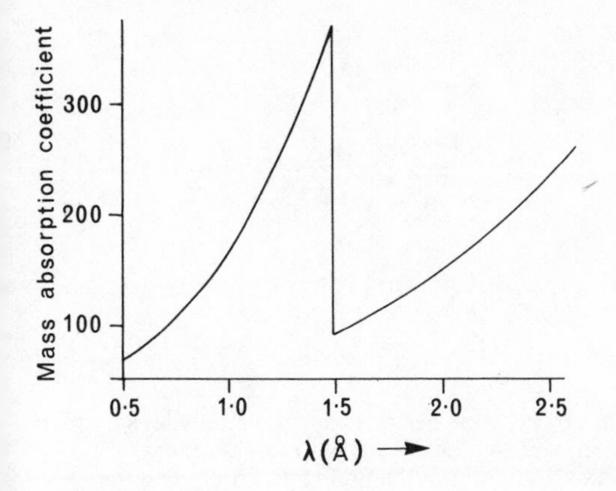

Fig 7.3 Dependence of the mass absorption coefficient of nickel on wavelength.

We conclude with a word of caution about the nature of the emission from commercial X-ray tubes. Tungsten slowly evaporates from the hot filament in use and eventually deposits as a thin film on the target and the windows. Then the radiation emitted by the target will be contaminated especially with the strongest line in the tungsten spectrum, $L\alpha_1$ ($\lambda = 1{\cdot}476$ Å), which represents an electronic transition from the M-shell to fill a vacancy in the L-shell. In old tubes also the intensity of emission is reduced by the presence of a strongly absorbent film of tungsten on the inner surface of the window.

Powder photographs

An essential difference between two and three-dimensional diffraction gratings is that the latter yield only a few diffracted beams for any one orientation of the grating relative to the incident beam. The number of diffracted beams that can be observed when X-radiation is diffracted by a crystalline substance can be increased in a variety of ways; several such ways that are applicable to a single crystal diffraction grating are discussed in the next chapter; here we remove the restriction that the grating be a single crystal and consider diffraction by a crystalline powder. It is assumed—and this is a practically valid assumption—that the powder has been so prepared that it consists of a very large number of minute crystal fragments in completely random orientation so that every possible lattice plane will be present in every possible orientation with respect to the incident X-ray beam. The Bragg Equation $\lambda = 2d \sin \theta$ will thus be satisfied for all planes (hkl) provided $d > \frac{1}{2}\lambda$. Since the only restriction placed by the Bragg Equation on the orientation of a reflecting plane is that it should make an angle θ with the incident X-ray beam, all planes with a given set of indices (hkl) whose normals lie on a cone of semiangle $90° - \theta$ about the direction of the incident beam will reflect. Figure 7.4 shows the normal N to a plane (hkl) with interplanar spacing d lying on a small circle of radius $90° - \theta$ about the direction X of the incident X-ray beam. Since the reflected beam R is required to be coplanar with X and N, R lies at the intersection of the great circle XN with a small circle of radius 2θ about the direction X′ of the emergent direct beam. Planes with indices (hkl) will thus give rise to a cone of diffracted beams with semiangle 2θ. The total diffraction pattern produced by a crystalline powder is thus a set of cones, each cone corresponding to a solution of the Bragg Equation.

The diffraction pattern produced by a powder is commonly recorded on a narrow strip of photographic film in a cylindrical camera whose axis is coincident with the

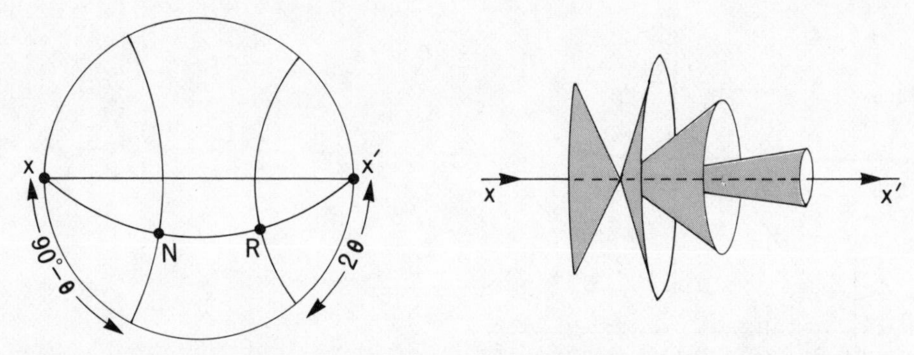

Fig 7.4 Generation of cones of diffracted X-rays by an X-ray beam incident on a powder specimen. The stereogram on the left shows the relationship between the normal N to a set of planes (*hkl*), the direction X of the incident X-rays, the direction R of the resulting diffracted beam and the Bragg angle θ. The drawing on the right shows cones of diffracted radiation emanating from a powder specimen for four solutions of the Bragg equation $\lambda = 2d \sin \theta$, the semi-angle of each cone being 2θ.

powder specimen. Each diffracted cone is recorded on the film strip as a pair of arcs, one on either side of the direction of the incident beam (Fig 7.5). The appearance of the diffraction pattern on the film strip after development depends on the way the film is mounted in the camera. There are three mountings in common use, which differ in the position of the free ends of the film relative to the incident beam. The *Bradley–Jay* mounting is such that the incident beam passes through the gap between the ends of the film and the undeviated beam leaves the camera through a hole punched in the centre of the film (Fig 7.5(a)). The *van Arkel* mounting simply has the positions of entry and exit reversed (Fig 7.5(b)). In the *Straumanis* mounting, both the incident and the undeviated beam pass through holes punched in the film, the gap between the ends of the film lying close to the radius of the camera normal to the incident beam (Fig 7.5(c)); most modern powder cameras use the Straumanis mounting.

Approximate Bragg angles can simply be determined by measuring the distance s between the midpoints of corresponding arcs on the film and assuming a value r equal to the radius of the camera for the radius of the film in the camera; then $\theta = s/4r$ if s is measured across the direction of the undeviated beam as in the Bradley–Jay mounting, $\frac{1}{2}\pi - \theta = s/4r$ if s is measured across the direction of the incident beam as in the van Arkel mounting. The Straumanis mounting yields concentric arcs about each of the two holes punched in the film so that measurements of s across one hole will give $\theta = s/4r$ and across the other $\frac{1}{2}\pi - \theta = s/4r$; the hole through which the incident beam enters the camera can easily be recognized by the splitting of the arcs closest to it into $\alpha_1\alpha_2$ doublets. Accurate measurement of Bragg angles requires film shrinkage during development, fixing, washing, and drying to be taken into account as well as the deviation of the mean radius of the film in the camera from the radius of the camera due to the finite thickness of the film. In cameras that use the Bradley–Jay or the van Arkel mounting this is achieved by constructing the camera so that a knife-edge casts a shadow just short of each end of the film, the angle between the knife-edges, $4\phi_k$, being determined by calibrating the camera with a substance whose unit-cell dimensions are very accurately known; if the measured distance between the shadows cast by the knife-edges is s_k, then $\theta = \phi_k s/s_k$ for the Bradley–Jay mounting and $\frac{1}{2}\pi - \theta = \phi_k s/s_k$ for the van Arkel mounting. The Straumanis mounting is however self-calibrating provided the powder pattern

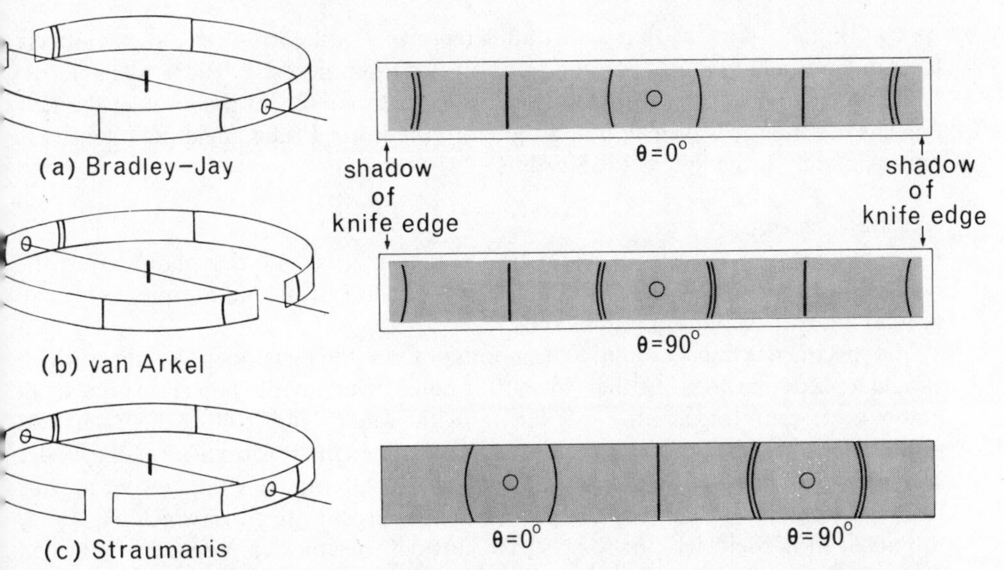

Fig 7.5 Film mountings for X-ray powder photography. The diagrams on the left show the arrangement of the film strip in the camera for three mountings that have been much used: the specimen is represented as a short thick line and the direction of the incident X-rays is arrowed. The diagrams on the right show the appearance of the film laid flat after development for each of the three mountings: exemplary low θ, moderate θ, and high θ powder rings are shown. Observation of $\alpha_1\alpha_2$ splitting serves to identify the high θ lines on the film.

concerned exhibits sharp high θ reflexions; then measurement of the positions of the midpoints of one pair of corresponding arcs, s_1 and s_1', across the exit hole and of another pair, s_2 and s_2', across the entrance hole yields a distance

$$s^* = \frac{s_1+s_1'}{2} - \frac{s_2+s_2'}{2}$$

equal to the distance between the centres of the two pairs; since one centre is at $\theta = 0$ and the other at $\theta = \frac{1}{2}\pi$, $s^* = \pi r$ so that $\theta = (\pi/4s^*)s$ if s is measured across the exit hole and $\theta = \frac{1}{2}\pi - (\pi/4s^*)s$ if s is measured across the entrance hole.

The Bradley–Jay mounting is suitable for recording powder patterns for comparative purposes and is capable of giving accurate d-spacings for low-θ lines; the van Arkel mounting is especially applicable to the accurate measurement of high angle lines; but both have been largely superseded in modern cameras by the Straumanis mounting, which combines the advantages of both with no intrinsic disadvantage.

It has already been said that high angle lines can be recognized immediately on inspection of a powder photograph because they display $\alpha_1\alpha_2$ splitting. That the Kα doublet will be resolved at high θ can simply be seen by differentiating the Bragg Equation

$$\lambda = 2d\sin\theta$$

Therefore $d\lambda = 2d\cos\theta\, d\theta$

whence $d\theta = \dfrac{\tan\theta}{\lambda}\, d\lambda$

As $\theta \to \frac{1}{2}\pi$, $\tan\theta \to \infty$ so that a small difference in λ will correspond to a relatively large difference in θ. For $CuK\alpha_1$ and $CuK\alpha_2$, for example, $\Delta\lambda = 0{\cdot}0038$ Å (Table 7.1) so that for $\theta = 80°$, $\Delta\theta = 0{\cdot}86°$ and for $\theta = 85°$, $\Delta\theta = 1{\cdot}72°$. In the doublet the $K\alpha_1$ line has the shorter wavelength and therefore the lower Bragg angle; its intensity is approximately twice that of the $K\alpha_2$ line.

Experimental procedure

For a full account of the practical details of taking X-ray powder photographs the reader is referred to Klug and Alexander (1954) or to Lipson and Steeple (1970). We present here an account in outline only.

The specimen is finely ground to a smooth powder, the constituent grains of which should have dimensions less than 45×10^{-3} mm. By mixing the powder with a small amount of gum tragacanth, moistening with water, and rolling between two microscope slides the powder specimen can be obtained in the form of a small cylinder of diameter 0·3 to 0·5 mm and about 1 cm in length. Alternatively the powder may be loaded into a thin-walled capillary tube made of borosilicate glass (which has a very low absorption coefficient for X-rays). The powder specimen so prepared is attached to a spindle that can be centred so that the length of the specimen can be brought into coincidence with the axis of the cylindrical camera (Fig 7.6). X-rays enter the camera through a *collimator* which is essentially a metal tube, of internal diameter ~0·5 mm, extending to within a few mm of the centre of the camera; the hole in the collimator is widened to about 1 mm diameter at the exit end to form a *guard tube* that serves to trap the radiation scattered from the end of the fine hole. The undeviated beam is led out of the camera through a similar tube, of rather larger internal diameter, into a *beam trap*. The beam trap is so constructed that it incorporates a fluorescent screen which is useful for alignment of the camera along the beam emitted from the

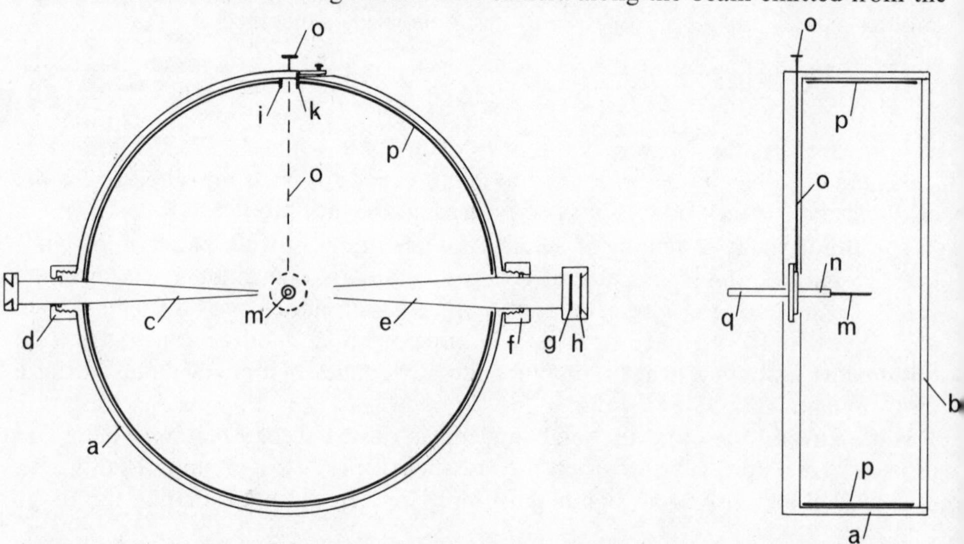

Fig 7.6 A type of powder camera of 57·3 mm radius shown about half-size in front and side elevation, a, cylindrical camera body. b, detachable lid. c, collimator. d, collimator locking screw. e, exit tube, f, exit tube locking screw. g, fluorescent screen. h, lead glass screen. i, fixed pin and k, movable pin with locking screw (movement of k away from i forces the film against the camera body; k is moved until resistance is felt and then locked). m, specimen. n, specimen holder. oo, centring plunger to move magnetic chuck (the spindle q is rotated manually and the plunger oo is applied until the specimen is centred throughout its rotation). p, film. q, drive spindle from detachable motor.

X-ray generator; the fluorescent screen is viewed through a lead-glass window. Centring of the specimen on the axis of the camera is achieved by removing the beam trap, illuminating the collimator with a light source, and adjusting the specimen mounting until the specimen is seen to remain stationary on rotation about its axis. The camera is in essence a light-tight box which must be loaded with film only in a darkroom. The various types of camera use different devices for holding the film strip firmly against a cylindrical metal former. Before loading the film is punched with one or two holes in the appropriate positions for the collimator and/or the beam trap to pass through. During exposure the specimen is rotated slowly about its axis by an electric motor; this serves greatly to increase the number of crystal orientations with respect to the incident beam that are present in the specimen. If the powder is too coarse or the specimen not rotated, complete randomness of orientation will not be achieved and the powder lines will appear 'spotty'.

Powder cameras of various diameters are available commercially, 57·3, 60, 114·6, 90, and 190 mm are diameters that have been used extensively. For collimation of the same quality a camera of large diameter will produce better resolution of the diffraction pattern; but this advantage is counterbalanced by the longer exposure required to produce lines of comparable intensity due to absorption and scattering of the diffracted X-rays by the air in the camera. Cameras of large radius are provided with a facility for evacuation, but this is usually only necessary when long wavelengths (e.g. $CrK\alpha$) are being used or when the specimen is being heated. For general purposes a camera of diameter 114·6 mm is very suitable and has the added advantage for preliminary work that 1 mm measured on the film corresponds to $1°$ if film shrinkage is neglected: thus if the distance between the mid-points of corresponding arcs is measured across the exit hole with a ruler as 64·0 mm, the Bragg angle of this reflexion is immediately determined as $16{\cdot}0°$ and this can simply be converted to a d-spacing by consulting tables giving $d-\theta$ relationships for commonly used radiations (e.g., Fang and Bloss, 1966). For accurate work of course this radius has no special advantage; measurement and calibration of the film will be required.

Powder diffractometry

In X-ray powder photography the whole diffraction pattern is recorded simultaneously on a photographic film; in X-ray powder diffractometry the diffraction pattern is scanned by a counter device which plots counter output against Bragg angle on a paper trace. The resolution obtainable in diffractometry under optimum working conditions is very much better than in photography and Bragg angles can be measured to much higher accuracy, but the apparatus is much more complex, very weak reflexions are difficult to distinguish from background noise, and a larger powder specimen is required. As will be exemplified later in this chapter the diffractometer is used mainly for problems that require highly accurate Bragg angles or high resolution, while the powder camera is used generally for identification.

A material is prepared for diffractometry by grinding to a smooth powder and sedimenting in a suitable volatile medium on to a microscope cover slip of diameter ~20 mm. It is usual to compress the sedimented specimen against a polished steel plate to ensure that its surface is flat. Randomness of orientation of crystallites in the powder specimen is further increased by rotating the specimen slowly during exposure about an axis normal to its plane. When crystallites tend to sediment with preferred orientation—that is when they are flakes or needles—special techniques of sample preparation have to be used.

The best operating systems currently in use make use of the parafocusing effect illustrated in Fig 7.7. The line focus of the X-ray tube and the entrance slit of the counter are constrained to lie on a circle, the *focusing circle*, so as to be equidistant from the specimen, the surface of which is tangential to the circle. This condition ensures that the diffracted X-rays to be measured are reflected from the surface of the specimen and so are effectively focused on the entrance slit of the counter. Theoretically the condition for the reflected X-rays to be focused is that the specimen should be an arc of the focusing circle, but it is practically more simple and found to be adequate if the specimen is tangential. The divergence of incident and diffracted beams parallel to the length of the line focus of the X-ray tube—that is, normal to the plane of Fig 7.7—is limited by passing each through *Soller slits*, a set of narrow slits formed by a pack of thin metal plates, each parallel to the plane of the diagram.

The diffraction pattern is scanned by rotating the counter at a steady speed about the centre of the surface of the specimen; the radius of this *scanning circle* varies from one make of instrument to another, but is often $\sim$200 mm. In order to maintain the geometry of the focusing circle the specimen is geared to rotate about the same axis at half the speed of the counter. A scale provides direct measurement of the scattering angle 2θ. Scanning speeds vary from $\frac{1}{8}$ to 2 degrees per minute. The counter, which may be a Geiger, proportional, or scintillation counter, outputs through electronic circuits, that we shall not describe in detail, to a pen recorder which plots counts per second, as a measure of intensity, on a continuous paper chart. The chart moves at a steady speed so that distances parallel to its length provide a measure of differences in 2θ. Chart speeds vary from 200 to 1600 mm per hour. Scanning and chart speeds are independently adjustable to suit the nature of the problem under investigation. On the chart one degree of 2θ may be represented by as little as 1·67 mm or as much as 213 mm. For very accurate measurements counter and specimen can alternatively be moved in angular steps as small as 0·01° and the diffracted X-radiation counted for a much longer time than would be possible in conditions of continuous scanning; the resultant counts are then plotted manually against 2θ to give a highly accurate peak profile.

The scanning range of diffractometers, except those built for special purposes, is limited to $\theta < 80°$ simply because at higher angles the counter would foul the X-ray tube; similarly they are limited to $\theta > 4°$ because at lower angles the intense undeviated beam would damage the counter. In practice these restrictions rarely matter.

It is worthy of note that since, as is evident from Fig 7.7, the area of the specimen irradiated by the incident beam varies with Bragg angle, it is important that the specimen should be homogeneous and large enough in area to catch the whole of the incident beam at the lowest Bragg angle to be used. If this second condition is not satisfied, peak heights at different Bragg angles will not be comparable.

It was pointed out at the beginning of this section that in powder photography the whole diffraction pattern is being recorded throughout the exposure whereas in diffractometry each part of the pattern is recorded at a different time. It becomes necessary therefore to stabilize the intensity of the incident X-ray beam; this is achieved by stabilizing the high voltage supply and the filament heating current or, less commonly, by continuous monitoring of the incident intensity.

A formal comparison of the advantages and drawbacks of powder photography and diffractometry would be misleading; the sort of use for which each technique is especially appropriate has already been pointed out in general terms and the reader

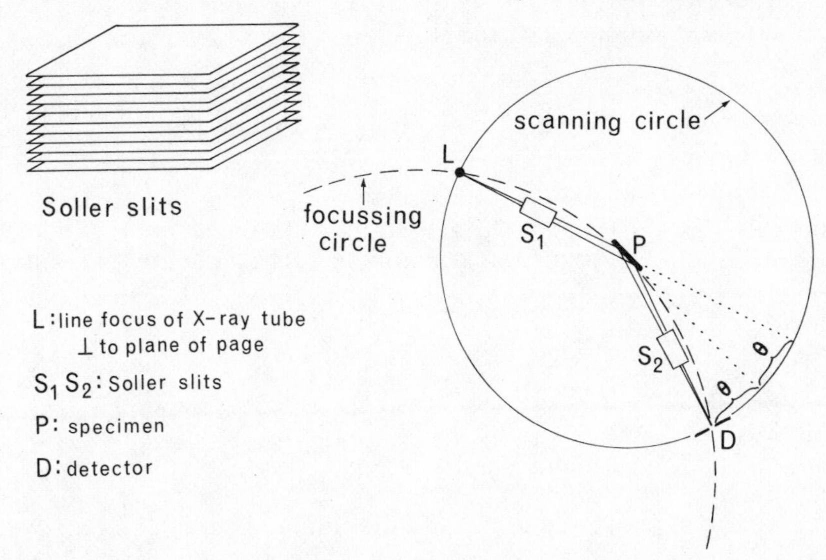

Fig 7.7 Geometry of a powder diffractometer utilizing the parafocusing effect. The construction of Soller slits is shown on a larger scale in the upper left-hand corner.

should draw his own conclusions from the examples discussed at the end of this chapter. It should be borne in mind that the area of overlap of the two methods is extensive. Most of the uses for which the diffractometer is superior stem from its better resolving power, as an indication of which we may compare the minimum Bragg angle at which the $CuK\alpha_1 - K\alpha_2$ doublet is resolved under optimum conditions, $\sim 20°$ by diffractometry compared with $\sim 55°$ by photography.

Interpretation of powder photographs

Measurement of a powder photograph provides a direct determination of the Bragg angle of every cone of diffracted radiation intersected by the film, that is of every *powder line* on the photograph. Solution of the Bragg Equation for each powder line yields the corresponding interplanar spacings, usually known as *d-spacings.* If the unit-cell dimensions of the substance are known it is then possible to determine the indices of the planes contributing to each powder line. If the unit-cell is unknown, it is generally possible with more or less certainty to find unit-cell dimensions consistent with the powder pattern and thence to index all the lines of the pattern. In either case it will be necessary to consider how many symmetry related planes contribute diffracted radiation to each powder line; we deal with this topic first.

Multiplicity Factors

All lattice planes of equal *d*-spacing of necessity give rise to reflexions at the same Bragg angle. Reflexions from each such plane will be independent of one another; consequently the intensity of a powder line is simply the sum of the intensities of all reflexions contributing to it. All lattice planes that are related by symmetry will have the same *d*-spacing so that the number of planes contributing to a powder line *hkl* will be the number of planes in the form $\{hkl\}$ of the Laue group. Thus the line of longest *d*-spacing in a powder photograph of a primitive cubic substance of Laue group $m3m$ is composed of reflexions from all the planes of the form $\{100\}$; the six planes $(100), (010), (001), (\bar{1}00), (0\bar{1}0), (00\bar{1})$ all contribute equally to the powder line, the

Table 7.2 Multiplicity factors for cubic powder lines

Indices of powder line	$h00$	$hh0$	hhh	hhl	$hk0$	hkl
Laue group $m3m$						
Multiplicity factor	6	12	8	24	24	48
Line intensity	6 I($h00$)	12 I($hh0$)	8 I(hhh)	24 I(hhl)	24 I($hk0$)	48 I(hkl)
Laue group $m3$						
Coincident lines unrelated by symmetry	—	—	—	—	$hk0$, $kh0$	hkl, lkh
Multiplicity factor	6	12	8	24	12, 12	24, 24
Line intensity	6 I($h00$)	12 I($hh0$)	8 I(hhh)	24 I(hhl)	12 [I($hk0$) +I($kh0$)]	24 [I(hkl) +I(lkh)]

multiplicity of which is then said to be 6. By way of example suppose that the intensities of the lines containing the 100 and 111 reflexions in a cubic powder pattern are respectively I and 2I; since {100} contains 6 planes and {111} contains 8 planes in both cubic Laue groups, the ratio of the intensities of the 100 and 111 reflexions is $\frac{1}{6}$I:$\frac{2}{8}$I, i.e. $\frac{2}{3}$. The lines on a powder pattern are conventionally indexed if possible with the indices of that contributing reflexion which has $h \geqslant k \geqslant l$; thus the powder line to which reflexions from all the planes of the form {100} contribute is known as the 100 line, rather than as 010 or 001 or $00\bar{1}$.

Multiplicity factors for the cubic Laue group $m3m$ are shown in Table 7.2. The geometry of the cubic unit-cell is such that certain values of $h^2+k^2+l^2$ can be obtained from quite different values of h, k, and l so that planes unrelated by symmetry have the same d-spacing $a(h^2+k^2+l^2)^{-\frac{1}{2}}$ and in consequence the corresponding powder lines coincide. Thus the {300} and {221} planes have the same d-spacing $\frac{1}{3}a$ so that the 300 and 221 powder lines are coincident. Measurement of the intensity of such a line provides information only about the combined intensities of the coincident lines 6I(300)+24I(221); I(300) and I(221) can be separately evaluated only by study of the single crystal diffraction pattern.

In the other cubic Laue group $m3$ coincidence of powder lines due to reflexion from planes unrelated by symmetry can arise in another way. For instance the forms {210} and {120} are distinct, but the geometry of the lattice is such that they have identical d-spacings $a/\sqrt{5}$; the intensity of reflexion from planes of the two forms will be different so that the intensity of the composite line will be 12I(210)+12I(120). This sort of coincidence affects all lines of the types $hk0$ and hkl in the powder patterns of substances of Laue group $m3$, but for all other types of reflexions multiplicity factors are the same as those for Laue group $m3m$. Multiplicity factors for Laue group $m3$ are shown alongside those for $m3m$ in Table 7.2, where pairs of reflexions that are coincident but independent in the lower symmetry group are designated as $hk0$ and $kh0$, hkl and lkh.

In summary the intensity of a powder line hkl is related to the intensity of the Bragg reflexion hkl by the multiplicity factor for the form $\{hkl\}$ in the appropriate Laue group provided no other form has the same interplanar spacing.

We have taken our examples of multiplicity factors entirely from the cubic system because we shall be mainly concerned with the interpretation of cubic powder patterns. What has been said here applies however in general to all other systems.

Interpretation when the unit-cell is known

From measurements of the Bragg angles of the lines of a powder pattern the d-spacings of the lines can simply be calculated by application of the Bragg Equation. It is convenient to arrange such a set of observed d-spacings in sequence of decreasing d. The observed d-spacings are then compared with the d-spacings of all planes that can give rise to reflexion calculated from the known unit-cell dimensions. In the cubic system the comparison invariably yields unambiguous indices for all low θ lines, but becomes progressively less certain as θ increases unless the unit-cell edge is very accurately known. The d-spacing of the highest θ line that can be unambiguously indexed is then used to recalculate a, from which a revised set of d-spacings is calculated; the indexing may then be carried to higher θ and the process of successive improvement of a continued. In systems of lower symmetry it is usual to calculate by computer from the unit-cell constants a set of d-spacings sorted in sequence of decreasing d. Ambiguity in the indexing of observed lines usually arises at a much lower Bragg angle than in the cubic system and it is advisable, whenever possible, to compare the intensities of observed powder lines with intensity data for reflexions obtained from single crystals of the substance, taking multiplicity factors into account. By comparison with single crystal intensity data it may be possible to index unambiguously a line that would otherwise have to be referred to two or more forms unrelated by symmetry but with equal, or nearly equal, d-spacings.

Interplanar spacings are most conveniently calculated from unit-cell constants by use of the fundamental reciprocal lattice relationship $d^*_{hkl} = 1/d_{hkl}$. In the triclinic system, the most general case, d^* can simply be shown by three-dimensional geometry to be related to the reciprocal lattice constants by

$$\frac{1}{d^2_{hkl}} = d^{*2}_{hkl} = h^2a^{*2} + k^2b^{*2} + l^2c^{*2} + 2klb^*c^* \cos\alpha^* + \\ + 2lhc^*a^* \cos\beta^* + 2hka^*b^* \cos\gamma^*.$$

When the reference axes are orthogonal this reduces to

$$\frac{1}{d^2_{hkl}} = d^{*2}_{hkl} = h^2a^{*2} + k^2b^{*2} + l^2c^{*2}$$

and further reduces for the cubic system to

$$\frac{1}{d^2_{hkl}} = d^{*2}_{hkl} = (h^2 + k^2 + l^2)a^{*2}$$

i.e. $$d_{hkl} = \frac{a}{\sqrt{(h^2 + k^2 + l^2)}}.$$

Interpretation when the unit-cell is unknown

In the cubic system unambiguous indexing can usually be achieved, otherwise we are on much less certain ground especially in systems of low symmetry; even with elaborate computer programmes available for selecting and adjusting unit-cell constants to produce a perfect fit, within the limits of error of measurement, between calculated and observed d-spacings there can be no certainty that the interpretation is correct unless additional information is available from single crystal studies. An instructive example of misinterpretation is provided by Christophe-Michel-Lévy and Sandrea (1953), who indexed a powder pattern of 15 lines given by the mineral högbomite on a tetragonal unit-cell with a 8·34, c 7·96 Å; McKie (1963) employed single crystal data to show that the symmetry was hexagonal and the true unit-cell dimensions were a 5·72, c 23·0 Å. Indexing of powder photographs in these circumstances should only be attempted when crystals of a size suitable for single crystal X-ray study cannot be isolated and electron diffraction studies are for one reason or another impossible. In general it is true to say that as symmetry decreases from cubic to triclinic the number of lines in powder patterns taken with X-radiation of the same wavelength increases, the number of adjustable unit-cell parameters increases also, as does the number of correspondences available between observed and calculated d-spacings. Although there is always a risk of misinterpretation, plenty of interpretations have been shown by subsequent single crystal studies to be substantially correct. A clear account of the methods of indexing appropriate to each system and when the system is unknown is given by Lipson and Steeple (1970). We shall here confine ourselves mainly to cubic patterns which can always be indexed satisfactorily except when very few lines are present.

In the three systems with orthogonal axes, cubic, tetragonal, and orthorhombic, there is a simple expression for the spacing d_{hkl} of (hkl) lattice planes. The first plane of the (hkl) set out from the origin makes intercepts a/h, b/k, c/l on the reference axes and the length of the normal from the origin to this plane is of course the interplanar spacing d_{hkl} (ON in Fig 5.9). The direction cosines of the normal are $d_{hkl}/(a/h)$, $d_{hkl}/(b/k)$, $d_{hkl}/(c/l)$ respectively. Now when the reference axes are orthogonal the sum of the squares of the direction cosines of the normal is unity. Therefore

$$d_{hkl} = \left\{\frac{h^2}{a^2}+\frac{k^2}{b^2}+\frac{l^2}{c^2}\right\}^{-\frac{1}{2}}.$$

In the cubic system $a = b = c$ and the expression for d_{hkl} reduces to

$$d_{hkl} = \frac{a}{\sqrt{(h^2+k^2+l^2)}}.$$

Substitution in the Bragg Equation yields, for a cubic substance

$$\sin^2\theta = \frac{\lambda^2}{4a^2}(h^2+k^2+l^2)$$

which can conveniently for our present purpose be rewritten as

$$\sin^2\theta = \frac{\lambda^2 N}{4a^2},$$

where N is an integer that can be expressed as the sum of three squares.

The equation at the end of the preceding paragraph provides a means of indexing the powder pattern of any substance that is known from other information or suspected from the simplicity of the pattern to be cubic. From each measured Bragg angle $\sin^2\theta$ is calculated. The common factor of the set of $\sin^2\theta$ values is found, either by inspection or graphically; the value of N for each measured line is then given by the ratio of $\sin^2\theta$ for the line to the common factor. The graphical method is particularly suitable when θ has not been measured particularly accurately and no computing facilities are available. On a sheet of rectangular graph paper the measured values of $\sin^2\theta$ are plotted along one axis and possible values of N, that is all integers that can be the sum of three squares, from 0 to at least 30 are plotted on the other axis. Lines are drawn through each $\sin^2\theta$ value parallel to the N axis right across the sheet of graph paper. A ruler is laid along the $\sin^2\theta$ axis and rotated slowly about the origin until all its intercepts with $\sin^2\theta$ lines are at integral values of N; a line is drawn through the origin at this inclination and the value of N corresponding to each measured line is read from the graph (Fig 7.8).

Once a set of values of N has been obtained each line on the pattern can be indexed from $N = h^2+k^2+l^2$, applying the convention $h \geqslant k \geqslant l$. Some values of N can, as has been remarked earlier, correspond to more than one set of indices, for example $N = 9$ yields 221 and 300, both of which have $d = \frac{1}{3}a$. Certain other values of N cannot be expressed as the sum of three squares, e.g. 7, 15, 23, 28; such integers are given by $m^2(8n-1)$ where m and n are integers (a proof of this expression is given in Lipson and Steeple, 1970, Appendix 3). If the cubic substance has a non-primitive lattice, systematic absences will appear in the sequence of values of N: powder lines

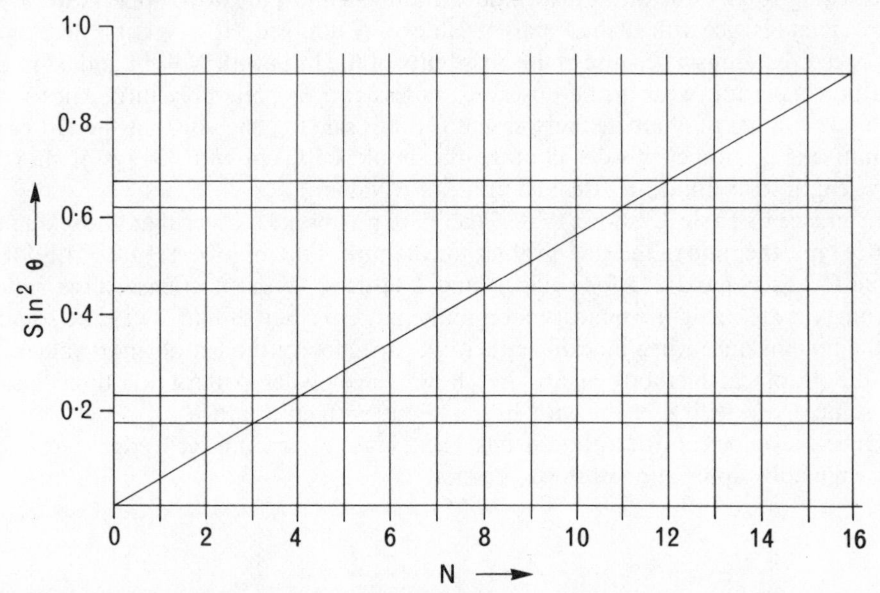

Fig 7.8 Indexing of a cubic powder pattern when the unit-cell dimension is not known, even approximately. Measured values of $\sin^2\theta$ are plotted on the vertical axis and drawn out as horizontal lines. The sequence of integers from zero is plotted on the horizontal axis and each is drawn out as a vertical line. A ruler is rotated from the $\sin^2\theta$ axis about the origin until all its intercepts with horizontal lines coincide with their intersections with the vertical set; a line is drawn through the origin at this inclination. For simplicity only values of N up to 16 are shown.

for which $h+k+l$ is odd will be absent when the lattice type is I, while lines for which h, k, l are neither all even nor all odd will be absent when the lattice type is F. The sequence of powder lines that may be observable in cubic substances with P, I, or F lattices is given in Table 7.3. Other lines may of course be absent due to the presence of translational symmetry elements or may be too weak to be observed due to the atomic arrangement associated with each lattice point. At first sight it might seem that, since N must always be even for an I lattice, P and I lattices would be indistinguishable, an I lattice yielding a spurious set of N values, each one half of the true N value; but

Table 7.3 Permissible reflexions for cubic lattice types
$N = h^2+k^2+l^2$

N	P	I	F	N	P	I	F
1	100	—	—	16	400	400	400
2	110	110	—	17	410, 322	—	—
3	111	—	111	18	330, 411	330, 411	—
4	200	200	200	19	331	—	331
5	210	—	—	20	420	420	420
6	211	211	—	21	421	—	—
7	—	—	—	22	332	332	—
8	220	220	220	23	—	—	—
9	300, 221	—	—	24	422	422	422
10	310	310	—	25	500, 430	—	—
11	311	—	311	26	510, 431	510, 431	—
12	222	222	222	27	333, 511	—	333, 511
13	320	—	—	28	—	—	—
14	321	321	—	29	520, 432	—	—
15	—	—	—	30	521	521	—

the forbidden N values, 7, 15, 23, etc, serve to make the distinction, i.e. if N appears to be equal to any one of these forbidden numbers and the lattice appears to be P, then it is an I lattice with all the apparent values of N doubled. However, it is necessary to proceed cautiously because of the possibility of the lines with $N = 14$ and 30 in an I lattice being too weak to be observed; unless two or preferably three lines with forbidden values of N are actually absent it is not safe to conclude that the lattice is primitive and another powder photograph should be taken with X-rays of shorter wavelength to examine an extended range of N values.

Reference to Table 7.3 shows that, if there are no absences other than those due to lattice type, the ratio of the $\sin^2\theta$ values for the three lines of lowest θ are 3:4:8 for a cubic F lattice and 1:2:3 for cubic P and I lattices. With experience these ratios become easy to recognize on inspection of the pattern; they should always be sought in the first instance when indexing either by inspection of the list of $\sin^2\theta$ values or by the graphical method. Figure 7.9 shows the powder patterns of three cubic substances, one with a P, one with an I, and one with an F lattice.

Once the cubic powder pattern has been indexed the unit-cell edge a can be determined by application of the expression

$$a = \frac{\lambda}{2\sin\theta}\sqrt{(h^2+k^2+l^2)}.$$

Maximum accuracy in the evaluation of a is obtained by using the measured value of θ for the line of highest θ as is evident from the following argument. Differentiation of the Bragg Equation $\lambda = 2d\sin\theta$ with respect to θ and d yields

$$2\sin\theta.dd+2d\cos\theta.d\theta = 0$$

Therefore $$\frac{dd}{d\theta} = -d\cot\theta.$$

Therefore for a fixed error in θ, the error in d will be least when $\cot\theta$ has its smallest value, that is as $\theta \to 90°$.

Examples of indexing and determination of a from cubic powder photographs are provided by Fig 7.8 and Table 7.4.

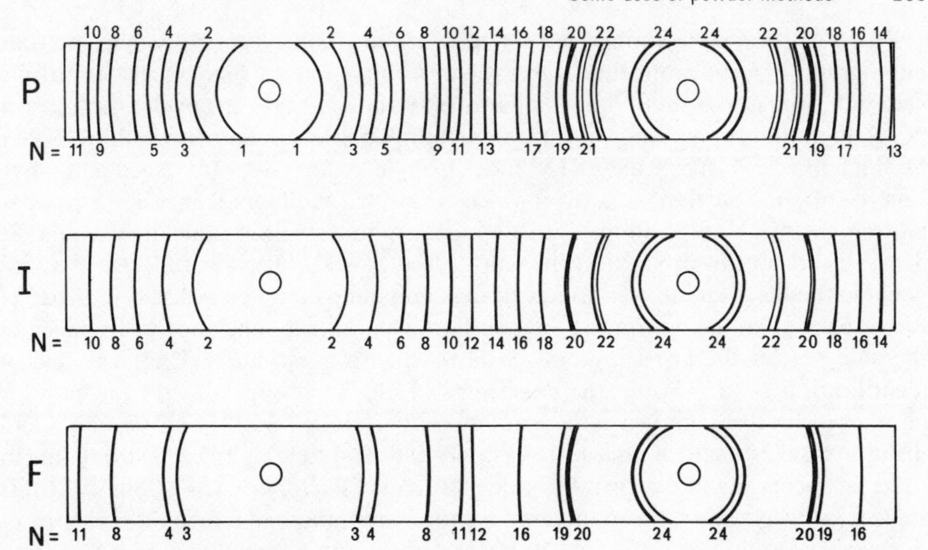

Fig 7.9 Diagrammatic representation of powder photographs, in the Straumanis setting, of three hypothetical cubic substances with $a = 3{\cdot}86$ Å, taken with CuKα radiation ($\lambda = 1{\cdot}542$ Å). The substance giving rise to the upper photograph has a primitive lattice so that the only absent lines are those for $N = (h^2+k^2+l^2) = 7, 15, 23$. In the middle photograph the substance has an I lattice so that there are systematic absences corresponding to odd values of $h+k+l$. In the lowermost photograph the substance has an F lattice so that reflexions are only present when h, k, l are all even or all odd.

Table 7.4 Indexing a cubic powder pattern

Procedure: (1) Measure film and evaluate θ for each line (column 3).
(2) calculate $\sin^2\theta$ for each line (column 4).
(3) *either* plot $\sin^2\theta$ for each line on graph (Fig 7.8) and draw a straight line through intersections of lines of calculated $\sin^2\theta$ and lines of constant N; then read off the value of $N = h^2+k^2+l^2$ for each line (column 5).
or determine highest common factor of $\sin^2\theta$ values by inspection of column 4 and thence deduce values of N (column 5).
(4) index each line from its N value (column 6).
(5) determine the lattice type.
(6) use the measured value of θ for the line of greatest N to calculate the unit-cell edge a.

line	intensity	$\theta°$	$\sin^2\theta$	N	hkl
a	w	24·3	0·169	3	111
b	m	28·3	0·225	4	200
c	s	42·1	0·450	8	220
d	s	51·9	0·619	11	311
e	vw	55·2	0·675	12	222
f	m	71·6	0·900	16	400

Lattice type = F

$$a = \frac{\lambda\sqrt{N}}{2\sin\theta} = \frac{1{\cdot}542\times 4}{2\sin 71{\cdot}6°} = 3{\cdot}25\ \text{Å}$$

λ(CuKα) = 1·542 Å

Intensity scale: s > m > w > vw

Some uses of powder methods

It is not proposed here to offer an exhaustive list of the uses to which X-ray powder diffraction studies have been put, but merely to indicate the scope of the methods by mention of a wide ranging variety of fruitful uses.

Probably the best known use of powder methods, especially the powder photograph, is for identification. Over the past thirty years a very extensive card file listing the *d*-spacings and relative intensities of the lines on the powder patterns of many thousands of substances has been built up; this is known as the Powder Diffraction File, formerly the ASTM Index (supplementary data are issued annually). Each compound on file has a card on which are given *d* and I for all lines in the powder pattern, chemical composition, unit-cell constants from single crystal data if available, density, and optical properties; in many cases powder lines are indexed. Accompanying the file is an index in which all the compounds on file are listed in order of the *d*-spacings of their strongest lines so that a particular substance will appear in the index under the *d*-spacing of each of its six strongest lines.[2] The technique of identification is to measure the *d*-spacings of all the strong lines on the powder photograph of the unknown, if there are more than six to select the six strongest and make a visual estimate of their relative intensities, and to look up each strong line in order of decreasing intensity in the index until a satisfactory match is found. At this stage it is always advisable to obtain a powder photograph of a reliable specimen of the substance identified taken on a camera of the same radius and with the same radiation as the photograph of the unknown. Most laboratories concerned with identification maintain a collection of powder photographs of well authenticated substances in their field. To ease the task of reducing a powder pattern to a set of *d*-spacings, there are commercially available rulers that give a direct measurement of *d*, for a camera of given radius and radiation of given λ, to sufficient accuracy for use with the File. For purposes of identification, powder photography is generally preferred to diffractometry because the whole range of θ can be sampled more quickly and the pattern converted to a list of *d*-spacings more easily; nevertheless in particular circumstances diffractometry may be selected for identification. There are of course some potential snags to be borne in mind when using the File: the unknown may not be on file, or it may be a member of a solid solution series only the end members of which are on file, or it may be a mixture. In general however the Powder Diffraction File is a powerful tool for the identification of unknown substances.

Identification of the several constituents of a mixture by powder methods is never a straightforward task unless the substances that may be present are restricted in number and not more than three, or at most four, of them are present in the mixture. For reliable identification it is essential to compare the powder pattern of the mixture with the superimposed patterns of the suspected constituents. The limit of detection of any constituent depends very much on whether its powder pattern contains a very strong line that is clearly resolved from the lines of the other constituents; even when this criterion is satisfied it is rarely possible to detect a substance present to the extent of <5 per cent. For the detection of substances present in small concentrations diffractometry and photography are generally balanced: the former gives better resolution while the latter exhibits very weak lines with less ambiguity. The limit of detection by optical examination (chapter 12) is generally very much lower, but identification of the impurity is less satisfactory than by powder methods; if the impurity is in the form of grains of manageable size, a sufficient number of grains to make a powder specimen can be picked out under the microscope and identification achieved by powder photography.

[2] This is the indexing system for inorganic compounds including minerals. For organic compounds there is a separate card file and a separate index in which a compound is listed under the *d*-spacing of each of its three strongest lines.

Powder diffractometry is particularly useful for accurate determination of the unit-cell dimensions of solid solutions. A simple example is provided by the Cu–Au system whose melts yield on rapid cooling (quenching) cubic solid solutions for which a varies linearly with atomic percentage from 3·608 Å for pure copper to 4·070 Å for pure gold. Measurement of the d-spacing of an indexed high-θ line by diffractometry will thus yield the composition of the quenched alloy. Precision can be greatly improved by admixture with a pure substance whose unit-cell dimensions are very precisely known and which serves as an internal standard of θ on the diffractometer trace; the standard is chosen to have a peak close to the peak whose d-spacing is to be determined and the proportions in the diffractometer sample are adjusted by trial and error until the two peaks are of comparable height so that uncertainties in scanning and chart speeds have minimal effect. An example of the use of internal standards is provided by the determination of composition of olivine in the binary solid solution series Mg_2SiO_4–Fe_2SiO_4 (Yoder and Sahama, 1957). For this purpose silicon is selected as internal standard because its 111 peak, at $2\theta = 28{\cdot}465°$ for CuKα radiation, is strong, precisely known, and close to the 130 peak in olivine at 2θ ranging from $\sim 32\frac{1}{2}°$ for Mg_2SiO_4 to $\sim 31°$ for Fe_2SiO_4; the greatest length to be measured on the diffractometer trace is thus only 4° of 2θ. The diffractometer is run at least six times over the pair of peaks and the average of their separation used to calculate d_{130} for the olivine specimen. Thence the composition of the solid solution can be obtained by substitution in the expression

$$\text{Mol per cent } Mg_2SiO_4 = 4233{\cdot}91 - 1494{\cdot}59\, d_{130},$$

which is based on measurements of d_{130} for olivines of known composition. The resultant composition determined by diffractometry is in this case subject to an error that may be as high as 4 per cent mainly because most naturally occurring olivines are not strictly binary solid solutions but contain small amounts of other components such as Mn_2SiO_4 in solid solution. In developing such a method an essential preliminary is of course to obtain and index diffractometer traces of compositions close to the end numbers because not only will d_{hkl} vary with composition but so will I_{hkl}; it is necessary to choose a peak that remains strong throughout the compositional range and to choose a standard peak that is not interfered with by peaks of the solid solution in any compositional range.

The techniques described in the preceding two paragraphs are of particular use in synthetic studies and in the determination of phase diagrams, topics that will be developed in Part II, especially chapter 16.

The method of diffractometry with an internal standard can be applied to the accurate determination of the unit-cell dimensions of a substance of high symmetry whose crystal system is known from other evidence or in general when approximate unit-cell dimensions only are available, for one reason or another, from single crystal X-ray or electron diffraction photographs.

Powder photography and diffractometry provide a satisfactory means of determining coefficients of thermal expansion, which will be anisotropic (chapter 11) for crystalline solids other than those belonging to the cubic system. Furnaces capable of heating a powder specimen in a camera or on a diffractometer to temperatures in excess of 2000 °C are available; in high temperature work it is usually necessary to evacuate the camera or diffractometer space and the furnace must be split so that it does not interfere with incident or diffracted beams. Low temperature cameras in which the specimen is cooled by a stream of coolant, liquid air or rarely liquid helium, are available and can, with special precautions, be operated at temperatures as low as 2 K. In both high and low temperature cameras and diffractometers the temperature of the specimen is recorded by a thermocouple placed as close to the specimen as possible without interference with the incident or selected diffracted beams; it is usually necessary to make a calibration correction for the temperature difference between specimen and thermocouple by replacing the specimen with substances that have accurately known melting or transformation temperatures.

So far we have considered only uses of powder methods that depend primarily on measurements of d-spacings, but the relative intensities of powder lines can be measured and, after appropriate corrections, utilized in crystal structure determination. For this purpose intensity data collected from single crystals are preferable, not only are there more data, but the problem of coincident reflexions does not arise; nevertheless many structures of substances, such as alloys, for which it is difficult to obtain single crystals have been based on powder intensity data.

Powder methods have also been used in kinetic studies of polymorphic transformations. If the intensities of adjacent lines, one belonging to the reactant and the other to the product phase, are measured in a series of specimens subjected to isothermal heating for various lengths of time, a plot of their ratio against time enables the time to be evaluated for a certain fraction of the reactant phase to be transformed. Similar measurements for other temperatures then yield a determination of the activation energy of the process.

8
Single crystal X-ray diffraction patterns

When a parallel beam of monochromatic X-radiation falls on a stationary single crystal very few lattice planes will be oriented so as to satisfy the Bragg Equation and in consequence very few reflexions will be observable. There are two ways in which the number of reflexions can be increased. One is to allow the crystal to oscillate or rotate during its exposure to monochromatic radiation; the other is to allow the wavelength of the incident radiation to be variable by using the total emission of the X-ray tube while keeping the crystal stationary. The first approach is the more productive and is employed in all but one of the experimental methods to be described in this chapter; the second approach gives rise to Laue photography which, although important, is generally less informative.

The apparatus necessary for each single crystal technique is described in the appropriate section of this chapter. The one experimental feature common to all, the mounting of the crystal on *arcs*, we shall deal with at this point. A small crystal whose dimensions should ideally be within the range 0·5 to 0·05 mm is selected under a binocular microscope. The crystal is then glued to a thin glass fibre (about 15 mm in length and less than 0·5 mm in diameter) so that a simple zone axis is approximately parallel to the fibre. The method of locating the zone axis in the crystal depends on the nature of the substance under investigation. If the crystal has well developed faces it will be possible to locate prominent zone axes by direct morphological inspection under the microscope. For instance if the substance is known to be tetragonal and to have {100} prism faces commonly well developed, [001] will lie parallel to the faces of this form, while [100] and [010], which are equivalent, will be normal to the prism faces. If however the crystal has no well developed faces but is transparent and optically anisotropic, the polarized light techniques described in chapter 12 may serve to locate one or more simple zone axes in the crystal. When both these approaches fail, trial and error X-ray methods have to be used.

For ease of handling the glass fibre, to which the crystal is to be attached, its opposite end is pushed into a pea-sized blob of plasticine. When the crystal has been firmly glued to the fibre, the plasticine serves to fasten the fibre quite rigidly to the arcs. Crystallographic arcs, illustrated in Fig 8.1, consist of a manually adjustable slide *a*, carrying a projection against which the plasticine blob is pressed, surmounting a pair of worm operated arcs. The upper arc *b* provides a movement of 30° in either a clockwise or anticlockwise sense along the circumference of a vertical circle parallel

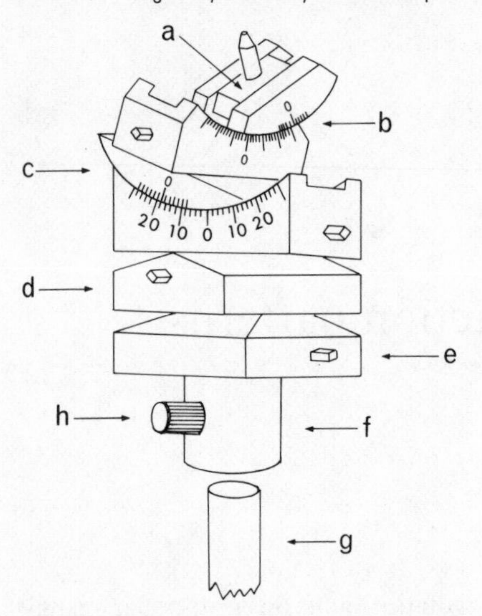

Fig 8.1 Crystallographic arcs. The labelling of the drawing is described in the text.

to the flat face of the arc. The lower arc *c* provides a similar movement in the vertical plane perpendicular to the plane of the upper arc *b*. Below the arcs are parallel worm operated slides *d*, *e*, to allow the crystal to be moved in a horizontal plane into coincidence with the axis of the spindle of the camera; the cap *f* at the base of the set of arcs fits over the camera spindle *g* and is clamped to it by tightening the knurled screw *h*.

The purpose of crystallographic arcs is to enable one to align the crystal so that it has a zone axis accurately parallel to a chosen direction in the X-ray camera. The techniques of crystal setting are described in Appendix F, which also provides details of crystal mounting techniques.

Oscillation photography

The oscillation camera, shown diagramatically in Fig 8.2, comprises a spindle *a* which is rigidly attached to the circular horizontal scale *b*, the arm *c* is clamped to the spindle and rests at its other extremity against a cam *d* driven by a synchronous electric motor *e* geared down to about one revolution per minute. Rotation of the cam causes the spindle, to which the arcs *f* are clamped, to oscillate through a definite angle; most cameras are equipped with alternative cams to give a choice of 5°, 10°, or 15° oscillation. A pin-hole collimator *g* is mounted so that its axis intersects the oscillation axis at right-angles; most cameras are supplied with fine and coarse collimators. The spindle can be raised or lowered so as to position the crystal in the X-ray beam. A cylindrical brass cassette whose axis is coincident with the oscillation axis constrains the photographic film *h*, which is in a light-tight envelope, to a cylindrical shape, the collimator protruding through the gap between the edges of the film; a typical film diameter is 60 mm. The undeviated X-ray beam passes through a hole in the cassette to be absorbed by a circular lead disc, the back-stop, attached to a removable cap (Fig 8.2(b)). By removing the cap it is a simple matter to test, with a fluorescent screen, whether the X-ray beam is passing correctly through the camera. Since the undeviated beam would produce serious fogging if it fell on the film a hole is punched in the film and a small tightly fitting brass collar is pressed through the hole in the film into the hole in the cassette to ensure correct alignment. The

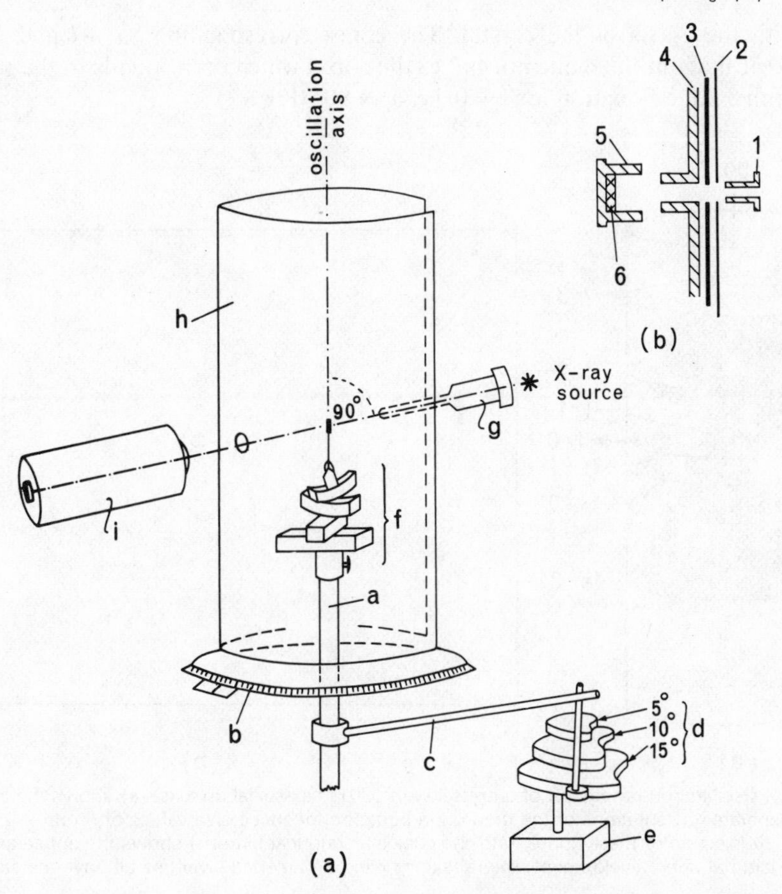

Fig 8.2 The oscillation camera. The essentials of the camera are shown in (a), the labelling of which is explained in the text. The detail of the backstop, or beam trap, is shown enlarged and exploded in (b): 1, collar; 2, opaque paper; 3, film; 4, wall of cassette; 5, cap; 6, lead disc.

oscillation camera is equipped with a telescope *i* whose axis is accurately aligned with that of the collimator to facilitate alignment and setting of the crystal. By swinging a lens into position in front of its objective the telescope *i* is simply converted into a microscope through which the crystal can be observed in a beam of light directed through the collimator; by focusing the microscope on the crystal and rotating the crystal, the crystal can be accurately centred on the oscillation axis of the camera and its height can be adjusted so that it lies precisely at the intersection of the oscillation axis with the incident beam from the collimator.

For oscillation photography the crystal is set so that a prominent zone axis is accurately parallel to the oscillation axis, that is the spindle axis, of the camera. Monochromatic X-radiation is incident on the crystal perpendicular to the oscillation axis. Suppose that the [001] axis of the crystal is parallel to the oscillation axis, then $i_c = 90°$ and the third Laue Equation reduces to

$$c \cos \delta_c = l\lambda.$$

Diffracted beams are consequently restricted to a series of cones, each cone being associated with a particular value of l, coaxial with the oscillation axis of the camera and with the z-axis of the crystal. The cones corresponding to $+l$ and $-l$ are symmetrical about the plane normal to the z-axis which corresponds to the solution of the third Laue Equation for $l = 0$, i.e. $\delta_c = 90°$ (Fig 8.3).

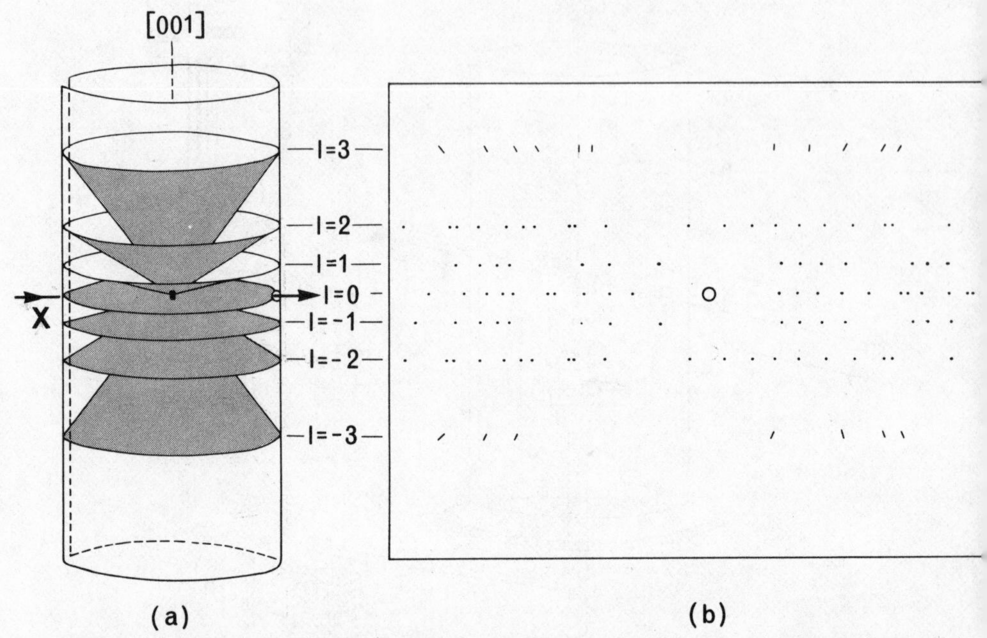

Fig 8.3 Oscillation photograph of a crystal with [001] as oscillation axis. (a) shows the coaxial cones generated by solutions of the third Laue Equation for successive values of l from -3 to $+3$ and the intersection of these cones with the coaxial cylindrical film. (b) shows the appearance of the film laid flat after development; the reflexions lying on any one layer line all have the same l index.

If the crystal is not oscillated, but kept stationary, very few diffraction maxima will occur because simultaneous solutions of the first and second Laue Equations,

$$a(\cos\delta_a - \cos i_a) = h\lambda$$

and

$$b(\cos\delta_b - \cos i_b) = k\lambda,$$

will be extremely rare. The number of diffraction maxima is substantially increased by oscillating the crystal about its z-axis so that the angles i_a and i_b can vary to some definite extent while i_c remains fixed at 90°. All the resultant diffracted beams lie on the surfaces of the set of cones that constitute the solutions to the third Laue Equation for different values of l. The set of cones intersects the cylindrical film in a set of circles, which appears, when the film is laid flat after development, as a set of parallel straight lines. All the reflexions recorded on the film lie on these straight lines, which are known as *layer lines*. Each layer corresponds to a solution of the third Laue Equation for a particular value of l and the layer lines are symmetrically disposed about that one of their set corresponding to $l = 0$, the *zero layer line*. For the lth layer line it is evident from Fig 8.4 that δ_c can be calculated by measuring the height H_l of this layer line above the zero layer line since the camera radius r is known.

$$\cot\delta_c = \frac{H_l}{r}.$$

Combination of this equation with the third Laue Equation yields the generally useful equation

$$c = \frac{l\lambda}{\cos \cot^{-1} H_l/r}$$

which enables the lattice spacing parallel to the oscillation axis to be evaluated.

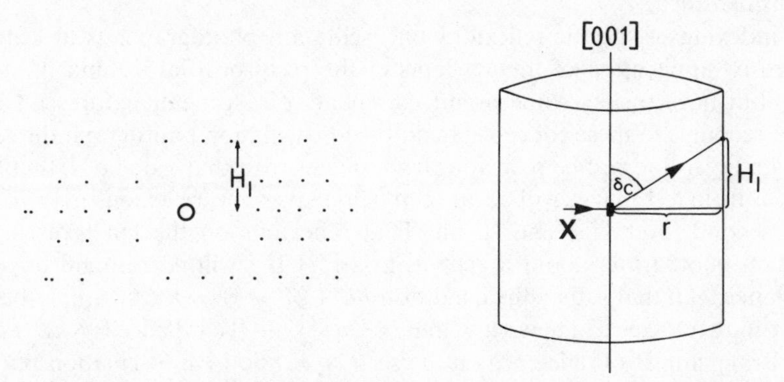

Fig 8.4 The geometry of layer line generation. The left-hand diagram is a portion of an oscillation photograph and the right-hand diagram illustrates the relationship between δ_c, H_l, and r.

Since the cotangent of δ_c is proportional to H_l, better accuracy would appear to be obtainable by measurement of the height of the highest observable layer line above the zero layer line, or better by measurement of the distance between the $+l$ and $-l$ layer lines for the greatest practicable value of l. However factors such as the divergence of the incident X-ray beam and the progressive reduction in angle between the diffracted beam and the surface of the film as l increases make the use of upper layer lines of no great advantage in improving the accuracy of δ_c. In general, measurement of layer line spacings yields unit-cell dimensions of accuracy no better than ± 1 per cent and can only be regarded as a preliminary means of determination of unit-cell dimensions; more sophisticated methods will be described later in this chapter.

The three unit-cell edges, a, b, and c, can be measured by mounting the crystal with its [100], [010], and [001] axes parallel in turn to the oscillation axis of the camera. If some other prominent zone axis can be located and the crystal mounted with this axis parallel to the oscillation axis of the camera, then the layer line spacing of the resultant oscillation photograph will provide a measurement of the spacing of lattice points in this direction by application of the general equation (p. 155) for $i = 90°$,

$$t \cos \delta = n\lambda$$

whence $$t = \frac{n\lambda}{\cos \cot^{-1} H_n/r}.$$

Thus for a monoclinic crystal measurement of the layer line spacing on a [110] oscillation photograph would serve to evaluate the lattice spacing parallel to the [110] zone axis and so to distinguish between a P lattice for which $t_{[110]} = \sqrt{(a^2+b^2)}$ and a C lattice for which $t_{[110]} = \frac{1}{2}\sqrt{(a^2+b^2)}$ if the magnitudes of a and b are already known. The determination of lattice type by this means is however fraught with danger because it is difficult to locate with certainty any but the most obvious zone axes even in a crystal with very well developed faces. In the early years of X-ray crystallography it was common practice to determine the β angle of monoclinic unit-cells by measuring the layer line spacing on [101] oscillation photographs and then applying the relationship $t^2_{[101]} = a^2 + c^2 + 2ac\cos\beta$; but later work using more sophisticated methods has in many cases shown the selected oscillation axis to have been misidentified.

The indexing of specific reflexions on oscillation photographs is in general best achieved by application of the concepts of the reciprocal lattice and the reflecting sphere, but nevertheless some useful comments of a general nature can be made without recourse to these concepts. In an [001] oscillation photograph the reflexions on the lth layer line will be reflexions from planes with that value of l, that is to say reflexions from $hk1$ planes will occur on the first layer line, reflexions from $hk2$ planes on the second layer line, and so on. That reflexions on the nth layer line of an oscillation photograph about a general axis $[UVW]$ will correspond to reflexions from planes (hkl) that satisfy the condition $hU + kV + lW = n$ can simply be seen by transforming the axes to a new set with $c' = Ua + Vb + Wc$, whence $l' = hU + kV + lW$.

The Bragg angles of reflexions on the zero layer line of an oscillation photograph can be determined directly in just the same way as for a powder photograph. For a cubic crystal the unit-cell dimension a can be determined from the layer line spacing of an oscillation photograph taken about an identifiable axis and the reflexions on the zero layer line can then be indexed by direct measurement of the Bragg angle θ and application of the expression

$$\lambda = \frac{2a\sin\theta}{\sqrt{(h^2 + k^2 + l^2)}}.$$

Consider for instance a [100] oscillation photograph which will yield a value for a whatever the lattice type; the reflexions on the zero layer line will be $0kl$ reflexions for which $\lambda = (2a\sin\theta)/\sqrt{(k^2 + l^2)}$. If the oscillation axis is parallel to a zone axis along which the spacing of lattice points is such that the unit-cell dimension is not deducible unless the lattice type is known, care must be exercised. Consider for instance a [111] oscillation photograph: if the lattice is P or F, $t_{[111]} = a\sqrt{3}$, but $t_{[111]} = \frac{1}{2}a\sqrt{3}$ for an I-lattice. In such a case as this, indexing of the zero layer reflexions on the [111] oscillation photograph should not be attempted until a has been determined directly from a [100] oscillation photograph.

Interpretation of oscillation photographs using the reciprocal lattice

The position of a reflexion on an oscillation photograph is specified by two coordinates (Fig 8.5), the height H of the reflexion above the zero layer line and M, which is defined as the distance of the reflexion measured along its layer line from the intersection of the plane containing the incident X-ray beam and the oscillation axis with the plane of the film. Our task now is to relate the measured coordinates of a reflexion to the coordinates of the reciprocal lattice point that gives rise to it on passing through the reflecting sphere. For this purpose it is convenient to define a system of cylindrical coordinates for the reciprocal lattice: ζ is defined as the perpendicular distance of the reciprocal lattice point from the plane which is perpendicular to the oscillation axis and contains the incident X-ray beam, and ξ is defined as the perpendicular distance of the reciprocal lattice point from the oscillation axis (Fig 8.6(a)). It is convenient, for reasons which will appear in due course, in

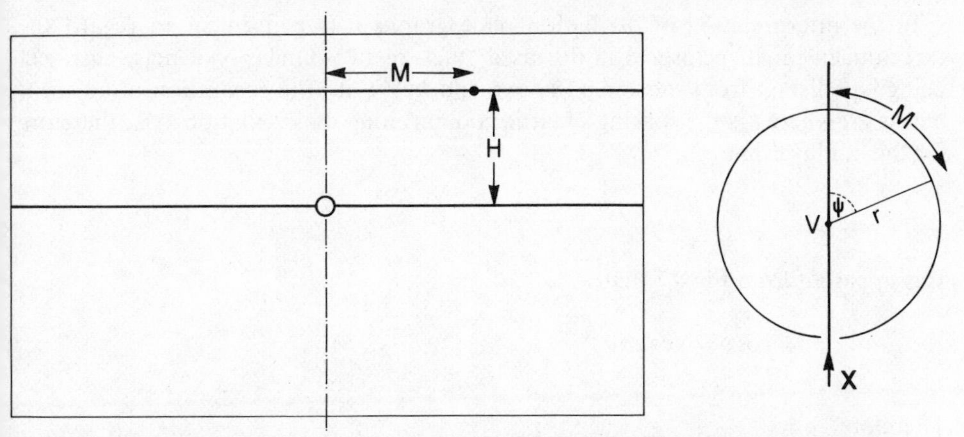

Fig 8.5 The coordinates H and M of a reflexion on an oscillation photograph. The selected reflexion is situated on the film at a perpendicular distance H from the zero layer line and at a distance M from the trace (shown as a dash-dot line) of the plane containing the incident beam and the oscillation axis. The right-hand diagram illustrates the relationship $M = r\psi$ of the film coordinate M to the angle ψ, which is the angle between two planes passing through the oscillation axis V, one containing the incident beam and the other the diffracted beam.

interpreting diffraction patterns to take the proportionality factor K in $d^* = K/d$ equal to the wavelength λ of the incident monochromatic radiation; this has the effect of making the radius of the reflecting sphere equal to one reciprocal lattice unit. One final preliminary statement remains to be made: that is the definition of the angle ψ as the angle made by the plane containing the incident X-ray beam and the oscillation axis with the plane containing the diffracted beam and the oscillation axis (Fig 8.6(b)); it is evident from Fig 8.5 that $\psi = M/n$.

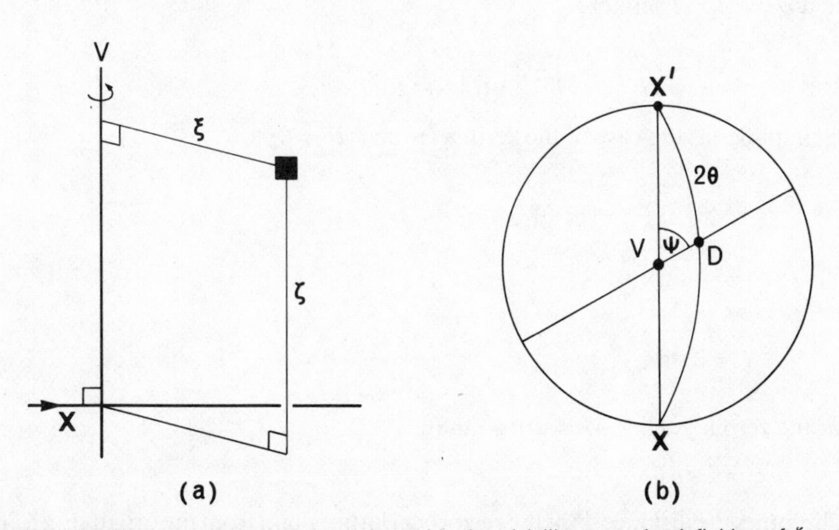

Fig 8.6 Cylindrical coordinates for the reciprocal lattice. (a) illustrates the definition of ζ as the perpendicular distance of the reciprocal lattice point (solid square) from the plane which is perpendicular to the oscillation axis V and contains the incident beam X, and ξ as the perpendicular distance of the reciprocal lattice point from the oscillation axis. (b) illustrates the definition of ψ as the angle between the plane containing the oscillation axis V and the incident beam XX′ and the plane containing the oscillation axis V and the diffracted beam D (cf. Fig 8.5).

In the interpretation of oscillation photographs it is convenient to regard the three-dimensional reciprocal lattice as a stack of two-dimensional nets, each net being equidistant from those next above and below it. The separation of adjacent nets is λ/t where t is the spacing of lattice points along the oscillation axis; therefore for the nth layer line

$$\zeta = \frac{n\lambda}{t}.$$

It is apparent from Fig 8.7 that

$$\zeta = \cos\delta = \cos\cot^{-1}\frac{H}{r}$$

$$\text{Therefore}\quad t = \frac{n\lambda}{\cos\cot^{-1}\dfrac{H}{r}}.$$

Thus t can be evaluated by measurement of H when r and λ are known.

Measurement of the reflexion coordinates M and H yields a direct determination of the reciprocal lattice coordinate ξ. It is evident from Fig 8.7 that the circular section of the reflecting sphere in a plane of constant ζ has radius $\sqrt{(1-\zeta^2)}$ and its centre is unit distance from the oscillation axis. For a reciprocal lattice point lying on the reflecting sphere (Fig 8.7(c))

$$\xi^2 = 2-\zeta^2-2\sqrt{(1-\zeta^2)}\,.\cos\psi.$$

We have already seen that $\zeta = \cos\cot^{-1} H/r$, which can alternatively be expressed as

$$\zeta^2 = \frac{H^2}{H^2+r^2}$$

and that $\psi = M/r$. Therefore

$$\xi^2 = 1+\left(\frac{r^2}{H^2+r^2}\right)-2\left(\frac{r^2}{H^2+r^2}\right)^{\frac{1}{2}}\cos\frac{M}{r},$$

which simplifies in the case of the zero layer line ($H = 0$) to

$$\xi^2 = 2-2\cos\frac{M}{r}$$

$$= 4\sin^2\frac{M}{2r}$$

i.e. $$\xi = 2\sin\frac{M}{2r};$$

but for the zero layer line $M = 2r\theta$ so that

$$\xi = 2\sin\theta$$

and ξ becomes the distance d^* of the reciprocal lattice point from the origin (Fig 8.7(d)).

Since the calculation of ξ from measurements of M and H is rather awkward for upper layer lines, charts, known as *Bernal charts*, are commercially available with lines of constant ζ and curves of constant ξ drawn at 0.05 intervals (Fig 8.8) for the common camera radii. By superimposing a Bernal chart on an oscillation photograph

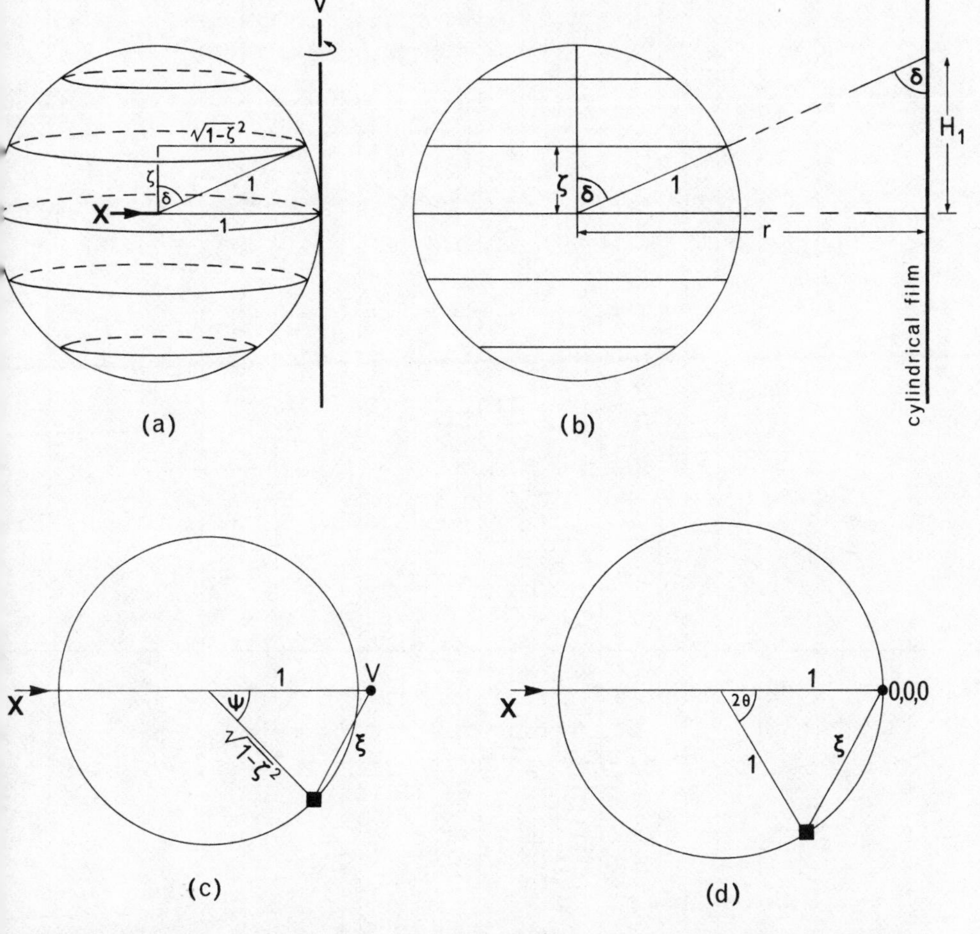

Fig 8.7 Interpretation of oscillation photographs in terms of the reciprocal lattice. (a) shows that a reciprocal lattice plane of constant ζ intersects the reflecting sphere in a circle of radius $\sqrt{(1-\zeta^2)}$ whose centre is unit distance from the oscillation axis V. (b) is a composite diagram in direct and reciprocal space to illustrate the relationship $\cot\delta = H_1/r$ for the first layer line on an oscillation photograph. (c) illustrates the relationship $\xi^2 = (1-\zeta^2)+1-2\sqrt{(1-\zeta^2)}\,.\cos\psi$ for a non-zero layer reciprocal lattice point lying on the reflecting sphere and (d) illustrates the simpler case of a zero layer reciprocal lattice point where $\xi = 2\sin\theta$.

so that its line of zero ζ coincides with the zero layer line and the origin of the chart coincides with the centre of the punched hole through which the undeviated X-ray beam passed, the coordinates ξ and ζ of each reflexion on the film can be read directly. It is to facilitate the use of the Bernal chart that the proportionality constant K for the reciprocal lattice is commonly taken as equal to λ in practical work; the radius of the reflecting circle is then independent of the wavelength of the radiation used so that the curves of constant ξ and ζ shown on the chart for a camera of given radius are generally applicable. To avoid confusion about the magnitude of K it is usual to refer to reciprocal units when $K = 1$ as Å^{-1}, but when $K = \lambda$ no reciprocal unit is specified. As will be apparent from the argument that follows only rarely does ξ need to be known to higher accuracy than is obtainable by use of the Bernal chart.

As the crystal oscillates the reciprocal lattice net oscillates correspondingly about

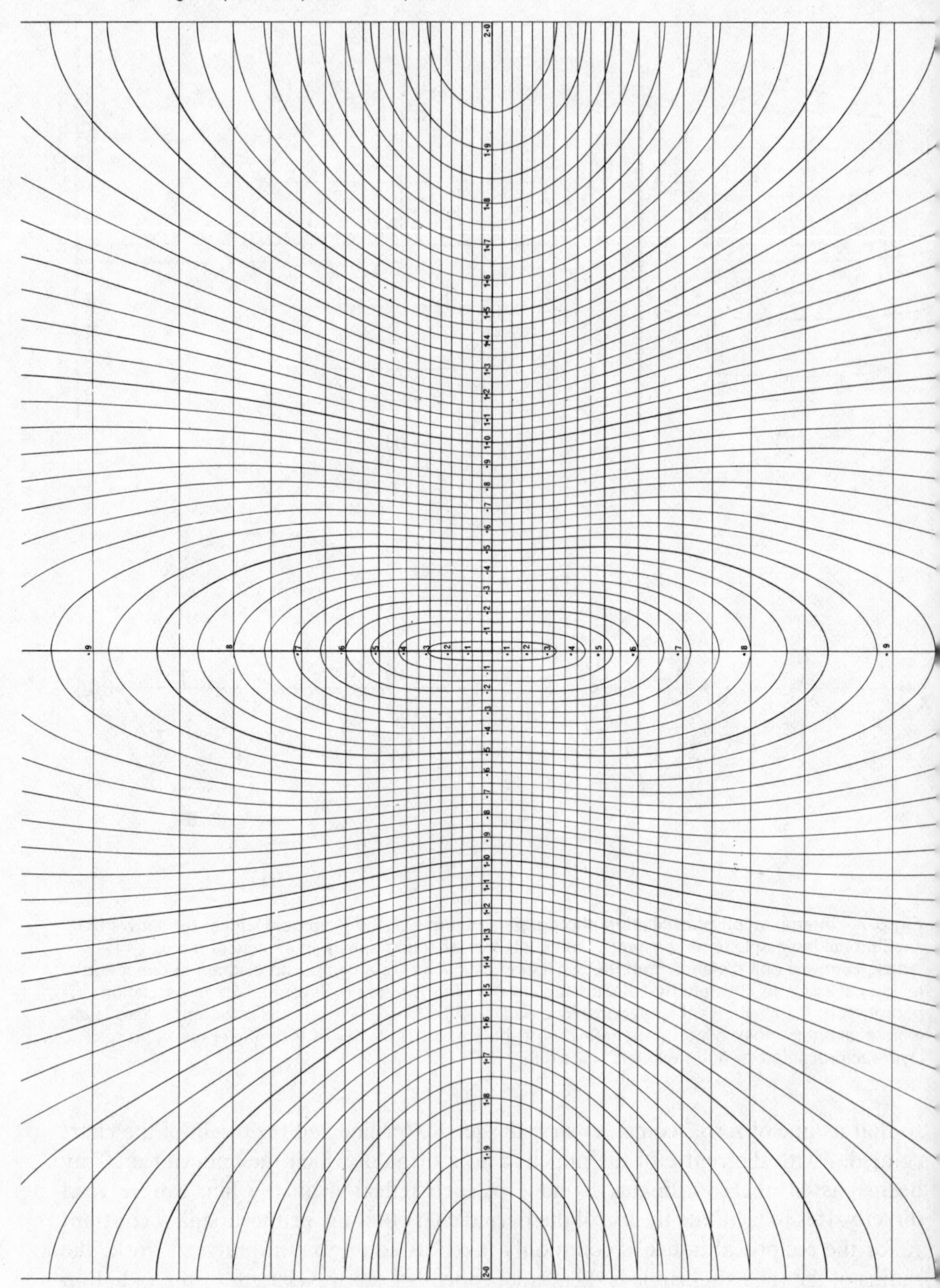

Fig 8.8 The Bernal chart. The parallel lines are lines of constant ζ in 0·05 intervals and the curves are curves of constant ξ in 0·05 intervals. The chart shown is approximately the size required for a camera of 30 mm radius. The chart is reproduced by courtesy of the Institute of Physics.

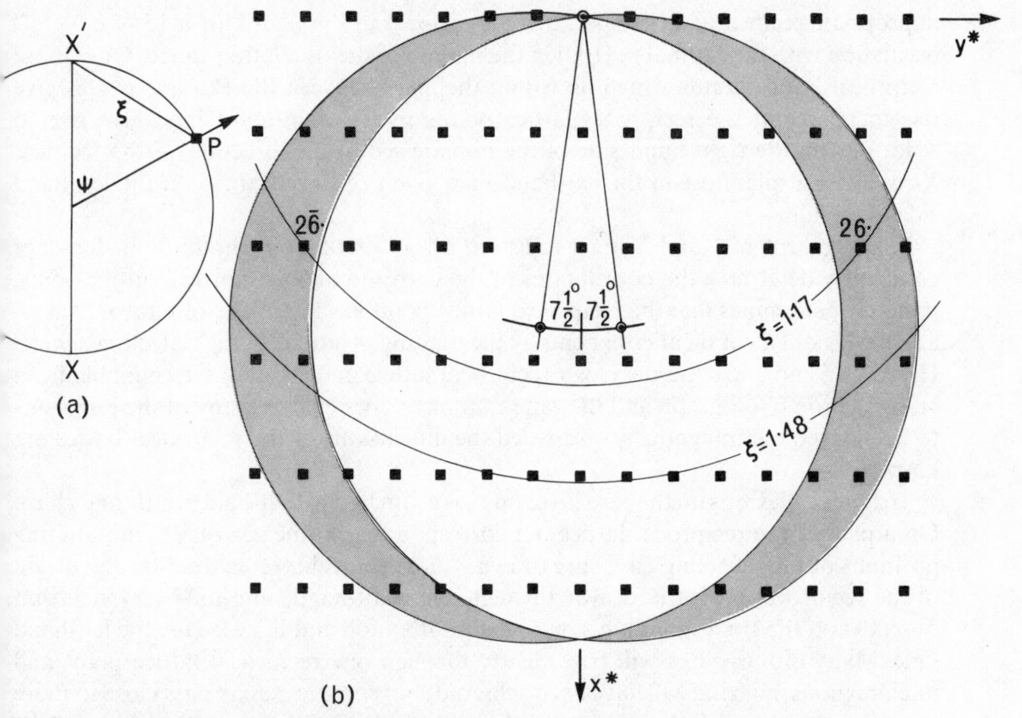

Fig 8.9 Interpretation of oscillation photographs in terms of the reciprocal lattice. (a) shows the general relationship between ξ and ψ for a reciprocal lattice point on the reflecting sphere. In (b) the x^*y^* reciprocal lattice net for a monoclinic crystal is taken to be stationary and the limits of oscillation of the reflecting sphere are shown as their circular sections of intersection with the net; the areas of the reciprocal lattice net intersected by the reflecting sphere are shown as shaded *lunes* for a 15° oscillation about x^*. The arc of $\xi = 1{\cdot}17$ passes through the reciprocal lattice points $2\bar{6}$. and 26., but the arc of $\xi = 1{\cdot}48$ does not pass through any reciprocal lattice points within the lunes; the resulting oscillation photograph may be expected to exhibit reflexions on the zero layer line with $\xi = 1{\cdot}17$, but no reflexions with $\xi = 1{\cdot}48$.

the vertical through its origin. Any reciprocal lattice point which passes through the reflecting sphere in the course of the oscillation gives rise to a diffracted beam in the direction of the radius of the reflecting sphere through the reciprocal lattice point[1] (Fig 8.9). Now and subsequently we shall treat the oscillation of the reciprocal lattice of the crystal relative to the reflecting sphere as a movement of the reflecting sphere through a stationary reciprocal lattice; to treat the relative motion in this way leads to some simplification of expression. Since the reflecting sphere oscillates about the vertical through the origin of the stationary reciprocal lattice, any reciprocal lattice point in the volume between the extreme positions of the reflecting sphere will give rise to a diffracted beam as the reflecting sphere passes through it. We illustrate this in Fig 8.9 for the zero layer of a monoclinic crystal in which the oscillation axis is parallel to [001]. The limiting positions of the reflecting sphere become for any one layer the limiting positions of the *reflecting circle* in which the sphere intersects the

[1] Since the origin of the reciprocal lattice lies on the oscillation axis, all diffracted beams pass through that point on the oscillation axis. But it is geometrically more convenient to suppose that all diffracted beams pass through the centre of the reflecting sphere when discussing the geometry of the oscillation photograph. It should be apparent to the reader that this geometrical simplification is generally acceptable even though it may occasionally make diagrams which combine reciprocal and direct space look inconsistent at first sight.

appropriate reciprocal lattice net; these are shown in the figure for a 15° (i.e. $\pm 7\frac{1}{2}°$) oscillation with the normal to (100) in the middle of the oscillation range. Only those reciprocal lattice points which lie within the *lunes* between the extreme circles give rise to reflexion; the reciprocal lattice points in the right-hand lune give rise to reflexions on the right-hand side of the film viewed in the direction of the incident X-ray beam while those in the left-hand lune give rise to reflexions on the left-hand side of the film.

Measurement of ζ and ξ for a reflexion on an oscillation photograph does not completely determine the coordinates of the corresponding reciprocal lattice point; it merely determines that the reciprocal lattice point lies in a plane of known ζ on an arc of constant ξ cut off at either end by the limiting positions of the reflecting sphere. However ζ and ξ are the only two reciprocal lattice coordinates determinable from an oscillation photograph and their measurement does enable many of the reflexions to be indexed unambiguously[2] provided the dimensions of the reciprocal lattice are known.

To index reflexions on the zero layer line ξ is estimated with the aid of a Bernal chart. On a plan of the reciprocal lattice net corresponding to the zero layer the limiting positions of the reflecting circle are drawn. An arc of radius ξ centred on the origin of the reciprocal lattice is drawn through the right-hand lune for every reflexion observed on the corresponding side of the photograph and likewise for the left-hand lune. Many of these arcs will pass clearly through one reciprocal lattice point and unambiguous indexing will have been achieved; but some arcs may pass close to more than one reciprocal lattice point so that the corresponding reflexions cannot be certainly indexed. If it is intended to determine the *diffraction symbol* of the substance the indices of reciprocal lattice points lying within the lunes but not giving rise to reflexion should be noted.

We take as our example a monoclinic crystal oscillating about its [001] axis with the incident X-ray beam normal to (100) in the middle of the 15° oscillation range. The zero layer line will contain only reflexions of the type *hk*0, the corresponding reciprocal lattice section will be the a^*b^* plane, which has a rectangular mesh, and the limiting positions of the reflecting circle will be symmetrically disposed about the x^*-axis. The procedure for indexing the zero layer line of the photograph is (i) draw out the reciprocal lattice net which passes through the origin and is normal to the oscillation axis [001], a convenient scale in most cases being 100 mm = 1 reciprocal unit, (ii) on this a^*b^* net draw an arc of unit radius about the origin of the net, (iii) mark on the arc two points $7\frac{1}{2}°$ on either side of the intersection of the arc with the x^*-axis, (iv) with these points as centres draw circles of unit radius to represent the limiting positions of the reflecting circle for the zero layer line, (v) draw an arc of radius ξ centred on the origin of the reciprocal lattice net for every observed reflexion to traverse the right-hand lune for reflexion on the right-hand side of the zero layer line (the film being viewed in the direction of the incident beam) and to traverse the left-hand lune for those on the left, (vi) passage of a ξ-arc through one reciprocal lattice point provides immediate indexing of the corresponding reflexion, (vii) note the indices of reciprocal lattice points lying within the lunes that do not give rise to reflexions.

In general the oscillation axis and the reciprocal axis corresponding to it will not be parallel so that in higher layers a reciprocal lattice point does not lie on the

[2] For unambiguous indexing of all the reflexions on a diffraction pattern recourse must be made to the moving film methods described later in this chapter.

oscillation axis. In the monoclinic example shown in Fig 8.10 the reciprocal axis z^* is inclined at $90° - \beta^*$ to [001] in the plane normal to y^*. The nth reciprocal lattice layer intersects the oscillation axis at a distance $\zeta = n\lambda/c$ from the origin and contains reciprocal lattice points whose l-index is equal to n. The reciprocal lattice point $00n$ is distant $\zeta \cot \beta^* = nc^* \cos \beta^*$ (Fig 8.10(b)) from the oscillation axis along the x^*-axis, that is along the reciprocal lattice row $h0n$. To index an upper layer line of an [001] oscillation photograph it is thus necessary to mark on the a^*b^* reciprocal lattice net the point of intersection O_n of the oscillation axis, which will lie for the nth layer at a distance $\zeta \cot \beta^*$ from the origin of the net, the reciprocal lattice point $00n$, in the $-x^*$ direction (i.e. between $00n$ and $\bar{h}0n$). The radius of the reflecting circle for this

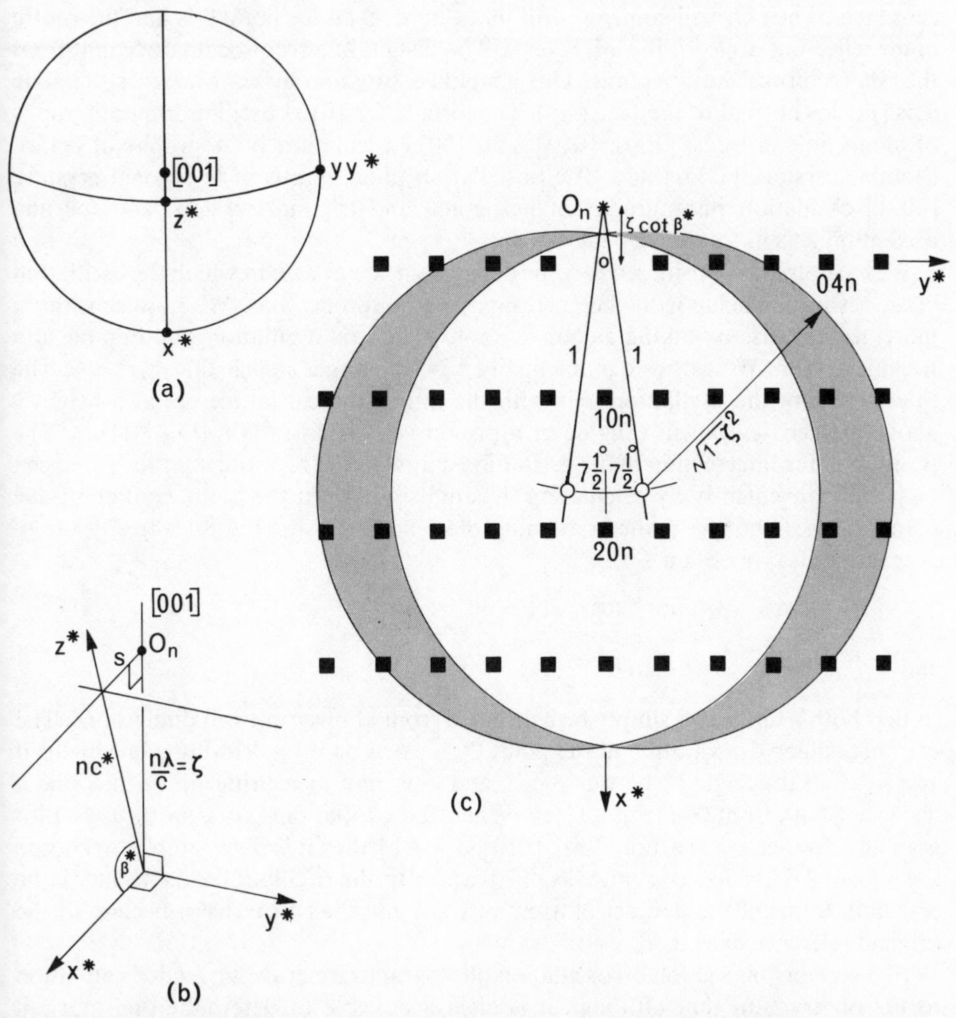

Fig 8.10 Indexing of reflexions on the nth layer line of the oscillation photograph of a monoclinic crystal whose oscillation axis is [001]. The stereogram (a) shows the relationship between direct and reciprocal lattice axes. The perspective drawing (b) shows the displacement $s = nc^* \cos \beta^* = \zeta \cot \beta^*$ of the point of intersection O_n of the oscillation axis with the a^*b^* reciprocal lattice net from the reciprocal lattice point $00n$; for n positive the displacement is in the direction $-x^*$. (c) shows the a^*b^* reciprocal lattice net with circles of radius $\sqrt{(1-\zeta^2)}$ drawn with centres distant one reciprocal lattice point from O_n, which is distant $\zeta \cot \beta^*$ along $-x^*$ from the origin $00n$ of the net.

layer is $\sqrt{(1-\zeta^2)}$ and its centre lies at a distance of 1 reciprocal lattice unit from this intersection O_n in a direction parallel to the incident beam, in this case parallel to x^* for the middle of the oscillation range (Fig 8.10(c)). The procedure for indexing the nth layer of the photograph is (i) on the a^*b^* reciprocal lattice net draw an arc of unit radius about a point O_n along $-x^*$ at a distance $\zeta \cot \beta^*$ from the reciprocal lattice point $00n$, (ii) mark the two points $7\frac{1}{2}°$ on either side of the intersection of the arc with x^*, (iii) with these points as centres draw the limiting reflecting circles of radius $\sqrt{(1-\zeta^2)}$, (iv) then proceed as for the zero layer line (v)–(vii).

If the oscillation axis and the corresponding reciprocal axis are coincident, the reciprocal lattice points $00l$ will lie on the oscillation axis and therefore the reciprocal lattice net for the nth layer will be superimposed without displacement on that for the zero layer. Then O_n will coincide with the reciprocal lattice point $00n$ and the centre of the reflecting circle for the nth layer will be distant one reciprocal lattice unit from the $00n$ reciprocal lattice point. This simplified situation arises whenever the zone axis $[pqr]$ is normal to the face (pqr). This occurs for [010] oscillation photographs of monoclinic crystals, [100], [010], and [001] oscillation photographs of orthorhombic crystals, $[UV0]$ and [001] oscillation photographs of tetragonal crystals, [0001] oscillation photographs of hexagonal and trigonal crystals, and for any oscillation axis in a cubic crystal.

We have already considered in some detail a particular case in which the oscillation axis does not coincide with the corresponding reciprocal axis. We now consider a more general case by taking as our example the z-axis oscillation photograph of a triclinic crystal. We earlier defined in Fig 6.24, the angle ε such that $z{:}z^* = \varepsilon$. The intersection of the oscillation axis with the nth reciprocal lattice net at a height ζ above the zero layer will thus lie at a distance $\zeta \tan \varepsilon$ from $00n$ (Fig 8.11(a)). The position of the intersection of the oscillation axis with the reciprocal lattice plane can be found conveniently by calculating the angle η between the plane containing the z and z^* axes and the plane containing the z and y^* axes (Fig 8.11(b)). By use of Napier's Rules we obtain

$$\tan \eta = -\tan \alpha^* \cos \beta$$

and $$\cos \varepsilon = \sin \alpha^* \sin \beta$$

so that both η and ε can simply be calculated from known unit-cell dimensions. The a^*b^* net is then drawn out and the point O_n located on it by drawing a line inclined at η to y^* in the angle γ^* between $-x^*$ and $-y^*$ and measuring off on this line a distance $\zeta \tan \varepsilon$ from $00n$ (Fig 8.11(c)). When the oscillation axis is not a crystallographic reference axis, i.e. not [100], [010], or [001], then it is often simpler to choose a new set of reference axes with, say z', parallel to the oscillation axis, to index the crystal in terms of the new set of axes, x', y', z', and then to transform back to the original reference axes, x, y, z.

To conclude this section on oscillation photographs we draw the reader's attention to the observation that although it is always possible to determine one unit-cell dimension from an oscillation photograph, that parallel to the oscillation axis, reflexions can only be indexed when the reciprocal lattice geometry of the crystal is known. The determination of reciprocal lattice geometry is most easily achieved by the use of Weissenberg and precession photographs; oscillation photographs are practically useful only as a preliminary to more thorough investigation by moving film methods and for some specialized applications outside the scope of this textbook.

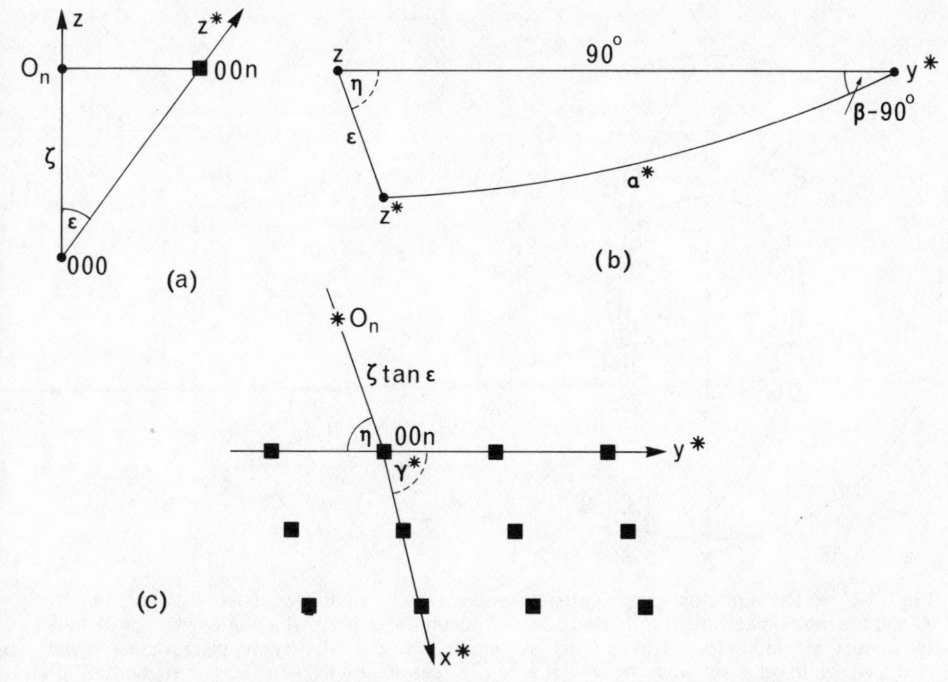

Fig 8.11 Indexing of a non-zero layer on the oscillation photograph of a triclinic crystal. (a) is a plane diagram showing the intersection of the *n*th reciprocal lattice net with z^* at 00*n* and with *z* (the oscillation axis) at O_n. The spherical triangle (b) illustrates the calculation of ε and η from α^* and β. The position of O_n on the a^*b^* reciprocal lattice net is shown in (c).

Rotation photography

The oscillation camera can simply be adapted to permit the crystal to rotate through 360° by removing the arm from the spindle to the cam and introducing a belt-drive from the camshaft to the spindle. The resultant rate of rotation of the crystal is of the order of one revolution per minute. The resultant rotation photograph is of course very similar to an oscillation photograph of the same substance taken about the same axis but displays many more reflexions. Measurement of layer line spacing enables the repeat distance between lattice points along the rotation axis to be determined in just the same way as from an oscillation photograph.

In the course of a complete rotation every lattice plane giving rise to a reflexion *hkl* will pass through an orientation which satisfies the Bragg Equation twice (Fig 8.12). The two reflexions produced will be symmetrically disposed about the vertical line representing the intersection of the film with the plane containing the incident X-ray beam and the rotation axis. Moreover the $(\bar{h}\bar{k}\bar{l})$ lattice plane will give rise to two similarly disposed reflexions of equal intensity and these will be positioned on the film so that they are related to the *hkl* reflexions by a line of symmetry coincident with the zero layer line. Rotation photographs thus always have the symmetry of the two-dimensional point group 2*mm*. In a case in which two or more lattice planes have the same *d*-spacing (and therefore identical Bragg angle) and give rise to reflexions on the same layer line these reflexions will be superimposed in the rotation photograph: for instance, on the [001] rotation photograph of a cubic crystal the 501, 051, $\bar{5}01$, and $0\bar{5}1$ reflexions will all be superimposed at a point on the first layer line.

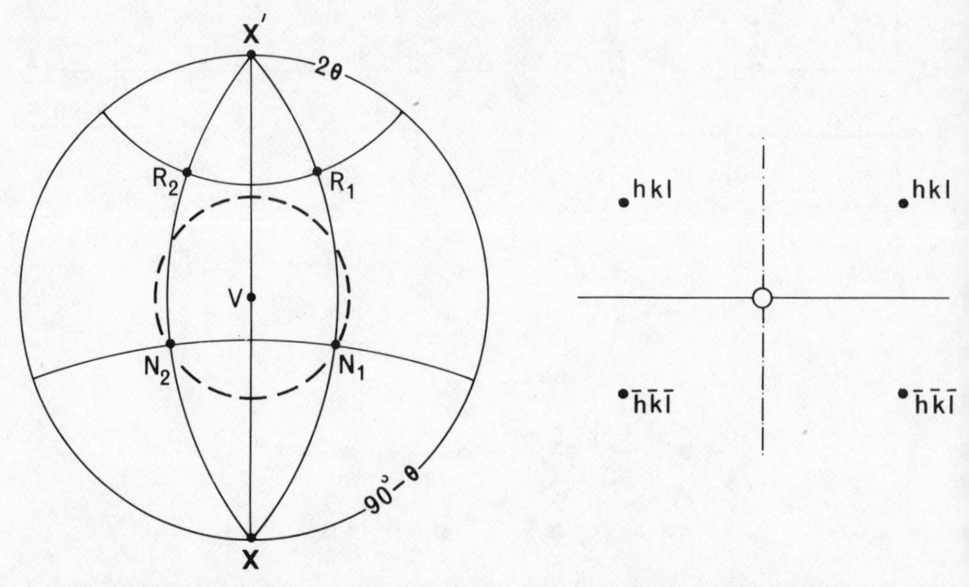

Fig 8.12 Rotation photography. On the stereogram the incident X-ray beam is denoted by X and X′ and the rotation axis by V. If in the course of rotation of the crystal a plane (*hkl*) comes into the correct orientation for reflexion, when its normal is N_1, then during complete rotation it will also give rise to reflexion when its normal is N_2. The resultant reflexions, R_1 and R_2 on the stereogram, lie at the same height from the zero layer line (the primitive of the stereogram and the solid line on the drawing of a rotation photograph on the right) and are symmetrically disposed about the plane containing X, X′ and V (shown as a dot-dash line in the right-hand diagram). The ($\bar{h}\bar{k}\bar{l}$) plane will give rise to two identical reflexions which are related on the rotation photograph to the *hkl* reflexions by a line of symmetry coincident with the zero layer line.

A rotation photograph is more useful than an oscillation photograph for the determination of the lattice type of a cubic or tetragonal crystal from an [001] photograph by measurement of θ for all reflexions on the zero layer line (the example discussed earlier) because it will display all reflexions of non-zero intensity for which θ is less than some angle little short of 90°. But exposure times for rotation photographs are very much longer than for oscillation photographs and, except for the sort of problem mentioned immediately above, rotation photographs are little used.

Interpretation of rotation photographs by use of the reciprocal lattice

As the crystal rotates about an axis, which for argument we take as [001], the reflecting sphere sweeps through the reciprocal lattice so that in the course of a complete 360° rotation it passes through all reciprocal lattice points lying on the zero layer reciprocal lattice net within a circle of radius two reciprocal units about the origin of the net. For an upper layer the reflecting circle cuts an annular swathe through the relevant reciprocal lattice net with internal and external radii $1 \pm \sqrt{(1-\zeta^2)}$ (Fig 8.13). Thus for one setting of the crystal all the reciprocal lattice points within the torus produced by the rotation of the reflecting sphere about the rotation axis pass through the reflecting sphere. Moreover each such reciprocal lattice point passes through the reflecting sphere twice in the course of complete rotation of the crystal and so gives rise to a reflexion on the right and on the left-hand side of the incident X-ray beam. Reciprocal lattice points with the same value of ξ in the same reciprocal lattice net will of course give rise to superimposed reflexions.

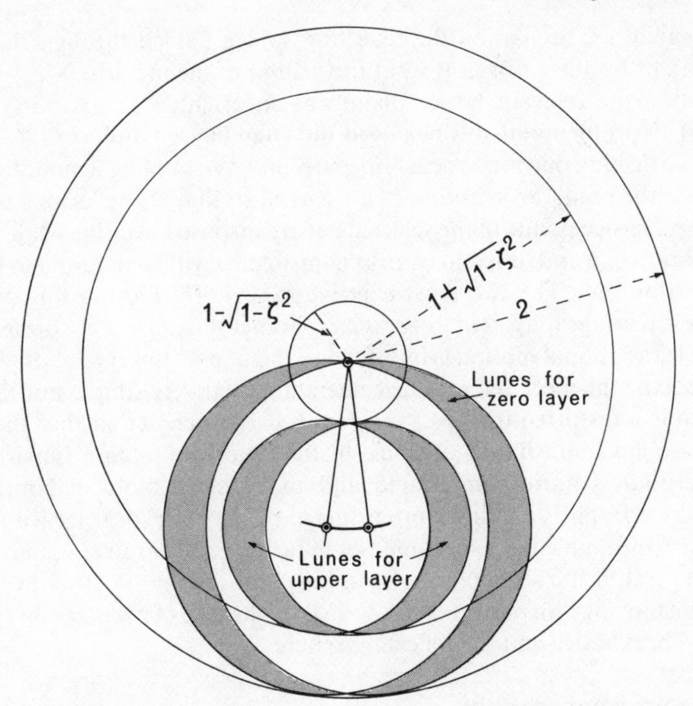

Fig 8.13 Rotation photography: the limiting torus. As the crystal describes a complete rotation the reflecting sphere sweeps through all reciprocal lattice points on the zero layer net within a circle of radius 2 reciprocal units and cuts an annular swathe with internal and external radii $1 \pm \sqrt{(1-\zeta^2)}$ through non-zero layer nets. The lunes for the zero layer and one non-zero layer of a 15° oscillation photograph are shown shaded.

Only very rarely, and then when the rotation axis is a symmetry axis of high order, is it profitable to attempt to index a rotation photograph. All the reciprocal space that can be sampled by a rotation photograph can be more informatively sampled by a series of 15° oscillation photographs taken at 15° intervals.

The *limiting torus* of a rotation photograph encloses all those reciprocal lattice points which may give rise to a reflexion in the course of a complete rotation and comprises all that volume of reciprocal space which can be studied with one rotation photograph or a series of oscillation photographs taken about the same axis. The only way in which reciprocal lattice points outside the limiting torus can be investigated is by changing the rotation or oscillation axis of the crystal. By use of a variety of rotation axes all the reciprocal lattice points within a sphere of radius equal to 2 reciprocal units centred on the origin of the reciprocal lattice can be brought into the reflecting position; this sphere is known as the *limiting sphere*. All reciprocal lattice points corresponding to lattice spacings greater than $\frac{1}{2}\lambda$ Å lie within the limiting sphere and so the smaller the magnitude of λ the greater the number of reflexions that may be obtained; in other words, since $\sin\theta \leqslant 1$ it follows from the Bragg Equation that reflexion can only take place for lattice planes whose spacing $d \geqslant \frac{1}{2}\lambda$.

Moving film methods

We have already seen that the coordinates of a reciprocal lattice point are not completely determinable from an oscillation photograph because one cannot know at what stage of the oscillation the reflecting sphere passed through the reciprocal lattice point, or in other words at what inclination of the incident X-ray beam to its mean position the relevant lattice plane was so oriented as to satisfy the Bragg Equation. This problem can only be solved and unambiguous indexing of all reflexions achieved by selecting one reciprocal lattice net and by coupling a smooth movement of the film to the oscillatory motion of the crystal so that the reflexions produced by the reciprocal lattice points of the selected net are disposed over the whole area of the film. A separate film and different camera adjustments will be required to record each reciprocal lattice net. The two most effective means of achieving this objective are *Weissenberg* photography and *precession* photography. In the former a simple oscillatory translational motion is imparted to the film as the crystal oscillates about the camera axis; the mechanics of the camera are relatively simple but the resultant photograph is a distorted image of the reciprocal lattice net so that the indices of reflexions are not immediately obvious. In the latter the motion imparted to both film and crystal is more complicated and the camera more elaborate; but the resultant photograph is an undistorted image of the reciprocal lattice net and the reflexions are indexable by inspection. Some further comparative comments will be made at the end of the section on precession photography. For the explanation of both types of moving film photograph[3] we shall make use, of necessity, of the concepts of the reciprocal lattice and the reflecting sphere.

Weissenberg photography

The essential feature of the Weissenberg camera is that it selects one layer line of an oscillation photograph and distributes the reflexions of the layer line over the whole area of the film so that the coordinates of each reciprocal lattice point giving rise to a reflexion can be unambiguously determined. The selection is achieved by the use of screens and the layer line is spread over the area of the film by moving the film backwards and forwards parallel to the oscillation axis of the crystal.

We now proceed to describe the Weissenberg camera. The arcs carrying the crystal are attached to a spindle driven by a synchronous electric motor as in the oscillation camera except that the spindle is horizontal (Fig 8.14). A worm drive from the motor moves the cylindrical film cassette parallel to its axis, which is coincident with the oscillation axis of the crystal, so that for every 1° rotation of the crystal the cassette moves 0·5 mm; at either end of the traverse of the cassette the motor is reversed by a micro-switch. The cassette has an axial slit about 5 mm wide to accommodate the collimator. Immediately inside the cassette and mounted independently of it is a cylindrical *screen* opaque to X-rays; the screen, which likewise has a slit through which the collimator projects, is in two halves, one attached to each end of the camera. The gap between the two halves of the screen can be varied in width (usually 2–4 mm) and its mean position set to coincide with the cone of diffracted beams of a selected layer line. A back-stop to absorb the undeviated X-ray beam is mounted inside one half-screen on an adjustable slide so that it can be set opposite the collimator. The radius of the cylindrical film is usually 28·65 mm and that of the screens 23·02 mm. The oscillation range, usually set to cover just over 180°, can be varied by adjusting the positions of the micro-switches that reverse the movement of the film cassette.

[3] The account of the theory of moving film methods and the interpretation of Weissenberg and precession photographs provided here is necessarily confined to essentials. Of the many excellent textbooks and monographs to which the reader might turn for more detailed information we draw attention particularly to Henry, Lipson, and Wooster (1960), Buerger (1942), Buerger (1964), Nuffield (1966), and Jeffery (1971).

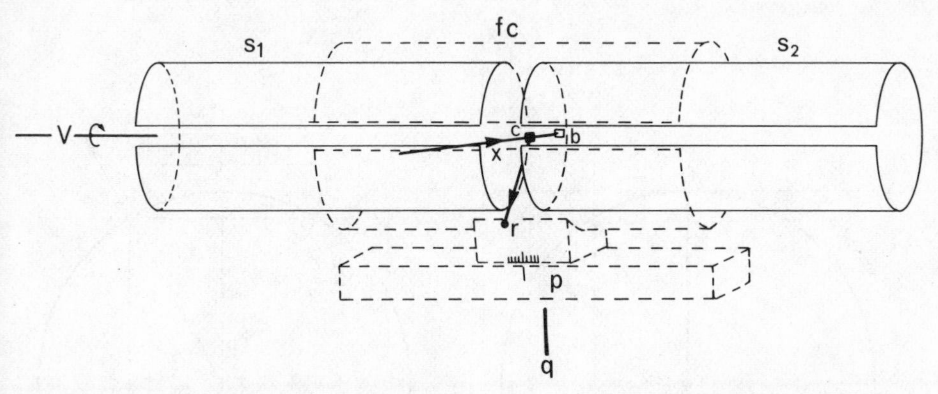

Fig 8.14 The Weissenberg camera. The perspective drawing shows the essential features of a Weissenberg camera. The crystal *c* is mounted on arcs on an oscillating spindle *V*. The split screens (s_1, s_2) are rigidly attached to the camera body in such a manner that the magnitude and position of their separation can be adjusted. The collimator, through which the incident X-ray beam, *X*, passes is rigidly mounted on the camera base. The back-stop, or beam trap, *b*, is attached to one screen. The film cassette (*fc*) is supported by the carriage *p* and can be set at a range of positions relative to a fiducial mark on the carriage. The carriage describes a linear motion parallel to the oscillation axis, its motion being geared to the oscillatory motion of the crystal. The bearings of the carriage are rigidly attached to the camera body. For zero layer photographs the oscillation axis is perpendicular to the incident X-ray beam; for non-zero layer photographs the camera body is inclined to the incident beam by moving the camera body through the appropriate angle about the vertical axis *q*, which passes through the intersection of the spindle axis *V* and the incident beam *X*.

The camera can be rotated through a selected angle about a vertical axis which passes through the centre of the crystal; this facility is necessary for recording upper layer diffraction patterns. The film cassette consists of two parts, the cylindrical cassette itself which holds the film and a *carriage* which is moved backwards and forwards on rails parallel to the oscillation axis of the crystal by a reversible motor. The cassette can be locked to the carriage at different distances from a fixed point on the carriage; we shall see that this is a useful facility when taking upper layer photographs.

We consider first the formation of zero-layer photographs. The axis of the collimator is perpendicular to the oscillation axis; the crystal is situated at their intersection and oriented so that a prominent zone axis is parallel to the oscillation axis. The reflexions of the zero layer are generated by reciprocal lattice points situated on the net containing the origin and perpendicular to the oscillation axis. The resultant diffracted beams therefore lie in the plane containing the incident beam and perpendicular to the oscillation axis (Fig 8.15(a)); if the centre of the gap in the screens is set to coincide with this plane, diffracted beams generated by reciprocal lattice points not situated on this net will be absorbed by the screens.

It is immediately apparent from inspection of a zero-layer Weissenberg photograph, such as that shown in Fig 8.16, that some reflexions lie on prominent straight lines running at a slant across the film. Another prominent feature of the photograph is the disposition of the reflexions on sets of curves. We now proceed to explain the geometry of a zero-layer photograph in general terms and then to discuss a specific example, a zero-layer photograph of a monoclinic crystal with [010] as oscillation axis.

Consider a line OP in reciprocal space lying in the plane normal to the oscillation axis and passing through the origin O of reciprocal space. Suppose that initially the line OP is tangential to the reflecting circle (Fig 8.17(a)) and that for this orientation of the crystal the extreme left-hand side of the film (Fig 8.17(b)) is opposite the gap

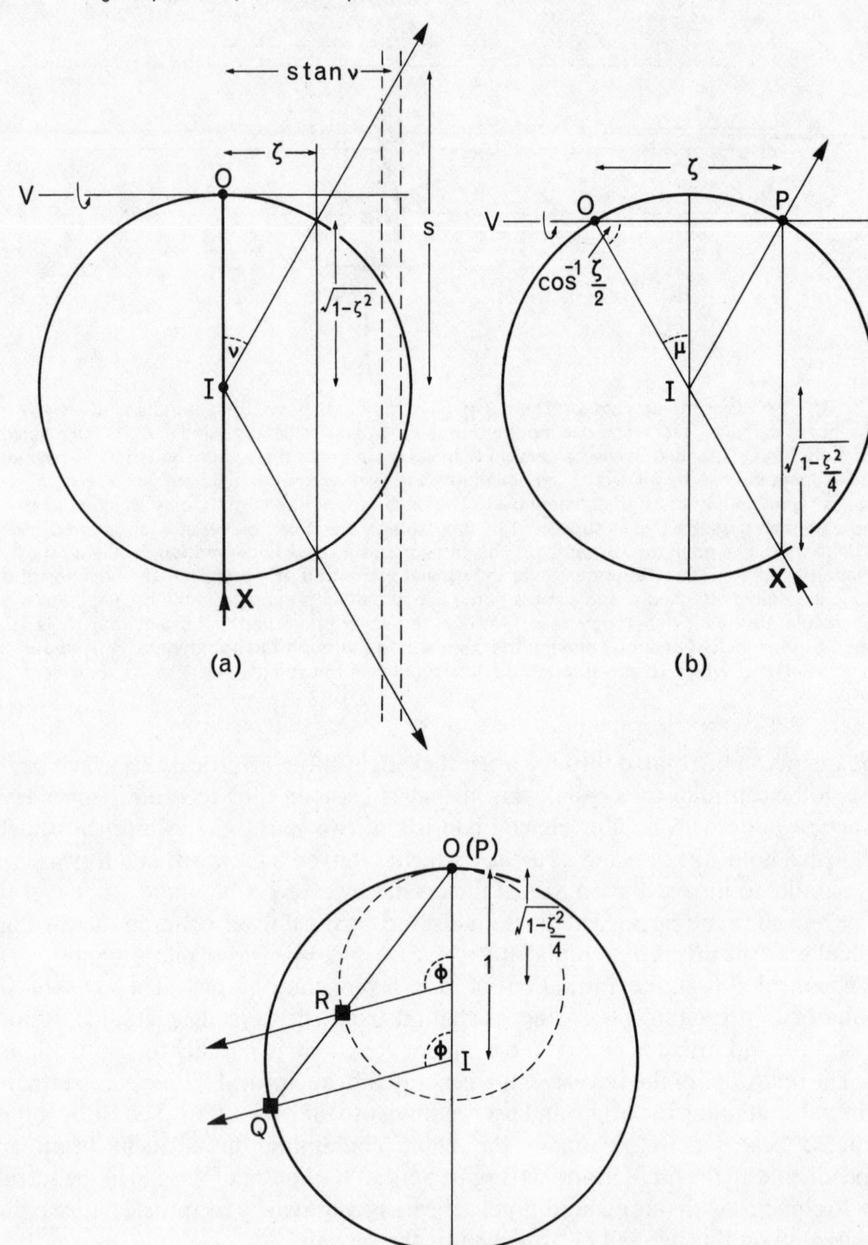

Fig 8.15 Weissenberg photography. (a) shows how a reciprocal lattice net normal to the oscillation axis V at a height ζ above the zero layer gives rise to a cone of diffracted rays of semi-angle $90° - \nu$ where $\nu = \sin^{-1} \zeta$ when the incident X-ray beam XO is perpendicular to the oscillation axis. For the zero layer, $\zeta = 0$ and the cone becomes a plane; for non-zero layers, reflexions for a layer of known ζ will pass through the gap in screens of radius s when the centre of the screen gap is separated by a distance $s \tan \nu$ from the plane normal to the oscillation axis and containing the incident beam (normal beam Weissenberg). (b) illustrates the essential geometry of the equi-inclination Weissenberg for non-zero layers: the incident beam XO is inclined to the oscillation axis V at the angle $\cos^{-1} \frac{1}{2}\zeta$. (c) illustrates the change of scale in non-zero layer equi-inclination Weissenbergs.

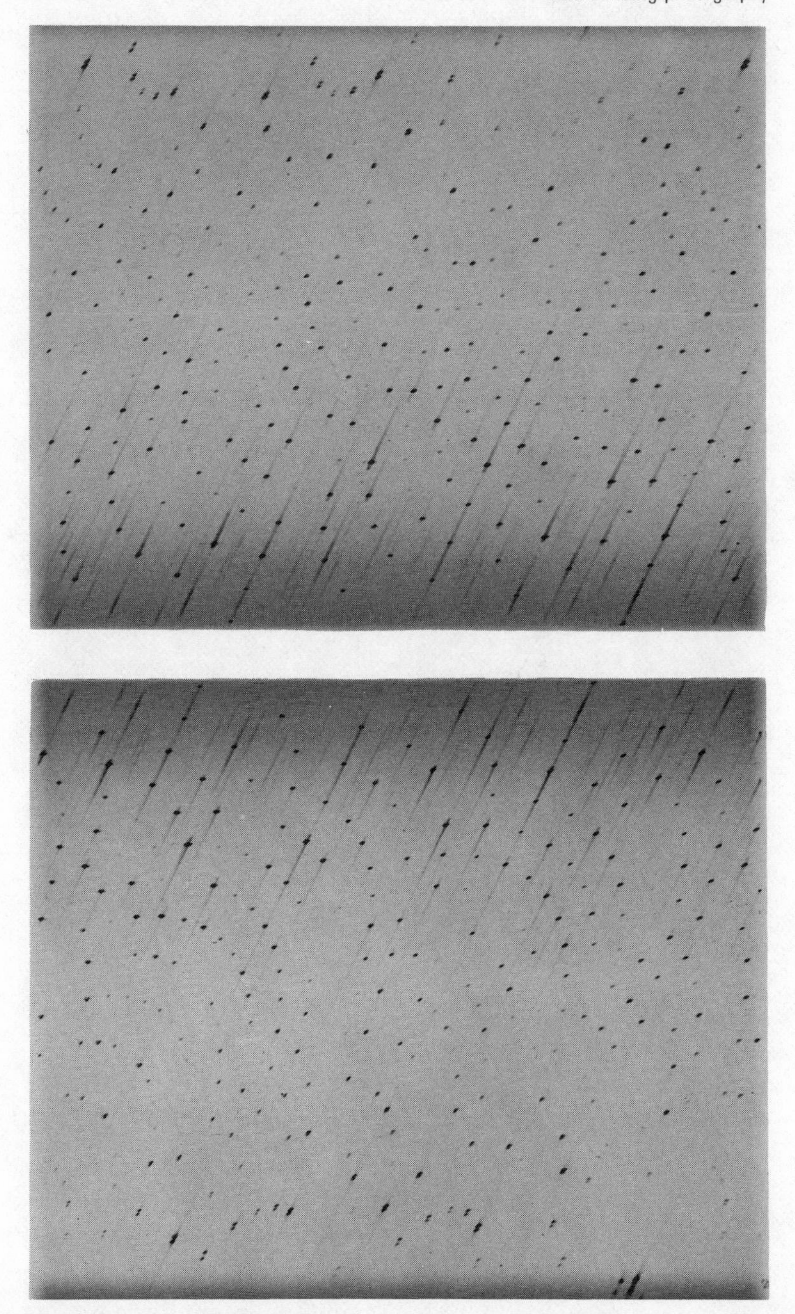

Fig 8.16 Zero layer y-axis Weissenberg photograph of the monoclinic mineral *latiumite* taken with CuKα radiation.

in the screen. For this position of the line OP any reciprocal lattice points that lie on OP cannot also lie on the reflecting circle and so cannot give rise to diffracted beams. The undeviated X-ray beam travelling along XO produced would pass through the gap in the screen to reach the film at O′ were it not trapped by the back-stop attached to the screens. The origin of coordinates on the film is taken as O′ and the reference axes X, Y are as shown in Fig 8.17(b). The X-axis is the median line of the

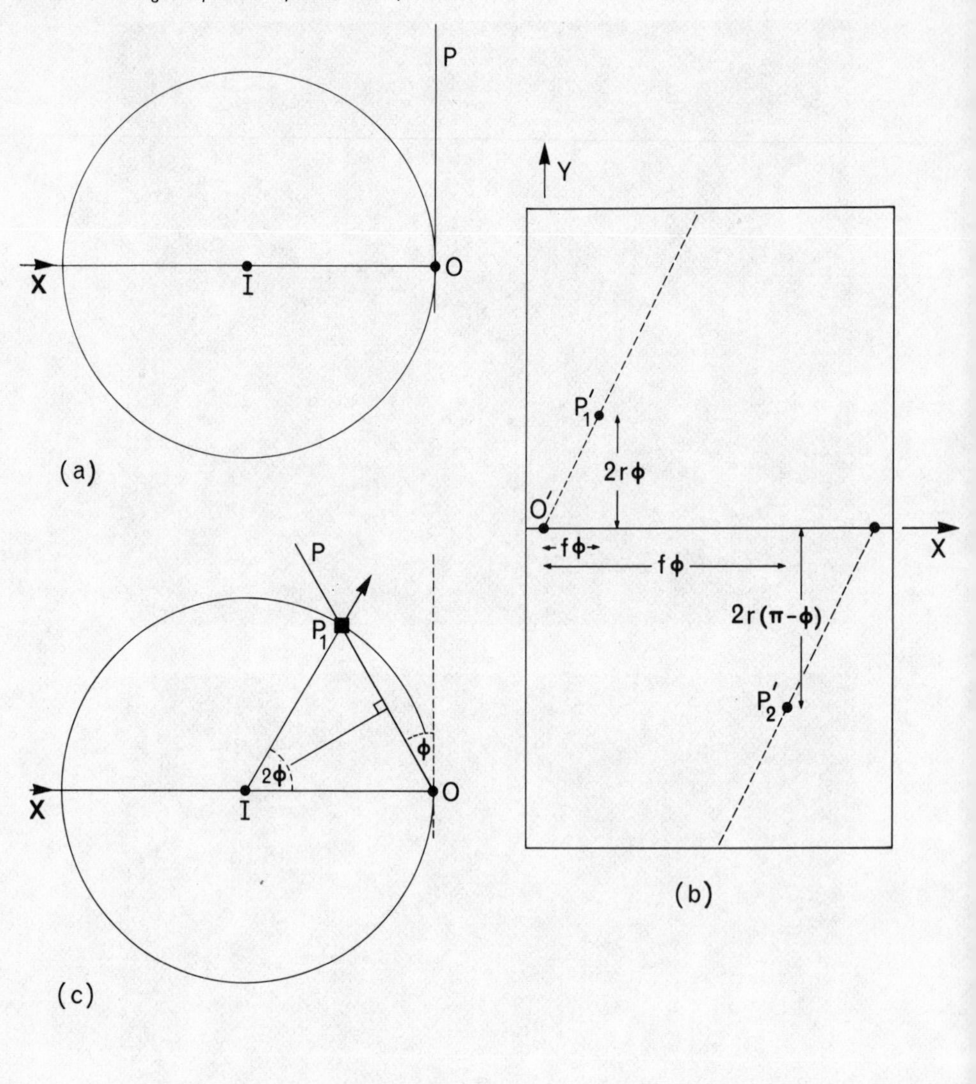
P
X
I
O
(a)
Y
P'1
2rφ
O'
fφ
fφ
X
2r(π-φ)
P'2
(b)
P
P1
φ
2φ
X
I
O
(c)

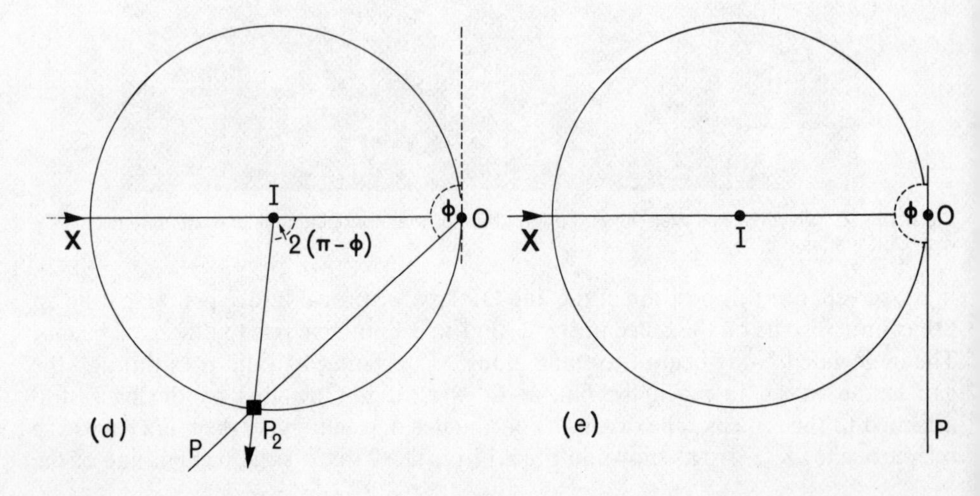
I
φ
O
X
2(π-φ)
P
P2
(d)
X
I
φ
O
(e)
P

film and corresponds to the locus of the point of intersection of the forward direction of the incident X-ray beam with the film as the crystal oscillates. The positive direction of the Y-axis is chosen so that the upper half of the film records diffracted beams produced by the passage of the reciprocal lattice net through the upper semicircle of the reflecting circle. As the line OP rotates anticlockwise the film moves from right to left so that the screen gap moves across the film from left to right. For a rotation of OP through $\phi°$ from its initial position the film moves through a distance $f\phi$, where f is for most instruments 0·5 mm per degree; at this stage (Fig 8.17(c)) OP intersects the reflecting circle in P_1 and if P_1 happens to be a reciprocal lattice point the reflected beam will make an angle 2ϕ with the forward direction of the incident beam. The reflexion P'_1 produced by the reciprocal lattice point P_1 will thus lie at a distance $2r\phi$ (for ϕ in radians, i.e. $\pi r\phi/90$ for ϕ in degrees) from the median line of the film, where r is the camera radius, so that the X, Y coordinates of this reflexion will be $f\phi$, $2r\phi$. Since both coordinates are proportional to ϕ, reflexions from reciprocal lattice points lying on OP will themselves lie on a straight line inclined at the angle $\eta = \tan^{-1}(\pi r/90f)$ to the median line of the film. When the camera radius $r = 57{\cdot}3/2 \simeq 90/\pi$ mm and $f = 0{\cdot}5$ mm per degree, $\eta = \tan^{-1} 2 = 63°26'$.

As the angle ϕ increases X and Y will increase until when ϕ is a little short of 90° Y will correspond to the edge of the film; reflexions with ϕ close to 90° are not recorded on the film but pass through the axial slit in the film cassette. For ϕ greater than 90° reflexions from reciprocal lattice points on the line OP lie on the lower half of the film (Fig 8.17(d)), the diffracted beam IP_2 making the angle $2(\pi-\phi)$ with the forward direction of the incident beam. The film coordinates of the reflexion P'_2 produced by the reciprocal lattice point P_2 will be $X = f\phi$, $Y = 2r(\phi-\pi)$. Reflexions due to lattice points on the line OP such that $\frac{1}{2}\pi < \phi < \pi$ will thus lie on a line of the same slope η as the line on the upper half of the film; this line will intersect the median line of the film at a point distant $180f$ mm from O′ corresponding to $\phi = 180°$ (Fig 8.17(e)). When $\phi = 180°$ PO produced bears the same relationship to the reflecting circle as OP for $\phi = 0°$ (Fig 8.17(a)); further rotation allows reciprocal lattice points on PO produced to come into the reflecting position.

We turn now to consideration of a line QR parallel to OP in the plane normal to the oscillation axis and at a perpendicular distance q from OP. We choose QR such that when OP is in its initial position tangential to the reflecting circle, QR does not intersect the reflecting circle (Fig 8.18(a)). As the crystal rotates anticlockwise from its initial position the line QR first touches the reflecting circle at a point S (Fig 8.18(c)) such that $\cos\phi = IT = IS-TS = 1-q$, IO and IS being radii of the reflecting circle. If there happens to be a reciprocal lattice point at S it will thus give rise to a reflexion S′ with film coordinates (Fig 8.18(b)) $X = f\phi$, $Y = r\phi$, where $\phi = \cos^{-1}(1-q)$.

Further increase in ϕ will cause QR to intersect the reflecting circle in two points Q_1 and R_1 (Fig 8.18(d)). The diffracted beams IQ_1 and IR_1 produced by reciprocal

Fig 8.17 The generation of lines of reflexions on a zero-layer Weissenberg photograph. Diagrams (a), (c), (d), and (e) show successive stages in the rotation of a line OP in reciprocal space normal to the oscillation axis and passing through the origin O. In (a) OP is tangential to the reflecting circle; at stage (c) it has rotated through the acute angle ϕ; at stage (d) ϕ has become obtuse; and in (e) ϕ is shown equal to 180°. Diagrams (c) and (d) illustrate values of ϕ for which reciprocal lattice points, P_1 and P_2, lie on the reflecting circle and give rise to the reflexions P'_1 and P'_2 shown on the drawing (b) of the resulting Weissenberg photograph; the coordinates of P'_1 and P'_2 with respect to the film axes X and Y are indicated. (b) is drawn half-size for a camera with $r = 28{\cdot}65$ mm and $f = 0{\cdot}5$ mm per degree.

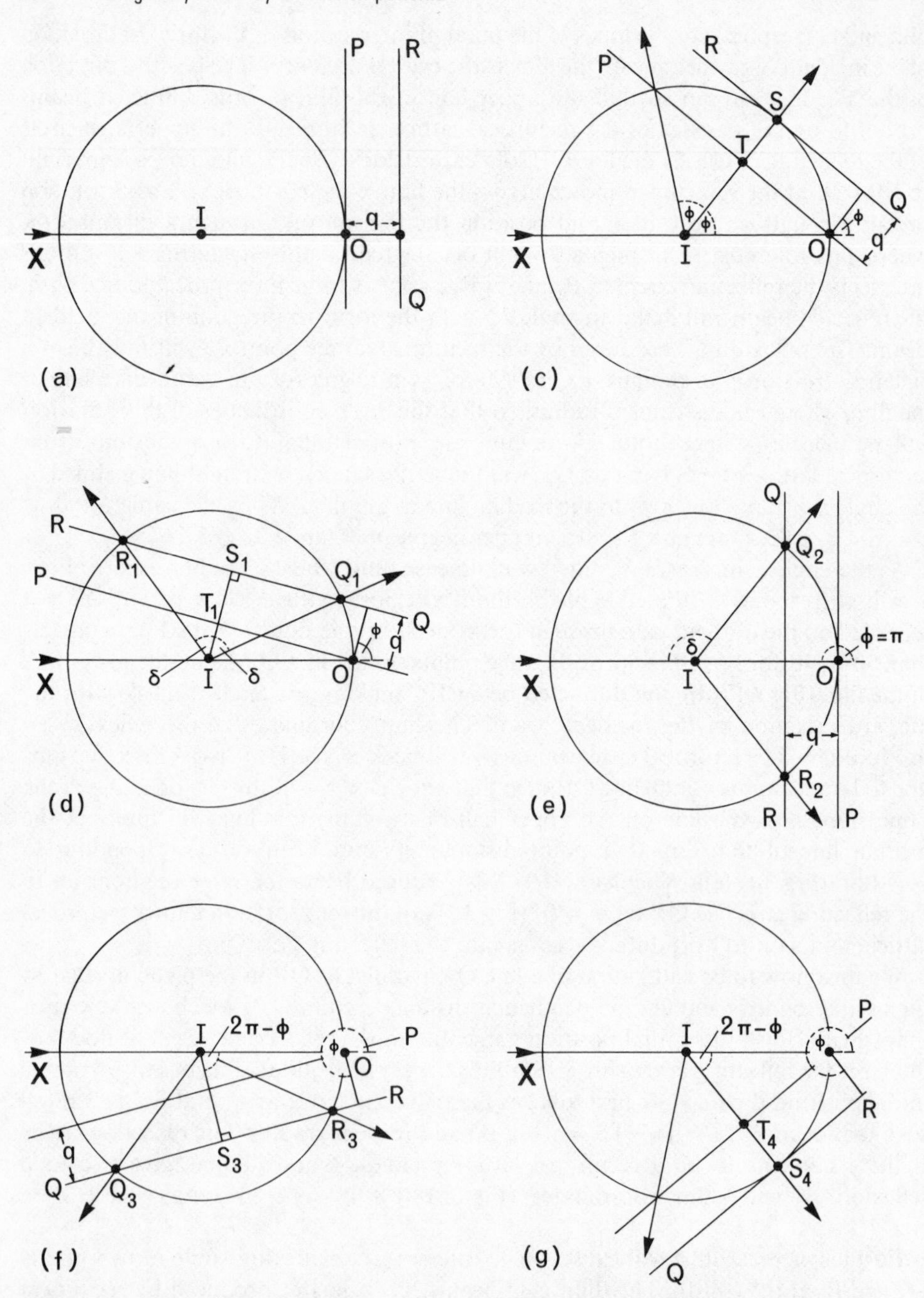

Fig 8.18 The generation of reflexions on a zero-layer Weissenberg photograph by a reciprocal lattice line which does not pass through the origin. The diagrams (a), (c), (d), (e), (f), (g) show successive stages of the rotation of the non-central line QR in the zero-layer reciprocal lattice net from $\phi = 0$ in (a) to $\phi = \pi$ in (e) and $\phi > \pi$ in (f) and (g). Diagram (b) is a drawing of the resultant Weissenberg photograph (half size for a camera with $r = 28{\cdot}65$ mm, f $= 0{\cdot}5$ mm per degree) with the hypothetical reflexions produced by a continuum of reciprocal lattice points on the line QR shown as broken curves.

lattice points at Q_1 and R_1 respectively make equal angles δ with IS_1 and are thus inclined at angles $\phi \pm \delta$ to the forward direction of the incident X-ray beam. It is

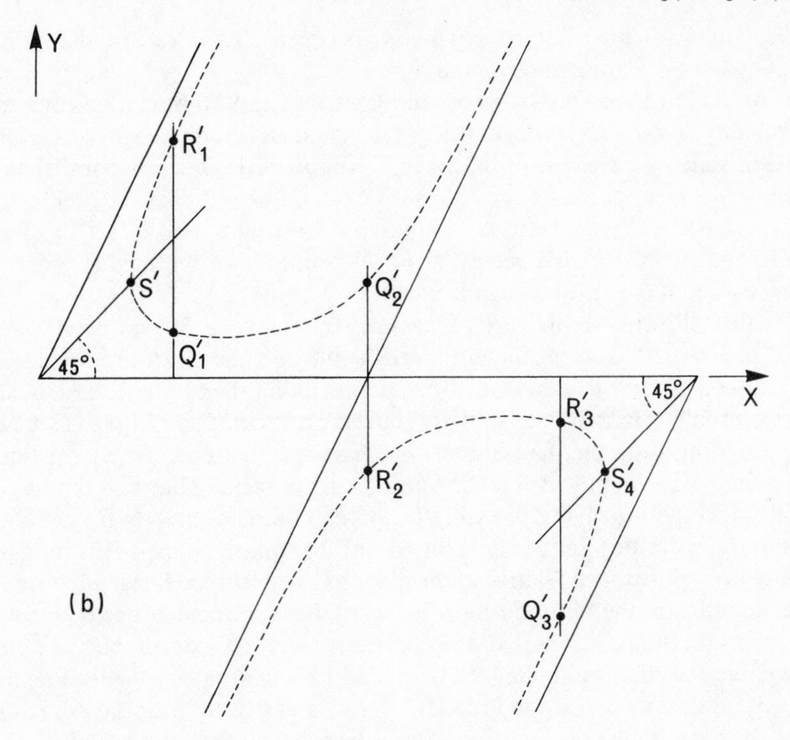

evident from the figure that $\cos\delta = IS_1 = IT_1 + T_1S_1 = \cos\phi + q$. The film co-ordinates of the reflexions Q_1' and R_1' produced by reciprocal lattice points at Q_1 and R_1 will thus be $X = f\phi$, $Y = r\phi - r\cos^{-1}(\cos\phi + q)$ and $X = f\phi$, $Y = r + r\cos^{-1}(\cos\phi + q)$ respectively. Reflexions corresponding to intersections of the line QR with the reflecting circle thus lie on the broken curve shown in Fig 8.18(b). Points on this curve, such as Q_1' and R_1', which correspond to simultaneous reflexion have their X-coordinates equal and the mean of their Y-coordinates lies on the line $X = f\phi$, $Y = r\phi$ which passes through O′ and S_1'; for $r = 57{\cdot}3/2$ mm and $f = 0{\cdot}5$ mm per degree, this line is inclined at $\tan^{-1}(\pi r/180f) = 45°$ to the median line of the film.

As ϕ increases from $\cos^{-1}(1-q)$ the Y-coordinate of the upper part of the curve increases until the top edge of the film is reached at Y just less than $r\pi$. Further increase in ϕ will cause the intersections of QR with the reflecting circle to be on opposite sides of the incident X-ray beam: a reciprocal lattice point which has crossed the incident beam will give rise to a reflexion whose coordinates are $X = f\phi$, $Y = r(\phi - 2\pi) + r\cos^{-1}(\cos\phi + q)$ so that the reflexion lies on the lower half of the film. When $\phi = \pi$ (Fig 8.18(e)) the two reflexions Q_2' and R_2' are equidistant from the median line of the film. When ϕ increases beyond π a stage will be reached at which both intersections of QR lie in the lower semicircle of the reflecting circle and then both the reflexions produced will be on the lower half of the film (Fig 8.18(f)). Further rotation eventually brings QR into an attitude where it is tangential to the reflecting circle at S_4 (Fig 8.18(g)); the diffracted beam from a reciprocal lattice point at S_4 makes an angle $2\pi - \phi$ with the forward direction of the incident X-ray beam so that the film coordinates of the reflexion S_4' will be $X = f\phi$, $Y = r(\phi - 2\pi) = -r\cos^{-1}(1-q)$. In practice the range of movement of the film cassette and the standard size of X-ray film limit the oscillation range of the Weissenberg camera to about 200° and so it may not be possible experimentally to record the complete curves of constant q on one photograph.

Curves similar to those shown as broken lines on Fig 8.18(b) can be constructed for any value of $q < 2$ reciprocal units. In particular if two reference axes, x and y, in reciprocal space are chosen so that x is normal to OP and y is parallel to OP, we can construct a set of curves for reflexions from points on lines parallel to y with $x = 0, \pm 0{\cdot}1, \pm 0{\cdot}2, \ldots \pm 1{\cdot}9$ reciprocal units. By drawing a second set of identical curves displaced from the first set by $X = -90f$ mm along the median line of the film we have a set of curves of constant x with $y = 0, \pm 0{\cdot}1, \pm 0{\cdot}2, \ldots \pm 1{\cdot}9$ reciprocal units. A chart showing both sets of curves, known as a *Weissenberg chart*, is illustrated in Fig 8.19. The prominent straight lines on the chart are separated by $90f$ mm and correspond to reflexions from points along the reference axes x and y. The chart is usually 135 mm long so that it covers an oscillation range of as much as 270°. By superimposing the base line of the chart (printed on transparent film) on the median line of the film, which is the line $Y = 0$, the rectangular coordinates of all reciprocal lattice points giving rise to reflexion can be read directly from the chart and the reciprocal lattice net can be plotted out on squared paper. It is immaterial where the chart is positioned relative to the film provided its base line is superimposed on the median line of the film; the coordinates of the reciprocal lattice points will of course depend on the positioning of the chart but when the coordinates are plotted on graph paper the effect of moving the base line of the chart along the median line of the film will be seen to correspond merely to a rotation of the zero layer reciprocal lattice net. It is always convenient to position the chart so that one prominent linear alignment of reflexions coincides with a diagonal line on the chart; then all reflexions will lie on one set of curves (the corresponding reciprocal lattice points will then lie on lines parallel to one of the reference axes) and, if the reciprocal lattice net is rectangular, at the intersections of the two sets. In the previous sentence the term 'set of curves' implies not only those curves actually drawn on the chart but also interpolated curves.

Before going on to consider upper layer Weissenberg photographs we illustrate the formation of a zero-layer photograph by showing how the $h0\bar{1}$, $h00$, and $h01$ reciprocal lattice points give rise to reflexions in the case of a monoclinic crystal oscillated about [010]. Figure 8.20(a) shows the orientation of the a^*c^* reciprocal lattice net when the gap in the screens exposes the point O′ on the film (Fig 8.20(d)). In this orientation x^* is tangential to the reflecting circle. As the crystal rotates anticlockwise the reciprocal lattice point $20\bar{1}$ passes through the reflecting circle to give rise to a reflexion on the upper half of the photograph. Figures 8.20(b) and (c) show the successive orientations of the reciprocal lattice net as the 200 reciprocal lattice point passes into and out of the reflecting circle to give rise to reflexions respectively on the upper and lower halves of the photograph. Figures 8.20(e) and (f) show the successive orientations of the reciprocal lattice net as 301 passes into and out of the reflecting circle to give rise to reflexions respectively on the upper and lower halves of the film. The reflexions $h0l$ lie on curves corresponding to $q = \lambda/c = c^* \sin\beta^*$ and the $h0\bar{l}$ reflexions lie on curves corresponding to $q = -c^* \sin\beta^*$. The straight line through O′ and the reflexions $h00$ are the expression on the film of the axis x^* of the monoclinic reciprocal lattice. A parallel straight line through $00\bar{1}$, O″, and 001 on the film corresponds to the z^*-axis of the reciprocal lattice, the distance O′O″ on the film being equal to $f\beta^*$. On Fig 8.20(d) we have drawn curves of constant l; we could just as well have drawn curves of constant h, which would be

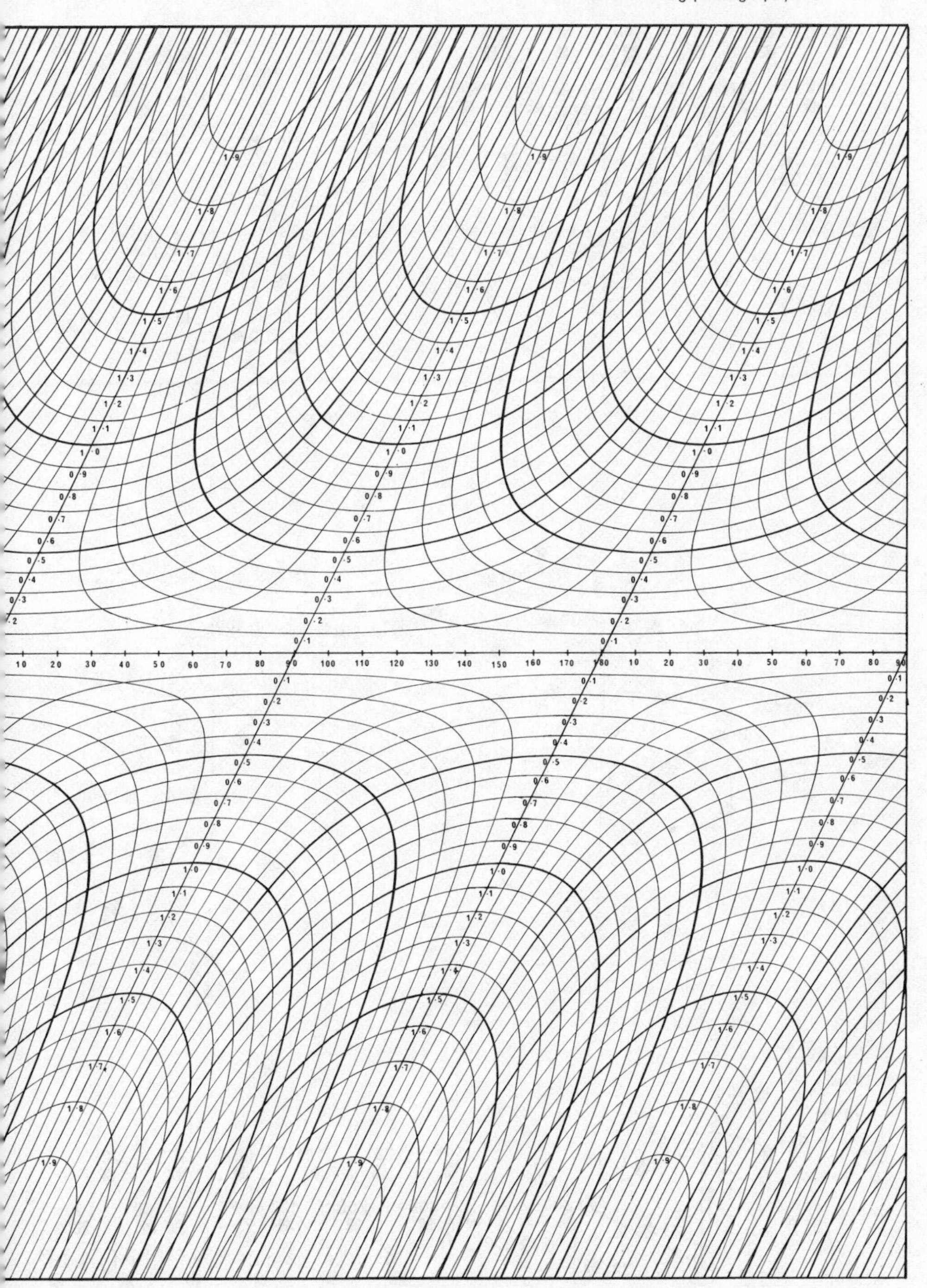

Fig 8.19 The Weissenberg chart. The chart shown is for a camera of diameter 57·3 mm, $f = 0{\cdot}5$ mm per degree; it is reproduced by courtesy of the Institute of Physics.

similar in form, but displaced by the distance O′O″ from the set actually shown. To plot a reciprocal lattice net from a zero-layer Weissenberg photograph one of the straight lines on the Weissenberg chart should be superimposed on one of the

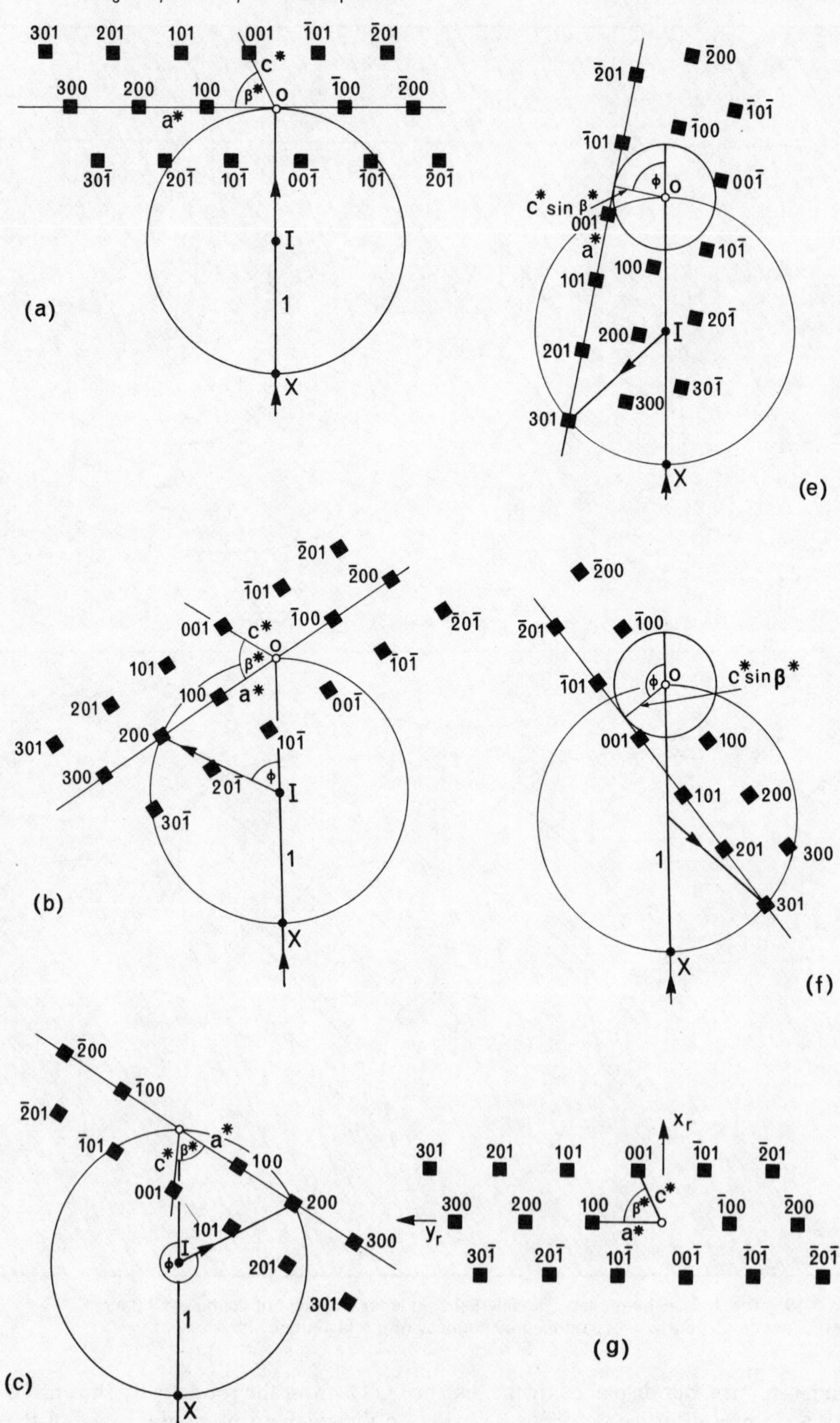

301 201 101 001 1̄01 2̄01
c*
300 200 100 β* O 1̄00 2̄00
a*
301̄ 201̄ 101̄ 001̄ 1̄01̄ 2̄01̄
I
1
X
(a)
2̄00
2̄01
1̄01̄
1̄00
1̄01
ϕ
O
001̄
c* sin β*
001
a*
101̄
100
101
201̄
200
I
201
301̄
300
301
X
(e)
2̄01
1̄01
2̄00
001
c*
1̄00
2̄01̄
O
β*
1̄01̄
101
100
a*
001̄
201
200
101̄
301
ϕ
300
201̄
I
301̄
1
X
(b)
2̄00
1̄00
2̄01
ϕ
O
c* sin β*
1̄01
001
100
101
200
201
300
1
301
X
(f)
2̄00
1̄00
2̄01
a*
1̄01
c*
β*
100
001
200
101
300
I
ϕ
201
301
1
X
(c)
x_r
301 201 101 001 1̄01 2̄01
c*
y_r
300 200 100 β* 1̄00 2̄00
a*
301̄ 201̄ 101̄ 001̄ 1̄01̄ 2̄01̄
(g)

prominent lines of reflexions, *h*00 or 00*l*, and then the coordinates of each reflexion should be read off in rectangular reciprocal coordinates x_r, y_r (Fig 8.20(g)).

We now turn to upper-layer Weissenberg photographs, restricting our discussion to the most commonly used type, *equi-inclination* photographs. In an oscillation photograph—and a Weissenberg is only a specialized sort of oscillation photograph—a reciprocal lattice net normal to the oscillation axis at a height ζ above the zero layer gives rise to a cone of diffracted rays (Fig 8.15(a)) of semiangle $90° - \upsilon$, where $\sin \upsilon = \zeta$. Therefore if the screen is moved through a distance $s \tan \upsilon$, where s is the radius of the screen, the reflexions from this layer will pass through the gap in the screen to be recorded on the film. But the geometry of such a photograph (a *normal beam Weissenberg*) is rather inconvenient and it is better to rotate the incident beam relative to the crystal so that the incident beam lies on the surface of the cone of diffracted beams generated by the reciprocal lattice net (Fig 8.15(b)). When this is done the origin O of the reciprocal lattice lies a perpendicular distance $\frac{1}{2}\zeta$ below the plane passing through the centre of the reflecting sphere and normal to the oscillation axis so that incident and diffracted beams make equal angles $90° - \mu$, where $\mu = \sin^{-1}\frac{1}{2}\zeta$, with the oscillation axis. The selected reciprocal lattice net intersects the reflecting sphere in a circle of radius $\sqrt{(1 - \frac{1}{4}\zeta^2)}$ passing through the oscillation axis at the point P (Fig 8.15(b)). With such an *equi-inclination* arrangement it is possible to record all the reciprocal lattice points within a circle of radius $2\sqrt{(1 - \frac{1}{4}\zeta^2)}$ centred on the oscillation axis and moreover the geometry of X-ray reflexion is identical with that for the zero-layer except that the radius of the reflecting circle is reduced from one to $\sqrt{(1 - \frac{1}{4}\zeta^2)}$ reciprocal unit.

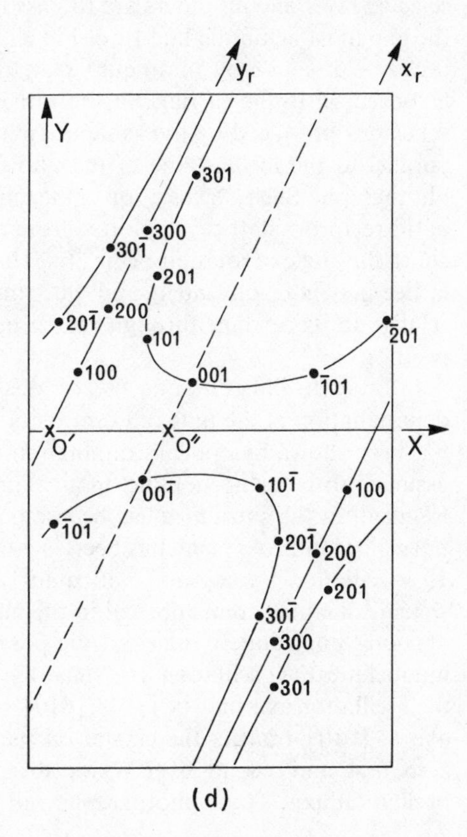

Fig 8.20 Generation of the zero-layer Weissenberg photograph of a monoclinic crystal oscillated about [010]. (a) illustrates the orientation in which x^* is tangential to the reflecting circle. (b) and (c) illustrate the passage of the 200 reciprocal lattice point into and out of the reflecting circle as the crystal rotates anticlockwise. (e) and (f) illustrate the passage of the 301 reciprocal lattice point into and out of the reflecting circle. (d) shows the disposition of the reflexions produced on the resulting Weissenberg photograph by the reciprocal lattice points shown in (a)–(f); the rectangular film axes are labelled *X* and *Y* as in Fig 8.17 and the reference axes of the Weissenberg chart are labelled x_r and y_r. (g) shows the reciprocal lattice net plotted from (d) by reading off values of x_r and y_r for each observed reflexion.

It is apparent from Fig 8.15(c) that a reciprocal lattice point Q lying in the plane through the origin perpendicular to the oscillation axis gives rise to a diffracted beam (which will be recorded as a reflexion on the zero-layer photograph) which makes an angle ϕ with the forward direction of the incident X-ray beam where the reciprocal lattice point is distant $OQ = 2\sin\frac{1}{2}\phi$ from the origin O. A reciprocal lattice point R in an upper layer giving rise to a diffracted beam inclined at the angle ϕ to the forward direction of the incident beam will lie at a distance $PR = 2\sqrt{(1-\frac{1}{4}\zeta^2)}.\sin\frac{1}{2}\phi$ from the intersection P of the oscillation axis with its reciprocal lattice net. Thus the Weissenberg chart constructed for the zero layer can be used to plot the reciprocal lattice net for an upper layer provided that all coordinates read from the chart are reduced by the factor $\sqrt{(1-\frac{1}{4}\zeta^2)}$. The reciprocal lattice net for an upper layer may not necessarily have a reciprocal lattice line passing through its origin and so the slanting lines which are so obvious on a zero layer photograph may be missing or, as in the case (Fig 8.10(a)) of a monoclinic crystal oscillating about [001], only one reciprocal lattice line (that parallel to x^* and containing the reciprocal lattice points $h0n$ in the nth layer) passes through the oscillation axis so that the nth layer equi-inclination Weissenberg photograph will show only $h0n$ reflexions lying on a straight diagonal line.

The essential requirement for an equi-inclination photograph, that the incident X-ray beam should lie on the cone of diffracted beams for the layer concerned, is achieved experimentally by turning the camera through the angle μ about the vertical axis through the crystal, where $\sin\mu = \frac{1}{2}\zeta$, and keeping the collimator, which is rigidly attached to the base of the camera, stationary. If the diffracted beams of the selected layer and no others are to pass through the gap in the screens, the centre of the gap must lie on the line PI of Fig 8.21. Therefore each screen must be moved in the same direction by an amount $s\tan\mu$ where s is the screen radius. If the cassette is locked on to the carriage in the same position as for the zero-layer photograph reflexions produced by the same orientation of the crystal will be displaced $r\tan\mu$ parallel to the median line of the photograph relative to those on the zero layer photographs. Such a translation of the reflexions on the film corresponds to a rotation of the reciprocal lattice net plotted from coordinates measured with the Weissenberg chart, the angle of rotation being $\{(r/f)\tan\mu\}^\circ$. This rotation of successive reciprocal lattice nets is inconvenient and can simply be eliminated by moving the cassette relative to its carriage through $r\tan\mu$ before exposure starts and locking it in this position.

In order to illustrate one use of Weissenberg photographs we now discuss the determination of the unit-cell dimensions and diffraction symbol of a crystal which has been shown by optical examination to be biaxial (chapter 12) so that it may be assumed to be orthorhombic, monoclinic, or triclinic. Suppose that a single crystal fragment of the substance has been mounted on a glass fibre and that a zone axis normal to a mirror plane has been located and set parallel to the oscillation axis. A 15° oscillation photograph will quickly yield a good approximate value for the spacing of lattice points normal to the mirror plane. The finding of one mirror plane of course immediately rules out the possibility that the substance is triclinic. If it is monoclinic, the oscillation axis must by convention be [010]; if it is orthorhombic, the oscillation axis may be [100], [010], or [001]. We tentatively index the oscillation axis as [010], transfer the crystal on its arcs to the Weissenberg camera, and take zero, first, and second layer Weissenberg photographs with the presumed [010] as oscillation axis. These photographs will enable us to plot out the $h0l$, $h1l$, and $h2l$

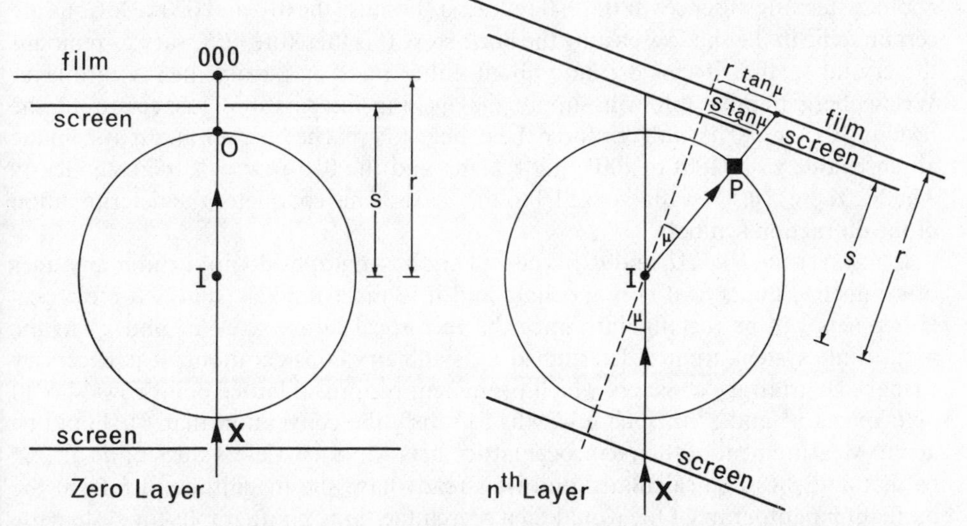

Fig 8.21 Upper layer equi-inclination Weissenberg photographs. The left-hand diagram shows the experimental arrangement for a zero-layer photograph contrasting with the arrangement for an upper-layer photograph shown on the right. For an *n*th layer photograph the camera axis is inclined relative to the fixed collimator through the angle $\mu = \sin^{-1}(\frac{1}{2}\zeta_n)$, the screen is translated through $s \tan\mu$ and the film carriage is translated through $r \tan\mu$, where *s* and *r* are screen and camera radii respectively.

reciprocal lattice nets. If these three nets each exhibit symmetry 2*mm* with their mutually perpendicular lines of symmetry normal to reciprocal lattice point rows and if the intensities of the pairs of the reflexions related positionally by the lines of symmetry are approximately equal, then the substance is orthorhombic. The geometry of the Weissenberg camera is such that lines of symmetry in reciprocal lattice nets do not obviously appear as such on the photograph; a line of symmetry in the net generates one of the parallel slanting straight lines on the film and symmetry related reflexions of equal intensity will lie on either side of the line but will not have symmetry related film coordinates. If our crystal turned out to be orthorhombic x^* and z^* would be chosen as the directions normal to the two lines of symmetry on the reciprocal net plots. The reciprocal lattice dimensions a^* and c^* could then be evaluated and the dimensions *a*, *b*, *c*, of the unit-cell could be calculated from them and from the direct determination of *b* from the oscillation photograph. The next stage in this investigation would be to search the three reciprocal lattice nets for systematically absent reflexions. Adequate sampling of general *hkl* reflexions should have been achieved to determine whether the lattice type is A, B, C, P, I, or F. The reciprocal lattice net *h*0*l* would then be searched first for independent systematic absences which would indicate the presence of (010) glide planes translating $\frac{1}{2}a$ or $\frac{1}{2}c$ or $\frac{1}{2}(a+c)$ and secondly for independent systematic absences in the *h*00 or 00*l* reflexions which would indicate the presence of screw diads parallel to the *x* or *z* axes. Systematic absences in the 0*kl* and *hk*0 reflexions, which would imply the presence of *b* or *c* or *n* glides parallel to (100) or *a* or *b* or *n* glides parallel to (001), can be investigated by looking at the 00*l*, 01*l*, 02*l*, and *h*00, *h*10, *h*20 rows in the three nets, although it must be borne in mind that the amount of information available may not be adequate for a conclusive statement. What we particularly lack is information about systematic absences in the $0k0$ reflexions because the 010 and 020 reflexions are certain to lie in the shadow cast by the back-stop. It is therefore necessary to remount the crystal so that it can oscillate about either its x or z axis; then a zero-layer Weissenberg photograph will supply the missing information. Inspection of the resulting $0kl$ or $hk0$ net will reinforce the conclusion reached earlier about systematic absences due to a (100) or (001) glide plane and the $0k0$ row will indicate clearly whether there is a screw diad parallel to the y-axis. This completes the determination of the diffraction symbol.

If however the $h0l$, $h1l$, and $h2l$ Weissenberg photographs do not exhibit any lines of symmetry, the crystal is monoclinic and it is most unlikely that the reciprocal lattice nets will be rectangular. Since the reciprocal lattice axes x^* and z^* in the monoclinic system are not determined by symmetry considerations, it is necessary to make an arbitrary choice of which prominent reciprocal lattice point rows should be taken as x^* and z^* in such a way as to satisfy the convention that β^* should be acute. Measurement of the reciprocal lattice nets will then yield values of a^*, c^*, β^* so that a, c, β can be calculated and we already have the magnitude of b from the oscillation photograph. One would then search the three photographs for systematic absences in general hkl reflexions to determine the lattice type, and for systematic absences in the $h0l$ reflexions for evidence of an a or c or n glide plane parallel to (010). Again it would be necessary to remount the crystal and to take a zero-layer photograph about either [100] or [001] to investigate systematic absences in the $0k0$ reflexions, which would indicate the presence of a screw diad parallel to [010]. The diffraction symbol will then be completely determined.

We conclude by noting that in both the orthorhombic and the monoclinic case we may have chosen the crystallographic reference axes unconventionally. If that is so, it may be necessary in the orthorhombic case to transform the axes so that they conform to the conventions laid down in the *International Tables for X-ray Crystallography*, or in the monoclinic case to choose alternative x and z axes. The reader will have noticed that in both cases it was necessary to remount the crystal; had a precession camera been available however, the crystal could have simply been transferred on its arcs from one camera to the other so that the supplementary data, the $hk0$ or $0kl$ layers, could be obtained without any necessity for remounting the crystal.

Precession photography

In essence the precession method differs from those previously described in that the movement of the crystal is not an axial oscillation but a precession. In consequence the camera motion is essentially three-dimensional and not easily described in terms of two-dimensional diagrams. The diffraction pattern is recorded on a plane film which provides an undistorted photograph of a selected reciprocal lattice plane; the provision of an undistorted representation of a reciprocal lattice plane coupled with uniform spot shape is the essential purpose of the method. We begin by discussing the geometry of zero-layer photographs in terms of the reciprocal lattice and reflecting sphere, pass on to consider upper layer photographs, and conclude with an outline description of the precession camera.

A crystal mounted on arcs, whose spindle axis is normal to the incident X-ray beam, is set initially so that a prominent zone axis is coincident with the X-ray beam and so perpendicular to the spindle axis. This contrasts with the requirement for oscillation and Weissenberg photography that a zone axis should be parallel to the spindle axis of the arcs and so perpendicular to the incident X-ray beam. In the initial setting of a crystal on the precession camera (Fig 8.22(a)) the reciprocal lattice layer through the origin and normal to the selected zone axis is thus tangential to the reflecting sphere; in consequence no zero-layer reflexions can occur in this orientation of the crystal. In order to record the zero-layer reflexions the zone axis is moved through the angle $\bar{\mu}$ so that the zero layer of the reciprocal lattice intersects the reflecting sphere in a small circle of radius $\sin\bar{\mu}$ (Fig 8.22(b)). The zone axis is then caused to *precess* about the direction of the incident X-ray beam so that it describes a conical surface whose apex is the centre of the crystal, whose axis is coincident with the incident X-ray beam, and whose semiangle is $\bar{\mu}$ (Fig 8.22(c)). The precession is achieved by coupling an oscillation of the spindle of the arcs about its own axis with an oscillation of the spindle about a second axis, which is normal to both the spindle and the incident X-ray beam and passes through their point of intersection (Fig 8.22(d)); this movement is such that the line in the crystal coincident with the spindle axis lies in the plane of the spindle axis and the incident X-ray beam throughout the precession.

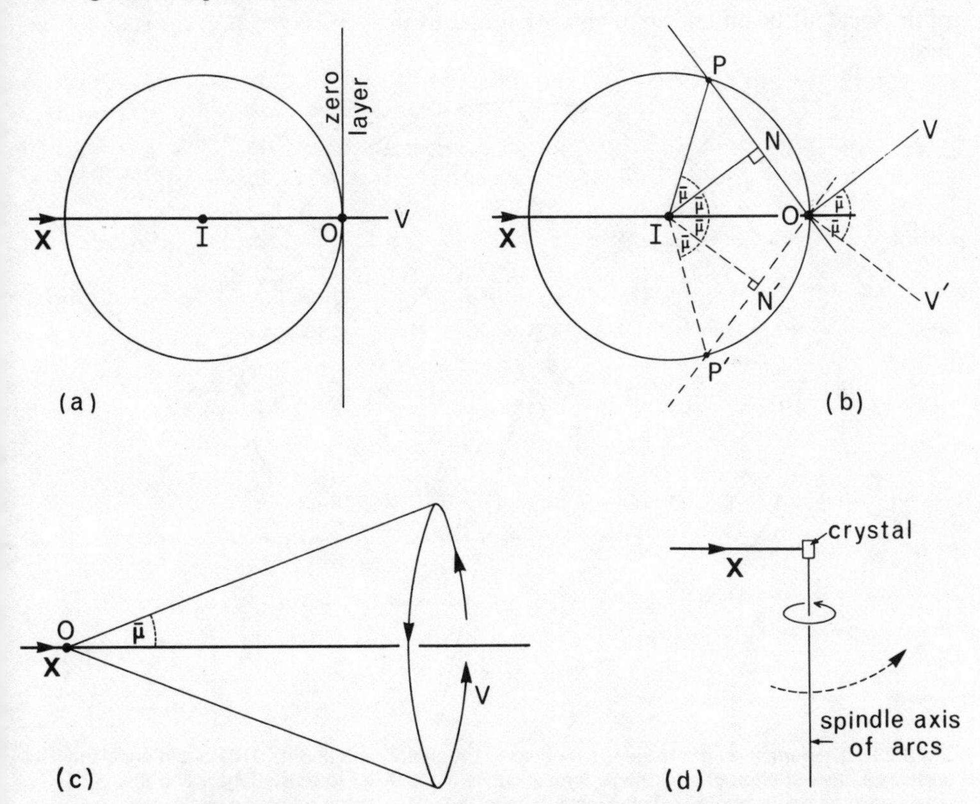

Fig 8.22 Geometry of zero-layer precession photography. (a) shows the initial setting of the crystal in which the zero-layer reciprocal lattice net normal to the selected zone axis OV is tangential to the reflecting sphere. In (b) the zone axis OV is inclined at the angle $\bar{\mu}$ to the incident X-ray beam so that the zero-layer reciprocal lattice net intersects the reflecting sphere in a small circle of radius NP $= \sin\bar{\mu}$; the two orientations in which the plane of the diagram contains the normals to the zero layer, IN and IN′, are shown. (c) shows the precession of the selected zone axis OV about the incident X-ray beam on the surface of a cone of semi-angle $\bar{\mu}$. (d) illustrates the precession of the selected zone axis by coupled oscillations about two axes, both of which are perpendicular to the incident X-ray beam: one oscillation axis is the spindle axis of the arcs and the other is perpendicular to the plane of the diagram.

Figure 8.22(b) shows two positions in the course of the motion: IN and IN′ are directions parallel to the selected zone axis, which has a constant inclination $\bar{\mu}$ to the incident beam direction XIO; OP, and OP′ are the corresponding positions of the zero reciprocal lattice layer normal to the selected zone axis. The zero layer thus intersects the reflecting sphere at all stages of the motion in a small circle of radius $\sin \bar{\mu}$ which always passes through the origin O. In the course of a complete precession the small circle sweeps through all the reciprocal lattice points of the zero layer lying within a circle of radius $2 \sin \bar{\mu}$ about the origin of the reciprocal lattice. In Fig 8.23 the relative motion of the reciprocal lattice plane and the reflecting sphere is illustrated in terms of a stationary reciprocal lattice and a moving reflecting sphere. At any given time the reflecting sphere intersects the zero layer in a circle passing through the origin of reciprocal space. During the precession this circle of intersection effectively rolls about O: three successive positions of the circle of intersection are shown, two of which correspond to the positions shown in Fig 8.22.

The film is caused to precess in a manner identical to that of the crystal so that its centre remains stationary at a point distant F from the crystal in the forward direction of the incident beam and its plane is parallel to the zero layer at every stage of the motion.

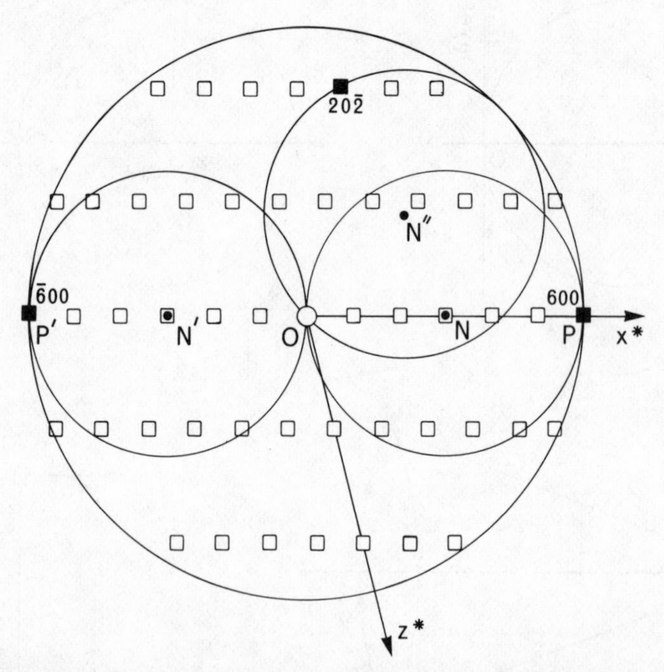

Fig 8.23 The formation of the zero-layer precession photograph about [010] of a monoclinic crystal, i.e. the *h*0*l* photograph. The reciprocal lattice net is taken to be stationary and the reflecting sphere rolls about O. Three instantaneous positions of the reflecting circle (radius = $\sin \bar{\mu}$) are shown; these circles (centres: N, N′, N″) give rise to the reflexions 600, $\bar{6}$00, and 20$\bar{2}$, the relevant reciprocal lattice points being shown as solid squares. The reflecting circle sweeps out in the course of its motion a circle of radius $2 \sin \bar{\mu}$ about O.

In illustrating the production of precession photographs it is usually convenient to superpose diagrams illustrating the geometry in both direct and reciprocal space. We assume that the crystal is situated at the centre I of the reflecting sphere and choose a scale such that the crystal to film distance, F mm, in direct space measured

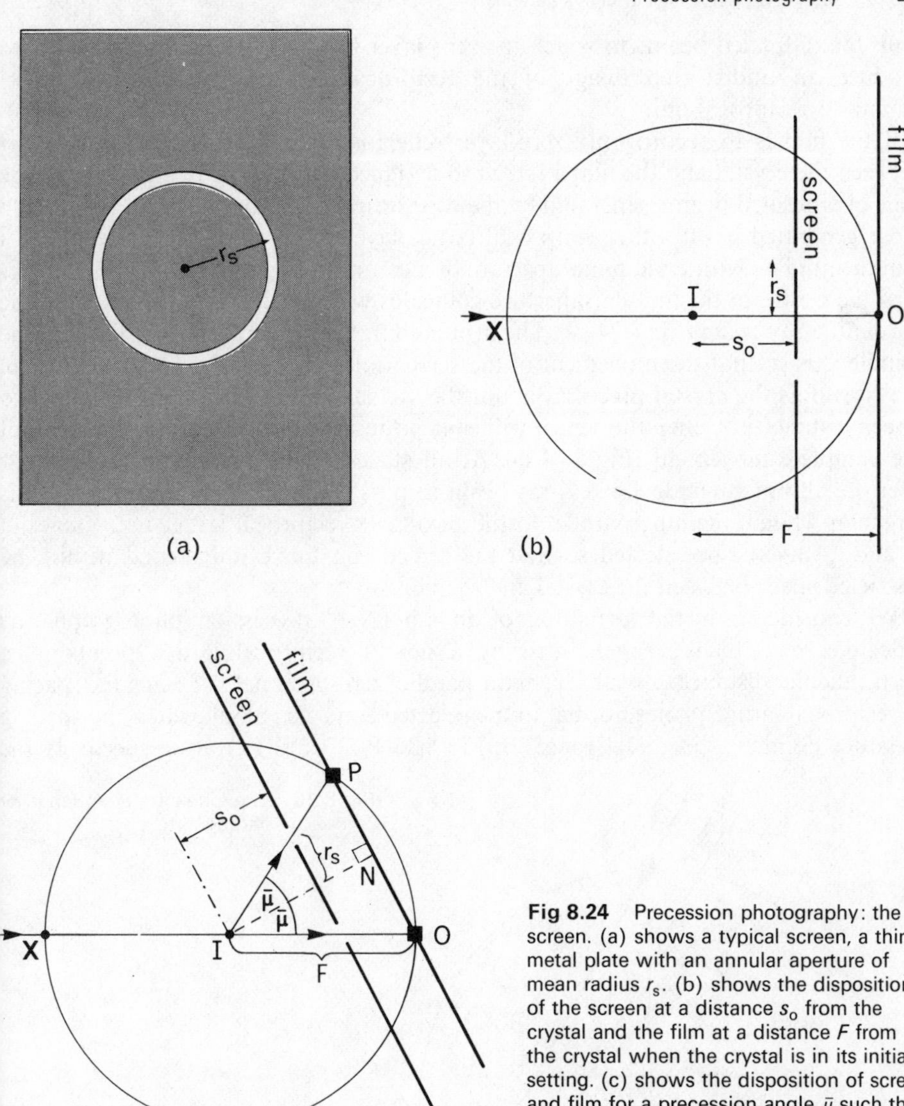

Fig 8.24 Precession photography: the screen. (a) shows a typical screen, a thin metal plate with an annular aperture of mean radius r_s. (b) shows the disposition of the screen at a distance s_o from the crystal and the film at a distance F from the crystal when the crystal is in its initial setting. (c) shows the disposition of screen and film for a precession angle $\bar{\mu}$ such that $s_o = r_s \cot \bar{\mu}$ so that the screen allows passage only of zero-layer reflexions throughout the movement.

in the direction of the undeviated X-ray beam is equal on the diagram to the radius of the reflecting sphere, one unit in reciprocal space. In Fig 8.24(c) then the centre of the film is at O and a reciprocal lattice point at P gives rise to a diffracted beam parallel to IP which intersects the film at P. The ratio OP/IO = $2 \sin \bar{\mu}$ is the same in direct and in reciprocal space; in direct space IO = F mm and in reciprocal space IO = 1 reciprocal unit. In direct space therefore OP = Fd^*, where d^* is equal to the distance represented by OP in reciprocal space, the distance of the reciprocal lattice point P from the origin O of reciprocal space. As the crystal precesses about IO the zero layer generates a cone of diffracted beams of semiangle $\bar{\mu}$ and this cone rolls about the line IO which remains on the surface of the cone throughout the motion. The film moves so that its plane is always parallel to the zero layer, the crystal to film distance is constant and equal to IN = $F \cos \bar{\mu}$, and the point O is fixed in position.

Thus the diffracted beams to which the zero layer gives rise strike the film so as to produce an undistorted image of the reciprocal lattice plane on a scale of F mm = 1 reciprocal unit.

If the film is to record only zero-layer reflexions it will be necessary to insert between the crystal and the film a *screen* so designed as to allow free passage of the cone of diffracted beams generated by the zero reciprocal lattice layer and to absorb those generated by all other reciprocal lattice layers. The form of the screen is a thin metal plate with an annular aperture of mean radius r_s (Fig 8.24(a)). When $\bar{\mu}$ is zero the centre of the annular aperture coincides with the forward direction of the incident X-ray beam (Fig 8.24(b)). The arm holding the screen is rigidly fixed to the spindle axis so that the movement of the screen follows precisely the movement of the crystal as the crystal precesses about the incident X-ray beam. If the crystal to screen distance is $r_s \cot \bar{\mu}$ the screen will isolate the zero-layer reflexions throughout the complete movement (Fig 8.24(c)). At all stages of the precession the annular aperture allows the undeviated X-ray beam to pass; the screen precesses round this direction while remaining parallel to the zero-layer reciprocal lattice net. Values of s_0 and r_s have to be selected so that the screen can move unimpeded within the restricted space between the crystal and the film.

We consider now the formation of an nth layer[4] precession photograph, the reflexions on which are generated by a net of reciprocal lattice points at a perpendicular distance $\zeta_n = n\lambda/t$ from the parallel zero-layer net, λ/t being the spacing of reciprocal lattice planes normal to the selected zone axis along which the spacing of lattice points is t in direct space. In Fig 8.25 P_nQ_n and PO are respectively the

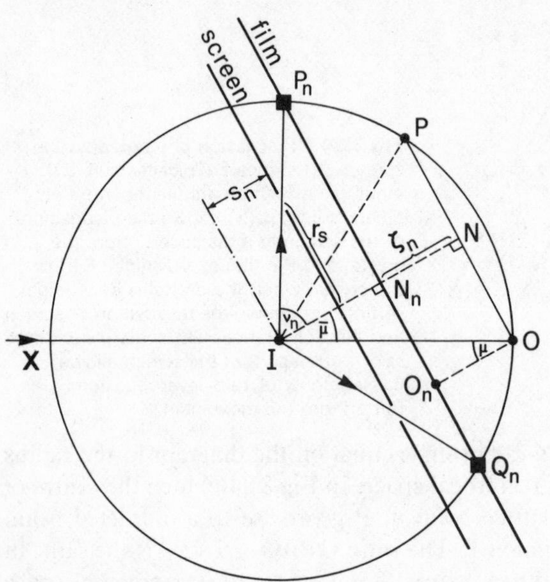

Fig 8.25 The formation of an nth layer precession photograph.

intersections of parallel nth and zero reciprocal lattice layers with the plane of the diagram. Again we imagine the crystal to be situated at the centre I of the reflecting sphere and the film to be coincident with the nth reciprocal lattice layer. IP_n and IQ_n lie on the surface of the cone of diffracted beams for the nth layer, the semiangle of this cone being υ_n such that $\cos \upsilon_n = IN_n = IN - N_nN = \cos \bar{\mu} - \zeta_n$. Just as for the

[4] For the nth layer to be an *upper* rather than a *lower* layer photograph the positive direction of $[UVW]$ must be directed back along the incident X-ray beam.

zero-layer photograph a screen has to be introduced between the crystal and the film to eliminate reflexions from all but the selected nth layer. It is apparent from Fig 8.25 that if the annular circular aperture in the screen is of mean radius r_s, then the screen has to be placed at a perpendicular distance s_n from the crystal such that $s_n = r_s \cot v_n = r_s \cot \cos^{-1}(\cos \bar{\mu} - \zeta_n)$. The magnitudes of r_s and s_n must be so chosen that the movement of the screen is not impeded by any other part of the camera. Since we imagine the film to be coincident with the nth reciprocal lattice layer it will have to be moved relative to its position for a zero-layer photograph by a distance $F\zeta_n$ so that the perpendicular crystal-to-film distance for an nth layer photograph is $F \cos \upsilon_n = F(\cos \bar{\mu} - \zeta_n)$ and the magnification factor remains equal to F. There is now no fixed point on the film; its centre O_n precesses about the forward direction of the incident X-ray beam in just the same way as does the intersection of the zone axis with the nth reciprocal lattice layer. In order to obtain the requisite movement of the film the film-cassette is mounted on an arm which can be brought forward in the direction normal to the plane of the film so that the cassette is displaced forwards relative to the mounting which controls the movement of the film; in terms of Fig 8.25 the film is positioned for recording the nth layer so that it moves about the stationary point O, which corresponds to the centre of the zero-layer film. As the crystal precesses the small circle (radius $\sin \upsilon_n$) in which the nth reciprocal lattice layer intersects the reflecting sphere sweeps through the reciprocal lattice plane in a circular area of radius $O_nP_n = O_nN_n + N_nP_n = ON + N_nP_n = \sin \bar{\mu} + \sin \upsilon_n$. Figure 8.26, which illustrates this point, shows the same instant intersection as Fig 8.25. It will be apparent from Fig 8.26 that nth layer photographs necessarily have a central blind spot of radius

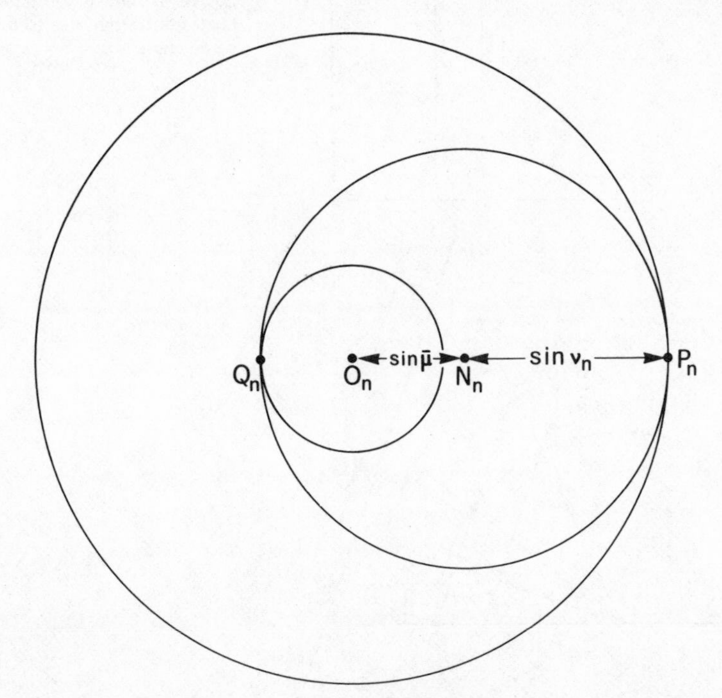

Fig 8.26 The central blind spot in nth layer precession photographs. The figure shows the same instant intersection of the reflecting circle with the nth reciprocal lattice net as Fig 8.25. Only reciprocal lattice points in the annular area centred on O_n between the limiting circles of radii O_nQ_n and O_nP_n give rise to reflexion. On the photograph $O_nQ_n = F(\sin \nu_n - \sin \bar{\mu})$ and $O_nP_n = F(\sin \nu_n + \sin \bar{\mu})$.

$O_nQ_n = N_nQ_n - N_nO_n = F(\sin \upsilon_n - \sin \bar{\mu})$, corresponding to a circular area of the reciprocal layer which remains throughout the motion within the reflecting sphere.

In order to take an nth layer precession photograph it is necessary to know the magnitude of ζ. If ζ is not already known from prior study with oscillation photographs, it can very easily be determined by inserting a film in a light-tight envelope in the screen holder and taking a precession photograph at a known precession angle $\bar{\mu}$ to yield what is known as a *cone-axis photograph*. Reciprocal lattice layers normal to the zone axis will produce diffracted beams lying on coaxial cones; the semiangle for the cone produced by the nth layer is ν_n and for the zero layer $\nu_0 = \bar{\mu}$. These cones of diffracted beams intersect the film in concentric circles of radius $r_n = s' \tan \upsilon_n = s' \tan \cos^{-1}(\cos \bar{\mu} - \zeta_n)$, where s' is the perpendicular distance from the crystal to the screen holder (Fig 8.27). By measurement of the radii of the circles so produced ζ can be evaluated with sufficient accuracy for first, second, etc layer photographs to be taken.

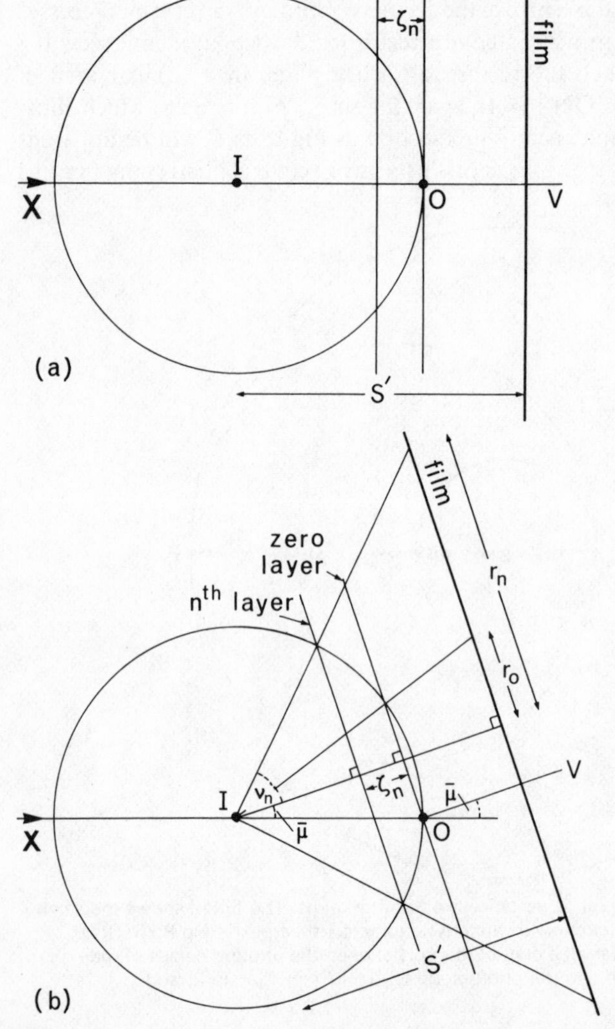

Fig 8.27 The formation of a cone-axis photograph. (a) and (b) show respectively the geometry for $\bar{\mu} = 0$ and for a non-zero value of $\bar{\mu}$. The film is positioned in the screen holder at a distance s' from the crystal at I. (b) shows an instant in the precession motion which causes the zero layer reciprocal lattice net to give rise to a cone of diffracted beams of semi-angle $\bar{\mu}$ and the nth layer net to give rise to a cone of semi-angle ν_n.

Our description of the precession camera will be confined to an outline of the essentials of the instrument. Being fundamentally a three-dimensional rather than an axial apparatus it is difficult to describe in terms of two-dimensional diagrams (Fig 8.28), but easy enough to understand when seen. The motor drives a spindle *a* coincident with the incident X-ray beam. Rigidly attached to the spindle is a graduated arc *b* whose centre lies at the point of intersection of the forward direction of the incident beam with the film. A bearing at the centre of the film holder *c* maintains a rod perpendicular to the plane of the film throughout the motion; the other end of the rod engages the arc and is clamped in position at the chosen precession angle $\bar{\mu}$. The film holder and the spindle *d*, to which the arcs *e* carrying the crystal *f* are attached, are mounted on gimbals, i.e. free moving mutually perpendicular bearings. The motion of film and crystal are linked so that each precesses identically about the direction of the incident beam. The screen *g* is rigidly attached to the spindle *d* which

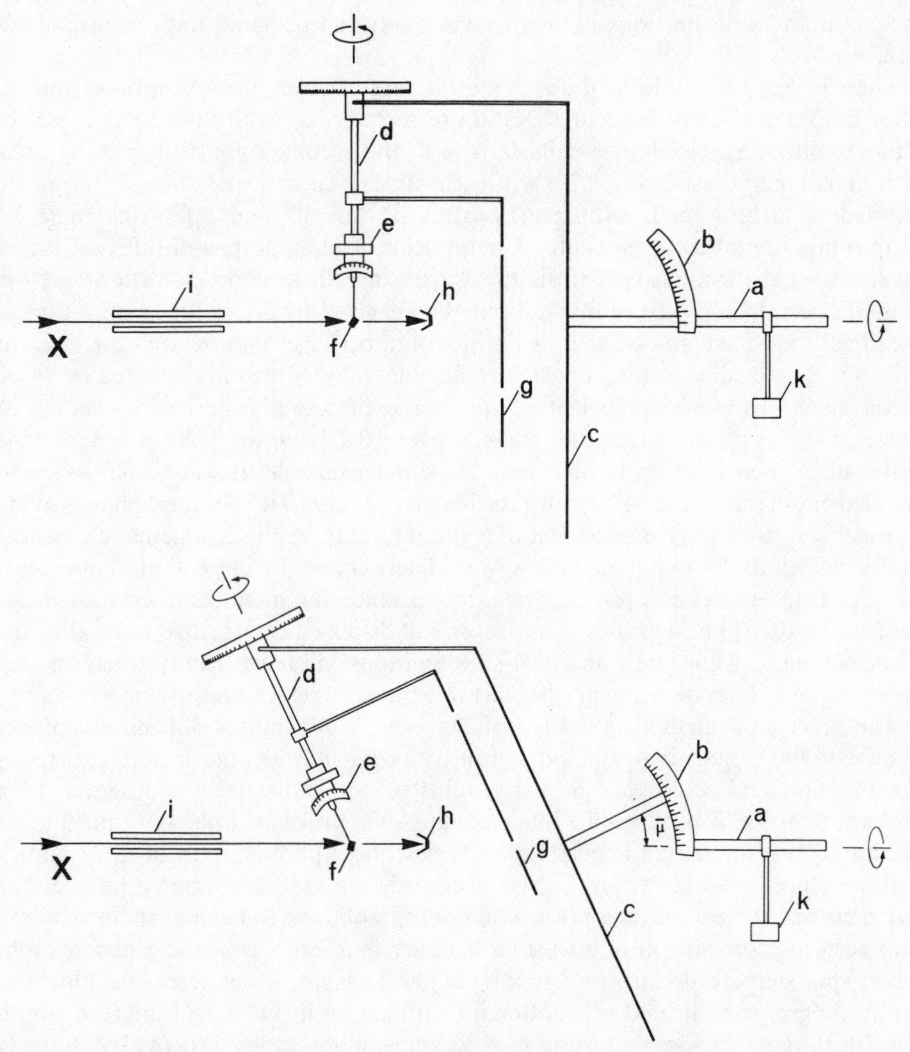

Fig 8.28 Schematic representation of a precession camera. The disposition of the parts of the camera are shown in the upper diagram for $\bar{\mu} = 0$ and in the lower diagram for a zero-layer photograph at $\bar{\mu} = 25°$. The labelling of the diagrams is explained in the text.

carries the arcs so that throughout its motion its plane is parallel to that of the film and its annulus is so placed as to transmit the selected cone of diffracted beams. A back-stop *h* which slips on to the collimator *i* has its absorbing cup situated immediately behind the crystal. Since smooth precession is essential if the intensities of the reflexions recorded on the film are to have significance, a counter weight, *k*, whose moment about the motor axis can be varied, is attached to the motor spindle to balance the moment of the various moving parts of the camera about this axis. For mechanical reasons the maximum attainable precession angle $\bar{\mu}$ is in most instruments not greater than 35°. The crystal-to-film distance is commonly fixed at 60 mm so that *F* is a constant equal to 60 mm; in some cameras the crystal-to-film distance is adjustable and then it can very simply be measured accurately by photographing a reciprocal lattice layer of a crystal whose unit-cell dimensions are known precisely. The back-plate of the film cassette is drilled with two pinholes which allow light to fall on the film and so serve to define the horizontal line through the centre of the film; the mid-point of the line joining the two black spots corresponds to the centre of the film.

There is very little to be said about the interpretation of precession photographs. The photograph is a direct representation of a reciprocal lattice net on the scale of *F* mm to one reciprocal lattice unit. Zero, first, and second layer [010] photographs of a monoclinic crystal (Fig 8.29) will each display an array of spots that can be indexed on a unit mesh with axial repeats a^* and c^* and interaxial angle β^* (conventionally taken to be acute). Comparison of photographs of different layers taken about the same axis is straightforward, since there is no distortion and the magnification factor is constant, and may be achieved by direct superimposition of the films. Since the blank circle in the centre of upper layer photographs increases in radius as ζ_n increases, some important reflexions may be missing and that may be inconvenient. However a sufficiently large sample of *hkl* and *h*0*l* reflexions should be obtainable from zero, first, and second layer [010] photographs to enable the systematic absences in these reflexions to be determinable. It will be necessary to supplement these photographs with a zero-layer [100] or [001] photograph to provide a reasonably accurate measurement of *b* and information about systematic absences in 0*k*0 reflexions. Laue symmetry is always determinable by inspection of appropriately oriented precession photographs: for instance for monoclinic crystals (Laue group 2/*m*) [010] photographs of any layer will display a central diad normal to the plane of the film while [100] and [001] photographs will have 2*mm* symmetry if zero layer and only lines of symmetry parallel to z^* and x^* respectively if upper layer.

The precession method is nicely balanced in its advantages and disadvantages relative to the Weissenberg method. Because it provides, when the crystal is correctly set and film and screen are properly adjusted, an undistorted photograph of a reciprocal lattice plane it enables interaxial angles in the plane to be very much more accurately measured than is possible on Weissenberg photographs. In the accuracy with which reciprocal cell edges can be measured there is little to choose between the two methods. Reflexions on a precession photograph tend to be uniform in shape so that accurate comparison of intensities is easier than on Weissenberg photographs where spot shape, especially on upper layer photographs, varies across the film. The unravelling of complicated orientational relationships in twins and intergrowths is simplified by the lack of distortion in precession photographs. Where exceptionally small or unstable crystals have to be used the shorter exposure time in which it is possible to obtain a satisfactory photograph by the precession method may be an

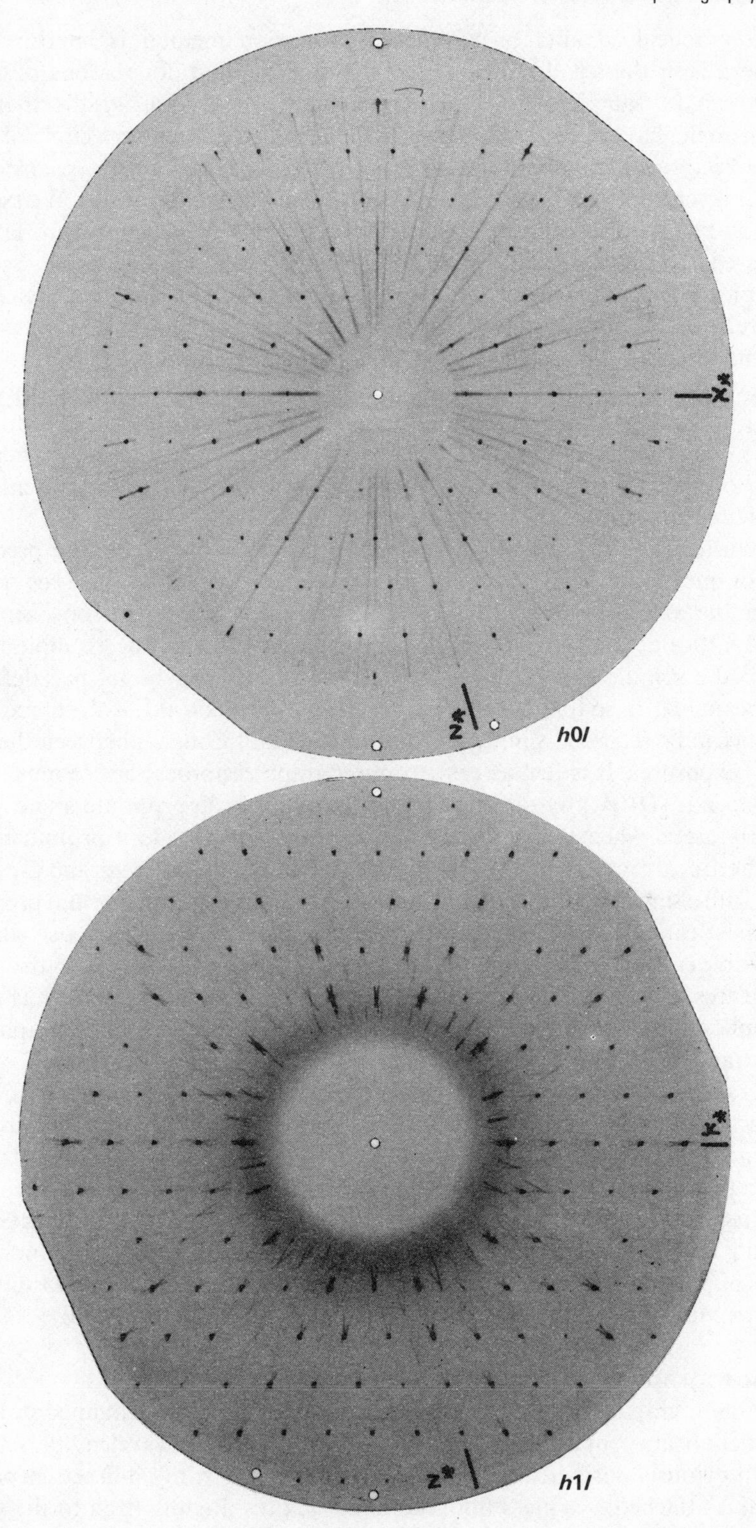

Fig 8.29 Zero and first layer [010] precession photographs of the monoclinic mineral *latiumite* taken with CuKα radiation, $\bar{\mu} = 25°$, $F = 60$ mm (three-quarters actual size).

important practical advantage. Where the precession method is inferior to the Weissenberg is in the smaller area of reciprocal space that, for reasons of camera geometry, can be sampled by a zero layer photograph. Usually the maximum precession angle that can conveniently be used is 30° and this allows a circle of radius $2 \sin \bar{\mu} = 1$ reciprocal unit about the origin of the zero layer net to be recorded; the area of nth layer nets that can be recorded is of course smaller. In the Weissenberg method however it is theoretically possible to record reflexions from a zero-layer net encompassed by a circle of radius two reciprocal units and on upper layer Weissenberg photographs, although the area that can be recorded decreases with increasing ζ_n, the area recorded is always substantially greater than on a precession photograph for the same value of ζ_n. Moreover it is possible with the Weissenberg camera to record layers of very much higher ζ value than with the precession camera; this is an important advantage when collecting intensity data for structural work and, moreover, occasionally symmetry elements that appear to be present when only reflexions of small Bragg angle are inspected may be seen to be absent when reflexions further out from the origin are considered.

In conclusion we draw attention to a useful facility provided by the precession camera: for one mounting of the crystal with a selected reciprocal lattice row parallel to the spindle axis of the arcs it is possible to record more than one zero-layer reciprocal lattice net and the corresponding upper layer nets. For example if x^* is parallel to the spindle axis of the arcs, [010] and [001] can be set parallel to the incident beam in turn so that the $h0l$, $h1l$, $h2l$, etc and the $hk0$, $hk1$, $hk2$, etc reciprocal lattice nets can be recorded simply by rotating the dial through α between the [010] and [001] exposures. It is just as easy to photograph reciprocal lattice nets normal to any zone axis $[OVW]$ by turning the dial through the appropriate angle. This is particularly useful when the reciprocal lattice row is parallel to a prominent zone axis and the arcs are interchangeable between oscillation, Weissenberg, and precession cameras; with a single mounting of the crystal oscillation, Weissenberg, and precession photographs can be taken and inter-related. For example suppose that an orthorhombic crystal has been mounted so that its [100] axis is parallel to the spindle axis of the arcs. This mounting would enable zero and upper layer [100] Weissenberg photographs and zero and upper layer [010] and [001] precession photographs to be taken by transferring the crystal on its arcs from one camera to the other.

In general the choice between Weissenberg and precession photography depends on the nature of the problem in hand and of the material under investigation. The principal advantage of the oscillation camera over the moving film cameras is that it provides a two-dimensional record of a truncated torus of reciprocal space, truncated because the film is of finite length, and this is the only way in which the regions between reciprocal lattice layers can conveniently be investigated; this facility is important in the study of phenomena, such as anti-phase domains, which give rise to diffracted intensity maxima that do not correspond to reciprocal lattice points.

Laue photography

For Laue photography the single crystal under investigation is maintained stationary in an incident beam containing a wide spectral range of X-ray wavelengths. As in the methods previously described for studying single crystal X-ray diffraction patterns the crystal is attached to a glass fibre mounted on arcs. Rigidly fixed to the spindle carrying the arcs is a circular scale, which permits the crystal to be rotated through a known angle between successive exposures; the incident X-ray beam is perpendicular to the spindle axis. In various circumstances it may be convenient to use either a cylindrical film coaxial with the spindle axis (Fig 8.30(a)) or a flat film with its plane perpendicular to the incident beam. Alternative positions for a flat film are in common use: in the *back-reflexion position* the film is situated between crystal and collimator (Fig 8.30(b)) to record only reflexions of high Bragg angle, whereas in the *front-reflexion position* the film is situated on the other side of the crystal (Fig 8.30(c)) and records only reflexions of low Bragg angle. The back-reflexion arrangement is particularly useful for crystals which absorb X-rays very strongly and for large crystals (especially metal crystals).

There is no specially designed camera for Laue photography: an oscillation or a Weissenberg camera can be used without adaptation for recording Laue patterns on cylindrical film, while a precession camera can be used directly for taking front-reflexion flat film photographs and an oscillation camera can be adapted for either front or back-reflexion photographs. The only camera conditions to be satisfied are that the crystal should remain stationary during exposure and that either the axis of a cylindrical film or the plane of a flat film should be perpendicular to the incident X-ray beam.

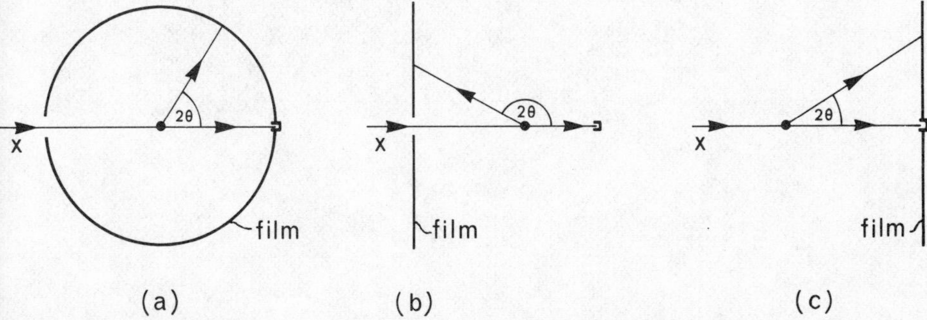

Fig 8.30 Laue photography. (a), (b), (c) show respectively the cylindrical film, back-reflexion flat film, and front-reflexion flat film arrangements. The crystal is shown as a solid circle and the beam trap as a section through a cup.

It was remarked at the beginning of this chapter that for a stationary crystal bathed in a parallel beam of monochromatic radiation few solutions of the Bragg Equation will occur. We have already explored ways in which the crystal can be moved in a regular manner while keeping the incident radiation monochromatic. In Laue photography we are concerned with increasing the number of solutions of the Bragg Equation for a stationary crystal and we do this by allowing the wavelength of the incident radiation to be variable. A broad X-ray spectrum in the incident beam is achieved by utilizing the unfiltered output of an X-ray tube; for this purpose the higher the atomic number of the target element the better, provided it is a sufficiently good thermal conductor to withstand a high current density, i.e. W is preferable to Mo, but Cu will do very well.

The reflexions on a Laue photograph cannot be easily indexed. The angle between the forward direction of the incident X-ray beam and the diffracted beam will be equal to 2θ so that θ is readily determinable. However a particular value of θ may correspond to more than one solution of the Bragg Equation $\lambda = 2d \sin \theta$ when λ is variable and d is unknown. For instance if the emission spectrum of the X-ray tube extends from $< \lambda'$ to $> 3\lambda'$, where λ' is a particular wavelength, the $2h, 2k, 2l$ and the $3h, 3k, 3l$ reflexions will have the same Bragg angle as the hkl reflexion and so all

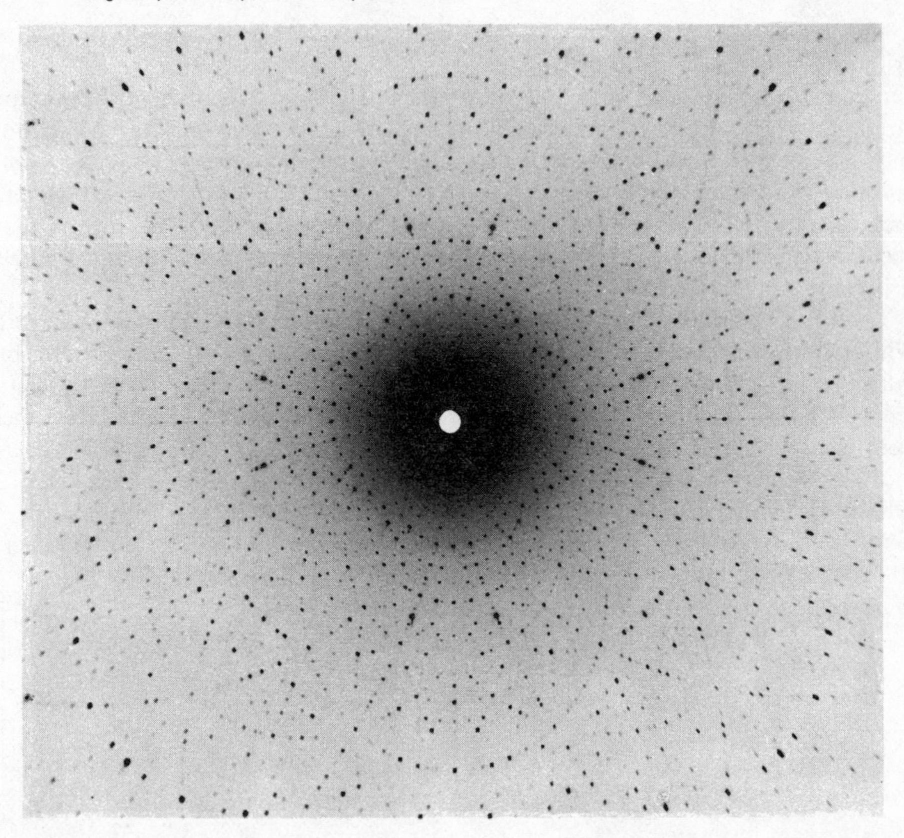

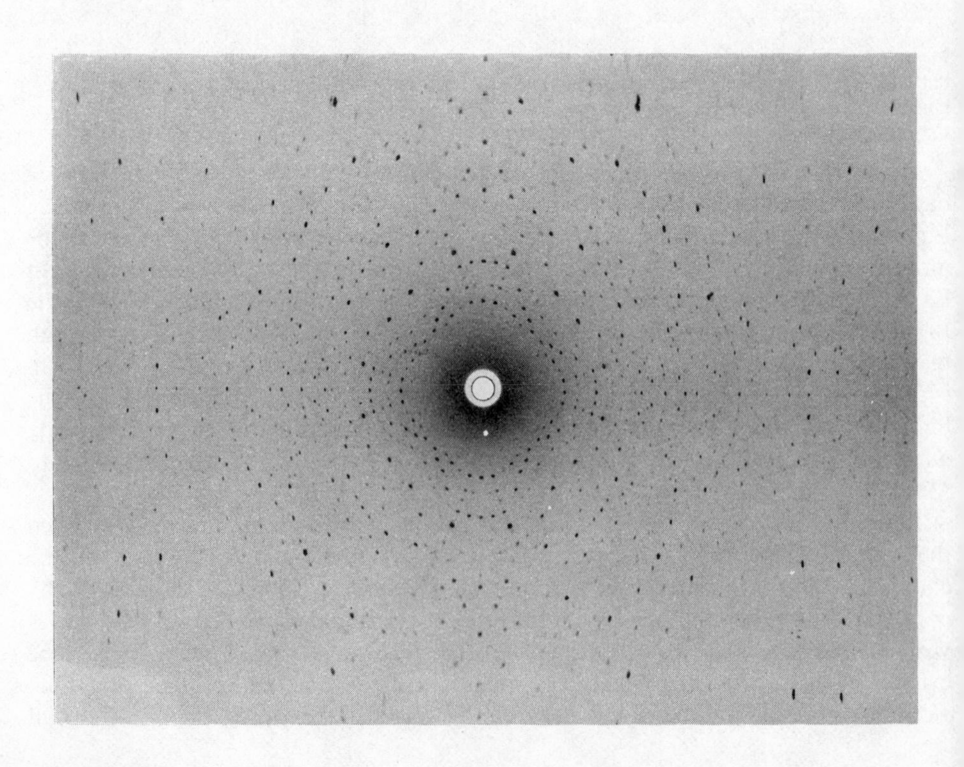

Fig 8.31 Laue photographs of a crystal of the tetragonal mineral *vesuvianite* (Laue group 4/*mmm*) taken with the incident beam parallel to the tetrad. The lower photograph was taken in a cylindrical camera; the upper photograph was taken with the front-reflexion flat film arrangement.

three reflexions will be superimposed. Indexing can be achieved by correlating the angular relationships between the normals to reflecting planes with the known axial ratios and interaxial angles. Indexed Laue photographs have some specialized uses, such as for the determination of the orientation of large single crystals of cubic metals, a topic which lies outside our scope. Except for such specialized applications it is rarely necessary to index Laue photographs; the commonest uses of the method do not involve indexing and are (i) the determination of the Laue symmetry of a crystal and (ii) the setting of a crystal with an identified zone axis in a particular direction relative to the camera geometry as a preliminary to oscillation or moving film photography.

We turn now to consider some of the general characteristics of Laue photographs. Simple inspection of a Laue photograph (Fig 8.31) reveals reflexions lying on curves such that each curve corresponds to the intersection with the film of a cone, the surface of which contains the forward direction of the incident beam. With flat films such intersections are conic sections, but with cylindrical films the nature of the curve is more complicated. Figure 8.32 illustrates the diffraction geometry for the generation of such a curve. Consider the plane represented by the great circle PZ whose pole is N. The reflected beam R will be coplanar with the incident X-ray beam XX′ and with the normal N to the plane PZ; it will therefore lie on the great circle X′PN and will be so placed that RP = PX′ = θ, where θ is the Bragg angle. Since the pole N of the great circle PZ lies on the great circle X′PR, these great circles intersect orthogonally so that $\widehat{ZPX'} = \widehat{ZPR} = 90°$. The spherical triangles ZPX′ and ZPR are thus congruent (with common side ZP, right-angles at P, X′P = PR) so that ZR = ZX′ = ψ. This result will be true for any plane which contains the direction Z and satisfies the Bragg Equation for this orientation of the incident X-ray beam. In

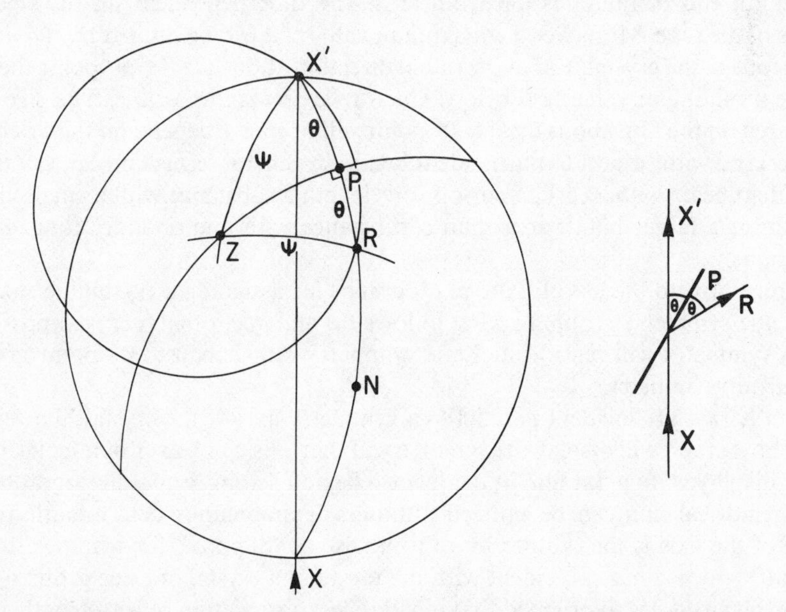

Fig 8.32 The diffraction geometry for the generation of a curve of reflexions on a Laue photograph. The diagram on the left is described in the text; that on the right illustrates the relationship of the incident and diffracted beams to the plane P.

particular if Z is a zone axis $[UVW]$ any plane containing the direction Z will lie in the zone $[UVW]$ and, if it is in the reflecting position, the direction of the diffracted beam arising from it will lie on the small circle whose stereographic centre is Z and which contains the forward direction of the incident X-ray beam. Reflexions generated by the planes of a zone $[UVW]$ will thus lie on a cone containing the incident direction, whose semiangle is the angle between the zone axis and the forward direction of the incident beam; on the film the cone will be represented by reflexions lying on a curve through the point of zero Bragg angle. If $[UVW]$ is a prominent zone axis in the crystal, there will be many reflexions on the curve and reflexions generated by planes of low indices will lie at the mutual intersection of several such curves, each related to a prominent zone axis. The shape of a zonal curve will depend very much on the angle ψ between the zone axis and the forward direction of the incident beam. For $\psi = 90°$ the cone of diffracted beams becomes a plane containing the incident X-ray beam; on a flat film this will be manifested as a radial line of reflexions through the $\theta = 0°$ (front-reflexion set-up) or $\theta = 90°$ (back-reflexion set-up) point. On a cylindrical film however the coplanar reflexions for which $\psi = 90°$ will only lie on a straight line on the film when the zone axis $[UVW]$ is either coaxial with the film cylinder or normal to the plane containing the incident beam and the axis of the cylindrical film. That is to say the only straight lines of reflexions on a cylindrical Laue photograph will be in the horizontal and vertical directions through the point on the film corresponding to $\theta = 0°$.

Another obvious feature of Laue photographs (Fig 8.31) that is worthy of comment is the absence of reflexions on the film over an area centred on the intersection of the forward direction of the incident beam with the film (this is of course not a feature of back-reflexion photographs). A reflexion close to the forward direction of the X-ray beam must have rather a small Bragg angle θ. Therefore for this reflexion $\lambda/2d$ must be small. As we have already seen (Fig 7.2) there is a sharp cut-off at the low wavelength end of the emission from an X-ray tube dependent on the operating voltage of the tube. Moreover the maximum value of d will be limited by the unit-cell dimensions of the crystal. For every radial direction about the $\theta = 0°$ point there will thus be a minimum value of θ below which reflexion is impossible. The size of the blank area on the film about the $\theta = 0°$ point will of course depend on the orientation of the crystal with respect to the incident beam; in general terms one can say that for an incident beam with a certain cut-off wavelength a substance with a small unit-cell will exhibit a larger blank area than a substance with one or more long unit-cell dimensions.

We turn now to the use of Laue photographs for assigning a crystalline substance to its Laue symmetry group. In what follows the statement that a crystal possesses a certain symmetry will refer to its Laue symmetry rather than to its point group or space group symmetry.

When X-rays are incident parallel to a symmetry axis of the crystal, the resultant Laue photograph will display the symmetry of that axis. Such axial symmetry is most clearly displayed on a flat film in the front-reflexion setting, but a photograph taken on a cylindrical film can be utilized although unambiguous determination of the nature of the axis is then rather more troublesome. Suppose, for instance, that the incident X-ray beam is coincident with the tetrad in a crystal of Laue group $4/m$ and that the plane (hkl) is so oriented that it will reflect X-radiation of wavelength λ. Then the symmetry related planes $(k\bar{h}l)$, $(\bar{h}\bar{k}l)$, and $(\bar{k}hl)$ will be similarly inclined to the direction of the incident beam and will also reflect X-rays of wavelength λ (Fig 8.33).

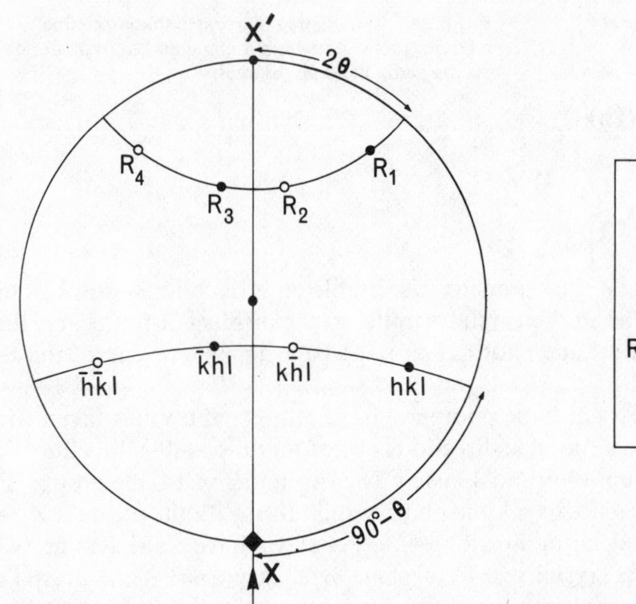

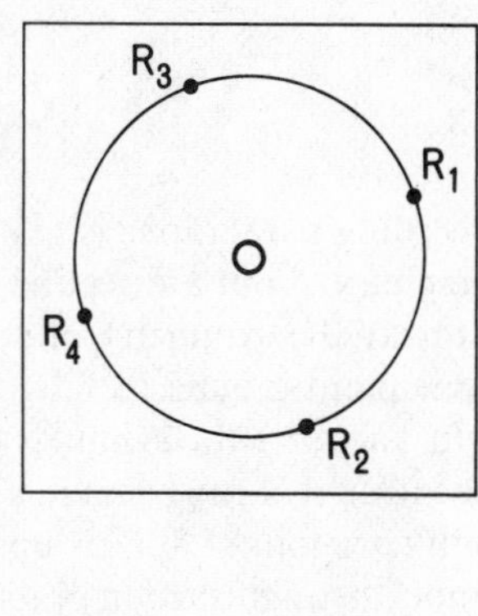

Fig 8.33 Laue photography of a crystal of Laue group $4/m$ with the incident X-ray beam parallel to the tetrad. The stereogram on the left shows the disposition of the normals to four planes related by the tetrad and the disposition of the resulting reflexions R_{1-4}. The portion of a flat film Laue photograph on the right shows the disposition of the four reflexions on a circle about the exit-hole in the centre of the film.

All four reflexions will be equal in intensity and will be disposed on the film about the point of intersection of the incident X-ray beam with the film in a manner consistent with tetragonal symmetry about that point. The Laue photograph taken as a whole will thus display tetragonal symmetry, every reflexion being equal in intensity to three others related to it spatially by successive rotation through 90° about the direction of the incident X-ray beam. Likewise when the incident beam is coincident with a hexad, triad, or diad axis in the crystal the corresponding symmetry will be apparent on a Laue photograph; and when the incident beam is coplanar with a mirror plane in the crystal, the resultant Laue photograph will display a line of symmetry parallel to the mirror plane and passing through the point of intersection of the incident beam with the film.

Since the Bragg Equation restricts θ to values between 0° and 90°, the X-ray beam must be incident on the same side of the (hkl) plane as the outward direction of the normal to the plane for the hkl reflexion to be produced; if the X-ray beam is incident, at the correct angle, on the other side of the (hkl) plane it will give rise to the $\bar{h}\bar{k}\bar{l}$ reflexion (Fig 8.34). It is thus impossible to record an hkl reflexion and a $\bar{h}\bar{k}\bar{l}$ reflexion without moving the crystal relative to the incident X-ray beam so that a Laue photograph never exhibits both hkl and $\bar{h}\bar{k}\bar{l}$ reflexions. For the tetragonal example that we have been considering this means that if reflexions are recorded from the planes (hkl), $(k\bar{h}l)$, $(\bar{h}\bar{k}l)$, $(\bar{k}hl)$, reflexions will not be recorded from their opposites $(\bar{h}\bar{k}\bar{l})$, $(\bar{k}h\bar{l})$, $(hk\bar{l})$, $(k\bar{h}\bar{l})$. Of course if the crystal is rotated through 180° about an axis normal to the tetrad so as to bring the opposite sense of the tetrad into coincidence with the forward direction of the incident beam, then all four opposites will reflect and reflexions from (hkl), etc will be absent.

The general conclusion to be drawn from the tetragonal example discussed in the

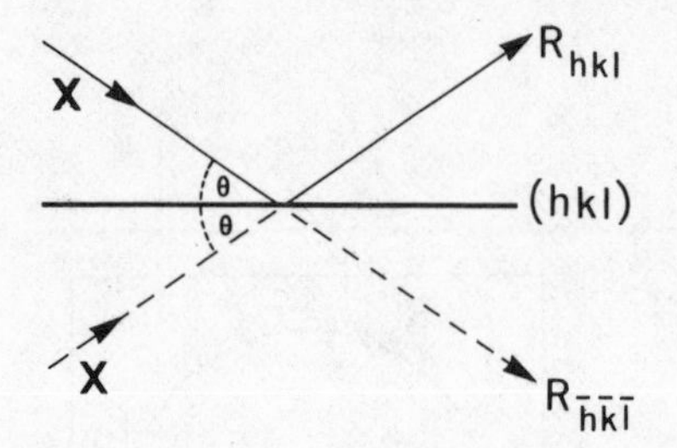

Fig 8.34 The diagram illustrates the impossibility of recording reflexions from a plane and from its opposite on the same Laue photograph.

preceding paragraphs is that the symmetry discernible on a Laue photograph is the symmetry about a direction in the crystal parallel to the incident X-ray beam; that is to say the symmetry of a Laue photograph must be assignable to one of the ten plane point groups.

In Table 8.1 the symmetry of Laue photographs of tetragonal crystals taken with the incident X-ray beam in a specified direction is listed for all possible directions for both tetragonal Laue groups (4/*m* and 4/*mmm*). The two tetragonal Laue groups are simply distinguished in practice by a Laue photograph taken with the incident X-ray beam parallel to [001]: the photograph for a 4/*m* crystal will have plane symmetry 4, whereas that for a 4/*mmm* crystal will have plane symmetry 4*mm*. If the crystal is already known to be tetragonal, its Laue group can be uniquely determined by taking just this one photograph. But if the possibility of the crystal being cubic has not been ruled out by other evidence, the observation that one Laue photograph has plane symmetry 4*mm* merely indicates that the Laue group of the crystal is either 4/*mmm* or *m*3*m*; Laue photographs in other orientations will have to be taken to distinguish between these two possibilities.

There is no standard procedure for determining the Laue symmetry of a crystal. The successive photographs necessary in a particular case will depend on the evidence provided by those already taken and on any reliable information that may happen to be available from prior study of certain physical properties of single crystals of the substance; for instance preliminary optical examination (chapter 12) may have given a clear indication of crystal system. Quite commonly the Laue symmetry of a crystal is determined incidentally by observation of intensity relationships of reflexions in the course of the investigation of its reciprocal lattice geometry by one or other of the moving-film methods.

Table 8.1
Symmetry of Laue photographs of tetragonal crystals

Laue group	Direction of incident X-rays	Symmetry of Laue photograph
4/*m*	[001]	4
	$\langle UV0\rangle$	*m*
	$\langle UVW\rangle$	1
4/*mmm*	[001]	4*mm*
	$\langle 100\rangle$, $\langle 110\rangle$	2*mm*
	$\langle UV0\rangle$, $\langle U0W\rangle$, $\langle UUW\rangle$	*m*
	$\langle UVW\rangle$	1

The point groups of each tetragonal Laue group are:

4/*m*: 4, $\bar{4}$, 4/*m*

4/*mmm*: 422, 4*mm*, $\bar{4}2m$, 4/*mmm*

A difficulty commonly encountered in the course of determination of Laue symmetry by means of Laue photographs is that only when the crystal is very precisely set with its symmetry axis parallel to the incident beam will the resultant photograph clearly display the symmetry of the axis. An error in setting of as little as 5 minutes of arc may substantially affect the appearance of the photograph. Let us suppose that two symmetry related planes are inclined at angles $\theta+\delta\theta$ and $\theta-\delta\theta$ to the incident beam; then the two reflexions produced will be recorded on a flat film in the front-reflexion setting, at a perpendicular distance R from the crystal (Fig 8.35), at distances $R\tan 2(\theta+\delta\theta)$ and $R\tan 2(\theta-\delta\theta)$ from the centre of the film. Moreover the two planes will reflect different wavelengths $\lambda+\delta\lambda$ and $\lambda-\delta\lambda$, where $\delta\lambda = 2d\cos\theta.\delta\theta$ and, as we have seen earlier, intensity varies quite rapidly with wavelength in certain parts of the spectral range emitted by an X-ray tube. Thus the two reflexions may be markedly different in intensity as well as being noticeably asymmetrically disposed on the photograph even though the mis-setting of the crystal is slight. But it is difficult to generalize and the experienced crystallographer may be able to discern a suspicion of the presence of a symmetry axis at a considerable inclination to the incident beam. When inspection of a Laue photograph reveals a suspicion of the presence of a symmetry axis, the crystal should be adjusted to bring that direction into closer alignment with the incident beam and another Laue photograph should be taken.

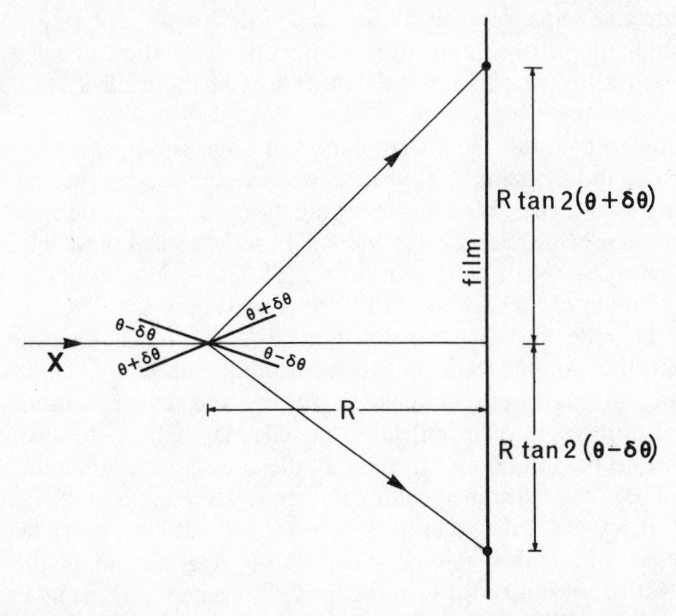

Fig 8.35 The figure illustrates the point that only when the incident X-ray beam is precisely parallel to a symmetry axis does the resultant Laue photograph clearly display the symmetry axis. The two lattice planes, shown as bold lines, are related by a diad axis inclined at the small angle $\delta\theta$, in the plane of the diagram, to the incident X-ray beam. The resultant reflexions will not be symmetrically disposed about the centre of the Laue photograph and will not be equal in intensity.

We deal generally with the setting of crystals on single crystal cameras in Appendix F, but it is appropriate to discuss here certain features of Laue photographs which are utilized for that purpose. A prominent zone in the crystal, because it contains a large number of lattice planes, will usually correspond to a curve with many closely spaced reflexions on the photograph and such curves are an obvious feature of most Laue photographs. The point of intersection of two or more such prominent zonal curves may be expected to correspond to the direction of the normal to a lattice plane of very simple indices. Symmetry axes are invariably parallel to such directions and it is, at least in principle, a simple matter to search for symmetry axes by bringing each such prominent zonal intersection in turn into coincidence with the incident beam.

It is always true to say that the presence of a mirror plane in a Laue group implies the presence of an axis of twofold or higher symmetry normal to it. Therefore if a mirror plane, which is necessarily perpendicular to a zone axis of simple indices, is located and the crystal is set with the mirror plane perpendicular to the spindle axis of the arcs on which the crystal is mounted, then a symmetry axis must be parallel to the spindle axis.

There is no generally applicable procedure for locating the crystallographic axes of a crystal of known or unknown symmetry. Each problem has to be tackled by the crystallographer in the light of what he knows at the start, or learns as he proceeds, about the Laue group of the crystal, in relation to the apparatus immediately available to him, and always bearing in mind the intensity of labour he is able to devote to the problem. The choice of procedure will depend very much on the experience and skill of the crystallographer. For a shapeless opaque crystal, the most difficult sort of subject, the present authors would usually choose to use Laue photographs taken on cylindrical film to locate symmetry directions; but other crystallographers might prefer to locate a principal zone by taking a series of precession photographs of small precession angle ($\bar{\mu} = 10°$) at appropriate intervals of rotation of the spindle axis. Either approach will succeed; one or the other may be more efficient in a particular case.

We turn now to consider the interpretation of Laue photographs in terms of the reciprocal lattice and the reflecting sphere:[5] we have the choice of adopting either of two alternative approaches. We can either take the constant K to be equal to unity so that the dimensions of the reciprocal lattice will be independent of wavelength, but the radius of the reflecting sphere will be variable (Fig 8.36(a)); or we can take the constant K equal to λ so that the dimensions of the reciprocal lattice vary with wavelength, but the radius of the reflecting sphere is constant. If $K = 1$, all those reciprocal lattice points within the volume between the reflecting sphere of minimum radius (corresponding to maximum wavelength in the incident radiation capable of generating an observable reflexion) and the reflecting sphere of maximum radius (corresponding to the minimum wavelength, the cut-off wavelength, in the incident radiation) will be in the reflecting position for some wavelengths in the incident beam. Overlapping reflexions will occur if the reciprocal lattice points hkl; $2h, 2k, 2l$; $3h, 3k, 3l$; etc lie within this volume (Fig 8.36(b)). The second of the alternative approaches, with $K = \lambda$, is however more fruitful in general for the interpretation of Laue photographs. With $K = \lambda$, the reflecting sphere has radius equal to one reciprocal unit and each lattice plane is represented in reciprocal space by a radial streak; the end of the streak nearer the origin corresponds to the minimum wavelength emitted by the X-ray tube (the magnitude of λ_{min} depends on the nature of the target and on the applied voltage as indicated in Fig 7.2) and the end of the streak away from the origin fades away at a wavelength that is greater for planes which reflect strongly than for those which give rise only to weak reflexion.

For a particular wavelength the reciprocal lattice points corresponding to lattice

[5] The reader who has not followed the introduction to the reciprocal lattice in chapter 6 and who has omitted the sections of this chapter concerned with moving film methods should pass on to the next chapter.

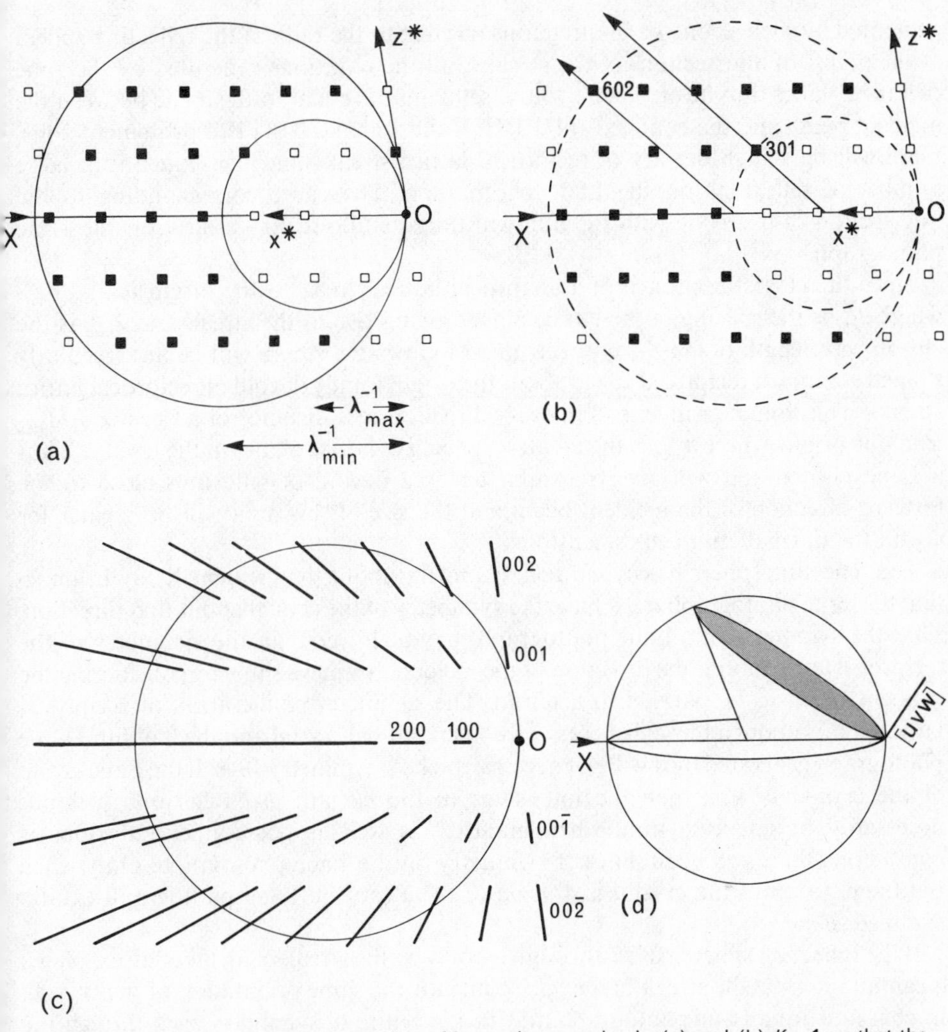

Fig 8.36 Reciprocal lattice interpretation of Laue photographs. In (a) and (b) $K = 1$ so that the reciprocal lattice dimensions are constant, but the radius of the reflecting sphere varies between λ_{min}^{-1} and λ_{max}^{-1} where λ_{min} is the cut-off wavelength of the incident spectrum and λ_{max} is arbitrarily taken at some wavelength where the intensity falls below a certain level; the ratio $\lambda_{max}/\lambda_{min}$ is taken as 2. (b) illustrates the parallelism of the reflected beams produced by the 301 and 602 reciprocal lattice points of a monoclinic crystal. In (c) and (d) $K = \lambda$ so that the reflecting sphere is of unit radius and each reciprocal lattice point becomes a streak radiating from the origin O as shown in (c); in (c) the ratio $\lambda_{max}/\lambda_{min}$ is taken at the lower value of 1·4 for clarity of the diagram and again a monoclinic lattice is exemplified. (d) serves to illustrate the point that the plane (shaded) in reciprocal space representing the zone $[UVW]$ intersects the reflecting sphere in a small circle; the Laue reflexions generated by such a zone lie on the surface of a cone whose apex is at the centre of the reflecting sphere and whose base is the small circle.

planes lying in a zone are coplanar and this plane passes through the origin of reciprocal space. When the incident X-ray beam is polychromatic, as it is in Laue photography, each such reciprocal lattice point is replaced by a radial streak lying in the plane (Fig 8.36(c)). Moreover the streaks from the planes (hkl), $(2h, 2k, 2l)$, $(3h, 3k, 3l)$, etc will overlap if the range of wavelengths in the incident beam is sufficiently large. In general the plane in reciprocal space which represents the zone $[UVW]$ cuts the reflecting sphere in a small circle (Fig 8.36(d)) and the reflexions generated by such a zone lie in directions parallel to the radii of the reflecting sphere at the points of intersections of the streaks; all the reflexions generated by the zone of lattice planes thus lie on a cone whose semiangle is equal to the angle between the incident beam and the zone axis $[UVW]$. If the zone $[UVW]$ is a prominent zone, there will be a high density of reciprocal lattice streaks and consequently a large number of reflexions on the Laue photograph. The curve corresponding to the intersection of this cone with the film will thus stand out very clearly on the Laue photograph.

Since the closest approach of a reciprocal lattice streak to the origin is λ_{min}/d_{hkl}, where d_{hkl} is the spacing of the lattice planes giving rise to the streak and λ_{min} is the cut-off wavelength of the X-ray spectrum, it follows that there will be an irregularly shaped volume of reciprocal space about the origin totally devoid of reciprocal lattice streaks. This volume will extend in every direction to a distance of at least λ_{min}/d_{max} from the origin, where d_{max} is the greatest spacing of lattice planes in the crystal. Thus a Laue photograph will always exhibit an area devoid of reflexions close to the forward direction of the incident beam and the size of this area will be greater for crystals with small unit-cell dimensions.

The reflecting sphere has cylindrical symmetry about the incident X-ray beam so that the Laue photograph must have the symmetry of the crystal about that direction. That the symmetry of a Laue photograph may be lower than the symmetry of the reciprocal lattice about the direction of the incident beam we show by considering the incident beam to be parallel to a tetrad. The symmetry of the array of reciprocal lattice points about a tetrad is necessarily $4mm$; but the crystal and the (flat film) Laue photograph generated by it will not necessarily show symmetry $4mm$. If the Laue group of the crystal is $4/m$, then the intensities of the hkl and $\bar{h}kl$ reflexions will not necessarily be equal so that, when intensity as well as position of reflexions is considered, there will be no lines of symmetry on the Laue photograph. Only when the Laue group of the crystal is $4/mmm$ or $m3m$ can the Laue photograph exhibit symmetry $4mm$.

If the reflecting sphere passes through a point on the streak of an (hkl) lattice plane, it cannot intersect the streak of the ($\bar{h}\bar{k}\bar{l}$) plane for the same orientation of the crystal. In consequence a Laue photograph may lack a centre of symmetry even though the diffraction pattern as a whole may be centrosymmetric. The symmetry of a Laue photograph is always that of one of the ten two-dimensional crystallographic point groups.

In the treatment of Laue photographs as far as we have taken it in this chapter the use of the reciprocal lattice and the reflecting sphere are not essential; but they do provide, as we have sought to show in the preceding paragraphs, an elegant way of explaining the diffraction pattern produced.

Determination of accurate unit-cell dimensions

It is often necessary to be able to determine unit-cell dimensions very much more accurately than is possible by measurement of layer line spacings on oscillation photographs or by direct measurement of Weissenberg or precession photographs. Obvious uses for accurate unit-cell dimensions are in the determination of thermal expansion coefficients (chapter 11) and for the conversion of the atomic coordinates which are the end result of a structure determination to accurate bond lengths and bond angles. Various methods for the accurate determination of unit-cell dimensions have been in general use over the past few decades; one obvious approach is to determine a^*, b^*, and c^* by very precise measurement of the Bragg angle for $h00$, $0k0$, and $00l$ reflexions. If the crystal is monoclinic, β^* can be measured directly on an [010] precession photograph and if it is triclinic the three interaxial angles α^*, β^*, and γ^* can be measured on the appropriate precession photographs. In these two systems, the only systems for which interaxial angles have to be measured, the form of the expressions for deriving unit-cell dimensions from reciprocal lattice dimensions (Table 6.7) are such that errors in the measurement of interaxial angles in reciprocal space may seriously affect the accuracy of both interaxial angles and unit-cell edges in direct space. We confine our discussion of the accurate measurement of unit-cell dimensions here to one very elegant, accurate, and generally applicable method, which makes use of the doublet splitting at high θ of reflexions on Weissenberg photographs.

As we have pointed out earlier (chapter 7) the characteristic X-radiation emitted from a crystallographic X-ray tube consists of a $K\beta$ line, which is filtered out, and the two closely spaced lines $K\alpha_1$ and $K\alpha_2$. The reflexions produced by the $K\alpha_1$ and $K\alpha_2$ wavelengths are resolved only at high Bragg angle. On a Weissenberg photograph all reflexions at low Bragg angles will appear to be single spots; but as θ increases the difference in Bragg angle for the $K\alpha_1$ and $K\alpha_2$ lines gradually increases until at high Bragg angle reflexions from a plane will be resolved into clearly separated pairs of reflexions, the inner and stronger of which is produced by the shorter α_1 wavelength in the incident beam and the outer by the longer, and weaker, α_2 wavelength.

The difference $\delta\lambda$ in wavelength between the α_1 and α_2 lines is accurately known for all X-ray sources in common use. Therefore the Bragg angle θ for the α_1 reflexion can be found by measuring the difference in Bragg angle $\delta\theta$ between pairs of reflexions produced by resolution of the α_1 and α_2 lines in the following manner. Suppose the wavelength of the α_1 radiation is λ, then for reflexion from a plane of spacing d,

$$\lambda = 2d\sin\theta \qquad \text{for the } K\alpha_1 \text{ wavelength}$$

and

$$\lambda+\delta\lambda = 2d\sin(\theta+\delta\theta) \qquad \text{for the } K\alpha_2 \text{ wavelength.}$$

Therefore

$$\sin(\theta+\delta\theta)-\sin\theta = \frac{\delta\lambda}{2d} = \frac{\delta\lambda}{\lambda}\sin\theta$$

$$\sin\theta\cos\delta\theta+\cos\theta\sin\delta\theta-\sin\theta = \frac{\delta\lambda}{\lambda}\sin\theta$$

$$\cos\theta\sin\delta\theta-\sin\theta(1-\cos\delta\theta) = \frac{\delta\lambda}{\lambda}\sin\theta$$

$$\cos\theta = \sin\theta\left(\frac{1-\cos\delta\theta}{\sin\delta\theta}+\frac{\delta\lambda}{\lambda\sin\delta\theta}\right)$$

$$\cos\theta = \sin\theta\left(\tan\frac{\delta\theta}{2}+\frac{\delta\lambda}{\lambda\sin\delta\theta}\right)$$

$$1-\sin^2\theta = \sin^2\theta\left(\tan\frac{\delta\theta}{2}+\frac{\delta\lambda}{\lambda\sin\delta\theta}\right)^2$$

and so

$$\sin^2\theta = \left\{1+\left(\tan\frac{\delta\theta}{2}+\frac{\delta\lambda}{\lambda\sin\delta\theta}\right)^2\right\}^{-1}.$$

For a zero-layer Weissenberg photograph, if the separation of the α_1 and α_2 reflexions from the same lattice plane is δs, measured perpendicular to the median line of the film, then $\delta\theta = \delta s/2r$, where r is the camera radius (Fig 8.37(a)). The magnitude of

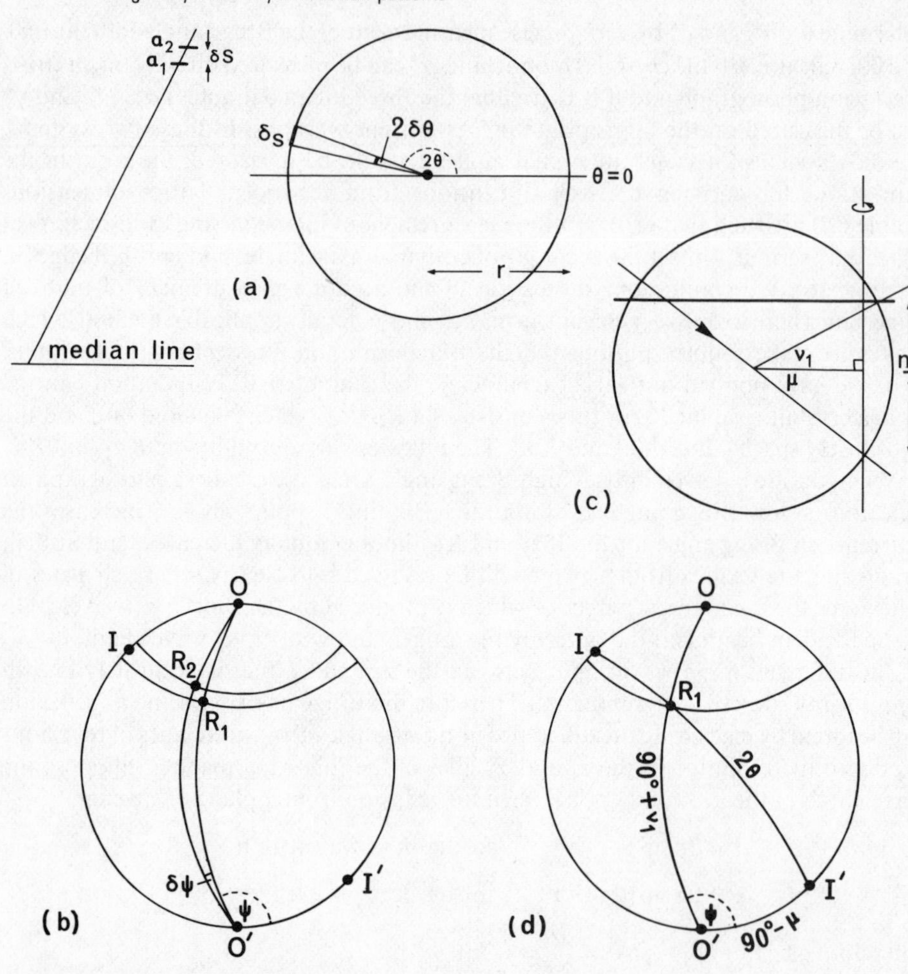

Fig 8.37 Accurate determination of unit-cell dimensions by measurement of $\alpha_1\alpha_2$ splitting. The measurement of the separation δs of the $\alpha_1\alpha_2$ doublet is made perpendicular to the median line of the Weissenberg photograph, as shown on the left in (a), so that $\delta s = 2r.\delta\theta$. The stereogram (b) illustrates the diffraction geometry for the α_1 reflexion R_1 and the α_2 reflexion R_2 in an *n*th layer Weissenberg photograph; OO′ is the oscillation axis and II′ is the incident beam direction so that $OI = 90°-\mu$, $OR_1 = 90°-\nu_1$, and $OR_2 = 90°-\nu_2$. (c) illustrates the generation of the α_1 reflexion in an *n*th layer equi-inclination Weissenberg; the intersection of the *n*th reciprocal lattice net with the plane of the diagram is shown as a bold line. The stereogram (d) serves to illustrate the relationship between θ, ν_1, μ, and ψ for the α_1 reflexion R_1.

$\sin^2\theta$ can thus be calculated from measurements of δs for the pair of resolved $\alpha_1-\alpha_2$ reflexions. The form of the relationship between δs and $\sin^2\theta$ is such that the value of $\sin^2\theta$ is rather insensitive to errors in measurement of δs.

For an *n*th layer equi-inclination Weissenberg photograph the relationship between the measured value of δs and $\sin^2\theta$ is more complicated. The Bragg angle θ of the α_1 reflexion cannot be deduced directly from measurement of δs. The distance s of a reflexion from the median line of the photograph gives the angle $\psi = s/r$ between the plane containing the forward direction of the incident X-rays and the oscillation axis and the plane containing the reflected beam and the oscillation axis; thus measurement of the separation of the resolved $\alpha_1-\alpha_2$ doublet, δs, gives a value of $\delta\psi$

for the reflexion (Fig 8.37(b)). The equi-inclination angle μ for the nth layer is such that $\sin\mu = n\lambda/2t$, where t is the periodicity of lattice points along the zone axis parallel to the oscillation axis. The angle μ at which the camera is set cannot be right for both the α_1 and α_2 wavelengths. Reflexions will lie on two cones, one of semiangle $90°-\upsilon_1$ and the other of semiangle $90°-\upsilon_2$, where υ_1 and υ_2 are related to the camera angle μ. It is apparent from Fig 8.37(c) that for the α_1 reflexion

$$\frac{n\lambda}{t} = \sin\upsilon_1 + \sin\mu \quad (1)$$

and correspondingly for the α_2 reflexion of the doublet

$$\frac{n(\lambda+\delta\lambda)}{t} = \sin\upsilon_2 + \sin\mu. \quad (2)$$

We can obtain an expression for θ by use of the spherical triangle $O'I'R_1$ (Fig 8.37(d)):

$$\cos 2\theta = \cos(90°+\upsilon_1)\cos(90°-\mu) + \sin(90°+\upsilon_1)\sin(90°-\mu)\cos\psi \quad (3)$$

i.e.

$$2\sin^2\theta = 1 + \sin\upsilon_1\sin\mu - \cos\upsilon_1\cos\mu\cos\psi \quad (4)$$

and correspondingly

$$2\sin^2(\theta+\delta\theta) = 1 + \sin v_2\sin\mu - \cos v_2\cos\mu\cos(\psi+\delta\psi). \quad (5)$$

From these relations, numbered (1) to (5) above, coupled with the Bragg Equation an expression can be obtained for relating θ to the measured separation δs of an $\alpha_1-\alpha_2$ doublet, the wavelengths λ and $\lambda+\delta\lambda$ of the relevant lines in the incident radiation, the camera angle μ, and the lattice spacing t. The lattice spacing t will be approximately known at the start and so each cycle of refinement in the computation will yield a more accurate value of t which is then used to calculate a better value of $\sin^2\theta$.

We now illustrate the way in which cell dimensions can be calculated from a set of values of $\sin^2\theta$ by considering a monoclinic example. The extension of the argument to the triclinic case is straightforward, but the expressions involved are of course more cumbersome. For a reciprocal lattice point in the monoclinic system d^* is given by

$$d^{*2} = h^2a^{*2} + k^2b^{*2} + l^2c^{*2} + 2hla^*c^*\cos\beta^*,$$

which becomes, on putting $q = d^*$ and differentiating,

$$qdq = (h^2a^* + hlc^*\cos\beta^*)da^* + k^2b^*db^* + (l^2c^* + hla^*\cos\beta^*)dc^* - hla^*c^*\sin\beta^*.d\beta^*$$

i.e.

$$qdq = A\,da^* + B\,db^* + C\,dc^* - D\,d\beta^*.$$

The coefficients A, B, C, D are calculated from the known approximate unit-cell dimensions; $q = \lambda/d = 2\sin\theta$ is obtained from measurements of $\alpha_1-\alpha_2$ doublet splitting in the way described earlier; and dq is taken to be the difference between this 'observed' value of q and that obtained from the known approximate unit-cell dimensions. Every measurement of an $\alpha_1-\alpha_2$ doublet thus gives rise to one such linear equation in the four unknowns da^*, db^*, dc^*, and $d\beta^*$. The set of linear equations derived from all the measurements made is soluble by least-squares methods to yield the more accurate reciprocal lattice dimensions a^*+da^*, b^*+db^*, c^*+dc^*, and $\beta^*+d\beta^*$; from these more accurate unit-cell dimensions and thence more accurate values of $\sin^2\theta$ for upper layer photographs can be calculated. A further cycle of refinement is then performed, and so on until by successive approximation unit-cell dimensions accurate to better than 1 in 10^3 are achieved.

The powerful technique outlined above, which was developed by Alcock and Sheldrick (1967) from a method restricted to zero-layer photographs (Main and Woolfson, 1963), has the advantages that all the measurements can be quickly and easily made with an ordinary travelling microscope (an accuracy of about 3 per cent in the measurements is all that is required) and that computer programmes for the least-squares calculations are readily available. The method is moreover found to be insensitive to even quite substantial errors in camera radius and in the equi-inclination angle μ. It is however observed that when data from Weissenberg photographs taken about only one oscillation axis are used, those cell dimensions which are not directly determinable from the zero-layer photograph are relatively less accurate. It is therefore advisable to include data from one or more Weissenberg photographs taken about a second axis. Of course one could merely use the three zero-layer photographs taken respectively about the x, y, and z axes; but much more data become available and higher accuracy is in consequence achieved if equi-inclination upper layer photographs about at least one axis are included.

We shall not discuss here other methods of determination of accurate unit-cell dimensions. For an excellent and thorough treatment of this topic the reader is referred to Woolfson (1970).

9
Principles of structure determination: the diffraction of X-rays, neutrons and electrons

We turn now to a brief description of the essential problem involved in the determination of the structure of a crystal from its diffraction pattern. This provides a convenient opportunity to introduce an elementary account of the diffraction of electrons and neutrons by crystals in the course of which we shall emphasize the differences between the diffraction of X-rays, electrons, and neutrons by crystals and point out the advantages and disadvantages of using one or another type of radiation to solve a particular problem. X-ray diffraction studies provide the primary means for the determination of crystal structures, as they have since the early days of the determination of the structures of such simple salts as NaCl and CsCl, soon after the discovery of the diffraction of X-rays by crystals. Structures of increasing complexity have become soluble as the techniques of X-ray structure determination have progressed until it is now possible to solve the crystal structures of the biologically very important proteins, the unit-cells of which contain several thousand atoms.

The techniques of crystal structure determination are outside the scope of this textbook. Here we are concerned simply with stating the essential problem and indicating, in general terms, the lines on which it may be soluble by X-ray methods with or without the assistance of complementary neutron diffraction studies. We aiso comment on the uses of electron microscopy to complement X-ray diffraction methods in the study of single crystals.

We take as our starting point the familiar phenomenon of optical diffraction and, in particular, consider how a parallel beam of monochromatic light forms an image of a one-dimensional grating when a lens is inserted in the path of the light waves emergent from the grating. In Fig 9.1 a monochromatic light beam is incident on a grating G of transparent lines; the transmitted light waves then pass through a lens L so placed that a real image of the grating is formed in the plane I parallel to the plane of the grating G. The Abbe theory of image formation assumes that the formation of the image I takes place in two stages. As the incident plane wave impinges on the grating it is diffracted so as to give rise to sets of parallel beams corresponding to the zero, first, second, etc orders of diffraction. The lens L focuses these sets of diffracted beams in the plane D; the Fraunhofer diffraction pattern[1] of the grating is formed in the plane D, which is effectively at an infinite distance from the grating. The

[1] In Fraunhofer diffraction both the source and the observed diffraction pattern are effectively infinitely distant from the object. In Fresnel diffraction, in contrast, either the source, or the observed diffraction pattern, or both are not effectively at infinity.

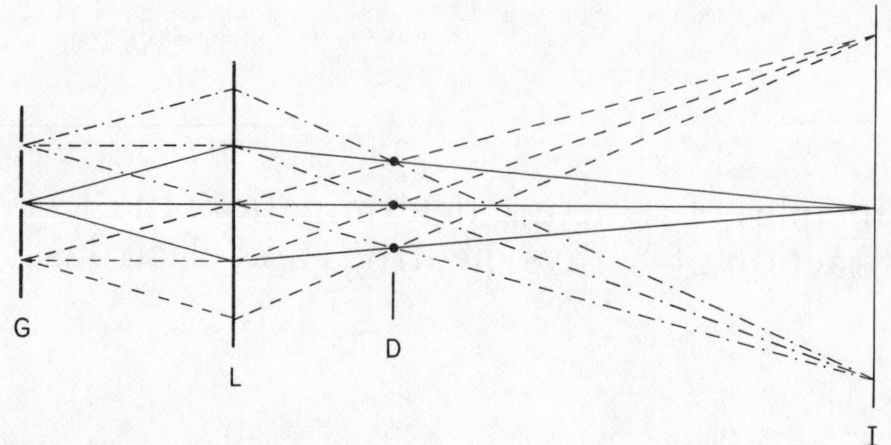

Fig 9.1 Abbe's theory of image formation. A parallel beam of monochromatic light is incident on the grating G. The diffracted rays are focused by the lens L to form a diffraction pattern in the plane D and a real image in the plane I.

waves travel on in such a manner that the diffracted beams produced by a particular grating element cross in the plane I so that a real image of the diffraction grating is produced in the plane I. The image of a two-dimensional grating is produced in a precisely analogous manner. But for a three-dimensional grating the complete diffraction pattern can only be recorded by allowing the angle of incidence of the light beam on the grating to be varied; it is thus impossible to focus the whole diffraction pattern simultaneously to form an image of the structure. With an optical lens only an image of a projection of the three-dimensional structure is obtainable.

The Abbe theory of image formation provides a means of calculating the form of the image produced by a grating of known size and shape. The direction, amplitude and phase of each diffracted beam can be calculated so as to give a complete description of the diffraction pattern formed in the plane D. The diffracted beams may then be imagined to be recombined so as to form the real image in the plane I and this can be calculated too. For a three-dimensional grating it is possible to calculate its complete diffraction pattern and to construct its image even though the image cannot be formed by an incident optical beam and a simple lens system.

If a comparable experimental arrangement could be used for X-rays it would then be possible, at least in principle, to form a real two-dimensional image of a crystal structure. But this cannot be done in practice because the refractive index of most solids for X-radiation differs so little from unity, by about 1 in 10^6, that it is impossible to construct a lens for X-rays. Without a lens there is no means of bending the diffracted X-rays to form an image of the crystal structure; but it remains possible to observe the diffraction pattern at a distance that is effectively infinite in comparison with the size of the grating elements of the crystal structure and from this Fraunhofer diffraction pattern it should be possible to calculate the image of the crystal structure. But again we are thwarted by a practical difficulty: there is no experimental means of measuring the phases of X-rays so that, except in certain special cases, the phases of the X-rays diffracted by a crystal structure are unknown. Measurement of an X-ray diffraction pattern yields the directions and the intensities of the diffracted beams; and from their measured intensities their amplitudes can be calculated. We can thus determine the shape and size of the unit-cell and calculate the structure amplitude from the measured intensity of every X-ray reflexion; but in the absence of any practical means of determining phase the description of the diffraction pattern remains incomplete.

The basic problem of X-ray crystal structure analysis is the completion of the description of the diffraction pattern by finding some way of evaluating the phase of every observed reflexion. In a few special cases symmetry considerations limit the number of possible structures to so small a number that it becomes feasible to compare the observed structure amplitudes with those calculated for each of the possible structures and so to determine which is the correct structure. For example a cubic structure of Laue group *m*3*m* with an F-lattice and four molecules of AX per unit-cell may have either the NaCl-type structure or the blende-type structure. From Table 9.1, in which the intensities of a few reflexions calculated for each of these possible structures are shown, it is apparent that comparison with the measured intensities of reflexions produced by the substance under consideration will rapidly show which of the two possible structures the substance actually has. In a simple case such as this a powder diffraction pattern would be adequate, care being taken to assign the appropriate multiplicity to each powder line and to take into account the several physical and geometrical factors which relate the measured intensity of any powder line to the intensity $I(hkl)$ of the corresponding diffracted beam produced by one unit-cell.

In most cases however the task of deducing the phases of the diffracted X-ray beams is very much more difficult. Various methods may be used to predict a probable structure for the substance and the correctness of the predicted structure can be tested by calculating structure factors from it. If there is reasonably good agreement between the amplitudes of the X-ray reflexions calculated from the proposed structure and the measured amplitudes, then the proposed structure may be assumed to be a good approximation to the real structure and it may be assumed that the phases of the X-ray reflexions from the crystal are the same as those calculated from the proposed structure. It is then possible to calculate an image of the structure from the measured amplitudes and the calculated phases. The positions of the atoms determined from this image should provide a better approximation to the real structure than that given by the initially proposed, or trial, structure. The process is

Table 9.1
Intensities of reflexions for AX structures of NaCl-type and blende-type.

Both structure types have cubic F-lattices with one formula unit per lattice point. The coordinates of the atoms associated with the lattice point at the origin are A 0, 0, 0; B 0, 0, $\frac{1}{2}$ for the NaCl structure and A 0, 0, 0; B $\frac{1}{4}, \frac{1}{4}, \frac{1}{4}$ for the blende (ZnS) structure. Intensities are given by

$$\mathrm{I}(hkl) = 16(f_\mathrm{A}+(-1)^l f_\mathrm{X})^2 \quad \text{for NaCl}$$

and

$$\mathrm{I}(hkl) = 16\left(\left[f_\mathrm{A}+f_\mathrm{X}\cos 2\pi\frac{h+k+l}{4}\right]^2 + f_\mathrm{X}^2\sin^2 2\pi\frac{h+k+l}{4}\right) \quad \text{for ZnS}$$

Reflexion	I(*hkl*) for NaCl		I(*hkl*) for blende	
111	$16(f_\mathrm{A}-f_\mathrm{X})^2$	w	$16(f_\mathrm{A}^2+f_\mathrm{X}^2)$	m
200	$16(f_\mathrm{A}+f_\mathrm{X})^2$	s	$16(f_\mathrm{A}-f_\mathrm{X})^2$	w
220	$16(f_\mathrm{A}+f_\mathrm{X})^2$	s	$16(f_\mathrm{A}+f_\mathrm{X})^2$	s
311	$16(f_\mathrm{A}-f_\mathrm{X})^2$	w	$16(f_\mathrm{A}^2+f_\mathrm{X}^2)$	m
222	$16(f_\mathrm{A}+f_\mathrm{X})^2$	s	$16(f_\mathrm{A}-f_\mathrm{X})^2$	w

Relative intensities: s = strong, m = medium, w = weak

repeated successively until the discrepancy between the structure amplitudes of the proposed and real structures is minimized. The details of the methods used in structure determination are outside our scope here; the reader interested in pursuing this topic is referred to Woolfson (1970).

Diffraction of neutrons by crystals

The interaction between a neutron beam and an atom gives rise to both coherent and incoherent scattering. This elementary treatment[2] is confined to coherent scattering, that is to scattering such that there is a definite phase relationship between incident and scattered beams so that beams scattered from different atoms in a crystal structure can interfere. Such scattering is analogous to the diffraction of X-rays.

The wavelength associated with a neutron is given by the de Broglie Equation $\lambda = h/mv$, where h is Planck's constant, m is the mass of the neutron, and v is its velocity. The range of velocities in the neutron beam extracted from a reactor is related to the temperature of the reactor by $\frac{1}{2}mv^2 = \frac{3}{2}kT$, where v is now the root mean square velocity, k is Boltzmann's constant, and T is in degrees Kelvin. In practice the temperature of a reactor is such that the emergent neutrons have root mean square velocities corresponding to wavelengths in the range 1·3 to 1·6 Å, an appropriate wavelength range for diffraction by crystals. In crystallography it is necessary to use a parallel neutron beam that is, at least approximately, monochromatic; this is achieved by collimating the neutron beam emergent from the reactor and allowing the collimated beam to be incident on a single crystal monochromator set at such an angle as to satisfy the Bragg Equation for reflexion of neutrons of the required wavelength from the face of the monochromator. However the neutron beam emergent from a reactor is generally much weaker than the X-ray beam emergent from an X-ray tube so that for diffraction studies it is necessary to use neutron beams of rather large cross-sectional area. Because of this, collimation cannot produce an accurately parallel beam and in consequence the monochromator does not produce a strictly monochromatic beam, but a beam with a wavelength spread of about 0·05 Å. Such a wavelength spread limits the resolution of the observable diffraction pattern. Since the cross-section of the neutron beam has to be large to provide adequate intensity, it is necessary to use larger crystal specimens than is usual in X-ray studies. This requirement poses no special problems where powder work is concerned because the absorption of neutrons is very much smaller than the absorption of X-rays by crystals; but it does preclude single crystal studies on substances which do not readily form large single crystals.

Neutron diffraction is governed by the same physical principles as X-ray diffraction but the mechanism of scattering is different. We have earlier (chapter 6) shown that for X-ray diffraction, when an unpolarized plane wave of unit amplitude is incident on an atom of atomic scattering factor f, the amplitude of the scattered radiation at a distance R from the atom is given by

$$\frac{1}{R}\cdot\frac{e^2}{mc^2}\cdot f\left(\frac{1+\cos^2 2\theta}{2}\right)^{\frac{1}{2}}$$

and there is a phase difference π between the scattered and incident X-radiation. The amplitude of the scattered radiation varies with direction because of the angular dependence of the polarization factor $\surd(\frac{1}{2}+\frac{1}{2}\cos^2 2\theta)$ and because the radius of the

[2] For a comprehensive account of neutron diffraction the reader is referred to Bacon (1962).

atom is comparable with the X-ray wavelength. Interference thus occurs between the X-radiation scattered from different parts of the atom so that the atomic scattering factor f decreases as $(\sin\theta)/\lambda$ increases. When neutrons are scattered by a non-magnetic atom however it is the nucleus that is solely responsible for the scattering, the electrons being too small to deflect the neutron beam. When a plane wave of unit amplitude is incident on the nucleus of an atom in a solid the nucleus is not free to recoil and the scattered neutron amplitude at a distance R from the nucleus is simply given by b/R, where b is a nuclear property known as *scattering length*. There is no polarization factor involved in neutron scattering and, since the radius of the nucleus is very small compared with the neutron wavelength, b is independent of scattering angle; b is also very nearly independent of neutron wavelength. The relationship however between scattering length for neutrons and atomic species is very much more complicated than the relationship between atomic scattering factor for X-rays and atomic species. The nuclei of different isotopes of the same element differ in their scattering lengths so that the coherent scattering length $\bar{b}$ for an element has to be taken as the average of $b \times$ (isotopic abundance) for all the isotopes of the element. Moreover if an isotope has non-zero nuclear spin, its scattering length has two possible values and this must be taken into account in calculating the coherent scattering length $\bar{b}$. Values of $\bar{b}$ cannot at present be calculated satisfactorily and so have to be determined experimentally. It is interesting to observe that $\bar{b}$ is of the same order of magnitude as $e^2 f/mc^2$ so that a crystal scatters neutrons and X-rays by about the same amount. It is simply because neutron beams of comparable intensity cannot be produced that it is necessary to use larger crystals for neutron diffraction than for X-ray diffraction studies.

Unlike atomic scattering factors, which vary regularly with atomic number, scattering lengths vary quite irregularly with atomic number. For a few atoms (e.g. hydrogen, titanium, Ni^{62}) the scattering length is negative, the sign of b corresponding to the phase change on scattering, π for b positive and zero for b negative.

We have so far confined our discussion of coherent neutron scattering to that produced by interaction with non-magnetic nuclei, but neutrons are also scattered by interaction of their magnetic moments with the permanent magnetic moments of atoms containing unpaired electrons. Here the scattering is due to neutron-electron interaction so that the scattered neutron amplitude, just like scattered X-ray amplitude, falls off with increasing $(\sin\theta)/\lambda$. It is only for substances in which the magnetic moments of the atoms are regularly arranged, that is for ferromagnetic, anti-ferromagnetic and ferrimagnetic substances, that such scattering is coherent. In paramagnetic substances the magnetic moments of the atoms are randomly orientated so that magnetic neutron scattering is incoherent and merely contributes to the background scattering. Since the magnetic moment of an atom does not affect its scattering of X-rays, it is necessary to use neutron diffraction to study the alignment of magnetic moments in ferro-, anti-ferro-, and ferrimagnetic materials.[3]

Neutron diffraction is usually used to complement, not to replace, X-ray diffraction studies. In the first place neutron sources are not very widely available. Moreover the experimental techniques are not as simple and the resolution of the diffraction patterns obtained is not as good as with X-rays. Before doing any structural work with neutrons therefore, all the information that can readily be obtained with X-rays should be collected. We now conclude this brief account of neutron diffraction with

[3] For this purpose *polarized neutrons*, that is neutrons with their spins all aligned in one direction, are normally employed.

two examples in which neutron studies have provided solutions to structural problems that could not be solved by X-rays alone.

A consequence of the irregular variation of scattering length with atomic number is that certain light elements scatter neutrons as effectively as some heavy elements, whereas for X-rays the atomic scattering factor increases regularly with atomic number so that the light elements consistently scatter least. It is thus often impossible to locate very light atoms in the unit-cell by X-ray diffraction because their scattering effects are swamped by the much stronger scattering produced by any heavy elements that may be present; in some cases the irregularity of neutron scattering lengths provides a means of locating such light atoms. For example in NaH X-ray studies show that the sodium atoms are in face-centred-cubic array but fail to locate the positions of the hydrogen atoms because their contribution to the total scattering is so slight. The scattering lengths of hydrogen and sodium are however similar in magnitude but opposite in sign: $-0{\cdot}38 \times 10^{-12}$ cm for H, $0{\cdot}35 \times 10^{-12}$ cm for Na. Examination of the neutron powder diffraction pattern readily enables the hydrogen atoms to be located in the unit-cell and the structure of NaH to be determined as NaCl-type. This example illustrates one of the more important uses of neutron diffraction, the location of hydrogen atoms in crystal structures. Another example, on which we shall comment at length in chapter 10, is the location of the positions of the deuterium atoms in heavy ice, D_2O, at $-50\,°C$.

Another way in which neutron diffraction data may be very useful is in distinguishing between the structural sites occupied by atoms of different elements whose atomic numbers are rather close, provided that their neutron scattering lengths are not, as may be the case, also similar. It is for example difficult to distinguish between the sites occupied by Mg and Al atoms in complex oxide structures by the use of X-ray methods because the number of extranuclear electrons in Mg^{2+} and Al^{3+} is the same; but the neutron scattering lengths for Mg $0{\cdot}54 \times 10^{-12}$ cm and for Al $0{\cdot}35 \times 10^{-12}$ cm are sufficiently different to make a clear distinction in neutron diffraction patterns between the sites occupied by these two elements. For instance neutron diffraction shows unambiguously that $MgAl_2O_4$ has the normal spinel structure (chapter 10). In a similar way neutron diffraction can be applied to the study of ordering in alloys such as CuZn (see chapters 10 and 13), the atomic numbers of Cu and Zn being respectively 29 and 30 so that X-ray diffraction cannot easily distinguish between Cu and Zn occupied sites. Contrariwise X-ray diffraction is better for the study of ordering in Cu–Au alloys since the atomic numbers of Cu and Au, respectively 29 and 79 are very far apart while their neutron scattering lengths, $0{\cdot}79 \times 10^{-12}$ cm and $0{\cdot}76 \times 10^{-12}$ cm respectively, are very close together.

Electron microscopy

We provide here only a very brief account of the theory and applications of electron microscopy, paying particular attention to applications in the field of mineralogy. For a comprehensive treatment of electron microscopy the reader is referred to Hirsch *et al* (1965) and for a good account of mineralogical applications to the chapter by McConnell in Zussman (1967).

In an electron microscope the electrons emitted from a hot filament (the electron 'gun') are accelerated through a high potential difference V, of the order of 10^5 volts, and so acquire kinetic energy eV, where e is the electron charge. The velocities of electrons so accelerated are not negligible in comparison with the velocity of light c

so that a relativistic correction has to be made. The wavelength of an electron is thus given by

$$\lambda = h\left\{2m_0 eV\left(1+\frac{eV}{2m_0c^2}\right)\right\}^{-\frac{1}{2}},$$

where m_0 is the rest mass of an electron. Thus when $V = 1 \times 10^5$ volts, λ is 0·037 Å.

Electrons are scattered by electric fields and so an electron beam may be diffracted by atoms whose size is comparable to the electron wavelength. Unlike beams of X-rays or neutrons, an electron beam can be brought to a focus by electrostatic or electromagnetic fields so that in principle electron beams can be used to form magnified images of objects in precisely the same way as light rays are commonly used; because electron beams have shorter wavelengths the resolution obtainable in an electron microscope, as little as 8 Å,[4] is very much better than that obtainable with an optical microscope, about 10^3 Å. The resolution of an electron microscope is not yet quite good enough to obtain an image of a crystal structure, but the instrument is nevertheless very valuable for the study of fine scale phenomena in crystals. It is particularly useful in that it can provide not only the image but also the diffraction pattern of the same small crystal. Its disadvantage, with respect to X-rays, is that only very thin crystals, a few hundred Ångstrom units thick in the direction parallel to the incident electron beam, can be examined and the orientation of such a crystal can only be varied by about $\pm 30°$ relative to the incident electron beam; and, further, the crystals must be stable in high vacuum and unaffected by the local heating (usually several hundred degrees Celsius) caused by the incident electron beam. Moreover, in the electron microscope inelastic scattering of the incident electron beam by the specimen may not be negligible; this has the effect of decreasing the resolution of the microscope and of producing some complications in electron diffraction patterns.

We now consider the diffraction patterns obtainable in the electron microscope and go on to discuss, in brief, the uses of electron microscopy in mineralogy.

Since electrons are scattered by potential fields their scattering by an atom involves both its nucleus and its electrons. When an electron beam of unit amplitude is incident on an atom of atomic number Z the amplitude of the scattered electrons at a distance R from the atom is

$$\frac{f_e}{R} = \frac{me^2}{2h^2R}\left(\frac{\lambda}{\sin\theta}\right)^2 (Z-f),$$

where f is the atomic scattering factor of the atom for X-rays and f_e is known as its atomic scattering amplitude for electrons. The scattered amplitude thus varies with wavelength λ and with scattering angle θ. It can be shown that for $(\sin\theta)/\lambda$ greater than about 0·4 Å^{-1} scattering amplitude increases regularly with atomic number, while for $(\sin\theta)/\lambda$ less than about 0·4 Å scattering amplitude does not vary in a regular manner. For elements of the first three periods of the periodic table there is actually a tendency for the scattering amplitude at small angles to decrease with increasing atomic number (Fig 9.2). On average one can say that at $(\sin\theta)/\lambda = 0$, f_e is proportional to $\sqrt[3]{Z}$ so that light elements are better scatterers relative to heavy elements for electrons than for X-rays for which $f \propto Z$. So it would appear that electron diffraction might provide a useful means of determining the positions of light atoms in crystal structures, but uncertainties about the way in which the measured intensities of

[4] Imperfections in the magnetic electron lenses, especially spherical aberration of the objective lens, make this practical limit of resolution significantly in excess of the theoretical limit.

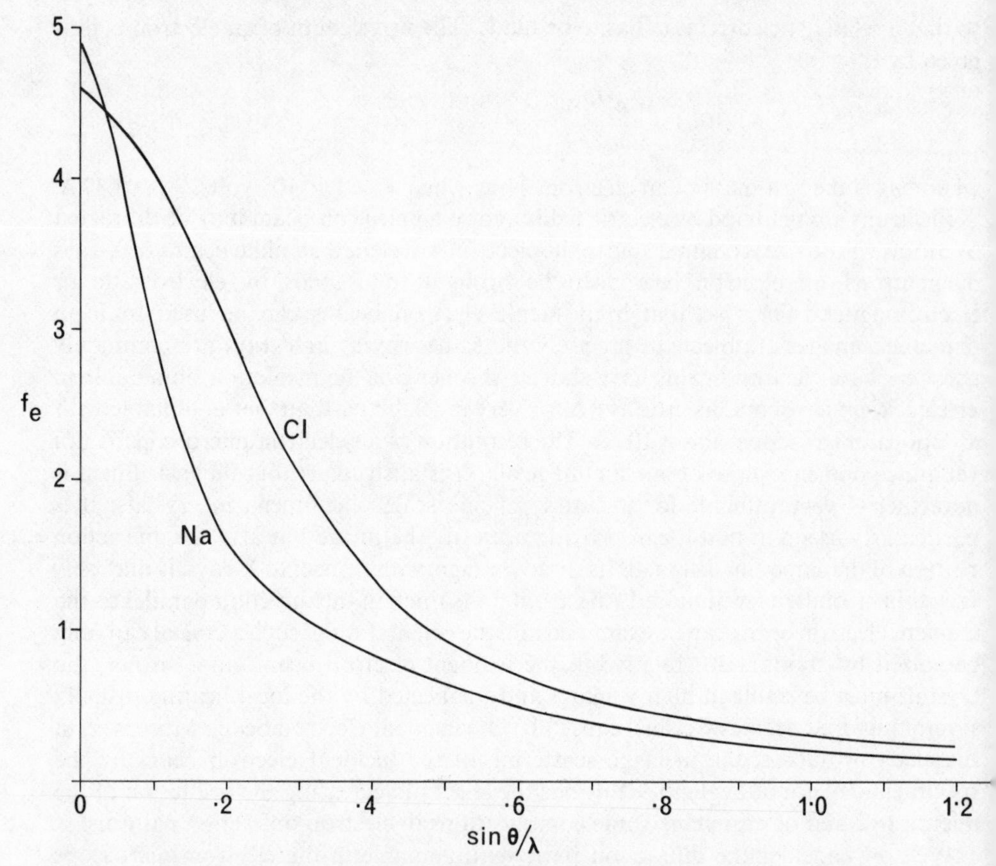

Fig 9.2 The variation of atomic scattering amplitude for electrons f_e with $(\sin\theta)/\lambda$ for Na and Cl; f_e is measured in Å. The data plotted refer to electrons at rest; for electrons of velocity v it is necessary to make a relativistic correction by multiplying by $(1-(v^2/c^2))^{-\frac{1}{2}}$.

electron reflexions are to be related to structure amplitudes limit the use of electron diffraction in structure determination.

Substitution of numerical values in the expressions for the amplitude of the electrons and of the X-rays scattered by a given atom leads to the expression $\{0{\cdot}85\lambda^2(Z-f)\,10^3\}/(\sin^2\theta)f$ for the ratio of the scattered electron amplitude to the scattered X-ray amplitude for the atom (λ is measured in Å). This ratio is of the order of 10^4 for small values of $(\sin\theta)/\lambda$ so that one can say that atoms scatter electrons very much more strongly than they scatter X-rays.

If a crystal is correctly oriented in an incident electron beam to produce a strong Bragg reflexion, the electron beam may be completely reflected after it has traversed as few as 25 lattice planes. In X-ray diffraction however the possibility of complete reflexion of the incident beam does not arise until at least 10^4 lattice planes have been traversed. Because atomic scattering amplitudes for electrons are so high the assumption made in chapter 6 that diffracted X-rays are not rescattered becomes invalid for electron diffraction; a scattered beam of electrons may be rescattered during its subsequent path through the crystal and multiple diffraction then occurs. It is this possibility of multiple diffraction which makes the relationship between observed intensity and structure amplitude uncertain. Very thin crystals may be used to reduce the uncertainty, but they are liable to bend and it is necessary to use a very narrow electron beam in order to ensure that diffraction occurs from a flat part of the crystal. Thus electron diffraction methods cannot easily be used for structure determination.

Electron diffraction is used extensively in the study of very small crystals so that it is worth while here to explore the special characteristics of electron diffraction patterns. In doing so we shall need to make use of the concepts of the reciprocal lattice and the reflecting sphere which were developed for X-ray diffraction in chapter 6. We shall suppose that the crystals under examination are very thin flakes or plates. The method of sample preparation is such that the smallest dimension of the crystal is usually normal to a lattice plane of simple indices and this direction is parallel to the incident electron beam. Since the crystals are very thin, only a few unit-cells thick, in the direction of the incident electron beam we are justified in assuming that electrons are scattered once only.

In deriving the conditions for diffraction by a crystal we assumed in chapter 6 that the number of unit-cells in the crystal was so very large that observable diffraction would only occur when waves scattered by each lattice point were in phase. In reciprocal space this condition is realized when the sphere of reflexion intersects a reciprocal lattice point. In a very thin crystal plate normal to [001] there are too few unit-cells in the z direction to produce complete destructive interference in all directions other than that for which there is a path difference of one wavelength for radiation scattered from adjacent unit-cells along the z direction. In consequence the condition imposed by the third Laue Equation is relaxed. Each reciprocal lattice point is drawn out into a spike normal to the plane of the plate, the spike extending a distance K/t in the [001] and [00$\bar{1}$] directions from each reciprocal lattice point (t = thickness of plate).

In electron diffraction not only are we concerned with very thin crystals, but with very short wavelengths: the wavelength of the electron beam will be about 0·05 Å so that, taking the reciprocal lattice constant K equal to unity, the radius of the reflecting sphere, K/λ, will be about 20 Å^{-1} and reciprocal lattice constants will be of the order of 0·1–0·2 Å^{-1}. Thus for a single orientation of the crystal the reflecting sphere is very likely to intersect several reciprocal lattice spikes and it becomes possible to record a diffraction pattern with monochromatic radiation and a stationary crystal. This point is illustrated in Fig 9.3, where an electron beam is shown incident along [001] of an orthorhombic crystal plate, the large faces of which are parallel to (001). In the resultant diffraction pattern $hk0$ reflexions lie in a circle at the centre of the photograph and $hk1$ reflexions lie in a concentric ring.

Since the wavelength of electrons is small compared with the spacing of reflecting planes in a crystal the Bragg angles of the lowest angle reflexions are of the order of 1°, very much less than for X-ray diffraction, and this has an interesting consequence. In Fig 9.4 the effective perpendicular distance from the specimen S to the film on which the electron diffraction pattern is recorded is L. Consider a reflexion at a distance R on the film from the intersection of the incident electron beam with the film. This reflexion is produced by the intersection of the reciprocal lattice spike through P with the reflecting sphere OQ. Since the Bragg angle θ is very small $\tan 2\theta = R/L$ becomes $2\theta = R/L$ and, for the same reason, OP = OQ.

But $\qquad \mathrm{OP} = d^*$

and $\qquad \mathrm{OQ} = \dfrac{2}{\lambda}.\sin\theta = \dfrac{2\theta}{\lambda}.$

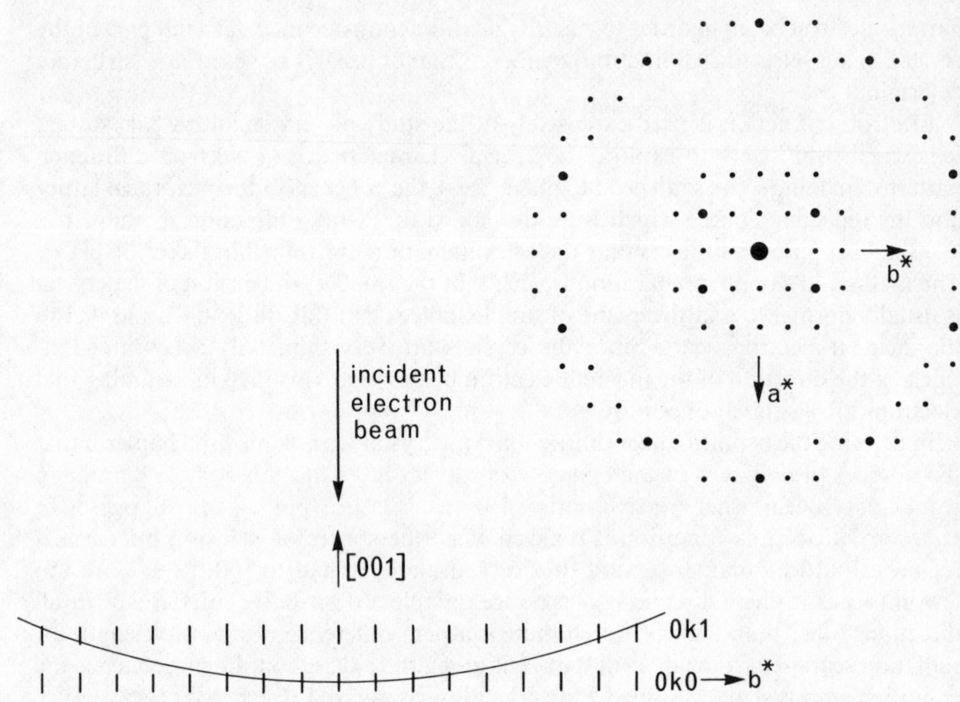

Fig 9.3 The formation of a diffraction pattern by a monochromatic electron beam incident on a stationary crystal. The figure shows the reciprocal lattice rows 0*k*0 and 0*k*1 of a very thin orthorhombic crystal whose smallest dimension is parallel to [001]. The electron beam is incident parallel to [001]. The reflecting sphere is seen to intersect three reciprocal lattice spikes on either side of the origin in the zero layer and the seventh and eighth spikes in the first layer. The diffraction pattern obtained is shown on the right: the central circular area containing *hk*0 reflexions is separated from the concentric ring of *hk*1 reflexions by an annular blank area.

Therefore $d^* = \dfrac{2\theta}{\lambda} = \dfrac{R}{L\lambda}$.

The observed diffraction pattern is thus an undistorted projection of the reciprocal lattice which gives rise to it. This property makes the interpretation of electron diffraction patterns rather easy once the effective crystal to film distance L has been determined; L is not a simple length but depends on the magnification of the electron lens system and may be evaluated from measurements of the diffraction pattern of a substance of known unit-cell dimensions.

Since the attitude of the crystal with respect to the incident electron beam can only be adjusted to a limited extent in the electron microscope the techniques of crystal orientation which we describe for X-ray diffraction in Appendix F are not available for electron diffraction and it becomes necessary to be able to index an electron diffraction pattern when the orientation of the crystal is unknown. This is quite straightforward provided the reciprocal lattice geometry of the crystal is already known.

Electron diffraction is obviously valuable for the study of fine-grained materials whose crystals are too small for study by single crystal X-ray diffraction methods. Other applications make use of electron diffraction in conjunction with *diffraction contrast*, a phenomenon that is due to the very strong diffraction of electrons by crystals. Suppose that the electron microscope is focused on the lower face of the

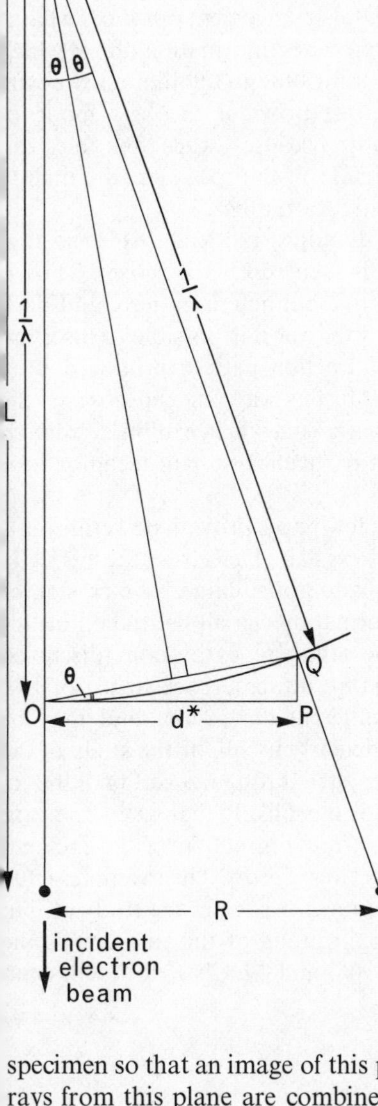

Fig 9.4 The undistorted nature of electron diffraction patterns. In the figure the Bragg angle θ is grossly exaggerated. The diagram is explained in the text.

specimen so that an image of this plane is formed. For a normal image the diffracted rays from this plane are combined to form the image (Fig 9.1). But if an aperture were placed in the lens system of the microscope in the plane D so as to allow only the zero-order diffracted beam to pass through it, no detail would be visible in the image which would appear as a uniformly illuminated area corresponding to the shape of the object. If however the object is oriented so as to produce a strong Bragg reflexion, much of the incident electron beam will be reflected and the intensity of the image will be correspondingly reduced. When the crystal contains defects of such a nature that some parts of it are in a reflecting position and others not, those parts of the image which correspond to little or no diffraction will appear bright while those which correspond to strong diffraction will appear dark. Such a *bright field image* is useful for the study of defects in the crystal specimen. Alternatively the aperture may be inserted in the lens system of the electron microscope so as to allow only the passage of a selected diffracted beam, say that corresponding to the 201 reflexion. In these circumstances the image of the crystal will appear bright only where the corresponding parts of the crystal are correctly oriented to diffract the incident electron beam into the 201 reflexion. This experimental arrangement produces what is known as a *dark field image*. The dark field image provides information about which parts of the crystal specimen are so oriented as to contribute to the diffracted beam, in our example 201. In practice diffraction contrast provides the best mode of observation in transmission electron microscopy because, when the electron microscope is simply used to obtain a magnified image of the specimen, the quality of the image is impoverished as the result of inelastic scattering.

In an electron microscope it is a simple matter to adjust the lens system so that either the image plane or the diffraction pattern is focused on the screen. Thus it becomes possible to obtain an electron micrograph, a diffraction pattern, and dark and bright field images of the same small crystal. Moreover it is possible to insert an aperture in the lens system so as to record the diffraction pattern produced by a selected area (diameter about 1μ) of the specimen. In this way one can observe the diffraction patterns produced by the various parts of a crystal which exhibit diffraction contrast in bright and dark field images. This is a particularly useful technique for the study of polyphase systems and twins.

The techniques that we have outlined in the last few pages provide powerful tools for the study, especially, of imperfections in small crystals. Defects in the perfectly regular arrangement of lattice points due to dislocations, bending of the crystal, or strains associated with the onset of a phase transformation can all be studied in this way. These techniques are also valuable for the study of exsolution (discussed thermodynamically in chapter 14), where there is an intimate coexistence of two phases, the lattices of which will be of slightly different dimensions and may be slightly differently oriented. They may also be particularly useful for the study of the fine scale twinning that results from cooling a crystal through a certain kind of polymorphic transformation; in such a specimen it is most likely that when one twin component is in the correct orientation for very strong diffraction of the incident electron beam the other twin component will diffract very feebly. The examples cited illustrate only a few of the many uses of the electron microscope in the study of thin crystals. For a comprehensive account of the applications of the instrument the reader is referred to the works of Hirsch *et al.* (1965) and McConnell in Zussman (1967).

10
Crystal chemistry

The diffraction methods explored in the preceding three chapters and especially the diffraction of X-radiation by single crystals provide the means by which the arrangement of atoms in a crystalline solid is determined. From such knowledge of atomic positions interatomic distances, i.e. bond lengths, and bond angles can simply be derived and may then provide information about the nature of the cohesive forces operative in the structure. Identification of the types of cohesive force operating in a particular structure then enable the physical and chemical properties of the solid substance to be understood.

There is, at least in theory, an alternative approach to understanding the physical and chemical properties of a crystalline solid substance; that is the calculation of the energy of the crystal structure for the composition concerned. But in practice even the simplest crystalline solids are too complex for precise quantum mechanical calculations to be possible. The concepts of quantum mechanics that have proved so fruitful for the understanding of atoms and simple molecules can yield no more than a qualitative interpretation of the nature of the cohesive forces operative in crystalline solids; but although only qualitative the quantum mechanical approach is nevertheless informative.

At the present stage of development of solid state chemistry the most fruitful way of approaching a crystal structure is to combine quantum mechanical concepts, necessarily qualitative, with the determined crystal structure and knowledge of observed physical and chemical properties; such an approach usually enables the factors determining the stability field of the crystalline solid and its reactivity to be identified. Identification of the atomic properties that determine the limits of stability and the reaction mechanisms of a crystalline solid are of prime concern to the solid state chemist and the mineralogist.

In this chapter we shall do no more than scratch the surface of solid state chemistry. We shall confine our discussion to simple structures, considering the range of compounds that adopt particular structures, investigating the relationship between structure type and type of cohesive force, and attempting to establish why a particular compound assumes the structure it does in the crystalline state. We shall assume that the reader has the depth of knowledge of the electronic structure of atoms and of valence theory that can be obtained by reading any of the many excellent introductory textbooks of inorganic or theoretical chemistry.

Description of crystal structures

In strictly formal terms a crystal structure is described completely by a statement of the dimensions of the unit-cell and the coordinates and nature of every atom in the unit-cell. For crystallographic purposes it is important to supplement this information by a statement of the space group and the types of position, whether special or general, occupied by atoms; the space group symbol of course includes information about the crystal system, the point group, and the lattice type. But for the crystal chemist it is more immediately useful to have some simple statement, whenever possible, about the way in which atoms are associated in the structure; such a statement involves the concepts of *close-packing* and *coordination polyhedra*, which we now develop.

It is convenient to begin our discussion of the description of crystal structures by considering the ways in which identical spherical atoms can be packed together so as to occupy the minimum possible volume, because so many actual structures can be described in terms of such *close-packing* of some or all of their constituent atoms. In taking this approach we are deliberately ignoring cohesive forces and treating the constituent atoms of the structure as inert rigid spheres. In the next few paragraphs we shall use the words *sphere* and *atom* without distinction.

There is only one way in which spheres can be arranged on a plane so as to minimize the area occupied, and that is illustrated in Fig 10.1(a). In such a *close-packed plane* of spheres each sphere has six spheres in contact with it; the arrangement is based on a hexagonal plane lattice with one atom associated with each lattice point. A three-dimensional close-packed structure is obtained when close-packed planes are stacked in such a manner that the total volume occupied is minimized. This is achieved when one close-packed plane is superimposed on another close-packed plane so that three spheres in one plane are in contact with the same sphere in the other plane (Fig 10.1(b)). It is immediately apparent from the figure that there are two ways in which the second plane can be superimposed on the first to satisfy this criterion.

It is convenient now to distinguish between close-packed planes with regard to the positions of their atoms in relation to the unit-mesh outlined in Fig 10.1(c), which has its x and y axes parallel to close-packed lines of atoms. A plane which has atoms with coordinates $x = 0$, $y = 0$, is designated a close-packed plane of type A. Planes with atoms at $\frac{2}{3}, \frac{1}{3}$ or $\frac{1}{3}, \frac{2}{3}$ with respect to the unit-mesh are designated respectively B and C. In Fig 10.1(b) the lower plane is taken to be an A plane and the upper plane a B plane. Such an AB sequence of close-packed planes is equivalent to an AC sequence, the two sequences being interconvertible merely by rotation of the reference axes, x and y, through 60°. But the distinction between B and C planes becomes effective when we consider the superimposition of a third plane. Three-dimensional close-packing can be achieved by superimposing on an AB pair either an A plane to give an ABA sequence or a C plane to give an ABC sequence; in either case the atoms in the third plane lie immediately above 'holes' in the plane immediately below it, a B plane. In the ABC sequence the atoms of the third plane lie above 'holes' common to the two lower planes, whereas in the ABA sequence the atoms of the third plane lie directly above those of the first plane.

Close-packed structures in which every sphere, or atom, is equivalent can be derived from the two infinite sequences ABABAB ... and ABCABC ... Many other sequences of close-packed planes, such as ȦBACȦBAC ..., ȦBABCȦ ..., are possible; but ABAB ... and ABCABC ... are the two simplest and moreover the only two in

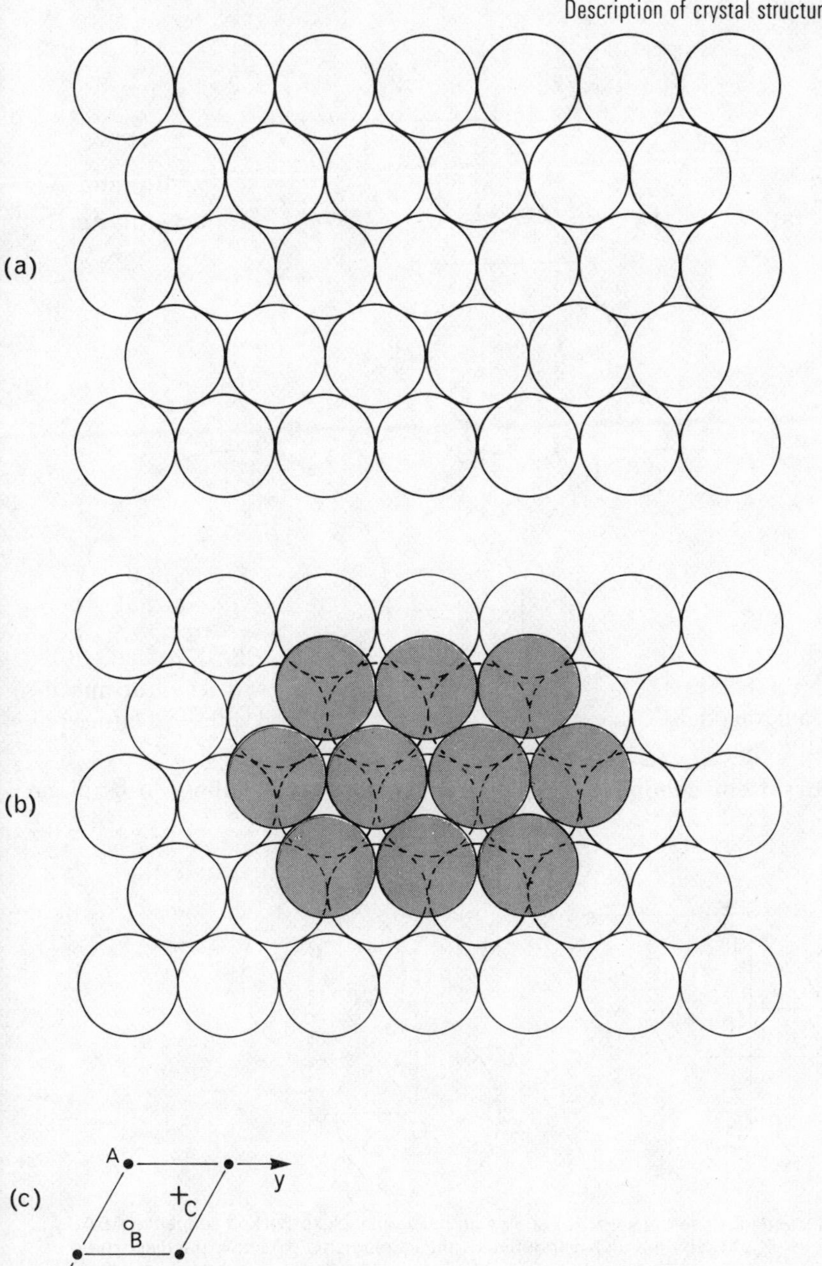

Fig 10.1 Close-packing of spheres: (a) is the plan of a close-packed plane of spheres and (b) shows the superposition of one close-packed plane of spheres (shaded) on another (unshaded). The nomenclature of close-packed planes is indicated in (c), from which it is apparent that the lower and upper planes shown in (b) are respectively in the A and B orientations.

which each atom is related to every other by either the lattice translation or a symmetry operator.

The sequence of close-packed planes ABABAB..., which repeats every two layers, is known as *hexagonal close-packing* (Fig 10.2(a)); its space group is $P6_3/mmc$ and its

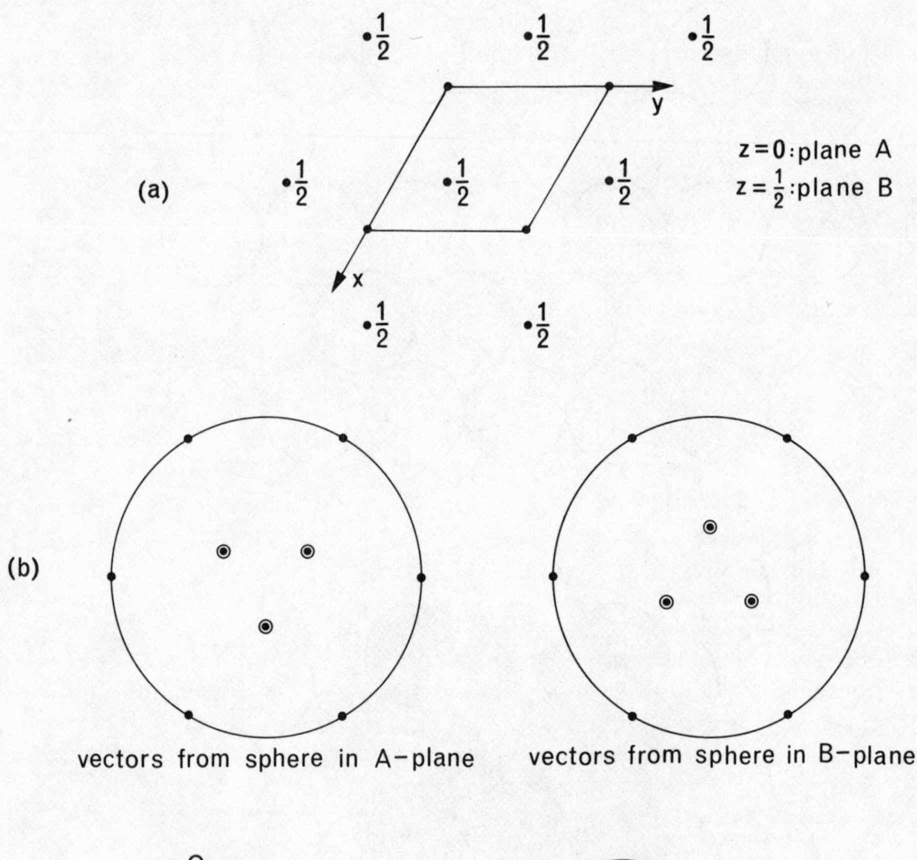

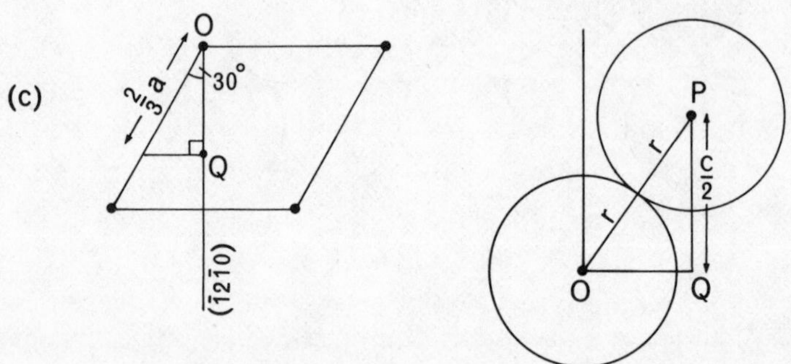

Fig 10.2 Hexagonal close-packing: (a) shows an hexagonal close-packed sequence ABAB . . . in projection on (0001) with the unit-cell outlined; the stereograms (b) show the disposition of nearest neighbours for a sphere in an A plane and a sphere in a B plane; and the sections (c) serve to demonstrate that $c/a = \sqrt{\frac{8}{3}}$.

point group 6/*mmm*. The close-packed planes are parallel to (0001) of the hexagonal unit-cell. Every sphere is in contact with twelve others, six in its own plane and three in each of the planes immediately above and below. Stereograms of the directions of the vectors joining any sphere to each of its twelve neighbours are shown in Fig 10.2(b); the disposition of the nearest neighbours to spheres in A and B planes differs only in orientation to the arbitrarily selected reference axes, *x* and *y*. In hexagonal close-packing the dimensions, *a* and *c*, of the hexagonal unit-cell are simply related to the radius, *r*, of the close-packed spheres. The *x* and *y* axes lie along close-packed lines so that $a = 2r$. A section (Fig 10.2(c)) through the origin O parallel to $(\bar{1}2\bar{1}0)$ shows the atom centred on P in contact with that centred on O so that $OP = 2r = a$. And by Pythagoras' Theorem $OP^2 = OQ^2 + QP^2$, where $OQ = \frac{2}{3}a\cos 30° = a/\sqrt{3}$ and $QP = \frac{1}{2}c$. Therefore $a^2 = \frac{1}{3}a^2 + \frac{1}{4}c^2$, i.e. $c/a = \sqrt{\frac{8}{3}} = 1{\cdot}6331$. A crystal in which the atoms are arranged precisely in hexagonal close-packed array thus has an hexagonal unit-cell, space group $P6_3/mmc$, and axial ratio $c/a = 1{\cdot}6331$.

The sequence of close-packed planes ABCABC..., which repeats every three layers (Fig 10.3(a)), is at first sight trigonal with a rhombohedral lattice. However this sequence gives rise to close-packed planes in other orientations so that there is additional symmetry and a special relationship between the dimensions a and c of the triple hexagonal unit-cell. This arrangement is based on a cubic F-lattice and has space group $Fm3m$, point group $m3m$; one atom is associated with each lattice point. The close-packed planes shown in Fig 10.3(a) are parallel to the (111) plane of the cubic unit-cell (Fig 10.3(b)) and they are repeated by its symmetry elements so that there are in all four orientations of close-packed planes, parallel to $\{111\}$. This sequence is known as *cubic close-packing*. Each atom is related to every other atom by a lattice translation of the cubic F-lattice and, as in hexagonal close-packing, has twelve nearest neighbours. But the vectors from an atom to its twelve nearest neighbours are differently disposed in cubic (Fig 10.3(c)) and hexagonal (Fig 10.2(b)) close-packing: in both there are six neighbours in the same close-packed plane, but the three in each of the adjacent planes are not directly superimposed in cubic as they are in hexagonal close-packing. It is evident from Fig 10.3(b) that atoms are in contact along the face diagonals of the cubic unit-cell, that is along $\langle 110 \rangle$ directions; the close-packed directions in the (111) plane are therefore $[1\bar{1}0]$, $[0\bar{1}1]$, $[\bar{1}01]$, and their opposites. This point is illustrated in the stereogram Fig 10.3(d) showing vectors to nearest neighbours; Fig 10.3(d) is drawn in the conventional orientation for cubic stereograms with z at the centre whereas Fig 10.3(c) is drawn with $[111]$ at the centre for ease of comparison with Fig 10.3(a). Since atoms are in contact along $\langle 110 \rangle$ directions the cubic unit-cell edge a is related to the atomic radius r by $a = 2\sqrt{2}r$. The four atoms in the cubic unit-cell occupy a volume $\frac{16}{3}\pi r^3 = \pi a^3/(3\sqrt{2})$. The percentage of the volume a^3 of the cubic unit-cell occupied by atoms is thus $100\pi/(3\sqrt{2}) = 74{\cdot}05$ per cent. In any sequence of close-packed planes, whether cubic, hexagonal, random, or any other, the percentage volume of space occupied by the close-packed spheres is of course the same. That is to say any close-packed array fills space as efficiently as any other: but only the sequences known as hexagonal and cubic close-packing have all spheres necessarily equivalent. Cubic and hexagonal close-packing and sequences that can be described as mixtures of these two are the only close-packed sequences commonly found in actual crystal structures.

In the description of crystal structures it is often possible to make the simplifying assumption (the validity of which we shall explore later in this chapter) that the atoms are all spheres in contact with one another. This is the assumption we have tacitly made in our discussion of close-packing where we have regarded the words atom and sphere as interchangeable. The strongest interactions between atoms in a crystalline solid will naturally be those between adjacent atoms so the nature of the distribution of atoms around one another will be one of the most significant features of the crystal structure. The term *coordination* is used to describe the number and arrangement of the near neighbours of an atom in a crystal structure, the number of near neighbours being known as the *coordination number*. In close-packed structures, as we have seen, each atom has twelve near neighbours: each, identical, atom can

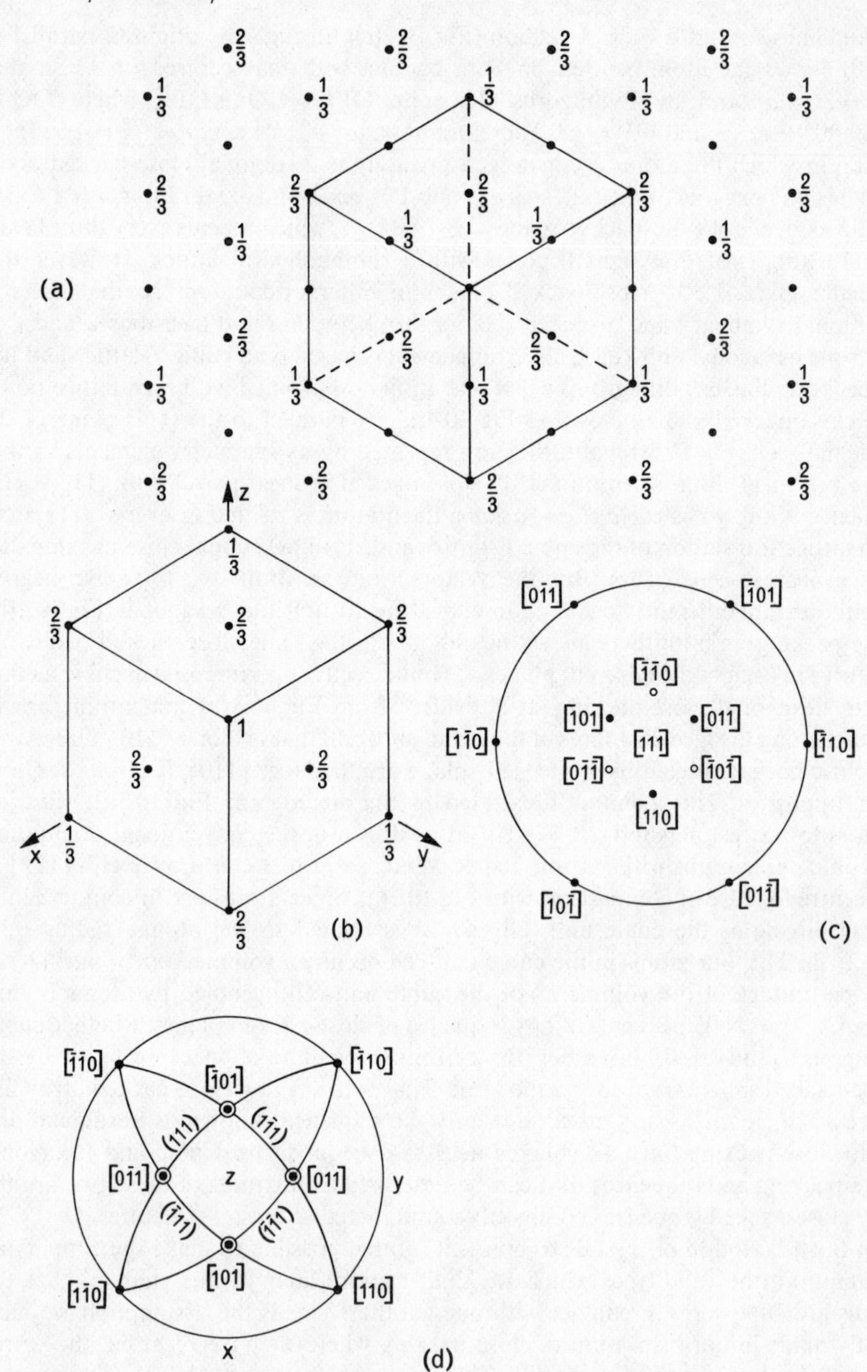

Fig 10.3 Cubic close-packing. The sequence of close-packed planes ABCABC . . . is shown in (a) in projection on a close-packed plane, the planes A, B, C being at heights $0, \frac{1}{3}, \frac{2}{3}$. The cubic F-cell is outlined in (a) and shown with lattice points on its underneath faces omitted in (b). The disposition of the vectors from an atom to its nearest neighbours is shown projected on a close-packed plane in the stereogram (c), where each vector is indexed on the cubic lattice. Close-packed directions and planes are shown in (d), which is oriented, unlike (a), (b) and (c) in the conventional orientation for cubic stereograms.

thus be said to have a coordination number twelve or to be twelvefold coordinated.

Its coordination number does not however completely describe the coordination of an atom. We have already seen in hexagonal and cubic close-packing that, although the coordination number is twelve in both, the arrangement of near neighbours is different in these two types of close-packing. For the close-packed structures we illustrated the spatial disposition of near neighbours by drawing stereograms of interatomic vectors from the central atom to its near neighbours (Figs 10.2(b), 10.3(c)); but although this is always a useful mode of representation, we can often do better by making use of the concept of the *coordination polyhedron*. The atom whose coordination is under discussion is taken to lie at the centre of a polyhedron and its near neighbours at the coigns of the polyhedron; by stating the nature of the polyhedron, the coordination number and the spatial disposition of interatomic vectors are completely specified. For example in NaCl (Fig 10.4) each sodium atom is surrounded by six chlorine atoms, each chlorine atom being situated at a distance $\frac{1}{2}a$ in a direction $\langle 100 \rangle$ from the sodium atom. The group of six chlorine atoms thus lie at the coigns of a regular octahedron centred on the sodium atom. To describe the coordination of sodium by chlorine in this structure as *octahedral* at once gives the coordination number as six and the disposition of the interatomic vectors between sodium and chlorine as mutually perpendicular. In this structure chlorine is likewise octahedrally coordinated by sodium.

Sodium chloride has cubic symmetry and in consequence the coordination octahedron is constrained to be a regular octahedron. But in many inorganic and mineral structures no such symmetry constraint operates so that coordination polyhedra may be distorted from regularity. The concept of the coordination polyhedron is only valuable when the shape of the polyhedron approximates to that of one of the regular solids. Fortunately this is quite commonly so. An example of extreme distortion is provided by the coordination of potassium by oxygen in the mineral sanidine $KAlSi_3O_8$: nine oxygen atoms lie at distances between 2·7 and 3·1 Å from the potassium atom and the potassium–oxygen vectors make various angles with one another. In a case such as this there is no advantage in visualizing a coordination polyhedron; it is better to illustrate the coordination of potassium by means of a stereogram showing the directions of the potassium–oxygen vectors, each pole being labelled with the length of the vector it represents.

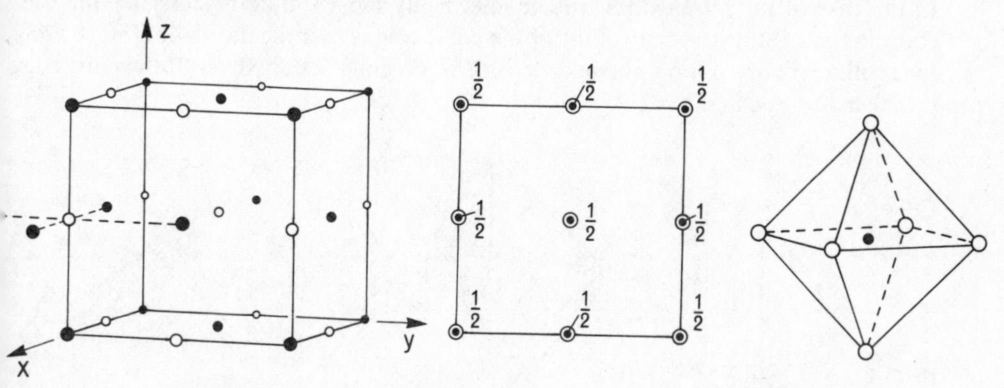

Fig 10.4 The NaCl structure: on the left is a perspective drawing of the structure, in the centre a plan of the structure on (001), and on the right a perspective drawing of the coordination octahedron of Cl^- about Na^+. Na^+: solid circles. Cl^-: open circles. The disposition of Na^+ ions about a Cl^- ion is indicated on the left-hand diagram.

We now proceed to investigate the relationship between the size of an atom and the size of the atoms coordinating it for each of the three most symmetrical coordination polyhedra, the regular cube, octahedron, and tetrahedron. We continue to assume that all atoms are rigid, non-interacting spheres and further assume that all the atoms coordinating the central atom are of the same kind. Let the radius of the central atom be r_A and the radius of a coordinating atom be r_X. If the structure is to be stable, all the coordinating atoms must be in contact with the central atom so that the length of any A–X interatomic vector will be $d = r_A + r_X$. Moreover coordinating atoms cannot overlap so that the X–X interatomic vectors cannot be less than $2r_X$, that is to say any edge of the coordination polyhedron must be $\geqslant 2r_X$ in length. The geometry of the coordination polyhedron is determined by the relationship between the distance from its centre to a coign, $d = r_A + r_X$, and the limiting length of any of its edges $t = 2r_X$. We consider first the *cube* (Fig 10.5(a)), where d is half the length of a body diagonal so that

$$d = \frac{\sqrt{3}}{2}t.$$

Therefore $r_A + r_X = \frac{\sqrt{3}}{2} 2r_X$

and hence $\frac{r_A}{r_X} = \sqrt{3} - 1 = 0{\cdot}732.$

In the *octahedron* (Fig 10.5(b)) a central section passing through four coordinating atoms is a square so that

$$d = \frac{\sqrt{2}}{2}t.$$

Hence $r_A + r_X = \frac{\sqrt{2}}{2} 2r_X$

and $\frac{r_A}{r_X} = \sqrt{2} - 1 = 0{\cdot}414.$

In the case of the *tetrahedron* it is geometrically most simple to consider the four coordinating atoms to occupy four of the eight coigns of a regular cube (Fig 10.5(c)); the resultant coordination polyhedron is then a regular tetrahedron. If the cube edge is a, then $d = \frac{\sqrt{3}}{2}a$ and $t = \sqrt{2}a$.

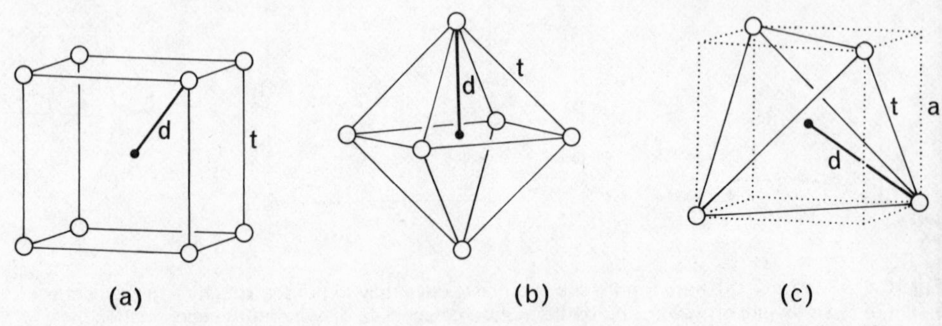

Fig 10.5 Coordination polyhedra, the cube [8], the octahedron [6], and the tetrahedron [4]. Interatomic vectors A–X and X–X are respectively *d* and *t*.

Hence $$d = \frac{\sqrt{3}}{2\sqrt{2}} t,$$

$$r_A + r_X = \frac{\sqrt{3}}{2\sqrt{2}} 2r_X$$

and $$\frac{r_A}{r_X} = \sqrt{\left(\frac{3}{2}\right)} - 1 = 0{\cdot}225.$$

In each case if the radius ratio r_A/r_X exceeds the limiting value the coordinating atoms will not be in contact along an edge of the coordination polyhedron but will be held apart because of the excessive size of the central atom. In the limiting case there is contact between the central atom and the coordinating atoms as well as between adjacent coordinating atoms. If however the radius ratio r_A/r_X falls short of the limiting value for the polyhedron concerned, adjacent coordinating atoms will be in contact with one another but not with the central atom; in other words the space available for the central atom is too large.

We return now to close-packed structures to consider in terms of coordination polyhedra the interstices between the atoms in close-packed array. Interstices are of two types, octahedral and tetrahedral. Consider for example the two close-packed planes in orientations A and B shown in Fig 10.6(a). Octahedral interstices with their centres mid-way between the two close-packed planes have x and y coordinates corresponding to those of atoms in a close-packed plane in orientation C. Tetrahedral interstices lie in two sets of positions: one, at a height of three-quarters of the separation of the close-packed planes, is in the A orientation and the other, at a height

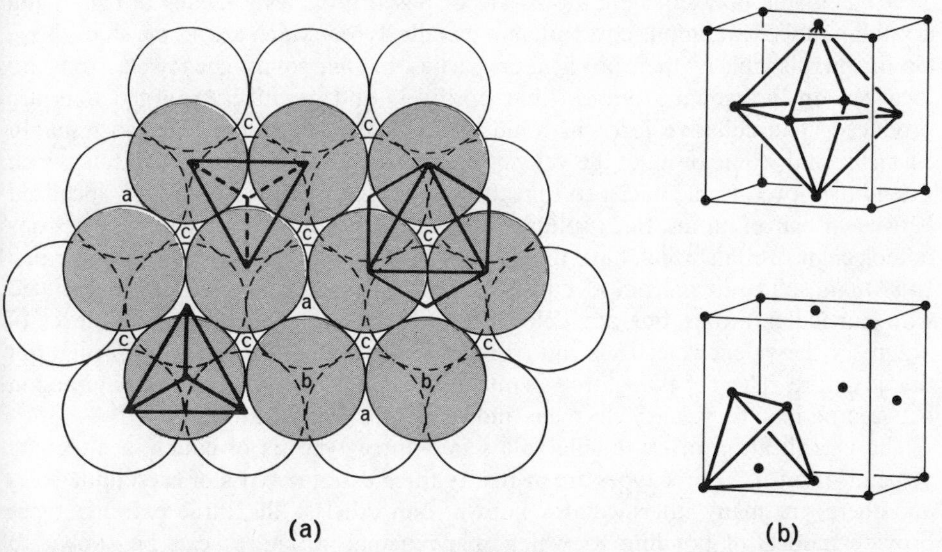

Fig 10.6 Interstices between atoms in close-packed array. (a) shows a close-packed plane of type B (shaded circles) superimposed on a close-packed plane of type A (open circles); coordination polyhedra are outlined for one of the octahedral interstices c at $z = \frac{1}{2}$ (where z is a fraction of the separation of the A and B planes) and for one of each of the two types of tetrahedral interstices, type a at $z = \frac{3}{4}$ (downward pointing tetrahedron) and type b at $z = \frac{1}{4}$ (upward pointing tetrahedron). (b) shows the coordination polyhedra in cubic close-packing for the octahedral interstice at $\frac{1}{2}, \frac{1}{2}, \frac{1}{2}$ and for the tetrahedral interstice at $\frac{3}{4}, \frac{1}{4}, \frac{1}{4}$ outlined within the cubic unit-cell.

of one-quarter of the interplanar separation, is in the B orientation. In the unit-cell of the hexagonal close-packed arrangement (Fig 10.2(a)) the octahedral interstices thus have coordinates $\frac{1}{3}, \frac{2}{3}, \frac{1}{4}$ and $\frac{1}{3}, \frac{2}{3}, \frac{3}{4}$, while the tetrahedral interstices have coordinates $0, 0, \frac{3}{8}$; $0, 0, \frac{5}{8}$; $\frac{2}{3}, \frac{1}{3}, \frac{1}{8}$; and $\frac{2}{3}, \frac{1}{3}, \frac{7}{8}$. In the whole structure there are one octahedral and two tetrahedral interstices per atom.

For cubic close-packing we refer the coordinates of interstices to the cubic unit-cell (Fig 10.6(b)). Octahedral interstices are centred on $\frac{1}{2}, \frac{1}{2}, \frac{1}{2}$ and positions related thereto by F-lattice translations, i.e. $0, 0, \frac{1}{2}$; $0, \frac{1}{2}, 0$; and $\frac{1}{2}, 0, 0$. Tetrahedral interstices are sited at $\pm\frac{1}{4}, \frac{1}{4}, \frac{1}{4}$ and positions related thereto by F-lattice translations, i.e. $\pm\frac{3}{4}, \frac{3}{4}, \frac{1}{4}$; $\pm\frac{3}{4}, \frac{1}{4}, \frac{3}{4}$; $\pm\frac{1}{4}, \frac{3}{4}, \frac{3}{4}$. Again the frequency of interstices per atom is one octahedral and two tetrahedral.

In many essentially ionic structures, both inorganic and mineral, the anions are in approximately close-packed array and the smaller cations occupy octahedral or tetrahedral interstices within the array. Usually the cations are too large to fit into interstitial sites without distorting the anion array from strict close-packing, but to describe such structures in terms of approximate close-packing of anions with cations on interstitial sites is generally worth while. For instance the sodium chloride structure may be regarded as a cubic close-packed array of chloride anions with the smaller sodium cations occupying all octahedral interstices.

Cohesive forces in crystals

Electrons in the complete electronic shells of an atom are relatively little affected by the presence of neighbouring atoms so that in any simple study of bond formation only the outer, or valency, electrons need be considered. The way in which the valency electrons of the bond-forming atoms are disposed provides a means of classifying crystalline solids into ionic, covalent, and metallic types, which are, as we shall show, also distinguishable by their physical properties. In *ionic* solids valency electrons are localized on individual atoms so that positively and negatively charged ions are produced. The cohesive force in ionic solids is then essentially electrostatic. In *covalent* solids some or all of the valency electrons occupy molecular orbitals which extend over two atomic nuclei so that the valency electrons are effectively localized between a pair of atoms. Such solids are held together by the presence of electrons in molecular orbitals which have the effect of binding atoms to one another in pairs. In *metallic* solids the valency electrons of the constituent atoms are not associated with particular atoms but are delocalized. At this first approximation level of argument the valency electrons can be regarded as able to move freely throughout the structure. The cohesive force in metallic crystals arises from the interaction between peripatetic valency electrons and positively charged atoms.

The classification of crystalline solids into three types is of course a sweeping generalization. The three types are in reality three extreme types of crystalline solid and there are many intermediates known. Nevertheless the three extreme types provide models of bonding to which many actual structures can be shown to approximate.

Before exploring in greater detail the cohesive forces operative in crystalline solids that can be classified as ionic, covalent, or metallic it is necessary to consider a mode of interatomic interaction that is present in all crystalline solids but dominant only in *molecular* crystals; this type of interaction is that produced by the ubiquitous *van der Waals* forces.

Van der Waals interaction

From the classification of crystalline solids outlined above one type has been omitted. This is the class of so-called *molecular crystals*, which are composed of atoms or discrete groups of atoms in which valency requirements are fully satisfied within the atom or group. Examples of molecular crystals are the noble gases, which can be crystallized at low temperatures and have their electrons only in completely filled electronic shells, and such substances as iodine and benzene which exist as free molecules outside the solid state. The great majority of substances that form molecular crystals are organic and not of primary concern to us here; but the relatively weak intermolecular forces that determine the configuration of such crystals are operative also in all other types of crystalline solid and cannot be ignored. These forces are known as *van der Waals* or *London* forces.

Van der Waals forces operate in all states of matter and are perhaps most familiar in the gaseous state where they are in part responsible for deviations of real gases from the ideal gas equation. In the van der Waals equation of state for a gas $[P+(a/V^2)](V-b) = RT$ the term a/V^2 is attributable to van der Waals forces. These forces are diverse in origin: in classical terms the principal contribution to them, where non-polar atoms or molecules are concerned, arises from the interaction of non-permanent dipoles. In, for example, the noble gas neon the K and L shells are filled and the electron density distribution is spherically symmetrical so that the atom is non-polar. But this statement is not strictly true because electron density distribution is a time-averaged property: at any given instant the electrons may not be symmetrically distributed about the nucleus so that the atom is at that instant dipolar. The dipole moment will vary continuously in magnitude and direction, its average over a period of time being zero. The instantaneous dipoles on each neon atom will however interact and it can be shown that the resultant interaction energy between a pair of atoms separated by a distance d is proportional to d^{-6}. In addition such atoms will also have instantaneous quadrupole moments giving rise to dipole-quadrupole and quadrupole-quadrupole interactions whose energies are proportional to d^{-8} and d^{-10} respectively. None of these instantaneous interactions gives rise to any directional effect when averaged over a period of time so the maximum interaction energy will be achieved simply when the atoms are packed as closely together as possible.

The atoms do not however approach one another infinitely closely because repulsive forces come into play when the electron clouds of adjacent atoms begin to overlap. Quantum mechanical arguments suggest that the energy of interaction due to such repulsive forces, the *overlap energy*, is proportional to $e^{-d/\beta}$, where β is a constant. However for the purposes of simple calculations it is adequate and usual to take the repulsive forces as proportional to d^{-n}, where $8 \leqslant n \leqslant 10$; the overlap energy is then proportional to d^{-n+1} with limits d^{-7} and d^{-9}. Figure 10.7 shows the variation of van der Waals energy and overlap energy with interatomic distance for a pair of atoms. It is apparent from the figure that the overlap energy increases very sharply when the interatomic distance falls below its equilibrium value. This of course means that a pair of atoms at their equilibrium separation will be strongly resistant to compression; they can be regarded effectively as rigid spheres so that in structures composed of like non-polar atoms, such as crystalline neon, the atoms can be considered as rigid spheres of radius equal to half the observed interatomic distance. Such atomic radii have no physical significance outside the crystalline state, where they represent half the equilibrium separation of atoms; the atom itself has no

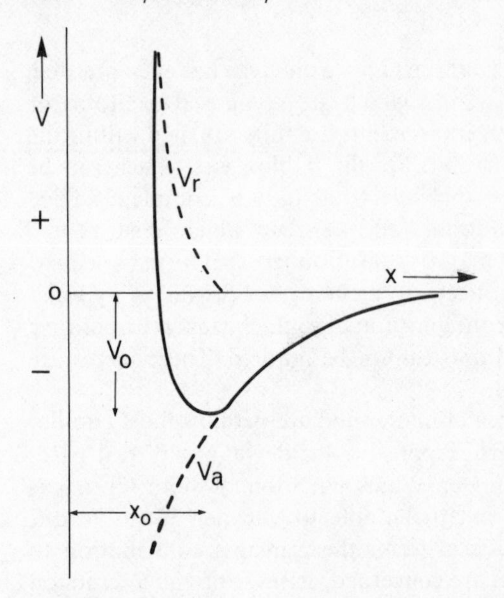

Fig 10.7 The potential energy, V (solid curve), of a pair of atoms plotted as a function of their separation, x. The potential energy due to long-range attractive forces (V_a) and to short-range repulsive forces (V_r) is plotted as broken curves. The equilibrium separation x_0 occurs when the resultant potential energy is a minimum (V_0).

definable boundary, its electron density falling off gradually to zero only at infinity. It is only when atoms are juxtaposed in equilibrium in a crystal structure that it is reasonable to assign them radii, the magnitudes of the radii being such that their sum is equal to the observed interatomic separation in the structure. To the crystallographer radius is thus an important atomic property.

Since van der Waals forces are non-directional, structures composed of atoms whose interactions are limited to van der Waals and overlap interactions will have their atoms packed together as closely as possible, that is to say in crystallographic terms, they will be close-packed structures. This is so for all the noble gas elements whose crystal structures have been determined. And in molecular crystals generally only van der Waals and overlap forces are operative so that the molecules will be packed together as closely as possible having regard to molecular shape. For instance in iodine the diatomic molecules all have their lengths parallel to a plane and are arranged in a characteristic herring-bone pattern (Fig 10.8); each molecule in a particular plane is in contact with four molecules in its own plane, four molecules in the adjacent plane above and four in the adjacent plane below so that the molecules are packed together as tightly as possible. The contrast between the intra-molecular I–I separation of 2·68 Å and the closest distance of approach of iodine atoms in neighbouring molecules, 3·54 Å, provides some indication of the relative magnitudes of covalent and van der Waals forces.

The three ideal types of bonding

We return now to consideration of the cohesive forces that are mainly responsible for the structures of inorganic and mineral crystals. We consider three idealized models in turn: the metallic, covalent, and ionic bonds. We shall in the ensuing paragraphs discuss the structures to which each model would be expected to give rise and later on examine a variety of simple crystal structures in order to explore the validity of the classification of bonds into these three types.

The simplest description of the model *metallic bond* is provided by the statement that in a metallic bonded crystal structure the valency electrons have complete

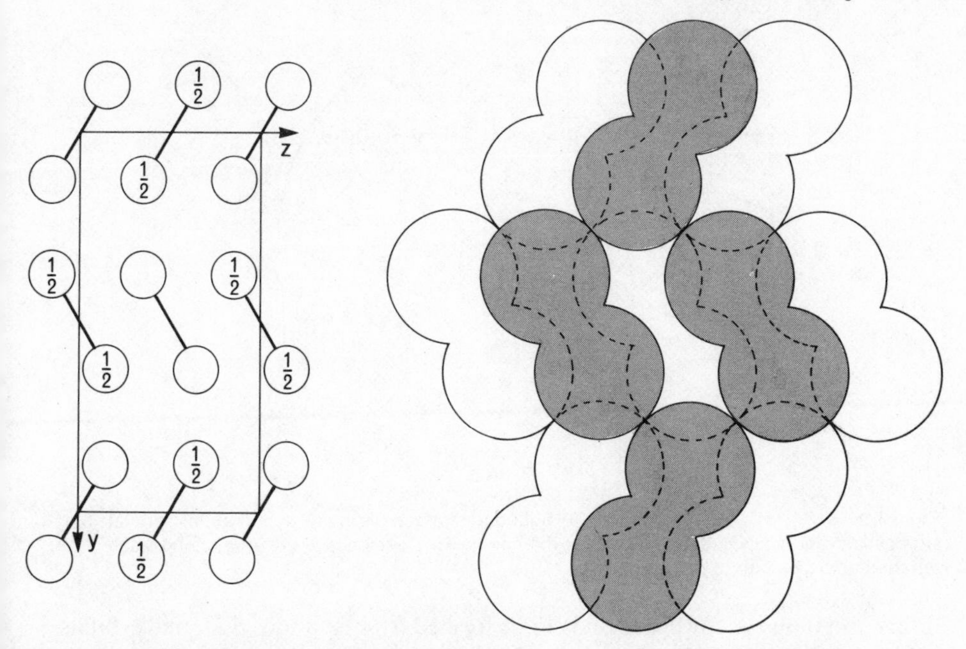

Fig 10.8 The crystal structure of iodine. The left-hand diagram is a plan on (100) of the unit-cell of I_2. The right-hand diagram shows the I_2 molecules at $x = \frac{1}{2}$ (shaded) superimposed on the sheet of I_2 molecules at $x = 0$ (unshaded). The unit-cell is an orthorhombic B-cell, an unconventional orientation.

mobility within the structure. This model has been developed to explain why metals crystallize in certain structures and to account for some of their physical properties, such as the characteristic mechanical behaviour of metals.

In this simple model metal atoms have inert gas configurations and are positively charged ions so that the cohesive forces will arise from interaction between the metal cations and the free roving electrons. Attractive forces arising in this way must be non-directional and give rise to maximum interaction energy when the metal occupies the minimum possible volume. This minimum is determined by equilibrium between the attractive forces and repulsive forces due to interaction between like charges and to overlap of the electron clouds of adjacent atoms. At equilibrium the metal structure can thus be regarded as a close-packed array of spherical atoms, the atomic radius being half the distance of closest approach of atoms in the structure.

The simple model of the metallic bond thus leads to the conclusion that metals should have close-packed structures. To establish whether a particular metallic element crystallizes in the hexagonal or cubic close-packed structure or with some more elaborate sequence of close-packed planes requires a more sophisticated model that is able to take account of interactions of at least second nearest neighbour atoms. Intuitively one might suppose that many metals will have the simple cubic or hexagonal close-packed structures and, as we shall see later in this chapter, this is found to be so.

In the ideal model of the *covalent bond* valency electrons occupy molecular orbitals so as to be effectively localized between atoms. In many structures the molecular orbitals extend over only two atomic nuclei and may be formed by linear combination of one atomic orbital of each of the two atoms. The strength of the resultant covalent bond will depend on the extent of overlap of the two atomic orbitals, so that it is

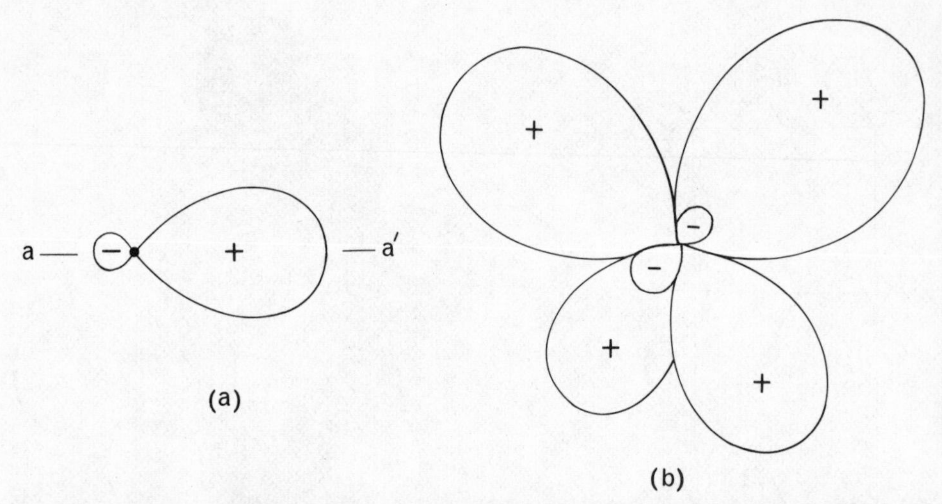

Fig 10.9 sp^3 hybrid atomic orbitals. (a) boundary surface of an sp^3 hybrid atomic orbital; the surface is cylindrically symmetrical about the line aa′. (b) perspective drawing to show the tetrahedral disposition of sp^3 orbitals.

rather common for covalent bonds to be formed from hybridized atomic orbitals.

The number of covalent bonds that an atom can form depends on its electronic configuration and the direction of the bonds is determined by which atomic orbitals are combined to form molecular orbitals. The covalent bonds that an atom can form are thus limited in number and restricted in direction. By way of example we take carbon, which has four valency electrons and can therefore form four covalent bonds. Maximum overlap of atomic orbitals is achieved when the four valency electrons are in sp^3 hybrid orbitals, which have their directions of maximum electron density disposed towards the corners of a regular tetrahedron centred on the carbon nucleus (Fig 10.9). When molecular orbitals are formed from these sp^3 atomic orbitals the carbon atom will be tetrahedrally coordinated; this is the situation in the covalent bonded crystalline form of carbon, the mineral *diamond*. In diamond the carbon atoms are disposed on a cubic F-lattice with two atoms associated with each lattice point (Fig 10.10). The coordination tetrahedron about each atom is oriented so that its

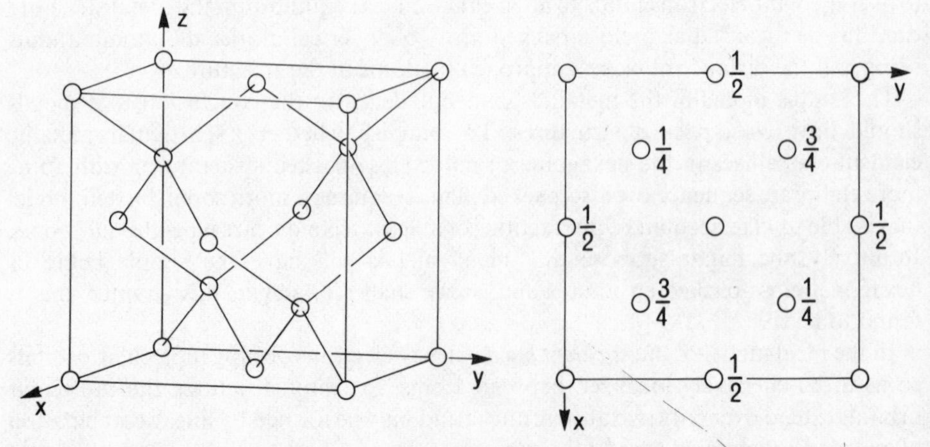

Fig 10.10 The crystal structure of diamond. On the left is a clinographic drawing of one unit-cell with interatomic vectors drawn in. On the right is a plan of one unit-cell on (001).

faces are parallel to {111} planes. The point group is *m*3*m*. A crystal of diamond is effectively a macro-molecule; all the bonds in the crystal are covalent bonds which are very strong and consequently diamond crystals are extremely hard. The requirement that the bonds from each carbon atom should be tetrahedrally disposed, because otherwise molecular orbital formation would not be energetically effective, leads to a very open structure, that is to a very low density; if the atoms are taken to be spherical with radii equal to half the equilibrium distance (1·544 Å) between nearest neighbours, then the proportion of the volume of the unit-cell occupied by carbon atoms can simply be calculated as 34 per cent, whereas the corresponding proportion for close-packed structures is 74 per cent.

A radius can be assigned unambiguously to covalently bonded atoms in crystals of the elements, such as carbon in diamond. But if such radii are summed to give an estimate of the length of the covalent bond between two different elements there may be considerable discrepancy between the calculated and observed bond lengths because covalent radii vary with the atomic orbitals used for bond formation and are moreover only precise for bonding between atoms of approximately equal *electronegativity* (Table 10.9). For example silicon crystallizes with the diamond structure, the equilibrium distance between silicon atoms being 2·352 Å and the radius of the silicon atom 1·176 Å. Silicon carbide, SiC, crystallizes in a variety of structures one of which has a structure simply related to that of diamond by substitution of Si for one of the C atoms of the repeat unit (this is identical with the structure of the mineral *blende*, ZnS, shown in Fig 10.19) so that every atom is tetrahedrally coordinated to four atoms of the other element. The sum of the covalent radii of C and Si determined from observed bond lengths in crystals of these elements is 1·948 Å; the observed C—Si bond length in the compound is 1·888 Å. The discrepancy arises because the C—Si bond is not purely covalent. In general one can say that the length of a bond between unlike elements is very closely equal to the sum of the covalent radii determined in the two elements only when the electronegativities of the two elements are of similar magnitude; when there is a substantial difference in electronegativity, as between C and Si, the bond will not be purely covalent and will not be equal in length to the sum of the covalent radii determined from crystals of the elements concerned.

In covalent bonds between like atoms the electrons involved in bond formation will be equally shared between the two atoms and the bond will have no polarity; such a bond is an ideal covalent bond. But when the two atoms are of different elements the atomic orbitals of each that are involved in bond formation will not play equal parts in molecular orbital formation and consequently the valency electrons in the bond will spend more time near one atom than the other. Such a bond will have at least a slight polarity; that is to say it will have some ionic character.

Purely covalent bonding is relatively rare in inorganic and mineral structures, although many such structures contain bonds that involve some proportion of covalent bonding. Covalent radii are consequently of little significance in the field with which we are concerned.

A selection of the hybrid orbitals that give rise to commonly observed coordination polyhedra are set out in Table 10.1. In reading this table it should be remembered that while either ionic or covalent bonding can give rise to tetrahedral or octahedral coordination, only covalent bonding can produce square planar or trigonal prismatic coordination.

In the model *ionic bond* each constituent atom in the structure has a noble gas

Table 10.1
Hybrid orbitals giving rise to commonly observed coordination polyhedra

Coordination number	Hybrid orbital	Coordination polyhedron
2	sp	linear
3	sp^2	triangular
4	sp^3	tetrahedral
	dsp^2	square planar
6	d^2sp^3	octahedral
	d^4sp	trigonal prism

configuration so that every cation and anion has a spherically symmetrical distribution of electron density. Positively and negatively charged ions attract each other with a force that is inversely proportional to the square of the distance between them; their interaction energy is therefore proportional to the distance d between their centres. As the two oppositely charged ions approach, their electron clouds increasingly overlap and repulsive forces become significant. The repulsive forces operative in ionic crystals are of the same nature as in molecular crystals; the energy of repulsive interaction is represented accurately as $e^{-d/\beta}$ and approximately as d^{-n} where $7 \leqslant n \leqslant 9$. Equilibrium is attained when the essentially electrostatic attractive forces are balanced by the repulsive overlap forces. Figure 10.7 illustrates the variation of energy with interionic separation. Since overlap energy increases rapidly as the interionic distance falls below its equilibrium value ions in ionic crystals behave rather like hard rubber balls.

The electrostatic energy of an ionic crystal depends on the arrangement of all the ions in the crystal so that any arrangement in which every ion is coordinated with as many as possible ions of opposite charge will be favoured. Overlap energy however merely depends on the closeness of approach of neighbouring ions. The equilibrium distance between the ions of two given elements may thus be expected to vary with the geometry of the several structures in which they occur. However the steepness of the increase in repulsive energy at short interionic distances may be expected to make the variation in length of a given ionic bond in different structures relatively small.

In any group of isostructural compounds equilibrium cation–anion distances vary regularly from compound to compound. This point is exemplified in Table 10.2 which shows equilibrium interionic distances in those alkali halides that crystallize with the NaCl structure. One can, for instance, see that interionic distances in the fluorides are systematically about 0·50 Å shorter than in the corresponding chlorides and similarly the potassium salts have interionic distances shorter by about 0·14 Å compared with corresponding rubidium salts. On the basis of such comparisons it is evidently reasonable to assign radii to ions in ionic crystals. Such ionic radii will not be precise because the difference in observed interionic distance in corresponding series of salts of any two selected ions is only approximately constant; nevertheless a set of ionic radii would be a useful guide for the prediction of interionic distances. Such ionic radii have no physical significance outside the crystalline state; in general an atom or ion cannot be assigned a definite size, but it is useful and, up to a point, valid to regard ions in crystal structures as rigid spheres of determinable radius.

X-ray diffraction techniques provide a means of determining crystal structures

Table 10.2
Observed interionic distances in alkali halides with the NaCl structure

	Li^+	Na^+	K^+	Rb^+	Cs^+
F^-	2·01	2·31	2·67	2·82	3·00
Cl^-	2·57	2·81	3·14	3·29	
Br^-	2·75	2·98	3·29	3·43	
I^-	3·00	3·23	3·53	3·66	

Differences:

	Li^+	Na^+	K^+	Rb^+
(R—Cl) − (R—F)	0·56	0·50	0·47	0·47
(R—Br) − (R—Cl)	0·18	0·17	0·15	0·14
(R—I) − (R—Br)	0·25	0·25	0·24	0·23

	(Na—X) − (Li—X)	(K—X) − (Na—X)	(Rb—X) − (K—X)	(Cs—X) − (Rb—X)
F^-	0·30	0·36	0·15	0·18
Cl^-	0·24	0·33	0·15	
Br^-	0·23	0·31	0·14	
I^-	0·23	0·30	0·13	

from which equilibrium distances between adjacent ions in the structure can simply be evaluated. There remains the problem of how to establish a set of ionic radii from such information about sums of radii taken in pairs. In the first effective attempt to solve the problem, Landé assumed that LiI was precisely at the limit of stability of octahedral coordination, with the lithium cations fitting exactly into octahedral interstices in an ideally close-packed array of iodide anions (c.f. the NaCl structure, Fig 10.4). The iodide anions will then be in contact along face diagonals of the unit-cell so that $r_{I^-} = \frac{1}{4}a\sqrt{2}$. Lithium and iodine ions are in contact along the unit-cell edges so that $r_{I^-} + r_{Li^+} = \frac{1}{2}a$, whence $r_{Li^+} = \frac{1}{4}a(2-\sqrt{2})$. Once the radius of one ion has been evaluated in this way the radii of other ions can be simply determined from measured interionic distances in crystals of appropriate compounds. More reliable determinations of ionic radii have subsequently been achieved by assuming that the size of an ion in an ionically bonded structure is directly related to some property of the free ion such as its polarizability. One of the most valuable sets of ionic radii evaluated by such means is that due to Goldschmidt (1926), whose cation radii were improved by Ahrens (1952). Table 10.3 is based on these two publications. Other sets of radii do not differ greatly and all give useful estimates of interionic distances. Comparison of observed interionic distances with those calculated from Ahrens–Goldschmidt radii

Table 10.3
Ionic radii for octahedral coordination according to Ahrens (1952) for cations and Goldschmidt (1926) for anions. Radii are in Angstrom units.

Ion	Radius	Ion	Radius	Ion	Radius	Ion	Radius
Ag^+	1·26	Cu^+	0·96	Mn^{3+}	0·66	Sn^{4+}	0·71
Al^{3+}	0·51	Cu^{2+}	0·72	Mo^{6+}	0·62	Sr^{2+}	1·12
Au^+	1·37	Fe^{2+}	0·74	Na^+	0·97	Ti^{4+}	0·68
Ba^{2+}	1·34	Fe^{3+}	0·64	Nb^{5+}	0·69	Tl^+	1·47
Be^{2+}	0·35	Ge^{4+}	0·53	Ni^{2+}	0·69	W^{6+}	0·62
Ca^{2+}	0·99	Hg^{2+}	1·10	P^{5+}	0·35	Zn^{2+}	0·74
Cd^{2+}	0·97	K^+	1·33	Pb^{2+}	1·20	Zr^{4+}	0·79
Co^{2+}	0·72	Li^+	0·68	Rb^+	1·47		
Cr^{3+}	0·63	Mg^{2+}	0·66	Si^{4+}	0·42		
Cs^+	1·67	Mn^{2+}	0·80	Sn^{2+}	0·93		
Br^-	1·96	F^-	1·33	O^{2-}	1·40		
Cl^-	1·81	I^-	2·20	S^{2-}	1·74		

Table 10.4
Comparison of observed interionic distances (Å) for alkali halides with the NaCl structure with Ahrens-Goldschmidt radius sums

		Li^+	Na^+	K^+	Rb^+	Cs^+
F^-	obs	2·01	2·31	2·67	2·82	3·01
	Σr	2·01	2·30	2·66	2·80	3·00
Cl^-	obs	2·57	2·81	3·14	3·29	—
	Σr	2·49	2·78	3·14	3·28	3·48
Br^-	obs	2·75	2·98	3·29	3·43	—
	Σr	2·64	2·93	3·29	3·43	3·63
I^-	obs	3·02	3·23	3·53	3·66	—
	Σr	2·88	3·17	3·53	3·67	3·87

show (Table 10.4) that, while the relevant radius sum rarely agrees precisely with the measured interionic distance, the agreement is usually sufficiently close to justify the concept of spherical ions of definite radius in the alkali halides and this is so generally in ionic crystal structures. The sources of the discrepancy are various: the equilibrium distance between any two oppositely charged ions will not be determined exclusively by their mutual interaction but must depend more or less on interactions with other constituent ions in the structure and it has to be borne in mind that van der Waals forces will generally make some contribution, perhaps only a small contribution, to the attractive forces operating.

Recently ionic radii of the alkali metal and halide ions have been evaluated from measurement of interionic distances, thermal expansion coefficients, and isothermal compressibilities of single crystals of the alkali halides making use of the theoretical Born model of ionic solids.[1] The resultant ionic radii (Table 10.5) differ significantly from the generally accepted values, cation radii being about 0·20 Å larger and anion radii correspondingly smaller. These radii yield interatomic distances in excellent agreement with those determined directly by X-ray structure analysis for alkali halides with the NaCl structure.

Another way of determining ionic radii is to make a very thorough X-ray diffraction study of a simple salt so that the electron density distribution over the unit-cell can be calculated very accurately. The point at which the electron density is at a minimum on the line joining two neighbouring ions can then be taken as the point of contact of the two ions and determines the radius of each in the substance under investigation. In a purely ionic compound the minimum of electron density would be expected to be very nearly zero and the ions to be demonstrably spherical.

Table 10.5
Comparison of Fumi–Tosi radii with Ahrens radii for alkali metal ions and Goldschmidt radii for halide ions in octahedral coordination. All radii are in Å.

	Li^+	Na^+	K^+	Rb^+	Cs^+
Ahrens	0·68	0·97	1·33	1·47	1·67
Fumi–Tosi	0·90	1·21	1·51	1·65	1·80

	F^-	Cl^-	Br^-	I^-
Goldschmidt	1·33	1·81	1·96	2·20
Fumi–Tosi	1·19	1·65	1·80	2·01

[1] For further information about this approach see M. P. Tosi and F. G. Fumi, *J. Phys. Chem. Solids*, **25** (1964), 45.

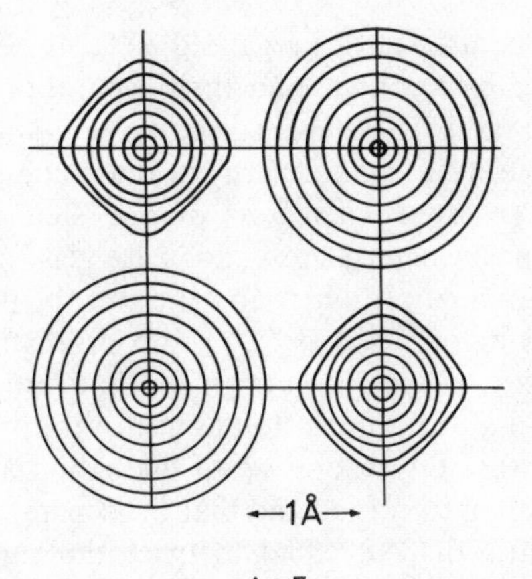

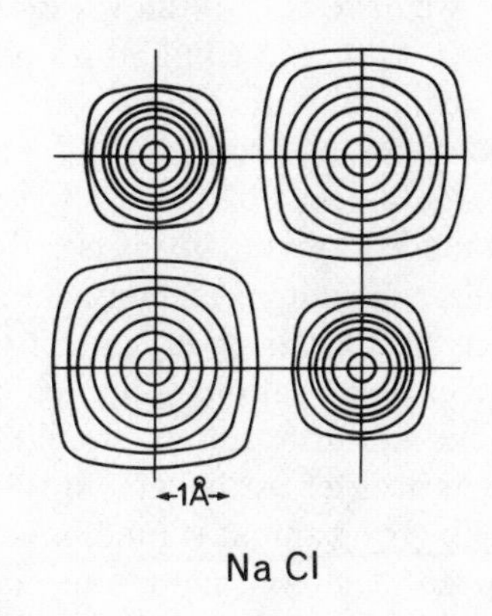

Fig 10.11 Electron density contours in *xy*0 sections of LiF and NaCl. Cations at 0, 0, 0; ½, ½, 0, and anions at ½, 0, 0; 0, ½, 0 only are shown. Contours are drawn at the same intervals in each section. In both salts electron density falls off from the centre of the ion more rapidly for the cation than for the anion. Such electron density maps yield ionic radii in agreement with those of Fumi and Tosi.

Few such studies have actually been made because it is experimentally difficult to achieve the requisite accuracy; but among the few are studies of NaCl and of LiF (Fig 10.11). The study of LiF is particularly interesting because it indicates that the cohesive forces operating are not purely ionic in character: neither type of ion is exactly spherical and the minimum electron density on the Li—F join is not even approximately zero. The conclusion that the Li—F bond has some covalent character is inescapable.

In crystal chemistry it is a convenience to be able to use ionic radii to interpret simple and complex structures in terms of packing considerations, but it must always be borne in mind that this is a grossly simplified approach. In the crystalline state ionic radii have precise significance only in the structural context in which they were determined. Ahrens–Goldschmidt radii and radii derived from the Born model are variously appropriate for the study of particular properties, each set being internally consistent at least to the extent that it gives approximate agreement with measured interionic distances. In the rest of this chapter we shall use Ahrens' set of ionic radii—which are more appropriate for simple packing considerations—except where we specify that other radii are being employed for a specific purpose.

All sets of ionic radii have been determined for octahedral coordination and have to be modified for other coordination polyhedra. The equilibrium separation of a given cation–anion pair will depend on cation–anion and anion–anion interactions in the structure and so on the coordination number of the cation. For example CsCl can crystallize under different physical conditions with the CsCl-type structure (Fig 10.17), in which the coordination number of Cs^+ is 8 and its coordination polyhedron a cube, and with the NaCl structure in which the coordination number of Cs^+ is 6 and its coordination polyhedron an octahedron; the Cs—Cl bond length in the former is 3·57 Å and in the latter 3·47 Å.

In general one can say that the radii of ions are 3 per cent larger in cubic coordination compared with octahedral coordination and 5–7 per cent smaller in tetrahedral compared with octahedral coordination. Tables of ionic radii in most works refer to octahedral coordination and it is necessary for the user to make the appropriate adjustments for other coordinations.

As we have already shown ions in ionic structures can be regarded as fairly rigid spheres so that at equilibrium adjacent ions are in contact and their separation is approximately equal to the sum of their ionic radii. The stable structure, that is the structure of lowest energy, will be such that the electrostatic energy of interaction is as large as possible; this situation will be achieved when ions of one sign are surrounded by as many as possible of the opposite sign. Since cations are generally smaller than anions the most stable structure for a purely ionic substance will be that which has its cations in contact with as many anions as possible. The coordination number of the cations is limited by the necessity to satisfy the criterion that cations and neighbouring anions should be in touch and the anions should not overlap; in geometrical terms this amounts to requiring that the separation of the cation and neighbouring anions should be equal to the sum of their radii and that the separation between neighbouring anions should be greater than or equal to twice the anion radius. In the structures of simple AX compounds these requirements are satisfied by cubic, octahedral, or tetrahedral coordination of cations by neighbouring anions. Ideally, as we showed earlier in our discussion of coordination polyhedra, a cation should adopt cubic (8-fold) coordination when $r_A/r_X \geqslant \sqrt{3}-1$, octahedral (6-fold) coordination when $\sqrt{3}-1 > r_A/r_X \geqslant \sqrt{2}-1$, and tetrahedral (4-fold) coordination when $\sqrt{2}-1 > r_A/r_X \geqslant \sqrt{\frac{3}{2}}-1$.

All the alkali–halides except CsCl, CsBr, and CsI crystallize with the NaCl structure at room temperature and it is to be expected that their radius ratios satisfy the criterion $\sqrt{3}-1 > r_A/r_X \geqslant \sqrt{2}-1$. Inspection of the relevant radius ratios listed in Table 10.6 shows that the criterion is not satisfied by this simple series of isostructural salts. The radius ratios of LiCl, LiBr, and LiI indicate tetrahedral coordination while those of KF, RbF, RbCl, RbBr, and CsF indicate cubic coordination. That the radius ratio criterion is not rigorously applicable is not surprising: equilibrium interionic separations do not depend solely on nearest neighbour cation–anion interaction, but also on interactions between ions of like charge. Such interactions between ions of like charge will become especially significant as the radius ratio approaches its lower limit for the postulated coordination; in these circumstances anions will be very nearly in contact so that anion–anion interactions may be expected to exercise a decisive effect. Anion–anion repulsion will tend to separate adjacent anions to a distance greater than the sum of their formal radii and so increase cation–anion separation; the resultant effect is in favour of a higher coordination number. This explains, at

Table 10.6
Radius ratios for the alkali halides

The ratios shown are based on Ahrens cation radii and Goldschmidt anion radii for octahedral coordination.

	Li^+	Na^+	K^+	Rb^+	Cs^+
F^-	0·511	0·729	1·000	1·106	1·255
Cl^-	0·376	0·536	0·736	0·812	0·923
Br^-	0·347	0·495	0·680	0·750	0·852
I^-	0·309	0·441	0·605	0·668	0·760

The alkali halides that fall between the solid lines have radius ratios $\sqrt{2}-1 \leqslant r_A/r_X \leqslant \sqrt{3}-1$ and should have the NaCl structure; those between the dashed lines crystallize in that structure at room temperature.

least qualitatively, why LiCl, LiBr, and LiI are found to have the 6-fold coordinated NaCl structure at room temperature and why the interionic distances in these salts are consistently greater than the sum of the relevant ionic radii for octahedral coordination. Indeed in general simple binary salts with radius ratios marginally below $\sqrt{2}-1$ adopt the NaCl structure.

To explain why the NaCl structure is adopted by alkali halides with radius ratios slightly in excess of $\sqrt{3}-1$ it is necessary to bear in mind that in the real crystal the cohesive forces are unlikely to be purely electrostatic. If there is any appreciable covalent content in the bonds simple geometrical arguments will not be applicable. In this context it is relevant to note that the lattice energies of the NaCl (octahedral coordination) and CsCl (cubic coordination) structures are closely similar so that it is reasonable to expect subsidiary interactions to play a critical role in determining which structure is adopted in alkali halides with radius ratios close to $\sqrt{3}-1$. Slight covalent character and van der Waals forces may well tip the balance, but one cannot be sure in the present state of knowledge precisely why KF, RbF, RbCl, RbBr, and CsF adopt the 6-fold coordinated NaCl structure. When the radius ratio is substantially greater than $\sqrt{3}-1$, as is the case in CsCl, CsBr, and CsI, the 8-fold coordinated CsCl structure is adopted in conformity with radius ratio predictions.

Crystal structures of the elements

We now proceed to describe the crystal structures of a selection of those chemical elements that form crystalline solids under atmospheric conditions and to relate the structures found to the bond models, van der Waals, metallic, and covalent. It will become apparent as we proceed that, while some elements have crystal structures typical of one of the bond models, others have structures that cannot be so simply classified.

Most *metals* adopt one of three simple structures: cubic close-packed, hexagonal close-packed, or body-centred cubic. The first and second of these we have already discussed in general terms; the third has a cubic I-lattice with one atom associated with each lattice point (Fig 10.12). Every atom thus has an identical environment with cubic coordination, the eight atoms of the coordination polyhedron being situated at distances $\frac{1}{2}(a\sqrt{3})$ in $\langle 111 \rangle$ directions from the selected atom. If each atom

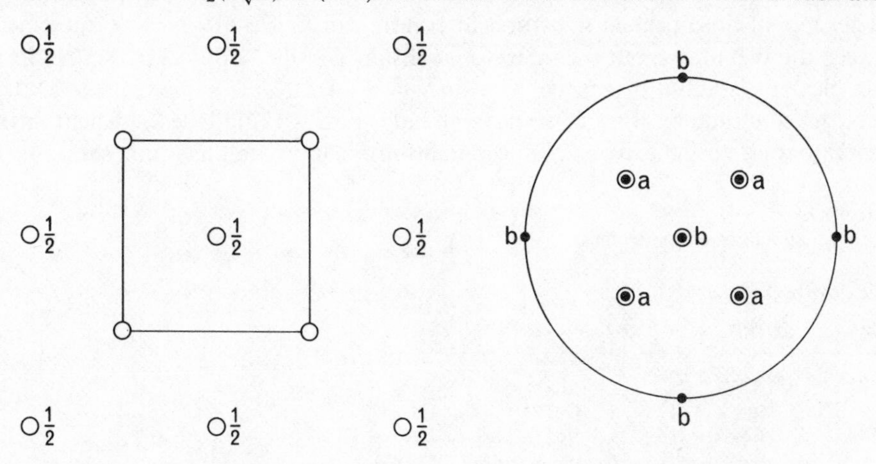

Fig 10.12 The body-centred cubic metal structure. On the left is a plan of the structure on (001) with the unit-cell outlined. The stereogram on the right shows the disposition of interatomic vectors from an atom to atoms a distant $\frac{1}{2}(a\sqrt{3})$ and to atoms b distant a.

is taken to be a sphere of radius $\frac{1}{4}(a\sqrt{3})$, that is half the distance of closest approach, then it can simply be shown that 68 per cent of the volume of the unit-cell is occupied by metal atoms. That this percentage falls so little short of the corresponding figure for close-packed structures, which we have shown to be 74 per cent, suggests that the body-centred cubic metal structure is very nearly close-packed; yet the coordination number is only eight compared with twelve in the close-packed structures. However in addition to its eight nearest neighbours every atom has six neighbours that are only slightly further removed: these atoms are disposed in $\langle 100 \rangle$ directions about the selected atom and distant from it one unit-cell edge *a*. The distance to the six second-nearest neighbours is thus only 15 per cent greater than that to the eight nearest neighbours so that one can say that every atom is close to fourteen others in body-centred cubic metals; in close-packed metals of course every atom is *equally* close to twelve others.

The crystal structures adopted by a selection of metals at room temperature and pressure are shown in Table 10.7. It is immediately apparent that some generalizations can be made. Metals of Group Ia crystallize in the body-centred cubic structure while those of Group Ib are cubic close-packed. The Group II metals beryllium and magnesium have the hexagonal close-packed structure with c/a close to the ideal value of 1·63; but the Group IIb metals, although they have the same arrangement of atoms, have axial ratios significantly different from 1·63. For the metallic elements in the remainder of the periodic table relationships are less obvious and moreover several are polymorphic. For example iron has the body-centred cubic structure from its melting point down to 1401 °C and again below 906 °C, while between 1401 °C and 906 °C it is cubic close-packed. Cobalt is cubic close-packed above 500 °C and at lower temperatures is hexagonal close-packed with stacking faults at irregular intervals which give rise to sequences such as ABABCACAC... In contrast nickel is cubic close-packed at all temperatures up to its melting point.

The energy of transformation between two polymorphs of a metallic element is usually quite small, for example 0·22 kcal mole^{-1} between body-centred cubic and cubic close-packed iron at 906 °C. The range of stability of a given structure is thus widely variable from one metal to another even when the electronic structures of their atoms are similar. When a metal crystallizes under different conditions in the cubic and hexagonal close-packed structures interatomic distances are in close agreement between the two forms; but the interatomic distance in the body-centred cubic form of an element is about 3 per cent less than in a form of the same element with a close-packed structure. But, as we have already seen, packing is less efficient in the former so that the effective volume per atom is approximately the same as in

Table 10.7
The crystal structures of some metals

Cubic close-packed:					
Ca, Sr;	Ni, Pt;	Cu, Ag, Au;	Al		
Hexagonal close-packed:					
	Be, Mg;	Ti;	Co;	Zn, Cd	
c/a	1·57 1·62	1·59	1·63	1·86 1·89	
Body-centred cubic:					
Li, Na, K, Rb;	Ba;	V;	Cr;	Fe	

close-packed metal structures. The *metallic radius* of an atom is taken as half the distance of closest approach in a close-packed form of the metal; if there is no stable close-packed form, then it is necessary either to take 1·03 times the radius derived from the body-centred cubic form, if there is one, or to derive the required metallic radius from interatomic distances in an alloy involving the element in question.

In cubic close-packing all the twelve nearest neighbours of an atom are constrained by symmetry to be equidistant from it. But in hexagonal close-packing symmetry does not relate interatomic distances between nearest neighbours when each atom is in the same close-packed sheet to interatomic distances between nearest neighbours when the relevant atoms are in different sheets. Of the metals listed in Table 10.7 as having the hexagonal close-packed structure type all except zinc and cadmium have c/a within 4 per cent of the ideal value of 1·63 and their structures can therefore be regarded as close approximations to hexagonal close-packing. In zinc however the separation of nearest neighbour atoms in the same close-packed plane is 2·66 Å while the separation of nearest neighbours in adjacent close-packed planes is 2·91 Å. Binding within close-packed sheets is therefore stronger than between the sheets. Consistent with this conclusion are certain directional physical properties of single crystals of zinc. For instance zinc is observed to have a 'good $\{0001\}$ cleavage', which implies that the structure fractures relatively easily along (0001) planes as one would expect if inter-sheet bonding were weaker than bonding within the sheets. Also the coefficient of thermal expansion normal to (0001) is greater than in directions parallel to (0001); for expansion from 20 °C to 100 °C 0·486 per cent and 0·115 per cent respectively. A similar situation obtains in cadmium. By way of conclusion all one can say is that these two elements do not have purely metallic structures; it is probable that bonding within close-packed sheets here has some covalent character.

The *non-metallic* elements might be expected from what we have already said about simple bond models to be covalently bonded, but this is not a valid generalization because only tetravalent atoms can form infinite three-dimensional structures linked by pure covalent bonds. The resultant structural type is that of diamond, in which every atom is tetrahedrally coordinated. The Group IV elements, carbon, silicon, germanium, and tin (below 13 °C) crystallize with the diamond structure (Fig 10.10). Diamond, although it is familiar both as a gemstone and as an abrasive, is not the thermodynamically stable form of carbon at room temperature and pressure; the stable structure for the element carbon under these conditions is graphite. In the graphite structure the carbon atoms form sp^2 hybrid orbitals which overlap to form σ molecular orbitals and are arranged in an infinite planar hexagonal ring structure (Fig 10.13). Each atom has a p orbital that is not involved in the formation of the sp^2 hybrid orbital; these atomic orbitals combine as π molecular orbitals extending over the whole planar sheet. The electrons in the π-orbitals are partially delocalized to the extent of being free to move in directions parallel to the sheet with which they are associated. There can thus be no covalent bonding between adjacent sheets in the graphite structure; the sheets are linked together to form three-dimensional crystals only by van der Waals forces. In the simplest of the several structures adopted by graphite there is a two-layer repeat with alternate atoms directly superimposed on atoms in adjacent sheets (Fig 10.13). The contrast between bond strength within (covalent bonding) and between (van der Waals bonding) sheets in the graphite structure is indicated by the separation of nearest neighbour carbon atoms: 1·4 Å for pairs of atoms in the same sheet compared with 3·35 Å for pairs in which one atom lies in the sheet immediately above or below the sheet to which the other atom of

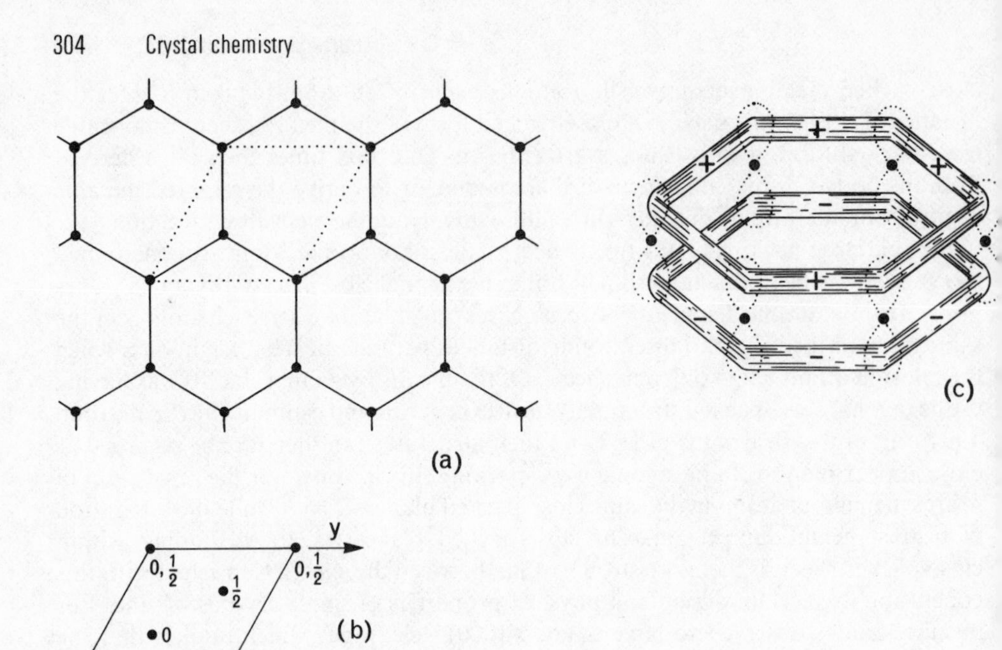

Fig 10.13 The crystal structure of graphite. (a) shows the arrangement of atoms in a single sheet of carbon atoms with the unit mesh dotted. (b) shows the unit-cell of the simplest structure adopted by graphite in plan on (0001). (c) shows the disposition of π-orbitals about a C_6 hexagon in graphite.

the pair belongs. Weak electrical conductivity parallel to (0001) and the strongly insulating nature of graphite perpendicular to (0001) are consequent on the existence of π-orbitals parallel to (0001).

Elements capable only of forming fewer than four covalent bonds do not crystallize in purely covalent bonded structures. Their structures are mostly molecular, the molecules being groups of atoms covalently bonded into infinite chains or sheets. Bonding between the chains or sheets is either attributable to van der Waals forces or to bonds that have some covalent and some metallic character. The Group V elements, arsenic, antimony, and bismuth are of this sort; all three have similar crystal structures in which each atom is closely bonded to three others so as to form an infinite puckered sheet (Fig 10.14). The sheets are stacked to form a three-dimensional structure in such a manner that every atom has three close neighbours in adjacent sheets (Fig 10.14(b)). In arsenic the closest approach of atoms where both belong to the same puckered sheet is 2·51 Å and where each belongs to a different sheet is 3·15 Å. The bonding within each sheet is evidently very much stronger than that between adjacent sheets. Arsenic is electrically an insulator, but it has a metallic lustre. It would therefore seem probable that while bonding within each sheet may be purely covalent, that between the sheets is partially metallic but has some covalent character. The structures of antimony and bismuth are similar; but as the atomic number increases in this Group the difference between the two bond lengths becomes less marked (Table 10.8) and the crystals become generally less covalent and more metallic in character.

Thermal expansion data, also set out in Table 10.8, are consistent with this conclusion: the greatest difference between the coefficients of linear expansion parallel

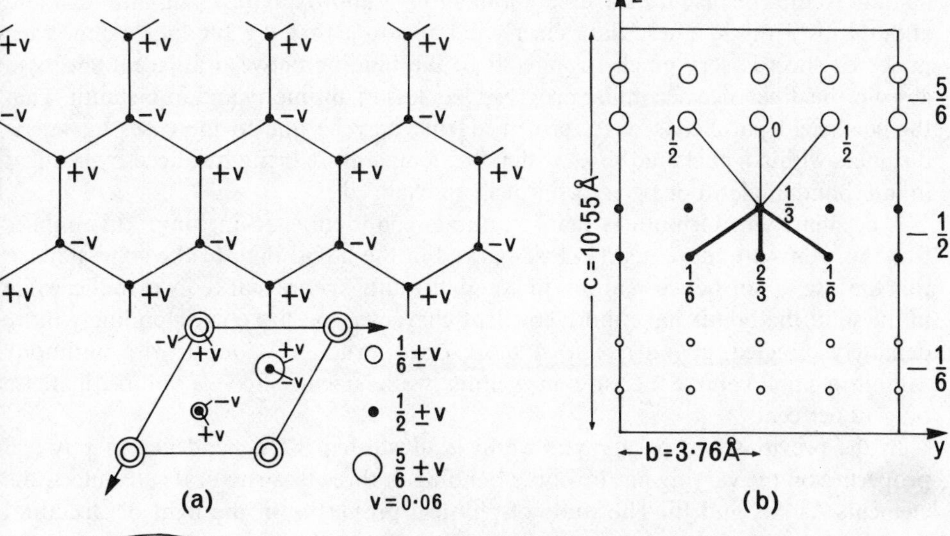

(c)

Fig 10.14 The crystal structure of arsenic. (a) is the projection of one puckered sheet of As atoms on (0001) with the contents of one unit-cell shown below; the puckered sheets are stacked with translation vector $\frac{2}{3}a+\frac{1}{3}b+\frac{1}{3}c$. (b) is the projection down [21$\bar{1}$0] on to the (10$\bar{1}$0) plane with coordinates out of the projection plane and interatomic vectors shown only for one atom and its near neighbours. Coordinates refer to the C-cell with x_c parallel to [21$\bar{1}$0] and y_c, z_c parallel respectively to [01$\bar{1}$0] and [0001] of the hexagonal unit-cell. (c) is a stereogram showing the disposition of vectors from an As atom at $z=\frac{1}{2}+v$ to its near neighbours, solid circles representing vectors to atoms 3·14 Å away and in the next sheet, open circles representing vectors to atoms 2·51 Å away and in the same sheet as the central As atom.

to x and to z is observed in arsenic and the coefficient measured parallel to z in arsenic is nearly three times the corresponding coefficients for antimony and bismuth. Moreover in arsenic the coefficient of thermal expansion parallel to the x-axis is effectively zero whereas for both antimony and bismuth it is appreciable in magnitude;

Table 10.8
Interatomic distances and coefficients of thermal expansion for the Group Vb elements

	As	Sb	Bi
X—X	2·51	2·91	3·10
X—X′	3·15	3·36	3·47
(X—X′)−(X—X)	0·64	0·45	0·37
$\alpha_x . 10^6$	~0	8	12
$\alpha_z . 10^6$	47	16	16

X—X is the shortest interatomic distance within the puckered sheets. X—X′ is the shortest interatomic distance between adjacent puckered sheets. Distances are measured in Ångstrom units. α_x and α_z denote coefficients of linear thermal expansion parallel to the x and z axes respectively; the units of both are deg^{-1}; the temperature range to which they refer is 20–400 °C.

its ratio to the coefficient measured parallel to z is approximately $\frac{1}{2}$ in antimony and about $\frac{3}{4}$ in bismuth. These data clearly demonstrate that the binding within each puckered sheet is very much stronger than the binding between adjacent sheets in arsenic, the difference becoming progressively less in antimony and in bismuth. That the coefficient parallel to x is zero in arsenic may be due to increases in As–As distances within a puckered sheet with rising temperature being balanced by changes in interbond angles to leave a fortuitously unchanged.

A peculiarity of bismuth is that it exhibits contraction on melting. This implies that the atoms are more nearly close-packed in the liquid than in the solid state so that the interaction between atoms in liquid bismuth is more markedly metallic, while in the solid the bonds have more covalent character and are correspondingly more definitely directed, giving rise to a more open structure. Contrariwise antimony exhibits a small volume increase on melting, while arsenic shows a volume increase of ~ 10 per cent.

In the previous two paragraphs we have illustrated the dependence of physical properties on the varying nature of the bonding in three isostructural substances, the elements As, Sb, and Bi. The study of physical properties in the light of structural knowledge can in appropriate circumstances provide a powerful tool for the elucidation of the nature, and especially the variation in nature of bonding in crystalline solids. We shall discuss the physical properties, especially the directional physical properties, of crystalline solids further in chapter 11.

The Group VI elements, sulphur, selenium, and tellurium are capable of forming only two covalent bonds per atom. Sulphur and selenium both form molecular structures with puckered rings linked together by van der Waals forces; in sulphur the rings may be six or eight membered and in selenium they are eight membered. More interesting crystallographically are the stable structures of selenium and tellurium; in each of these structures there is a helical chain of atoms disposed about a screw triad (Fig 10.15) with bond angles between atoms of the same helix of 105° in Se and 102° in Te, angles consistent with the hypothesis that bonding within each helix is essentially covalent. In the stable structures of both elements the helices are packed

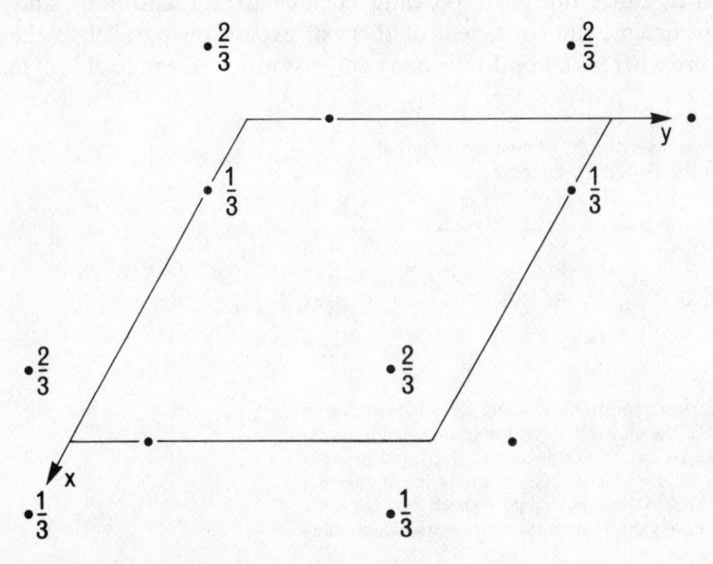

Fig 10.15 The crystal structure of hexagonal selenium.

so that each helix has six near neighbours suggesting that bonding between helices is due to non-directional forces. The closest separation of pairs of selenium atoms where one of the pair belongs to one spiral and the other to an adjacent spiral is 3·49 Å, rather less than the interatomic separation between the puckered rings of the unstable modification 3·53 Å. This discrepancy taken together with the metallic lustre and semiconducting properties of the modification of selenium stable at room temperature and pressure suggests that the bonding between spirals has some metallic character. Bond lengths and physical properties indicate that inter-spiral bonding in tellurium is more metallic. As in Group V metallic character increases with atomic number.

The atoms of the Group VII elements are capable of forming only a single covalent bond. The only element of the Group that is crystalline at room temperature and pressure is iodine, which has a molecular structure in which I_2 molecules are linked by van der Waals forces. The non-directional nature of van der Waals forces is clearly demonstrated by the manner in which the I_2 molecules are packed together in the orthorhombic structure as tightly as is consistent with their shape (Fig 10.8). In the orthorhombic B-cell of iodine the linear I_2 molecules are all parallel to (100) and aligned in the herring-bone pattern that is typical of the packing of linear molecules in molecular crystals. The same sort of packing is typical too of planar molecules, such as benzene (Fig 10.16) where the plane of every molecule in the structure is parallel to the *z*-axis of the orthorhombic unit-cell and the planes of the molecules are symmetrically inclined to the *x*-axis; the whole structure represents a close-packing of molecules having regard to their shape.

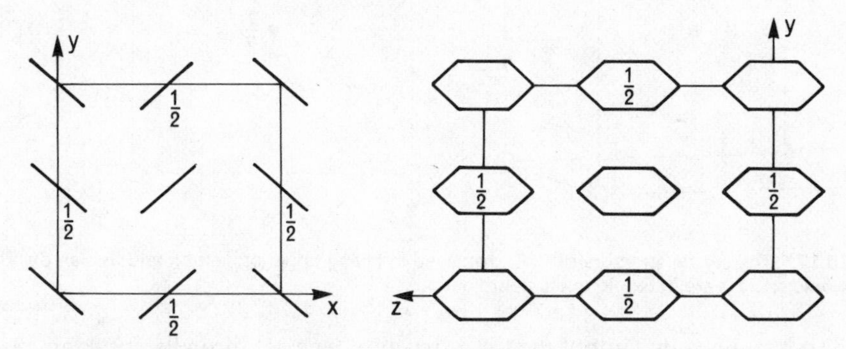

Fig 10.16 The crystal structure of benzene. In both diagrams coordinates refer to the centres of molecules. In the right-hand diagram it should be noted that the molecules are not parallel to the plane of projection (100); those centred on $x = \frac{1}{2}$ are inclined in the opposite sense to those centred on $x = 0$.

In iodine the interatomic distance within each molecule in the crystal is 2·68 Å, closely similar to the I–I distance in gaseous iodine, 2·66 Å. The van der Waals intermolecular bonding in the crystal leads to a closest distance of approach between iodine atoms of different molecules of 3·54 Å, a clear demonstration of the relative weakness of van der Waals forces.

We have not attempted to discuss the structures of all the elements that exist as stable crystalline solids under atmospheric conditions, but those we have discussed provide a survey of typical elementary structures. The reader in search of more comprehensive information is referred to the useful compilation of Wyckoff (1963).

AX and AX_2 structures

We turn now to description of the structures commonly formed when one element is bonded to another either in equal proportions, the AX structures, or in the ratio 1:2, the AX_2 structures. There is, as one would expect, great diversity in the structures adopted by AX and AX_2 compounds in the crystalline state, and many of the structures are known in only one or two compounds. Each of the types that we discuss in the following pages is important to the extent that it is adopted by a substantial number of compounds. We shall concentrate immediately on the description of each structure and defer discussion of A—X bond type to the next section.

CsCl type

The lattice is primitive cubic and there is one formula unit in the repeat unit. The two atoms are disposed in the unit-cell in such a manner that if, in CsCl itself, the chlorine atom is situated at the origin then the caesium atom lies at the centre of the unit-cell, at $\frac{1}{2},\frac{1}{2},\frac{1}{2}$ (Fig 10.17). Both atoms are in 8-fold coordination and their coordination polyhedra are cubes. The structure can conveniently be described with reference to the coordination cubes about the caesium atoms: every cube shares each of its six faces with another coordination cube.

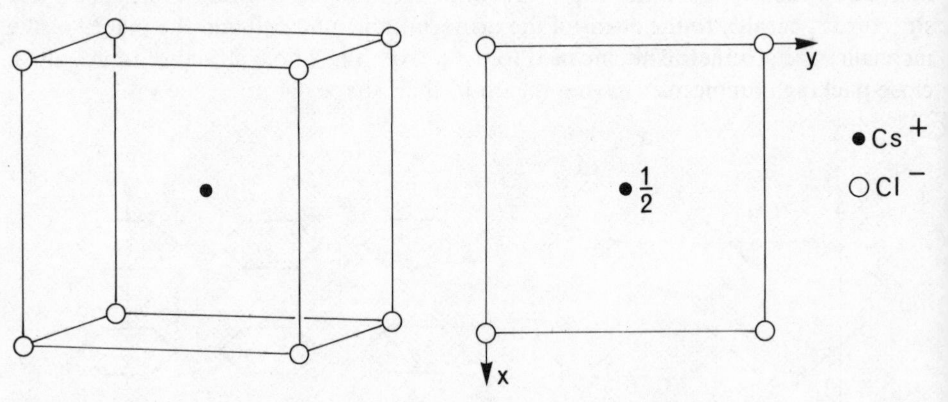

Fig 10.17 The crystal structure of CsCl displayed in clinographic projection and in plan on (001). Both ionic species are in 8-fold cubic coordination.

The compounds that adopt the CsCl structure type are, broadly speaking, of two kinds. First, there are the halides of the larger univalent cations such as the chlorides, bromides, and iodides of caesium and thallium. Second, there are intermetallic phases such as AlNi, CuZn, CrAl.

NaCl type

We have already dealt in passing with this structure type (see Fig 10.4). The structure is based on the cubic F-lattice and has one formula unit in the repeat unit. Every atom is thus related to atoms of its own kind by lattice translation. If, in NaCl itself, a sodium atom is situated at the origin, then a chlorine atom will lie at the centre of an edge of the unit-cell, e.g. at $0,0,\frac{1}{2}$. Both atomic species are in octahedral coordination, the octahedra being constrained by symmetry to be regular. The structure can simply be described with reference to the coordination octahedra about the sodium atoms: every octahedron shares each of its twelve edges with another coordination octahedron.

If the chlorine atoms in the type structure are to be regarded as rigid spheres, then they are in cubic close-packed array and the array has a sodium atom in each of its octahedral interstices. In most of the compounds that adopt the NaCl structure it is however likely that the atoms of the larger species are held apart by the smaller atoms and are not strictly close-packed. It may nevertheless be useful for comparative purposes to regard the NaCl structure type as an approximately cubic close-packed array of atoms of the larger sort with all its octahedral interstices occupied by those of the smaller species.

The variety of compounds that adopt the NaCl structure type is remarkable. They fall into four classes: (i) the majority of all alkali halides and some related compounds such as KCN; (ii) most of the oxides, sulphides, selenides, and tellurides of the alkaline earths, such as MgO, CaS, BaSe; (iii) the nitrides, phosphides and hydrides of various metals, such as ZrN, TiC, NaH; (iv) compounds of the Group Vb elements, phosphorus, arsenic, and antimony, with various trivalent metals, especially the lanthanide elements, such as CeP, GdAs, SnSb.

NiAs type

In this structure the lattice type is hexagonal and the unit-cell contains two formula units. Each atomic species occupies a set of special equivalent positions in space group $P6_3/mmc$. A plan showing four unit-cells of the structure in projection down the z-axis is shown in Fig 10.18(a). The arrangement of arsenic atoms is reminiscent of hexagonal close-packing; but the axial ratio $c/a = 1{\cdot}39$ is significantly less than the value, 1·63, for

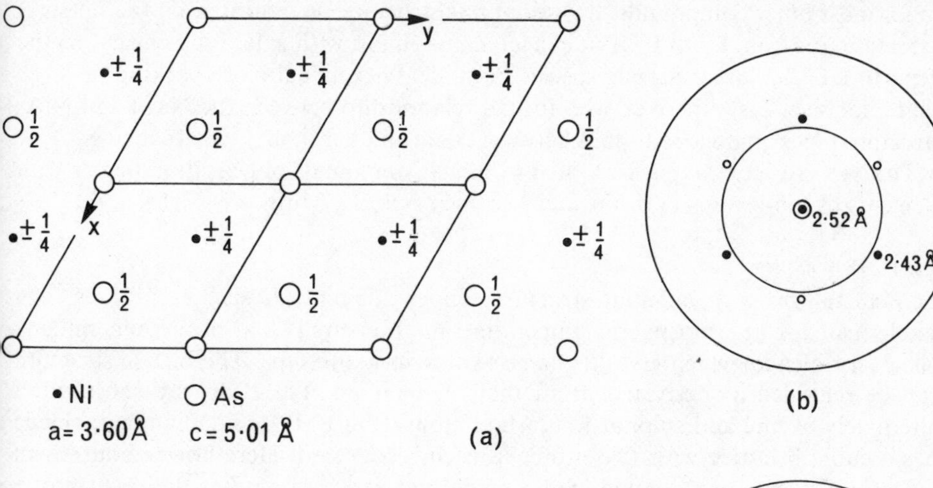

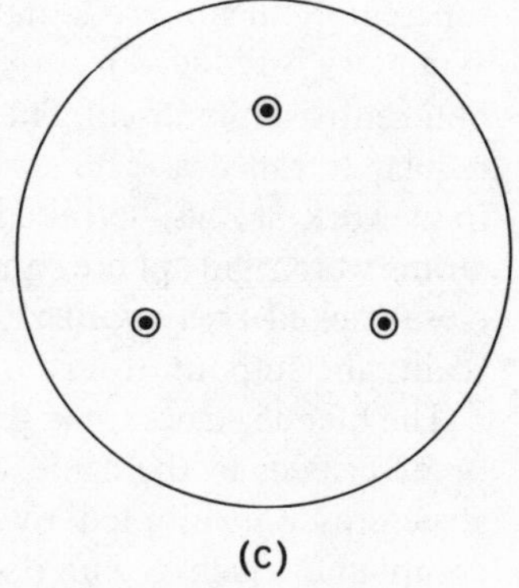

Fig 10.18 The NiAs structure. (a) is a plan of the structure on (0001) with four unit-cells outlined. The stereogram (b) shows the disposition of vectors from a nickel atom at $z = \frac{1}{4}$ to its near neighbours, of which six are arsenic atoms distant 2·43 Å and two are nickel atoms distant 2·52 Å; if the poles of the six Ni–As vectors lay on the small circle shown the coordination would be regular octahedral. The stereogram (c) shows the disposition of vectors from the arsenic atom at the origin to its near neighbours which are all nickel atoms. The plan and the stereograms are all oriented similarly.

ideal hexagonal close-packing. If the arsenic atoms were in ideal hexagonal close-packing, then the nickel atoms would lie in the octahedral interstices of the close-packed array. The coordination polyhedron of arsenic about nickel is in reality a distorted octahedron, the distortion being a compression of the octahedron normal to the pair of opposite faces lying parallel to (0001). Each octahedron shares this pair of opposite faces with those of adjacent coordination octahedra so that the Ni—Ni separation parallel to the *z*-axis is rather short; at 2·52 Å it is only marginally longer than the distance between Ni and its six nearest As neighbours, 2·43 Å. These two neighbouring nickel atoms cannot be omitted from the complete statement of the coordination of nickel in the structure; thus the coordination of nickel is 6-fold by arsenic and 2-fold by nickel. It is convenient to represent such a distorted coordination polyhedron by a stereogram showing the directions of the vectors from the selected atom to its near neighbours. To those familiar with the stereographic projection such a stereogram provides a clearer impression of the departure from regular octahedron geometry than could quickly be obtained from a list of angles between interatomic vectors. Figures 10.18(b) and (c) are stereograms showing respectively the disposition of near neighbours about nickel and about arsenic. It is apparent from the latter that each arsenic atom is surrounded by six nickel atoms situated at the corners of a trigonal prism.

We have seen that if the arsenic atoms were in ideal hexagonal close-packing, the nickel atoms would occupy all the octahedral interstices of the close-packed array. In NiAs itself there is a marked deviation from ideal close-packing and this is so for most other compounds that adopt this structure, as indicated by the range of axial ratio c/a from 1·2 to 1·7. But in a few compounds, with axial ratios near 1·63 the departure of the larger atomic species from ideal hexagonal close-packing must be slight; for such cases there exists a formal relationship between the NaCl and NiAs structure types analogous to that between cubic and hexagonal close-packing.

The NiAs structure type is adopted by many compounds of transition metals with elements of sub-groups IVb, Vb, and VIb such as PtSn, MnBi, VS, NiTe.

Zinc blende (ZnS) type

ZnS crystallizes with different structures under different physical conditions. The stable structure at atmospheric temperature and pressure is that of the mineral zinc blende (or blende) which gives its name to this structure type. The blende structure can be regarded as derivative from that of diamond (Fig 10.10) by substitution alternately of zinc and sulphur for carbon atoms (Fig 10.19). Like diamond, blende has a cubic F-lattice with two atoms in each repeat unit. Here however atoms of different elements are situated at the origin and at $\frac{1}{4}, \frac{1}{4}, \frac{1}{4}$ so that the point group symmetry is reduced from $m3m$ in diamond to $\bar{4}3m$ in blende, which is consequently non-centrosymmetrical. The coordination polyhedra about both zinc and sulphur are regular tetrahedra. The structure can simply be described as a three-dimensional framework of ZnS_4 tetrahedra linked to one another by their corners. If the sulphur atoms were rigid spheres and in contact with one another, they would form a cubic close-packed array with zinc atoms occupying alternate tetrahedral interstices. In reality the sulphur atoms are not close-packed but are held apart by zinc atoms.

The blende structure is geometrically related to the wurtzite structure (the next to be described) in the same way as cubic is to hexagonal close-packing. These two structures are adopted by much the same range of compounds; indeed several compounds, such as ZnS itself, crystallize with either structure. The blende structure

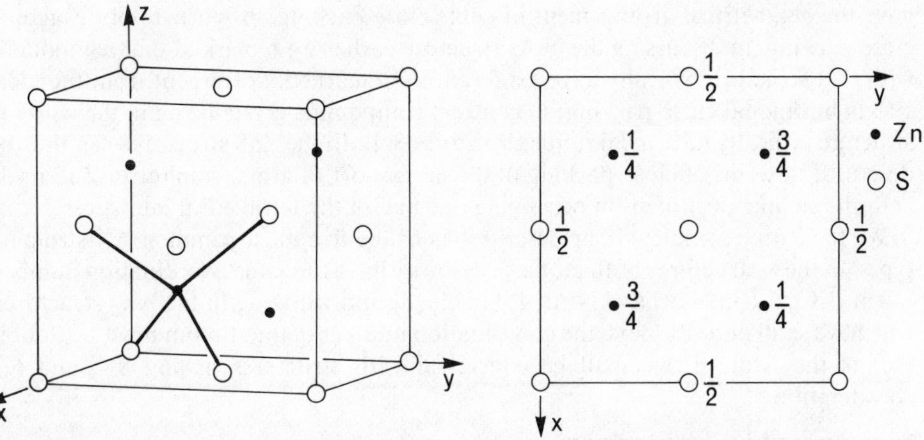

Fig 10.19 The crystal structure of blende, ZnS. On the left is one unit-cell in clinographic projection with bonds from the zinc atom at $\frac{1}{4}, \frac{1}{4}, \frac{1}{4}$ indicated. On the right is a plan of the structure on (001).

is adopted by the cuprous halides; by the sulphides, selenides, and tellurides of Group IIIb; and by compounds of Group IIIb with Group Vb elements such as AlSb, GaP.

Wurtzite (ZnS) type

Wurtzite is the polymorph of ZnS stable at atmospheric pressure above 1020 °C. It exists metastably at room temperature, persisting for long periods of time, but transforming to blende when ground in a mortar or otherwise subjected to shearing stresses. The lattice type is primitive hexagonal and the unit-cell contains two formula units (Fig 10.20). As in blende each atomic species is tetrahedrally coordinated to the other. Every coordination tetrahedron has one of its faces parallel to (0001) and is linked at each corner to three other tetrahedra. The distinction between the blende and wurtzite structures lies in the disposition of second nearest neighbours, that is in the arrangement of atoms of the same element. In blende the sulphur (or zinc) atoms

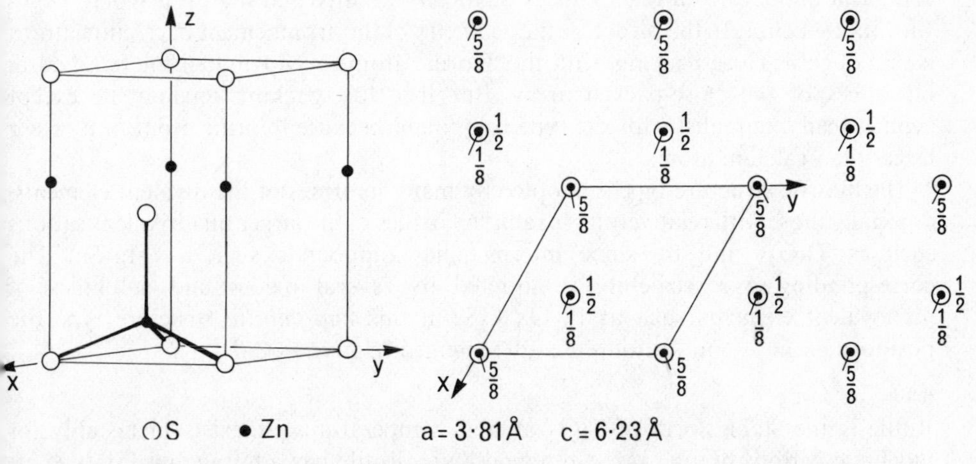

Fig 10.20 The crystal structure of wurtzite, ZnS. The left-hand diagram is a clinographic projection of one unit-cell with bonds from the zinc atom at $\frac{2}{3}, \frac{1}{3}, \frac{1}{8}$ indicated. The right-hand diagram is a plan of the structure on (0001) with one unit-cell outlined.

have the geometrical arrangement of cubic close-packing, in wurtzite of hexagonal close-packing. In discussing the NiAs structure earlier we remarked that compounds with that structure typically have axial ratios *c*/*a* markedly divergent from the ideal 1·63 of hexagonal close-packing; in contrast compounds crystallizing in the wurtzite structure typically have axial ratios close to 1·63. Both the ZnS structures can thus be described in terms of close-packing of the larger sort of atom, sulphur in ZnS itself, with the smaller sort of atom occupying one half of the tetrahedral interstices.

We have now completed our descriptions of the five most common AX structure types. In these structures both atoms necessarily have the same coordination number, 8:8 in CsCl, 6:6 in NaCl and NiAs, 4:4 in blende and wurtzite. In the AX_2 structures, which we shall describe next, the coordination number of the A atom must be double that of the X atom; we shall be concerned with structures having 8:4, and 6:3 coordination.

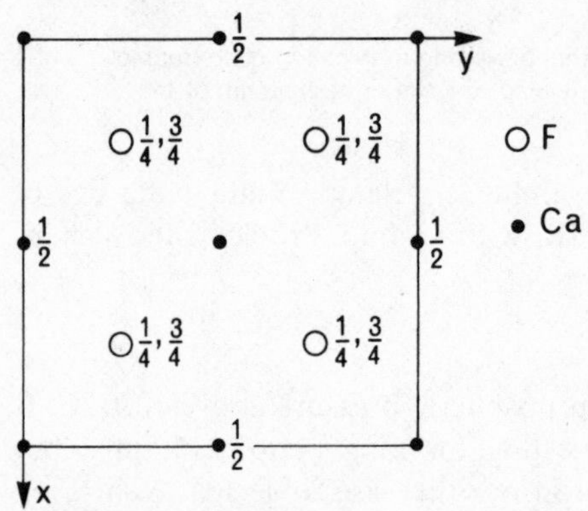

Fig 10.21 The crystal structure of fluorite, CaF_2 shown in plan on (001).

Fluorite (CaF_2) type

This structure type has a cubic F-lattice with one formula unit in the repeat unit. A plan of the structure on (001) is shown in Fig 10.21. Each calcium atom is 8-fold coordinated by fluorine, the coordination polyhedron being a cube. Each fluorine atom is coordinated to four calcium atoms arranged at the corners of a regular tetrahedron. The structure can simply be described in terms of the coordination cubes of fluorine about calcium: each cube is identically oriented and shares each of its edges with another cube. In this structure the geometry of the arrangement of calcium atoms is that of cubic close-packing, with the fluorine atoms occupying all the tetrahedral interstices of the close-packed array. But the close-packing analogy is, except geometrically, unrealistic for the type compound because fluorine atoms are rather larger than calcium atoms.

The fluorite structure type is adopted by many fluorides of the divalent elements, especially those with relatively large radii; by oxides of the larger quadrivalent cations such as ThO_2; and by some intermetallic compounds such as Mg_2Sn. The corresponding A_2X structure is adopted by several oxides and sulphides of monovalent elements such as Li_2O, K_2S; in this *anti-fluorite* structure type the positions of the A and X atoms are interchanged relative to CaF_2.

Rutile (TiO_2) type

Rutile is the stable form of TiO_2 at high temperatures; it exists metastably for indefinite periods of time at room temperature. Rutile has a tetragonal P-lattice, its space group being $P4_2/mnm$ and its point group $4/mmm$. There are two formula units in the unit-cell, all the atoms of the same element being related by the symmetry elements of the space group. A plan of the structure is shown in Fig 10.22(a). Each titanium atom is in 6-fold coordination, the coordination polyhedron being a slightly distorted octahedron. Each oxygen atom is coordinated to three titanium atoms disposed at the corners of a plane triangle parallel to (110) or ($\bar{1}10$). The coordination polyhedron about each titanium atom shares two of its edges with adjacent polyhedra, one about the titanium atom immediately above and the other about the titanium atom immediately below it in the [001] direction. There are thus infinite

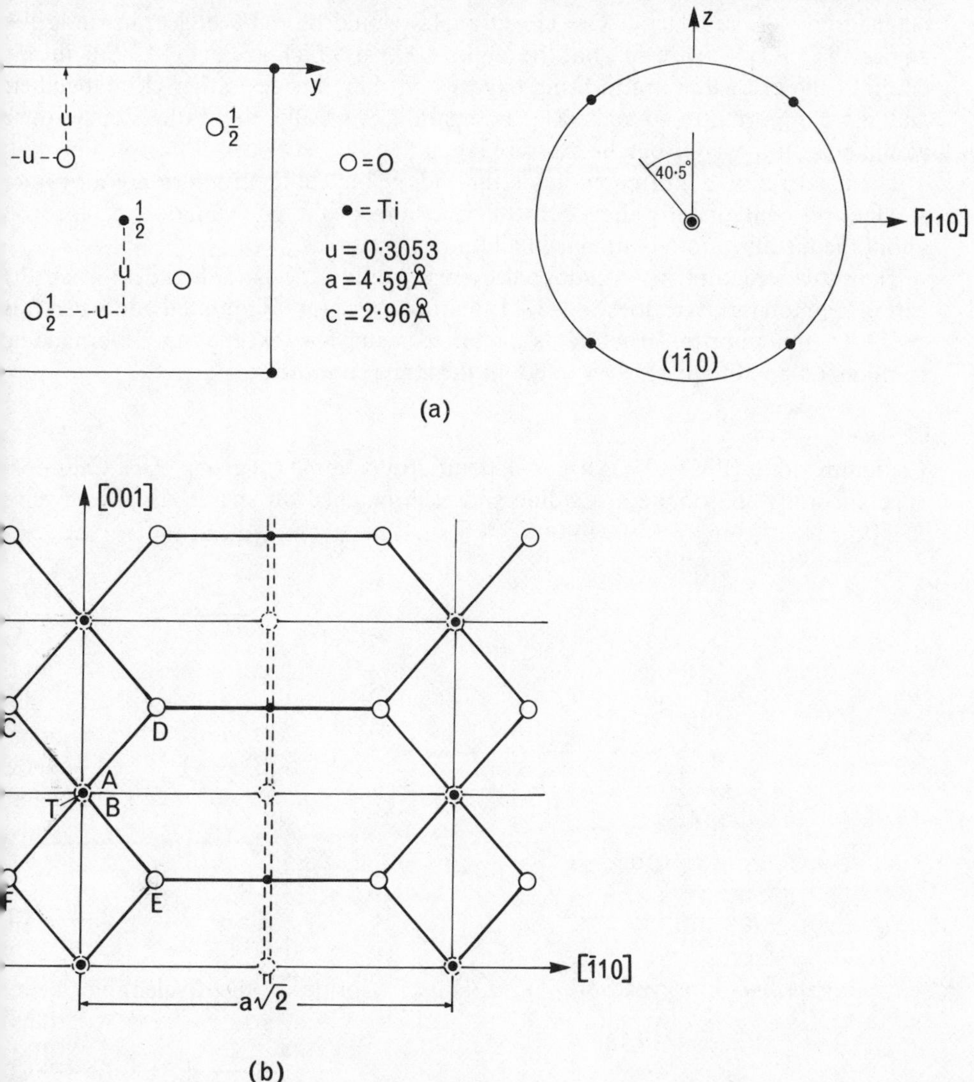

Fig 10.22 The crystal structure of rutile, TiO_2. (a) shows, on the left, a plan of the structure on (001) and, on the right, a stereogram of the disposition of Ti—O vectors from the titanium atom at $\frac{1}{2}, \frac{1}{2}, \frac{1}{2}$; the stereogram is oriented so that its primitive is the plane (1$\bar{1}$0) in order to display the distorted octahedral coordination. (b) is a section of the rutile structure through the origin and parallel to (110); the broken circles represent oxygen atoms $(\frac{1}{2}-u)\sqrt{2}a$ above and below the plane of the section; bold lines representing vectors from titanium atoms to their near oxygen neighbours are shown broken when they lie out of the plane of the section; thin lines indicate the intersection of unit-cell faces with the plane of the section. Diagram (b) displays the structure as a set of [001] chains of edge-sharing octahedra, alternate chains having their octahedra rotated through 90° about [001].

chains of TiO_6 distorted octahedra running parallel to the z-axis. Each oxygen atom in the structure is involved in edge sharing between adjacent octahedra of one chain and forms the apex of one octahedron of an adjacent chain (Fig 10.22(b)). The distortion of the TiO_6 octahedra is quite symmetrical: the distance from the titanium atom to each of the four oxygens (C, D, E, F in Fig 10.22(b)) that are involved in edge sharing in its own chain is 1·92 Å, slightly less than the distance, 2·01 Å, to the two oxygens, A and B, which are involved in edge-sharing in adjacent chains. If the octahedron were regular all O—Ti—O angles would be right-angles; in rutile the angles $\widehat{ATC}$, $\widehat{BTC}$ . . . are 90°, but the angles $\widehat{CTD}$ and $\widehat{ETF}$ are only 81°. Significant effects of the distortion are to bring oxygens on shared edges rather close together, and, more importantly, to make Ti—Ti separations parallel to z rather longer than would be so if the TiO_6 octahedra were regular.

The axial ratios c/a of compounds that adopt the rutile structure are all rather similar and consequently the extent of distortion of the coordination octahedron about the metallic atom is similar in all known examples.

The rutile structure type is adopted by many difluorides and dioxides; when the cation to anion radius ratio is $< \sqrt{3}-1$ the rutile structure is adopted and when it is $> \sqrt{3}-1$ the fluorite structure is preferred with few exceptions. Intermetallic compounds do not appear ever to adopt the rutile structure.

CdI_2 type

Cadmium iodide (Fig 10.23) is trigonal, point group $\bar{3}m$, space group $P\bar{3}m$. Cadmium is octahedrally coordinated to iodine and each octahedron shares six of its twelve

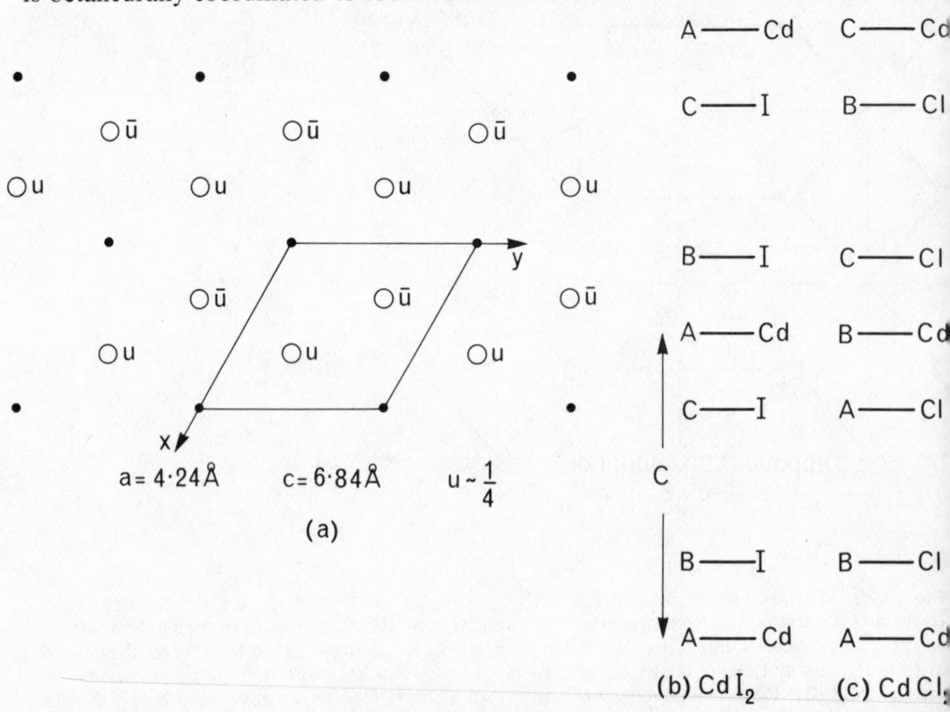

Fig 10.23 The CdI_2 and $CdCl_2$ structure types. (a) is a plan of CdI_2 on (0001), solid circles representing cadmium atoms and open circles halogen atoms. (b) shows the sequence of (0001) planes of Cd and I atoms in CdI_2 and (c) shows the corresponding sequence in $CdCl_2$, the relative orientation of planes being indicated in close-packing nomenclature A, B, C.

edges with other octahedra so as to form infinite sheets of octahedra parallel to (0001). All such sheets are identical and are related by the c lattice repeat. If the (0001) planes of iodine atoms were precisely $\frac{1}{4}c$ above and below the planes of cadmium atoms and if the axial ratio c/a were precisely 1·63, then the iodine atoms in this structure would be geometrically in hexagonal close-packed array and the cadmium atoms would be situated in octahedral interstices between alternate iodine planes. But in the real structure iodine atoms are not exactly in hexagonal close-packed array; each iodine atom is coordinated to three cadmium atoms in the same (0001) plane and the separation between cadmium and iodine planes is less than $c/4$. Each iodine atom has twelve iodine neighbours, nine of which are coordinated to one or more of the same three cadmium atoms as itself, while its other three iodine neighbours lie in the adjacent (0001) iodine plane and are coordinated to cadmium atoms in the next cadmium plane. Interaction between iodine atoms in adjacent (0001) planes is the only cohesive force linking adjacent sheets of cadmium–iodine octahedra. This structure is thus significantly different from the AX and AX_2 structures previously described in that it is not a three-dimensional framework of bonds between unlike atoms: in this structure there must be effective forces between adjacent iodine planes.

The CdI_2 structure type is adopted by the iodides, bromides, and hydroxides of several metals, mostly transition elements, as well as by the sulphides, selenides, and tellurides of certain quadrivalent elements such as Sn, Ti, Zr.

$CdCl_2$ type

This structure type is closely related to CdI_2. Here too there are sheets of cadmium atoms in octahedral coordination, but equivalent $CdCl_2$ layers are stacked on a rhombohedral lattice. The $CdCl_2$ structure has the point group $\bar{3}m$ and space group $R\bar{3}m$. If the layer shown in plan in Fig 10.23(a) is taken to be that associated with the lattice point at the origin of the triple hexagonal unit-cell so that cadmium atoms have coordinates $0,0,0$; $\frac{2}{3},\frac{1}{3},\frac{1}{3}$; and $\frac{1}{3},\frac{2}{3},\frac{2}{3}$ (that is the sequence of cadmium layers is, in close-packing notation, ABCA...), the sequence of chlorine atoms in the complex layer repeated by the rhombohedral lattice is CBACBACB.... The sequence of (0001) chlorine sheets is thus geometrically that of cubic close packing (with reference axes at 60° to those used elsewhere in this text). The relationship between the stacking sequences in CdI_2 and $CdCl_2$ are shown diagramatically in Fig 10.23(b), (c); the c-repeat of the hexagonal unit-cell in CdI_2 corresponds to two iodine planes whereas that of the triple hexagonal unit-cell in $CdCl_2$ corresponds to six chlorine planes.

The $CdCl_2$ structure type is rather less common than the CdI_2 type but is found in a variety of chemically similar compounds. No significant distinction can be made between compounds that adopt one rather than the other of these two structures.

Bonding in AX and AX_2 structures

In concluding the description of each simple structure type in the last section we listed the sorts of compounds which crystallize with the structure concerned. Inspection of these brief lists in the light of the most elementary chemical knowledge indicates that a particular structure type is not usually characteristic of one particular bond model.

Before investigating further the nature of the relationship between structure type and bond type it is necessary to develop the bond models that we set up earlier to allow for the existence of bonds intermediate in character between the simple types, ionic, covalent, and metallic. Consider for instance the blende structure in which both the A and the X atoms are in tetrahedral coordination. Tetrahedral coordination would be expected for an ionic structure with a radius ratio between $\sqrt{2}-1 = 0{\cdot}414$ and $\sqrt{\frac{3}{2}}-1 = 0{\cdot}225$. Alternatively tetrahedral coordination is characteristic of covalent bonding achieved by the formation of molecular orbitals from sp^3 hybrid atomic orbitals. In the case of ZnS itself the radius ratio for fully ionized atoms $r(Zn^{2+})/r(S^{2-}) = 0{\cdot}4$, consistent with the tetrahedral coordination of the blende structure. However tetrahedral sp^3 hybrid orbitals can be formed from the $4s$ and $4p$ atomic orbitals of Zn and the $3s$ and $3p$ atomic orbitals of S. Thus both the purely ionic and purely covalent models can account for the observation of tetrahedral coordination in blende. To determine the real nature of the bond between Zn and S in blende is not an easy task: detailed quantum mechanical calculations can be made in theory but may prove to be too difficult in practice even for such a simple structure; alternatively an exceptionally accurate diffraction study should enable the distribution of electrons within the unit-cell, and particularly the number associated with each atom, to be determined. In ZnS covalent bonding requires that Zn should contribute its two $4s$ electrons and that S should contribute its two $3s$ and four $3p$ electrons to form molecular orbitals so that if these eight electrons spend equal amounts of time in the vicinity of each atom there will always be four electrons associated with the Zn atom and four with the S atom. The resulting average configuration $Zn^{2-}S^{2+}$ is improbable on general chemical grounds, because zinc and sulphur are known to be respectively cationic and anionic whenever they display ionic character. Moreover quantum mechanical calculations have indicated that there is an excess negative charge of about $\frac{1}{3}$ on sulphur. The Zn—S bond in blende is therefore likely to be about 60 per cent ionic in character.

We have from the start assumed the simple bond models, ionic, covalent, and metallic, to be ideals or extremes. We now point to reasons why a more sophisticated approach to bonding in the crystalline state is necessary for full understanding and outline such an approach in terms of the model bonds. For a more thorough treatment the reader is referred to textbooks on valency theory such as Murrell, Kettle, and Tedder (1971).

In the *ionic* bond model each ion was assumed to be spherical. But as two ions approach they interact so as to *polarize* each other (Fig 10.24); that is to say the electron cloud about each ion is distorted from the spherical symmetry of the free ion by the presence of the other ion. The *polarizing power* of an ion depends on its charge and size. Polarizing power increases in general with increasing positive charge and decreasing size, that is to say with the magnitude and concentration of the net charge on the ion. The effect of the polarizing power of one ion on another ion is measured by the *polarizability* of the second ion. Polarizability increases with increasing negative charge and increasing size; the larger and more highly charged an anion is, the more weakly will its outer electrons be bound to its nucleus and the more easily will it be distorted from the spherical symmetry of the free ion. Strong polarization

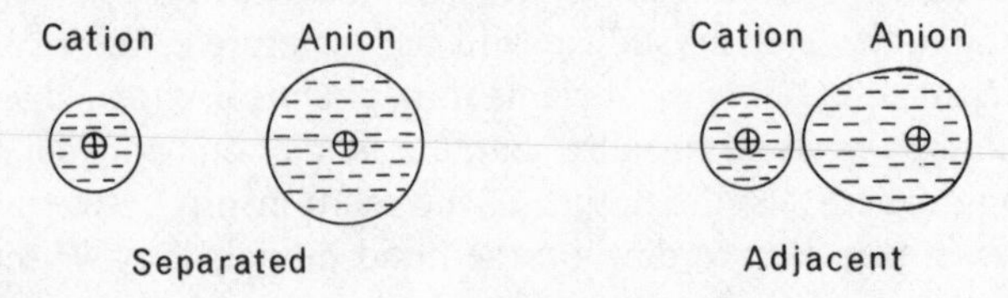

Fig 10.24 The polarization of the electron cloud of a large anion (e.g. F^-) by the approach of a small cation (e.g. Li^+).

is therefore to be expected in structures composed of cations of high polarizing power and anions of high polarizability; that is small, highly charged cations in combination with large highly charged anions. The effect of polarization is illustrated diagrammatically for LiF in Fig 10.24. In polarization the outer electrons of an anion, being less strongly bound to the nucleus, will be more affected than the electrons of its inner shells so that the valency electrons of the anion are concentrated to some extent in the space between the cation and the anion; polarization is therefore equivalent to the introduction of some covalent character to the essentially ionic bond. Polarization will tend to decrease the equilibrium distance between cations and anions in the structure and it will of necessity be more pronounced in structures with low coordination numbers such as tetrahedrally coordinated structures.

In *covalent* bonds the electrons involved in bond formation will spend equal amounts of time in the vicinity of each atom of the bond. But when the two bonding atoms are of different elements, it is improbable that the bonding electrons will be equally shared. If on average the bonding electrons spend more time in the vicinity of one atom, that atom will have an excess negative charge and the other atom an excess positive charge. The bond will thus have partial ionic character.

Our comments on ionic and covalent bonds indicate that each may have some of the character of the other. It would not be unrealistic to suppose that a smooth gradation of bond character from pure ionic to pure covalent could be established in an appropriate selection of compounds. What primarily concerns us here is the nature of the change in structure that may occur in the course of this gradual change in bond type.

For a substance AX the normalized wave functions for its covalent and ionic structures, A—X and $A^+—X^-$ respectively, can be formally represented as ψ_{cov} and ψ_{ionic}. Putting λ equal to the ratio ψ_{ionic}/ψ_{cov}, the relative weightings to be assigned to the two wave functions in the electron distribution will be $\psi^2_{ionic}:\psi^2_{cov} = \lambda^2:1$. The *percentage ionic character* of the bond is now defined as $100\lambda^2/(1+\lambda^2)$. This ratio is not generally susceptible to direct calculation and it is necessary to seek some means of estimating it. Of the several means that have been employed, we shall confine ourselves to the one that makes use of the concept of *electronegativity*.[2] Pauling (1960) rather loosely defined the electronegativity of an atom in a molecule as its power to attract electrons to itself. If two atoms have equal electronegativities, they will attract electrons to the same extent and a purely covalent bond will result. But if one atom has a higher electronegativity than the other, it will attract the electrons involved in bond formation more than the other and the bond will be, at least partially, ionic. In general the greater the electronegativity difference between the two atoms, the greater the ionic character of the bond between them.

Various ways of determining electronegativities with precision have been proposed, some experimental and others theoretical. We shall deal here only with the method of Allred and Rochow (1958). It is assumed that the electronegativity of an atom is proportional to the force exerted by the nucleus of the atom on the electrons in the bond and that this force is equal to Z^*e^2/r^2, where Z^* is the effective nuclear charge (i.e. the total nuclear charge less the amount by which the other electrons of the atom screen the bonding electrons from the nuclear charge), r is the mean distance of the bonding electrons from the nucleus, and e is the electronic charge. The covalent radius of the atom is taken to be equal to r. This approach places electronegativities on the

[2] For a detailed treatment of this topic the reader is referred to Coulson (1963), Cotton and Wilkinson (1966), or Phillips and Williams (1965–6).

same arbitrary scale (which conveniently uses the scale 0–4) as that used by Pauling who estimated electronegativities less precisely from ionic resonance energies. On this scale the electronegativity x of an atom is given by

$$x = 0{\cdot}359\frac{Z^*}{r^2} + 0{\cdot}744$$

Once the electronegativities of two elements involved in bond formation have been established, the percentage ionic character of the bond may be estimated from their difference. Several curves have been proposed to relate electronegativity difference to percentage ionic character. We shall not here explore their quantitative differences, but merely state that it is generally agreed that when the electronegativity difference between two atoms A and X, $|x_A - x_X|$, is about 1·7 the A—X bond has about 50 per cent ionic character.

Qualitatively electronegativity is a useful concept for collating the properties of ionic compounds in the absence of any direct determination of the percentage ionic character of their bonds. But the interpretation of electronegativity differences in terms of current theories of bonding is obscure so that it is not generally possible to use the concept quantitatively. As well as varying from element to element the electronegativity of different valence states of the same element will be different because the screening effect of the bonding electrons and the atomic radius will be different. Table 10.9 lists the electronegativities of some elements calculated in each case for the covalent radius appropriate to the common electronic configuration.

In the simple covalent bond model which was our starting point we assumed that the electrons involved in bond formation were localized between two atomic nuclei, the bonding electrons occupying molecular orbitals formed by the overlap of atomic orbitals of the two atoms. In a crystalline solid however molecular orbitals may be more extensive in that they may extend over more than two atomic nuclei or even throughout the crystal structure.

If neighbouring atoms are sufficiently close together in a crystal structure, the outermost of their occupied atomic orbitals will overlap to form molecular orbitals. Thus when two atoms of lithium are in close proximity the 2*s* atomic orbitals of each will overlap to form two molecular orbitals, one of lower and the other of higher energy than an isolated 2*s* atomic orbital. In a crystal of lithium containing N atoms $2N$ molecular orbitals will be formed. None of these molecular orbitals will be associated with any particular pair of Li atoms, but every Li atom will contribute by overlap of its 2*s* atomic orbital to every molecular orbital. The energies of the

Table 10.9
Electronegativities of some elements according to Allred and Rochow (1958)

Ag	1·42	Co	1·70	Mg	1·23	Sb	1·82
Al	1·47	Cr	1·56	Mn	1·60	Se	2·48
As	2·20	Cs	0·86	Mo	1·30	Si	1·74
Au	1·42	Cu	1·75	N	3·07	Sn	1·72
Ba	0·97	F	4·10	Na	1·01	Sr	0·99
Be	1·47	Fe	1·64	Nb	1·23	Te	2·01
Bi	1·67	Ge	2·02	Ni	1·75	Ti	1·32
Br	2·74	H	2·20	O	3·50	Tl	1·44
C	2·50	Hg	1·44	P	2·06	W	1·40
Ca	1·04	I	2·21	Pb	1·55	Zn	1·66
Cd	1·46	K	0·91	Rb	0·89	Zr	1·22
Cl	2·83	Li	0·97	S	2·44		

resultant molecular orbitals will be symmetrically disposed above and below the energy of the 2*s* atomic orbital of an isolated Li atom. The molecular orbitals can thus accommodate 4*N* electrons in a *band* of closely spaced energy levels. As *N* increases the energy levels of the band become closer together and the energy difference between the highest and lowest energy levels, the band width, increases. When *N* is very large, as it will be in a real crystal, the energy levels of the band are very closely spaced, almost a continuum. Moreover band width becomes sufficiently large in a crystal for the possibility of overlap between bands due to different atomic orbitals to arise. Consider for instance the case of an atom of an element that has two electrons in 2*s* atomic orbitals and its 2*p* orbitals unoccupied; in a crystal of the element the energy of the top of the band due to the combination of 2*s* atomic orbitals may be greater than the bottom of the band due to the combination of 2*p* atomic orbitals, which can be used to form molecular orbitals even though they are not occupied in isolated atoms of the element. In general if the energies of the atomic orbitals are well separated and the atoms are sufficiently far apart in the crystal, there will be no overlap of energy bands (Figs 10.25(a) and (b)). The valence electrons will then occupy the lowest available energy levels in the lowest band, which they may or may not fill. If however the energies of the atomic orbitals are closer together, energy bands will overlap; then the valence electrons will again occupy the lowest available energy levels (Fig 10.25(c)) but these will be distributed among two or more bands.

When the energy bands in a solid are fully occupied, the substance is an insulator and is, in terms of the simple bond model, a covalent substance. In such circumstances the simple model of molecular orbitals associated with a pair of atoms is equivalent to the delocalized molecular orbital model; in crystal chemistry the former generally provides the more useful approach. In contrast the delocalized molecular orbital model applied to metals is mostly more productive than the simple metallic bond model of electrons moving freely within the solid. If electrons are free to move in a metal crystal, then there must be ionization of the metal atoms so that the electrons move in a periodic field. In consequence certain energies are forbidden and the electrons occupy permitted energy bands analogous to those derived by delocalized molecular orbital theory. If one or more energy bands are only partially occupied by electrons the substance has metallic properties. Conduction of electricity takes place

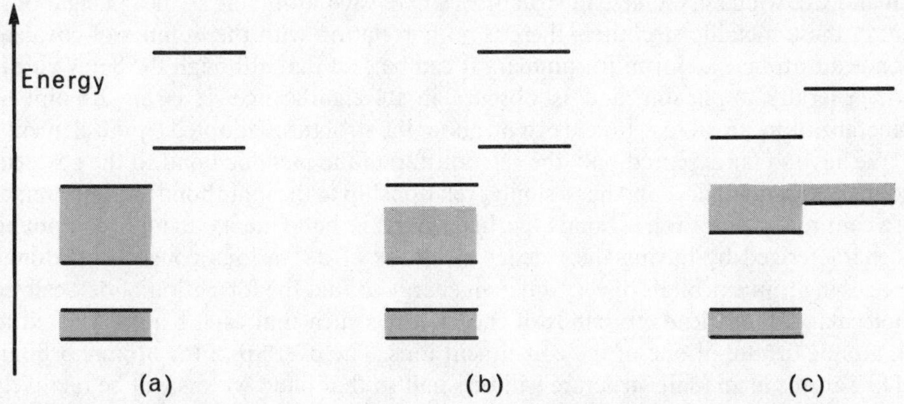

Fig 10.25 Electron distribution in a covalent compound (a) and in metals (b) and (c). The shaded regions are occupied by electrons. In (a) two energy bands are fully occupied and the next higher band is empty. In (b) one energy band is filled, the next partially occupied and the highest empty. In (c) the lowest energy band is again fully occupied while the next two above overlap and are both partially occupied.

by excitation of electrons to higher, unoccupied, energy levels of the band leading to a net flow of electrons. Absorption of light excites some electrons into the higher, unoccupied, energy levels of the band; radiation emitted as the electrons fall back to lower energy levels gives rise to the characteristic *lustre* of metals. Neither of these properties is displayed by compounds which have only fully occupied bands and have their highest occupied band substantially separated in energy from their lowest unoccupied band.[3]

Metallic properties such as electrical conductivity and magnetic susceptibility can be explained in some detail in terms of the width and relative positions of their energy bands and the extent of electron occupation of their several bands. For metals with cubic or hexagonal close-packed or body-centred cubic structures the band structure has been satisfactorily investigated theoretically; but distorted versions of these structures and more complex structures have not been amenable to theoretical study. We have here introduced this very brief account of band theory primarily to demonstrate that in such a relatively sophisticated treatment there is only a distinction of degree between metallic and covalent bonds with every possible variation between these extremes. For our limited crystallo-chemical purpose it is sufficient to think in terms of two extreme structures, one metallic and the other covalent: in metallic bonds the valence electrons are delocalized and in covalent bonds the valence electrons are associated with pairs of atoms. Structurally metallic bonding gives rise to close-packed and nearly close-packed (e.g. body-centred cubic) structures, whereas covalent bonding gives rise to structures with directed bonds.

It is appropriate at this point to introduce what is commonly known as the *8–N rule*, which states that 'a *b*-subgroup element in the *N*th group of the Periodic Table tends to crystallize in a structure with $8-N$ nearest neighbours'. Thus the carbon ($N = 4$) atoms in diamond have four nearest neighbours; the Group Vb elements As, Sb, Bi crystallize in structures where each atom has three nearest neighbours; the Group VIb elements S, Se, Te have structures in which each atom has two nearest neighbours; and in iodine ($N = 7$), which forms molecular crystals, each atom has only one near neighbour. In all these structures of elements with $N \geqslant 4$ the number of near neighbours is to be correlated with the number of covalent bonds the atom can form. But the $8-N$ rule also holds for the Group IIb metals, Zn and Cd, which crystallize in structures where each atom has six near neighbours but in these metallic structures there is no correlation with the number of covalent bonds the atoms can form. In summary it can be said that although the $8-N$ rule is not generally applicable and is obscure in its significance, it is an attempt at generalization, albeit to a limited extent, about the structures adopted by the elements.

We have so far explored only the relationship of the metallic bond to the covalent bond; but the metallic bond has a similar relationship to the ionic bond as the extremes of a continuous variation of bond type. In terms of the band theory an ionic compound is characterized by having the valency electrons of its various constituent atomic species in atomic orbitals of very different energy so that the formation of delocalized molecular orbitals leads to bands of energy levels such that each band is related to an atomic orbital of one of the constituent ions. The overlap of the atomic orbitals of like atoms in an ionic structure will be small so that band widths will be relatively narrow and bands will rarely overlap in energy. As the structure changes from that

[3] In making this statement and throughout this chapter we neglect the special case of the class of compounds known as *semiconductors*. The reader who desires to explore such compounds is referred to textbooks of crystal physics, such as Kittel (1971).

appropriate to the ionic configuration A^+X^- towards that of the metallic (alloy) configuration AX the bands broaden and become less closely related to the atomic orbitals of the isolated atoms A and X; the bond simultaneously decreases in ionic and increases in metallic character.

In conclusion it can be said that the three simple bond models are extremes and that bonds of intermediate character, ionic-covalent, covalent-metallic, metallic-ionic, are to be expected to occur quite commonly.

We turn now specifically to discuss bonding in the simple AX and AX_2 structures of the previous section. In doing so we shall ignore the intermetallic compounds that crystallize in several of the typical simple structures; the crystal chemistry of alloy systems is a highly specialized field outside our scope and the reader is referred for a brief account to Phillips and Williams (1965) or for a thorough treatment to Hume-Rothery, Smallman, and Haworth (1969). Here we shall be concerned essentially with the compounds of the transition metals and the metals of Groups I and II with the Vb, VIb, and VIIb elements, which provide examples of all the points that need to be made.

Of the AX structures the CsCl type is the least common. It is adopted by ionic compounds where the radius ratio is greater than $\sqrt{3}-1 = 0{\cdot}732$, the minimum ratio permissible for structures with cubic coordination. However some alkali halides with radius ratios appropriate to the CsCl-type actually crystallize with the NaCl-type structure as do some other alkali halides with radius ratios less than the minimum limit for octahedral coordination. The NaCl-type structure is found to occur in a great many predominantly ionic compounds in which the cation has an inert gas configuration. Thus alkali and alkaline-earth salts adopt this structure regularly, while transition metal salts, other than oxides, prefer other structures. In other compounds which adopt the NaCl-type structure however bonding must be predominantly covalent; and in some of these it is necessary to postulate partial metallic character, for example the mineral galena, PbS, which has characteristically metallic lustre.

The blende and wurtzite structures are closely related and ZnS is not unusual in crystallizing in both structures, each stable under different physical conditions. Most of the compounds that crystallize in one or both of these structures are dominantly covalent in character, such as the chalconides of the less electronegative elements. Very few oxides or halides crystallize in either of these structures.

It is difficult to generalize about the nature of the bonding in compounds that crystallize in the NiAs-type structure. In the type compound the coordination polyhedron of nickel about arsenic is a trigonal prism and therefore the Ni—As bonds cannot be dominantly ionic in character. The close approach of nickel nuclei in directions parallel to [0001] is also inconsistent with ionization of the nickel atoms. It is generally true that the axial ratio c/a in this structure type decreases with increasing metallic character of the bonding, so that one would expect metallic bonding, Ni—Ni bonding in this case, in directions parallel to [0001]. One might expect bonding between nickel and arsenic on general grounds to be dominantly covalent, but the properties of NiAs are intermediate between those of metallic alloys and ionic compounds. Most substances that adopt this structure tend to have variable composition, that is to be non-stoichiometric; they are to be regarded not as compounds in the strict sense, but rather as thermodynamic phases.

Of the AX_2 structures the fluorite and rutile types are adopted by dominantly ionic structures, especially by fluorides and oxides. Most of the compounds that crystallize in these structures have radius ratios within the limits for cubic and octahedral coordination respectively. In the case of the rutile structure there are some compounds that adopt it which have radius ratios just less than $\sqrt{2}-1 = 0{\cdot}414$; the reason for crystallization in the rutile structure in these cases is thought to be analogous to that for the adoption of the NaCl structure by the lithium halides.

The oxides and chalconides of the alkali metals crystallize in the anti-fluorite structure with the alkali metal ions in tetrahedral coordination and the anions in cubic coordination. This is the highest regular coordination possible because octahedral coordination of the cations in an A_2X compound would imply 12-fold coordination of the anions, that is close packing of the anions and a close-packed structure only has one octahedral interstice per close-packed atom. The radius ratio criterion consequently does not apply for this structure type: tetrahedral coordination is observed for cations in compounds with the anti-fluorite structure at radius ratios ranging from 0·31 in Li_2Te to 1·1 in Rb_2O.

Following the sequence of our descriptive section we come now to the cadmium iodide structure type, which must involve two distinct bond types: in CdI_2 itself there will be I—I bonding between the iodine atoms of adjacent (0001) sheets as well as Cd—I bonding. Compounds having this structure are characterized by easy cleavage parallel to (0001), an observation which suggests that the only cohesive forces operating between adjacent iodine sheets are van der Waals forces. Bonding between cadmium and iodine is unlikely to be dominantly ionic $Cd^{2+}I_2^-$ because this would lead to rather strong electrostatic repulsive forces between necessarily similarly charged iodine ions in adjacent sheets which are attracted only by the weak van der Waals forces. Covalent bonding between cadmium and iodine is possible by utilizing the sp^3d^2 octahedral hybrid orbitals of Cd and the $5p$ orbitals of iodine to give the formal configuration $Cd^{4-}I_2^{2+}$. Various pieces of experimental evidence suggest that the Cd—I bonds are intermediate between covalent and ionic, perhaps so balanced that the iodine atoms are effectively electrostatically neutral so that there is no significant electrostatic force between adjacent iodine layers.

It is apparent from the data presented in Table 10.10 that in AX_2 compounds, where X is a halogen, decreasing electronegativity and increasing polarizability of the halogen are to be correlated with change in structure from the fluorite or rutile types through the cadmium chloride type to the cadmium iodide type. The cadmium

Table 10.10
The structures of some AX_2 compounds where X is a halogen

The electronegativities are denoted by $x(A)$ and $x(X)$. Structure types: F = fluorite, R = rutile, Cc = cadmium chloride, Ci = cadmium iodide.

		F	Cl	Br	I
	$x(X)$ =	4·10	2·83	2·74	2·21
	$x(A)$				
Mg	1·23	R	Cc	Ci	Ci
Ti	1·32	R	Ci	Ci	Ci
Cd	1·46	F	Cc	Cc	Ci
Mn	1·60	R	Cc	Ci	Ci
Fe	1·64	R	Cc	Ci	Ci
Co	1·70	R	Cc	Ci	Ci
Ni	1·75	R	Cc	Cc	Cc

chloride structure is evidently adopted by compounds that are rather more ionic in character than those crystallizing with the cadmium iodide structure. This conclusion is consistent with the greater separation of cadmium atoms in adjacent layers in the $CdCl_2$ structure where the shortest Cd—Cd vector is $\frac{2}{3}\mathbf{a}+\frac{1}{3}\mathbf{b}+\frac{1}{3}\mathbf{c}$ compared with $\mathbf{c}$ in the CdI_2 structure ($\mathbf{c}$ for $CdI_2 \simeq \frac{1}{3}\mathbf{c}$ for $CdCl_2$). In this set of halides there would appear to be no correlation between structure type and either the electronegativity or the polarizing power of the metallic atoms concerned.

In the preceding pages we have discussed a few of the structures in which compounds whose formulae can be written as AX or AX_2 crystallize. Many other such compounds have structures simply related to those we have discussed, while others have totally different structures. We have seen that it is usually difficult and often impossible to determine the nature of the binding forces from knowledge of the structure alone. Structure analysis provides information essentially about bond lengths and inter-bond angles. To obtain information about bond strengths supplementary studies of such physical properties as thermal expansion and compressibility and the determination of melting or sublimation temperatures are required. In general one can say that small coefficients of thermal expansion, low compressibilities and high melting points or sublimation temperatures are indicative of the presence of strong binding forces in the crystal structure. Properties such as high electrical conductivity and metallic lustre are especially indicative of the presence of some degree of metallic bonding in the structure. In short a thorough physical as well as crystallographic study of a compound is necessary before the binding forces operative in it can be classified. Classification in terms of the simple ionic, covalent, and metallic bond models, even allowing for the existence of intermediates between these three extreme types, is not ultimately adequate but it has the merit of simplicity and is appropriate in its degree of sophistication to the sort of data available for most crystalline compounds.

In conclusion we return briefly to close-packing. Of the AX and AX_2 structures that we have discussed only caesium chloride, fluorite (but not anti-fluorite) and rutile have anions arranged in a manner that bears no geometrical resemblance to close-packing (Table 10.11). In all the other structures the anions have close-packed arrangements without being actually close-packed and the cations are situated in one or other type of interstice in the anion array. A peculiarity of the fluorite structure is that the cations have the cubic close-packed arrangement with the anions occupying

Table 10.11
Apparent close-packing in some AX and AX_2 structures

Where both atomic species are in close-packed array, the species that usually has the smaller radius is shown in parenthesis.

	Structure type	Close-packed atom	Type of close-packing	Interstices occupied: Coord. no.	Number
AX	CsCl	—	—	—	—
	NaCl	X(A)	cubic	[6]	all
	NiAs	X	hexagonal	[6]	all
	blende	X(A)	cubic	[4]	half
	wurtzite	X(A)	hexagonal	[4]	half
AX_2	fluorite	A	cubic	[4]	all
	rutile	—	—	—	—
	CdI_2	X	hexagonal	[6]	half
	$CdCl_2$	X	cubic	[6]	half

sites corresponding to the tetrahedral interstices in cubic close packing; but the resemblance to cubic close-packing can be no more than formal because fluorine anions are significantly larger than calcium cations. There is however a remarkable similarity between the unit-cell edges of fluorite 5·46 Å and elementary calcium 5·58 Å; in both structures the arrangement of calcium atoms is that of cubic close packing and in metallic calcium the atoms are actually close-packed. The significance, if any, of this relationship remains obscure.

Complex ionic compounds

For many of the more complicated inorganic oxides and silicates the ionic bond model is found to provide an adequate basis for understanding the structure even when some, or all, of the bonds have as much as 50 per cent covalent character. In compounds that contain complex ions such as $(CO_3)^{2-}$ or $(SiO_4)^{4-}$ even the C—O or Si—O bonds, in reality dominantly covalent, can safely be treated as ionic bonds if no more than a general understanding of the structure is required. This over-simplification fails however to explain some observed structural features, especially inter-bond angles. In such structures the anions together with any large cations, such as Na^+, Ca^{2+}, K^+, that may be present constitute a framework which is often geometrically related to close-packing; all the smaller cations occupy tetrahedral or octahedral interstices in the framework. Very small cations such as Si^{4+}, Fe^{3+}, etc tend to occupy tetrahedral interstices, intermediate sized cations such as Mg^{2+}, Ti^{4+}, Fe^{2+}, etc tend to occupy octahedral interstices, and Al^{3+} is quite commonly situated on both octahedral and tetrahedral sites.

Many structural features of complex ionic compounds can be accounted for in terms of the simple electrostatic considerations epitomized in *Pauling's Rules.* The four Rules[4] are:

(1) *The nature of the coordinated polyhedra:* A coordinated polyhedron of anions is formed about each cation, the cation–anion distance being determined by the radius sum and the ligancy of the cation by the radius ratio.

(2) *The electrostatic valency rule:* In a stable ionic structure the valence of each anion, with changed sign, is exactly or nearly equal to the sum of the strengths of the electrostatic bonds to it from the adjacent cations.

(3, 4) *The sharing of polyhedron corners, edges, and faces:*

(3) The presence of shared edges and especially of shared faces in a coordinated structure decreases its stability; this effect is large for cations with large valence and small ligancy.

(4) In a crystal containing different cations those with large valence and small coordination number tend not to share polyhedron elements with each other.

Discussion of the first rule must begin by defining the term ligancy as equivalent in crystallographic terms to coordination number. This rule is simply a concise statement of the principles that we have already applied to the interpretation of simple ionic crystals and extends the application of those principles to more complex structures. The dependence of the cation–anion distance on the radius sum implies that for this limited objective ions can be regarded as rigid spheres. The cation is usually smaller than the anions to which it is bonded so that the nature of its

[4] We quote the Rules without change of wording from Pauling (1960).

coordination polyhedron is determined by the maximum number of anions that can be in contact with the cation without overlapping one another.

The second rule expresses the principle of local charge balance, that is to say in a stable ionic structure the charge on an ion is neutralized by the presence of neighbouring ions. Electrostatic bond strength is simply defined as the valency of a cation divided by its coordination number and represents the amount of positive charge contributed by the cation to each bond. Thus an ion of valency z in n-fold coordination has bonds radiating from it, each of bond strength z/n. The rule requires that the sum of the electrostatic strengths of the bonds to an anion of valency $-y$ should be given by $y \simeq \sum_i (z_i/n_i)$ where the anion is bonded to i cations. For instance in rutile Ti^{4+} is in octahedral coordination so that the electrostatic strength of each Ti—O bond is $\frac{4}{6} = \frac{2}{3}$. Each oxygen atom is bonded to three titanium atoms so that the sum of the electrostatic strengths of the bonds to each oxygen atom is $3(\frac{2}{3}) = 2$ and is equal to the charge on the oxygen anion with reversed sign. The electrostatic valency rule is not obeyed rigorously in all complex structures, but in silicates at least it is rare for the sum of the electrostatic bond strengths to an oxygen atom to lie outside the range $2 \pm \frac{1}{6}$.

The third and fourth rules embody the obvious principle that electrostatic energy is minimized when cations are as far apart as can be consistent with the existence of a crystal structure, that is to say when polyhedra share corners. Figure 10.26 illustrates this point: the separation of the cations at the centres of a pair of octahedra sharing a common corner has a maximum value of $2d_{AX}$ but if the octahedra have a common edge the maximum separation of the cations falls to $\sqrt{2}d_{AX}$ if the octahedra remain regular. In reality coordination octahedra usually distort in such a manner that cation–anion distances are little affected, but the length of the shared edge tends to decrease so that interbond angles depart from the ideal value of 90°, and cation–cation distances tend to be increased. For instance in rutile (Fig 10.26(c)) the shared

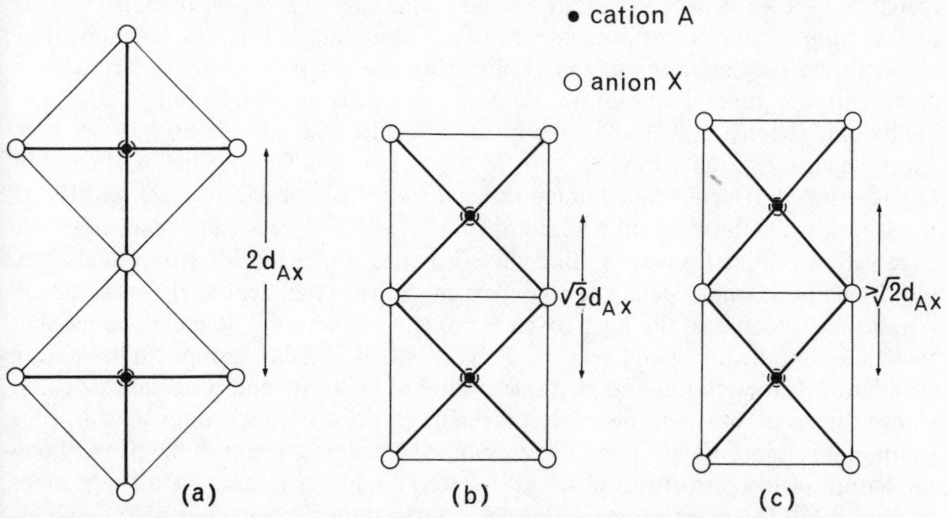

Fig 10.26 Distortion of coordination octahedra due to edge sharing. Bold lines represent bonds from cations A to anions X and light lines outline coordination octahedra. In (a) only a corner is shared by the adjacent pair of octahedra so that the shortest approach of cations is $2d_{AX}$. In (b) an edge is shared and adjacent cations are only $\sqrt{2}d_{AX}$ apart. The distortion of coordination octahedra in rutile (TiO_2), shown in (c), is such as to increase the separation of adjacent cations to more than $\sqrt{2}d_{AX}$; the O—O distances in the plane of the diagram are 2·53 Å on the shared edge and 2·96 Å on the other edge.

octahedral edges, of which there are two, have a length of 2·53 Å; while the lengths of the unshared edges are 2·78 Å and 2·96 Å, eight of the former and two of the latter (Fig 10.22).

It is obviously disadvantageous for cations of relatively large charge and small radius and consequently of small coordination number to share polyhedron elements; if they do so it is likely to be a corner rather than an edge and most unlikely to be a face. For instance in all the SiO_2 structures in which silicon is tetrahedrally coordinated to oxygen the tetrahedra share corners with other SiO_4 tetrahedra and in none of the known structures are edges or faces shared.

We now proceed to discuss two examples of complex ionic structures, the minerals forsterite Mg_2SiO_4 and common spinel $MgAl_2O_4$ in the light of Pauling's Rules.

Forsterite is orthorhombic with point group *mmm*, and space group P*bnm*. This space group is a non-standard setting of P*nma*, which was explored in some detail in another non-standard setting P*mcn* in chapter 4; coordinates of general and special equivalent positions for P*mcn* are listed in Table 4.4. Figure 10.27 shows a plan of the forsterite structure on (100) based on the recent refinement of Birle *et al.* (1968). Magnesium atoms are octahedrally coordinated by oxygen and lie on two sets of special equivalent positions: one half of the magnesium atoms lie on mirror planes in positions with coordinates $x, y, \frac{1}{4}$, etc, where $x = 0{\cdot}99$, $y = 0{\cdot}28$, and the other half lie on centres of symmetry at 0, 0, 0 etc. Silicon atoms are tetrahedrally coordinated by oxygen and occupy one set of special equivalent positions on mirror planes; they are situated at $x, y, \frac{1}{4}$, etc, where $x = 0{\cdot}43$, $y = 0{\cdot}09$. Oxygen atoms lie on one set of general equivalent positions, x, y, z etc, where $x = 0{\cdot}28$, $y = 0{\cdot}16$, $z = 0{\cdot}03$, and two sets of positions on mirror planes, $x, y, \frac{1}{4}$, etc, where $x = 0{\cdot}77$, $y = 0{\cdot}09$, and $x = 0{\cdot}22$, $y = 0{\cdot}45$. Each oxygen atom is bonded to one silicon and three magnesium atoms.

The electrostatic strength of a Si—O bond is $\frac{4}{4} = 1$ and of a Mg—O bond is $\frac{2}{6} = \frac{1}{3}$. The sum of the electrostatic strengths of the bonds to any oxygen atom is therefore $1 + 3.\frac{1}{3} = 2$, which exactly balances the charge of -2 on the fully ionized oxygen atom. The electrostatic valency rule is thus satisfied in this structure. The coordination polyhedra about the small highly charged Si^{4+} cation have the low coordination number 4 and do not share any elements with one another, that is to say they are *isolated* SiO_4 tetrahedra. Each tetrahedron does however share its three edges which are parallel to (100) with MgO_6 octahedra. The octahedra about Mg atoms situated at centres of symmetry each share two opposite edges with SiO_4 tetrahedra while those about Mg atoms on mirror plane sites each share only one edge with a SiO_4 tetrahedron. Since Si—O bonds are relatively strong and since O—Si—O bond angles tend not to depart much from the ideal angle for a regular tetrahedron because of the high covalent content of the bond, it might be thought that SiO_4 tetrahedra would tend to be very nearly regular tetrahedra in silicate structures. However recent very precise studies of forsterite and other silicates have shown that even SiO_4 tetrahedra become markedly distorted when involved in edge sharing; it is interesting to note that despite the covalent content of the Si—O bond the nature of the distortions observed is interpretable in terms of Pauling's Rules (cf. Fig 10.26). The magnesium sites on centres of symmetry, being linked to two SiO_4 tetrahedra, are consequently more stringently constrained than the other type of Mg site. Therefore the isostructural compound $CaMgSiO_4$ has the relatively large Ca^{2+} ions situated on mirror planes, Mg^{2+} continuing in occupation of the centre of symmetry sites.

In the forsterite structure the arrangement of oxygen atoms is approximately that

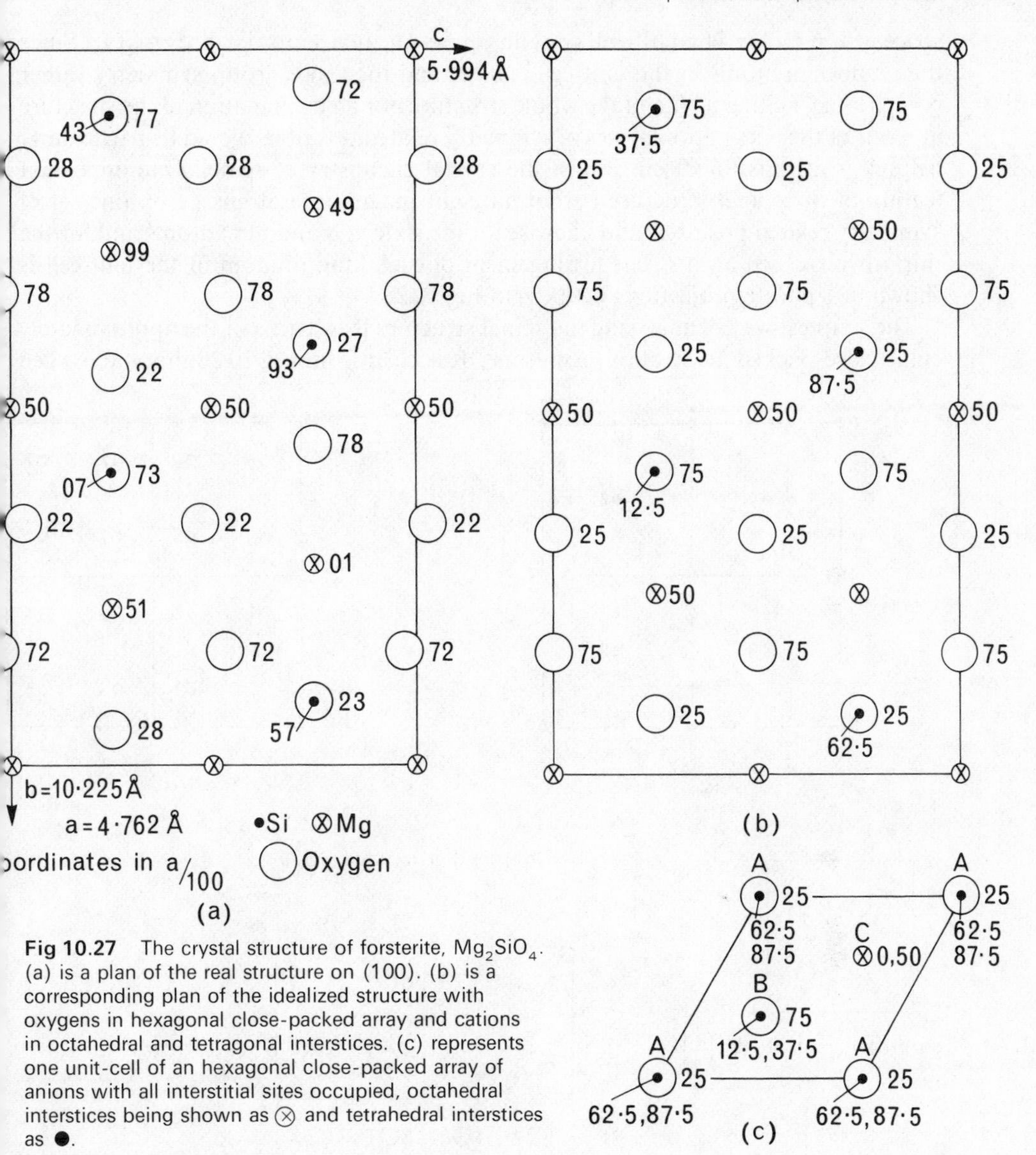

Fig 10.27 The crystal structure of forsterite, Mg_2SiO_4. (a) is a plan of the real structure on (100). (b) is a corresponding plan of the idealized structure with oxygens in hexagonal close-packed array and cations in octahedral and tetragonal interstices. (c) represents one unit-cell of an hexagonal close-packed array of anions with all interstitial sites occupied, octahedral interstices being shown as ⊗ and tetrahedral interstices as ●.

of hexagonal close-packing, with close-packed sheets parallel to (100) and close-packed rows of atoms parallel to [001]. The axial ratio $a/c = 3{\cdot}18$, rather smaller than the corresponding ratio for close-packing of rigid spheres $2 \times 1{\cdot}63 = 3{\cdot}26$. The separation of oxygen atoms along the z-axis is $c/2 = 2{\cdot}99$ Å, which is rather larger than the diameter of an O^{2-} ion 2·80 Å. Moreover what we have regarded as (100) sheets of oxygen atoms are not strictly planar, but regularly puckered with atoms about 0·20 Å above and below the mean position of the plane.

We shall return to the forsterite structure later in this chapter to consider its response to the substitution of other cations for Mg^{2+} and Si^{4+}.

The other structure that we use to exemplify Pauling's Rules is that of the mineral *spinel*, $MgAl_2O_4$. The unit-cell of spinel is cubic, its point group being $m3m$ and its space group Fd3m. The unit-cell contains eight formula units, i.e. $8MgAl_2O_4$. Since the number of atoms in the unit-cell is large and the space group symmetry rather complicated, neither a plan of the whole structure nor an examination of the structure in terms of the occupation of sets of symmetry related positions would be particularly helpful as a basis for discussion of the crystal chemistry of spinel. One important feature of the spinel structure is that all eight magnesium atoms lie on one set of symmetry related positions and likewise for the sixteen aluminium atoms and for the thirty-two oxygen atoms. The arrangement of each kind of atom in the unit-cell is shown in separate projections on (001) in Fig 10.28.

The simplest way of analysing the spinel structure is in terms of the approximately cubic close-packed array of oxygen atoms. It is clear from Fig 10.28 that the oxygen

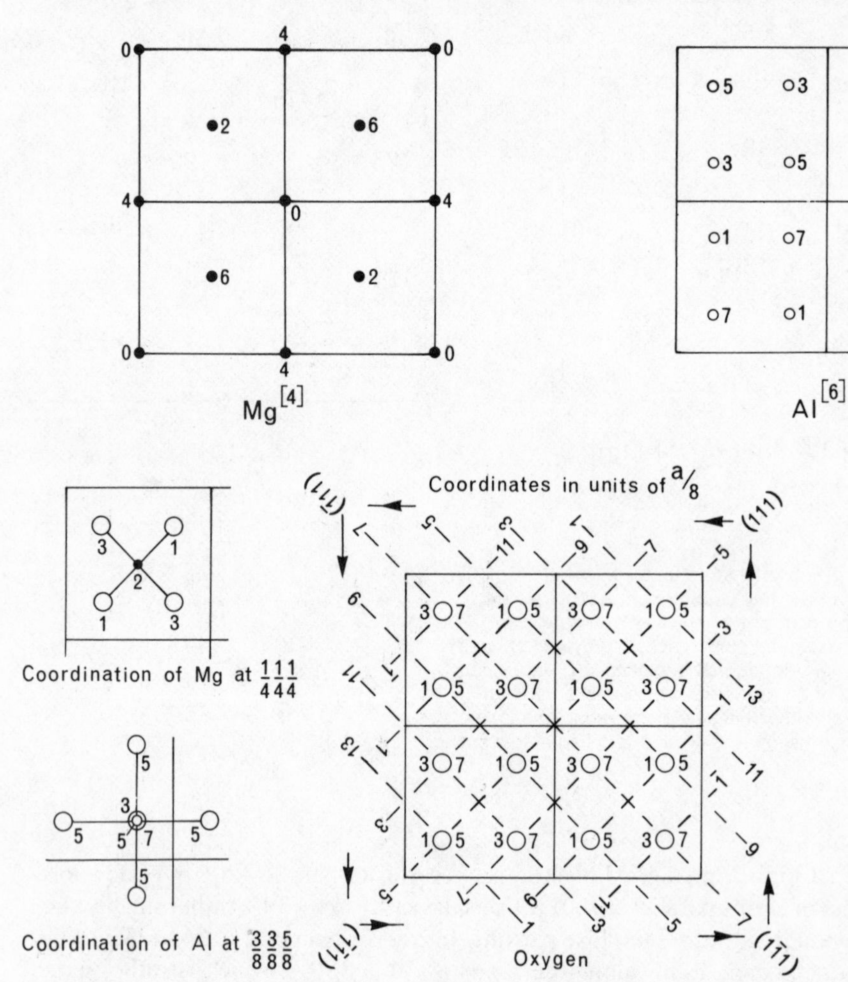

Fig 10.28 The crystal structure of spinel, $MgAl_2O_4$. Each diagram is a plan on (001) with one unit-cell shown divided into quarters. The two upper diagrams show the Mg and Al cation sites. The lower diagram on the right shows the positions of oxygen anions in the ideal structure. In the oxygen diagram dashed lines represent contours on {111} planes, the heights of the contour lines being shown in units of $\frac{1}{8}a$ on adjacent sides of the unit-cell outline and the indices of the relevant plane at the corner between these two sides (e.g. the (111) close-packed oxygen plane has contour heights $-\frac{1}{8}, \frac{1}{8}, \frac{3}{8}$ indicated along the right-hand side of the figure, $\frac{5}{8}$ at the top right-hand corner, and $\frac{7}{8}, \frac{9}{8}, \frac{11}{8}$ along the top of the diagram). The lower left-hand diagrams show the coordination of oxygen about Mg and Al cations respectively.

array, assuming perfect cubic close-packing, has a unit-cell edge equal to half that of the spinel unit-cell. There are therefore 8 close-packed 'oxygen sub-cells' in the spinel unit-cell and consequently 32 octahedral interstices and 64 tetrahedral interstices in the unit-cell. One half of the octahedral sites are occupied by aluminium ions and one eighth of the tetrahedral sites are occupied by magnesium ions. The occupied sites are regularly disposed in such a manner that tetrahedra share corners only with octahedra and the octahedra share edges with one another. The manner in which the occupied cation sites are disposed is most clearly to be seen by considering the sequence of cations normal to one set of $\{111\}$ close-packed oxygen planes. There are two types of cation arrangement between adjacent close-packed oxygen sheets; the two arrangements, which we define as types I and II, alternate through the whole structure (Fig 10.29(c)). Consider a type I cation layer between oxygen sheets in the A and B orientations of close-packing and a type II cation layer between B and C oxygen sheets. In the type I cation layer aluminium atoms occupy three of the four octahedral sites which, as is apparent from Fig 10.29(a), are in the C orientation in close-packing terminology. In the type II cation layer only one of the four octahedral sites is occupied by Al and two of the eight tetrahedral sites are occupied by Mg.

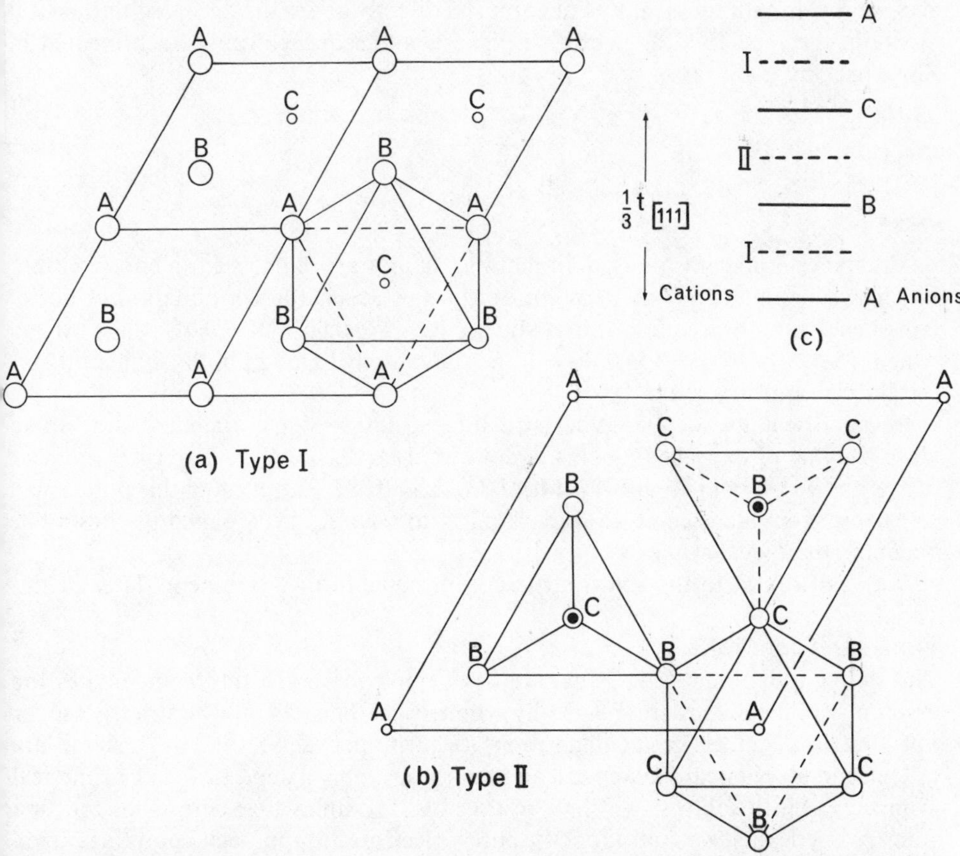

Fig 10.29 The crystal structure of spinel, $MgAl_2O_4$, shown in plan on (111). (a) shows a type I layer containing Al cations in three quarters of the octahedral interstices between A and B close-packed oxygen planes. (b) shows a type II layer containing Al cations in one quarter of the octahedral interstices and Mg cations in two eighths of the tetrahedral interstices between B and C close-packed oxygen planes. (c) shows the stacking sequence of (111) anion and cation layers along [111].

The aluminium atom necessarily occupies an A site; one magnesium atom is on a B site and the other on a C site (Fig 10.29(b)). There is only one C site that can be occupied by Mg if its tetrahedron is not to share a face with the Al octahedron in the type I cation layer immediately below. And there is only one B site that can be occupied by Mg if its tetrahedron is not to share any elements with other MgO_4 tetrahedra in this type II layer. The Al then enters the only octahedral site in which sharing with Mg tetrahedra of the same layer is restricted to sharing of corners. The sequence of type I and type II cation layers is repeated in such a manner that Mg tetrahedra face one another across the vacant octahedral site of the intervening type I layer.

If Fig 10.29(b) is imagined to be superimposed on Fig 10.29(a) it becomes apparent that every oxygen atom is coordinated to one magnesium atom and three aluminium atoms. The sum of the electrostatic strengths of the bonds to any oxygen atom is then $\frac{2}{4}+3.\frac{3}{6}=2$ which exactly balances the charge on a fully ionized oxygen atom. The spinel structure thus satisfies exactly the electrostatic valency rule.

This sequence of oxygen and cation sheets is repeated so that the oxygen atoms are in cubic close-packing. The next type I cation layer then falls between C and A oxygen sheets with its vacant octahedral site directly above the occupied tetrahedral B-site in the type II layer beneath it. The whole sequence can be represented in close-packing notation as:

Oxygen	A		B		C		A		B		C		A		B
Al(I)		C				B				A				C	
Al(II)				A				C				B			
Mg(II)				BC				AB				CA			

The spinel structure is unusual in that its Mg ions are in tetrahedral coordination. In silicates and in many other structures Mg is in octahedral coordination, which is consistent with the radius ratio $r(Mg^{2+})/r(O^{2-}) = 0{\cdot}66/1{\cdot}40 = 0{\cdot}47$, while Al for which $r(Al^{3+})/r(O^{2-}) = 0{\cdot}51/1{\cdot}40 = 0{\cdot}36$ is commonly found in both octahedral and tetrahedral coordination.

In our discussion of the spinel structure so far we have assumed ideal cubic close-packing of oxygens, but this is not strictly true. Ideally the oxygen atom at x, x, x has $x = \frac{3}{8}$; but in reality for $MgAl_2O_4$, $x = 0{\cdot}387$. The effect of this quite small departure from ideal cubic close-packing is to enlarge the tetrahedral sites while retaining their ideal symmetry.

We shall return to the spinel structure for some further comments later in this chapter.

Non-spherical ions

The simple ionic bond model that we have employed up to this point makes the assumption that all ions are spherically symmetrical. This can only be strictly true for ions that have an inert gas configuration; ions with partially filled d or f orbitals are not in general spherically symmetrical. In our discussion of departures from spherical symmetry, which follows, we shall restrict our examples to elements of the first transition series and we shall adopt the purely electrostatic approach known as *crystal field theory*.

Only when the $3d$ orbitals of a free ion are occupied by either five or ten electrons is the electron density distribution in the ion spherically symmetrical. When electron density distribution is non-spherical the equilibrium distance between the ion and an ion to which it is bonded will vary with the direction of the bond because the short range repulsive forces are determined by the overlap of the electron clouds of the two ions and the $3d$ orbitals are the outermost orbitals in the case of ions of the first transition metal series.

A transition metal ion in octahedral coordination has its $3d$ orbitals in the orientation corresponding to minimum energy when the electron density maxima of the d_{z^2} and $d_{x^2-y^2}$ orbitals (Fig 10.30) are directed towards the six coordinating ions. The remaining three $3d$ orbitals, d_{xy}, d_{yz}, d_{zx}, then have their directions of maximum electron density disposed symmetrically between groups of four coplanar coordinating ions. In such a coordination environment the d_{z^2} and $d_{x^2-y^2}$ orbitals retain their equivalence but become energetically distinct from the other three orbitals d_{xy}, d_{yz}, d_{zx}. An electron in a d_{z^2} or a $d_{x^2-y^2}$ orbital will have a higher energy than an electron in a d_{xy}, d_{yz}, or d_{zx} orbital because the electrons in the former set of orbitals are closer to the coordinating anions. The effect of the coordinating anions is represented in the energy level diagram of Fig 10.31(a). The total energy of the electrons associated with the cation is raised when the cation is surrounded by six anions; the five $3d$ orbitals, degenerate in the free cation, become split into two levels with energy difference Δ_0 such that there is a triply degenerate energy level at $\frac{2}{5}\Delta_0$ below the mean energy, comprising the d_{xy}, d_{yz}, d_{zx} orbitals, and a doubly degenerate energy level at $\frac{3}{5}\Delta_0$ above the mean energy, comprising the d_{z^2} and $d_{x^2-y^2}$ orbitals. The orbitals of these two energy levels are commonly denoted, in notation borrowed from group theory, as t_{2g} and e_g respectively. The difference between the energy levels t_{2g} and e_g is Δ_0 which is known as the *crystal field splitting*. The magnitude of Δ_0 depends on the nature of the coordinating anions; it may be sufficient to play a significant part in determining at least some of the physical and chemical properties of the cation in this environment.

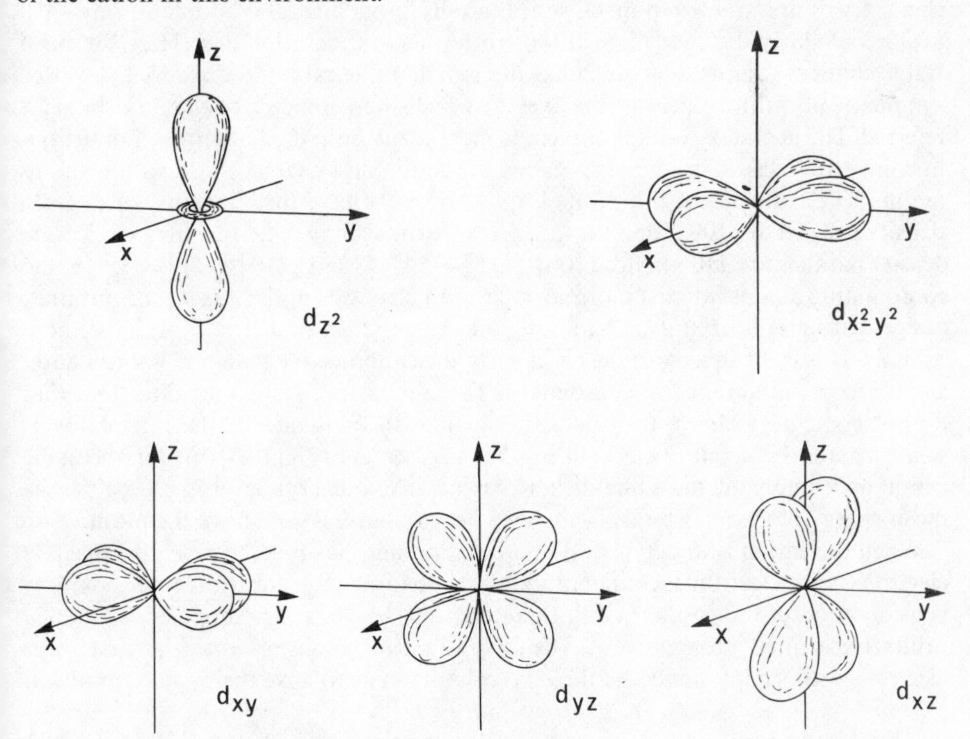

Fig 10.30 Boundary surfaces for $3d$ electrons.

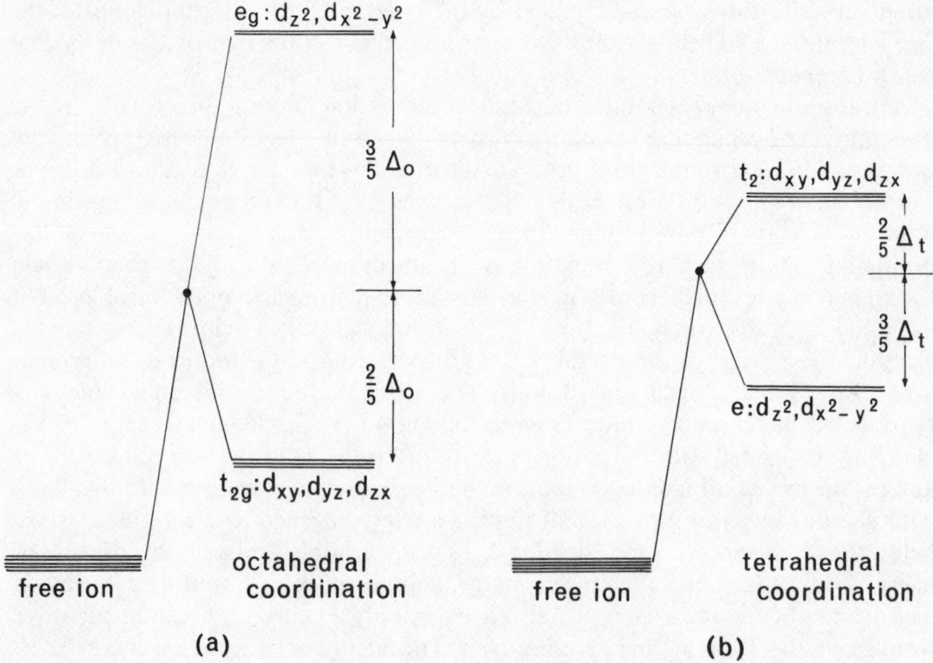

Fig 10.31 Energy level diagrams to illustrate crystal field splitting of the 3*d* orbitals of ions in (a) octahedral and (b) tetrahedral coordination.

Before going on to consider the effect of crystal field splitting on the crystal chemistry of first transition metal series cations in octahedral coordination we shall explore similarly the case of such ions in tetrahedral coordination. Here the most stable configuration is that which has the four tetrahedral bond directions disposed symmetrically with respect to the axes to which the geometry of the 3*d* orbitals is referred. The geometry of the problem is most easily described in terms of directions in point group $\bar{4}3m$. The vectors between the cation and its four coordinating anions are in $\langle 111 \rangle$ directions; the d_{z^2} and $d_{x^2-y^2}$ orbitals have their maxima of electron density parallel to $\langle 100 \rangle$; and the d_{xy}, d_{yz}, d_{zx} orbitals have their maxima of electron density parallel to $\langle 110 \rangle$. Since $[100]{:}[111] = 54°44'$ and $[110]{:}[111] = 35°16'$ the coordinating anions interact more strongly with electrons in the d_{xy}, d_{yz}, d_{zx} orbitals, designated here t_2 orbitals,[5] than with those in the d_{z^2} and $d_{x^2-y^2}$ orbitals, the *e* orbitals. This is the inverse of the effect when the cation is in octahedral coordination and moreover in tetrahedral coordination the anions approach along directions that do not coincide with electron density maxima. In consequence the interaction is weaker here, the crystal field splitting being $\Delta_t = \frac{4}{9}\Delta_0$ (Fig 10.31(b)) for the same cation and anions at the same distance apart. The *e* energy level is $\frac{3}{5}\Delta_t$ below the mean energy of the 3*d* orbitals, while the t_2 energy level is $\frac{2}{5}\Delta_t$ above the mean.

When the cation is in octahedral coordination and has four, five, six, or seven 3*d* electrons, two electronic configurations are possible. We illustrate this point by consideration of the case in which there are four 3*d* electrons. Suppose that the 3*d* orbitals are filled progressively. Then Hund's Rule indicates that the first three electrons will occupy singly the three t_{2g} orbitals so as to have their spins parallel. If

[5] The difference in designation of groups of orbitals in octahedral and tetrahedral environments follows from the symmetry requirements of group theory notation, which we shall not go into.

the $3d$ orbitals were not subjected to crystal field splitting, the fourth electron would enter one of the e_g orbitals. But in the crystal this would mean that the fourth electron has excess energy Δ relative to each of the first three electrons. If however the fourth electron were to enter one of the t_{2g} orbitals its spin would have to be anti-parallel to that of the electron already in this orbital and the energy of the ion would be increased by an amount P due to pairing of electron spins and to the presence of a second electron which must increase the electrostatic repulsive energy of the orbital. If $\Delta > P$, the fourth electron will enter a t_{2g} orbital; but if $\Delta < P$ it will enter an e_g orbital. The former situation is described as the *low spin configuration* or, since it is associated with a high value of Δ, as a *high crystal field splitting.* The latter situation, corresponding to a low value of Δ is known as the *high spin configuration* or as a *low crystal field splitting.* High and low spin configurations occur in the first transition metal series when there are four, five, six, or seven $3d$ electrons (Table 10.12), but with other numbers of d electrons only one configuration is possible.

For an ion in tetrahedral coordination the possibility of either the high or low spin configuration arises when it has three, four, five, or six $3d$ electrons. But here the crystal field splitting Δ is much smaller than for ions in octahedral coordination; indeed no example has yet been reported of a case where Δ is large enough to force spin pairing in the e orbitals in preference to occupation of the t_2 orbitals. Thus in all known examples the electrons are in the *high spin configuration*, every $3d$ orbital being singly occupied before spin pairing begins.

When the electron density distribution in an ion is not spherically symmetrical there can be no justification for regarding the ion as even approximately a rigid sphere of fixed radius. The equilibrium distance of the ion from another to which it is coordinated is determined by the overlap of the electron clouds of the two ions. In ions of the first transition metal series the $3d$ orbitals extend further from the nucleus than do the fully occupied inner orbitals so that the $3d$ electrons play a critical part in determining equilibrium distances. Take for example an ion in octahedral coordination with four $3d$ electrons in the high spin configuration. If the fourth electron is in the d_{z^2} orbital, then the electron density will be greater along the z-axis than along the x or y axes (the axial notation referring to the interatomic vectors between the transition metal ion and its coordinating ion which will not of course in general be parallel to the crystallographic reference axes). The equilibrium distance apart of the ions along the z-axis will consequently be greater than along the x and y axes so that the coordination polyhedron will be distorted from regular octahedral shape by elongation along one axis. On the other hand if the fourth electron enters

Table 10.12
High spin and low spin configurations of ions in octahedral coordination

Number of $3d$ electrons	High spin configuration $\Delta < P$		Low spin configuration $\Delta > P$	
	t_{2g}	e_g	t_{2g}	e_g
4	↑ ↑ ↑	↑	↑↓ ↑ ↑	
5	↑ ↑ ↑	↑ ↑	↑↓ ↑↓ ↑	
6	↑↓ ↑ ↑	↑ ↑	↑↓ ↑↓ ↑↓	
7	↑↓ ↑↓ ↑	↑ ↑	↑↓ ↑↓ ↑↓	↑

the $d_{x^2-y^2}$ orbital, the electron density will be least along the z-axis so that the coordination octahedron will be distorted from regularity by having one axis shorter than the other two, equal, axes. It is however found that the situation we considered first, with $z > x = y$, is generally energetically more favourable. Distortion of a coordination polyhedron from regularity produced in this sort of way is known as the *Jahn–Teller effect.*[6]

Jahn–Teller distortion occurs in octahedral coordination whenever the e_g orbitals are unequally occupied. This occurs when there are four $3d$ electrons in high spin configuration as in Cr^{2+}, Mn^{3+}, when there are seven $3d$ electrons in low spin configuration as in Co^{2+}, and when there are nine $3d$ electrons as in Cu^{2+} (see Table 10.13). Thus $CrCl_2$ has an orthorhombic structure related to the rutile structure (Fig 10.32) in which all the chromium ions are equivalent and in similarly distorted octahedral coordination to chlorine; of the six Cl^- ions surrounding any Cr^{2+} ion two are distant 2·92 Å and four are distant 2·37 Å from the cation. In $CrCl_2$ therefore the e_g electron must be in the d_{z^2} orbital.

A similar effect is to be expected when the t_{2g} orbitals are unequally occupied; but these orbitals do not have their electron density maxima directed along interatomic vectors so that the effect will be very much smaller. There is little or no experimental evidence for distortion of coordination octahedra due to unequal occupation of t_{2g} orbitals.

Table 10.13
The Jahn–Teller effect for octahedrally and tetrahedrally coordinated ions

Coordination	Distortion	Number of 3d electrons	Configuration	t_{2g} d_{xy}	d_{yz}	d_{zx}	e_g $d_{x^2-y^2}$	d_{z^2}
Octahedral	$c < a = b$	4	high spin	↑	↑	↑	↑	
		7	low spin	↑↓	↑↓	↑↓	↑	
		9		↑↓	↑↓	↑↓	↑↓	↑
	$c > a = b$	4	high spin	↑	↑	↑		↑
		7	low spin	↑↓	↑↓	↑↓		↑
		9		↑↓	↑↓	↑↓	↑	↑↓
				e $d_{x^2-y^2}$	d_{z^2}	t_2 d_{xy}	d_{yz}	d_{zx}
Tetrahedral	$c < a = b$	4	high spin	↑	↑		↑	↑
		9		↑↓	↑↓	↑	↑↓	↑↓
	$c > a = b$	3	high spin	↑	↑	↑		
		8		↑↓	↑↓	↑↓	↑	↑

In tetrahedral coordination unequal occupation of t_2 orbitals leads to distortion of inter-bond angles. Here neither the t_2 nor the e orbitals have their electron density concentrated along the directions in which the four coordinating ions approach the transition metal ion. However the t_2 orbitals, that is d_{xy}, d_{yz}, d_{zx}, have their electron density maxima closer to the interatomic vectors than do the e orbitals so that, when they are unequally occupied, the coordinating ions will be repelled by those orbitals that contain additional electrons. Inter-bond angles will then deviate from the 109°28′ appropriate to regular tetrahedral coordination. If there is only one electron in a t_2 orbital and it is in the d_{xy} orbital, then the coordinating ions will be repelled towards

[6] For a more extended discussion of the Jahn-Teller effect in molecules and in ionic crystals the reader is referred to Cotton and Wilkinson (1966) or Phillips and Williams (1965).

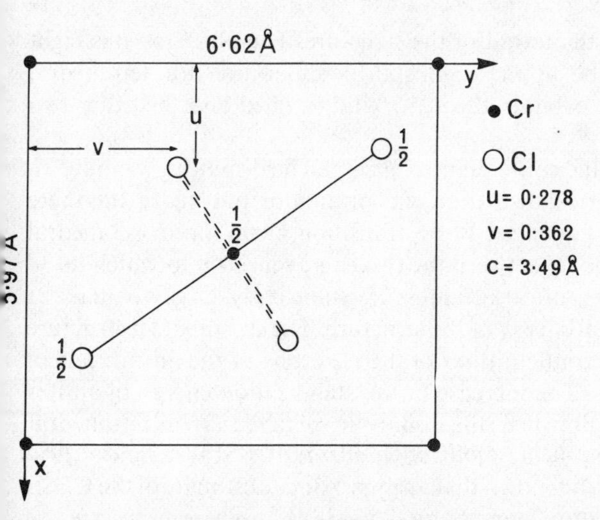

Fig 10.32 The crystal structure of $CrCl_2$. The structure is orthorhombic but related to that of rutile (TiO_2, tetragonal, cf. Fig 10.22(a)). The coordination octahedron about each Cr^{2+} cation displays Jahn-Teller distortion indicative of the presence of one electron in the $3d_{z^2}$ orbital. Bonds from the Cr^{2+} ion at $\frac{1}{2}, \frac{1}{2}, \frac{1}{2}$ are shown in the figure.

the z-axis so that the two inter-bond angles measured across z and $-z$ will be decreased (Fig 10.33(a)) and the z-axis of the tetrahedron will be increased in length. When however there are two electrons in t_2 orbitals, one in d_{yz} and one in d_{zx}, the coordinating ions will be repelled towards the xy plane so that the same two inter-bond angles will now be increased (Fig 10.33(b)) and the length of the z-axis will be decreased. Jahn–Teller distortion will thus be expected to occur in tetrahedral coordination, high spin configurations only being considered, when there are three, four, eight, or nine $3d$ electrons (Table 10.13) as in Ni^{2+}, Cu^{2+}. As for octahedral coordination, the effect is sufficiently strong to be observable only when the orbitals of higher energy, the t_2 orbitals, are unequally occupied; it has not been reported for cases of unequal occupation of e orbitals, which occur when an ion has one or six $3d$ electrons.

By way of example we take $NiCr_2O_4$ which adopts a modification of the spinel structure with symmetry reduced from cubic to tetragonal. The Cr^{3+} ion is in regular octahedral coordination to oxygen. The coordination polyhedron about the Ni^{2+} ion is distorted from regular tetrahedral shape, being elongated 14 per cent along that one

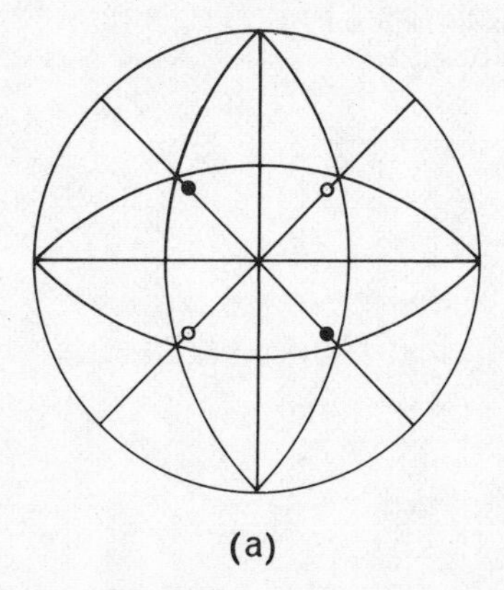

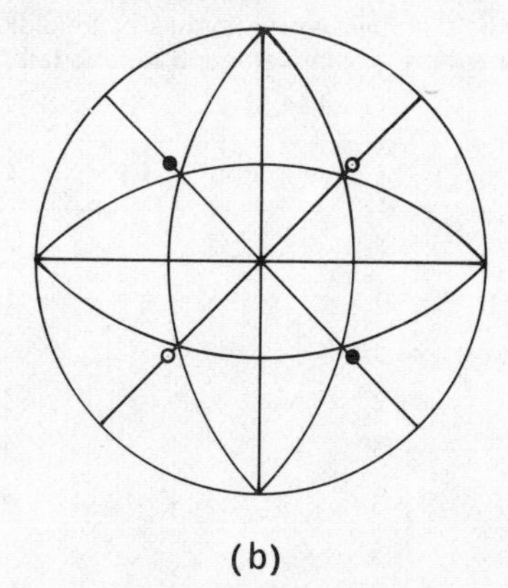

(a) (b)

Fig 10.33 Stereograms showing inter-atomic vectors to illustrate Jahn–Teller distortion of coordination tetrahedra. In (a) the d_{xy} orbital is occupied by one electron and the other t_2 orbitals are unoccupied: the coordination tetrahedron is distorted from cubic symmetry (indicated by the plotted $\langle 110 \rangle$ zones) to tetragonal symmetry with $c/a > 1$. In (b) the d_{xy} orbital is unoccupied and there is one electron in each of the other t_2 orbitals, d_{yz} and d_{zx}: again the coordination tetrahedron is distorted to tetragonal symmetry, but now with $c/a < 1$.

of its $\bar{4}$ axes which is parallel to the tetrad of the structure. The Ni^{2+} ion has eight $3d$ electrons, which must here be in the high spin configuration for tetrahedral coordination so that one of its t_2 orbitals, the d_{xy} orbital, is filled and the other two contain one electron each.

In discussing the crystallographic consequences of crystal field splitting we have so far restricted ourselves to geometrical distortion of coordination polyhedra, but there may also be direct energetic consequences. For a transition metal ion in octahedral coordination the energy difference Δ_0 between the two energy levels into which its $3d$ orbitals are split is of the order of 50 kcal mole^{-1}, sufficiently large to make a significant contribution to the total energy of the structure. The stability of a structure will therefore be affected by the configuration of the electrons in the $3d$ orbitals of constituent transition metal ions. The increase in the stabilization energy of an ion due to its $3d$ orbitals being split when the ion is in octahedral or tetrahedral coordination is known as *crystal field stabilization energy* (CFSE) or *ligand field stabilization energy* (LFSE). It can be shown that an approximate estimate of the CFSE of an ion in octahedral coordination is given by $(2n_t - 3n_e)\frac{1}{5}\Delta_0$ where n_t, n_e are the number of electrons in the t_{2g}, e_g orbitals respectively. The corresponding expression for an ion in tetrahedral coordination is $(3n_e - 2n_t)\frac{1}{5}\Delta_t$, where $\Delta_t \simeq \frac{4}{9}\Delta_0$. It is apparent from Table 10.14, which shows the CFSE calculated from these expressions for various numbers of $3d$ electrons, that tetrahedral coordination is not generally stabilized relative to octahedral coordination. Table 10.15 lists the CFSE's of some common ions; the data in this table, based on determinations of Δ_0 and Δ_t by spectroscopic studies of oxides, indicate that some ions have a strong preference for octahedral coordination.

Some of the structural consequences of CFSE are well illustrated by those AB_2O_4 compounds which contain transition metal ions and crystallize in the *spinel* structure. Consider first the common mineral *magnetite*, Fe_3O_4. If the formula were rewritten as AB_2O_4, that is as $Fe^{2+}Fe_2^{3+}O_4$, one might expect by analogy with $MgAl_2O_4$, which we discussed earlier, that the ferrous ions would be in tetrahedral coordination and the ferric ions in octahedral coordination. However Fe^{2+} has six $3d$ electrons

Table 10.14
Crystal field stabilization energies of transition metal ions in high spin configuration for octahedral and tetrahedral coordination. $\Delta_t \simeq \frac{4}{9}\Delta_0$.

Number of 3d electrons	Octahedral Configuration t_{2g}	e_g	Octahedral CFSE	Tetrahedral Configuration e	t_2	Tetrahedral CFSE
1	1		$\frac{2}{5}\Delta_0$	1		$\frac{3}{5}\Delta_t$
6	4	2		3	3	$\simeq 0{\cdot}27\Delta_0$
2	2		$\frac{4}{5}\Delta_0$	2		$\frac{6}{5}\Delta_t$
7	5	2		4	3	$\simeq 0{\cdot}53\Delta_0$
3	3		$\frac{6}{5}\Delta_0$	2	1	$\frac{4}{5}\Delta_t$
8	6	2		4	4	$\simeq 0{\cdot}36\Delta_0$
4	3	1	$\frac{3}{5}\Delta_0$	2	2	$\frac{2}{5}\Delta_t$
9	6	3		4	5	$\simeq 0{\cdot}18\Delta_0$
5	3	2	0	2	3	0
10	6	4		4	6	

Table 10.15
Spectroscopically determined crystal field stabilization energies for some transition metal ions in high spin configurations

Energies are in kcal mole^{-1}.

	Number of 3*d* electrons	CFSE	
		Octahedral	Tetrahedral
Ti^{3+}	1	20·9	14·0
Fe^{2+}	6	11·9	7·9
Co^{2+}	7	22·2	14·8
Ni^{2+}	8	29·2	8·6
Mn^{3+}	4	32·4	9·6
Cu^{2+}	9	21·6	6·4
Fe^{3+}	5	0	0
Mn^{2+}	5	0	0

so that its CFSE would be expected to be greater for octahedral than for tetrahedral coordination (Table 10.14) and this is borne out by the experimental data of Table 10.15. In short Fe^{2+} displays a *preference* for octahedral coordination. The CFSE of Fe^{3+} however is zero for both octahedral and tetrahedral coordination so that it exhibits no *site preference*. In magnetite therefore CFSE determines that the ferrous ions occupy randomly half the octahedral sites in the spinel structure and the ferric ions occupy the remainder of the octahedral sites and all the tetrahedral sites. This structure, which is not restricted to the composition Fe_3O_4, is known as the *inverse spinel* structure and formulated as $B^{[4]}[AB]^{[6]}O_4$ to distinguish it from the *normal spinel* structure $A^{[4]}B_2^{[6]}O_4$.

Crystal field stabilization energy is only one of the factors that determine whether an AB_2O_4 compound adopts the normal or the inverse spinel structure and, taken in isolation, it does not always predict correctly. For instance $CoAl_2O_4$ would be expected to be inverse because the spherical Al^{3+} atom has zero CFSE and consequently no site preference, while Co^{2+} would be expected on the basis of the data of Table 10.15 to have a clear preference for octahedral coordination. However the structure of $CoAl_2O_4$ is found to be normal, $Co^{[4]}Al_2^{[6]}O_4$. Detailed examination of the factors, other than CFSE, that determine whether the inverse or normal spinel structure is adopted by a particular compound is outside our scope.[7] Certainly the CFSE of non-spherical ions is always a significant and often the controlling factor. A more general and often important factor is the electrostatic energy of the structure which will be dependent on the precise positions of the oxygen atoms and so may be significantly different for the normal and inverse structures of the same compound. It seems likely that this factor is principally responsible for $CoAl_2O_4$ having the normal spinel structure. One factor that at first sight might be thought to be significant is the relative size of the A and B cations, especially when neither is a transition metal ion; there appears however in the spinels to be no relationship between relative cation size and the adoption of one or other form of the structure.

Ideal and defect structures

So far we have assumed every crystalline solid to be a regular three-dimensional arrangement of atoms, an arrangement that has an identifiable unit-cell which, by repetition on a lattice, gives rise to the structure of the whole crystal. In such a crystal

[7] The interested reader is referred to Greenwood (1968).

of an *ideal structure* every crystallographically equivalent site, whether equivalent by lattice translation or by operation of the symmetry elements of its space group, is occupied by a chemically identical atom. Thus in common spinel $MgAl_2O_4$ (Fig 10.28) every magnesium atom occupies a site identical in every respect to the sites occupied by all other magnesium atoms; and likewise for aluminium and for oxygen. In this structure all the atoms of one chemical element occupy a set of equivalent positions and those of each other element occupy a distinct set of equivalent positions; in short the structure of common spinel is an *ideal structure*. However an inverse spinel, such as $Fe[NiFe]O_4$, has a random distribution of nickel and ferric ions on its octahedral sites with specific occupation of its tetrahedral sites by ferric ions and its anion sites by oxygen. That the point group of crystals of this substance is the same as that of crystals of common spinel, *m3m*, indicates that point group symmetry is only the symmetry of the *average structure* of the substance concerned. The contents of a unit-cell of $Fe[NiFe]O_4$ cannot be described precisely: for each octahedral site there is a probability of $\frac{1}{2}$ of occupation by Ni^{2+} and an equal probability of occupation by Fe^{3+}. In such a *defect structure* each octahedral site can conveniently be said to be occupied by a 'half-atom' of nickel and a 'half-atom' of iron, a statement that has only statistical significance. In such a structure electrical neutrality is maintained over the whole structure, but not necessarily for each unit-cell. Likewise the electrostatic valency principle holds for the average structure, but not necessarily for every oxygen atom in the structure. For instance in the spinel structure every oxygen anion is bonded to one cation in a tetrahedral site and to three cations in octahedral sites; in $Fe(NiFe)O_4$ the tetrahedrally coordinated Fe^{3+} cation gives rise to bond strengths of $\frac{3}{4}$, the octahedrally coordinated Fe^{3+} to $\frac{1}{2}$ and the octahedrally coordinated Ni^{2+} to $\frac{1}{3}$ so that the sum of the bond strengths reaching an O^{2-} anion may be $2 \pm \frac{1}{4}$ or $2 \pm \frac{1}{12}$, always slightly greater or less than the ideal value 2. However the electrostatic valency rule is statistically obeyed by the average structure: $\frac{1}{2}Ni^{2+} + \frac{1}{2}Fe^{3+}$ represents a total charge of $2\frac{1}{2}+$ on each octahedral site so that the sum of the bond strengths reaching each oxygen atom from its three neighbouring octahedral sites is $3 \times \frac{5}{12} = 1\frac{1}{4}$ which, added to $\frac{3}{4}$ from the neighbouring tetrahedral site, yields a total bond strength of 2 exactly balancing the charge of -2 on the oxygen anion.

Confining our argument still to the spinel structure, there are spinels which exhibit randomness of occupation of both tetrahedral and octahedral sites. For instance $MgGa_2O_4$ has its Mg^{2+} and Ga^{3+} cations distributed statistically according to the scheme expressed by the structural formula $[Mg_{0\cdot33}Ga_{0\cdot67}]^{[4]}[Mg_{0\cdot67}Ga_{1\cdot33}]^{[6]}O_4$ so that the Mg/Ga ratio is identical on tetrahedral and octahedral sites. In $MgFe_2O_4$ however the two types of site are occupied in different proportions; thus $[Mg_{0\cdot1}Fe_{0\cdot9}]^{[4]}[Mg_{0\cdot9}Fe_{1\cdot1}]^{[6]}O_4$ represents the equilibrium distribution of the two cations between the two types of site at a certain temperature; this sort of distribution is usually temperature-dependent.

In the case of the inverse spinel *magnetite* Fe_3O_4, which can be represented as $Fe^{3+}[Fe^{2+}Fe^{3+}]O_4^{2-}$, the ferrous and ferric ions that formally occupy the octahedral sites cannot be distinguished by Mössbauer spectroscopy (chapter 15) because the transfer of the valency electron by the process $Fe^{3+} + e^- \rightleftarrows Fe^{2+}$ is very rapid. All the iron atoms on octahedral sites are continuously changing their oxidation state so that averaged over a finite time the occupation of every octahedral site is identical. The occurrence of facile electron transfer between iron atoms on octahedral sites incidentally accounts for the high electrical conductivity and for the optical opacity of magnetite.

In general it is not experimentally easy to determine site occupations in compounds with spinel structures, some of the more sophisticated diffraction techniques being usually necessary. Nevertheless a very considerable body of structural research has been done on the spinels, the objective of most of which has been the interpretation of their strongly structure-dependent magnetic and electrical properties.

Each of the defect structures that we have considered so far, $NiFe_2O_4$, $MgGa_2O_4$, $MgFe_2O_4$, Fe_3O_4 has had a chemical composition that could be expressed by a simple formula even though some or all of its cations may be distributed randomly over one or more sets of equivalent sites in its structure. But much the same sort of defect structure can occur in compounds that do not have such simple *stoichiometric* compositions. For instance it is known experimentally that Fe^{2+} can substitute for Mg^{2+} in $MgAl_2O_4$ to give a continuous range of *solid solution*[8] between the extremes or *end-members* $MgAl_2O_4$ and $FeAl_2O_4$; in short all compositions $[Mg_xFe_{1-x}]Al_2O_4$, where $0 \leqslant x \leqslant 1$, have the spinel structure. In this *solid solution series* Mg^{2+} and Fe^{2+} are randomly distributed on the tetrahedral sites of the spinel structure. Solid solution is also possible between a normal spinel at one end of the series and an inverse spinel at the other extreme composition; for example $ZnFe_2^{3+}O_4$ is normal, $Fe^{3+}[Fe^{2+}Fe^{3+}]O_4$ is inverse and compositions intermediate between the two can be represented as $[Zn_{1-x}Fe_x^{3+}]^{[4]}[Fe_x^{2+}Fe_{2-x}^{3+}]^{[6]}O_4$. In general in discussing solid solutions it is of prime importance to know which sites in the structure are available for occupation by the atoms of each element concerned.

Atoms of various elements are able to substitute for one another to a greater or smaller extent in solid solutions provided they form bonds of similar type and are of similar size. In general a continuous range of solid solution between two end-members occurs only when the ions that substitute for each other differ in radius by not more than 15 per cent of the radius of the smaller ion; moreover it is essential that the end-members should be structurally similar. Whether solid-solution is continuous over the whole range or restricted to within a few per cent of each end-member then depends on the ability of the structure to tolerate the presence of ions of different radii on structurally equivalent sites. Thus in the olivine structure there is continuous solid solution between forsterite Mg_2SiO_4 and fayalite Fe_2SiO_4, where the radii of the large cations are $r(Mg^{2+}) = 0{\cdot}66$ and $r(Fe^{2+}) = 0{\cdot}74$ Å. However monticellite $CaMgSiO_4$, although isostructural, has its calcium atoms completely filling one set of equivalent (mirror plane) sites and there is no appreciable solid solution between monticellite and forsterite. The reason for this is that $r(Ca^{2+})$ at 0·99 Å is substantially more than 15 per cent larger than $r(Mg^{2+})$ at 0·66 Å and the structure cannot tolerate the excessive local distortion produced by random occupation of 6-fold coordinated mirror plane sites by cations so diverse in size; only when the mirror plane sites are fully occupied by the large cation is the structure stable. At intermediate compositions $[Ca_xMg_{2-x}]SiO_4$, where $0 < x < 1$, it is energetically more favourable for forsterite and monticellite to coexist as separate phases than to form a solid solution.

In some structures continuous solid solution between a particular pair of end-member compositions occurs only at high temperatures where the increased amplitude of thermal vibrations makes it possible for the structure to tolerate the local strains generated by random occupation of equivalent sites by ions of diverse size. Such solid solutions respond to decrease in temperature under equilibrium

[8] We treat solid solutions here in simple structural terms and postpone a fuller discussion in terms of thermodynamic stability to chapter 13.

conditions either by *ordering* of cations of different kinds on to distinct sets of equivalent sites, or by *exsolution*. It is however usually possible to *quench* the high-temperature equilibrium state by very rapid cooling and so preserve the solid solution for more convenient structural study at room temperature. Experimentally in silicates and many other complex ionic structures it is very easy to quench high-temperature equilibrium states (see chapter 16) and correspondingly difficult to achieve equilibrium site occupations at lower temperatures; this is mainly because the rate of diffusion of ions through structures falls off very rapidly with decreasing temperature so as to be negligible well above room temperature.

It is appropriate at this point to draw attention to the analogous behaviour of many complex ionic, especially mineral, structures on the one hand and alloy systems on the other. Solid solution, exsolution, and order–disorder phenomena are well known and in many respects similar in both; but the experimental study of simple alloy systems is generally easier because equilibrium is very much more rapidly attainable. In much of what follows we could equally well take our examples from alloys as from complex ionic structures; but, as we are primarily concerned with minerals, we shall take most of our examples from the latter field and have recourse to alloy studies only for the provision of certain particularly simple illustrations of principle.

It is very common for crystals of naturally occurring minerals to have random occupation of equivalent structural sites by a variety of cations; that is to say minerals are very commonly solid solutions and it is most unusual to find a naturally occurring end-member. It is convenient to describe mineral compositions in terms of the appropriate percentages of end-member compositions in the solid solution; the actual composition is imagined to be produced by *isomorphous replacement* of one or more ions of its dominant end-member by various other ions in appropriate amounts. Some common replacements of cations bonded to oxygen are shown in Table 10.16. It is apparent from the table that ions of different valency may replace one another; but when that occurs there must be a compensating replacement to maintain overall electrical neutrality. Thus if Al^{3+} replaces Si^{4+} there must elsewhere in the structure be replacement of cations of valency m (e.g. Na^{+}) by cations of valency $m+1$ (e.g. Ca^{2+}) to exactly the same extent; the total replacement can be expressed as $Na^{+}Si^{4+} \rightarrow Ca^{2+}Al^{3+}$. In isomorphous replacement most cations show a clear preference for sites with the coordination number appropriate to their radius; but Al^{3+} is unusual in being able to replace either on tetrahedral or on octahedral sites. By way of a very simple example of the representation of the composition of a solid solution in terms of percentages of real or hypothetical end-members we take the natural olivine $Mg_{1\cdot828}Ni_{0\cdot005}Fe^{II}_{0\cdot160}Mn^{II}_{0\cdot003}Ca_{0\cdot004}Si_{0\cdot996}O_4$, which may be represented as 91·4 per cent Mg_2SiO_4 (forsterite), 0·25 per cent Ni_2SiO_4, 8·0 per cent Fe_2SiO_4 (fayalite), 0·15 per cent Mn_2SiO_4 (tephroite), 0·2 per cent Ca_2SiO_4. If only the principal constituents are of interest the olivine can be represented compositionally as $Fo_{92}Fa_8$ where Fo = the forsterite end-member and Fa = the fayalite end-member.

It is the case of isomorphous replacement that makes the chemical compositions of so many silicate minerals appear at first sight to be highly complex. But when the crystal structure is known it becomes possible to assign the cations to sites of appropriate coordination number making use of the general rule that vacancies are, for energetic reasons, more likely to occur on sites of highest coordination number. Where there is more than one set of equivalent sites of the same coordination number, as in the case for octahedral sites in the olivine structure, the occupation of the two

Table 10.16
Cations which commonly replace one another when bonded to oxygen

Coordination	Cations
Tetrahedral	Si^{4+}, Al^{3+}
Octahedral	Al^{3+}, Fe^{3+}, Mg^{2+}, Ti^{4+}, Ni^{2+}, Fe^{2+}, Mn^{2+}
>6	Na^{+}, Ca^{2+}, K^{+}

or more sets cannot be distinguished on the basis of composition alone. Collation of all the reliable analyses that have been made of any mineral species usually indicates that the structure is stable over quite a wide range of composition, the observed range being restricted in certain directions, not by structural instability, but by low concentration of the rarer cations in the medium in which crystallization took place; it is rare however for the observed range of naturally occurring compositions to extend quite to pure end-member compositions of the solid solution, although the pure end-members can often be prepared synthetically.

In the defect structures that we have discussed so far crystallographically equivalent sites are occupied randomly by two or more ionic species. There is however another sort of defect structure in which there is only partial occupation of one or more sets of crystallographically equivalent sites, or, in other words, a random distribution of atoms of one species and vacancies on one or more sets of equivalent sites. Before going on to consider such defect structures it is necessary to devote some attention to the general question of vacancy defects in crystal structures.

It is thermodynamically impossible for any crystal at temperatures above absolute zero to have a perfectly regular three-dimensional array of atoms. In *real* crystals, as distinct from the *ideal* crystals that we have considered up to this point, there must be defects at all attainable temperatures and such essential defects are of two kinds. The *Schottky defect* is such that the proportion of vacancies on each set of equivalent sites is the same, the proportion increasing with temperature from zero at zero K. Schottky vacancies occur by migration of atoms to the surface of the crystal. The *Frenkel defect* is such that a vacancy is created by the migration of an atom from its proper structural site to an 'interstitial' position, that is to a site not occupied in the ideal structure. Here too the number of defects per unit volume increases with temperature from zero at zero K; but, in contrast, Frenkel defects do not necessarily involve all sets of equivalent sites. Sometimes only one atomic species is involved and, since the displaced atom is retained in the structure, no charge imbalance is produced. Frenkel defects are more likely to involve the smaller atoms of the structure, in ionic structures the smaller cations, because room must be available interstitially to accommodate the displaced atom. Thus in a structure with a close-packed anion framework and cations on tetrahedral sites interstitial to the close-packed array it is obviously energetically more probable that a cation rather than an anion will migrate to an unoccupied tetrahedral or octahedral interstitial site to create a Frenkel defect. Real crystals at room temperature usually contain both types of defect, but one type is usually predominant. These two types of inevitable defect play an important part in determining the magnitude of certain physical properties of single crystals, such as electrical conductivity and ionic diffusion.

We turn now from the ubiquitous Schottky and Frenkel defects to consider a group of compounds, known as the *berthollides*, some of which exhibit in addition vacancy defects of quite a different kind. The berthollides, alternatively known as *non-*

stoichiometric inorganic compounds, are a chemically diverse group characterized at their simplest by formulae of the type AX_x where x is non-integral and variable over quite a wide range without change of structural type.[9] One of the simplest examples of such a compound is ferrous oxide, which is known to be variable in composition in the range FeO to $Fe_{0{\cdot}84}O$, the precisely stoichiometric composition being thermodynamically unstable except perhaps at very low temperatures. Electrical neutrality is preserved in compositions other than FeO by oxidation of the appropriate amount of ferrous iron to the ferric state so that the formula of a non-stoichiometric ferrous oxide Fe_xO should strictly be written as $Fe^{3+}_{2-2x}Fe^{2+}_{3x-2}O$. Physical measurements show that in Fe_xO, where $x < 1$, a proportion of the iron sites in the structure are vacant. It used to be supposed that throughout the range to $Fe_{0{\cdot}84}O$ the vacant iron sites were randomly distributed, that the approximately cubic close-packed oxygen arrangement remained unperturbed even when the proportion of unoccupied octahedral cation sites approached 16 per cent; more recent work has shown that, though this may be so for small departures from stoichiometry, larger departures are accommodated by adjustment of the whole structure.

The structural study of non-stoichiometric compounds is currently highly active and we can do no more here than take some outstanding examples to illustrate the sorts of ways in which various simple structures adapt to departures from the ideal composition. The field is wide, complex, and inadequately explored; but one can generalize to the extent of saying that most non-stoichiometric compounds have one set of equivalent sites not fully occupied. There are exceptions and one of the simplest is titanium monoxide, ideally TiO but in reality compositions in the range $Ti_{0{\cdot}8}O$ to $TiO_{0{\cdot}7}$; in this compound, which has the NaCl structure, both cation and anion sites may have vacancies. At the extreme composition $Ti_{0{\cdot}8}O$ the anion sites are fully occupied and the cation sites are only partially occupied. At the other extreme $TiO_{0{\cdot}7}$ the anion sites are partially and the cation sites fully occupied. At intermediate compositions both cation and anion sites are only partially occupied; indeed at the composition TiO only 85 per cent of each type of site is occupied.

We return now to ferrous oxide. When Fe_xO is crystallized at high temperature and rapidly cooled (quenched), it develops small volumes or *micro-domains* of the inverse spinel Fe_3O_4 in which two thirds of the iron atoms are in the ferric state. The micro-domains are too small to be regarded as a second phase and moreover they are in continual movement within the host structure, changing in shape and size as they move. In both the ideal FeO structure and the Fe_3O_4 structure the oxygen atoms are in approximately cubic close-packing and in both structures Fe^{2+} lies on octahedral sites. In FeO all the octahedral interstices of the cubic close-packed anion array are occupied by Fe^{2+}; in Fe_3O_4 half the octahedral interstices are necessarily vacant, one quarter are occupied by Fe^{2+} and another quarter by Fe^{3+}, the remaining Fe^{3+} anions lying on tetrahedral sites. The proportion of the volume of the crystal of Fe_xO that must be assigned to the Fe_3O_4 micro-domains increases as x decreases such that the structure can be represented as $(4x-3)FeO.(1-x)Fe_3O_4$. In this first example of a berthollide not only are some of the octahedral cation sites of ideal FeO vacant, but some of the ideally unoccupied tetrahedral sites are occupied by Fe^{3+}.

Titanium monoxide responds quite differently to departures from the ideal composition TiO. In TiO_x vacancies are ordered in such a manner as to give rise to

[9] In general terms we may define as *daltonides* compounds to which Dalton's laws of constant and multiple proportions apply and as *berthollides* compounds to which these laws do not apply.

a *superlattice*.[10] A superlattice can be defined in general terms as a structure in which the atoms are in positions that deviate only slightly from those of the simple parent structure, at least one set of equivalent sites of the parent structure being split into two or more sets of equivalent sites in the superlattice. We shall give point to this definition by considering the simple alloy system Cu–Au.

The pure elements copper and gold both crystallize with the cubic close-packed structure. At high temperatures there is continuous solid solution between copper and gold, solid solutions of all compositions having the cubic close-packed structure with random occupation of all structural sites by Cu and Au in the appropriate proportions. At room temperature however the equilibrium structures of compositions close to CuAu and to Cu_3Au are ordered, Cu and Au atoms being disposed in an ordered manner on the sites of the cubic close-packed parent structure (Fig 10.34). In the ordered phase CuAu copper and gold atoms segregate into alternate (001) planes; the symmetry of the structure drops from cubic to tetragonal; the four equivalent sites of the cubic close-packed unit-cell become two sets of two equivalent sites, one occupied by Cu (at $z = \frac{1}{2}$, say) and the other by Au (at $z = 0$); and, if reference axes are maintained in the same orientation, the axial ratio c/a drops from unity for the cubic close-packed high-temperature phase to 0·932 for the tetragonal phase stable at lower temperatures. In the ordered phase Cu_3Au cubic symmetry is retained; the unit-cell still contains four atoms disposed in precisely the same manner as in the high-temperature cubic close-packed phase; but one site is occupied exclusively by Au and the other three by Cu so that the lattice type becomes primitive.

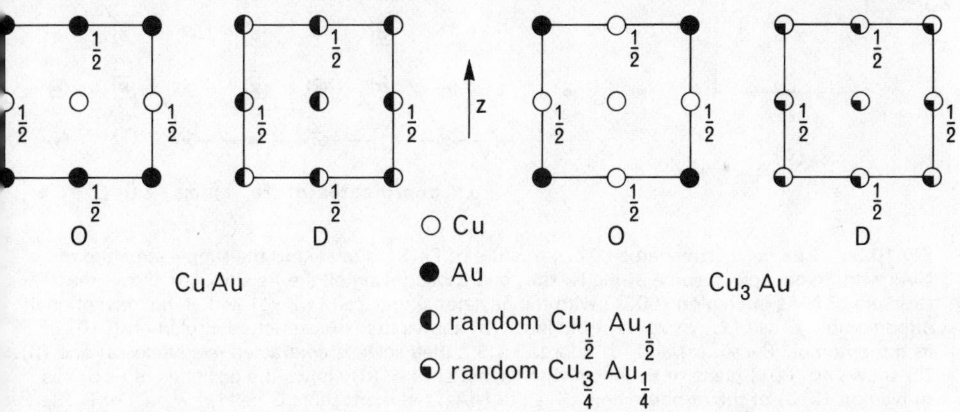

Fig 10.34 The copper-gold alloys. The four structural plans on (100) show in sequence from left to right the ordered (O) structure of CuAu, the disordered (D) structure of CuAu, the ordered (O) structure of Cu_3Au, and the disordered (D) structure of Cu_3Au. The ordered structure of CuAu is tetragonal; the other structures are cubic.

These two ordered phases in the Cu–Au system are in their different ways examples of superlattice structures; in CuAu the symmetry is lower than and derivative from that of the parent structure, while Cu_3Au undergoes a change of lattice type from F to P on ordering with falling temperature.

We return now to the berthollide compounds: the several superlattice structures reported for TiO_x are complex and, although intrinsically of great interest,

[10] In strictly crystallographic terms *superlattice* is a misnomer; *superstructure* would be the more apt term. However the description of this sort of structural modification as a superlattice is firmly embedded in the literature and it would be simply pedantic to insist on using the nomenclatorially proper term superstructure here.

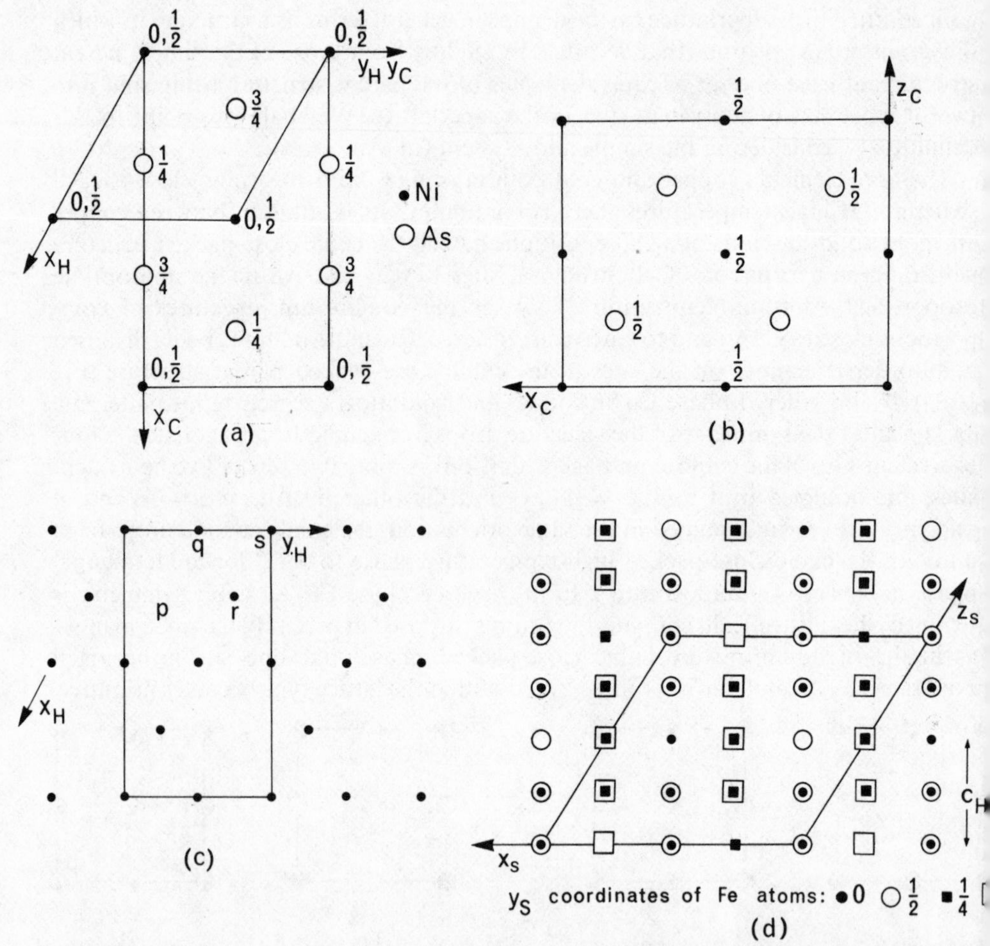

Fig 10.35 The Fe_7S_8 superlattice. The structure of Fe_7S_8 is related to the simple structure of NiAs with Fe occupying some of the Ni sites and S occupying all the As sites. (a) shows the structure of NiAs in plan on (0001) with the hexagonal unit-cell (x_H, y_H) and an unconventional orthorhombic C-cell (x_c, y_c, $z_c = z_H$) outlined. (b) shows the NiAs structure in plan on (010) of its orthorhombic C-cell. (c) and (d) refer to Fe_7S_8; their scale is contracted relative to (a) and (b). (c) shows an (001) plane of Fe atoms with vacant sites p. (d) shows the positions of Fe atoms in plan on (010) of the orthorhombic C-cell of NiAs; the monoclinic C-cell (x_s, y_s, z_s) of Fe_7S_8 is outlined.

inappropriate to an introduction to the subject[11] so we turn for illustration to the geometrically simpler superlattices of FeS_x. Several superlattices have been reported for FeS_x; all are based on the nickel arsenide structure (Fig 10.35(a), (b)). For example in Fe_7S_8 all the sulphur sites are occupied, but one in four of the iron sites is vacant in alternate (001) planes of iron atoms (Fig 10.35(c), (d)). The vacancies are regularly arranged on a plane hexagonal mesh of side $2a_H$, where a_H is the vector corresponding to the unit-cell edge of the parent NiAs structure; and in successive defect planes the vacancies occur in the positions p, q, r, s, p,... as shown in Fig 10.35(c). All atoms are displaced from the corresponding positions in the parent NiAs type unit-cell so

[11] The reader who desires to explore the TiO_x structures is advised to read the chapter by A. D. Wadsley in Mandelcorn (1964) and references cited therein.

hat a unit-cell of the superlattice is monoclinic-C related to the parent unit-cell by he matrix $420/020/\overline{2}\overline{1}2$. On heating above 360 °C the Fe_7S_8 superlattice is destroyed ınd the disposition of vacancies becomes random on a nickel arsenide type structure. n Fe_7S_8 therefore there is an order–disorder transformation generally analogous vith the transformations we have mentioned previously in the alloy structures CuAu ınd Cu_3Au, and in the alloy CuZn which will be discussed shortly.

Not all berthollides form superlattices: some crystallize in what are known as *shear structures*. One of the simplest examples of a shear structure, which will serve to define he type, is $(Mo, W)O_x$ where $2{\cdot}875 < x < 3$. The crystal structure of WO_3 is a listortion of the simple cubic ReO_3 structure. In ReO_3 rhenium atoms are sited at he corners and oxygens at the mid-points of the edges of the cubic unit-cell (Fig .0.36(a)), each Re atom being coordinated octahedrally by oxygen. These octahedra ıre linked by their corners so that each oxygen is shared by two Re atoms; each ›ctahedron is thus an ReO_3 unit. It is thus convenient to illustrate the structure by howing only ReO_3 octahedra: in Fig 10.36(b) Re atoms are represented by solid ircles and the coordination octahedra of oxygens are indicated in projection. The tructure of WO_3 is similar to that of ReO_3, but distorted so that its symmetry is nonoclinic. Based on this structure are the structures of a series of oxygen deficient hases of molybdenum-tungsten oxides whose compositions lie in the range $BO_{2{\cdot}875}$ o $BO_{2{\cdot}929}$, where B = Mo, W. Within this compositional range phases of definite omposition occur, which can be represented as B_nO_{3n-1} where $n = 8, 9, 10, 11, 12,$ 14. In the structures of these phases the oxygen deficiency is taken up by the formation of planes of discontinuity in the distorted ReO_3 structure where groups of four ›ctahedra share edges. The four oxygen atoms involved in edge sharing in any such group, marked a, b, c, d in Fig 10.36(c), are linked only to cations within the group; he other 14 oxygen atoms shown in the figure (6 in the plane of the diagram, 4 above ınd 4 below) are bonded to one cation of the group and to one other cation. Each uch group can thus be regarded as containing four cations and $4 + \frac{14}{2}$ oxygens so that t has the composition B_4O_{11} and is deficient relative to the stoichiometric omposition $4BO_3$ by one oxygen atom. By regular disposition of such B_4O_{11} groups hrough the distorted ReO_3 structure the structures of the phases B_nO_{3n-1} are roduced. Figure 10.36(d) illustrates the structure of one such phase, Mo_9O_{26}. This tructure can be regarded as being built of strings of nine octahedra, all of which hare corners with adjacent octahedra except for the first and last pairs of octahedra of the string which share edges with the extreme pairs of octahedra of an adjacent tring as Mo_4O_{11} groups.

Such B_nO_{3n-1} phases are strictly stoichiometric, continuously variable composition n the range $8 \leqslant n \leqslant 14$ being obtained by the crystallization of compositionally ıdjacent phases in appropriate proportions. Structurally distinct BO_x phases also ›ccur and some have cations in 4-fold or 7-fold coordination. The common feature of all BO_x phases is that electrical neutrality is maintained by the metal ion adjusting ts valence state; thus an oxide Mo_nO_{3n-1} can be represented as $Mo_2^{5+}Mo_{n-2}^{6+}O_{3n-1}$. It has not however so far proved possible to distinguish the sites occupied by the ations of higher and lower valence state. The electrical and magnetic properties of uch phases indicate that the extra electron formally associated with Mo^{5+} is lelocalized and so has the characteristics of a valency electron in a metallic structure.

Shear structures are also found in the titanium oxides Ti_nO_{2n-1}, where $4 \leqslant n \leqslant 10$. In such oxides the rutile structure is regularly interrupted by face-sharing octahedra. The principle of formation of these shear structures is essentially the same as for those

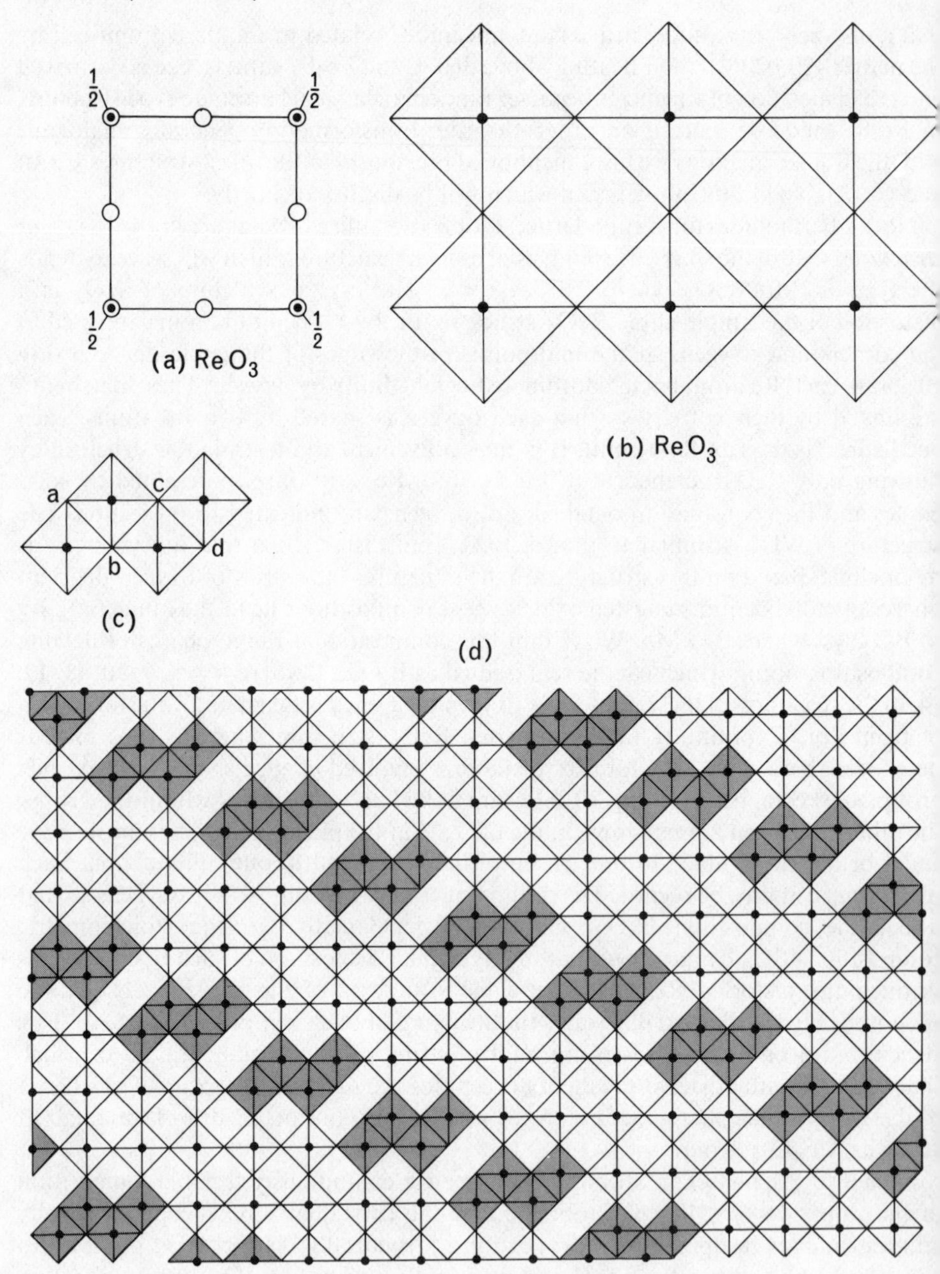

Fig 10.36 The shear structure Mo_9O_{26}. The structure of molybdenum and tungsten oxides are related to the simple cubic structure ReO_3 shown in plan on (001) in (a). In the same orientation and on the same scale the arrangement of corner-sharing octahedra in ReO_3 is shown in (b). The formation of a group of four octahedra sharing edges to accommodate oxygen deficiency in an Mo_4O_{11} group is illustrated on a smaller scale in (c); the oxygen atoms a, b, c, d are linked only to cations within the group. The plan (d) illustrates the disposition of such Mo_4O_{11} groups (shaded) in the structure of Mo_9O_{26}: the structure is composed of strings of nine octahedra which share corners, except for the first and last pairs of octahedra which share edges with those of adjacent strings. In (b), (c), and (d) solid circles represent cations and oxygen atoms are situated at intersections.

of the molybdenum-tungsten oxides, but their geometry is very much more difficult and we shall not attempt here to explore the Ti_nO_{2n-1} structures.

A considerable body of evidence has accumulated in recent years to suggest that structures with any appreciable proportion of randomly vacant sites tend to be thermodynamically unstable. Greater stability seems usually to be achieved when the defects are ordered either by the formation of micro-domains of another phase, as in Fe_xO, or by superlattice formation, as in FeS_x, or alternatively when the vacancies are eliminated by formation of shear structures, as in $(Mo, W)_nO_{3n-1}$. A substantial field remains to be explored and there are many difficulties not only in the diffraction analysis of such complex structures, but also in the preparation and chemical characterization of the phases concerned. Much of the impetus which has advanced such studies so rapidly in recent years springs from commercial interest in the electrical and magnetic properties of non-stoichiometric compounds.

In conclusion we deal briefly with a rather different sort of defect structure; these are structures in which some of the atoms do not lie on definite structural sites. The extreme situation in which certain atoms are totally mobile is most unusual; very much more common is the presence in the structure of groups of atoms that are free to rotate about one or more axes.

The classic example of free ionic mobility is the polymorph of AgI stable between 146 °C and the melting point at 555 °C. In this structure iodine atoms are situated at the corners and body centre of the cubic unit-cell; the silver atoms are not fixed in position. Apparently Ag^+ ions can migrate freely through the fixed structure of large, easily deformed I^- anions. The easy movement of Ag^+ ions imparts to this structure an anomalously high electrical conductivity.

Free rotation of a group of atoms within a structure is simply exemplified by the alkali cyanides such as KCN. At room temperature the structure of KCN is simply related to the NaCl type: the K^+ ion lies on the cation sites of the type structure and the CN^- ion is free to rotate about the anion sites, neither carbon nor nitrogen atoms having fixed positions in the structure. In free rotation the statically linear CN^- ion becomes spherically symmetrical with an effective radius of 1·92 Å so the effective radius ratio $r_{K^+}/r_{CN^-} = 1{\cdot}33/1{\cdot}92 = 0{\cdot}693$ and is within the limits appropriate to the NaCl structure type for strictly ionic bonding. The energy required for free rotation of the CN^- ion is supplied by the thermal energy of the crystal, which at −106 °C becomes inadequate to maintain the CN^- ion in free rotation; at this temperature the structure transforms to a closely related orthorhombic structure which has its cyanide ions aligned parallel to the [110] direction of the high-temperature structure.

Although quite a number of examples are known of the KCN type of defect structure and of structures in which planar ions such as CO_3^{2-} and NO_3^- are free to rotate about the axis normal to their planes, this sort of defect structure is necessarily more restricted in its occurrence and of less general interest than the defect structures discussed earlier.

Order-disorder transformations

In this chapter we have mainly been concerned with the description and explanation of static structures. We have occasionally mentioned the different structures or *polymorphs* of certain substances stable in different ranges of temperature and pressure. We are now concerned with the way in which certain structures respond to changes in temperature, that is with a certain type of *phase change*. Order–disorder transformations are a particularly interesting type of phase change because the structure changes gradually with changing temperature, rather than abruptly as in melting or in some other types of solid to solid transformation; moreover the change from an ordered to a disordered structure is related generally to changes from ideal to defect structures as we have already indicated in the cases of CuAu and Cu_3Au. We shall here confine our discussion of this extensively studied phenomenon to a simple alloy example, the order–disorder transformation in β-brass, CuZn.

Above 460 °C the structure of CuZn is the body-centred cubic metal structure (Fig 10.12) with Cu and Zn atoms randomly distributed over all structural sites. When CuZn is annealed below 460 °C a superlattice is formed which has the CsCl type of structure in which each atom is surrounded by eight nearest neighbours of the other kind. In simple structural terms the atomic radii of Cu (1·24 Å) and Zn (1·33 Å) are such that at high temperatures no distinction is necessary between atoms of the two elements. But at lower temperatures the strain associated with random distribution of atoms of the two elements on the two sites, 0, 0, 0 and $\frac{1}{2}, \frac{1}{2}, \frac{1}{2}$, of the cubic I-cell causes the structure to become unstable relative to the ordered structure in which one kind of atom is concentrated on the 0, 0, 0 site and the other on the $\frac{1}{2}, \frac{1}{2}, \frac{1}{2}$ site; the cubic lattice type changes from I to P without any very great change in unit-cell dimensions.

The order–disorder transformation in β-brass is usually discussed in terms of the *Bragg–Williams model* of order–disorder in AB alloys. We shall deal in some detail with the Bragg–Williams treatment after *entropy* has been introduced in chapter 13. Here we shall confine ourselves to definition of an order parameter and to consideration of some of the results obtainable from this simple model of order–disorder.

We suppose that the crystal contains N atoms of element A and an equal number N atoms of element B. We further suppose that the ordered structure of the alloy has two types of site, α and β, and that in the ordered structure all N atoms of A lie on α-sites and all N atoms of B lie on β-sites. In the case of β-brass this amounts to supposing that the copper atoms all lie on unit-cell corner sites, 0, 0, 0 etc, and that all the zinc atoms lie on body centre sites, $\frac{1}{2}, \frac{1}{2}, \frac{1}{2}$ etc. In a partially ordered structure the N α-sites will be occupied by a number $R < N$ of A atoms and by $N-R$ B atoms, the β-sites being occupied by $N-R$ A atoms and R B atoms. The *degree of order s* of such a structure is defined as the difference between the proportion of atoms correctly and incorrectly placed relative to the fully ordered structure, that is

$$s = \frac{2R}{2N} - \frac{2(N-R)}{2N}$$

$$= \frac{2R}{N} - 1.$$

In the fully ordered structure $R = N$ so that $s = 1$; and in the completely disordered structure A atoms are equally distributed between α- and β-sites so that $R = N/2$ and $s = 0$. We have supposed that in the fully ordered structure all the A atoms lie on α-sites; an identical ordered structure would be produced by ordering A atoms on to β-sites and for this structure $R = 0$ and $s = -1$; the two structures are simply brought into coincidence by translation of the origin of coordinates from 0, 0, 0 to $\frac{1}{2}, \frac{1}{2}, \frac{1}{2}$. It is necessary to consider only the tendency of A atoms to lie on one type of site, which we take to be α; that is to say we are concerned only with $N/2 \leqslant R \leqslant N$ and $0 \leqslant s \leqslant 1$.

In a structure with degree of order s the proportion of A atoms occupying α-sites, the sites occupied by A atoms in the fully ordered structure, is $R/N = \frac{1}{2}(1+s)$; and the proportion of A atoms on β-sites, that is wrongly situated relative to the chosen ordered structure, is $(N-R)/N = \frac{1}{2}(1-s)$. The ratio of rightly to wrongly placed A atoms is thus $R/(N-R) = (1+s)/(1-s)$. We shall make some use of this ratio in succeeding paragraphs.

Central to the study of order–disorder transformations is an understanding of how the degree of order varies with temperature. The Bragg–Williams model provides a most important theoretical account of the temperature dependence of s, to which the behaviour of a number of alloy systems approximates. We shall show in chapter 13 that the ratio of rightly to wrongly situated A atoms $R/(N-R)$ varies with absolute temperature T according to the equation

$$\ln\frac{1+s}{1-s} = \frac{zws}{kT},$$

where k is Boltzmann's constant and z is the coordination number of the α- or β-sites. The energy term w needs more explanation. The statement that the bond energies of A—A, B—B, and A—B bonds are respectively w_{AA}, w_{BB}, and w_{AB} means that the energy of an assemblage of A and B atoms is decreased by the amount w_{AA} when one A—A bond is formed, by w_{BB} when one B—B bond is formed, and by w_{AB} when one A—B bond is formed. Now the energy term in the equilibrium expression with which we are concerned, w, is defined as $w = w_{AB} - \frac{1}{2}(w_{AA}+w_{BB})$ and illustrated in Fig 10.37. In the fully ordered structure ($s = 1$) all the bonds between nearest neighbours are A—B bonds; whereas in the completely disordered structure, as will be shown in chapter 13, one half of the bonds between nearest neighbours are A—B bonds, one quarter A—A, and one quarter B—B. The amount by which the energy of the fully ordered structure is lower than that of the completely disordered structure is thus $\frac{1}{2}\{w_{AB} - \frac{1}{2}(w_{AA}+w_{BB})\} = \frac{1}{2}w$ per bond. In a structure with z-fold coordination of atoms on α-sites by atoms on β-sites (and vice versa) and N atoms of each kind there will be Nz bonds so that the energy of the fully ordered structure will be lower than that of the completely disordered structure by $\Delta E = \frac{1}{2}Nzw$.

We shall show in chapter 13 that the stable solution of the equilibrium equation

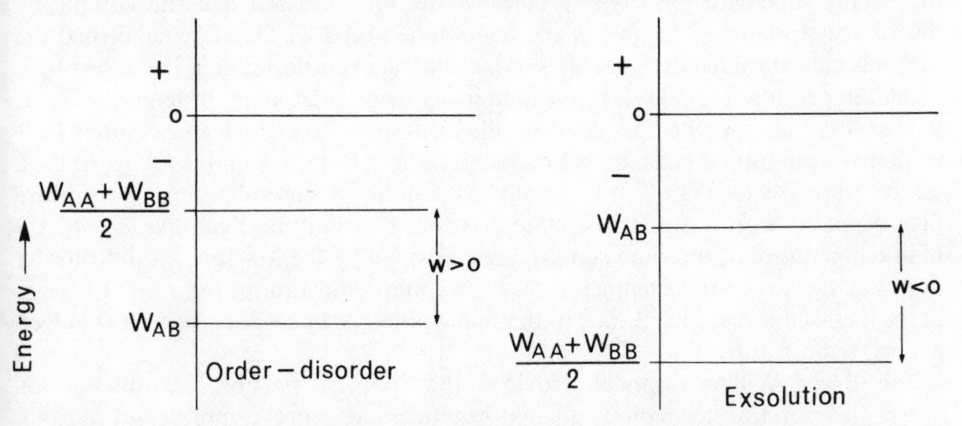

Fig 10.37 Energy diagrams for order-disorder and exsolution. If the decrease in the energy of an assemblage of A and B atoms is w_{AA}, w_{BB}, w_{AB} when one A—A, one B—B, one A—B bond is formed, then $w = w_{AB} - \frac{1}{2}(w_{AA}+w_{BB})$ is positive in an order-disorder system and negative for exsolution into distinct A and B phases.

for $w > 0$ is $s = 0$, corresponding to complete disorder in the crystal, for all temperatures greater than the *critical temperature* $T_c = zW/2k$ and $0 < s \leqslant 1$ for all temperatures below the critical temperature, s ranging from unity at zero K to zero at T_c and having an equilibrium value appropriate to each intermediate temperature. It will be shown that as the temperature of a fully ordered crystal at zero K is increased the degree of order decreases at an increasing rate which becomes catastrophic as the critical temperature is approached (Fig 10.38). Moreover the energy required (Fig 13.9), to produce the same amount of change in the degree of order, δs, gets less as the temperature rises from zero K towards T_c and is in the limit negligible. In short the process of disordering can be regarded as self-catalytic or, in Sir Lawrence Bragg's most expressive words 'demoralization sets in and there is a complete collapse of the ordered state'. Such processes, which become increasingly rapid as they approach completion, are known as *cooperative processes.* This particular cooperative process yields a steadily increasing rate of decrease of s from unity at zero K to zero at the critical temperature T_c. Above T_c the magnitude of s is uniformly zero.

The stable solution of the equilibrium equation for $w < 0$ corresponds to *exsolution* of the alloy AB into a mixture of two phases, the pure metals A and B. The phenomenon of exsolution, with which we are not specifically concerned here, is discussed in chapter 14.

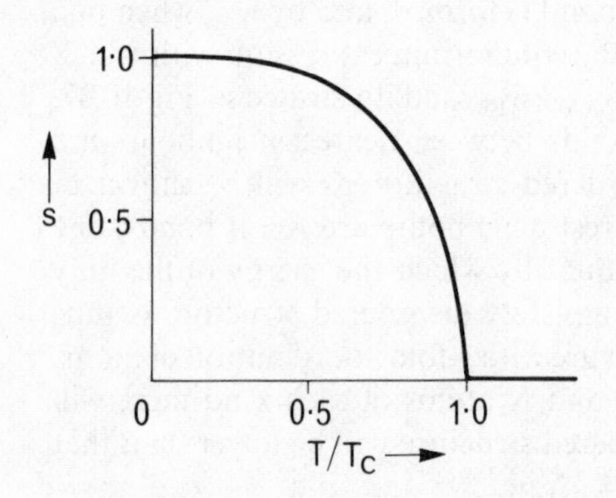

Fig 10.38 The temperature dependence of the order parameter s in the Bragg–Williams model for AB alloys. T_c is the critical temperature for disordering.

The magnitude of the critical temperature, given by $T_c = zw/2k$, is proportional to the energy difference $\Delta E = \frac{1}{2}Nzw$ between the fully ordered and the completely disordered structures. The greater the magnitude of ΔE for the alloy concerned the higher is the critical temperature of its order–disorder transformation. If T_c is less than room temperature the alloy will be completely disordered at room temperature. If T_c is greater than room temperature the alloy will show a certain degree of order s if it is in its equilibrium state at room temperature; but by rapid cooling from a temperature greater than T_c it is usually possible to preserve the high-temperature disordered state as a metastable state at room temperature. Such *quenching* of a high-temperature equilibrium state is made possible by the low rates of diffusion of atoms in the structure at temperatures near room temperature; the rate of change of the metastable disordered state to the stable ordered state will be very slow indeed at such temperatures.

The Bragg–Williams model provides the simplest possible account of an order–disorder transformation; all real examples are more complex and various other theoretical models have been proposed to take some of the complexities into account. We concern ourselves now with one particular feature of real order–disorder transformations which is not built into the Bragg–Williams model. In the

Bragg–Williams model degree of ordering is determined by the occupancy of two types of structural site that occur regularly throughout the structure. This sort of order, known as *long range order*, can be produced only if A—B bonds are stronger than the mean bond strength of A—A and B—B bonds so that an A atom will have a tendency to surround itself with B atoms. In the Bragg–Williams model this tendency becomes zero at a certain critical temperature above which an A atom does not discriminate between A and B neighbours. In reality the tendency persists above the critical temperature so that there are still small volumes of the ordered structure with A atoms preferentially surrounded by B atoms even though there is no regular ordering throughout the whole structure, that is to say there is no long range order but some degree of *short range order*. The degree of short range order σ can simply be defined in terms of the probability ρ_{AB} of a particular bond being an A—B bond as $\rho_{AB} = \frac{1}{2}(1+\sigma)$ by analogy with the definition of degree of long range order s in terms of the probability R/N of an α-site being occupied by an A atom, $R/N = \frac{1}{2}(1+s)$. The one-dimensional structure illustrated in Fig 10.39 provides an example of a situation in which the long range order parameter $s = 0$ and a high degree of short-range order is present. The hypothetical structure shown can be divided into four *domains* in each of which there is perfect order; the whole structure has equal numbers of each kind of atom on each type of site and so has zero long range order.

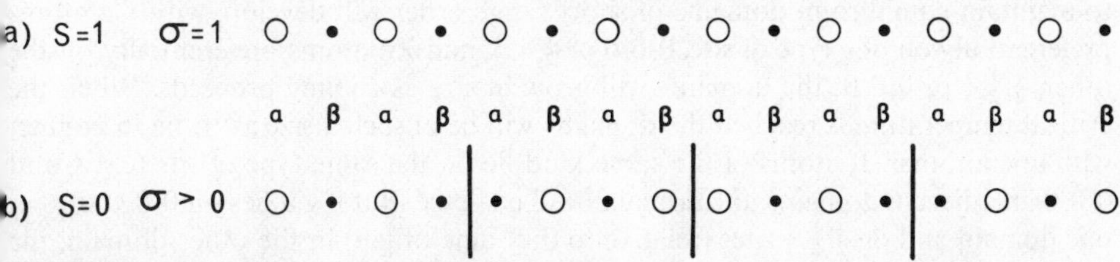

Fig 10.39 Short- and long-range order in a one-dimensional structure. (a) illustrates a state of complete long-range and short-range order; α and β sites alternate in the structure, each α-site is occupied by an A atom (open circle) and each β-site by a B atom (solid circle). In (b) α-sites are equally occupied by A and B atoms and β-sites likewise so that the long-range order parameter s is zero; but the occupation of α and β sites is not random and the structure is divided into domains in which A atoms lie on α-sites with B atoms on β-sites in one domain and vice versa in the adjacent domain.

The persistence of short range order above the critical temperature for long range ordering is illustrated in Fig 10.40 where σ is plotted against T/T_c. It is apparent that, although σ falls quite sharply in the neighbourhood of the critical temperature, it does not reach zero until a temperature well in excess of T_c has been attained. Domains of short range order persist in the structure above the critical temperature. For example in β-brass at high temperatures equal numbers of Cu and Zn atoms will lie on each

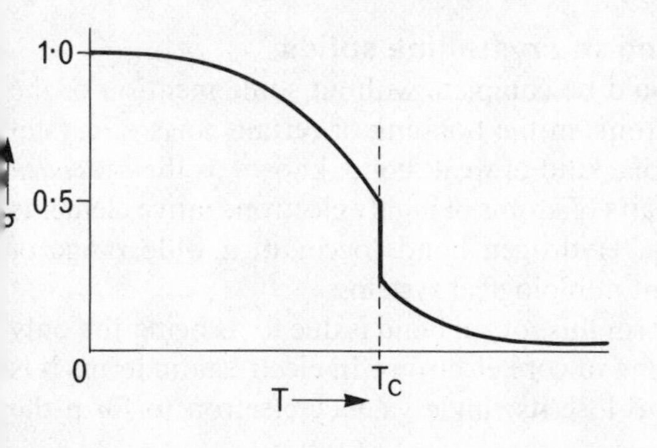

Fig 10.40 The temperature dependence of the short-range order parameter σ.

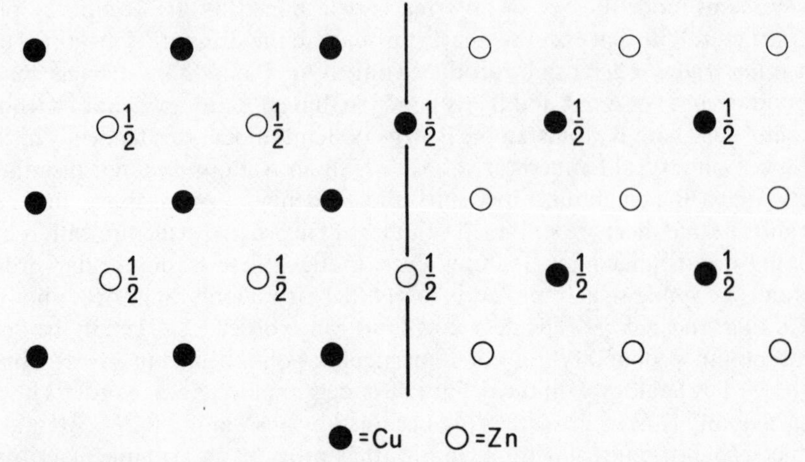

Fig 10.41 An anti-phase domain boundary (solid line) in β-brass. On one side of the boundary α-sites and β-sites are respectively occupied by Cu and by Zn atoms; across the boundary site occupation is reversed.

type of structural site, 0, 0, 0 and $\frac{1}{2}, \frac{1}{2}, \frac{1}{2}$, indiscriminately. As the crystal is slowly cooled to maintain equilibrium domains of short range order will develop, with Cu atoms preferentially on one type of site, 0, 0, 0 or $\frac{1}{2}, \frac{1}{2}, \frac{1}{2}$, and Zn atoms preferentially on the other, $\frac{1}{2}, \frac{1}{2}, \frac{1}{2}$ or 0, 0, 0; the domains will grow in size as cooling proceeds. When the critical temperature is reached the domains will be of such a size as to be in contact with one another. If atoms of the same kind lie on the same type of site (e.g. Cu at 0, 0, 0) in adjacent domains, the domains will coalesce. But if Cu lies on 0, 0, 0 sites in one domain and on $\frac{1}{2}, \frac{1}{2}, \frac{1}{2}$ sites (relative to the same origin) in the other domain, the domains will be exactly out of step with their α- and β-sites interchanged (Fig 10.41); adjacent domains related in this way are known as *antiphase domains.* Increase in the degree of long range order of an antiphase domain structure can only be achieved by the growth of one domain at the expense of its neighbours, which is a very slow process, much slower than ordering of a randomly disordered structure. Indeed in certain alloys, such as Cu_3Au where the gold atom can occupy one of four distinct types of site, antiphase domain structures are very stable.

We have discussed order–disorder transformations exclusively in terms of metal solid solutions, that is alloys. But they are well known too in other fields of solid solution as well as in defect structures with vacancies, such as Fe_7S_8. Order–disorder transformations also play an important role in the explanation of certain physical properties of crystalline solids, such as ferromagnetism and ferroelectricity.

Hydrogen and hydrogen-bonding in crystalline solids

No discussion of crystal chemistry could be complete without some mention of the important role played by hydrogen atoms in the bonding of certain sorts of crystal structure, in particular the formation of a kind of weak bond, known as the *hydrogen bond,* which serves as a link between pairs of atoms of highly electronegative elements such as fluorine, oxygen, or nitrogen. Hydrogen bonds occur in a wide range of structures and are especially important in biological systems.

The unique ability of hydrogen to form this sort of bond is due to its being the only element whose atoms are wholly lacking in core electrons. In electrostatic terms it is possible for a hydrogen atom either to lose its single valency electron to form the cation H^+ or, by capturing an electron, to fill its 1*s* orbital and so form the anion H^-. The hydrogen anion, having a complete outer shell, is effectively spherical and its radius, 1·54 Å, is intermediate between those of the anions F^- and Cl^-. The H^- anion thus behaves structurally in much the same way as halide anions; for instance the alkali hydrides all have structures of NaCl type. The H^+ cation in contrast has no electrons; it is a positively charged nucleus and consequently able to penetrate the electron clouds of neighbouring anions such as O^{2-}. The equilibrium distance between the hydrogen nucleus H^+ and the nucleus of the host anion will be determined by the balance between the electrostatic attractive forces between the H^+ nucleus and the electrons of the host anion on the one hand and the repulsive forces between the H^+ nucleus and the nucleus of the host anion on the other. However hydrogen–oxygen bonds are not dominantly ionic in character; it is more realistic to give emphasis to the formation of a covalent bond by overlap of the 1*s* atomic orbital of hydrogen with an sp^3 hybrid orbital of the oxygen atom.

The formation of a hydrogen bond between a pair of highly electronegative atoms is due to interaction between the hydrogen atom, which is covalently bonded to one of the atoms, and a lone pair of electrons of the other atom. The interaction must be essentially electrostatic because hydrogen is capable of forming only one covalent bond. Such interaction, necessarily always rather weak, is only significant when the hydrogen atom is able to approach very close to the second atom: this can only happen when the covalent bond between hydrogen and the first atom is strongly polar and when the second atom has rather few core electrons, so that overlap forces are minimal. Thus hydrogen bonding occurs only with strongly electronegative elements, mainly those of the first period such as N, O, and F. A pair of atoms linked by a hydrogen bond approach each other more closely than they would do in the absence of the intervening hydrogen atom. For instance in inorganic compounds the separation of oxygen atoms not linked to the same cation is usually not less than about 3·3 Å, while the separation of a hydrogen-bonded pair of oxygen atoms is only 2·5–2·8 Å.

The direct location of hydrogen atoms in a crystal structure is, at least, difficult by X-ray diffraction methods because the atomic scattering factor of hydrogen is exceedingly small. In structures composed entirely of atoms of low atomic number, such as carbon, nitrogen, oxygen, precise intensity measurements may make possible the location of hydrogen atoms from X-ray diffraction data, but in general it is necessary to infer the positions of hydrogen atoms from the recognition of short interatomic distances which may be taken to imply the occurrence of hydrogen bonds. In very few inorganic compounds have the positions of hydrogen atoms been determined directly, by neutron or X-ray diffraction or by spectroscopic methods.

The consequences of hydrogen bond formation are most clearly explained by reference to specific examples. We select examples which involve the ammonium ion, NH_4^+, the hydroxyl ion, OH^-, and crystalline H_2O, ice. In each case the covalent bonds between hydrogen and nitrogen or oxygen involve sp^3 hybrid orbitals. In NH_4^+ the hydrogens are disposed at the apices of a regular tetrahedron, but in many of the structures in which it occurs the NH_4^+ ion can apparently be treated, at least to a first approximation, as a spherical ion of radius 1·48 Å with a central positive charge. Thus the halides NH_4Cl and NH_4Br have NaCl-type structures while NH_4I has the CsCl-type structure; here NH_4^+ resembles Rb^+ which is of similar radius. However the fluorides of NH_4^+ and Rb^+ are significantly different: RbF has a NaCl-type structure, while NH_4F has a wurtzite-type structure and moreover in NH_4F the N—F separation is substantially less, at 2·66 Å, than the radius sum, $r_{NH_4^+} + r_{F^-} = 1{\cdot}48 + 1{\cdot}36 = 2{\cdot}84$ Å, calculated from the effective radius of NH_4^+ in its other halide structures. It is inferred that there is hydrogen-bonding between nitrogen and fluorine in NH_4F, the tetrahedral disposition of F^- about NH_4^+, and vice versa, being strongly favoured by the interaction of hydrogen atoms, covalently bonded to nitrogen, with lone pairs of electrons in tetrahedral sp^3 orbitals of fluorine atoms. Of the halogens only fluorine is sufficiently electronegative to form hydrogen bonds with nitrogen in its simple ammonium salts.

In minerals hydrogen bonding usually occurs as a link between two oxygen atoms, the oxygen–oxygen distance in O—H—O being about 2·7 Å. The hydrogen atom is asymmetrically situated between the two oxygen atoms at about 1 Å from the nearer of the two. We take as our first example *ice* H_2O, which has a structure related to that of wurtzite with all Zn and S sites occupied by oxygen. Each oxygen atom is surrounded by four hydrogen atoms situated on oxygen–oxygen joins, each hydrogen atom being distant 1·01 Å from one of its oxygen neighbours and 1·75 Å from the other. The water molecule is polar with two centres of positive charge corresponding to the two hydrogen atoms covalently bonded to the oxygen atom and two centres of negative charge corresponding to the other two sp^3 hybrid orbitals of the oxygen atom which are occupied by lone pairs of electrons; the disposition of the centres of positive and negative charge is tetrahedral. Hydrogen bonding arises in ice by the interaction of a hydrogen atom of one water molecule with a lone pair of an adjacent molecule. Each oxygen atom is thus tetrahedrally coordinated to its four nearest oxygen neighbours with one hydrogen atom situated on each O—O join. Each oxygen atom might be expected to have two of its four associated hydrogen atoms close to itself and the other two rather further away. However neutron diffraction studies of the deuterium analogue of ice, D_2O, indicate that each oxygen atom is associated with four 'half-atoms' of deuterium at a distance of 1·01 Å and four 'half-atoms' of deuterium at a distance of 1·75 Å from the oxygen nucleus, the 'half-atoms' being so disposed that there is one deuterium atom between each pair of oxygen atoms. The diffracted neutrons effectively record a time average of the equal occupation of two alternative sites by each deuterium atom (Fig 10.42). All the available evidence points to the occurrence of precisely the same situation in ice, H_2O. The diffraction evidence is supported by thermochemical studies of the residual entropy of ice (and D_2O) which are consistent with the existence of eight possible hydrogen sites, four close and four distant, associated with each oxygen atom being occupied at any instant in such a manner that two hydrogen atoms lie close to that oxygen atom and two are closer to

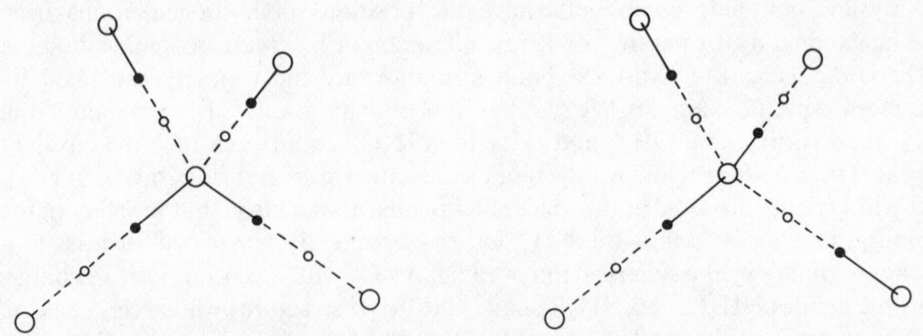

Fig 10.42 Two of the six possible configurations of hydrogen atoms about an oxygen atom in ice with retention of the H_2O molecule. Large open circles represent oxygen atoms; small circles represent hydrogen sites, solid when occupied, open when vacant.

its nearest neighbour oxygens. The H_2O molecule is thus retained in the structure, but the hydrogen atoms associated with each oxygen atom are constantly changing so that over a finite period of time each possible hydrogen site is effectively occupied by a 'half-atom'.

The hydroxyl ion, OH^-, is likewise polar and would be expected to have a tetrahedral charge distribution with its single centre of positive charge corresponding to the attached hydrogen atom and its three centres of negative charge each corresponding to an sp^3 hybrid orbital of the oxygen atom occupied by a lone pair of electrons. In some structures the hydroxyl ion appears at first sight to behave as a spherical anion of radius 1·53 Å. For instance both CaI_2 and $Ca(OH)_2$ (the mineral *portlandite*) crystallize with the CdI_2 structure type suggesting that the OH^- ion, just like the truly spherical I^- ion, imposes no directional constraints. However, precise X-ray diffraction studies have shown that in portlandite the hydrogen atoms lie between adjacent layers of CaO_6 octahedra; each is immediately above or below the oxygen atom with which it is associated and disposed so as to lie as far as possible from the nearest calcium atoms. The separation of oxygen and hydrogen nuclei is 0·79 Å.[12] The negative end of the polar hydroxyl ion thus points towards the adjacent calcium cations.

In other structures, in contrast, the hydroxyl ion may be associated with hydrogen bond formation. Then the hydrogen atom of the hydroxyl ion is involved in the 'short contact' of a hydrogen bond and the three centres of negative charge on the OH^- ion are either directed towards cations or involved in the 'long contacts' of hydrogen bonds. In $Zn(OH)_2$ for instance each Zn^{2+} cation is tetrahedrally coordinated by OH^- and each hydroxyl ion is linked to two cations and hydrogen bonded to two other hydroxyl ions, one hydrogen bond presumably being a 'short contact' and the other a 'long contact'. In $Zn(OH)_2$ the four nearest neighbours of each oxygen atom, Zn, Zn, H (close), H (distant), are approximately tetrahedrally disposed.

We take as our final example of a structure with hydrogen bonds the mineral *diaspore*, AlO(OH), which is shown in Fig 10.43. Diaspore is orthorhombic, space group P*bnm*, and all its atoms lie on (001) mirror planes. All the aluminium atoms are octahedrally coordinated and equivalent to one another. Each coordination octahedron has four shared edges; the two shared edges parallel to (001) give rise to chains of octahedra parallel to [001] and the other two serve to link adjacent chains in pairs. Oxygen atoms are of two kinds, which we denote O_I and O_{II}; all are bonded to three aluminium atoms. The three aluminium atoms to which each O_I atom is bonded are approximately coplanar and distant from the oxygen atom 1·85 Å, 1·85 Å, and 1·86 Å (Fig 10.43). The other kind of oxygen atom O_{II} is bonded to three aluminium atoms at a distance of 1·98 Å and to one hydrogen atom 1·01 Å away. Thus the O_{II} atom is the oxygen of the hydroxyl ion in AlO(OH); the approximately tetrahedral disposition of its four near neighbours, three aluminium atoms and one hydrogen atom, is consistent with the disposition of charges on an hydroxyl ion. The hydroxyl ion forms a hydrogen bond with an O_I atom which does not belong to any of the three coordination octahedra in which the OH^- ion is involved. The oxygen–oxygen separation in the O_{II}—H ... O_I hydrogen bond is 2·65 Å, but neutron diffraction studies indicate that the bond is not linear, the O_{II}—H vector being inclined at 12° to the O_I—O_{II} vector. Oxygen–oxygen distances on shared edges, 2·46 Å or 2·54 Å, are short, shorter even than the separation of hydrogen bonded oxygens,

[12] Neutron diffraction studies however yield the value 0·94 Å. Such discrepancies are not unusual and are attributable to the different modes of scattering of X-rays and neutrons (chapter 9).

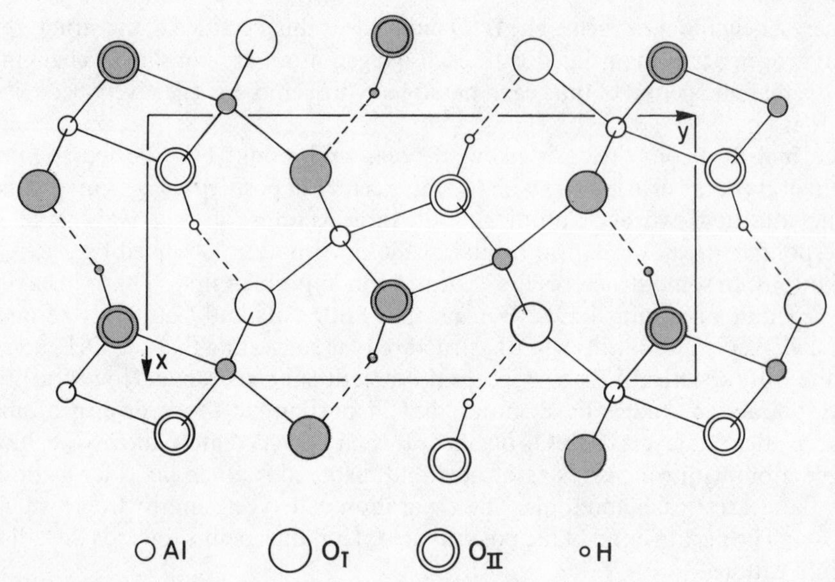

Fig 10.43 The crystal structure of disapore, AlO.OH. All atoms lie on mirror planes at $z=\frac{1}{4}$ (open circles) and $z=\frac{3}{4}$ (shaded). The space group is *Pbnm* and the unit-cell dimensions are *a* 4·401, *b* 9·421, *c* 2·845 Å.

because repulsion between adjacent cations brings the anions on the edge shared between their polyhedra closer together. The discrepancy between Al—O_I and Al—O_{II} separations in this structure may be correlated with the difference in electrostatic interaction of O^{2-} and OH^- anions with an Al^{3+} cation.

The application of Pauling's electrostatic valency principle to the essentially ionic structure of diaspore is instructive. Since Al^{3+} is octahedrally coordinated each Al—O bond should have a strength of $\frac{3}{6}=\frac{1}{2}$. The sum of the electrostatic strengths of the bonds from aluminium to both O_I and O_{II} atoms should thus be $3\times\frac{1}{2}=1\frac{1}{2}$. If the electrostatic valency principle were to hold precisely the hydrogen atom would have to be shared equally between an O_I and an O_{II} atom, contributing a bond strength of $\frac{1}{2}$ to each. But the hydrogen atom is more closely associated with the O_{II} atom so that it should contribute bond strengths of $>\frac{1}{2}$ to O_{II} and $<\frac{1}{2}$ to O_I. Moreover Al—O_{II} bonds are longer than Al—O_I bonds so that the sum of the strengths of the three bonds from Al^{3+} to O_{II} would be expected to be $<1\frac{1}{2}$ and the corresponding sum at O_I to be $>1\frac{1}{2}$. If the inequalities, $>\frac{1}{2}+<1\frac{1}{2}$ and $<\frac{1}{2}+>1\frac{1}{2}$, sum to 2 for each sort of oxygen atom, then the electrostatic valency principle holds precisely for this structure, but it can never be demonstrated that this is so. Direct application of Pauling's Rules does however, in this case, demonstrate the essential correctness of the structure determination.

In conclusion and by way of summary, it can be said that hydrogen-bonding influences significantly the type of structure adopted by a compound. The hydrogen bond is at once a directed bond and essentially ionic. Ionic compounds in which hydrogen-bonding occurs thus tend to adopt relatively open structures compared with the close-packed or approximately close-packed anion frameworks favoured by so many other ionic compounds. It remains to be said that, although in the examples of hydrogen bonding we have cited above the hydrogen atom is closer to one of the atoms involved in hydrogen-bonding than to the other, examples do exist for which all the available evidence indicates that the hydrogen atom is equidistant from the two atoms to which it is linked. For instance the F—H—F bond appears to be symmetrical in crystals of KHF_2.

11
Crystal physics

In the last chapter we were concerned with the structures of a variety of crystalline solids. We turn now to consideration of the effect of regular atomic structural arrangement on physical properties. Because a crystal has a regular atomic structural arrangement its physical properties will not, in general, be identical in every direction; such crystals are said to be *anisotropic*. The physical properties of gases, liquids, glasses, polycrystalline solids and certain single crystals (which we shall specify in due course) are independent of direction; such substances are described as *isotropic*. In other words when a particular physical property of a substance in a given state can be described in terms of a single coefficient the substance in that state is said to be isotropic for that property and it may be isotropic for all properties; but when complete description of the physical property of the substance in the given state requires the statement of several coefficients the substance in that state is said to be anisotropic for that property.

For the full mathematical treatment of the physical properties of single crystals the reader is referred to Nye (1960) and to Wooster (1949). Here we shall keep mathematics to a minimum and concentrate on the physics of the anisotropy of physical properties. We shall discuss in detail only two physical properties, diffusion and thermal expansion; in the course of the discussion we shall develop points of general physical significance applicable to properties other than the one under discussion and we shall develop some generally useful techniques for the analysis of anisotropic physical properties. The two properties selected for detailed study are not without interest: diffusion is important for the understanding of solid state reaction processes and studies of thermal expansion can provide significant information about the relative strengths of interatomic binding forces.

Diffusion

Diffusion is, like electrical and thermal conductivity, a transport property; the existence of a concentration gradient causes a flow of matter. It is found experimentally that in polycrystalline solids the flux **J** of an atomic species across unit area of a plane normal to the concentration gradient $\partial c/\partial r$ of that species is directly proportional to the concentration gradient,

i.e. $$\mathbf{J} = -D\frac{\partial c}{\partial r}.$$

The coefficient D is known as the *diffusion coefficient* and is found experimentally to be commonly independent of the concentration gradient. This simple relationship of widespread validity is often called *Fick's first law.* The solution of the differential equation for particular experimental conditions is outside the scope of this text; the interested reader is referred to Shewmon (1963) for a thorough account of the solutions of Fick's first law. We note that for polycrystalline solids the 'resultant' vector $\mathbf{J}$ is found experimentally to be parallel to the 'applied' vector $\partial c/\partial r$ and that in identical experimental conditions the ratio $\mathbf{J}/(\partial c/\partial r)$ is independent of the directions of the two vectors: the diffusion coefficient D is thus not direction-dependent in polycrystalline solids.

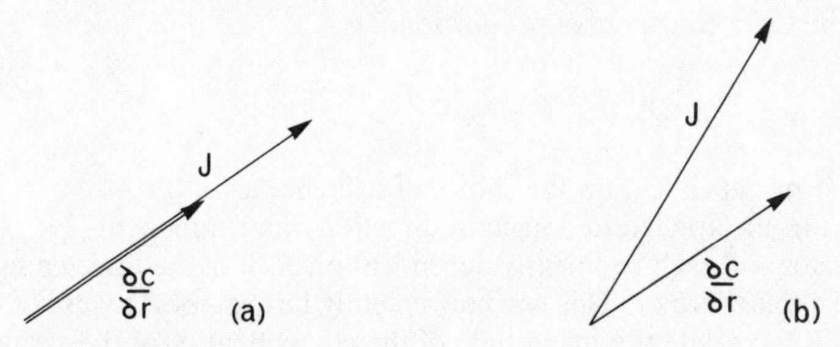

Fig 11.1 The relationship of the resultant flux **J** to the applied concentration gradient $\partial c/\partial r$ for diffusion in (a) a polycrystalline solid, (b) an anisotropic single crystal.

In single crystals however diffusion presents a more complex problem because the resultant flux $\mathbf{J}$ is not in general parallel to the applied concentration gradient $\partial c/\partial r$ (Fig 11.1) and moreover the angle between the two vectors varies with direction. It is the formal relationship between such applied and resultant vectors that is the essence of crystal physics. The mathematics of crystal physics is simple but rather cumbersome unless the so-called 'dummy suffix notation' is employed. We shall not here introduce this elegant notation, but use more familiar mathematics, sacrificing elegance to ease and speed of understanding. We begin by showing in general that in a single crystal there are at least three mutually perpendicular directions in which the applied and resultant vectors are parallel to each other; we consider the proposition first in two dimensions and then its extension to three dimensions.

Second rank tensors

In order to describe an anisotropic physical property it is necessary to choose a set of reference axes and it is convenient that they should be orthogonal axes. The reference axes for the description of the physical property do not need in the first instance to bear any special relationship to the crystallographic reference axes of the crystalline substance concerned, although at a later stage it will be necessary to establish the relationship between the two axial systems.

Suppose that in general an applied vector $\mathbf{p}$ gives rise to the resultant vector $\mathbf{q}$. In diffusion $\mathbf{p}$ is the concentration gradient $\partial c/\partial r$ and $\mathbf{q}$ is the flux $\mathbf{J}$. The components of $\mathbf{p}$ along the reference axes, x, y, z are denoted p_x, p_y, p_z respectively and similarly the components of $\mathbf{q}$ are q_x, q_y, q_z. If $\mathbf{p}$ lies along the x-axis (Fig 11.2), then its components

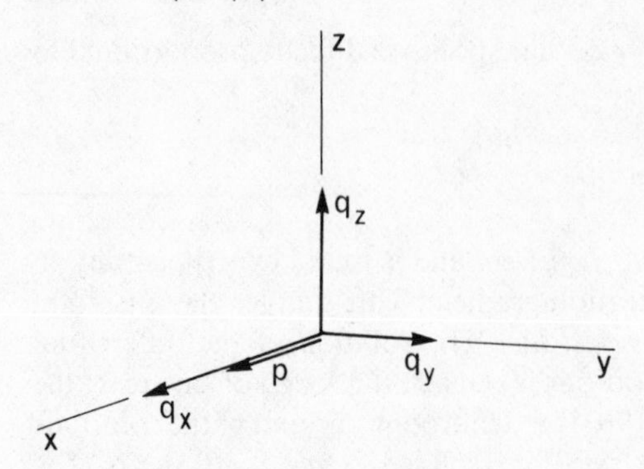

Fig 11.2 The components of the resultant vector **q** when the applied vector **p** lies along the *x*-axis.

are $p_x = p$, $p_y = p_z = 0$. The resultant vector **q** however will have in general three non-zero components, each of which is linearly proportional to $\mathbf{p} = p_x$. The magnitude of each of the three constants of proportionality

$$D_{xx} = \frac{q_x}{p_x}, \quad D_{yx} = \frac{q_y}{p_x}, \quad D_{zx} = \frac{q_z}{p_x},$$

will of course be dependent on the choice of reference axes.

If the vector **p** is applied in a general direction, the components q_x, q_y, q_z of the resultant vector will each be linearly dependent on each of the three components p_x, p_y, p_z of the applied vector. This can conveniently be expressed by use of coefficients of the form D_{kl} to relate the magnitude of the component q_k of the resultant vector to the magnitude of the component p_l of the applied vector; thus D_{zx} relates q_z to p_x and so on, so that the total relationship may be stated as

$$q_x = D_{xx}p_x + D_{xy}p_y + D_{xz}p_z$$
$$q_y = D_{yx}p_x + D_{yy}p_y + D_{yz}p_z$$
$$q_z = D_{zx}p_x + D_{zy}p_y + D_{zz}p_z.$$

The array of nine coefficients, D_{xx}, D_{xy}, etc (or in general D_{kl}) is known as a *second rank tensor.* Such a second rank (or second order) tensor expresses the linear relationship of the three components of the resultant vector **q** to the three components of the applied vector **p**. For the majority of physical properties, including diffusion, which can be represented by second rank tensors, thermodynamic arguments lead to the conclusion that $D_{lk} = D_{kl}$; when this is so the tensor is said to be *symmetrical.* If a second rank tensor is symmetrical it follows that the constant of proportionality has the same magnitude for p applied along the x-axis and q measured along the y-axis as for p applied along the y-axis and q measured along the x-axis, that is

$$D_{yx} = D_{xy}$$

i.e. $$\left(\frac{q_y}{p_x}\right)_{p=p_x} = \left(\frac{q_x}{p_y}\right)_{p=p_y}$$

The physical basis for such a relationship is not immediately obvious; we shall not attempt any explanation here, but simply assert that diffusion is a symmetrical second rank tensor property.

Principal axes

In general then the description of a physical property in a crystalline solid requires the specifications of an axial system and of the six coefficients D_{xx}, D_{yy}, D_{zz}, $D_{xy} = D_{yx}$, $D_{yz} = D_{zy}$, $D_{zx} = D_{xz}$ which relate the resultant vector components q_x, q_y, q_z to the applied vector components p_x, p_y, p_z. We now proceed to show that the number of independent coefficients can be further reduced to three because there is always a set of three orthogonal directions along which the resultant vector is parallel to the applied vector. If these three directions coincide with the chosen reference axes, then $D_{xy} = D_{yz} = D_{zx} = 0$ and

$$q_x = D_{xx}p_x$$

$$q_y = D_{yy}p_y$$

$$q_z = D_{zz}p_z.$$

Reference axes that have this property are known as *principal axes.*

Because a physical property, such as diffusion, is a property of the crystalline substance under discussion, the arbitrary choice of reference axes cannot affect the magnitude of the property in any direction, but only the way of describing it. Thus the coefficients of the second rank tensor necessary for complete quantitative description of the property will vary in magnitude with the choice of reference axes, but for a given applied vector **p** the same resultant vector **q** must always be obtained. The second rank tensor for any set of reference axes must therefore be related to that for any other axial system and in particular to that for the principal axes. We now explore the nature of this relationship in two-dimensions.

Suppose that the vector **q** has components q_x and q_y along the reference axes x and y respectively so that in Fig 11.3 **q** is represented by OQ, q_x by OX and q_y by OY. Then the component of **q** in a direction which has direction cosines l and m with respect to the reference axes will be

$$\begin{aligned}
\mathrm{OQ'} &= \mathrm{OR} + \mathrm{RQ'} \\
&= \mathrm{OR} + \mathrm{XS} \\
&= \mathrm{OX}\cos\widehat{\mathrm{ROX}} + \mathrm{XQ}\cos\widehat{\mathrm{QXS}} \\
&= \mathrm{OX}\cos\widehat{\mathrm{ROX}} + \mathrm{OY}\cos\widehat{\mathrm{YOR}} \\
&= q_x l + q_y m.
\end{aligned}$$

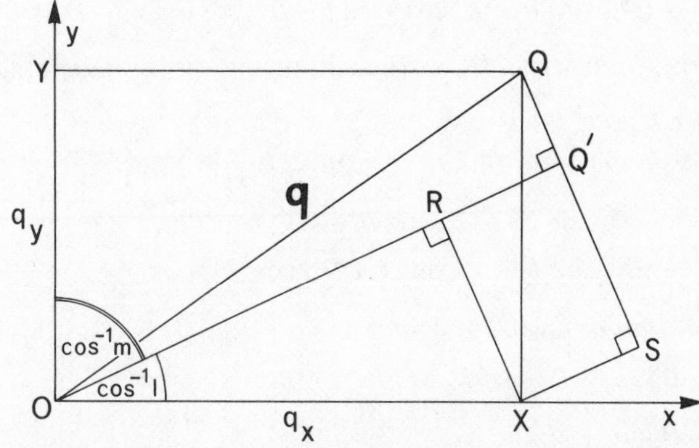

Fig 11.3 The component of a vector **q** in a direction whose direction cosines with respect to the reference axes x, y are l, m is $q_x l + q_y m$.

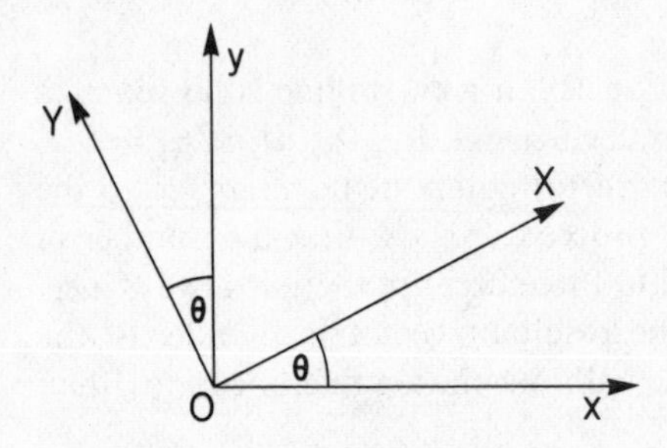

Fig 11.4 The orthogonal axes X, Y have direction cosines $(\cos\theta, \sin\theta)$, $(-\sin\theta, \cos\theta)$ with respect to the orthogonal axes x, y and the direction cosines of x, y referred to X, Y are $(\cos\theta, -\sin\theta)$, $(\sin\theta, \cos\theta)$.

Thus the components of the vector $\mathbf{q}$ along any set of reference axes can be calculated from its components along any other set of axes provided the direction cosines of the second axial system with respect to the first axial system are known. For instance the axes X, Y in Fig 11.4 have direction cosines $\cos\theta$, $\sin\theta$ and $-\sin\theta$, $\cos\theta$ with respect to the axes x, y.

Therefore $q_X = q_x \cos\theta + q_y \sin\theta$

and $q_Y = -q_x \sin\theta + q_y \cos\theta.$

But we have already seen that

$$q_x = D_{xx}p_x + D_{xy}p_y$$

and $$q_y = D_{yx}p_x + D_{yy}p_y.$$

Hence $$q_X = (D_{xx}\cos\theta + D_{yx}\sin\theta)p_x + (D_{xy}\cos\theta + D_{yy}\sin\theta)p_y$$

and $$q_Y = (-D_{xx}\sin\theta + D_{yx}\cos\theta)p_x + (-D_{xy}\sin\theta + D_{yy}\cos\theta)p_y.$$

Now the direction cosines of the x and y axes with respect to the X and Y axes are $\cos\theta$, $-\sin\theta$ and $\sin\theta$, $\cos\theta$ respectively so that

$$p_x = p_X\cos\theta - p_Y\sin\theta$$

and $$p_y = p_X\sin\theta + p_Y\cos\theta.$$

Therefore $$q_X = (D_{xx}\cos^2\theta + D_{yx}\cos\theta\sin\theta + D_{xy}\cos\theta\sin\theta + D_{yy}\sin^2\theta)p_X + (-D_{xx}\cos\theta\sin\theta - D_{yx}\sin^2\theta + D_{xy}\cos^2\theta + D_{yy}\cos\theta\sin\theta)p_Y$$

and $$q_Y = (-D_{xx}\cos\theta\sin\theta + D_{yx}\cos^2\theta - D_{xy}\sin^2\theta + D_{yy}\cos\theta\sin\theta)p_X + (D_{xx}\sin^2\theta - D_{yz}\cos\theta\sin\theta - D_{xy}\cos\theta\sin\theta + D_{yy}\cos^2\theta)p_Y.$$

Since the tensor is symmetrical, $D_{yx} = D_{xy}$ and these expressions simplify to

$$q_X = (D_{xx}\cos^2\theta + D_{xy}\sin 2\theta + D_{yy}\sin^2\theta)p_X + (-\tfrac{1}{2}D_{xx}\sin 2\theta + D_{xy}\cos 2\theta + \tfrac{1}{2}D_{yy}\sin 2\theta)p_Y$$

$$q_Y = (-\tfrac{1}{2}D_{xx}\sin 2\theta + D_{xy}\cos 2\theta + \tfrac{1}{2}D_{yy}\sin 2\theta)p_X + (D_{xx}\sin^2\theta - D_{xy}\sin 2\theta + D_{yy}\cos^2\theta)p_Y.$$

But the tensor for the X, Y axial system is

$$q_X = D_{XX}p_X + D_{XY}p_Y$$

$$q_Y = D_{YX}p_X + D_{YY}p_Y$$

with $D_{YX} = D_{XY}$.

Hence $$D_{XX} = D_{xx}\cos^2\theta + D_{xy}\sin 2\theta + D_{yy}\sin^2\theta$$
$$D_{XY} = \tfrac{1}{2}(D_{yy} - D_{xx})\sin 2\theta + D_{xy}\cos 2\theta$$
$$D_{YY} = D_{xx}\sin^2\theta - D_{xy}\sin 2\theta + D_{yy}\cos^2\theta.$$

Now if X and Y are to be principal axes, $D_{XY} = 0$ and the principal axes must be inclined to the reference axes x, y in such a manner that

$$\tan 2\theta = \frac{2D_{xy}}{D_{xx} - D_{yy}}.$$

Thus if the three coefficients D_{xx}, D_{xy}, D_{yy} referred to reference axes x, y are known, the directions of the principal axes X, Y can be found and the coefficients D_{XX}, D_{YY} along them can be determined.

Likewise in three dimensions if the tensor describing a physical property has been determined for one set of axes, the orientation of the principal axes can be found and the three coefficients necessary to describe the property on these axes can be evaluated. These three coefficients are usually denoted by the simplified symbols D_X, D_Y, D_Z so that for the principal axes we have

$$q_X = D_X p_X$$
$$q_Y = D_Y p_Y$$
$$q_Z = D_Z p_Z.$$

Clearly then the use of principal axes greatly simplifies the description of a physical property. We shall in what follows take principal axes as our reference axes.

Magnitude of a physical property in any direction in a crystal

In the direction of a principal axis the applied vector gives rise to a resultant vector parallel to itself. Thus a concentration gradient $\partial c/\partial X$ along the X-axis gives rise to a flux J of the diffusing atomic species along the X-axis and the diffusion coefficient along the X-axis is simply $D_X = J/(\partial c/\partial X)$. But when the applied vector is in some general direction the resultant vector is not necessarily parallel to it. In such a situation it is often convenient to measure the component $q_{\parallel}$ (Fig 11.5) of the resultant vector $\mathbf{q}$ in the direction of the applied vector $\mathbf{p}$. The ratio $q_{\parallel}/\mathbf{p}$, which we denote $D_{\parallel}$, is then the magnitude of the physical property in the direction of $\mathbf{p}$. In a single crystal $D_{\parallel}$ will usually vary with direction; and its magnitude in the direction whose direction

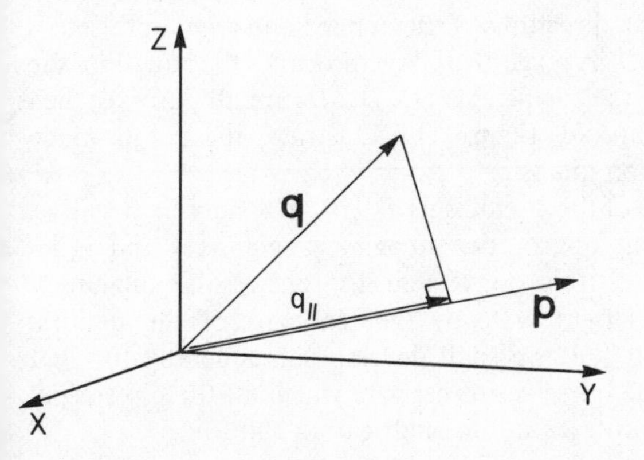

Fig 11.5 The component of the resultant vector **q** in the direction of the applied vector **p** is $q_{\parallel}$.

cosines relative to the principal axes are l, m, n can be related to the principal diffusion coefficients D_X, D_Y, D_Z by the following simple argument.

$$q_{\parallel} = lq_X + mq_Y + nq_Z,$$

but
$$q_X = D_X p_X = D_X l\mathbf{p}$$

$$q_Y = D_Y p_Y = D_Y m\mathbf{p}$$

$$q_Z = D_Z p_Z = D_Z n\mathbf{p}$$

Hence
$$D_{\parallel} = q_{\parallel}/\mathbf{p} = l^2 D_X + m^2 D_Y + n^2 D_Z.$$

The Representation Quadric

It is convenient to rewrite the equation

$$D_{\parallel} = l^2 D_X + m^2 D_Y + n^2 D_Z$$

as

$$D_X X^2 + D_Y Y^2 + D_Z Z^2 = D_{\parallel} r^2,$$

where $l = X/r$, $m = Y/r$, $n = Z/r$, so that the new equation can be compared directly with the equation to a quadric (the three-dimensional extension of the familiar two-dimensional conic)

$$\frac{X^2}{a^2} + \frac{Y^2}{b^2} + \frac{Z^2}{c^2} = 1.$$

Thus the quadric whose semi-axes are $a = D_X^{-\frac{1}{2}}$, $b = D_Y^{-\frac{1}{2}}$, $c = D_Z^{-\frac{1}{2}}$ has radius $r = D_{\parallel}^{-\frac{1}{2}}$ in the direction from the origin through the point with coordinates X, Y, Z, that is in the direction whose direction cosines are l, m, n. Such a quadric is known as the *representation quadric* and provides a very simple means of displaying the variation of $D_{\parallel}$ with direction in an anisotropic crystal. It is a general property of quadrics that at a point where the radius has direction cosines l, m, n the normal to the quadric surface has direction cosines proportional to l/a^2, m/b^2, n/c^2. Therefore at a point where the radius has direction cosines l, m, n the normal to the representation quadric has direction cosines proportional to lD_X, mD_Y, nD_Z. We have seen that if the applied vector $\mathbf{p}$ has direction cosines l, m, n, the resultant vector $\mathbf{q}$ has components $lD_X\mathbf{p}$, $mD_Y\mathbf{p}$, $nD_Z\mathbf{p}$. Therefore the radius and the normal to the representation quadric at any point represent respectively the directions of the applied and resultant vectors and the magnitude of the coefficient $D_{\parallel}$ is given by the reciprocal of the square of the radius at the point. Very commonly the coefficients D_X, D_Y, D_Z are all positive; then the representation quadric is an ellipsoid. Figure 11.6 illustrates the points made above for an ellipsoidal representation quadric.

Occasionally however one or two of the coefficients D_X, D_Y, D_Z may be negative. The representation quadric then has one or two imaginary semi-axes and is an hyperboloid. In such circumstances it is convenient to replace the imaginary parts of the quadric $D_X l^2 + D_Y m^2 + D_Z n^2 = D_{\parallel}$ by the real part of the quadric $D_X l^2 + D_Y m^2 + D_Z n^2 = -D_{\parallel}$, a point that we shall develop subsequently. In what follows immediately we shall confine ourselves to cases in which all three principal coefficients are positive so that the representation quadric is an ellipsoid.

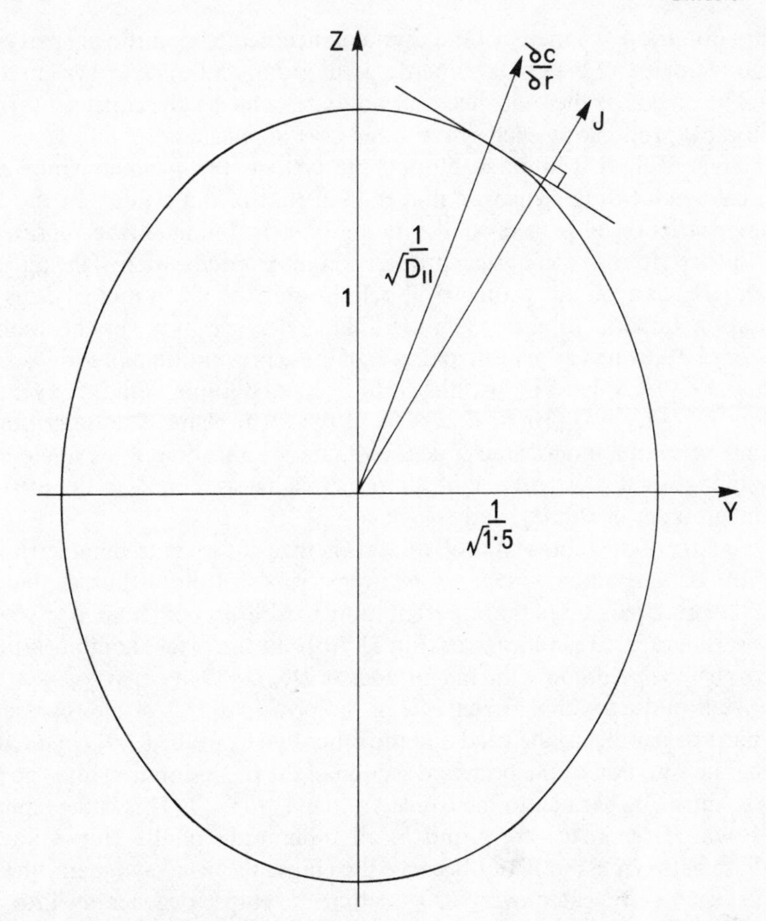

Fig 11.6 The representation quadric for self-diffusion in magnesium. The figure shows a section through the quadric in the plane of the two principal axes Y and Z and is drawn to scale. In magnesium (hexagonal) Z is parallel to [0001] and at the experimental temperature $D_X = D_Y = 1{\cdot}5 \times 10^{-8}\ \mathrm{cm^2\,s^{-1}}$, $D_Z = 1{\cdot}0 \times 10^{-8}\ \mathrm{cm^2\,s^{-1}}$.

Symmetry control of physical properties[1]

A quadric $X^2/a^2 + Y^2/b^2 + Z^2/c^2 = 1$ has the symmetry of the point group *mmm* with the diads of the point group parallel to the principal axes of the quadric. The directional variation of a physical property represented by a symmetrical second rank tensor thus has symmetry *mmm* and, in particular, is centrosymmetric. Moreover since the physical properties of a single crystal are consequences of its structure one would expect the symmetry of the directional variation of any physical property to be consistent with the symmetry of the structure. The relationship between the directional variation of a physical property and the point group symmetry of the crystal structure involved is expressed by *Neumann's Principle*, which states that the symmetry of a physical property must include the symmetry of the point group of the crystal. In the case of second rank tensor properties the symmetry inherent in the property combines

[1] In this section we use the convention that X, Y, Z are the principal axes of the representation quadric and x, y, z are the crystallographic reference axes. If one or more of the principal axes of the quadric is parallel to one or more of the crystallographic reference axes it is conventional to put $X \parallel x$, $Y \parallel y$, $Z \parallel z$. This simple convention does not apply in crystal optics, as we shall see in chapter 12.

with the point group symmetry of the crystal concerned to give the property in that crystal the symmetry of the holosymmetric point group of the crystal system to which the crystal belongs. It is therefore necessary only to consider the constraints imposed on the physical property by each of the seven crystal systems.

For crystals of the triclinic and monoclinic systems the symmetry inherent in a symmetrical second order tensor is higher than that of the crystal. In the triclinic system no restriction is placed on the shape or orientation of the representation quadric; the description of a physical property requires specification of the magnitudes of D_X, D_Y, D_Z and of the geometrical relationship of the principal axes of the representation quadric to the crystallographic reference axes. In the monoclinic system (Fig 11.7(a)) one of the principal axes of the representation quadric is required by Neumann's Principle to be parallel to the crystallographic diad, conventionally [010], and the other two principal axes lie in the (010) plane. The description of a physical property of a monoclinic crystal requires specification of the angle between the z crystallographic axis and one of the principle axes, X or Z, in the (010) plane, and the magnitudes of D_X, D_Y, and D_Z.

The symmetry of the representation quadric is that of the centrosymmetrical point group of the orthorhombic system, *mmm*. For crystals of the orthorhombic system therefore the principal axes of the representation quadric are constrained to be parallel to the crystallographic reference axes (Fig 11.7(b)). In the orthorhombic system it is thus necessary to specify only the magnitudes of D_X, D_Y, D_Z.

For crystals of the trigonal system one of the principal axes of the representation quadric must be parallel to the triad and the other two lie in the (0001) plane through the origin. The equation to the principal section of the representation quadric parallel to (0001) is, since Z is parallel to the triad, $X^2/a^2 + Y^2/b^2 = 1$. This is the equation to an ellipse whose semi-axes are a and b; all other radii of the ellipse have radii intermediate between a and b, but because the ellipse has triad symmetry the radii a and b must each be repeated every 120° so that the ellipse degenerates into a circle with $a = b$. Therefore in the trigonal system Z is parallel to the triad and $D_X = D_Y$ so that the representation quadric is an ellipsoid of revolution about the triad axis. For the complete description of a physical property of a trigonal crystal it is necessary to specify only two coefficients $D_X = D_Y$ and D_Z.

Likewise in the tetragonal and hexagonal systems the representation quadric is an

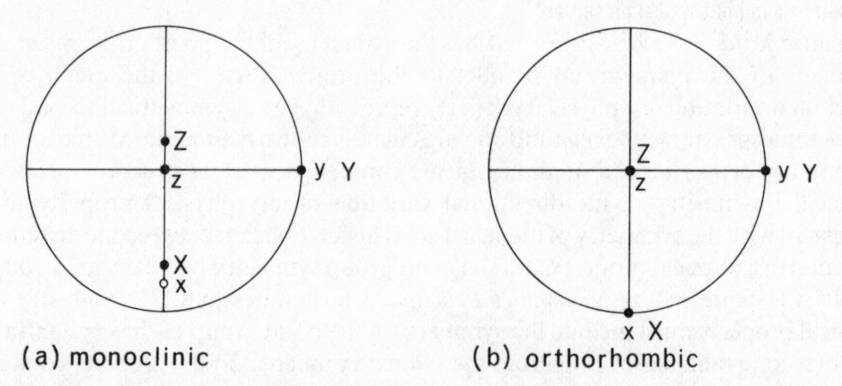

Fig 11.7 Stereograms illustrating Neumann's Principle. In the monoclinic system one of the principal axes, Y, of the representation quadric is constrained to be parallel to the diad, the other two may lie anywhere in the (010) plane. In the orthorhombic system the principal axes are parallel to the crystallographic axes.

ellipsoid of revolution about the axis of high symmetry. Again only two coefficients need to be specified D_Z and $D_X = D_Y$.

When the representation quadric is an ellipsoid of revolution the resultant vector **q** necessarily lies in the plane containing the applied vector **p** and the axis of high symmetry, the triad, tetrad, or hexad. All cases thus become two-dimensional and in particular the expression for $D_{\|}$ in a direction making an angle θ with the axis of high symmetry reduces to

$$D_{\|} = D_Z \cos^2\theta + D_X \sin^2\theta,$$

where Z is parallel to the axis of high symmetry z. The central section of the representation quadric perpendicular to the axis of high symmetry is, as we have shown, circular so that for all directions perpendicular to the axis of high symmetry the normal to the quadric coincides with its radius; therefore $\mathbf{q} \| \mathbf{p}$ for all directions perpendicular to the axis of high symmetry and for such directions $D_{\|} = D_X$.

Application of the argument used previously for the trigonal system to all four triads of the cubic system leads to the conclusion that the representation quadric for any cubic crystal is a sphere. The resultant vector **q** is parallel to the applied vector **p** for *all* directions in a cubic crystal. A second rank tensor property of a cubic single crystal thus requires only the specification of one coefficient D, which represents the ratio q/p for all directions, that is to say a cubic crystal is *isotropic* for second rank tensor properties just like a polycrystalline solid.

The constraints exerted by the symmetry of the crystal structure on the orientation and geometry of the representation quadric for a second rank tensor property are summarized in Table 11.1.

Diffusion coefficients

We return now to diffusion as an example of a second rank tensor property and give brief consideration to experimentally determined diffusion coefficients in single crystals. Surprisingly few determinations of high accuracy are available. Measured diffusion coefficients display marked anisotropy in general and strong temperature dependence. For instance the coefficients of self-diffusion of Mg through magnesium single crystals (hexagonal) are $D_Z = 1{\cdot}0 \exp(-32{,}200/RT)$ and $D_X = D_Y = 1{\cdot}5 \exp(-32{,}500/RT)$ where Z is parallel to the hexad and T is in kelvins. Thus at 300 K $D_X = D_Y = 4{\cdot}0 \times 10^{-24}\,\text{cm}^2\,\text{s}^{-1}$ and $D_Z = 4{\cdot}5 \times 10^{-24}\,\text{cm}^2\,\text{s}^{-1}$, while at 700 K $D_X = D_Y = 1{\cdot}85 \times 10^{-10}\,\text{cm}^2\,\text{s}^{-1}$ and $D_Z = 1{\cdot}0 \times 10^{-10}\,\text{cm}^2\,\text{s}^{-1}$. The anisotropy of the diffusion coefficients at both the selected temperatures is quite marked, but more

Table 11.1
The relationship of the magnitude and orientation of the principal axes of the representation quadric for second rank tensor properties to the symmetry of the crystal

Crystal system	Relationship between principal coefficients	Orientation of principal axes with respect to crystallographic axes
Triclinic	$D_X \neq D_Y \neq D_Z$	No special relation
Monoclinic	$D_X \neq D_Y \neq D_Z$	$Y \| y$
Orthorhombic	$D_X \neq D_Y \neq D_Z$	$X \| x$, $Y \| y$, $Z \| z$
Trigonal	$D_X = D_Y \neq D_Z$	Ellipsoid of revolution about z, $Z \| z$
Tetragonal	$D_X = D_Y \neq D_Z$	Ellipsoid of revolution about z, $Z \| z$
Hexagonal	$D_X = D_Y \neq D_Z$	Ellipsoid of revolution about z, $Z \| z$
Cubic	$D_X = D_Y = D_Z$	Sphere

remarkable is the very large temperature coefficient of D, which is very nearly independent of direction. In this example the relationship between the anisotropy of the principal diffusion coefficients and the crystal structure of Mg is not immediately apparent.

But in PbI_2, which has a structure similar to that of CdI_2 (Fig 10.23) the relationship between structure and anisotropy of diffusion coefficient for the self-diffusion of Pb is more clearly apparent. At 316 °C $D_Z = 2{\cdot}52 \times 10^{-6}$ cm^2 day^{-1} and $D_X = D_Y = 5{\cdot}02 \times 10^{-6}$ cm^2 day^{-1}. In the structure of this substance cation sites are much closer together in directions perpendicular to [0001] than in the direction [0001] and moreover it would be energetically difficult for a Pb ion to jump across from one PbI_2 layer to another.

In the mineral olivine there is marked anisotropy of diffusion coefficients for Ni. In an olivine of composition $(Mg_{0{\cdot}93}Fe_{0{\cdot}07})_2SiO_4$ experiments at high temperature yield $D_z = 25D_y$, $D_x \simeq D_y$ so that it would appear that the diffusion of cations is relatively facile in the [001] direction. Inspection of the plan of the olivine structure (Fig 10.27) shows that in the [001] direction there are chains of edge-sharing octahedra; we did not draw attention to these chains in chapter 10 because they were not relevant to our argument there, but the diffusion evidence taken together with recent structural studies on compositional derivatives of the olivine structure show the importance of this hitherto neglected structural feature.

In conclusion it is appropriate to say that the paucity of high quality data at present makes it impossible to provide adequate exemplification of the essential structural basis of diffusion anisotropy. Nevertheless diffusion must be one of the easiest of the anisotropic properties of single crystals to interpret in structural terms. It is for this reason and in spite of the shortage of good data that we have taken diffusion as one of our two exemplary physical properties.

Thermal expansion

We have earlier had occasion to comment on the nature of the interaction between an isolated pair of atoms and in particular on the form of the dependence of their interaction energy on interatomic distance (Fig 10.7). The minimum of the potential energy curve corresponds to the equilibrium distance between a pair of atoms at rest. This is an unattainable situation; at all temperatures the atoms oscillate relative to each other so that what one means by the equilibrium distance apart of a pair of atoms at any temperature is the separation of their mean positions. As temperature increases from absolute zero the potential energy of the atom pair increases and the separation of their mean positions (Fig 11.8) increases from x_0 along the broken line shown in the figure. If the potential energy trough were symmetrical, the broken line would be vertical and the 'coefficient of thermal expansion' of the atom pair would be zero: generalizing, it is because the broken line bends outwards that the coefficients of thermal expansion of crystalline solids are non-zero.

In a crystalline solid the effect of increasing temperature on atomic positions is more complex. The relative increase in the separation of a pair of adjacent atoms of the elements A and B may be markedly different from that of a pair of adjacent atoms of the elements A and C in a crystal of the elements A, B, and C. Even in a crystal of a chemical element there may be more than one bond type, as in graphite, and one would expect in such a case a marked difference in the rate of increase of the lengths of the different types of bonds. Moreover differential rates of increase of the separation of different pairs of atoms may cause coordination polyhedra of anions about cations

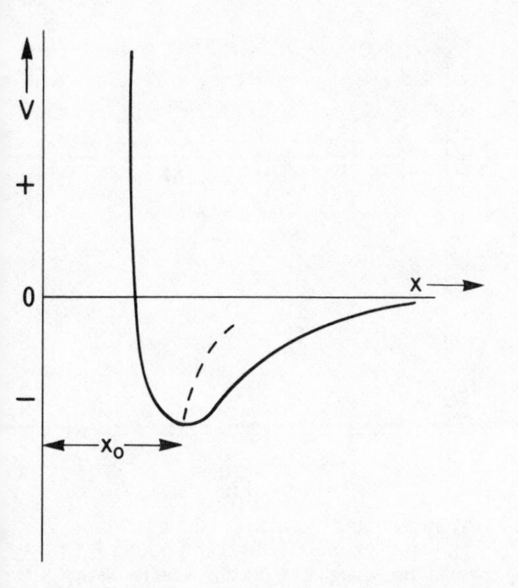

Fig 11.8 The variation of potential energy (V) with interatomic separation (x) for an isolated pair of atoms. The broken line represents the relation between the mean position of the atom pair and their potential energy; as temperature increases so will the potential energy of the atom pair and so will their mean separation.

to tilt relative to one another. In short the positions of all atoms in the structure will change with temperature, but the magnitude of such changes of position will be quite small. The overall effect can be described in terms of small changes in the dimensions of the unit-cell coupled with small changes in the coordinates of the atoms within each unit-cell.

Since all unit-cells are equivalent they must change in response to rising temperature in precisely the same way, the dimensions of the lattice change correspondingly by amounts usually of the order of 10^{-5} Å per °C. Since the lattice changes only in its dimensions, lattice planes persist as such, as do lattice directions, but with some small change in angular relationships.

The dimensions of any polycrystalline body change in response to changing temperature; the change is homogeneous, that is to say if a line of length l expands by an amount δl for a given temperature rise then $\delta l/l$ is independent of the magnitude of l. Thus a length l in any orientation in the polycrystalline body at T °C becomes $l+\delta l$ at $(T+1)$ °C. The increment δl is commonly expressed as $l\alpha$, where α is the change in length per unit length per degree and is known as the coefficient of linear expansion; for homogeneous expansion α is a constant independent of direction so that any length l at T °C becomes $l(1+\alpha)$ at $(T+1)$ °C. If α is independent of temperature in the temperature range under consideration, then the increase in a length l at T °C when the temperature is raised by t °C is $l\alpha t$. But for most polycrystalline substances α is temperature dependent so that the increase in length between the temperatures t_1 and t_2 is given by $l_{t_1}\int_{t_1}^{t_2} \alpha_t \, dt$.

When a polycrystalline body of volume V is heated through 1 °C its volume increases by an amount $\delta V = \alpha_v V$, where α_v is its coefficient of volume expansion. If the body is a cube of side l, then $V = l^3$ and

$$V+\delta V = (l+\delta l)^3$$

i.e. $$V(1+\alpha_v) = l^3(1+\alpha)^3$$

i.e. $$\alpha_v = (1+\alpha)^3 - 1$$
$$= 3\alpha$$

if powers higher than the first of the small coefficient α are neglected. If α_v is

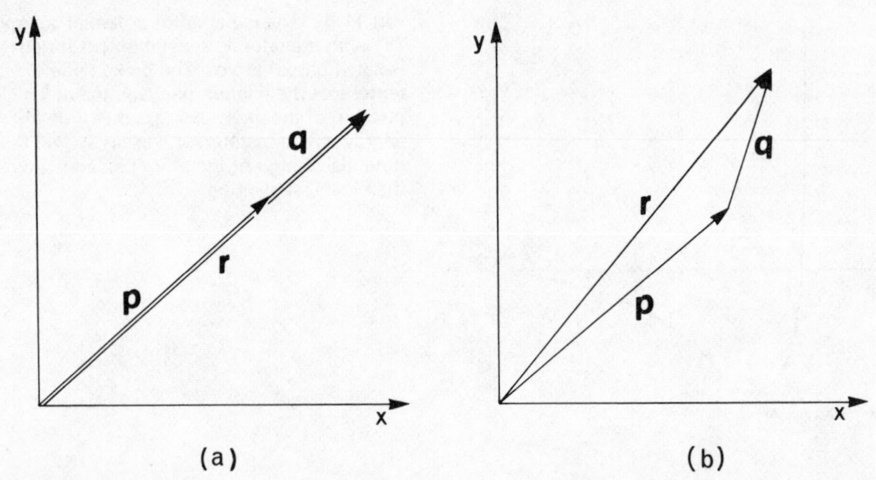

Fig 11.9 Two-dimensional thermal expansion in (a) an isotropic material, (b) an anisotropic material. A vector of length p expands by an amount q when the crystal is heated through T °C; in (a) the vectors **p** and **q** are parallel and in (b) they are non-parallel in general. The resultant vector $\mathbf{r} = \mathbf{p}+\mathbf{q}$; $r = p+q$ in (a) but not in (b).

independent of temperature in the temperature range under consideration, then the increase in volume, V at T °C, when the temperature is raised by t °C is $V\alpha_v t = 3V\alpha t$. In chapter 13 we shall be concerned only with coefficients of volume (or bulk) expansion of polycrystalline solids and in that context it is convenient to drop the subscript v (cf. Table 13.3).

When the temperature of a single crystal is raised the dimensions of the crystal increase. The expansion is, as for polycrystalline bodies, homogeneous but it is not necessarily of the same magnitude in all directions. In an isotropic crystal a vector of length **p** expands by a length **q** when the crystal is heated through 1 °C to become the vector $\mathbf{r} = \mathbf{p}+\mathbf{q}$, the vectors, **p**, **q**, **r** being parallel to one another (Fig 11.9(a)) so that $r = p+q$. If the coefficient of linear expansion α is defined[2] here, as for the polycrystalline case, as the change in length per unit length when the temperature of the crystal is raised through 1 °C, then $q = \alpha p$ and $r = (1+\alpha)p$. But many crystals are anisotropic for the property of thermal expansion; the vector **q** is not in general parallel to the vector **p** (Fig 11.9(b)), but is related to it by the coefficients α_{kl} of a symmetrical second rank tensor. The description of the thermal expansion of an anisotropic crystal is most easily obtained by the measurement of the three coefficients of linear expansion parallel to the principal axes $\alpha_X, \alpha_Y, \alpha_Z$. The change q in any other direction p when the temperature of the crystal is raised by 1 °C is given by

$$q_X = \alpha_X p_X$$

$$q_Y = \alpha_Y p_Y$$

$$q_Z = \alpha_Z p_Z,$$

where p_X, q_X, etc are the components of **p** and **q** on the principal axes.

The thermal expansion of a crystal must of course be consistent with the crystal's symmetry. The same relationships between principal axes and symmetry directions as we worked out for diffusion and listed in Table 11.1 apply in thermal expansion. It

[2] This definition of coefficient of linear thermal expansion agrees with that of Wooster (1949) and differs from that used by Nye (1960).

is perhaps necessary to make the obvious proviso that the temperature interval under consideration should not embrace a polymorphic transformation; at a polymorphic transformation the magnitudes and orientation of the principal axes of the representation quadric for thermal expansion are often profoundly altered.

We have defined coefficients of linear thermal expansion in terms of a temperature increment of 1 °C. However the coefficients are usually so small that in practice it is necessary to use much larger temperature increments, as much as 100 °C, in order to obtain measurements of acceptable accuracy. It is then necessary to assume constancy of α over the experimental temperature range. But it is always necessary to bear in mind that coefficients of thermal expansion are temperature dependent and vary particularly rapidly at temperatures close to phase transformations, including the melting point. Experimentally it is necessary to choose experimental temperature intervals with this sort of consideration in mind.

The formal description of thermal expansion is very much like that of any other symmetrical second rank tensor property. Although similar in form the diffusion tensor and the thermal expansion tensor are physically rather different. Both relate one vector to another: the diffusion tensor relates a resultant flux to an applied concentration gradient, but the thermal expansion tensor relates the orientation and distance from the origin of a point in the crystal after a temperature rise to the orientation and distance from the origin of the corresponding point before the temperature rise. In general terms the diffusion tensor relates vectors that represent distinct properties or processes while the thermal expansion tensor relates vectors of the same sort before and after an event.

We now proceed to explore the physical significance of the coefficients α_{kl} in the thermal expansion tensor by considering in the first instance a two-dimensional case; extension to three dimensions is mathematically straightforward, but more cumbersome and less easy to visualize physically. In terms of any axial system x, y the extension **q** of a vector **p** consequent on a change in the temperature of the crystal is given by

$$q_x = \alpha_{xx} p_x + \alpha_{xy} p_y$$
$$q_y = \alpha_{yx} p_x + \alpha_{yy} p_y.$$

Since the thermal expansion tensor is symmetrical $\alpha_{xy} = \alpha_{yx}$. A line initially parallel to the x-axis and of the length p_x (Fig 11.10) expands by an amount $q_x = \alpha_{xx} p_x$ and $q_y = \alpha_{yx} p_x$. Thus α_{xx} represents the extension per unit length of the initial line parallel to the x-axis and, since the magnitudes of all the coefficients α_{kl} are very small, α_{yx} represents the angle of rotation of the line from its initial orientation parallel to the x-axis towards the y-axis. Likewise a line of length p_y initially parallel to the y-axis expands by an amount $q_x = \alpha_{xy} p_y$ and $q_y = \alpha_{yy} p_y$ so that α_{yy} represents its extension per unit length parallel to the y-axis and α_{xy} represents its angle of rotation from the y-axis towards the x-axis. The rectangle with sides p_x and p_y thus distorts to a parallelogram and, since $\alpha_{xy} = \alpha_{yx}$, the resultant parallelogram is symmetrically disposed between the x and y axes, the angles between adjacent sides being $90° \pm 2 \tan^{-1} \alpha_{xy} \simeq 90° \pm 2\alpha_{xy}$ since all three coefficients α_{xx}, α_{xy}, α_{yy} are small. The line of length p which was initially the diagonal of the rectangle becomes a diagonal of the parallelogram. In general any line which is not parallel to a principal axis of the representation quadric rotates relative to a fixed axial system and changes its length as the temperature of the crystal changes. Only along the principal axes is the effect of temperature change restricted to change in length and not associated with rotation.

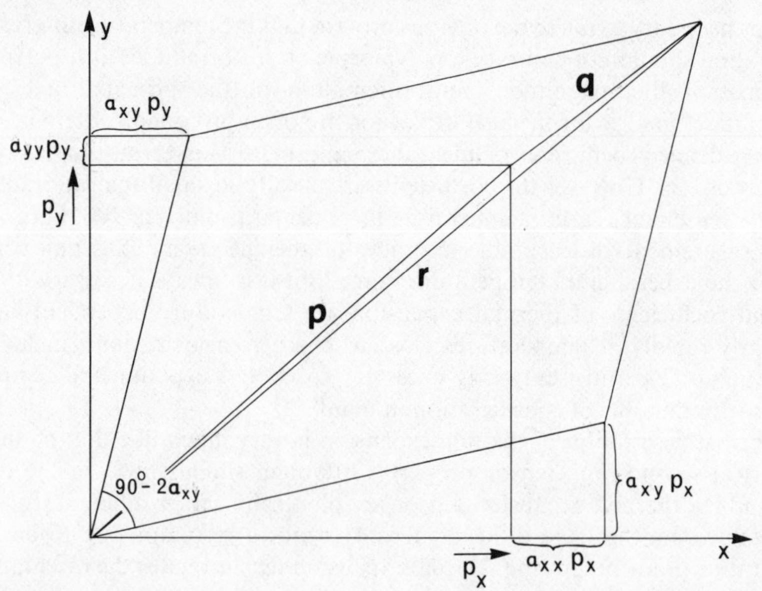

Fig 11.10 Two-dimensional thermal expansion. The vector **p** expands by **q** to become **r**. Its component p_x expands by $\alpha_{xx}p_x$ parallel to the x axis and by $\alpha_{xy}p_x$ parallel to the y axis; and its component p_y expands by $\alpha_{yy}p_y \| y$ and by $\alpha_{xy}p_y \| x$. The resultant parallelogram has angles between adjacent sides equal to $90° \pm 2\alpha_{xy}$, since α_{xx}, α_{xy}, and α_{yy} are very small. The diagram is greatly exaggerated.

It must however always be remembered that thermal expansion coefficients are usually no larger than about $10^{-5}\,\text{deg}^{-1}$ so that the greatest change in orientation of any line in a crystal amounts to no more than a few seconds of arc per degree Celsius. In practice therefore the distinction between the expansion of a line and its increase in length in its initial direction is negligible.

The insignificance of rotational effects is important when X-ray diffraction techniques are used to determine thermal expansion coefficients. Differentiation of the Bragg Equation

$$\lambda = 2d \sin \theta$$

yields

$$\frac{dd}{d} = -\cot \theta . d\theta.$$

In the context of thermal expansion we can put

$$\Delta d = \alpha\, d . \Delta t$$

where α is the change in spacing per unit length of the set of lattice planes giving rise to the measured reflexion and Δt is the temperature difference between successive measurements so that

$$\alpha = -\cot \theta \frac{\Delta\theta}{\Delta t}.$$

In order to obtain sufficiently large values of $\Delta\theta$ to be accurately measurable it is usually necessary to make use of reflexions of high Bragg angle and it may be necessary for Δt to be as much as 100 °C. It has of course to be assumed that α is effectively

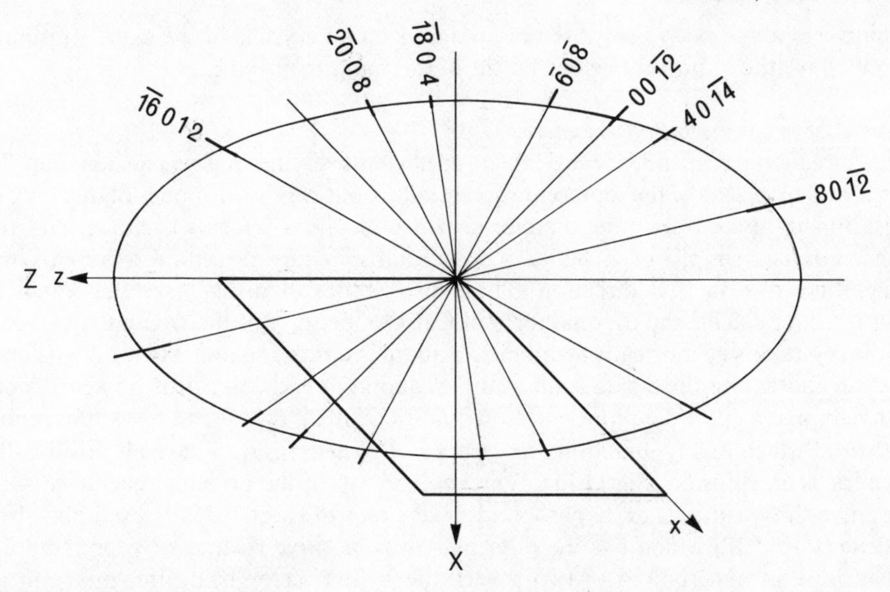

Fig 11.11 The (010) section of the representation quadric for thermal expansion of afwillite ($Ca_3(SiO_3OH)_2.2H_2O$; monoclinic; $a = 16{\cdot}21$, $b = 5{\cdot}63$, $c = 13{\cdot}12$ Å, $\beta = 134°48'$). The Z axis of the ellipsoid is effectively coincident with [001] in this monoclinic substance. This section of the thermal expansion quadric $\alpha_X X^2 + \alpha_Z Z^2 = 1$ has been determined by measurement of the variation of the d-spacings of $h0l$ reflexions with temperature: the directions of the normals to some of the planes used are shown on the diagrams, the thick part of each line representing the experimental error in $\alpha_{\|}^{-\frac{1}{2}}$. Analysis of all the experimental results has yielded $\alpha_X = 27{\cdot}8 \times 10^{-6}$, $\alpha_Z = 7{\cdot}1 \times 10^{-6}$ per deg. C.

constant even over such a large temperature range. The value of α obtained in this way may yield one of the principal coefficients of thermal expansion directly. For instance the determination of α from the change in the (0001) d-spacing of a trigonal crystal gives α_Z directly. But for a monoclinic crystal measurements of the change in d-spacing of several $h0l$ reflexions will usually have to be made in order to establish the orientation of the two principal axes in the (010) plane and the magnitude of the corresponding principal coefficients (Fig 11.11); in such a case the rotation of the normal to the ($h0l$) planes is ignored and α is taken to refer to the initial direction of the normal, an approximation which usually introduces a quite insignificant degree of error.

Coefficients of volume expansion for single crystals

Suppose that a cube of side l is cut out of a single crystal so that its edges are parallel to the principal axes of the representation quadric for thermal expansion. If the cube is heated through 1 °C it will deform to a parallelepiped with edges of length $l(1+\alpha_X)$, $l(1+\alpha_Y)$, $l(1+\alpha_Z)$. The increase in volume will be

$$\begin{aligned}\delta V &= l^3(1+\alpha_X)(1+\alpha_Y)(1+\alpha_Z) - l^3 \\ &= V(\alpha_X+\alpha_Y+\alpha_Z)\end{aligned}$$

if second and higher powers of the thermal expansion coefficients are deemed to be negligible. The coefficient of volume expansion $\alpha_v = \delta V/V$ is thus equal to the sum of the three principal coefficients of linear expansion,

$$\alpha_v = \alpha_X + \alpha_Y + \alpha_Z.$$

A polycrystalline body composed of an aggregate of crystals of the same substance would have the same volume coefficient of thermal expansion α_v.

Representation surfaces for thermal expansion

The directional dependence of linear coefficients of thermal expansion can be displayed by means of the representation quadric, but two other modes of display are occasionally used. In one the manner in which a sphere deforms is shown and the other consists simply of a figure whose radius in any direction represents the magnitude of α in that direction. These two modes of display are illustrated in Fig 11.12 for calcite, the trigonal form of $CaCO_3$; because of the trigonal symmetry of the crystal α is cylindrically symmetrical about the z-axis so that a two-dimensional section containing the z-axis is all that is required in each case. It is however more conventional to plot the representation quadric $\alpha_X X^2 + \alpha_Y Y^2 + \alpha_Z Z^2 = 1$, the radius vector of which in any direction represents $\alpha_{||}^{-\frac{1}{2}}$. When α_X, α_Y, α_Z are all positive the quadric is an ellipsoid (Fig 11.11). When one or two of the principal coefficients are negative the quadric is an hyperboloid of one or two sheets; this is exemplified by calcite (Fig 11.12) which has α_X, α_Y negative, α_Z positive so that its representation quadric is an hyperboloid of two sheets, the radius vector being imaginary in all directions for which $\alpha_{||}$ is negative. In order to display the variation of $\alpha_{||}$ with direction in such a case the real parts of two hyperboloids are drawn: $\alpha_X X^2 + \alpha_Y Y^2 + \alpha_Z Z^2 = 1$ is drawn out for directions in which $\alpha_{||}$ is positive and $\alpha_X X^2 + \alpha_Y Y^2 + \alpha_Z Z^2 = -1$ for directions in which $\alpha_{||}$ is negative, the positive and negative parts of the composite figure being clearly labelled.

When the representation quadric is an hyperboloid there will be a cone of directions asymptotic to the hyperboloid; all such directions have $\alpha_{||}^{-\frac{1}{2}} = \infty$ and are thus directions of zero thermal expansion. In the case of calcite for example the semi-angle θ of the cone, which is the angle between the surface of the cone and the [0001] axis, can simply be calculated from the expression

$$\alpha_{||} = l^2\alpha_X + m^2\alpha_Y + n^2\alpha_Z$$

by setting $\alpha_{||} = 0$ and noting that $\alpha_X = \alpha_Y$ to give

$$0 = (1-n^2)\alpha_X + n^2\alpha_Z$$

i.e.
$$0 = \alpha_X \sin^2\theta + \alpha_Z \cos^2\theta$$

and so
$$\tan\theta = \sqrt{\frac{-\alpha_Z}{\alpha_X}}.$$

Although we have devoted particular attention in considering representation surfaces for thermal expansion to the case of a substance with two principal coefficients negative because it exemplifies some points of interest, it must be emphasized that for the majority of crystals all three principal coefficients are positive and the representation quadric is an ellipsoid.

Relationship of thermal expansion coefficients to crystal structure

The relationship of linear thermal expansion coefficients to the differential strength of bonds in zinc and in arsenic, antimony and bismuth was commented on in chapter 10. In general it would seem reasonable to expect that directions in which bonds are relatively strong should have relatively small coefficients of linear expansion and that

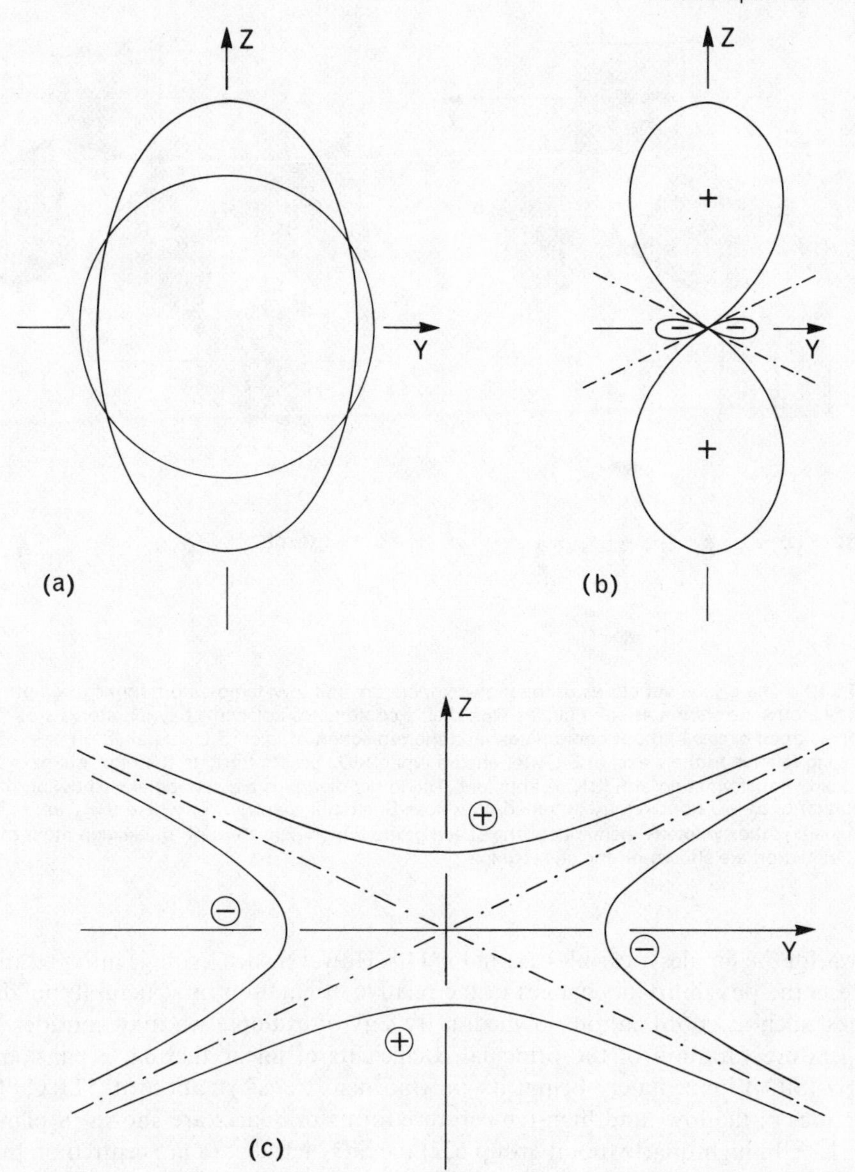

Fig 11.12 Thermal expansion of calcite, which has $\alpha_X = \alpha_Y = -5{\cdot}3 \times 10^{-6}$ (°C)$^{-1}$ and $\alpha_Z = 23{\cdot}6 \times 10^{-6}$ (°C)$^{-1}$. (a) shows the deformation of a sphere drawn in a calcite crystal, the diagram is exaggerated about 200 times for a 100 °C temperature increase. (b) is a plot of $\alpha_{\parallel}$ with the positive and negative loops of the figure marked. (c) is the representation quadric; the hyperboloid $\alpha_X X^2 + \alpha_Y Y^2 + \alpha_Z Z^2 = 1$ (labelled +) cuts the Z axis (the triad) and the hyperboloid $\alpha_X X^2 + \alpha_Y Y^2 + \alpha_Z Z^2 = -1$ (labelled −) cuts the Y axis. In (b) and (c) the directions of zero expansion are shown as dash-dot lines. All three diagrams are sections in the Z–Y plane; the figures shown are cylindrically symmetrical about the Z axis.

weak bond directions should have larger expansion coefficients. Moreover one might expect linear expansion coefficients to vary with temperature in much the same way as interatomic distances vary with temperature, that is to say to show the steadily increasing increase with temperature of the median line of the potential energy curve

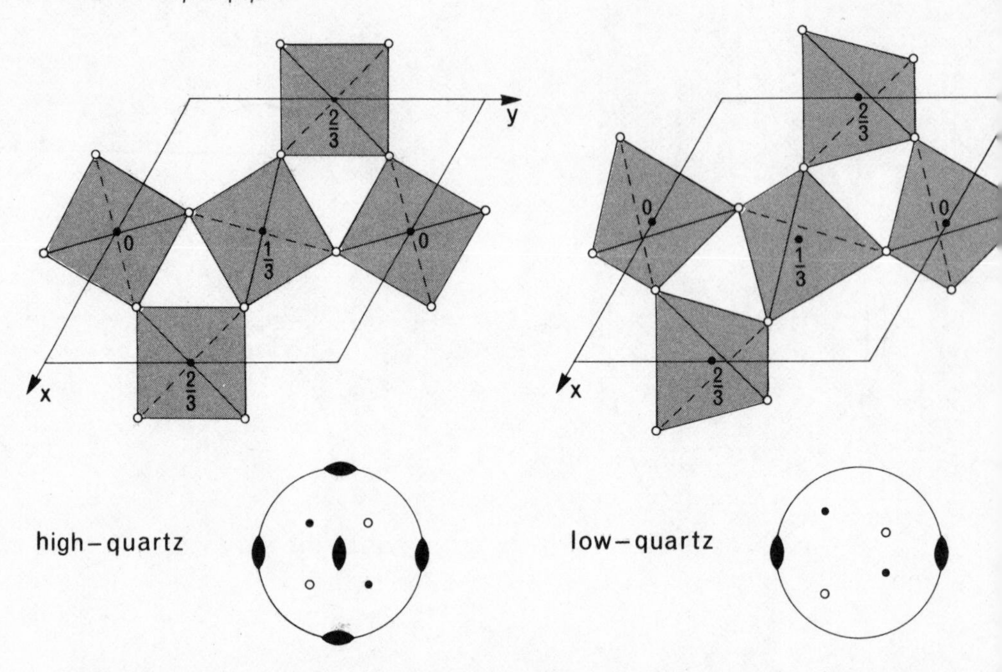

Fig 11.13 The crystal structures of the high-temperature and low-temperature forms of quartz SiO_2. Si atoms are shown as solid circles with their *z* coordinates adjacent; oxygen atoms are shown as open circles without coordinates; and the projection of each SiO_4 tetrahedron is shaded. The tilting relative to the *z* axis of SiO_4 tetrahedra when SiO_2 passes through the high-quartz → low-quartz transformation at 573 °C is apparent. The lower diagrams are stereograms showing the orientation of Si—O bonds in the tetrahedron whose Si atom lies at 0, u, $\frac{2}{3}$ (where $u = \frac{1}{2}$ in high-quartz); the symmetry elements of the space group which pass through the silicon atom of this tetrahedron are shown on the stereograms.

shown, for the simplest possible case, in Fig 11.8. However such a simple interpretation neglects the possibility of changes in the relative orientation of structural 'building blocks' such as coordination polyhedra. By way of example we may consider the temperature variation of the principal coefficients of linear thermal expansion of quartz (SiO_2) immediately below its polymorphic transformation at 573 °C. The structures of the low- and high-temperature forms of quartz are shown in plan in Fig 11.13. In high-quartz (point group 622) the SiO_4 tetrahedra are centred on three mutually perpendicular diads so that their position is rigidly fixed by the space group and all four Si—O bonds in any tetrahedron are required by symmetry to be equal in length although the bond angles O—Si—O need not all be equal. In low-quartz (point group 32) however only one diad ($\parallel x$, y, or u) passes through the centre of each SiO_4 tetrahedron, two pairs of distinct Si—O bond lengths are permitted in each tetrahedron and each tetrahedron may be tilted about its diad axis. Careful structural studies have shown that below the transformation temperature the angle of tilt of each SiO_4 tetrahedron with respect to the [0001] axis increases from 0° at 573 °C to 8·5° at 570 °C, 11·3° at 450 °C, and 15·6° at room temperature. The essential reasonableness of this conclusion can very easily be shown by calculating the angle of tilt of the tetrahedra from the measured unit-cell dimensions at various temperatures. Figure 11.14 shows the results of such calculations based on the assumptions that the tetrahedra remain regular below 573 °C with the same Si—O bond length as in high-

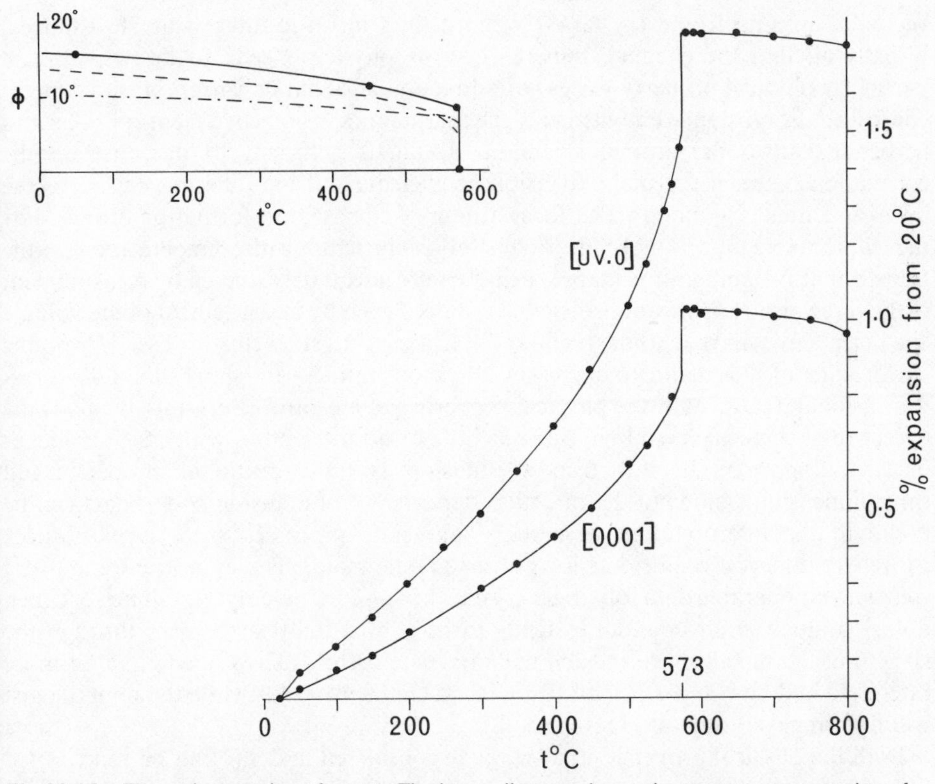

Fig 11.14 Thermal expansion of quartz. The lower diagram shows the percentage expansion of unit length at 20 °C parallel and perpendicular to the triad of low-quartz, which becomes the hexad of high-quartz. The upper diagram shows the angle of tilt, ϕ, of SiO_4 tetrahedra at various temperatures; ϕ has the same magnitude for all tetrahedra in the unit-cell and is defined for any tetrahedron as the angle between the diad of the tetrahedron parallel to [001] in high-quartz and the corresponding direction in low-quartz. The broken and dash-dot curves represent respectively the values of ϕ calculated from thermal expansion data $\|a$ and $\|c$ in the approximate manner described in the text.
Sources: thermal expansion data of Kozen and Takane quoted by Skinner in Clark (1966) and tilt angles of Young quoted by Megaw (1971).

quartz right down to 20 °C. The agreement between the angles of tilt derived from structure determinations and those calculated on this simple hypothesis is good enough to provide convincing evidence of the general correctness of the proposition that tilting of SiO_4 tetrahedra in a regularly progressive manner is the principal factor determining the temperature dependence of the thermal expansion of low-quartz. That the agreement is not perfect is scarcely surprising: it is known that as temperature falls from 573 °C towards room temperature the SiO_4 tetrahedra of low-quartz distort from regularity of shape and the Si—O bond lengths decrease in magnitude. The accurate structure refinement of low-quartz by Smith and Alexander (1963) has shown that at room temperature the two Si—O bond lengths of each tetrahedron are $1{\cdot}597 \pm 0{\cdot}003$ Å and $1{\cdot}617 \pm 0{\cdot}003$ Å.

Tensor properties: some generalities

Crystal physics is often regarded as a backwater of crystallography. The occasions when a detailed knowledge of the relationship between a resultant and an applied vector in an anisotropic crystal is of more than intrinsic interest are few indeed. Usually all that the chemist, mineralogist, or physicist needs to know is that a particular physical property varies with direction and that its variation is symmetry controlled. For instance knowledge of the thermal expansion coefficients parallel and perpendicular to the atomic layers in arsenic is sufficient to demonstrate the correlation between thermal expansion coefficients and the differing nature of the intra- and inter-layer bonds. The 'forewarning' of a phase transformation provided by the rapid increase of thermal expansion coefficients with temperature as the transformation temperature is approached can be adequately studied by measurement of the three principal linear coefficients or indeed even by measurement of the volume coefficient. We have earlier remarked on the almost exclusive use of volume coefficients of thermal expansion in thermodynamics. In short the full three-dimensional treatment of the physical properties of anisotropic crystals, which is the objective of classical crystal physics, has little or no interaction with other branches of crystallography. In this respect diffusion is an exceptional property: full three-dimensional study of the direction dependence of diffusion coefficients can be useful in the interpretation of certain solid-state processes such as exsolution. Moreover diffusion coefficients for polycrystalline materials may not correlate very well with comparable data for single crystals because in the polycrystalline specimen diffusion along grain boundaries tends to be substantially faster than through the structure; the metallurgist and the metamorphic petrologist will generally be more interested in diffusion coefficients for polycrystalline materials while the mineralogist will find single crystal data more useful.

Both the physical properties that we have considered in detail can be represented by second rank tensors, but there are many familiar anisotropic properties which cannot be so represented. For instance cleavage and rate of crystal growth are not tensor properties at all, while refractive index, although not itself a tensor property, is related to dielectric susceptibility which is a second rank tensor property. In the next chapter we shall develop this relationship and go on to give a fairly full account of the optical properties of anisotropic crystals; for the mineralogist optical properties are by far the most interesting and instructive of the physical properties of crystals.

The formal definition of tensors lies outside our scope, but it is appropriate at this point to make some simple general statements about tensors. Tensors of the zeroth, first, second, third, and fourth ranks may be involved in the formal description of physical properties. A zeroth rank tensor, or *scalar*, has magnitude but no direction; properties such as density and heat capacity are scalars. A first rank tensor, or *vector*, has both magnitude and direction; pyro-electricity is an example of a vector crystal property. A second rank tensor is a property that relates two vectors; diffusion, thermal expansion, thermal conductivity, electrical conductivity, dielectric susceptibility and magnetic susceptibility are familiar examples. Stress imposed on a crystal and the resultant strain are likewise second rank tensors, mechanical strain being analogous to thermal expansion which can be regarded as the strain induced in the crystal by a temperature change. A third rank tensor relates a vector and a second rank tensor; the best known third rank tensor property of crystals is piezoelectricity. A fourth rank tensor relates two second rank tensors, the most important being that which relates strain to stress. For polycrystalline solids the induced strain ε is related to the applied tensile stress σ by Hooke's Law, $\varepsilon = s\sigma$, where s is a constant known as the elastic modulus. But for an anisotropic crystal the relationship between ε and σ is more elaborate, requiring a fourth rank tensor, a 9×9 matrix, with as many as

36 independent constants, for its complete description.

Not only does crystal physics yield results that, as we have already said, are in detail of little interest in other fields of solid-state study, but the subject has, in its higher development, an abstruse mathematical formalism which sets it apart from other fields of solid-state study; moreover experimentation in this field presents exceptional difficulties. The sort of experiments described by Wooster and Breton (1970) require well developed and, usually, quite large single crystals; this requirement cannot be met for very many substances so that we remain ignorant of most of the physical properties in full three-dimensional detail of all but a few crystalline substances. The only anisotropic physical properties that are readily susceptible to detailed study over almost the whole range of the crystalline state are optical properties, which constitute the subject of the next chapter.

Point group determination

Although, as we have indicated in the preceding section of this chapter, the full three-dimensional study of the physical properties of single crystals is generally of no more than limited interest, there is one essential crystallographic task for which observations of anisotropic physical properties may be very useful. This task is the determination of the space group or the point group of a crystalline substance and in particular the establishment of whether the structure is centrosymmetric or non-centrosymmetric. We therefore conclude this chapter with an account of methods of point group determination. There is no generally applicable procedure; how one proceeds in any particular case will depend on what information is already available and on one's objective, which may be a complete structure determination or the preliminary description of a newly discovered substance.

If the substance is transparent and its crystals are not too small, optical examination (to be described in chapter 12) will quickly show whether it is isotropic and therefore to be assigned to the cubic system; or uniaxial and therefore belonging to the trigonal, tetragonal or hexagonal systems; or biaxial and therefore triclinic, monoclinic or orthorhombic. Observations of the orientation of optical properties with respect to the morphology of the crystals may enable one tentatively (but no more than that) to assign the substance to a particular crystal system, other than the cubic system where the assignment is unambiguous. The next step is, usually, to make a single crystal X-ray diffraction study of the substance in order to determine its Laue symmetry and then, by observation of systematic absences or moving film photographs, to determine its diffraction symbol.

In general the restriction imposed by Friedel's Law prevents this sort of study going beyond the determination of the diffraction symbol. It is not usually possible to determine whether the substance is centrosymmetric or non-centrosymmetric. Some diffraction symbols do however lead to a unique determination of space group and thus of point group. For instance a substance with the diffraction symbol *mmm*P*bcn* must be assigned to the point group *mmm* and to the space group P*bcn*. But a substance whose diffraction symbol is determined as *mmm*P.*cn* may have point group *mmm* or 2*mm* and space group P*mcn* or $P2_1cn$. In this second example the two possible space groups could be distinguished by determining whether or not the structure is centrosymmetric. If, in these circumstances, intensity data have been collected for a full structure analysis, then the most convenient way of determining whether the substance is centrosymmetric is by application of the *N*(*z*) *test*, which we now describe in outline. Reflexions of similar Bragg angle are grouped together (for instance, the range of sin θ from 0·3 to 0·5 might constitute one group, 0·4 to 0·6 the next, and so on). The assumption is then made that each atom in the structure scatters the same amplitude of X-radiation into every reflexion of a particular group. If the assumption is valid, it can be shown that the probability that the intensity of a given reflexion lies between I and $I+\delta I$ is independent of the actual atomic positions but is dependent on the symmetry of the structure. We now define $N(z)$ as the percentage number of reflexions in a particular group with intensities less than or equal to $z\bar{I}$, where z is a fraction and $\bar{I}$ is the mean intensity of the group. The relationship between $N(z)$ and z has been calculated for centrosymmetric and non-centrosymmetric structures. Curves of $N(z)$ plotted against z are shown in Fig 11.15 for the two cases; it is noticeable that the two curves follow markedly different courses especially for z between zero and 0·4. The $N(z)$ test consists of plotting $N(z)$ averaged over all the selected groups of reflexions against z for the measured intensities of the unknown structure and comparing the resultant plot with the standard centro- and non-centrosymmetric curves; good agreement with either standard curve yields a clear determination of whether or not the structure is centrosymmetric.

In the course of a structure analysis it may be possible to detect at an early stage the presence or absence of other symmetry elements such as diads and mirror planes which, being non-translational, do not give rise to systematic absences; if this can be done space groups with the same diffraction symbol, such as P2/*m*, P*m*, and P2, can be distinguished. The methods used however for making such distinctions lie outside our scope here.

We turn now to a property of X-ray diffraction which may be applied to point group determination in favourable circumstances; this is the property known as *anomalous scattering*. The X-radiation scattered by an atom is normally π out of phase

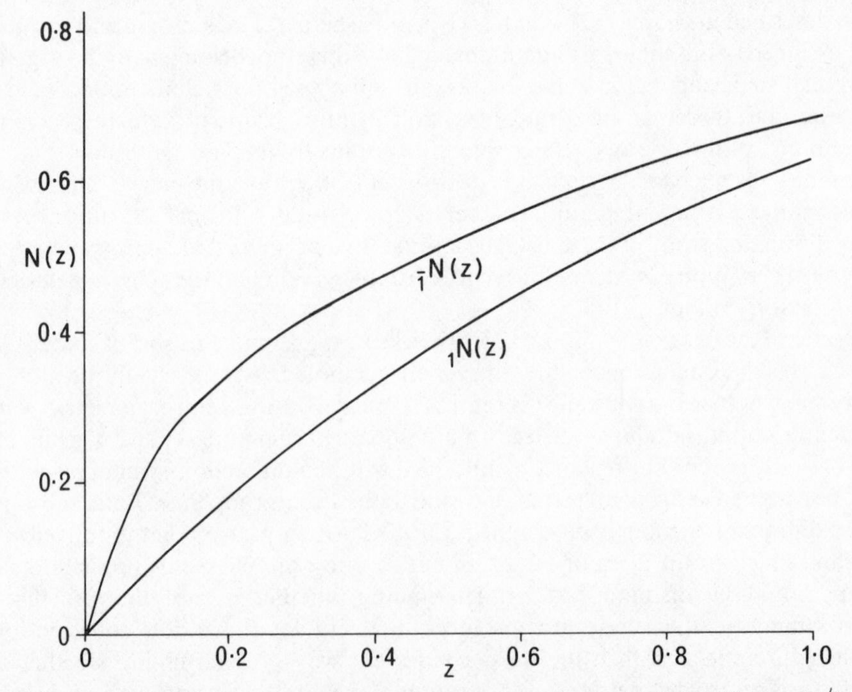

Fig 11.15 The $N(z)$ test. The theoretical curves for the centrosymmetric case $_{\bar{1}}N(z) = \mathrm{erf}\sqrt{(\frac{1}{2}z)}$ and for the non-centrosymmetric case $_{1}N(z) = 1 - \exp(-z)$ are shown plotted against z.

relative to the incident beam. But if the incident X-rays are of a wavelength just short of the absorption edge of one of the constituent atomic species of the structure concerned, then the X-rays scattered by atoms of that element do not have a phase difference π and are said to be anomalously scattered; the phase of such diffracted X-rays is advanced so that their effective path difference is less than that for X-rays scattered normally. Such anomalous scattering leads to a breakdown of Friedel's Law, as may be seen by considering the 111 and $\bar{1}\bar{1}\bar{1}$ reflexions of blende (ZnS). It was shown in chapter 6 that although the blende structure is non-centrosymmetric, its 111 and $\bar{1}\bar{1}\bar{1}$ reflexions are of equal intensity (Fig 6.21). But when the incident X-radiation is AuLα_1 which is scattered anomalously by Zn atoms, the diffraction pattern is modified in such a manner as though the Zn atoms were displaced from their actual positions so as to decrease the path travelled by the X-rays scattered by them. Thus for the 111 reflexion the zinc atoms behave as though they were displaced from their actual positions by a small distance in the AB direction and for the $\bar{1}\bar{1}\bar{1}$ reflexion as though they were displaced by an equal distance in the opposite direction BA (Fig 11.16(a)). The phase amplitude diagrams (Fig 11.16(b), (c)) indicate that the $\bar{1}\bar{1}\bar{1}$ reflexion is stronger than the 111 reflexion. Such a breakdown of Friedel's Law may enable the

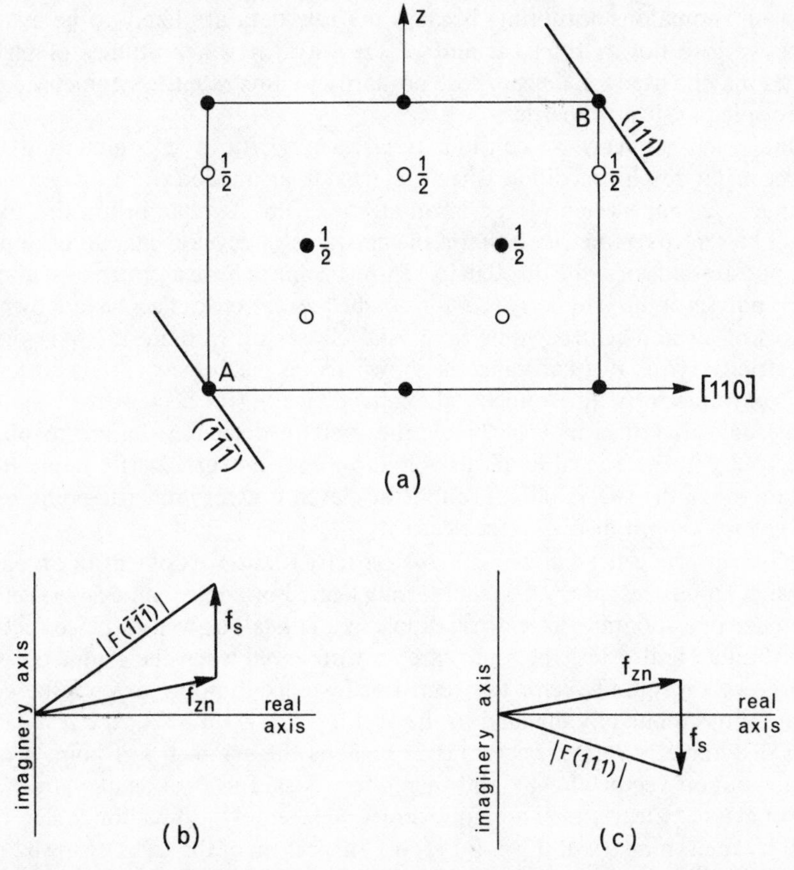

Fig 11.16 Anomalous scattering of AuLα_1 radiation by Zn atoms in blende (ZnS). (a) shows a projection of the blende structure on (1$\bar{1}$0) (cf. Fig 6.21); solid circles Zn, open circles S; the trace of the plane labelled (111) refers to the second plane out from the origin. (b) and (c) are respectively phase amplitude diagrams for the $\bar{1}\bar{1}\bar{1}$ and 111 reflexions; in each case the f_{Zn} vector makes an angle of $10\frac{1}{2}°$ with the real axis.

point group of a substance to be determined. For instance the Laue group of blende is $m3m$, which embraces the crystal classes 432, $\bar{4}3m$, and $m3m$; the only one of the three classes in which (111) and ($\bar{1}\bar{1}\bar{1}$) are not symmetry related is $\bar{4}3m$. The observation of unequal 111 and $\bar{1}\bar{1}\bar{1}$ intensities with X-radiation of suitable wavelength thus determines the point group of blende as $\bar{4}3m$. Since such intensity differences are likely to be very small, it would be unwise to rely on the observation of the intensities of only one pair of reflexions for point group determination. The observation of several such inequalities is necessary for unambiguous point group determination; for instance in Laue group $m3m$ the observation of systematic inequality in pairs of hkl and $h\bar{l}k$ reflexions leads unambiguously to the determination of the point group of the substance as $\bar{4}3m$. The success of this approach depends in the first place on the availability of X-radiation of the appropriate wavelength and secondly on the observation of intensity differences which can clearly be attributed to the breakdown of Friedel's Law and cannot be due to some other cause such as the differential absorption of X-rays by a crystal of irregular shape.

The quantitative, or even qualitative, study of the anisotropy of certain physical properties may be utilized for point group determination. Although it is usually more convenient to make use of X-ray methods for point group determination (such as the $N(z)$ test or anomalous scattering, because the raw data are likely to be available anyway), we give now a brief account of the ways in which studies of physical properties may be used for this purpose primarily to illustrate the symmetry control of anisotropic physical properties.

The first such property we consider is *piezoelectricity*. A crystal is said to be piezoelectric if it develops a dipole when subjected to an applied stress or, conversely, if it changes its shape when placed in an electric field. At equilibrium the applied stress will be centrosymmetric so that if the crystal is to develop charges of opposite sign at opposite ends of a line through its centre, it cannot have a centre of symmetry. Detailed analysis of the symmetry relations of the piezoelectric effect have shown that substances of all non-centrosymmetric crystal classes other than 432 may display piezoelectricity. Thus if a substance is shown to be piezoelectric, it must have a non-centrosymmetric point group. But the converse is not true because the magnitude of the piezoelectric effect may be below the limit of detection; failure to observe piezoelectricity in the crystal under examination does not necessarily imply that it belongs to one of the twelve point groups (the eleven centrosymmetric point groups and 432) which cannot display piezoelectricity.

The physical properties known as *pyroelectricity* (the development of an electric dipole when an unstressed crystal is uniformly heated or cooled) and *ferroelectricity* (the presence of a spontaneous electric dipole in a crystal) as well as piezoelectricity developed under hydrostatic pressure[3] are only observed when the symmetry of the crystal allows a resultant vector to occur. For instance in point group 2 the vector $U\mathbf{a}+V\mathbf{b}+W\mathbf{c}$ is related by the diad to the vector $-U\mathbf{a}+V\mathbf{b}-W\mathbf{c}$; the resultant of this pair of symmetry related vectors is $2V\mathbf{b}$. Thus the symmetry of point group 2 allows a resultant vector along y and therefore y is said to be a *unique direction* in crystals of class 2. Clearly then a unique direction is simply a direction which is not repeated by the symmetry of the point group. Inspection of the chart of the 32 point groups (Fig 3.20) shows immediately that the rotation axes in the point groups 2,

[3] Piezoelectricity may be developed by hydrostatic, compressive, or torsional stress systems. We are concerned here only with piezoelectricity developed under hydrostatic stress, which imposes the most stringent symmetry constraints of the three types of stress system.

2*mm*, 3, 3*m*, 4, 4*mm*, 6, and 6*mm* are unique directions, that all directions in point group 1 are unique, and that all directions parallel to the mirror plane in point group *m* are unique. Therefore if a crystal is found to be piezoelectric under hydrostatic stress or ferroelectric it must belong to one of these ten crystal classes which contain one or more unique directions and are known as the *polar classes.*[4] If piezoelectricity under hydrostatic stress or ferroelectricity is not observed in a crystal it is unsafe to conclude that the substance cannot belong to one of the ten polar classes; failure to observe either property may merely be due to its being too weak to be detected.

If a change in the temperature of a crystal causes an electric dipole to develop, the crystal is pyroelectric and in consequence must belong to one of the ten polar classes. But in practice it is difficult to be sure that the observed dipole moment is due to the pyroelectric effect rather than to a piezoelectric effect induced by the strains set up due to temperature gradients in the crystal during heating or cooling. For instance quartz, which belongs to class 32 and may therefore exhibit piezoelectricity under non-hydrostatic stress, commonly exhibits such 'false' pyroelectricity. In general it is wise to interpret the observation of pyroelectricity as indicating that the crystal belongs to one of the twenty crystal classes in which piezoelectricity may be observable, but does not necessarily belong to one of the ten polar classes in which true pyroelectricity may be observable.

Another physical property which is dependent on point group symmetry is *optical activity.*[5] A crystal is optically active if it rotates the plane of polarization of a beam of plane polarized light passing in certain directions through the crystal. For example crystals of quartz, class 32, are observed to rotate the plane of polarization in a clockwise sense in some crystals and anticlockwise in others. It can be shown that optical activity is restricted to crystals of those classes which contain no inversion axis of symmetry (including the mirror plane). These eleven *enantiomorphous* classes are listed in Table 11.2. In addition it can be shown that crystals of the classes *m*, *mm*, $\bar{4}$, and $\bar{4}2m$ may theoretically exhibit optical activity. However most crystals that are theoretically capable of exhibiting optical activity do not do so to any marked extent and moreover the observation of the property is difficult except for light travelling parallel to the principal axes of uniaxial crystals. Optical activity is thus not a property which can be utilized generally for point group determination.

The experimental methods of determination of the properties we have been

Table 11.2
Various groupings of non-centrosymmetric point groups

Crystal System	Piezoelectric classes	Polar classes	Enantiomorphous classes	Optically active classes
Triclinic	1	1	1	1
Monoclinic	2, *m*	2, *m*	2	2, *m*
Orthorhombic	222, *mm*2	*mm*2	222	222, *mm*2
Trigonal	3, 3*m*, 32	3, 3*m*	3, 32	3, 32
Tetragonal	4, $\bar{4}$, 422, 4*mm*, $\bar{4}2m$	4, 4*mm*	4, 422	4, $\bar{4}$, 422, $\bar{4}2m$
Hexagonal	6, $\bar{6}$, 622, 6*mm*, $\bar{6}m2$	6, 6*mm*	6, 622	6, 622
Cubic	23, $\bar{4}3m$	—	23, 432	23, 432

[4] The nomenclature is potentially confusing. *Polar directions* are directions whose opposite ends are not related by symmetry. Any non-centrosymmetric crystal has polar directions, but one polar direction may be related to other polar directions by symmetry. Thus while all unique directions are necessarily polar, polar directions are not necessarily unique.

[5] This property is discussed in chapter 12.

concerned with in preceding paragraphs are thoroughly discussed by Wooster and Breton (1970).

We conclude this brief survey of point group determination with some comments on the relevance of observations of crystal morphology. Before the advent of X-ray diffraction, morphological studies provided the principal means of point group determination, but such studies have for long been only of pedagogical and historical interest. Morphological studies are of course applicable only to substances which form well developed crystals. A well developed crystal of the substance under consideration is measured goniometrically and its faces are plotted on a stereogram. The stereogram is examined and the classes with which it is consistent are recorded. Other crystals of different habit, if available, are measured and the new faces added to the stereogram until as many faces as can be observed in crystals of the substance have been measured and plotted. In nineteen of the thirty-two crystal classes the general form is characteristic of the class; but in the remaining thirteen classes the general form is a special form[6] in one or more other classes so that, in these classes, the observation of one general form is insufficient to determine the point group symmetry. For example in the trigonal system the general form of class 32 is unique to that class; but the general form of class $\bar{3}$, a rhombohedron, is a special form in classes 32 and $\bar{3}m$ so that before a substance could be confidently assigned to class $\bar{3}$ crystals exhibiting other forms would have to be examined. Another commonly encountered difficulty is that the special forms of some point groups have higher symmetry than the point group: for instance the cube and the rhombic dodecahedron, which have symmetry $m3m$, are special forms in all the cubic point groups. Another potential source of error is that in non-holosymmetric point groups two forms of the same type may be developed so as to look as though they are a single form of higher symmetry. For example in classes 23 and $\bar{4}3m$ if the tetrahedra $\{111\}$ and $\{11\bar{1}\}$ are both present, they may present the appearance of an octahedron $\{111\}$ which is a special form in all the other cubic classes. Another potential source of error arises from the possibility that crystals may be twinned in such a manner that the twinning is not readily discernible by morphological examination; such crystals will appear to be of higher symmetry than their true point group symmetry. With so many chances of going inadvertently wrong it is fair to say that it is quite remarkable that the early crystallographers succeeded in assigning so many substances to their correct point group.

The valuable, but all too often indecisive, information about the probable point group of a substance derived from morphological observations can in some cases be supplemented—and may then become decisive—by the study of *etch figures*. Etch figures are produced on the natural faces of a crystal by the brief application of an appropriate solvent. In the early stages of the interaction between the solvent and the crystal small pits appear randomly disposed on the crystal faces. Under correct experimental conditions—and it may require some experimental trial and error to

[6] In this morphological context we define a *special form* as a form which bears some specialized relationship to the symmetry elements of the crystal class. The relationship may be that the normals to the faces of the form are parallel or perpendicular to a symmetry element (either an axis or a plane) or equally inclined to two symmetry axes. Special forms, defined in this way, are readily distinguishable by their appearance from the general form of the same crystal class. For example in class $\bar{4}$ the general form $\{hkl\}$ is a tetragonal sphenoid (that is a sort of tetrahedron elongated or shortened in the direction through the mid-points of one pair of opposite edges) while the special forms $\{hk0\}$ are tetragonal prisms. Likewise in class 23 the general form $\{hkl\}$ looks quite different from the special form $\{110\}$ although both have twelve faces; the former is a tetrahedral pentagonal dodecahedron and the latter is the familiar rhombic dodecahedron. In terms of the different definition of special form used in chapter 3 $\{hk0\}$ in class $\bar{4}$ and $\{110\}$ in class 23 would be general forms.

achieve this—the etch pits have demonstrably plane faces indicative of symmetry control of rate of solution. The shape and orientation of the etch pits on a particular crystal face corresponds to the projection of the point group symmetry of the crystal structure on the plane of the face concerned and so must conform to one of the ten two-dimensional point groups. Caution must however be exercised in the interpretation of etch figures (that is, the shape and orientation of etch pits) because the rate of solution in directions that are not symmetry related may be fortuitously similar so that etch figures of apparently higher symmetry than is consistent with the point group symmetry of the crystal may be observed. It is therefore safe only to derive from the study of etch figures a statement of the maximum possible point group symmetry of the crystal. A thorough account of the utilization of etch figure studies in point group determination is given in Buerger (1956).

12
Crystal optics

The interaction of light with crystalline matter has for long been a subject of fruitful study and it is appropriate that a substantial chapter in this book should be devoted to this topic. The microscopic study of crystal morphology is as old as microscopy itself, but it was not until the incorporation of polarizing devices into the, by then highly developed, compound microscope in the mid-nineteenth century that the study of crystal optics really began. By 1863 Zirkel had shown the polarizing microscope to be an invaluable tool for the characterization of minerals and for the elucidation of their mutual relations in rocks. It maintained its position as the principal instrument of the mineralogist and of the petrologist for many decades; and it proved particularly useful in certain branches of organic chemistry. Although now superseded in many of its functions by other physical instruments, it remains the most useful reconnaissance instrument for the study of transparent substances. A by no means insignificant feature of polarized light microscopy is the large amount of information that can be obtained in a short time; some of this information will be unambiguous, some will do no more than serve to suggest the need for further crystallographic or chemical investigation.

The study of opaque crystals in reflected polarized light has developed rapidly in recent years and may soon be expected to reach a level of sophistication that will make polarized light microscopy a powerful reconnaissance technique for all crystalline matter, whether opaque or transparent.

The nature of light

Light is electromagnetic radiation in the wavelength range 3800 Å (extreme violet) to 7800 Å (deep red). The theory of electromagnetic radiation was formulated by James Clerk Maxwell in 1864. The rigorous mathematical treatment of its application to light radiation is to be found in modern terminology in many standard textbooks of optics.[1] It suffices here merely to state certain of the conclusions that can be drawn from the theory. The first of these is that light is a transverse wave motion. Light passing through a vacuum or a material medium can be regarded as a periodic variation of the light vector in the plane perpendicular to the direction of propagation

[1] Preston (1928), Jenkins and White (1957), Ditchburn (1963), Longhurst (1957), Born and Wolf (1964), Lipson and Lipson (1969).

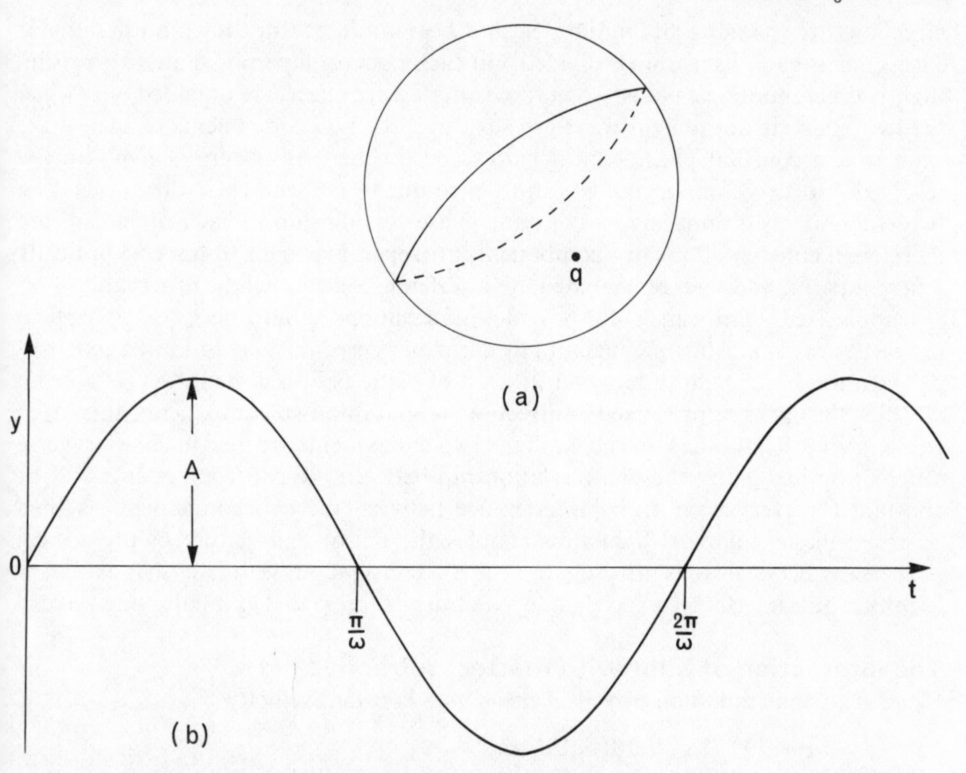

Fig 12.1 The transverse wave nature of light. In the stereogram (a) a direction q and the great circle of which it is the pole are shown; when q is the direction of propagation of light the light vector vibrates in directions parallel to the plane of the great circle. (b) shows the simple harmonic variation of the light vector $y = A \sin \omega t$ for a monochromatic wave.

of the wave motion (Fig 12.1). A ray of monochromatic light, that is light of a fixed and infinitesimally thin wavelength band, is accurately represented by simple harmonic variation of the light vector, $y = A \sin \omega t$, where y is the *light vector*, A the *amplitude*, ω the *frequency*, and t represents time.

The electromagnetic theory indicates that light has dual electric and magnetic properties represented by the complementary electric and magnetic vectors **E** and **H**. It is the electric vector that produces the photochemical reaction in a photographic emulsion and affects the retina of the eye; the electric vector **E** is therefore described loosely as the light vector, designated y.

The light emitted by an atom in a source such as a sodium lamp cannot however strictly be represented by a sine wave stretching from $-\infty$ to $+\infty$ because that would imply that the atom was radiating continuously and that cannot be so. Such a source contains very many atoms, each of which emits a succession of wave trains of finite length as electrons fall from excited states to the ground state; a large number, of the order of 10^8, of wave trains are emitted per second. The phase relationship between different wave trains is essentially random so that the beam of light emitted from the whole source consists of a large number of wave trains of finite length and random phase, each with its light vector vibrating in some direction in the plane normal to the direction of propagation. Such a beam is said to be unpolarized and incoherent; unpolarized because all directions of vibration are possible, and incoherent because the phase relationships between waves with different vibration directions are changing all the time. Such a beam of light will show no interference effects unless each wave train is divided and each portion superposed after traversing slightly different optical paths; observable interference effects are obtained only when the two portions of the light wave are such that the two components of each wave train bear a constant phase relationship to each other. Any beam of light can be resolved into two components vibrating in mutually perpendicular directions. The two components do not have a constant phase relationship to each other and are therefore incoherent. If the two components are separated, made to traverse optically different paths, and then recombined, no interference effects will be observable.

Unpolarized light can only provide information about the average optical properties of an anisotropic medium in the plane perpendicular to the direction of propagation of the light. If however a beam of plane polarized light, that is a beam in which the light vector is fixed in direction, is split into two components, then each component will have the same phase. If the two components are then made to traverse different optical paths, the phase relationship between the two components will be constant for every wave train. Interference between the two components is then possible. Plane polarized light thus enables the anisotropic nature of the optical properties of crystals to be investigated. Since a constant phase difference is necessary for observable interference, such light waves can be represented as infinite sine waves.[2]

The interaction of light with matter: refractive index

Electromagnetic radiation travels *in vacuo* with constant velocity

$$c = 299{\cdot}773 \pm 0{\cdot}010 \times 10^8 \text{ cm s}^{-1}$$

irrespective of frequency. In material media electromagnetic radiation travels with a slower velocity $v = c/n$, where the frequency dependent constant n is the *refractive index*. Refractive index may be regarded as a measure of the retardation of light by the medium. For gases n is only just greater than unity, for liquids and solids n lies in the range 1·3 to 2·1.

A beam of plane polarized light has its electric vector vibrating in a particular direction in the medium it is traversing and it will travel with a velocity given by the ratio of the velocity of light *in vacuo* to the refractive index of the medium for vibrations in that direction.

Gases, liquids, and glasses have no directional structure and therefore their refractive indices for any particular frequency of electromagnetic radiation will be independent of vibration direction. Such media are said to be *optically isotropic.* For crystalline substances on the other hand refractive index will in general vary with vibration direction for any given frequency: such media are said to be *optically anisotropic.*

The indicatrix

Electromagnetic wave theory leads to the conclusion that the dielectric properties of an isotropic medium for frequencies in the optical range are given by

$$\mathbf{D} = \kappa_0 K \mathbf{E},$$

where $\mathbf{D}$ is the electric flux density produced by the electric field strength, or 'light vector' $\mathbf{E}$, K is the *dielectric constant* of the medium, and κ_0 is the permittivity of a

[2] A more rigorous discussion can be found in any of the textbooks listed in footnote 1.

vacuum. It follows from Maxwell's equations that the velocity v of propagation of light waves through the medium can be expressed as

$$v = \frac{c}{\sqrt{K}},$$

where c is the velocity of light *in vacuo*. We have already defined the *refractive index* n of an isotropic medium as $n = c/v$, so that we now have

$$n = \sqrt{K}.$$

For the propagation of light through an anisotropic medium the component D_k of the electric flux density **D** is related to the component E_l of the electric field strength **E** by the equation

$$D_k = \kappa_0 K_{kl} E_l.$$

The dielectricity of an anisotropic crystal is thus, like diffusion and thermal expansion which we discussed in chapter 11, a symmetrical second rank tensor property. The dielectric properties of a crystal are characterized by the magnitudes and directions of the three principal dielectric constants K_X, K_Y, K_Z. The representation quadric for the tensor is the ellipsoid

$$K_X X^2 + K_Y Y^2 + K_Z Z^2 = 1,$$

which has semi-axes of magnitude $K_X^{-\frac{1}{2}}$, $K_Y^{-\frac{1}{2}}$, $K_Z^{-\frac{1}{2}}$ in the directions of the principal axes X, Y, Z respectively.

It can be shown by application of Maxwell's equations to the dielectric constant tensor that in general two waves, each plane polarized, may be propagated through an anisotropic crystal with a given wave normal.[3] The velocity v of each of the two waves will in general be different so that there will be two values of refractive index $n = c/v$ for every wave normal. The refractive indices of the two waves can conveniently be displayed as functions of the direction of their common wave normal by drawing the ellipsoid

$$\frac{X^2}{n_X^2} + \frac{Y^2}{n_Y^2} + \frac{Z^2}{n_Z^2} = 1,$$

where X, Y, Z are the principal axes of the dielectric constant tensor. The semi-axes of the ellipsoid, n_X, n_Y, n_Z are respectively equal to $\sqrt{K_X}$, $\sqrt{K_Y}$, $\sqrt{K_Z}$. This ellipsoid is known as the *indicatrix*, occasionally as the *Fletcher indicatrix*. We can rewrite the equation to the indicatrix as

$$B_X X^2 + B_Y Y^2 + B_Z Z^2 = 1,$$

where $B_X = 1/n_X^2 = K_X^{-1}$ and similarly $B_Y = K_Y^{-1}$, $B_Z = K_Z^{-1}$. B_X, B_Y, B_Z are the three principal *relative dielectric impermeabilities* of the medium and the indicatrix is the representation quadric for this property.

We consider now the propagation of light in any general direction OP in terms of the indicatrix (Fig 12.2). The two wave motions with the common wave normal OP will then have vibrations parallel to the plane perpendicular to OP through the centre O of the indicatrix. This plane will intersect the indicatrix in an ellipse with semi-axes OA and OB. It can be shown from Maxwell's equations (Appendix H of Nye, 1960)

[3] For a proof of this important result the reader is referred to Appendix H of Nye (1960).

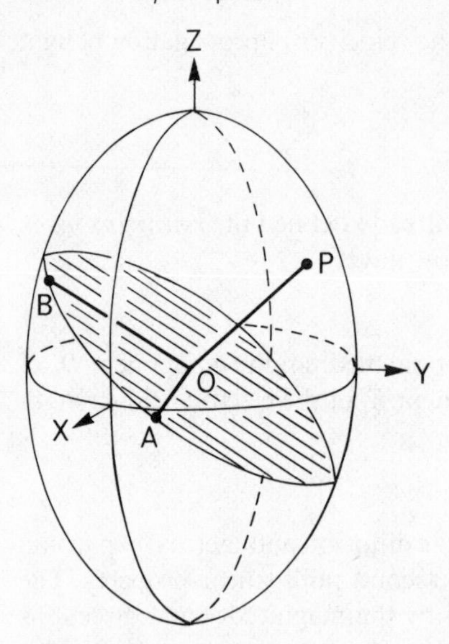

Fig 12.2 The indicatrix is shown in perspective as a drawing of its principal planes. The two wave motions which have the general direction OP as their common wave normal have their vibration directions respectively parallel to OA and OB, which are the semi-minor and semi-major axes of the shaded elliptical central section of the indicatrix normal to OP.

that the refractive indices of the two wave motions are given by OA and OB and further that the vibration direction of the wave motion with refractive index OA is parallel to OA and likewise the vibration direction of the wave motion with refractive index OB is parallel to OB. In the special case where the wave normal is parallel to X the two possible waves have refractive indices n_Y and n_Z; similarly when the wave normal is parallel to Y the refractive indices are n_Z and n_X, and when it is parallel to Z, n_X and n_Y. For this reason it is appropriate to describe the semi-axes of the indicatrix n_X, n_Y, n_Z as the *principal refractive indices.* It should be stressed that refractive index is not a tensor property although, as we have seen, the indicatrix is the representation quadric for another dielectric property.

In order to be consistent with our practice in chapter 11 we have used a rather cumbersome nomenclature. We now rewrite the equation to the indicatrix as

$$\frac{x^2}{\alpha^2}+\frac{y^2}{\beta^2}+\frac{z^2}{\gamma^2} = 1,$$

where α, β, γ are the principal refractive indices and x, y, z are the principal axes of the indicatrix. We add the condition $\alpha < \beta < \gamma$ (the reader will recall that we did not apply such a condition when considering representation quadrics in chapter 11), in consequence of which we can describe γ as the *semi-major axis*, α as the *semi-minor axis*, and β as the *third mean line* of the indicatrix (Fig 12.3). In crystal optics it is common, loosely, to use the same symbols, α, β, γ, to represent the directions of the *principal semi-axes* (strictly x, y, z) as to represent their magnitudes, the *principal refractive indices.*

In general a central plane section of the indicatrix will be an ellipse whose semi-axes are restricted only by the condition that they must both be $\geqslant \alpha$ and $\leqslant \gamma$. But there will always be two central sections that are circular as can readily be seen (Fig 12.3) if it is borne in mind that there must be a radius of length β in the $\alpha\gamma$ plane which is perpendicular to the semi-axis β; and further two such radii will be symmetrically disposed about the semi-major (or semi-minor) axis. The normal to each such *circular section* is known as an *optic axis.* Both optic axes will lie in the $\alpha\gamma$ plane which is thence known as the *optic axial plane* (abbreviation: OAP). The angle between the

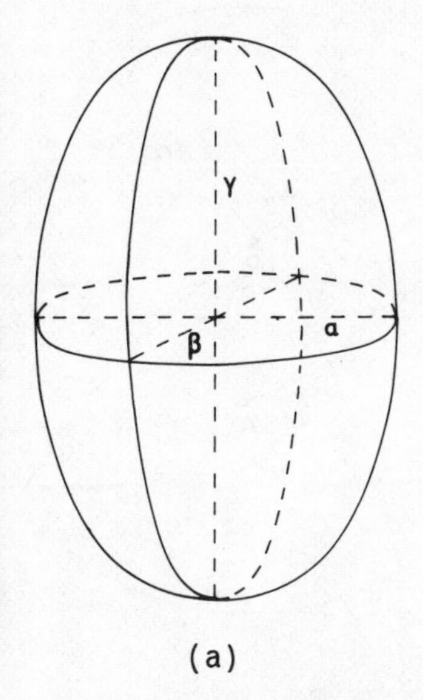

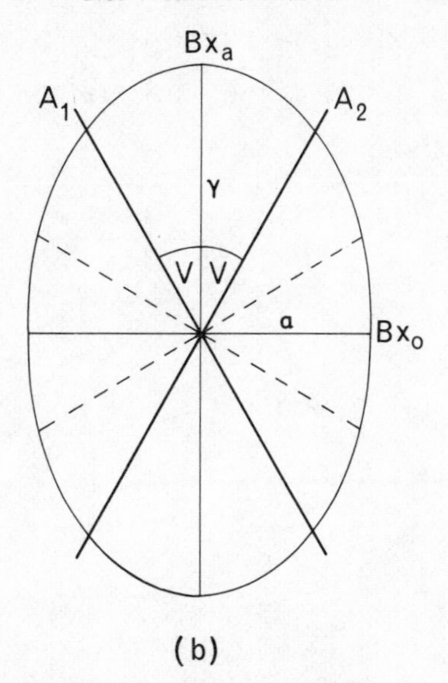

Fig 12.3 (a) is a three-dimensional drawing of the indicatrix to explain the axial nomenclature. (b) is the central section of the indicatrix in the $\alpha\gamma$ plane, the optic axial plane, with the optic axes A_1, A_2 shown as bold lines; perpendicular to each optic axis is a dashed line representing the trace of a circular section, of radius β, on this section of the indicatrix. The indicatrix shown is optically positive so that γ is the acute bisectrix (Bx_a) and α is the obtuse bisectrix (Bx_o); $2V_\gamma < 90°$.

optic axes is known as the *optic axial angle*, denoted by $2V$. Either α or γ may in different substances be the *acute bisectrix* (denoted Bx_a) of the optic axial angle, the other being the *obtuse bisectrix* (denoted Bx_o). The optic axial angle measured over the γ direction is designated $2V_\gamma$ and correspondingly when measured over the semi-minor axis of the indicatrix as $2V_\alpha$. If $2V_\gamma$ is acute the substance is said to be *optically positive*; if on the other hand $2V_\alpha$ is acute the substance is *optically negative*. In optically positive crystals the acute bisectrix is γ, while in optically negative crystals $Bx_a = \alpha$.

In developing the indicatrix we made use of, but did not prove, the following consequence of Maxwell's equations: light travelling in any general direction through an anisotropic substance has vibration directions parallel to the major and minor axes of the elliptical central section of the indicatrix perpendicular to the direction of propagation of the light wave motion and the refractive indices of the two waves are given by the lengths of the two semi-axes of the ellipse.[4] It is this property which makes the indicatrix so powerful a tool in the field of polarized light microscopy.

Biot–Fresnel construction

This extremely useful stereographic construction is based on a property of the indicatrix: that in an anisotropic medium light vibrates in the directions of the major and minor axes of the elliptical central section of the indicatrix perpendicular to the direction of propagation of the light. The construction consists of three steps and is illustrated in Fig 12.4(a).

[4] Except that when the direction of propagation is an optic axis the vibrations are not constrained to lie in two particular perpendicular planes; light travelling in such a direction has a single refractive index of magnitude β.

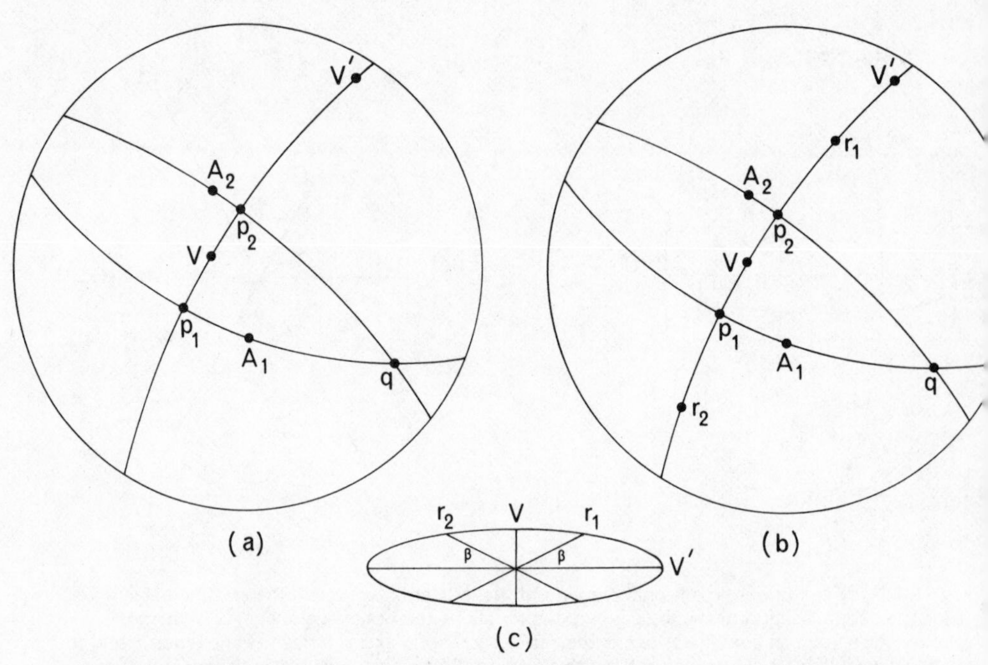

Fig 12.4 The Biot–Fresnel construction. The stereogram (a) shows the three steps in the construction for a direction of propagation q and optic axes A_1, A_2; V and V′ are the internal and external bisectors of p_1p_2. The stereogram (b) illustrates the proof of the construction: r_1 and r_2 are respectively the intersections of the great circle of pole q with the great circles representing the circular sections of the indicatrix, whose poles are A_1 and A_2. The plan (c) of the elliptical central section of the indicatrix perpendicular to the direction of propagation q illustrates the point that the intersection of the circular sections with this plane have radius β and are equally inclined to the principal axes of the ellipse.

(1) Construct the great circle whose pole is q, the direction of propagation of light. Vibrations will lie in two mutually perpendicular directions in this plane.
(2) Plot on the stereogram the directions A_1 and A_2 of the optic axes and construct the great circles through qA_1 and qA_2. Let these great circles intersect the great circle whose pole is q in p_1 and p_2 respectively.
(3) Construct the internal and external bisectors of p_1p_2 and label them V and V′ respectively. The required vibration directions are then V and V′.

A simple proof of this construction is due to Wooster (1949). Suppose the circular section of the indicatrix whose normal is A_1 intersects the plane whose normal is q in the direction represented by the pole r_1 on the stereogram shown in Fig 12.4(b); and suppose that the other circular section, normal A_2, intersects the same plane in the direction represented by the pole r_2. Then the elliptical central section of the indicatrix normal to the direction of propagation q will have radii equal to β in the directions represented by r_1 and r_2. But equal radii of an ellipse must by symmetry be equally inclined to the principal axes of the ellipse, which are the vibration directions. Therefore $r_2V = Vr_1$. Now r_1 lies on the great circle whose pole is q and on the great circle (not shown on the figure) whose pole is A_1. Therefore r_1 is the pole of the great circle qA_1p_1 and consequently $p_1r_1 = 90°$. Similarly $r_2p_2 = 90°$. Therefore $p_1V = p_1r_1 - Vr_1 = r_2p_2 - r_2V = Vp_2$. The vibration directions are therefore given by the internal and external bisectors, V and V′, of the angle p_1p_2.

Shape and orientation of the indicatrix in the seven crystal systems

For a medium wholly devoid of symmetry the indicatrix is a triaxial ellipsoid, whose semi-axes are conventionally denoted α, β, and γ such that $\gamma > \beta > \alpha$. Such a figure has holosymmetric orthorhombic symmetry *mmm*, one mirror plane being perpendicular to and one diad parallel to each semi-axis. If the medium has symmetry, as it usually does in the crystalline state, the shape of the indicatrix must be consistent with the point group symmetry of the medium. If the medium has symmetry lower than or equal to that of the indicatrix, that is if it belongs to the triclinic, monoclinic, or orthorhombic systems, the shape of the indicatrix will not be affected; its orthorhombic symmetry is then purely a property of electromagnetic radiation. If however the medium has higher symmetry, that is if it belongs to the trigonal, tetragonal, hexagonal, or cubic systems, the shape of the indicatrix will be modified. As we have seen in chapter 11 it is a general principle that the symmetry and orientation of a representation quadric is controlled by the point group symmetry of the crystal medium to which it refers: the indicatrix is peculiar in that it is the representation quadric of the rather uninteresting tensor property relative dielectric impermeability and the representation *surface* of the practically important non-tensor property refractive index, and further it is subject by convention to the restraint $\alpha < \beta < \gamma$ which is not normally applied to representation quadrics. We shall pursue this point by discussing the symmetry controls on the shape and orientation of the indicatrix in each of the crystal systems; the reader should compare this with Table 11.1.

In the *triclinic system* the indicatrix may be oriented with respect to the crystallographic reference axes in any manner. That the indicatrix has more symmetry than the triclinic crystal is a property of light, not of the crystal. However it is oriented the indicatrix always conforms to the maximum symmetry of the triclinic system, the presence of a centre of symmetry.

The *monoclinic system* is characterized by a diad parallel to [010] or a mirror plane (010) or both. One of the principal axes of the indicatrix must therefore be parallel to the crystallographic y axis or, in other words, α, β, or γ is parallel to [010]. It is convenient on practical grounds to distinguish between monoclinic substances in which the OAP is (010) (i.e. $\beta \parallel [010]$) and those in which the OAP is $\perp$ (010) (i.e. α or $\gamma \parallel [010]$). The two cases are illustrated in Fig 12.5.

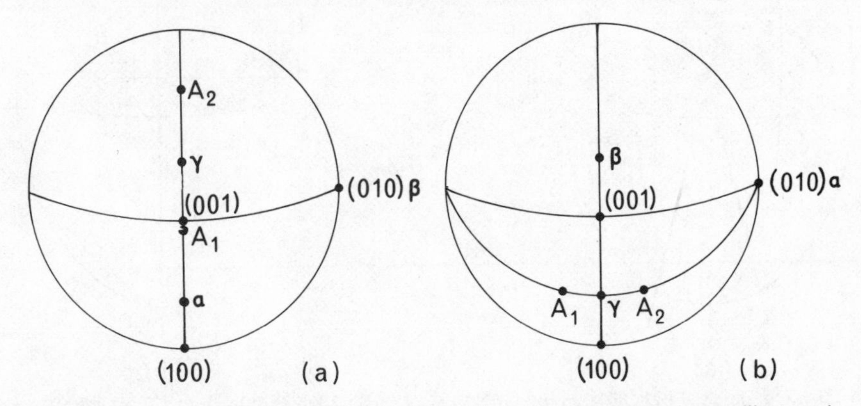

Fig 12.5 The two types of orientation of the indicatrix in the monoclinic system illustrated stereographically: (a) shows the case in which (010) is the optic axial plane and $\beta \parallel [010]$, (b) shows the case in which the optic axial plane is perpendicular to (010) and α or γ (in the example shown α) is parallel to [010].

In the *orthorhombic system* the symmetry axes, rotation or inversion diads, correspond to the symmetry axes of the indicatrix. The principal axes of the indicatrix, α, β, and γ, are therefore parallel to the crystallographic axes, x, y, and z, but not necessarily in that order.

In the *trigonal system* symmetry control of the shape of the indicatrix takes effect. Let the (0001) central section of the indicatrix have semi-axes n_1 and n_2. The equation to this elliptical section in polar coordinates will then be

$$\frac{r^2 \cos^2 \theta}{n_1^2} + \frac{r^2 \sin^2 \theta}{n_2^2} = 1,$$

where r is the radius at an angle θ to n_1, or in a more convenient form

$$\frac{r^2}{n_1^2} + r^2\left(\frac{1}{n_2^2} - \frac{1}{n_1^2}\right) \sin^2 \theta = 1.$$

In order to satisfy the trigonal symmetry of the medium, the ellipse must have equal radii every 120° and since the indicatrix is centrosymmetrical this implies that $r = \mu$ for $\theta = \phi$, $\phi + 60°$, and $\phi - 60°$ for all values of ϕ.

Hence $$\left(\frac{1}{n_2^2} - \frac{1}{n_1^2}\right) \sin^2 \phi = \left(\frac{1}{n_2^2} - \frac{1}{n_1^2}\right) \sin^2 (\phi + 60°) = \left(\frac{1}{n_2^2} - \frac{1}{n_1^2}\right) \sin^2 (\phi - 60°)$$

It is easily seen that these equations are only satisfied when $1/n_2^2 - 1/n_1^2 = 0$, that is when $n_1 = n_2$. The (0001) central section of the indicatrix is therefore a circular section and so likewise are all non-central (0001) sections of the indicatrix; the indicatrix has degenerated from a triaxial ellipsoid to an ellipsoid of revolution about the crystallographic triad. In other words the two optic axes A_1 and A_2 have coalesced into a single optic axis A (Fig 12.6) parallel to the triad and two of the semi-axes

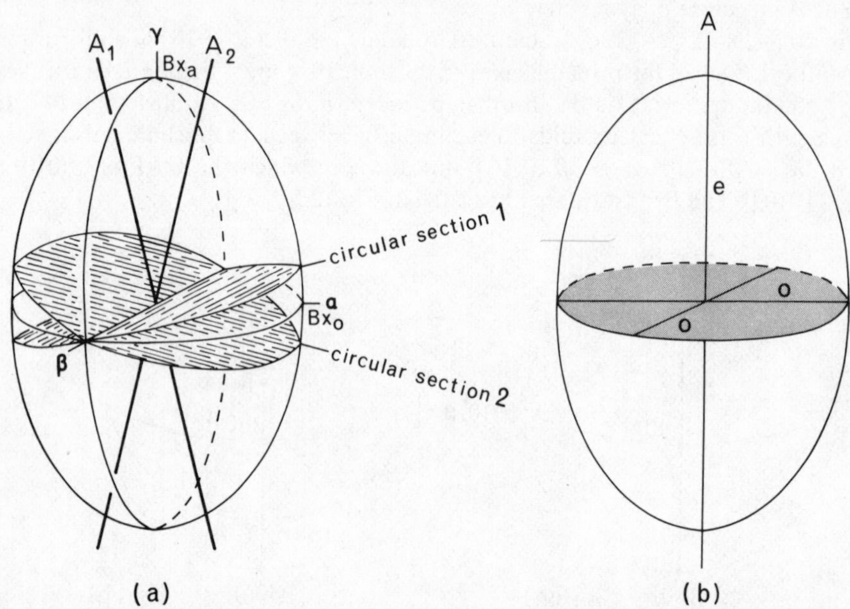

Fig 12.6 The uniaxial indicatrix as the limiting case of the biaxial indicatrix: (a) shows the circular sections (shaded) and the optic axes A_1, A_2, as bold lines in the biaxial indicatrix; (b) shows the single circular section (shaded) and the single optic axis A of the uniaxial indicatrix. In the uniaxial indicatrix the principal semi-axis in the direction of the optic axis is *e* and the radius of the circular section is *o*.

have become equal either as $\gamma = \beta$ or as $\alpha = \beta$. There is here a change of nomenclature: the principal refractive index corresponding to vibrations parallel to the single optic axis, the triad, is described as the *extraordinary* refractive index and denoted e (or ε) while the refractive index corresponding to vibrations in the single central circular section (0001) is described as the *ordinary* refractive index and denoted o (or ω). The concepts of the acute and obtuse bisectrices and the optic axial plane disappear and are replaced by a single optic axis parallel to the axis of high symmetry.

In the *tetragonal system* the indicatrix is likewise an ellipsoid of revolution about the axis of high symmetry, the tetrad. It is easily verified that an (001) central section of a triaxial ellipsoid constrained by symmetry to have mutually perpendicular equal radii must be circular and of course one of the principal axes of the indicatrix must by symmetry be parallel to the tetrad. As in the trigonal system the indicatrix has a single optic axis parallel to the crystallographic axis of high symmetry, the tetrad, and a single circular section parallel to (001).

In the *hexagonal system* the same argument holds as was used for the trigonal system.

The indicatrix in the *cubic system* is yet further degenerate. The four $\langle 111 \rangle$ triads give rise to four circular sections and in consequence every section of the indicatrix is circular: the indicatrix has degenerated into a sphere. In a cubic crystal refractive index, usually designated μ, is independent of vibration direction.

The symmetry control of the geometry of the indicatrix leads to a threefold classification of substances on the basis of their optical properties. Crystalline materials belonging to the triclinic, monoclinic, and orthorhombic systems are described as optically *biaxial*: in these systems the indicatrix is geometrically a triaxial ellipsoid with two optic axes and two circular sections. Crystalline materials belonging to the trigonal, tetragonal, and hexagonal systems are described as optically *uniaxial*: in these systems the indicatrix is an ellipsoid of revolution with a single optic axis, the axis of revolution of radius e, and a single central circular section of radius o. Cubic crystalline materials are optically *isotropic*: the indicatrix is a sphere of radius μ.

The *optic sign* of biaxial crystals has already been defined: for positive crystals $2V_\gamma < 90°$ and for negative crystals $2V_\alpha < 90°$. This convention is simply extended to uniaxial crystals: if the optic axes coalesce as γ, $2V_\gamma \to 0$ and γ becomes e (Fig 12.7). Therefore for positive uniaxial crystals $e > o$. If contrariwise the optic axes coalesce as α, $2V_\alpha \to 0$, α becomes e and for a negative uniaxial crystal $e < o$. Thus the extraordinary refractive index is relatively slow in a positive uniaxial substance; this constitutes the basis of the useful mnemonic POF = positive ordinary fast.

The Biot–Fresnel construction for locating the vibration directions for light travelling in a particular direction through a biaxial substance has already been described. The extension to uniaxial substances is a simplification: construct the great circle whose pole is the direction of propagation q (Fig 12.8) and the great circle through q and the optic axis A. The extraordinary vibration direction e' lies at the intersection of the two great circles, while the ordinary vibration direction o lies 90° away and therefore on the great circle whose pole is A. One of the refractive indices of a general section of a uniaxial crystal is thus always equal to o and the other, designated e', lies between the limiting principal refractive indices o and e in magnitude.

The polarizing microscope

It is appropriate at this stage to digress from the development of the properties of

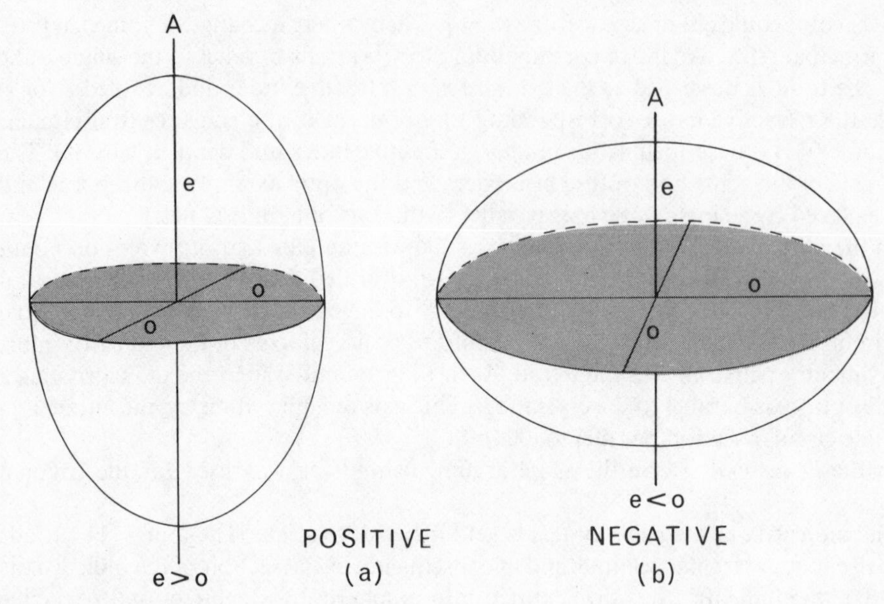

Fig 12.7 The sign convention for uniaxial crystals: (a) shows the indicatrix for a positive crystal with $e > o$ and (b) for a negative crystal with $e < o$.

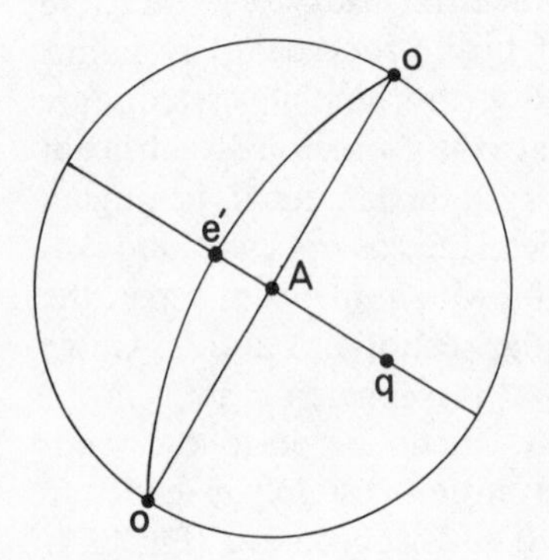

Fig 12.8 The Biot–Fresnel construction for uniaxial crystals. The stereogram shows a wave normal direction q and the optic axis A at the north pole. The vibration directions are e' at 90° from q in the plane qA and o at 90° from q in the plane (here the primitive) whose pole is A.

the indicatrix into a brief description of the principal instrument of crystal optics, the polarizing microscope, so that the practical significance of the theoretical work yet to be done can be made immediately clear. The polarizing microscope is in essence a compound microscope fitted with two polarizing devices, one above and one below the specimen stage, with a slot in the barrel to take accessory test plates, and with a variety of subsidiary lenses and diaphragms to provide the extreme versatility of illumination of the specimen that is requisite for adequate examination of transparent crystalline materials. There are on the market a great many different models of polarizing microscopes that differ substantially in the details of their construction, some having one particularly advantageous or convenient feature and others another. It would be unnecessarily restricting to describe one particular model and so this description will be confined to the essential features of the instrument.

The usual source of illumination is a tungsten filament lamp which may be provided with a focusing system and may be operated through a variable transformer so that its intensity can be varied. If a very high intensity of illumination is required, as is occasionally the case, a quartz-iodine lamp may be used alternatively. In most of the

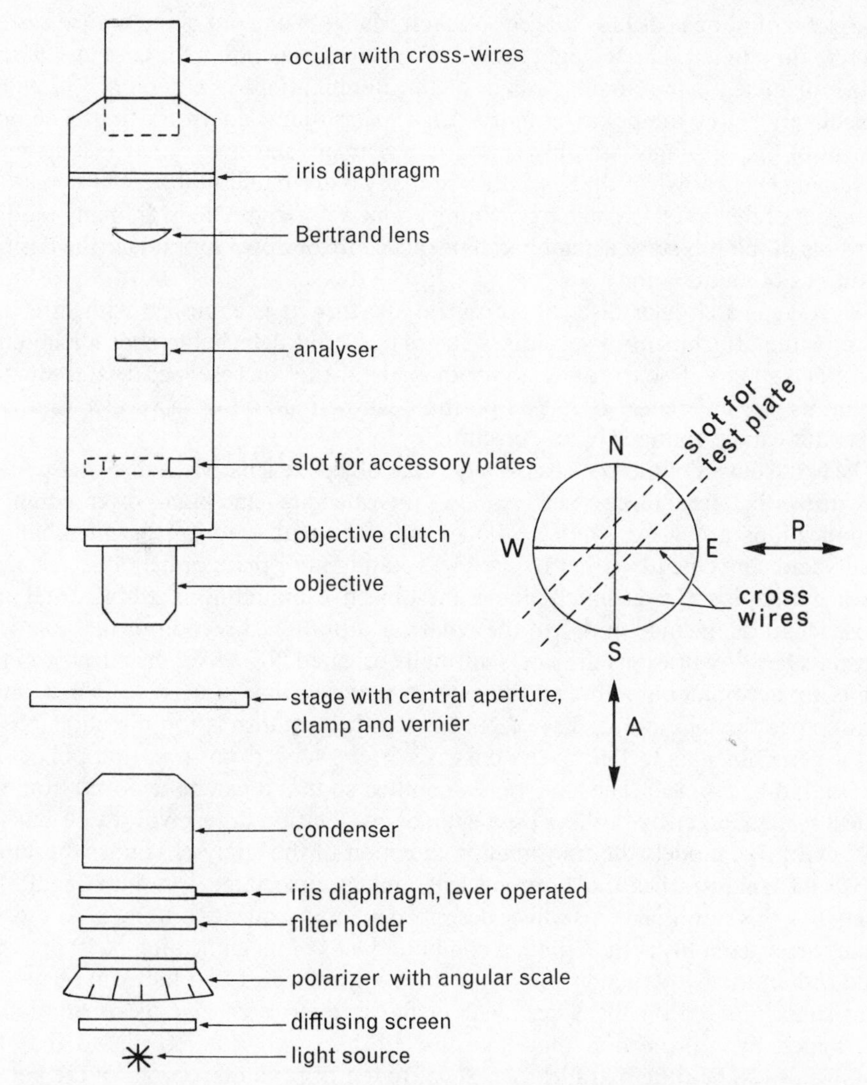

Fig 12.9 The polarizing microscope. The diagram on the left shows the construction of the microscope. The diagram on the right shows the stage orientation as it appears to the observer: the polarizer transmits vibrations from side to side, i.e. East–West, the analyser transmits vibrations from front to back of the stage, i.e. North–South, and the slot for test-plates is inclined at 45° to these directions in the NE–SW direction.

better class of modern microscopes the lamp is built in to the base of the stand (Fig 12.9).

Light enters the optical system of the microscope through the *sub-stage assembly*, the first unit of which is the *polarizer*, a circular disc of polaroid film about 3 cm in diameter mounted so that it can be rotated about the axis of the microscope. The polarizer is normally retained in a fixed position in use so that it transmits only light vibrating parallel to one of the cross-wires in the ocular; this fixed position is located by a click-stop. In the sequel it will be assumed that the vibration direction of light transmitted by the polarizer is from left to right of the field of view (or east–west if directions in the field of view are referred to the points of the compass $\text{W}\overset{\text{N}}{\underset{\text{S}}{}}\text{E}$). The polarizer mounting is usually hinged so that it can be swung out of the optical system if observation in unpolarized light should be required as is rarely the case. A *diffusing screen*, to increase the angular range of the illumination, is commonly provided immediately below the polarizer and a *filter holder* immediately above it. The next element of the substage assembly is an *iris diaphragm* and above that is a strongly converging lens known as the *condenser;* in many older models the condenser can be swung out of the optical system by pivoting about a horizontal axis. In many models the whole of the sub-stage assembly can be racked up or down to produce the desired conditions of illumination.

The stage is a circular disc with a central aperture. It is equipped with threaded holes for the attachment of subsidiary stages (*q.v.*) and detachable clips for holding slides. The stage is free to rotate about the axis of the microscope; it is graduated around its circumference so that its position can be read off on a vernier to about 0·1° and it can be clamped in any position.

The first element of the *barrel assembly* is the objective lens. Most microscopes are provided with at least three objectives, low-, intermediate-, and high-power (common magnifications are × 6·5, × 14, × 45) so mounted that they can be interchanged rapidly and conveniently either by use of a rotating nose-piece or better by a simple clutch mechanism. Immediately above the objective mounting is a horizontal *slot* whose length is inclined at 45° to the polarizer vibration direction, if the polarizer transmits E—W vibrations the slot is normally oriented NE—SW; the purpose of the slot is to accommodate the accessory test plates whose use is described later. Next comes the *analyser*, a disc of polaroid with its transmission direction N—S, that is perpendicular to that of the polarizer in the locked position (the polars are then said to be *crossed*). The analyser is mounted so that it can be removed from the optical system either by pushing on a slide or by rotation on a pivot. In all but the most expensive models the transmission direction of the analyser is not adjustable. Above the analyser lies the *Bertrand* lens and its associated iris diaphragm: the function of this subsidiary lens will be described in a subsequent paragraph. At the top of the barrel assembly is the *ocular*, a compound lens of magnification × 10 or × 15, fitted with mutually perpendicular cross-wires so positioned that they can be seen in focus superimposed on the object. The ocular is removable and its orientation is determined by a projection that fits into either of two slots so placed that the cross-wires will be either parallel to or at 45° to the vibration directions of the polars. The whole barrel assembly can be racked up or down to achieve focusing; commonly a coarse-adjustment and a calibrated fine-adjustment of the racking mechanism are provided. On some models the barrel is immovable and it is the stage and sub-stage assemblies that are together racked up or down to focus.

It is important that the axis of each optical element of the microscope should be coincident with the axis of rotation of the stage. This condition should be achieved by the makers for all except the interchangeable or movable elements, that is to say except for the objectives and on some models the condenser, and the Bertrand lens. The alignment of the objective is particularly important. If the objective is not aligned an object will not remain at the centre of the field on rotation of the stage. Models that have their objectives mounted on a revolving nose-piece are fitted with centring screws in the SE and SW positions on the upper part of the nose-piece, while those with an objective clutch have similarly disposed centring screws on each objective; in the former only the high-power objective need be centred. The procedure for centring is the same in each case (Fig 12.10). Select some detail of the object (a

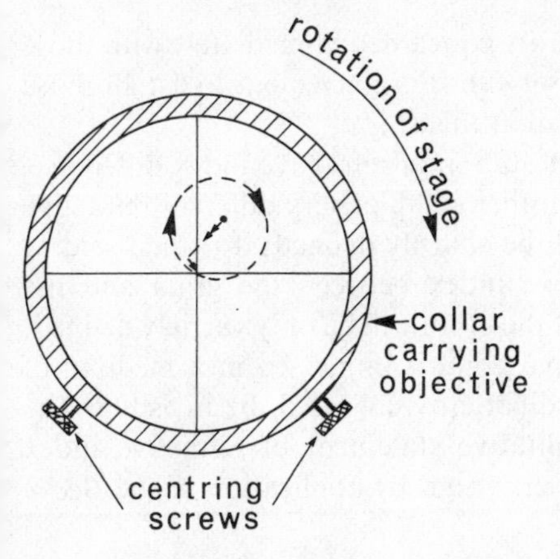

Fig 12.10 Centring the objective of a microscope. A recognizable point on a slide placed on the stage is moved to the intersection of the cross-wires and the centre of its circle of rotation is brought into coincidence with the cross-wires by adjustment of the centring screws; repetition of this procedure enables the high-power objective to be precisely centred.

speck of dust, the corner of a crystal, or some such *point*) and move it on to the intersection of the cross-wires. Rotate the stage and estimate the position of the centre of rotation of the point object. Bring the centre of rotation on to the intersection of the cross-wires by adjusting the centring screws which traverse in NW—SE and NE—SW directions. Repeat the procedure until the point object remains at the intersection of the cross-wires throughout a complete rotation of the stage. Several cycles of adjustment and testing are usually necessary. The centring of the objectives should be checked from time to time.

Microscopic examination of crystalline materials in parallel plane polarized light

Before considering what observations can be made with the polarizing microscope in parallel plane polarized light with the analyser withdrawn from the optical system, it is necessary to say a few words about the form in which the material can conveniently be examined. Small crystals ($\leqslant 1$ mm across) can be placed on a microscope slide, immersed in a liquid in which they are insoluble, the whole being covered with a glass 'cover slip' (a thin glass circle of about 1 cm diameter). Immersion in a liquid reduces light scattering from the crystal faces. A crystalline aggregate may be examined in either of two ways. It may be crushed with a pestle and mortar or in a mechanical grinder, the fine dust being removed by decantation in a liquid in which all the phases are insoluble and the large grains being removed by sieving through a sieve of suitable mesh (60 or 90 mesh, that is with holes of about 0·04 or 0·03 cm); the resulting powder can be mounted in a liquid on a glass slide under a cover slip. Alternatively a permanent thin-section of a coherent aggregate may be prepared; this is a technique that requires great skill if the resulting section is to retain all the textural features of the material. A slice about 0·5 cm thick and 2 cm square is cut from the specimen with a diamond impregnated saw; it is ground on a diamond- or carborundum-impregnated rotary lap until one surface is flat and polished; that surface is then cemented to a glass slide with Canada balsam or a plastic cement. The other surface of the slice is then ground on a rotary lap and by hand on successively finer grades of carborundum paste until a parallel sided section about 3×10^{-3} cm thick results; a cover slip is then cemented with Canada balsam or plastic cement on to the uncovered surface; the result is a thin section of standard thickness that should not deteriorate in many years.

Having got the crystalline material into a suitable form, what significant properties can be observed with the polarizing microscope? The answer to that question is the subject of the remainder of this chapter, but we are concerned immediately with those properties that can be observed with the analyser *out*. It is convenient to list all those properties and to discuss some of them in detail at this stage.

(1) *Relief.* This is the striking qualitative manifestation of refractive index difference. If a solid grain is immersed in a liquid, or in another solid, of very different refractive index, whether higher or lower, its margin will be sharply defined; it is then said to be in high relief. If the difference in refractive index between the grain and the immersion medium is slight, then the margin of the grain will be only vaguely defined; it is then said to be in low relief. If an isotropic grain is immersed in a medium of the same refractive index it will, unless it is differently coloured, be invisible. The observation of relief merely provides a qualitative statement of refractive index difference, but the difference can simply be given a sign by application of the *Becke line test.*

Consider a grain of an isotropic substance immersed in a liquid of higher refractive index. As the barrel assembly is racked up from the position of sharp focus a bright line, the Becke line, concentric with the margin of the grain will be seen to move outwards from the dark crystal/liquid boundary into the liquid. If the grain is of higher refractive index than the liquid the Becke line will move inwards from the grain margin as the barrel assembly is racked up. In general as the barrel assembly is racked *upwards* through the position of sharp focus the Becke line will move from the medium of lower refractive index through the grain boundary *into* the medium of *higher* refractive index.

The Becke line is usually attributed to either refraction at the tapering edge of the grain or to total reflexion at the grain/liquid boundary or to both. The refraction hypothesis implies that a grain of higher refractive index than the immersion medium acts as a converging lens (Fig 12.11 (a)) while a grain of lower index than the immersion medium acts as a diverging lens (Fig 12.11(b)). Total reflexion at the grain/liquid boundary will likewise deflect near parallel incident light towards the medium of higher refractive index (Figs 12.11(c), (d)). Both refraction and total reflexion are probably operative in practice in the production of the Becke line, one or other becoming dominant as the nature of the crystal/liquid boundary changes.

The sensitivity of the Becke line test is improved by stopping down the sub-stage iris diaphragm until the Becke line becomes distinct and by using an objective of intermediate- or high-power.

The meaningful extension of the Becke line test to anisotropic substances must be left to a subsequent paragraph. The Becke line test then becomes the basis of the most generally applicable method for the determination of the refractive indices of anisotropic solids.

(2) *Habit.* The shapes of any euhedral crystals that can be seen may be informative. By racking the barrel assembly up and down it is usually possible to obtain a three-dimensional view of the crystal grain. Measurement of the plane angles of faces lying parallel to the stage can be made by use of the angular graduations on the periphery of the stage and the vernier. In some cases, especially in the cubic system, it is possible to describe habit in some detail; in others such general descriptions as *prismatic* or *tabular* may be all that can be achieved. Preparations of grains obtained by crushing will of course usually yield no information about habit. Thin sections of crystal aggregates may, for example, display both long- and cross-sections of prisms or tablets.

(3) *Cleavage.* Crystals of many substances tend to fracture smoothly along structurally

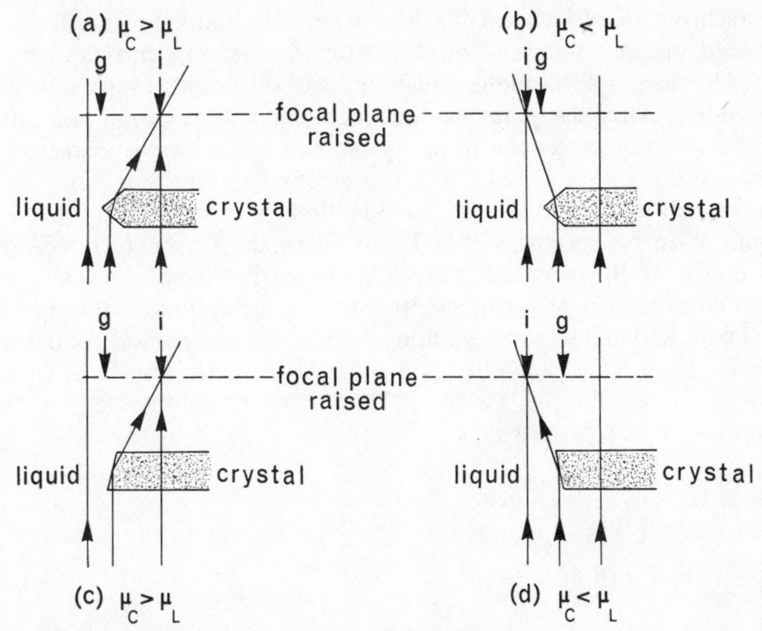

Fig 12.11 The Becke line test for a crystal immersed in a liquid.

controlled planes of simple indices. This special kind of fracture is known as *cleavage*. Cleavage is not usually observable microscopically in preparations of small crystals, but in preparations of crushed material, grains will usually tend to lie on cleavage planes if the substance possesses a good cleavage. In thin sections of crystalline aggregates *cleavage traces* show up as fine black, sometimes broken, lines which represent the intersection of the cleavage plane with the plane of the section. In a section of a crystalline substance for which the orientation of the cleavage is known, the cleavage trace on a section of identifiable orientation becomes an important reference direction; if the cleavage is (*hkl*) and the plane of the section (*pqr*), the cleavage trace will be the direction [(*hkl*), (*pqr*)]. Cleavages are made apparent in thin-sections by mechanical strains in grinding and strains induced by heating and cooling during the preparation of the section. Cleavage traces are only visible when the cleavage plane is nearly normal to the plane of the section; usually the angle between the two planes must be greater than 70°.

(4) *Colour*. Although the colour of a crystalline solid in transmitted light may be distinctive, it is very commonly due to quite small amounts of substituent cations or chromophore groups. The colour of anisotropic substances in plane polarized light will be dealt with further in a subsequent paragraph.

This is as far as the examination of crystalline materials can be taken without inserting the analyser into the optical system. The additional information obtainable by examination with the analyser inserted, that is examination between *crossed polars*, is discussed in the immediately following paragraphs.

Interference effects between crossed polars in parallel plane polarized light

Consider a thin parallel-sided crystal plate lying on the microscope stage between crossed polars, that is with the analyser in the optical system and the analyser vibration direction perpendicular to the polarizer vibration direction. Suppose the illumination to be monochromatic of wavelength λ. The plane polarized light transmitted by the polarizer and incident normally on the lower face of the crystal plate can be represented by the simple harmonic equation $y = A \sin \omega t$ where y is the light vector, A is amplitude, ω is frequency, and t is time measured with respect to some arbitrary zero. Let the vibration directions in the crystal plate correspond to refractive indices μ_1 and μ_2 and make angles θ and $90°—\theta$ respectively with the polarizer vibration direction. The amplitudes of the two wave motions within the crystal plate will be $A \cos \theta$ and $A \sin \theta$ respectively (Fig 12.12). Since the crystal plate has different refractive indices for the two wave motions, the time taken to pass through the crystal plate will not be the same for each. Suppose that the zero of time is taken at entry to the crystal plate and that the wave motion corresponding to refractive index μ_1 takes time t to pass through the plate while that corresponding to refractive index μ_2 takes time $t+(\alpha/\omega)$ to pass through the plate. Then immediately on emergence the two wave motions can be represented as

$$y_1 = A \cos \theta \sin \omega t$$

and

$$y_2 = A \sin \theta \sin (\omega t + \alpha)$$

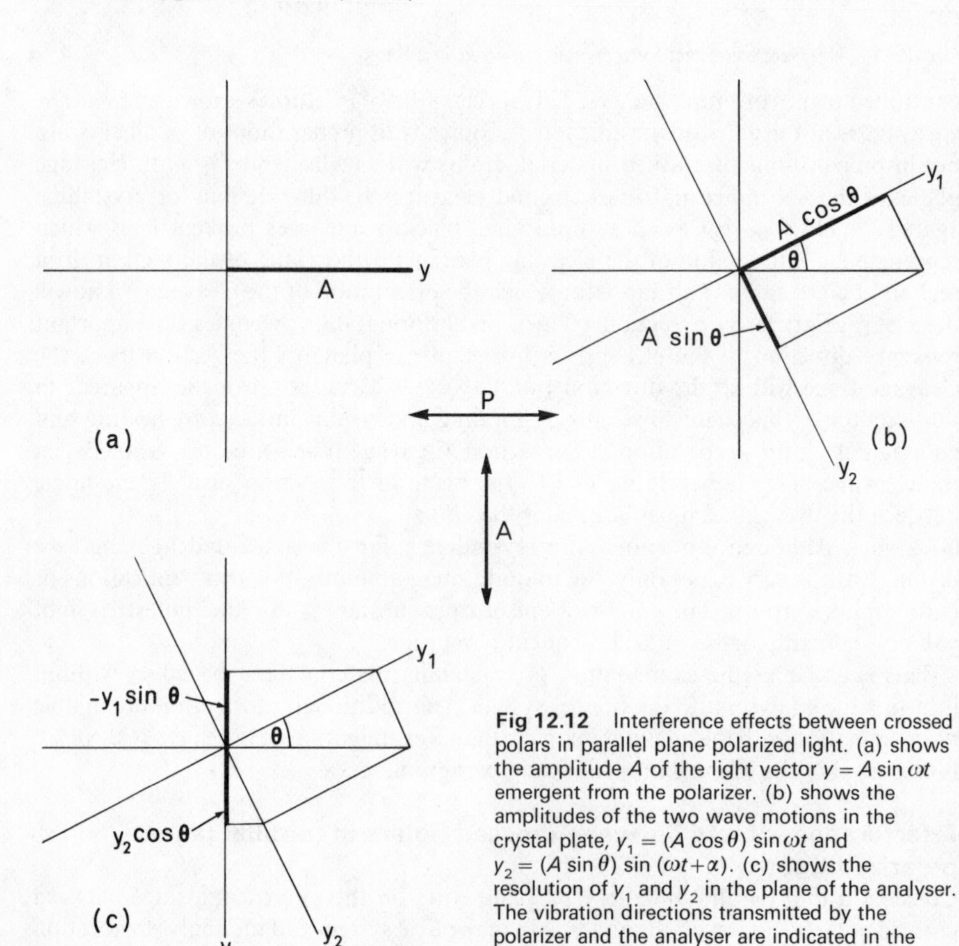

Fig 12.12 Interference effects between crossed polars in parallel plane polarized light. (a) shows the amplitude A of the light vector $y = A \sin \omega t$ emergent from the polarizer. (b) shows the amplitudes of the two wave motions in the crystal plate, $y_1 = (A \cos \theta) \sin \omega t$ and $y_2 = (A \sin \theta) \sin (\omega t + \alpha)$. (c) shows the resolution of y_1 and y_2 in the plane of the analyser. The vibration directions transmitted by the polarizer and the analyser are indicated in the centre of the figure.

The two wave motions will remain with constant phase difference α until they reach the front surface of the analyser. There they will interfere on resolution into the vibration plane of the analyser, which is perpendicular to that of the polarizer. The resultant light vector y is given (Fig 12.12) by

$$\begin{aligned} y &= y_2 \cos\theta - y_1 \sin\theta \\ &= A \sin\theta \cos\theta \{\sin(\omega t+\alpha) - \sin\omega t\} \\ &= A \sin 2\theta \sin\tfrac{1}{2}\alpha \cos(\omega t+\tfrac{1}{2}\alpha). \end{aligned}$$

The magnitude of the phase difference α can be evaluated quite simply. The velocities of the two wave motions will be respectively c/μ_1 and c/μ_2, where c is the velocity of light *in vacuo*. If the thickness of the crystal plate is d, then

$$t = \frac{d\mu_1}{c} \quad \text{and} \quad t + \frac{\alpha}{\omega} = \frac{d\mu_2}{c},$$

therefore $$\alpha = \frac{\omega d}{c}(\mu_2 - \mu_1).$$

Now the wavelength of the monochromatic source is λ in air and it is permissible to approximate the velocity of light in air to c (since the refractive index of air at atmospheric pressure is only 1·0003), so that $c = (\lambda\omega/2\pi)$ and the expression for the phase difference becomes

$$\alpha = \frac{2\pi d}{\lambda}(\mu_2 - \mu_1)$$

and for the resultant light vector transmitted through the analyser,

$$y = A \sin 2\theta \sin\left\{\frac{\pi d}{\lambda}(\mu_2 - \mu_1)\right\} \cos\left\{\omega t + \frac{\pi d}{\lambda}(\mu_2 - \mu_1)\right\}.$$

The intensity of light transmitted through the analyser is given by the square of the amplitude of the light vector y. The proportion of the incident intensity A^2 transmitted is therefore $\sin^2 2\theta \sin^2\{(\pi d/\lambda)(\mu_2 - \mu_1)\}$. As the stage carrying the crystal plate is rotated the intensity of light passing the analyser will vary as $\sin^2 2\theta$ and in particular will be zero for $\theta = \frac{1}{2}n\pi$ (Fig 12.13). The field of view will appear dark whenever θ is a multiple of 90°. There will be four such *extinction positions*, each corresponding to an orientation of the crystal plate in which its vibration directions are parallel and perpendicular to the polarizer vibration direction.

Extinction is a most important phenomenon. It makes possible the identification

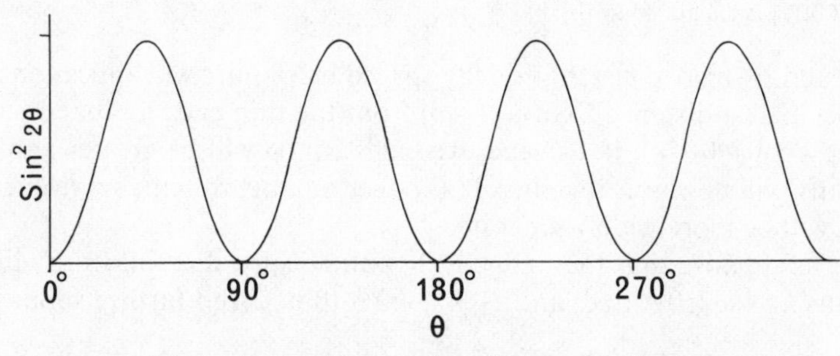

Fig 12.13 Plot of $\sin^2 2\theta$ against θ to illustrate the existence of extinction positions.

of the vibration directions of light travelling through a crystal parallel to the axis of the microscope. It is consequently a property central to the determination of refractive index in anisotropic materials. If the crystal plate is immersed in liquid of known refractive index and rotated into that extinction position for which the vibration direction μ_1 is parallel to the polarizer vibration direction, and if the analyser is then put out of the system and the Becke line test applied to the crystal/liquid interface, then it can be determined whether μ_1 is greater or less than the refractive index of the liquid. Rotation to the extinction position 90° away enables the same information to be obtained for μ_2.

The other factor in the expression for the amplitude of the light vector passing through the analyser, $\sin(\pi d/\lambda)(\mu_2-\mu_1)$, is just as important. This factor will be zero when $(\pi d/\lambda)(\mu_2-\mu_1) = n\pi$, where n is an integer; that is to say no light passes through the analyser when the condition $d(\mu_2-\mu_1) = n\lambda$ is satisfied at some general value of $\theta \neq \frac{1}{2}n\pi$.

The significance of this condition can be most clearly seen by considering a gently tapering wedge cut from a crystal of quartz (SiO_2, trigonal) in such a manner that the length of the wedge is parallel to the triad (Fig. 12.14). The length of the wedge is therefore parallel to the optic axis and the vibration directions in the wedge will be parallel and perpendicular to [0001], corresponding to refractive indices e and o. Suppose the wedge has an angle ϕ and x is measured along the base from the vanishingly thin end; then $d = x \tan\phi$.[5] The condition for zero intensity becomes

$$x \tan\phi.(e-o) = n\lambda$$

i.e.
$$x = \frac{n\lambda}{e-o}\cot\phi.$$

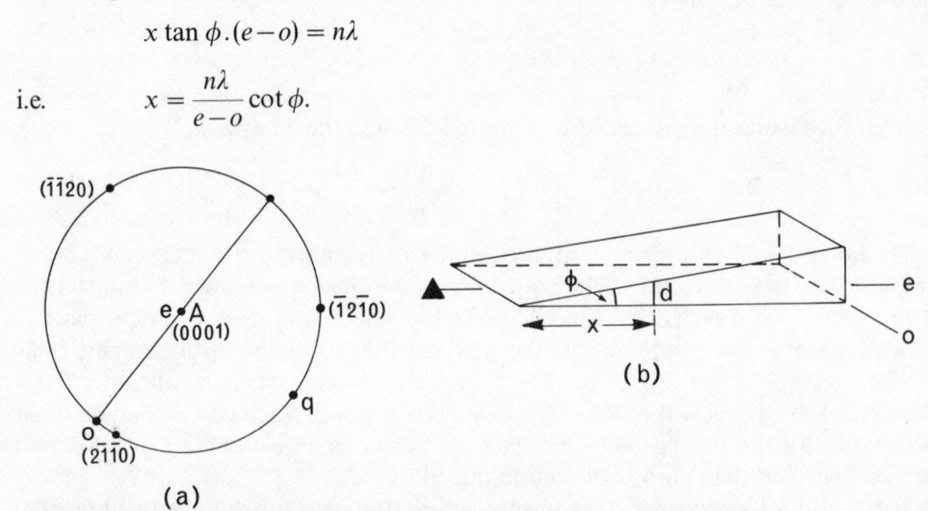

Fig 12.14 The quartz wedge. The stereogram (a) shows the vibration directions, e and o, for light propagated in a direction q parallel to the (0001) plane. The geometry of a thin wedge cut parallel to [0001] is illustrated in (b).

Since λ, ϕ, and $e-o$ are constant, equally spaced black lines will appear on the wedge, the first being at a distance $\lambda \cot\phi/(e-o)$ from the thin end; the band spacing will likewise be $\lambda \cot\phi/(e-o)$. If a steeper wedge is cut, ϕ will be greater and the bands consequently will be closer together. If a longer monochromatic wavelength is used, the bands will be more widely separated.

Now let us suppose that the wedge is, hypothetically, illuminated by light of two wavelengths $\lambda_1 \sim 6750$ Å (red) and $\lambda_2 \sim 4500$ Å (blue). And further suppose that the

[5] Refraction at the upper surface of the wedge may be neglected as ϕ is very small.

Fig 12.15 Diagrammatic representation of the appearance of a quartz wedge (thin end on the left) between crossed polars in (a) red light, (b) blue light, and (c) a combination of red and blue light. R = red, B = blue, W = white. The shaded areas represent low transmitted intensity adjacent to each black band.

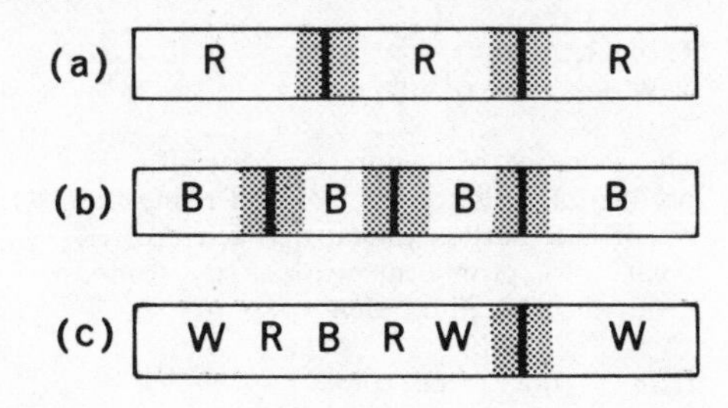

wedge angle $\phi = \tan^{-1} 0{\cdot}01$. The *birefringence* $e-o = 0{\cdot}009$ (for quartz $e = 1{\cdot}553$ and $o = 1{\cdot}544$). For the red wavelength the separation of dark bands $\lambda \cot\phi/(e-o) = 0{\cdot}75$ cm and for the blue wavelength 0·50 cm. The appearance of the wedge in red light alone, in blue light alone, and in red + blue light is illustrated diagrammatically in Fig 12.15: where both wavelengths are cut out a dark band will appear, where blue light is cut out a red band will appear, where red light is cut out a blue band will appear, and where both wavelengths are present in more or less equal intensity the wedge will appear whitish.

If the wedge is illuminated in truly white light, that is to say in a continuous spectrum ranging from deep blue ~ 3800 Å to extreme red ~ 7800 Å, black lines will not be produced, but instead a sequence of colour bands which represent the residue after wavelengths satisfying the condition $d(e-o) = n\lambda$ have been removed by interference as the thickness d gradually increases along the wedge. These colours are known as *interference colours* and the sequence in which they appear on a wedge between crossed polars is similar to that in Newton's rings and is known as Newton's scale of colours (Table 12.1).[6] The human eye is most sensitive to wavelengths in the yellow-green region and therefore the bands on the wedge that stand out most strongly are those from which yellow-green wavelengths have been removed by interference; these are known as the *sensitive tint* bands and appear as a distinctive reddish-violet colour. The wavelength eliminated at the first sensitive tint band is approximately 5600 Å. The sequence of sensitive tint bands corresponds exactly to the sequence of equally spaced black bands that would be seen in monochromatic light of wavelength 5600 Å. The sequence of colours apparent on the wedge is divided into *orders* by the sensitive tint (ST) bands. The first order, running from the vanishingly thin end of the wedge to the first order sensitive tint band is distinct from the others; its sequence of colours—grey, white, yellow, orange, red, sensitive tint—is not repeated in higher orders. In the second order the normal sequence—ST, blue, green, yellow, orange, red, ST—appears and is repeated in subsequent orders. The low order colours are relatively strongly coloured, while the high order colours are progressively less bold, until in the sixth order a pale whitish colour known as *high white* appears and persists throughout higher orders which are consequently indistinguishable.

Where the wedge is relatively thin, that is where low order colours appear between crossed polars, the condition for destructive interference $d(e-o) = n\lambda$ is satisfied by a single wavelength in the visible spectrum. When $d(e-o) = 7600$ Å, the condition is satisfied for a dark red wavelength 7600 Å with $n = 1$ and for the deepest blue

[6] In some textbooks of crystal optics Newton's scale of colours is shown as a colour-plate. Such plates always fail to give a true representation of the colours, because interference colours are necessarily viewed in transmitted light whereas the plate is printed for viewing in reflected light. We would advise the beginner in the use of the polarizing microscope to remind himself from time to time of the colours in Newton's scale by observing with the naked eye a quartz wedge placed between crossed sheets of polaroid.

Table 12.1
Newton's scale of colours

The sequence of colours is related to the product of the thickness t and birefringence $\Delta\mu$ of a colourless crystal between crossed polars. The prominent *sensitive tint* bands divide the scale into orders.

$t.\Delta\mu$ (Å)	Interference colour	Order
0	black	1st order
	grey	
	white	
	pale yellow	
	yellow	
	orange	
	red	
5600	sensitive tint	
	blue	2nd order
	green	
	yellow	
	orange	
	red	
11,200	sensitive tint	
	blue	3rd order
	green	
	yellow	
	orange	
	red	
16,800	sensitive tint	
	⋮	
23,000	high white	

wavelength 3800 Å for $n = 2$. And so as the thickness increases the condition for destructive interference is satisfied by an increasing number of wavelengths within the visible spectrum; wavelengths in different parts of the spectrum are eliminated and what passes the analyser gradually approximates, in so far as the human eye is concerned, to white light of reduced intensity (Fig 12.16). The point may be emphasized by considering a particular case in which $d(e-o)$ is large and equal to $5{\cdot}6 \times 10^{-4}$ cm: the following seven wavelengths will be eliminated 4000 Å (violet) for $n = 14$, 4308 Å (violet) for $n = 13$, 4667 Å (indigo) for $n = 12$, 5091 Å (blue-green) for $n = 11$, 5600 Å (yellow-green) for $n = 10$, 6222 Å (orange-red) for $n = 9$, and 7000 Å (red) for $n = 8$. Since these wavelengths are distributed across the range of the visible spectrum the wedge will appear whitish at this point.

A crystal plate of uniform thickness between crossed polars will display the interference colour appropriate to the value of $d(\mu_2 - \mu_1)$. If $d(\mu_2 - \mu_1) = 5600$ Å the interference colour observed will be the characteristic strong red-violet of first order sensitive tint. As the crystal plate is rotated by rotation of the stage extinction will be observed every 90° when either the μ_1 or μ_2 vibration direction is parallel to the polarizer vibration direction; in between extinction positions the intensity of the first order sensitive tint will vary as $\sin^2 2\theta$ reaching maximum intensity for $\theta = 45°$.

The beginner may find it difficult to recognize from memory a particular colour in Newton's scale, but natural crystals often have tapering edges on which it may be possible to trace the sequence of colours through the first and second orders and so

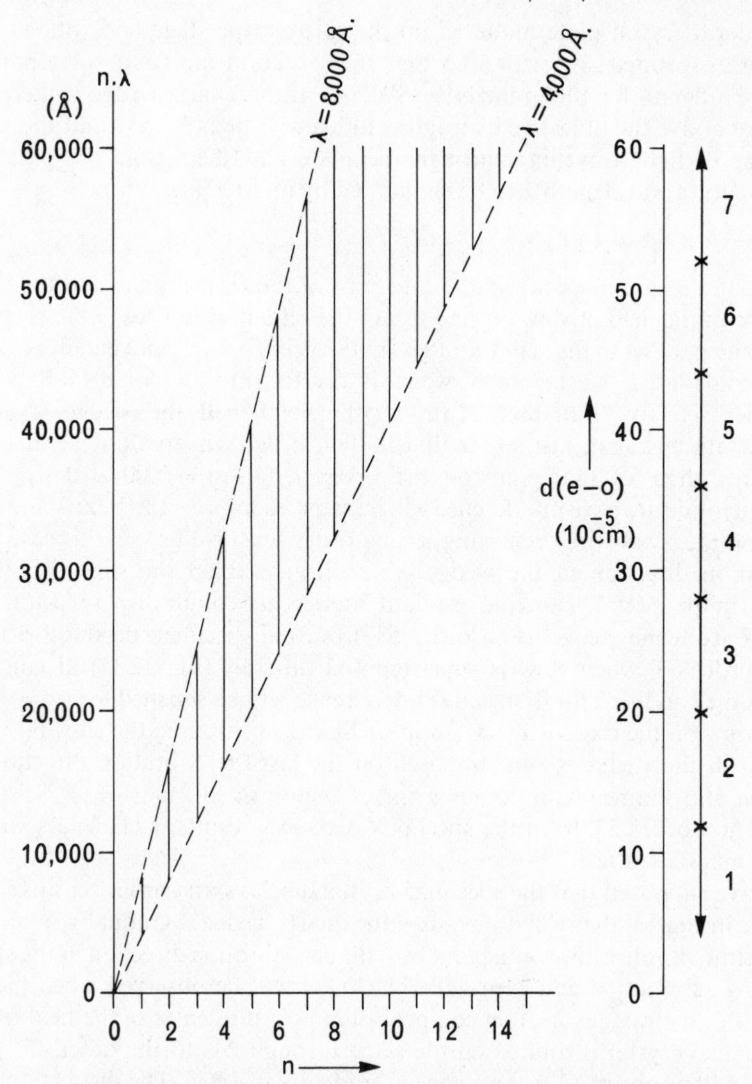

Fig 12.16 Wavelengths cut out by interference in a quartz wedge in white light between crossed polars. The range of wavelengths present in the light source is taken as 4000–8000 Å. Each vertical line shows the range of values of $n\lambda$ for wavelengths eliminated for a particular value of n. A horizontal line on the diagram thus gives the wavelengths eliminated by a plate of thickness d and birefringence $|e-o|$. The vertical line on the extreme right gives the number of wavelengths eliminated for the corresponding value of $d\,(e-o)$.

enable an unambiguous identification to be made of the interference colour exhibited by the uniformly thick body of the crystal. It is often useful to refresh the memory by inspection of a quartz wedge between crossed polars from time to time.

Accessory test plates and their use

(1) *The quartz wedge.* It is a simple matter for an experienced section maker to cut a wedge from a large quartz crystal so that its length is parallel, within a degree or so, to the triad. Since quartz is uniaxial positive, the vibration direction corresponding to the maximum refractive index e and consequently to the lowest velocity c/e will be parallel to the length of the wedge; such a wedge is described as *length slow.*

Consider a crystal plate mounted on the microscope stage and rotated through 45° from an extinction position so that the vibration direction corresponding to refractive index μ_1 lies along the NE—SW line. If the quartz wedge is then inserted in the slot above the objective its length will likewise lie NE—SW and the vibration directions of light travelling through the plate and the wedge are parallel. The condition for a wavelength λ to be eliminated by interference is now

$$d_p(\mu_1 - \mu_2) + d_w(e - o) = n\lambda,$$

where d_p and d_w are respectively the thickness of the crystal plate and the wedge at the centre of the field of view, μ_1 and μ_2 are the refractive indices of the crystal plate in the plane parallel to the stage, e and o are the principal refractive indices of quartz. If the crystal plate alone, before the wedge is inserted into the slot, exhibits a uniform first order sensitive tint, then $|d_p(\mu_1 - \mu_2)| = 5600$ Å. If the wedge is gradually inserted, thin end first, just so far that its first order sensitive tint band is on the cross-wires, then for that point on the wedge $d_w(e-o) = 5600$ Å. If $\mu_1 > \mu_2$, the condition for destructive interference will be satisfied for $n = 2 \times 5600$ Å and the field of view on the cross-wires will show second order sensitive tint; in this case the slow (e) vibration direction in the wedge is superimposed on the slow (μ_1) vibration direction in the specimen and the resultant interference colour displays *addition*—first order ST from the wedge + first order ST from the specimen produces a resultant second order ST when slow is superimposed on slow (Fig 12.17). If contrariwise $\mu_1 < \mu_2$, the condition for destructive interference will be satisfied for $n\lambda = 0$ and the field of view on the cross-wires will appear black; in this case the slow (e) vibration direction in the wedge is superimposed on the fast (μ_1) vibration direction in the specimen and the resultant colour displays *compensation*: first order ST from the wedge—first order ST from the specimen produces resultant blackness when slow is superimposed on fast.

We have supposed that the specimen crystal displays first order sensitive tint, but of course in general that will not be so. If the quartz wedge is gradually superimposed with its slow vibration direction parallel to the fast vibration direction in the specimen, increasing subtraction and eventually blackness will be observed when the correct thickness of the wedge for exact compensation is in the centre of the field of view. If the specimen crystal is rotated on the stage through 90° to the other 45° position, addition will be observed as the wedge is gradually inserted. The distinction between subtraction and addition can often be made more certain in the case of a small specimen crystal by using the high-power objective. If the specimen crystal shows an interference colour of high order it may be difficult to distinguish between addition

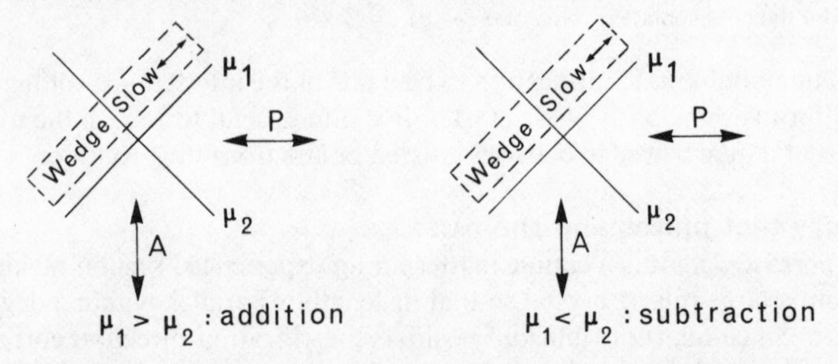

Fig 12.17 Addition and subtraction on superimposition of a quartz wedge in the 45° position NE–SW.

and subtraction as the wedge is inserted, although the latter is usually more easily recognized because it leads to the strong distinctive colours of the first and second orders; in such cases a thin part of the crystal may be found near its edge under high-power and the lower colours there displayed may be used to give an unambiguous determination. If a crystal is of variable thickness, attention should be concentrated on that part of it which displays first order sensitive tint; the distinction between second order sensitive tint and exact compensation yields the most unmistakable criterion of the relative slowness of the two vibration directions.

The property of extinction enables the vibration directions for light incident normally on a crystal plate to be identified. Use of the quartz wedge enables each vibration direction to be identified as slow or fast relative to the other. If, instead of a cut crystal plate, a uniaxial crystal from whose shape the direction of the optic axis can be determined, e.g. an elongated tetragonal prism, is the specimen on the stage, the optic sign of the substance can be established; with the optic axis lying NE—SW, that is parallel to the slow direction of the wedge, addition will indicate $e > o$ and the substance must be optically positive, while compensation will indicate a negative sign (Fig 12.18).

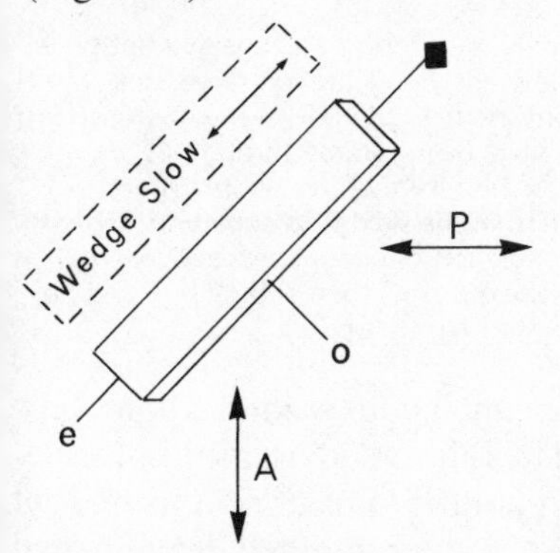

Fig 12.18 Superimposition of a quartz wedge parallel to the tetrad of a tetragonal crystal to determine optic sign.

If a crystal plate of known thickness is rotated into the 45° position that gives rise to compensation and the wedge is inserted until blackness is observed on the cross-wires, then $|\mu_1 - \mu_2|$ can be evaluated by noting the interference colour of the wedge on the cross-wires when the crystal plate is withdrawn from the optical system.

(2) *The sensitive tint plate.* This is the most generally useful testing instrument in white light; in monochromatic light it has no application since a change of one order cannot then be distinguished. It is constructed by sandwiching between glass a parallel-sided cleavage flake of gypsum, or more commonly mica, of such thickness that it displays first order sensitive tint. The cleavage flake is oriented so that its slow vibration direction lies parallel to the length of the mounted plate, that is to say the plate is length slow.

Suppose the specimen on the stage is rotated through 45° from extinction and displays first order sensitive tint. Insertion of the sensitive tint plate into the slot above the objective will then produce either addition, that is second order sensitive tint, or exact compensation. Addition will indicate that the relatively slow vibration direction in the specimen lies NE—SW on the stage. If the specimen displays a low white or grey interference colour, the insertion of the sensitive tint plate will produce either second order blue by addition or first order yellow by subtraction. The observation of a blue interference colour when the specimen is in one extinction position and a

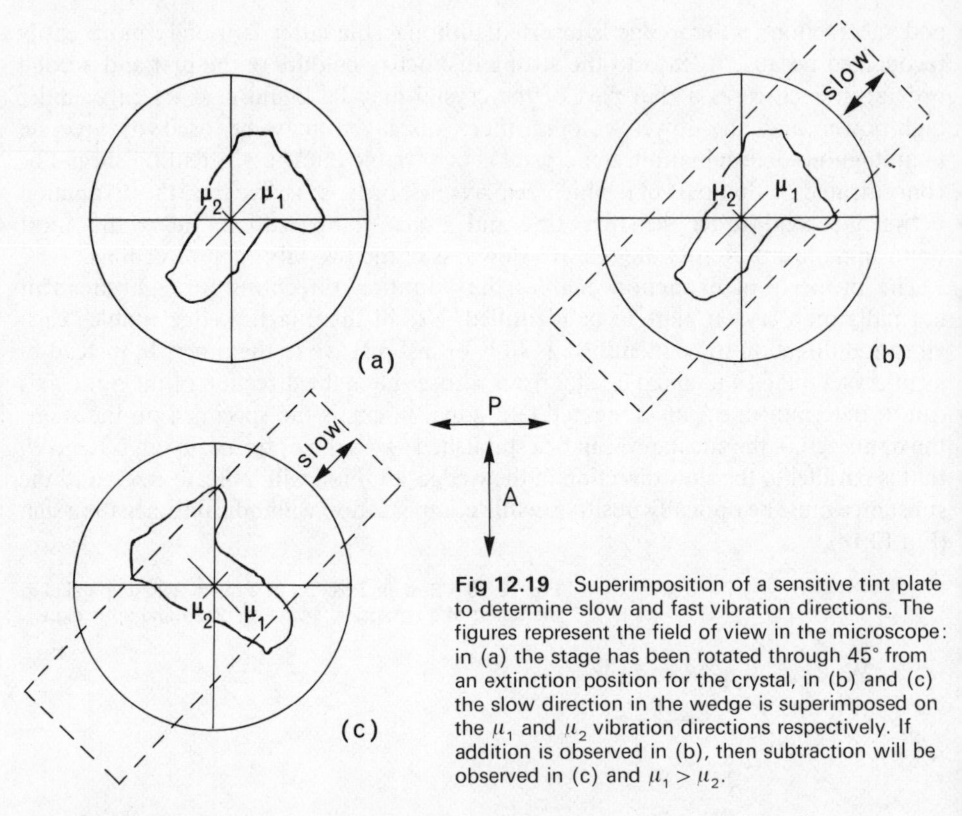

Fig 12.19 Superimposition of a sensitive tint plate to determine slow and fast vibration directions. The figures represent the field of view in the microscope: in (a) the stage has been rotated through 45° from an extinction position for the crystal, in (b) and (c) the slow direction in the wedge is superimposed on the μ_1 and μ_2 vibration directions respectively. If addition is observed in (b), then subtraction will be observed in (c) and $\mu_1 > \mu_2$.

yellow when it is in the other leads to a completely unambiguous identification of its slow and fast vibration directions (Fig 12.19). Indeed it is always advisable to observe the specimen, with the sensitive tint plate inserted, in both 45° positions: this merely involves rotation of the stage from one 45° position to the other and requires the relatively easy distinction to be made between interference colours that differ by exactly two orders. The advantage of the sensitive tint plate over the quartz wedge for this purpose is that it retards the whole beam of light reaching the eye by the same amount.

(3) *The quarter-wave plate.* This is constructed in precisely the same way as the sensitive tint plate, but the cleavage flake of mica is thinner and produces a retardation of only about 1600 Å (approximately $\lambda/4$ for Na_D). Its use is identical with that of the sensitive tint plate. It is only occasionally useful when dealing with specimens that exhibit first order interference colours.

Interference in strongly convergent light: the uniaxial interference figure

The preceding paragraphs have been concerned with the use of the polarizing microscope in its *orthoscopic* arrangement, in which the specimen is between crossed polars and illuminated by a parallel beam of plane polarized light. The microscope can alternatively be set up in the *conoscopic* arrangement so that the highly distinctive and informative interference effects when the anisotropic specimen is illuminated, between crossed polars, by a strongly convergent beam of plane polarized light can be observed.

The microscope is set up for conoscopic use by (i) inserting the analyser, (ii) inserting the sub-stage condenser, (iii) racking the sub-stage assembly up or down until the convergent beam is focused within the specimen crystal lying on the stage, (iv) opening the sub-stage iris diaphragm wide, (v) inserting the high-power objective and focusing it on the specimen crystal, (vi) *either* inserting the Bertrand lens *or* replacing the ocular by a *pin-hole*.

The function of the condenser is to provide a wide-angle cone of light focused on the specimen crystal on the stage. The divergent cone of rays emitted from the crystal is collected by the objective and in general the higher the power of the objective the wider the angle of the cone that can be collected. The rays passing through the objective form a small real image near the curved focal surface immediately above the objective; each point in this so-called *directions image* is a focal point of light that has passed in a definite direction through the crystal. On viewing the directions image through the analyser certain directionally dependent optical properties are revealed and a characteristic *interference figure* is displayed. The directions image is however too far away from the focal plane of the ocular to be visible without modification of the optical system of the eyepiece. The simplest modification is merely to remove the ocular so that the directions image can be viewed direct through the analyser. Alternatively the *Bertrand lens*, situated between the analyser and the ocular (Fig 12.9), may be inserted to produce a magnified directions image in the focal plane of the ocular. When dealing with small crystals, and in general to sharpen the focus of the interference figure, it is important to stop down the light near the top of the barrel assembly: if the Bertrand lens is used this can be done by stopping down the iris diaphragm immediately above it, while if observation is made directly a pin-hole can be inserted in place of the ocular. It is essential, especially when dealing with small crystals, to have the objective accurately centred.

Let us suppose that the specimen on the stage is a plate of uniform thickness t cut perpendicular to the optic axis of a uniaxial substance and that the incident light is monochromatic of wavelength λ. A beam of light incident on the specimen in a direction inclined to the optic axis, will be split into an ordinary and an extraordinary wave within the crystal. Since these have different refractive indices they will be refracted differently at the crystal surface and therefore will travel along slightly different paths within the crystal, becoming parallel again when they emerge from the crystal (Fig 12.20(a)). The difference between the angles which these two waves make with the optic axis is negligible. Thus the two wave motions can be considered to have a direction of propagation of light (p) within the crystal plate at an angle ϕ to the optic axis.[7] The effective thickness of the plate for this inclined direction of propagation will be $t \sec \phi$. The vibration directions of the extraordinary (e') and ordinary (o) wave motions for this direction of propagation can be determined by the Biot–Fresnel construction (Fig 12.20(b)). The magnitude of the directionally dependent refractive index e' can be evaluated from the geometry of the indicatrix in terms of e and o, which are constants for the substance. The equation to the section of the indicatrix in the plane defined by the optic axis and the direction of propagation (Fig 12.20(c)) is

$$\frac{x^2}{o^2}+\frac{y^2}{e^2} = 1.$$

[7] Ditchburn (1963, p. 512) gives a full explanation.

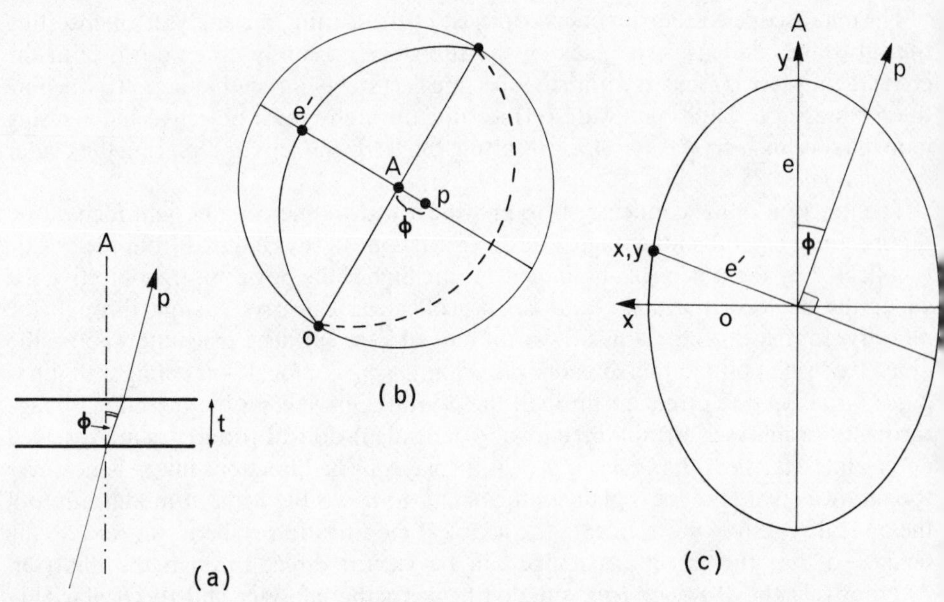

Fig 12.20 The formation of a uniaxial interference figure. (a) shows a light ray p passing through a crystal plate of thickness t at an angle ϕ to the optic axis A. (b) is the Biot–Fresnel construction to determine the vibration directions o and e'. (c) is the section of the indicatrix to illustrate the calculation of e'.

The coordinates of the intersection of the e' vibration direction with the indicatrix are $x = e' \cos\phi$, $y = e' \sin\phi$.

Therefore $$(e')^2\left(\frac{\cos^2\phi}{o^2}+\frac{\sin^2\phi}{e^2}\right)=1$$

and so $$e'=\frac{eo}{\sqrt{\{(e^2-o^2)\cos^2\phi+o^2\}}}.$$

Now the condition for destructive interference between crossed polars is $d(e'-o)=n\lambda$, where d is the actual distance traversed through the crystal plate. In this case $d = t\sec\phi$ and the condition becomes, on substitution of the expression for e' and rearrangement,

$$\frac{to}{\cos\phi}\left(\frac{e}{\{(e^2-o^2)\cos^2\phi+o^2\}}-1\right)=n\lambda.$$

The only variable in this expression is ϕ. Destructive interference will thus occur in monochromatic light when ϕ has certain values and a sequence of black rings will result, each ring corresponding to a particular value of n. The first ring out from the centre of the field of view will have $n = 1$, the next $n = 2$, and so on (Fig 12.22).

There are certain directions of propagation which must be considered separately. The parallel beam of plane polarized light emergent from the polarizer is made strongly convergent by passage through the condenser; the vibration direction of a particular wave will be changed since vibrations must be transverse but in the case of a direction of propagation q in the crystal lying in the plane defined by the optic axis and the polarizer vibration direction the o vibrations will have zero amplitude, because the o vibration direction is perpendicular to the polarizer vibration direction,

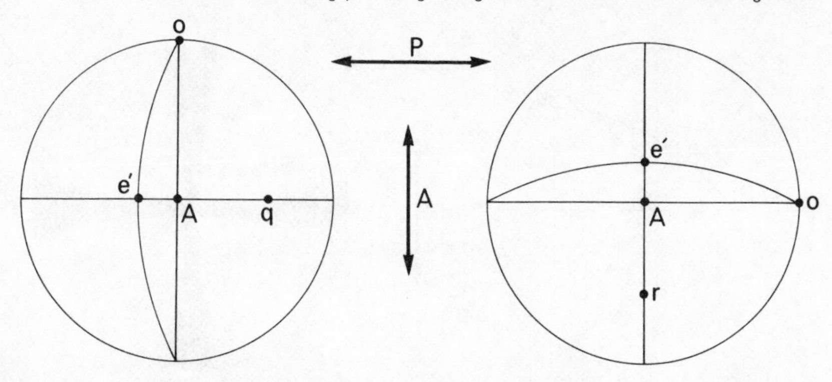

Fig 12.21 Biot–Fresnel constructions to illustrate the formation of the axial cross in uniaxial interference figures. The directions of propagation q and r lie respectively in the polarizer and analyser vibration planes. Their divergence from the optic axis A is grossly exaggerated for clarity.

while the e' vibrations will have finite amplitude (Fig 12.21); it is a general principle that periodic vibrations can be resolved into components in mutually perpendicular directions only so long as neither direction is perpendicular to the original vibration direction. The e' vibration direction lies in the plane perpendicular to the analyser vibration direction and therefore light propagated in directions such as q fails to pass the analyser; a black band running centrally along an E—W line appears in the field of view. Likewise directions of propagation such as r lying in the plane defined by the optic axis and the analyser vibration direction (Fig 12.21) will give rise to a central N—S black band. A direction of propagation such as r will have e' vibrations of zero amplitude, because the polarizer vibration direction is perpendicular to e'; and the o-vibrations, although of finite amplitude, are perpendicular to the analyser vibration direction.

What is seen in the field of view is a centred *uniaxial interference figure* consisting of a black cross, the arms of which run N—S and E—W, and a set of concentric black rings (Fig 12.22).

Reference to the condition for destructive interference $t(e'-o)\sec\phi = n\lambda$ shows that if the specimen crystal plate is replaced by one of greater thickness, $(e'-o)\sec\phi$ must decrease for a given value of n, therefore ϕ must decrease and the rings will move inwards towards the centre of the field of view so that more rings will be visible. If the wavelength λ is increased, $(e'-o)\sec\phi$ must increase and the rings will appear more widely spaced.

Precisely the same arguments apply to the formation of uniaxial interference figures

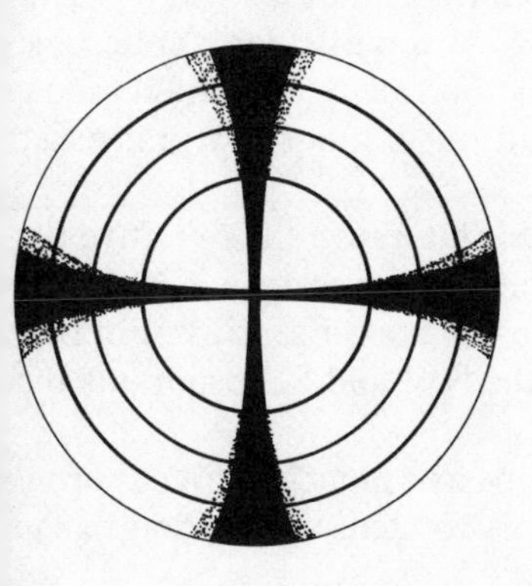

Fig 12.22 The appearance of a uniaxial interference figure in monochromatic light.

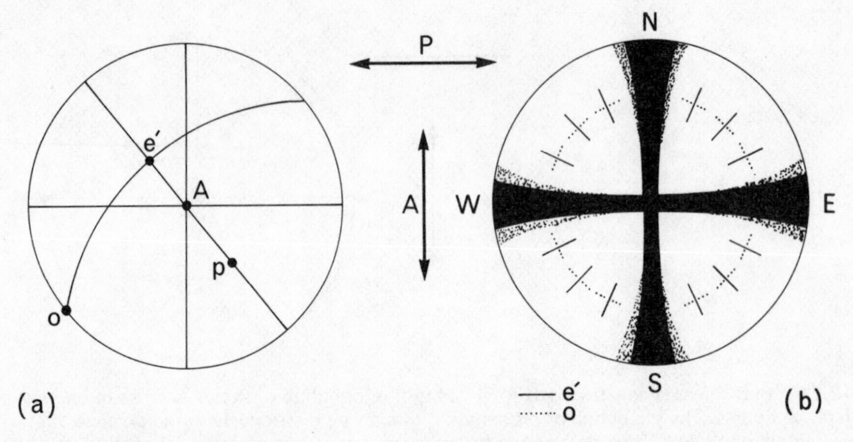

Fig 12.23 Vibration directions in the uniaxial interference figure. The Biot–Fresnel construction (a) illustrates the vibration directions *o* and *e′* for a general direction of propagation *p*. (b) shows the vibration directions in the uniaxial figure for a selection of directions of propagation.

in strongly convergent white light as were used for the production of interference colours by a quartz wedge in parallel plane polarized white light. Complete destructive interference occurs for certain wavelengths at certain angles of propagation through the crystal and a sequence of coloured rings is produced. The sequence of colours is that of Newton's scale (Table 12.1), the sensitive tint rings being the most prominent. The axial cross remains black because its formation is independent of wavelength.

The interference figure provides one of the simplest and most generally applicable means of determining *optic sign* in uniaxial substances. Consider a general direction of propagation *p* (Fig 12.23(a)). Light propagated in the direction *p* has vibration directions *e′* and *o* as shown in the figure. The direction *e′* lies in the plane defined by *p* and the optic axis (at the centre of the stereogram) and is therefore *radial* with respect to the field of view. The ordinary vibration direction *o* is *tangential* at the point represented by *p* in the directions image. Figure 12.23(b) displays the radial *e′* vibration directions and the tangential *o* vibrations for selected directions of propagation. If the specimen crystal is positive the slow (*e′*) vibration direction will always be radial while the fast (*o*) vibration direction will be tangential. When a length-slow sensitive tint plate is inserted in the 45° slot above the objective addition will be observed in the NE and SW quadrants (e' + slow in ST plate) and subtraction in the NW and SE quadrants (o + slow in ST plate). In white light sensitive tint rings of nth order will then become $(n+1)$th order in the top right and bottom left quadrants and $(n-1)$th order in the top left and bottom right quadrants (Fig 12.24); a particularly noticeable feature is that in the angle of the cross the top right and bottom left quadrants show second order blue while the top left and bottom right quadrants show first order yellow.

If a length-slow quartz wedge is pushed into the 45° slot above the objective the first order sensitive tint ring in the NE and SW quadrants will move inwards and its place will be taken by the second order and then by the third order sensitive tint ring and so on if the specimen crystal is positive, while in the NW and SE quadrants the rings will appear to move outwards.

If the optic axis is not quite parallel to the microscope axis an *off-centre* or *partial* figure (Fig 12.25) will be produced. As the specimen is rotated, by rotation of the

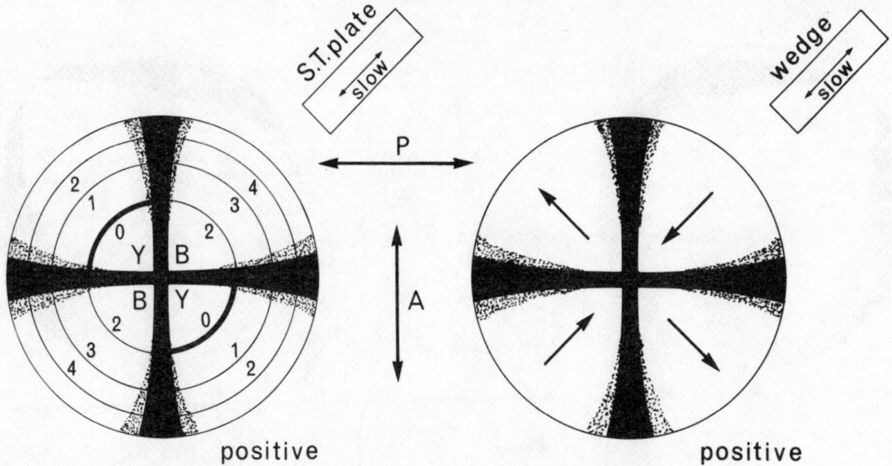

Fig 12.24 The use of accessory plates to determine the sign of a uniaxial figure.

Fig 12.25 An off-centre uniaxial figure. As the stage is rotated the optic axis describes a circle in the field of view centred on the cross-wires.

stage, the *brushes* or *isogyres* that constitute the axial cross will sweep across the field of view in such a manner that they always remain parallel to the polarizer and analyser vibration directions respectively, that is their attitude will always be N—S and E—W. The optic sign of the specimen crystal which produces such a figure can be determined by the use of accessory test plates with no greater difficulty than if the figure were precisely centred.

But if the axial cross, that is the direction of the optic axis, lies outside the field of view the determination of sign is more difficult. The angular radius of the field of view of most polarizing microscopes with the high-power objective inserted is about 55°. If the optic axis lies on the margin of the field of view of the directions image, the light propagated along the optic axis in the crystal must make an angle of 55° with the microscope axis after refraction at the crystal/air interface at the upper surface of the parallel sided crystal plate. Therefore the angle ϕ between the optic axis and the normal to the crystal plate will be given by $\sin 55° = o \sin \phi$ since light travelling along the optic axis has refractive index o. If $\phi > \sin^{-1}(\sin 55°/o)$, the brushes will still sweep across the field of view in such a manner as to be visible as either N—S or E—W dark lines and the quadrant observed at any point during the rotation of the stage can be identified by comparison of the movement of the brushes with the sequence of observations illustrated in Fig 12.26. The optic axis rotates on the surface of a cone whose axis is the microscope axis.

We have so far restricted our discussion of uniaxial interference figures to positive

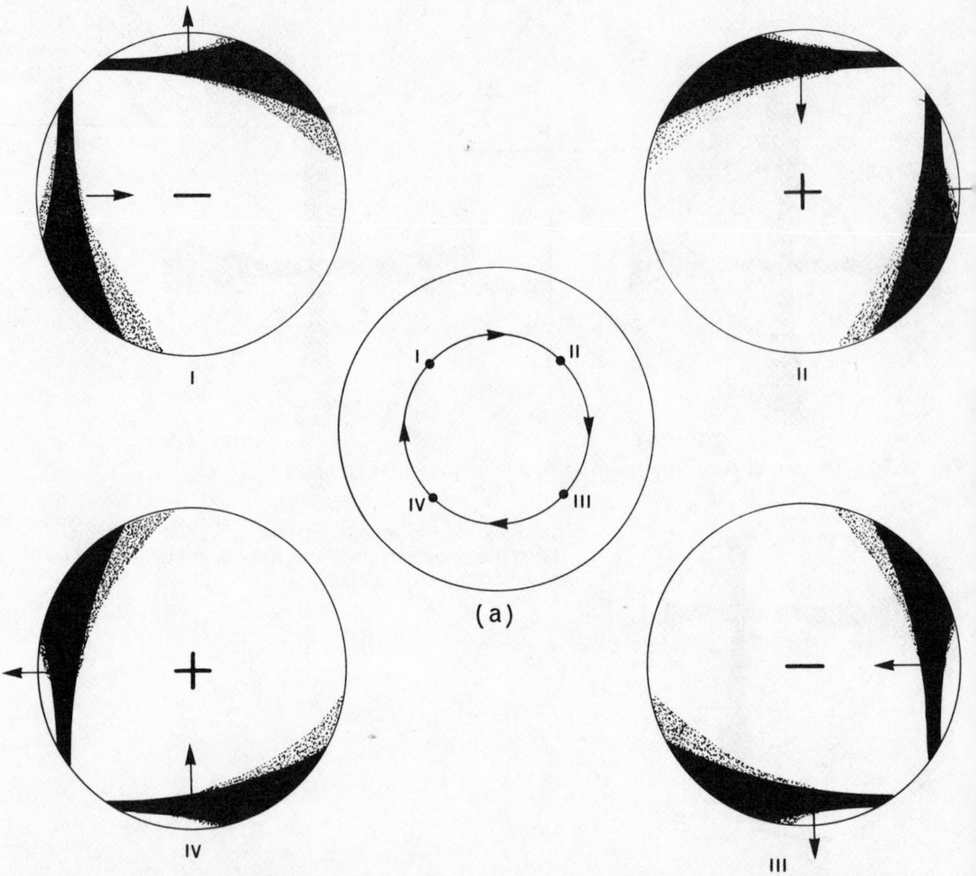

Fig 12.26 An off-centre uniaxial figure in which the axial cross lies outside the field of view. The stereogram (a) shows the precession of the optic axis about the microscope axis with the clockwise succession of positions I–IV shown. In each of the four partial figures the direction of movement of each isogyre for clockwise rotation of the stage is indicated by an arrow. The sign at the centre of each partial figure indicates for a positive crystal whether addition or subtraction occurs when a test plate is inserted.

crystals. The description of those for negative crystals is the same with fast and slow, addition and subtraction interchanged.

Interference effects in strongly convergent light: biaxial interference figures

The most practically useful figure is that produced when the acute bisectrix of the specimen crystal on the stage is parallel to the microscope axis. If the specimen is optically positive the γ vibration direction will then be parallel to the microscope axis. The orientation of the indicatrix and the appearance of the interference figure are shown in Fig 12.27 for the optic axial plane lying (a) in the so-called 90° position, i.e. E—W, and (b) in the 45° position, i.e. NE—SW.

The acute bisectrix figure is characterized by three features: (i) the *eyes* which correspond to propagation along the optic axes in the crystal, (ii) distorted rings about the eyes, black in monochromatic light and following Newton's scale in white light, which represent directions of equal retardation, and (iii) the *isogyres* which are black arcs corresponding to directions of propagation in the crystal which produce vibration

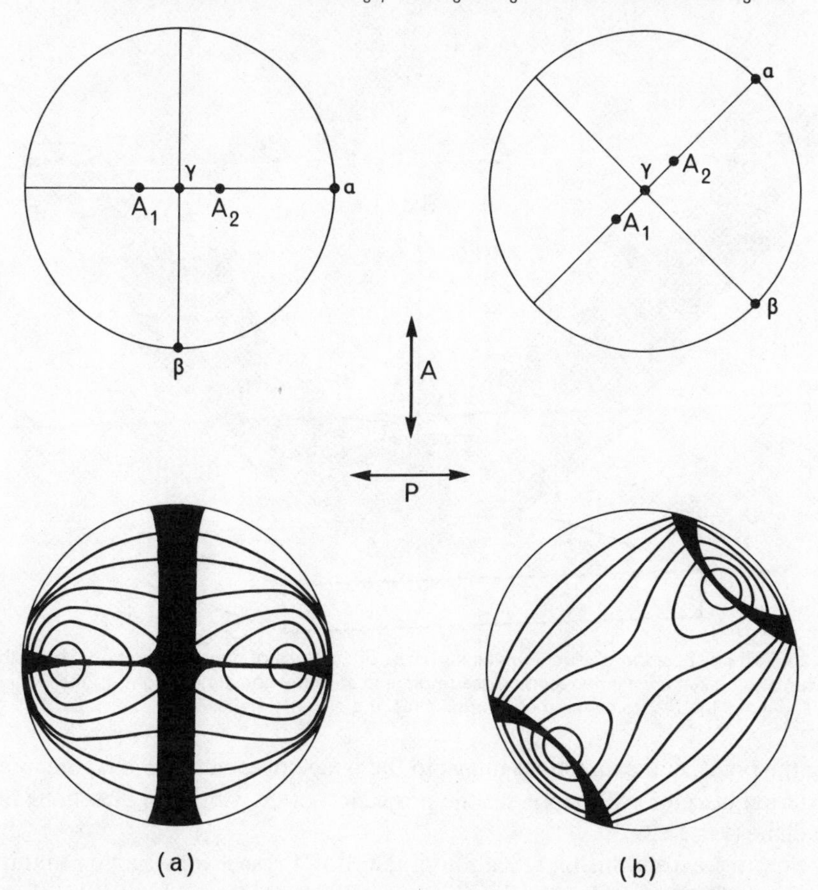

Fig 12.27 The biaxial interference figure. The two upper diagrams show the orientation of the indicatrix corresponding to the interference figures shown in the two lower diagrams. In (a), representing the 90° position, the optic axial plane is parallel to the plane defined by the microscope axis and the polarizer vibration direction, while in (b), representing the 45° position, the optic axial plane is parallel to the plane defined by the microscope axis and the bisector of the polarizer and analyser vibration directions.

directions parallel and perpendicular to the polarizer vibration direction in the directions image.

Propagation of light along an optic axis implies vibration directions in a circular section of the indicatrix. Interference effects will be nil and the eyes will appear black. Directions of propagation for which the retardation is the same lie on conical surfaces surrounding each optic axis. The cones become increasingly distorted as the angle between the direction of propagation and the optic axis increases until they merge to surround both optic axes symmetrically about the optic axial plane (Fig 12.28). The reader is referred for a full treatment of such *isochromatic surfaces* in terms of indicatrix geometry to Ditchburn (1963, p. 514). That the isogyres will be symmetrical about the optic axial plane in the 90° and 45° positions is self-evident. Their angular disposition with respect to the optic axes and the vibration directions of the polars is determinable stereographically by the Biot–Fresnel construction.

The vibration directions associated with any point on the directions image of the interference figure can simply be determined by a construction based on the Biot–Fresnel rule. The construction consists simply of bisecting the angle between the lines

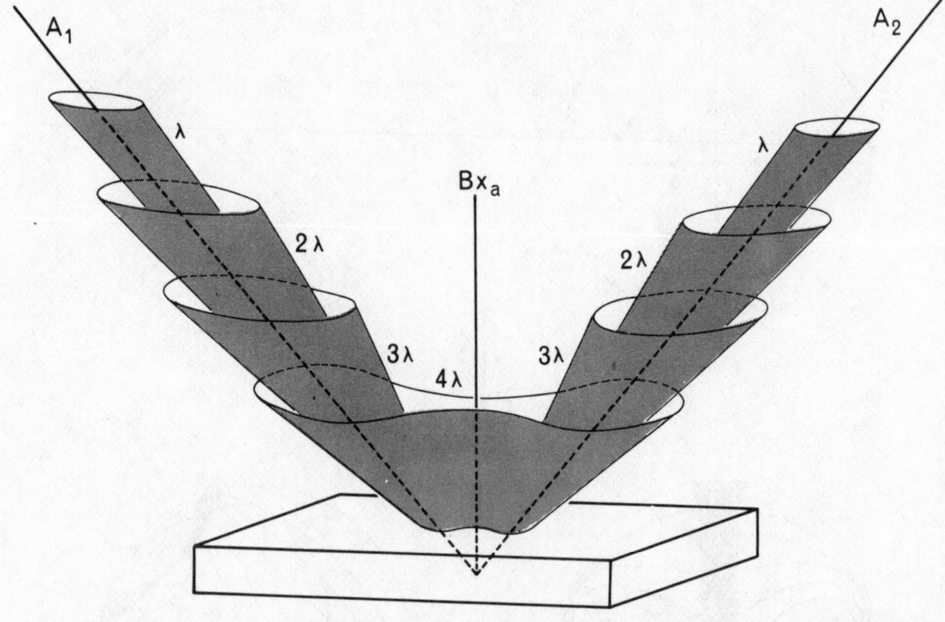

Fig 12.28 The disposition of isochromatic surfaces, or surfaces of equal retardation, about the optic axes of a biaxial crystal between crossed polars in strongly convergent light. In the figure the 4λ surface is the first to envelop both optic axes in a single surface.

joining the point on the directions image to the image of each optic axis; the internal and external bisectors will represent the projection of the vibration directions in the image plane (Fig 12.29).

It is easy to see from this modification of the Biot–Fresnel construction that in the 90° position (OAP lying E—W) the isogyres will constitute an E—W, N—S cross. Application of the construction will show that in the 45° position the isogyres are hyperbolas passing through the eyes and asymptotic to the polarizer and analyser vibration directions (Fig 12.31). On rotating the specimen crystal on the stage from the 90° to the 45° position the crossed isogyres separate to a maximum separation in the 45° position and on further rotation come together again until they cross when the OAP is N—S. In the crossed position the orientation of the OAP can readily be identified from the disposition of the rings of constant retardation about the eyes and

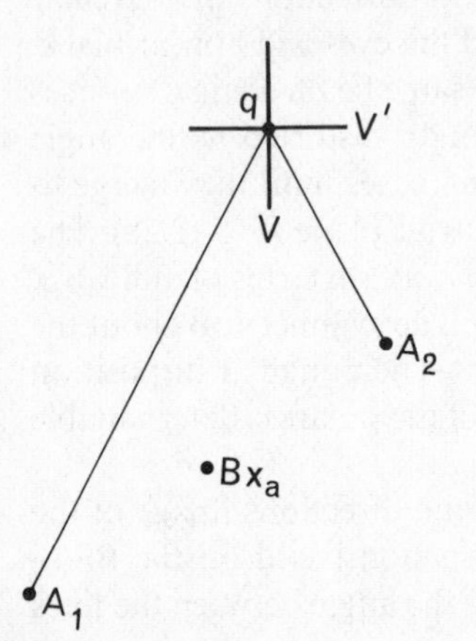

Fig 12.29 Vibration directions at a point in the directions image of a biaxial figure. The vibration directions, V and V′, are the internal and external bisectors of the lines joining the point q to the optic axes A_1 and A_2.

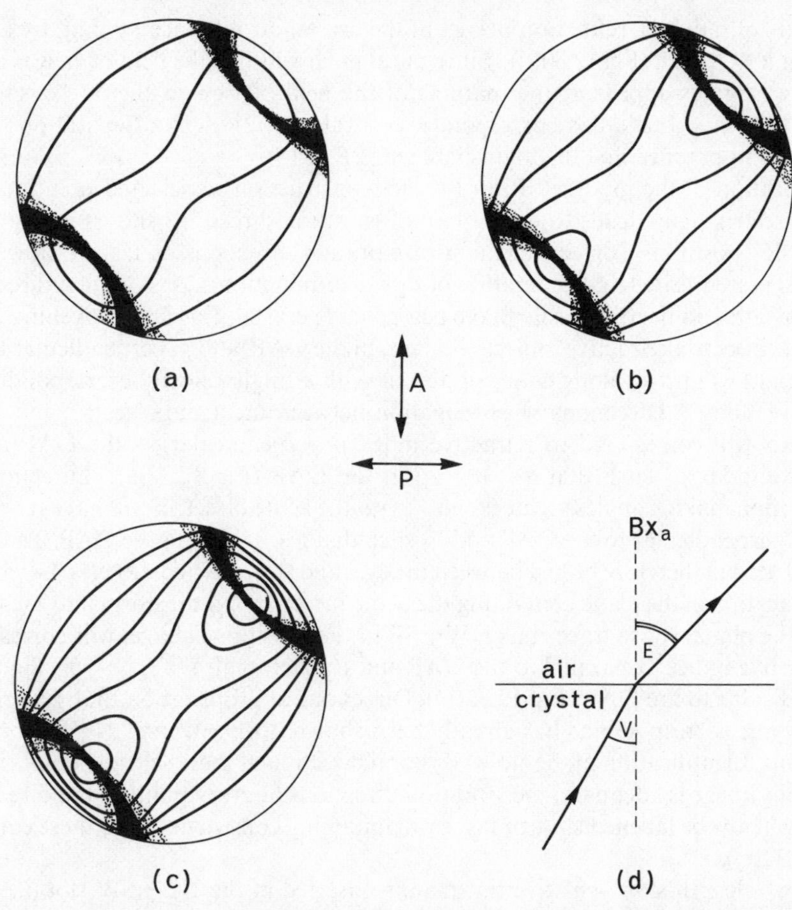

Fig 12.30 The acute bisectrix figures (a), (b), (c) show the change in aspect of the figure with increasing thickness for the same substance. If the substance is optically positive the interference colour at the centre of the figure is given by $d(\beta-\alpha)=n\lambda$. The diagram (d) illustrates the relationship of the angle V, between the acute bisectrix and an optic axis, and the angle E, between the acute bisectrix and the air path of a ray which has travelled along the optic axis in the crystal: $\sin E=\beta \sin V$.

moreover the brush perpendicular to the OAP is invariably more fuzzy than that representing the trace of the OAP on the directions image.

For a given substance the appearance of the acute bisectrix figure will depend, as in the uniaxial case, on the thickness of the specimen crystal on the stage (Fig 12.30). Light travelling along the microscope axis, that is parallel to the γ vibration direction if the crystal is positive, will have vibration directions corresponding to refractive indices α and β. The condition for destructive interference in the centre of the field of view will therefore be $d(\beta-\alpha)=n\lambda$. The magnitude of $\beta-\alpha$ can simply be determined by identification of the interference colour in white light at the centre of the field of view of an acute bisectrix figure in the 45° position when the thickness of the crystal is known.

The appearance of the acute bisectrix figure is also much dependent on $2E$, the angular separation in air of directions of propagation along the optic axes in the crystal. Light propagated along an optic axis in the crystal is inclined at an angle V to the acute bisectrix and therefore to the normal to the surface of a parallel-sided crystal plate. Since each optic axis is perpendicular to a circular section of the indicatrix of radius β, refraction occurs at the crystal/air interface so that, by Snell's Law, $\sin E = \beta \sin V$ (Fig 12.30(d)). Since the angular width of the field of view is about 110° the isogyres appear at the margins of the field of view in the 45° position if $E \sim 55°$; for $\beta = 1{\cdot}600$, this corresponds to a value of $2V \sim 60°$. Separation of the isogyres is in practice first distinguishable for $2E \sim 15°$.

Application of the modified Biot–Fresnel construction to selected points on the acute bisectrix figure leads to the pattern of vibration directions shown in Fig 12.31 for the 45° position. Correct labelling of vibration directions as fast or slow is an essential prerequisite to consideration of sign determination. Consider first directions of propagation in the optic axial plane of a positive crystal. For light travelling along the acute bisectrix refractive indices will be α, in the OAP, and β, perpendicular to the OAP. Light will travel along either optic axis with a single velocity corresponding to refractive index β. Directions of propagation between the acute bisectrix and either optic axis will correspond to refractive index β perpendicular to the OAP and a refractive index μ_1 such that $\alpha < \mu_1 < \beta$ in the OAP (Fig 12.32(a)). Directions of propagation making angles greater than V with the acute bisectrix will have refractive index β perpendicular to the OAP and μ_2 such that $\beta < \mu_2 < \gamma$ in the OAP. The trace of the OAP will therefore be fast between the eyes and slow outside the eyes. Directions of propagation in the plane containing the acute bisectrix and the normal to the OAP, that is the plane whose trace runs NW—SE in the directions image, will correspond to refractive indices α parallel to the OAP and μ_3 such that $\beta < \mu_3 < \gamma$ in the plane perpendicular to the OAP (Fig 12.32(b)). Directions of propagation that give rise to the isogyres correspond, as has already been shown, to E—W and N—S vibration directions. Identification of the slow vibration direction at these selected points in the directions image is adequate; the vibration directions at every point in the directions image can now be labelled slow or fast by maintaining consistency with these controls (Fig 12.32(c)).

If now a length-slow sensitive tint plate is inserted in the NE—SW slot (i.e. slow

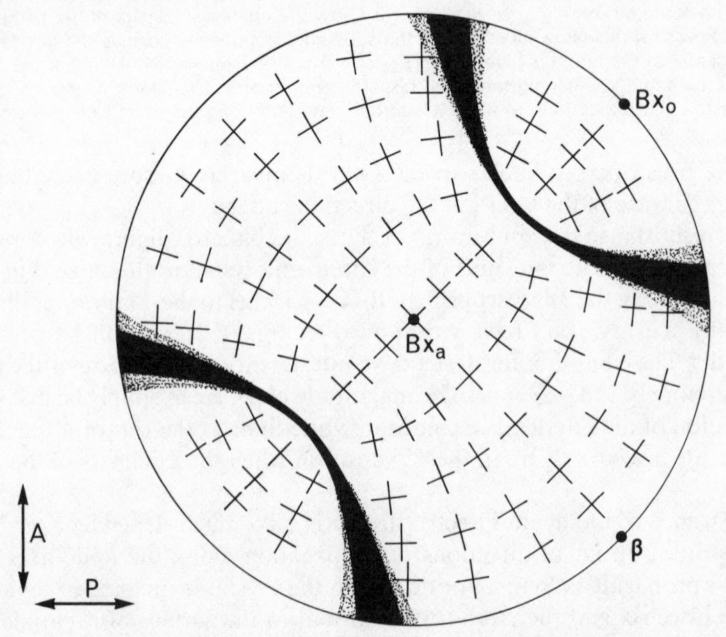

Fig 12.31 An acute bisectrix figure in the 45° position showing vibration directions.

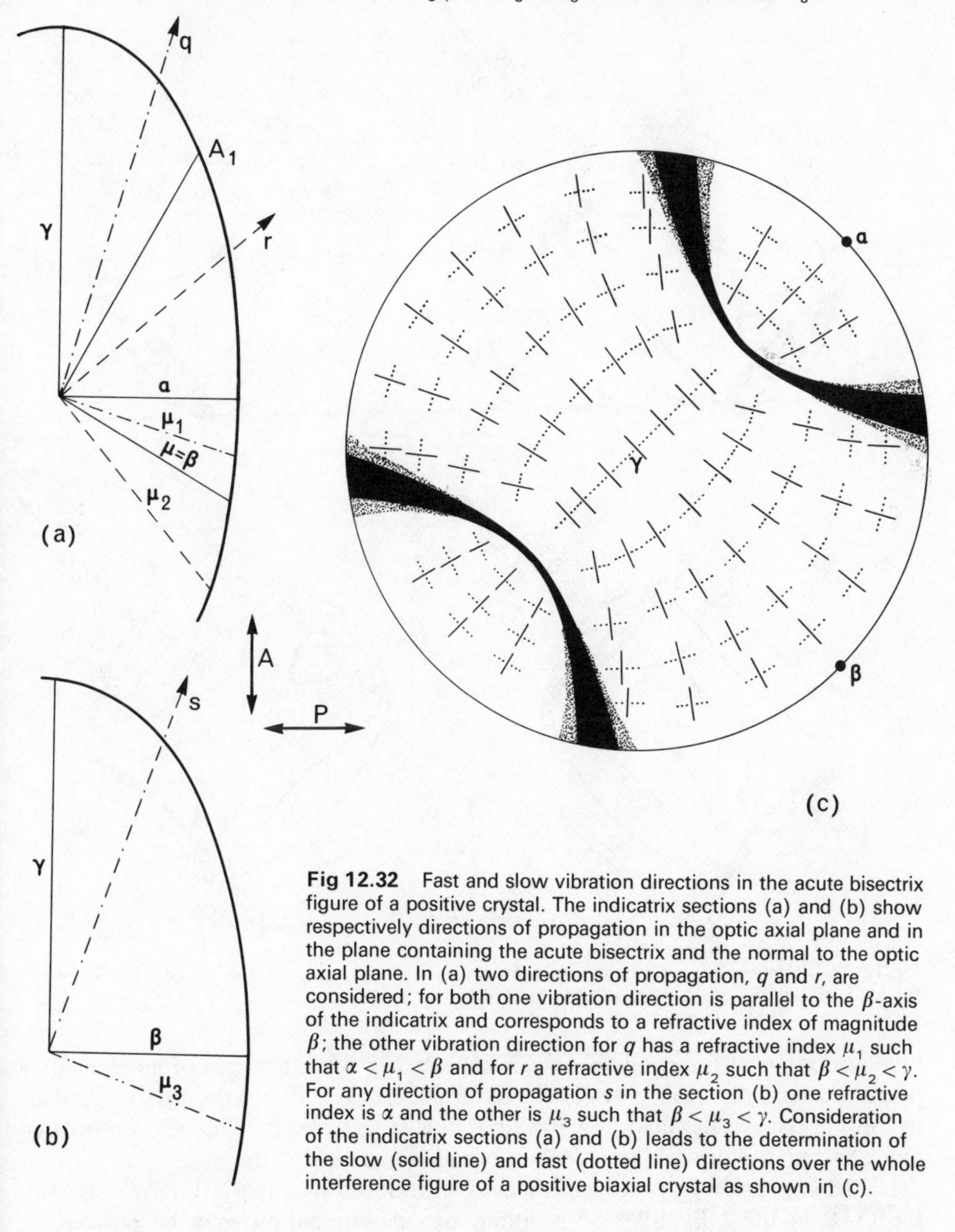

Fig 12.32 Fast and slow vibration directions in the acute bisectrix figure of a positive crystal. The indicatrix sections (a) and (b) show respectively directions of propagation in the optic axial plane and in the plane containing the acute bisectrix and the normal to the optic axial plane. In (a) two directions of propagation, *q* and *r*, are considered; for both one vibration direction is parallel to the β-axis of the indicatrix and corresponds to a refractive index of magnitude β; the other vibration direction for *q* has a refractive index μ_1 such that $\alpha < \mu_1 < \beta$ and for *r* a refractive index μ_2 such that $\beta < \mu_2 < \gamma$. For any direction of propagation *s* in the section (b) one refractive index is α and the other is μ_3 such that $\beta < \mu_3 < \gamma$. Consideration of the indicatrix sections (a) and (b) leads to the determination of the slow (solid line) and fast (dotted line) directions over the whole interference figure of a positive biaxial crystal as shown in (c).

in plate ∥ trace of OAP) addition will occur along the trace of the optic axial plane outside the eyes for a positive crystal and subtraction along the trace between the eyes. Generalizing, addition will be observed outside the isogyres and subtraction between the isogyres. In particular in white light the first order sensitive tint ring will become second order sensitive tint in the field outside the isogyres while it will be reduced to zero order in the central part of the field between the isogyres (Fig 12.33). The change in order, by two orders, of a sensitive tint ring at the isogyre is very easily recognizable and provides a foolproof means of sign determination. If $d(\beta - \alpha)$ is small for the specimen crystal the sign may be determined conveniently and with certainty

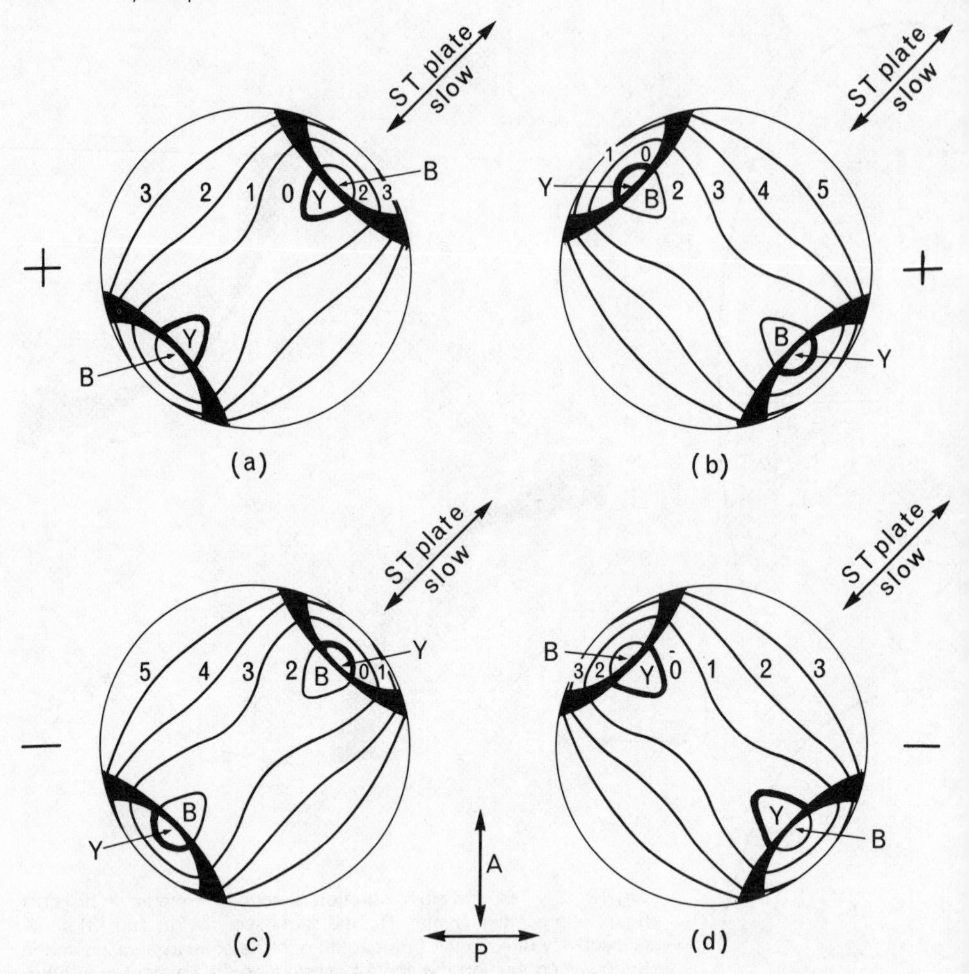

Fig 12.33 Superimposition of a sensitive tint plate on biaxial figures in the 45° position. (a) and (b) are for a positive crystal, (c) and (d) for a negative crystal.

by observation of the resultant interference colour on either side of an eye: for a positive crystal the field immediately out from the eye will be coloured blue, while just inside the eye the field will appear yellow (Fig 12.33(a)). The behaviour of an optically negative crystal is also shown in the same figure (c. d).

If a length-slow quartz wedge is slowly introduced into the NE—SW slot, with the OAP in the NE—SW 45° position, sign determination may be effected by observation of the progressive changes produced in the acute bisectrix figure. If the crystal is positive there will be addition in the areas outside the isogyres and this will be displayed dynamically by the inward movement of sensitive tint rings towards the eyes, an nth order ring being successively replaced by an $(n+1)$th, an $(n+2)$th and so on. In the central area between the isogyres there will be subtraction and the rings will move out, away from the trace of the OAP towards the edge of the field of view as nth order rings are replaced by $(n-1)$th, $(n-2)$th, and so on (Fig 12.34). In short the sensitive tint rings appear to flow from the areas outside the isogyres, through the eyes, and then out towards the NW and SE extremities of the field of view. The behaviour of the black rings in a positive acute bisectrix figure in

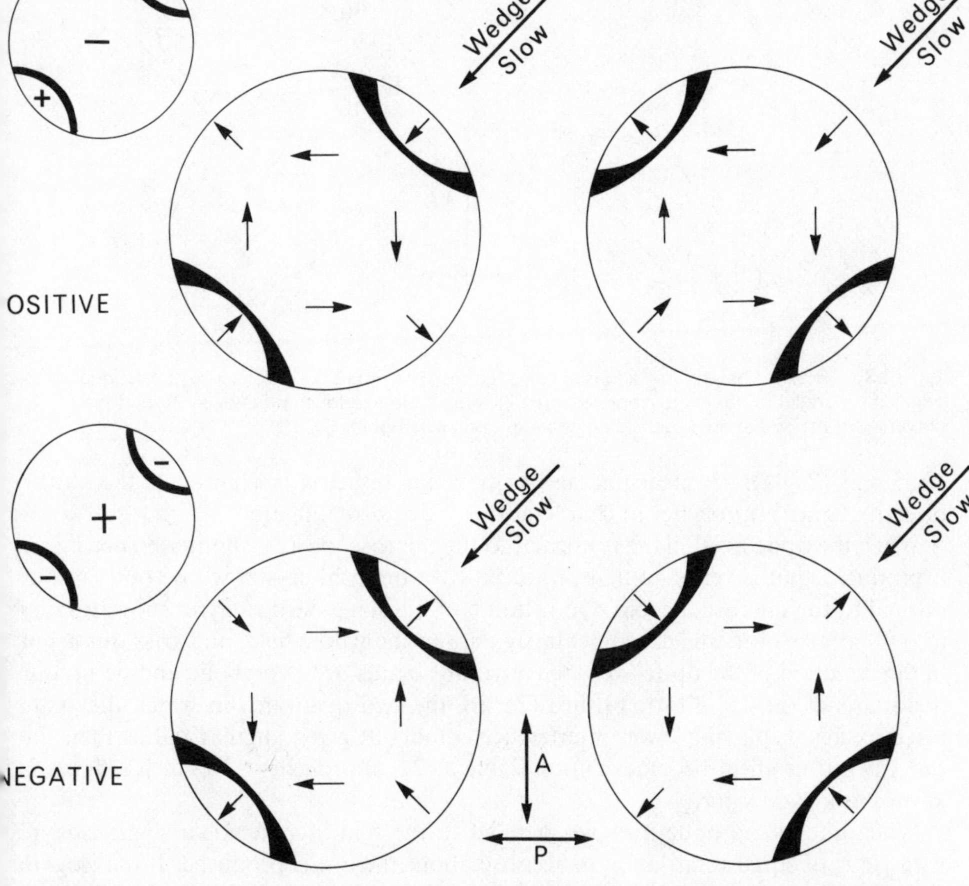

Fig 12.34 Superimposition of a quartz wedge moved from NE to SW on biaxial figures in the 45° position. The arrows show the direction of movement of the interference rings as the wedge is moved inwards. The two small diagrams on the left summarize the effect of superimposing a quartz wedge or a sensitive tint plate on crystals oriented with their optic axial planes NE–SW, the standard orientation.

monochromatic light will of course be identical. The behaviour of an optically negative crystal, for which the direction of movement of the rings is reversed, is shown also in Fig 12.34.

So far we have dealt exclusively with the acute bisectrix figure, that is with the interference figure produced conoscopically when the acute bisectrix of the specimen crystal lies parallel to the microscope axis. We have now to consider the interference figures produced by other orientations of the indicatrix.

Obtuse bisectrix figures are generally similar to acute bisectrix figures. In the 90° position the eyes are of course outside the field of view as is the case in acute bisectrix figures where $2V$ is large. On rotating the stage towards the 45° position the isogyres move out much more rapidly towards the edge of the field and this is characteristic. In the 45° position the isogyres will lie outside the field of view, but it is still possible to determine the optic sign of the substance by the use of accessory plates; a positive crystal will behave in the manner shown for a negative acute bisectrix figure in Figs

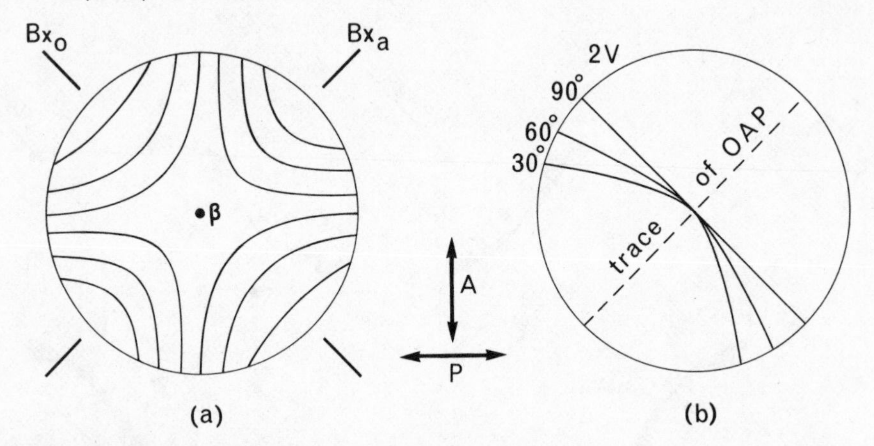

Fig 12.35 Biaxial flash figures and optic axis figures. The figure (a) is a flash figure produced when β is parallel to the microscope axis. (b) illustrates the relationship between 2V and the curvature of the single observable isogyre in an optic axis figure.

12.33 and 12.34. If $2V$ measured across the acute bisectrix is very small, the obtuse bisectrix figure approaches in character the optic normal figure.

When the optic axial plane is normal to the microscope axis an interference figure is produced that is very like that produced by a uniaxial crystal whose optic axis is normal to the microscope axis. On rotation of the stage dark isogyres move rapidly into the centre of the field, momentarily darken the whole field, and pass out again in the direction of the optic axes. Sensitive tint bands are hyperbolic and lie in four quadrants in the 45° position (Fig 12.35(a)), the two quadrants in which the acute bisectrix lies displaying lower interference colours at given angular radius than the other two (this effect becomes unnoticeable as $2V$ approaches 90°). Such a figure is known as a *flash figure.*

When one of the optic axes lies parallel to the microscope axis a single isogyre with rings of equal retardation concentric about the eye is produced. If the isogyre has marked curvature, optic sign can be determined in the normal way (Fig 12.35(b)), but if $2V$ is near 90° the isogyre will appear almost straight and sign will be indeterminable from the optic axis figure.

Crystals lying so that some general radius of the indicatrix lies parallel to the microscope axis may be such that a reliable sign determination can be made, but such orientations can never provide as much information as a centred acute bisectrix figure. If the acute bisectrix lies within about 40° of the microscope axis an amenable *off-centre figure* of one of the sorts shown in Fig 12.36 will be observed, but in general an off-centre figure yields no reliable information. An off-centre biaxial figure can be distinguished from an off-centre unaxial figure because in the latter the isogyres are bound to move across the field of view in such a manner that they always lie either N—S or E—W whereas in the former they can lie in any attitude.

A centred acute bisectrix figure yields information about the partial birefringence $\beta-\alpha$ if the specimen crystal is optically positive provided the thickness of the specimen crystal is known. If the colour displayed at the centre of the field of view, that is the

Fig 12.36 Biaxial off-centre figures. Successive stages are shown in the rotation of the figure produced by a crystal which has (a) Bx_a inclined to the microscope axis and Bx_o parallel to the plane of the microscope stage, (b) Bx_a inclined to the microscope axis and β parallel to the plane of the stage, and (c) Bx_a, Bx_o, and β in no special orientation.

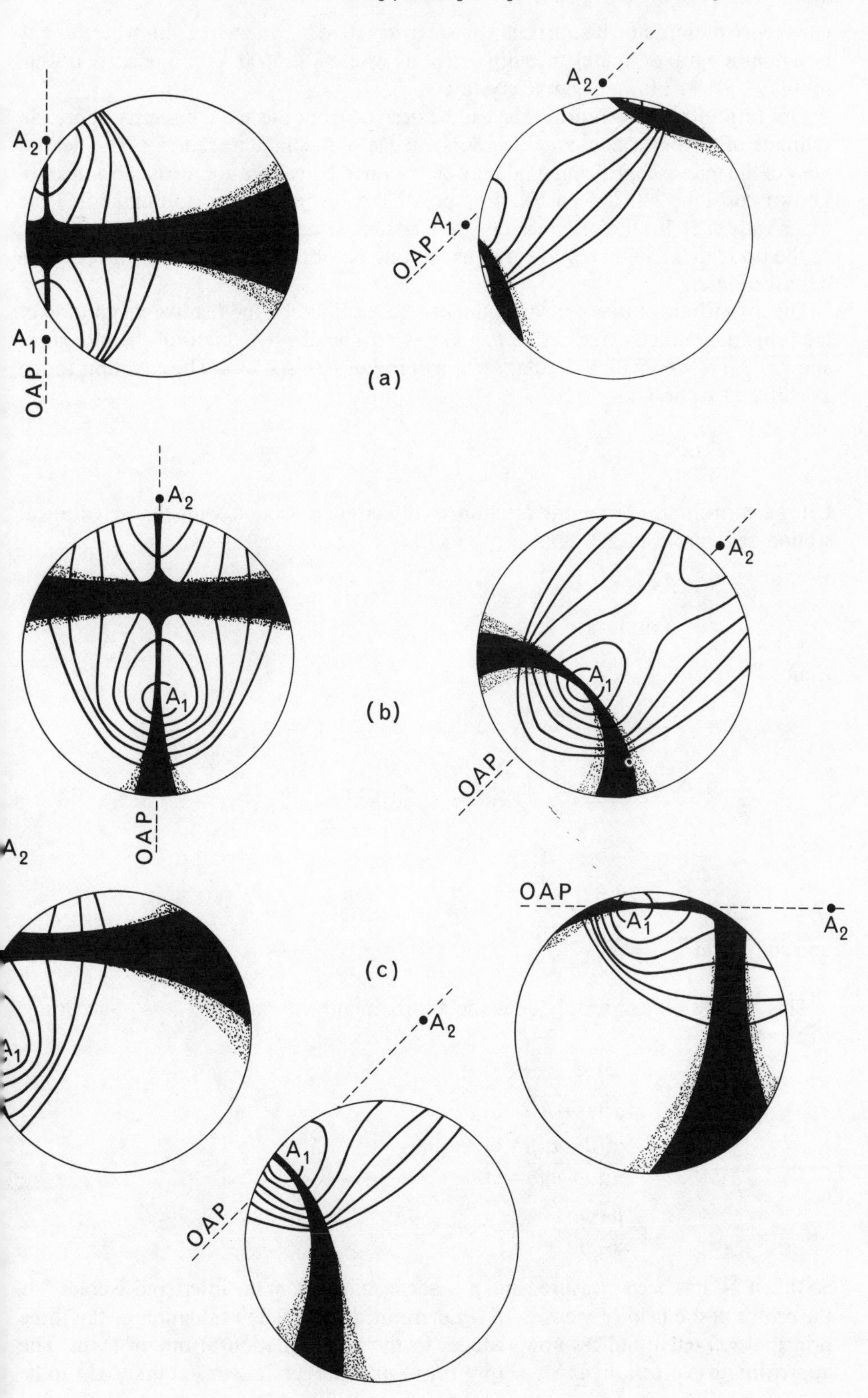
A2
A1
OAP
(a)
(b)
(c)

emergence of the acute bisectrix, is nth order sensitive tint in white light (where n may be a non-integral estimate), then $d(\beta-\alpha) = n\lambda$ where $\lambda \simeq 5600$ Å for a positive crystal (but $d(\gamma-\beta) = n\lambda$ for a negative crystal).

One further piece of information can be derived from the acute bisectrix figure: an estimate of the optic axial angle $2V$. Knowledge of the angular radius of the field of view of the microscope being used may be obtained by calibration with substances of known constant $2V$; it then becomes possible to estimate very approximately the magnitude of $2V$ for the specimen crystal. The precise determination of the magnitude of the optic axial angle requires the use of one or other of the special stages to be described later.

The magnitude of the optic axial angle is determined by the relative magnitude of the principal refractive indices, α, β, and γ. Consider the $\alpha\gamma$ section of the indicatrix and its intersection with a circular section of radius β (Fig 12.37). The equation to the $\alpha\gamma$ elliptical section is

$$\frac{x^2}{\alpha^2}+\frac{z^2}{\gamma^2} = 1.$$

Let the coordinates of the intersection of the circular section with the $\alpha\gamma$ elliptical section be x and z, then

$$x = \beta \cos V_\gamma$$

$$z = \beta \sin V_\gamma.$$

Hence

$$\frac{\beta^2 \cos^2 V_\gamma}{\alpha^2}+\frac{\beta^2 \sin^2 V_\gamma}{\gamma^2} = 1$$

$$\frac{\beta^2 \cos^2 V_\gamma}{\alpha^2}+\frac{\beta^2 \sin^2 V_\gamma}{\gamma^2} = \cos^2 V_\gamma + \sin^2 V_\gamma$$

$$\tan^2 V_\gamma = \frac{\beta^2/\alpha^2 - 1}{1-\beta^2/\gamma^2}$$

and so $$\tan^2 V_\gamma = \frac{1/\alpha^2 - 1/\beta^2}{1/\beta^2 - 1/\gamma^2}.$$

This expression can simply be made approximate without great loss of accuracy. Rearranging,

$$\tan^2 V_\gamma = \frac{\gamma^2(\beta^2-\alpha^2)}{\alpha^2(\gamma^2-\beta^2)}$$

$$= \frac{\gamma^2(\beta+\alpha)(\beta-\alpha)}{\alpha^2(\gamma+\beta)(\gamma-\beta)}$$

$$\simeq \frac{\beta-\alpha}{\gamma-\beta}.$$

So that if $2V$ has been measured and $\beta-\alpha$ determined from the interference colour at the centre of the field of view, $\gamma-\beta$ is determined; complete evaluation of the three principal refractive indices now reduces to the determination of one of them. The approximate expression $\tan^2 V_\gamma = (\beta-\alpha)/(\gamma-\beta)$, although it seems at first sight to be

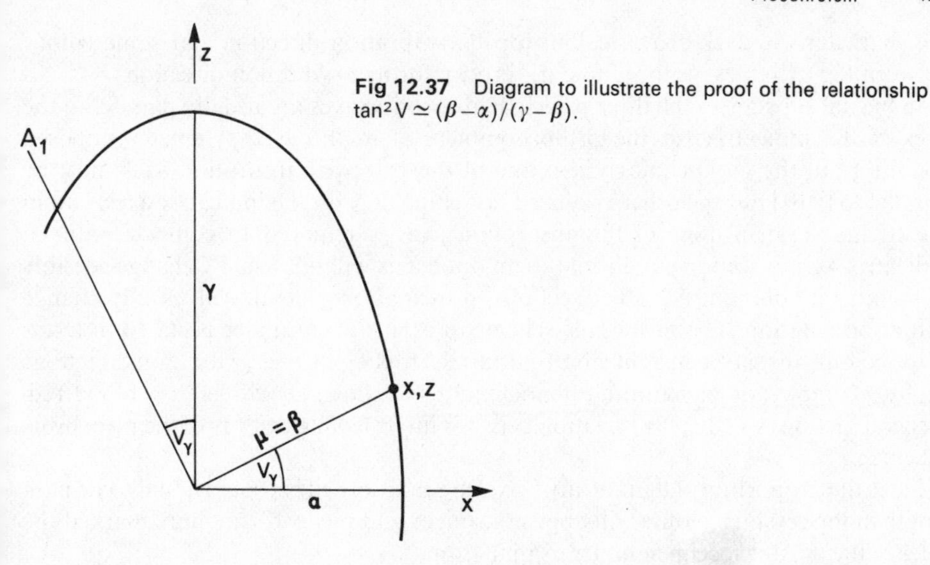

Fig 12.37 Diagram to illustrate the proof of the relationship $\tan^2 V \simeq (\beta-\alpha)/(\gamma-\beta)$.

a crude approximation, is usually adequate since V cannot readily be measured to better than $\pm 2°$.

It follows from the approximate expression $\tan^2 V_\gamma \simeq (\beta-\alpha)/(\gamma-\beta)$ that if

$\beta-\alpha < \gamma-\beta$, then $2V_\gamma < 90°$ and the crystal is positive; while if

$\beta-\alpha > \gamma-\beta$, then $2V_\alpha = 180° - 2V_\gamma < 90°$ and the crystal is negative.

Pleochroism

The transmission coefficient of light through most transparent substances varies with wavelength. When the transmission coefficient is very low over a certain wavelength band or bands the substance appears coloured in transmitted light. In isotropic (i.e. non-crystalline or cubic) substances such selective absorption of light is, in common with other optical properties, independent of vibration direction. Isotropic crystals viewed in white parallel plane polarized light with the analyser out will show no change of colour on rotation of the stage. In anisotropic crystals however the wavelength range, as well as the magnitude, of selective absorption is dependent on vibration direction: this property is known as *pleochroism.* Uniaxial and biaxial crystals viewed in white parallel plane polarized light with the analyser out will, if they are pleochroic, change colour on rotation of the stage.

Pleochroism in uniaxial substances can simply be summarized by a statement of the wavelength range or ranges strongly absorbed by light vibrating in the e and o vibration directions. Since absorption spectra have not been investigated for many minerals it is usual merely to state as the *pleochroic scheme* the colours displayed in transmitted light vibrating parallel and perpendicular to the optic axis. For instance the *pleochroic scheme* of a typical dravitic tourmaline (trigonal) is o = dark brown, e = yellow. An (0001) section of such a tourmaline would display in plane polarized light a dark brown colour invariant on rotation of the stage while a section parallel to [0001] would appear yellow when the optic axis was parallel to the polarizer vibration direction and would darken on rotation of the stage until maximum absorption corresponding to a dark brown colour was apparent with the optic axis perpendicular to the polarizer vibration direction. A general section would display the characteristic dark brown colour for its o-vibration direction and some colour between the extremes, yellow and dark brown, for its e'-vibration direction.

In biaxial substances the three principal absorption axes are usually parallel to the axes of the indicatrix; in the orthorhombic system this is a symmetry imposed condition; in the monoclinic system one of the principal absorption axes must be parallel to [010] but the other two need not coincide with the indicatrix axes, and in the triclinic system none of the absorption axes is required to coincide with an indicatrix axis. A section cut normal to an optic axis will not usually change sensibly in colour on rotation. All other sections, principal or general, will usually change colour on rotation. The pleochroic scheme of a biaxial substance is stated in terms of the colours displayed in light vibrating parallel to the α, β, and γ vibration directions as, for example, for piemontite (monoclinic) α = yellow, β = violet, γ = blood red. General sections will display variation between intermediate colours of the pleochroic scheme.

Since the proportion of light of any wavelength absorbed is proportional to its path length in the crystal, colours will appear stronger and pleochroism more marked the thicker the crystal specimen under examination.

Dispersion

In isotropic substances refractive index varies slightly with wavelength according to the empirical Cauchy Equation $\mu = A + B\lambda^{-2} + C\lambda^{-4} + \ldots$; usually the first two terms suffice for an adequate description of *normal dispersion* (Fig 12.38(a)). If white light is incident on a glass prism and the emergent light allowed to fall on a white screen, a spectrum will be seen because the refractive index of the glass, and consequently the angle of refraction, varies with wavelength. If a substance displays strong selective absorption, that is if it is coloured, the dispersion curve will be anomalous over the corresponding wavelength range (Fig 12.38(b)).

In a uniaxial substance both the principal refractive indices will be wavelength dependent. The orientation of the indicatrix, being symmetry controlled, will of course be independent of wavelength, but its shape will in general vary with wavelength. In the majority of substances dispersion of both o and e is small and dispersion of the birefringence $|e-o|$ is consequently negligible (Fig 12.38(c)). In some substances however the dispersion of birefringence is anomalous. For example in the tetragonal mineral apophyllite at certain compositions $o-e$ decreases smoothly with increasing λ (Fig 12.38(d)) and in consequence normal interference colours (Newton's scale) will not be displayed between crossed polars. The sequence of *anomalous interference colours* observable in a wedge of substance between crossed polars in parallel light or in an optic axis figure in strongly convergent light will depend on the precise nature of the dependence of $|e-o|$ on λ. The condition for destructive interference by an (001) section of the tetragonal mineral meta-torbernite where $e-o=0$ for $\lambda = 5120$ Å, is illustrated in Fig 12.39. It must be stressed that anomalous dispersion will not only change the sequence of interference colours but will characteristically give rise to colours different from those of Newton's scale.

An extreme case of anomalous dispersion in a uniaxial substance arises when the curves for dispersion of e and o cross (Fig 12.38(e)). This is the case in the tetragonal mineral meta-torbernite which is negative for short wavelengths, isotropic at 5120 Å (green), and positive at longer wavelengths. The interference colour displayed by an anisotropic section of meta-torbernite in white light between crossed polars will always be lacking in the green wavelength for which it is isotropic as well as in shorter

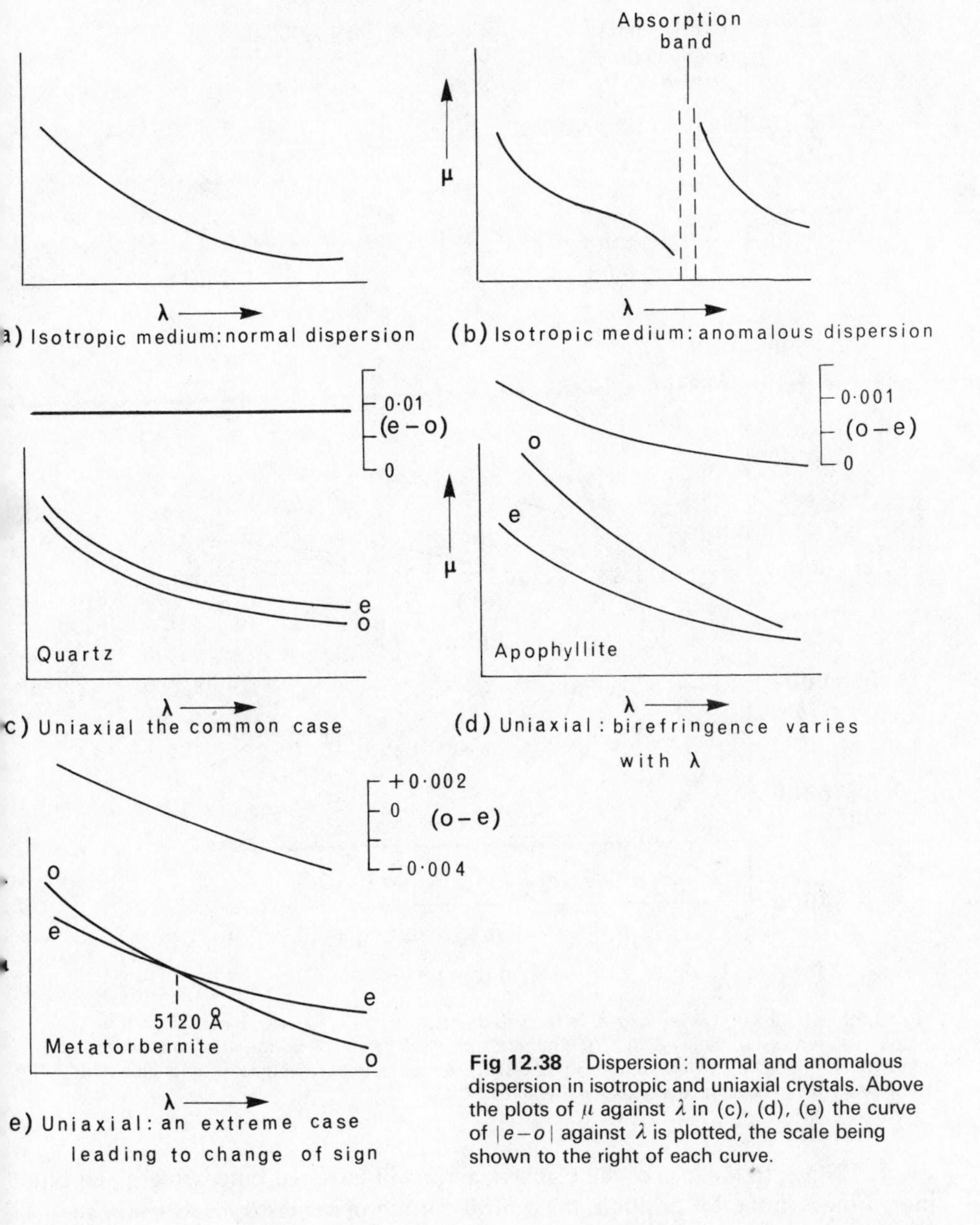

Fig 12.38 Dispersion: normal and anomalous dispersion in isotropic and uniaxial crystals. Above the plots of μ against λ in (c), (d), (e) the curve of $|e-o|$ against λ is plotted, the scale being shown to the right of each curve.

wavelengths such that $d(o-e') = n\lambda$ or longer wavelengths such that $d(e'-o) = n\lambda$; the interference colours produced by thin sections and small crystals will be characteristically anomalous.

In the orthorhombic system the orientation of the indicatrix remains symmetry controlled; only the shape of the indicatrix can be wavelength dependent. Dispersion in those orthorhombic substances that exhibit it is most clearly displayed in the acute bisectrix figure. Normal dispersion involves no crossing of the curves of the three principal refractive indices, α, β, and γ, plotted against λ; such wavelength dependence is known as *dispersion of the optic axes* because it is most clearly displayed as a variation of $2V$ with λ. Two cases may be distinguished, with $2V_{\text{red}} < 2V_{\text{blue}}$ or

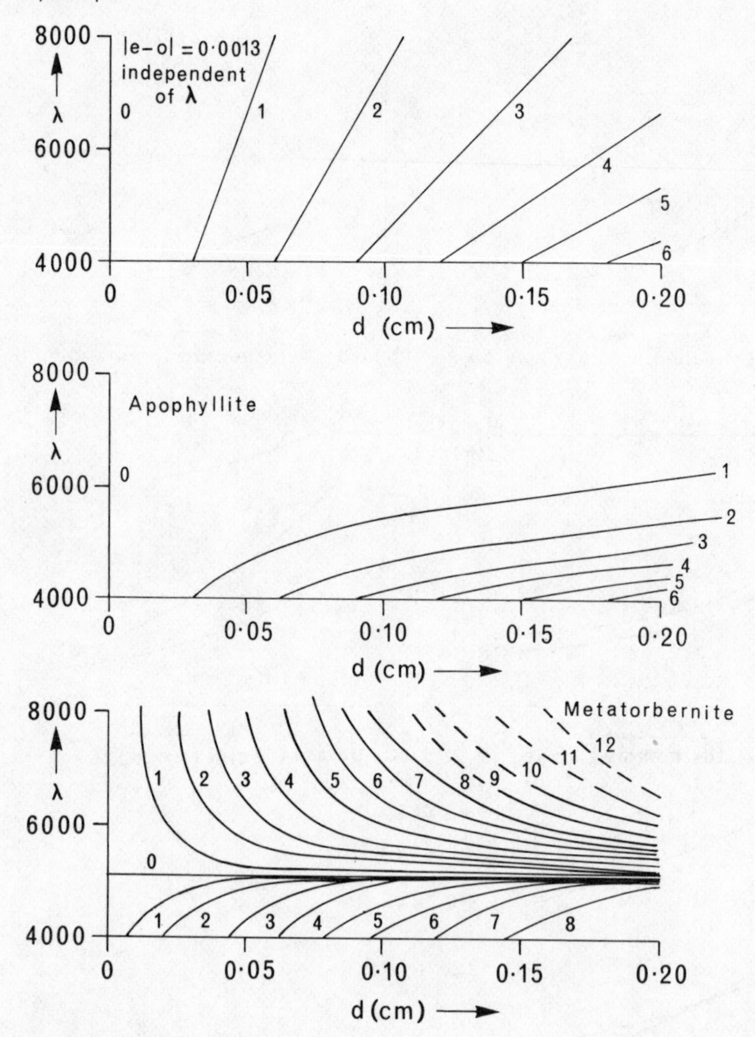

Fig 12.39 Graphs of $\lambda = |e-o|d/n$. In the top diagram $|e-o|$ is independent of λ. In the middle diagram $|e-o|$ varies with λ (the mineral *apophyllite*). In the bottom diagram, which refers to the mineral *meta-torbernite* (Fig 12.38(e)), an extreme case of dispersion is illustrated. The figure shown against each curve is the value of n.

$2V_{red} > 2V_{blue}$. In the former the black isogyres will have red outer fringes and blue inner fringes in the 45° position; the central portion of each isogyre representing the overlap of areas of extinction for various wavelengths (Fig 12.40). If on the other hand $2V_{red} > 2V_{blue}$, the isogyre for blue light, which will appear red, will lie inside the isogyre for red light, which will look blue.

In the extreme case, where two of the curves of the principal refractive indices plotted against wavelength cross, spectacular dispersion effects are produced; this is known as *crossed axial plane dispersion.* The best known example is the orthorhombic polymorph of TiO_2, *brookite.* The curves of refractive index against λ for each of the three principal vibration directions in brookite are shown in Fig 12.41(a). As λ increases from the blue end of the visible spectrum the optic axial angle in the (100) plane decreases from about 30° to zero at $\lambda = 5550$ Å (yellow-green) and then increases

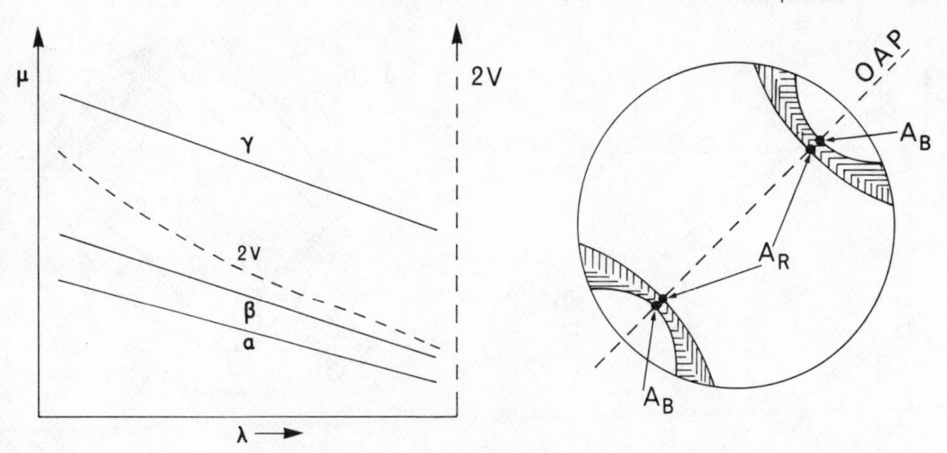

Fig 12.40 Dispersion of the optic axes in the orthorhombic mineral *barite*, $BaSO_4$. The diagram on the left shows the variation of α, β, γ and of 2V with λ; that on the right shows the appearance of the acute bisectrix figure in which the outer edges of the isogyres are coloured red and their inner edges are blue because $2V_{red} < 2V_{blue}$. The isogyres for red and blue light are shown respectively with vertical and horizontal hachures; in white light the former appears blue and the latter red. The positions of the optic axes for red and blue light are marked respectively A_R and A_B.

in the (001) plane (Fig 12.41(b)), the direction of the acute bisectrix remaining constant as [010] and the optic sign remaining positive. The positions of the isogyres in the acute bisectrix figure at three different wavelengths are shown in Fig 12.41 for an (010) plate with its *z*-axis lying NE—SW. If the light incident on the polarizer is passed through a monochromator of a type which permits rapid and smooth variation of the transmitted wavelength band, the movement of the isogyres in towards the centre of the field and outwards again at right-angles can be clearly demonstrated. The acute bisectrix figure of brookite in white light displays anomalous interference colours and is, not surprisingly, devoid of isogyres.

In the monoclinic system the orientation of the indicatrix is controlled by symmetry only to the extent that one of its diads must be parallel to [010]. Dispersion of the optic axes is still possible and in addition certain kinds of dispersion which are not permitted in systems of higher symmetry. In what follows it will be assumed that the monoclinic crystal is optically positive, so that γ is the acute bisectrix. It will be further assumed that we are dealing in each case with only one kind of dispersion.

If the acute bisectrix is parallel to the diad axis of symmetry, the only way in which the orientation of the indicatrix can vary with wavelength is shown in the stereogram Fig 12.42(a). As λ changes the optic axial plane rotates a few degrees about the monoclinic diad, that is about the acute bisectrix. The effect of this angular variation of OAP on the Bx_a figure in the 90° position is also shown in Fig 12.42(a). The 90° position provides a clearer distinction between the various types of dispersion peculiar to the monoclinic system than the more generally meaningful 45° position. This type of dispersion is known as *crossed dispersion*; the nomenclature is obvious from Fig 12.42(a). In crossed dispersion the isogyre through the eyes is fringed with red in the top left and bottom right quadrants, with blue in the bottom left and top right quadrants or vice versa. An example is borax, $Na_2B_4O_7.10H_2O$.

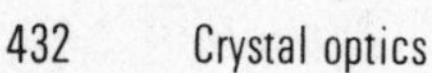

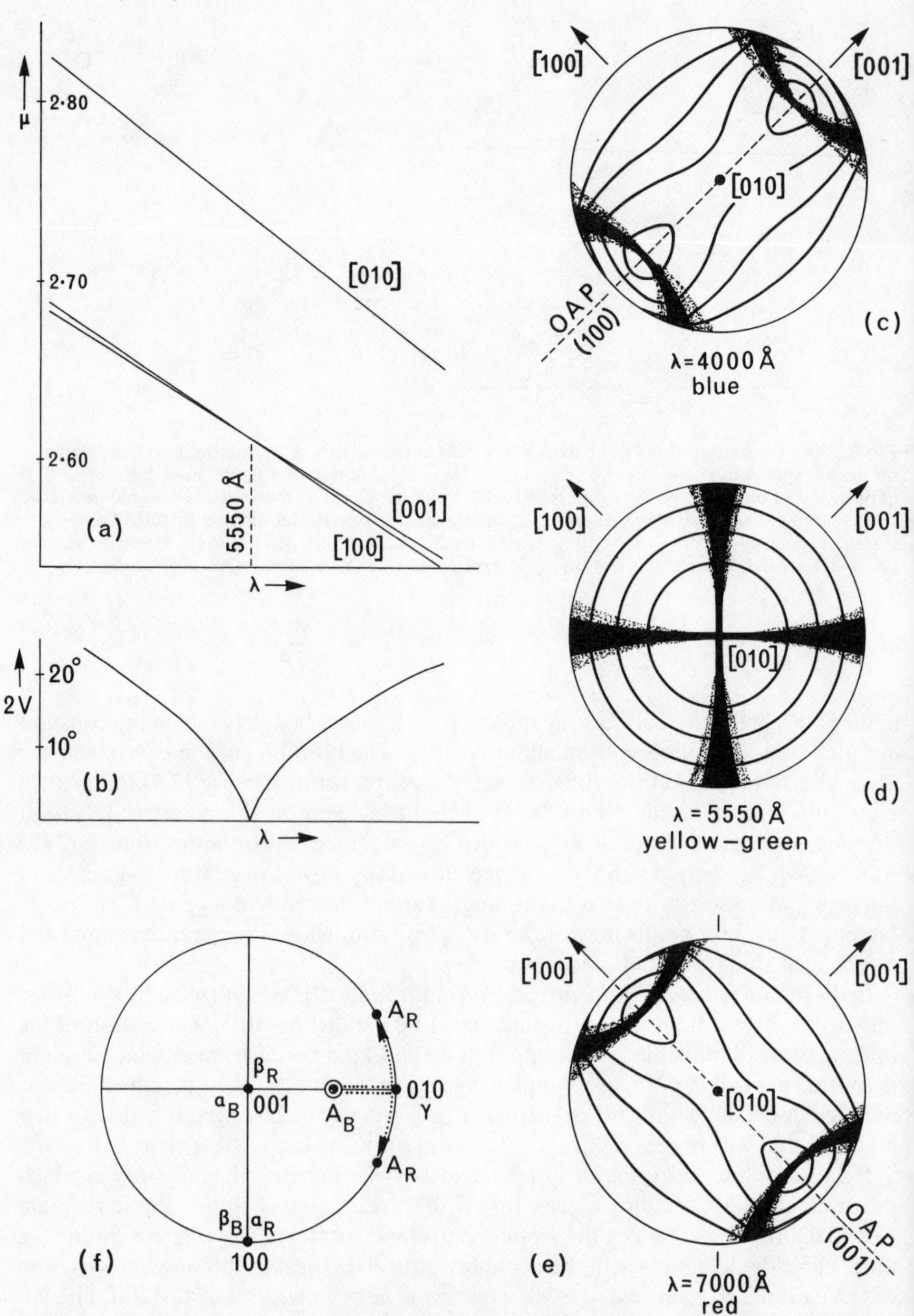

Fig 12.41 Crossed axial plane dispersion in the orthorhombic mineral *brookite*, TiO_2. The graph (a) shows the variation of the three principal refractive indices, corresponding to vibration directions parallel to the orthorhombic diads, with wavelength and (b) shows the relation between optic axial angle and wavelength. Brookite becomes optically isotropic at $\lambda = 5550$ Å. Drawings of the acute bisectrix figures, (c), (d), (e), are for a short wavelength, for $\lambda = 5550$ Å and for a long wavelength. The stereogram (f) illustrates the movement of the optic axes relative to the crystallographic axes as λ increases. In diagrams (c)–(f) the optic axial angles at low and high λ are exaggerated for clarity.

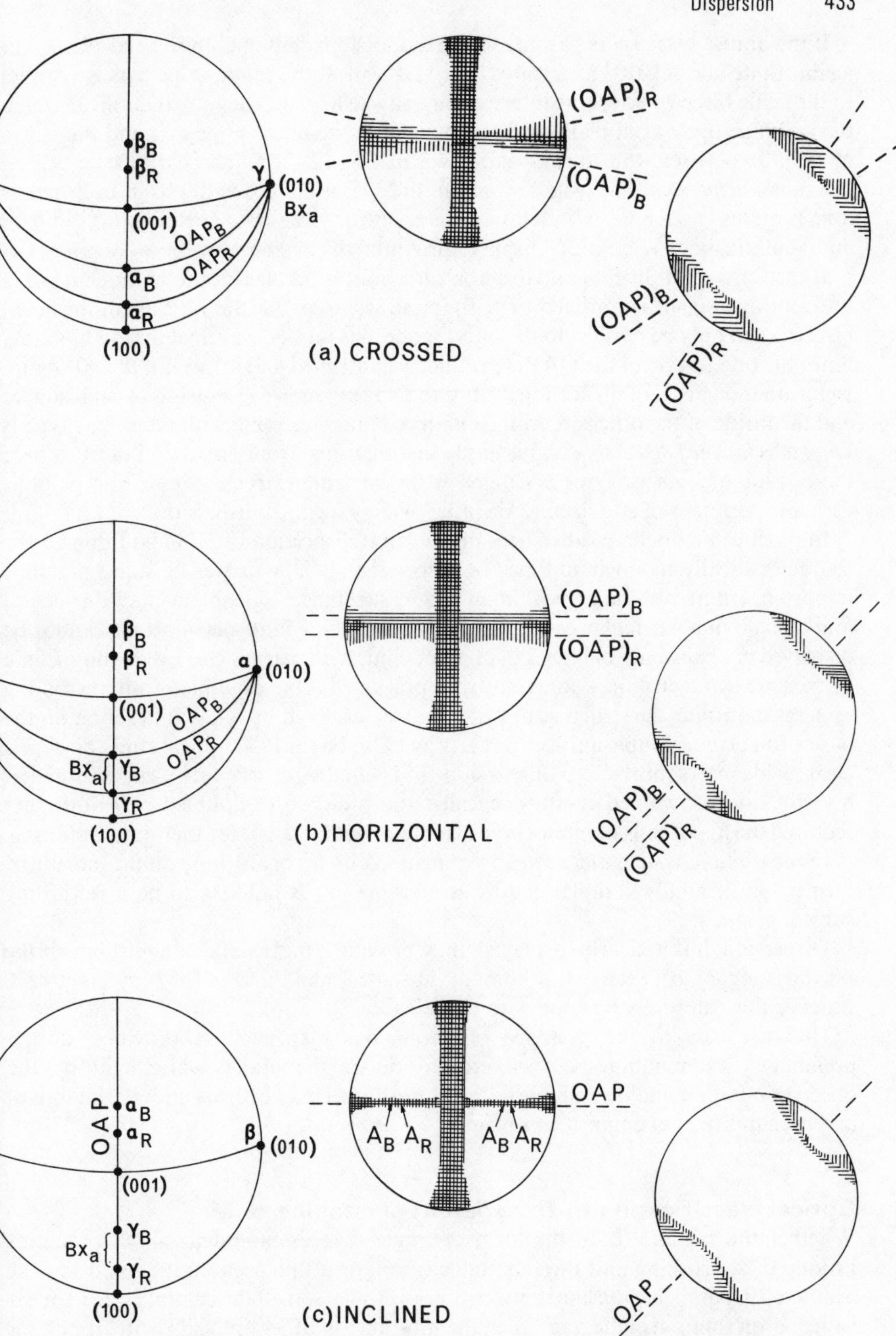

Fig 12.42 Dispersion in the monoclinic system. In each case illustrated the crystal is optically positive. (a) illustrates *crossed dispersion*, where $Bx_a \parallel [010]$. (b) illustrates *horizontal dispersion*, where $Bx_o \parallel [010]$. (c) illustrates *inclined dispersion*, where $\beta \parallel [010]$. In the examples shown $(2V)_{blue} = (2V)_{red}$ in each case. The isogyres for red and blue light are shown respectively with vertical and horizontal hachures; in white light the former appears blue and the latter red.

If the obtuse bisectrix is parallel to the monoclinic diad, the optic axial plane will again rotate about [010] as λ varies (Fig 12.42(b)). If the microscope axis is parallel to the acute bisectrix for some intermediate, say yellow, wavelength, then the extreme red and blue optic axial planes will be inclined to the microscope axis and the figure in the 90° position will appear as shown in Fig 12.42(b). This type is known as *horizontal dispersion*. The isogyre through the eyes is fringed with red in both upper quadrants and with blue in both lower quadrants or vice versa. The minerals sanidine and adularia, both $KAlSi_3O_8$, display horizontal dispersion.

If the third mean line, the β vibration direction, is parallel to the monoclinic diad the optic axial plane is confined to (010) for all wavelengths, but here again the acute bisectrix can only be parallel to the microscope axis for one wavelength (Fig 12.42(c)). Since the orientation of the OAP is symmetry controlled and invariant, the 90° figure yields no indication of dispersion, but in the 45° position the outside of one isogyre and the inside of the other are fringed with red and vice versa with blue. This type is known as *inclined dispersion* and is easily distinguished from dispersion of $2V$, where the outside of both isogyres is fringed with the same extreme colour, red or blue. Certain specimens of adularia, $KAlSi_3O_8$, display inclined dispersion.

In practice monoclinic substances may exhibit dispersion of $2V$ in addition to the characteristically monoclinic types of dispersion. It is a matter of some practical importance that, although most monoclinic substances do not display dispersion markedly, if in a particular case the evidence for a type of dispersion that cannot be displayed in a substance of higher than monoclinic symmetry is clear, then monoclinic or triclinic symmetry may confidently be presumed. One example of interest is the mineral staurolite, the structure of which was determined by X-ray diffraction on the basis of an orthorhombic unit-cell as early as 1929; but in 1956 optical study disclosed clear evidence of horizontal dispersion and simultaneously an X-ray diffraction investigation of twinned crystals revealed the presence of doubled reflexions that required the hypothesis of monoclinic symmetry with $\beta \simeq 90°$ for their interpretation.

In the triclinic system there are no symmetry controls of any kind on the indicatrix. Dispersion of highly complex nature is possible but is unlikely to be a rewarding subject of study.

Dispersion, if it is clearly displayed, may provide a means of distinguishing, in the biaxial category, between orthorhombic substances and those of lower symmetry. In practice this can rarely be done with confidence.

The first clue to the presence of anomalous dispersion is provided during preliminary examination between crossed polars in parallel white light by the observation of anomalous interference colours (especially browns and 'Berlin' blues) and of inability to extinguish completely.

Optical classification of transparent crystalline solids

Whether the material is in the form of single crystals mounted in oil, or grains produced by crushing and then mounted in oil, or a thin section, it should first be examined in parallel white light between crossed polars. If all the crystal grains appear isotropic on rotation of the stage, then the substance is either optically isotropic or, for one reason or another, all the grains are oriented so as to have an optic axis parallel to the microscope axis. If such grains yield no interference figure then the substance is optically isotropic, that is it is either non-crystalline or cubic.

If some crystals or grains appear isotropic while others exhibit interference colours, the interference figures produced by isotropic grains should be examined. This will yield a clear distinction between uniaxial and biaxial optic axis figures. If the substance is uniaxial, the sign of the figure may be determined by the use of one or other accessory test plates. If the substance turns out to be biaxial, a search of other grains should be made until one that yields a centred or nearly centred acute bisectrix figure is found. The sign can then be determined by use of one or other accessory test-plate.

Transparent crystalline solids can thus quite rapidly be assigned to the five categories: isotropic, uniaxial positive, uniaxial negative, biaxial positive, biaxial negative.

If the crystals or grains display any recognizable shape some comment on the crystal morphology and on the orientation of the indicatrix with respect to habit can be made at this stage. For instance it may be possible to describe the morphology of a uniaxial substance as tabular or prismatic and by noting the shape of isotropic sections to distinguish tetragonal from trigonal and hexagonal substances. If the substance is biaxial and prismatic it may be possible to demonstrate that one of the principal vibration directions is parallel to the prism axis; if that is so the substance is more likely to be orthorhombic or monoclinic than triclinic.

If the substance is biaxial, inspection of the acute bisectrix figure will yield a semi-quantitative estimate of the magnitude of the optic axial angle. If the isogyres remain in the field of view in the 45° position, then $2E < 110°$ and their separation will provide a semi-quantitative estimate of $2E$. If the isogyres disappear from the field before the 45° position is reached the angle relative to the 90° position at which they go out of the field is some guide to the magnitude of $2E$. If it is thought that $2V \sim 90°$, confirmation may be obtainable from the straightness of the single isogyre in the optic axis figure. In making such semi-quantitative estimates recourse should be had to oriented sections of substances of known $2V$.

If the approximate thickness of the crystal grains is known, observation of the closeness of the rings in a centred uniaxial figure enables the birefringence $|e-o|$ to be recognized as small, moderate, or large. If a uniaxial crystal can be recognized from its habit as lying with its optic axis parallel to the microscope stage and if its thickness is known approximately, then a more precise estimate of $|e-o|$ can be made from its interference colour between crossed polars.

If the approximate thickness of a crystal or grain of a biaxial substance lying with its acute bisectrix parallel to the microscope axis is known, then the partial birefringence, $\beta-\alpha$ if positive or $\gamma-\beta$ if negative, is determinable from the interference colour at the centre of the acute bisectrix figure.

Apart from such occasionally useful properties as dispersion or pleochroism we cannot take the optical characterization of a transparent crystalline solid further without actual measurement of one or more refractive indices, the subject of the next section.

Measurement of refractive index

For measurement of refractive index the material must be available either as small crystals or as grains produced by crushing. A portion of the specimen should be mounted in oil of some intermediate refractive index, say 1·600, unless there is evidence that the refractive indices are either very high or very low.

For isotropic solids there is only one refractive index to be measured and that is independent of vibration direction. For anisotropic solids only the principal refractive indices, o and e if uniaxial, α, β, and γ if biaxial, are in general worthy of precise measurement. The first task is therefore to identify the principal vibration directions.

If we are dealing with a mount of small crystals, most of the crystals will have the same crystallographic direction parallel to the microscope axis. If the substance is uniaxial with tabular habit all the crystals will lie on the basal plane, which will yield a centred optic axis figure, and only o will be easily measurable; if it is prismatic e and o will both be measurable and the optic axis should be recognizable from the habit; if the dominant form is $\{hkl\}$, o will always be measurable and it may be possible, in particular cases, to measure e' in an identifiable direction. If the substance is biaxial, a preparation of small crystals is less likely to provide orientations favourable for the determination of all three principal refractive indices than a preparation of crushed grains. A crystal that lies with its acute bisectrix parallel to the microscope axis will be suitable for the measurement of β and either α or γ (depending on whether it is positive or negative); if two refractive indices are known the third can be estimated by measurement of $2V$. Obtuse bisectrix and optic normal figures are not always reliably identifiable.

In a preparation of grains produced by crushing, the situation will be substantially the same as for a preparation of small crystals if the substance has a single perfect cleavage. Otherwise a search should be made for grains that lie in recognizable orientations; few may be found, but they will be invaluable.

The selected grain or grains should be worked towards the edge of the pool of oil with a fine needle mounted in a holder. Suitable steel needle holders are commercially available from biological equipment suppliers. But rather better, because it is lighter, is an artist's fine paint brush from which the camel hairs have been removed by a solvent; the needle can then be pushed down into the soft wood and glued into place with a strong adhesive. The selected grains should be pushed out beyond the edge of the pool of oil so that they lie together in an easily recognizable salient of oil. Place a drop of amyl acetate with a dropper close to the salient of oil. The oil will be repelled and the grains can be worked into the amyl acetate further from the residue of the pool of oil. When the amyl acetate surrounding the selected grains has dried, after five minutes or less, they can be transferred to another slide by being picked up on the point of the needle and immersed in a drop of oil of approximately known refractive index.

If the selected grain lies with a circular section of its indicatrix parallel to the microscope stage, that is if it appears isotropic, then it is immaterial in what direction the light passing through the grain is vibrating. If however it is anisotropic, it is important to isolate a single vibration direction for refractive index measurement. If we are dealing with a uniaxial crystal lying on a prism face, the crystal should be rotated into an extinction position between crossed polars, the analyser should be put out, and the Becke line test applied to yield information about the relative magnitude of either e or o with respect to the approximately known refractive index of the immersion medium. The grain should then be rotated to the other extinction position, with polars crossed; the analyser should be put out and the Becke line test applied again to give information about the other principal refractive index. If we are dealing with a biaxial crystal lying with its acute bisectrix parallel to the microscope axis, the two principal vibration directions should be identified by observation of the interference figure, and the Becke line test applied when each in turn lies parallel to the polarizer vibration direction. There is obviously no point in determining refractive indices for grains not in extinction positions.

The next step is to place a drop of amyl acetate close to the drop of refractive index oil, work the crystals into the amyl acetate, remove the residual oil with the edge of a piece of filter paper, wait until the amyl acetate has evaporated completely, and then immerse the crystals in a drop of oil of different refractive index. One proceeds in this manner until the Becke line test enables one of the principal refractive indices to be sandwiched between two oils of approximately known refractive index. Sets of standard oils are usually made up in refractive index intervals of 0·005.

When the refractive index of the crystalline solid is close to that of the immersion medium differential dispersion is usually evident. Refractive index changes more rapidly with wavelength for liquids than for crystalline solids in general, and when the refractive indices are close a blue and a red Becke line can usually be observed to move in opposite directions on racking the microscope barrel assembly up. This is the stage at which the accuracy of the comparison of refractive indices can only be improved by changing to a monochromatic light source, that is to a sodium lamp. A closer match can be achieved by mixing adjacent oils in the standard set in various proportions. When the match between one principal refractive index of the crystal and the refractive index of the immersion medium, as revealed by the total absence of relief, is as good as can be achieved, the oil should be transferred to a refractometer and its refractive index measured. Since the null point is often difficult to recognize, oils which are just higher and just lower in refractive index than the selected crystal vibration direction should also be measured on the refractometer. In this way refractive indices are easily measurable to 0·005 or better; there is, except in special cases, little point in attempting greater accuracy than $\pm 0{\cdot}001$.

All that remains is to measure the refractive index of the immersion medium and this is a simple task. For approximate work the *Leitz–Jelley Refractometer* is convenient, cheap, and requires only a small drop of liquid; it has the advantage of being usable in the high refractive index range up to 1·90. The instrument consists in essence of a slit B in a vertical steel plate and a slit C in another vertical steel plate at the same height as, and at a fixed horizontal distance from the first; a glass prism, precisely ground to a definite angle and cemented to a glass slide, is clipped to the second vertical plate behind the slit C (Fig 12.43). Light from the source A passing normally through the slide/liquid interface will be refracted upwards or downwards

Fig 12.43 The Leitz–Jelley refractometer. The instrument is shown schematically on the left. (a) and (b) show ray directions for a liquid of higher and lower refractive index respectively relative to the glass of the refractometer prism.

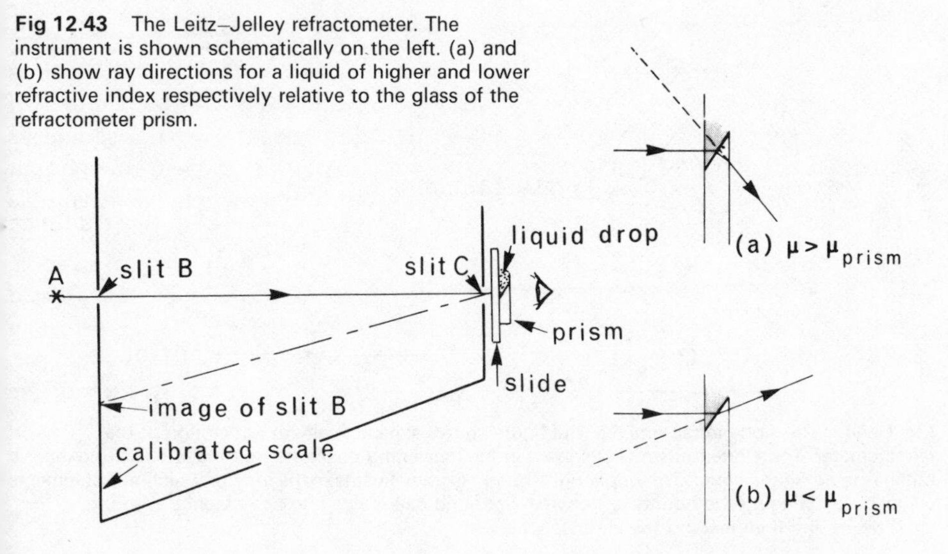

at the liquid/prism interface and the image of the slit B will be seen superimposed on a calibrated scale fixed on the first vertical plate. If a monochromatic light source is employed, usually a sodium lamp, the image of the slit B on the scale will provide a measure of the refractive index of the liquid in the angle of the prism. If white light is used the image of the slit B will be drawn out into a spectrum, but insertion of a filter that is strongly absorbing for Na_D in front of the slit B produces a black band in the image corresponding to the refractive index of the liquid for $\lambda(Na_D)$. The scale should be calibrated from time to time against pure liquids of known refractive index.

More commonly used and more accurate is the *Abbe Refractometer* in its various modifications. Light is reflected from a mirror on to the lower prism Q (Fig 12.44), the upper surface of which is roughly ground so as to act as a diffuse source of light entering the liquid film. Both prisms P and Q are of high refractive index glass. Light will then impinge on the interface between the liquid film and the upper prism P at all angles of incidence. Since the liquid must be of lower refractive index than the glass of the prism P for the instrument to be effective, light incident on the interface at grazing incidence will represent the limit of illumination of the exit face of the prism P, which is observed through the telescope T mounted on an arm attached to a calibrated scale. The telescope T is rotated until the intersection of its cross-wires coincides with the light/dark boundary in the image of the exit face of prism P. The light source must be monochromatic and is usually a sodium lamp. The use of the Abbe refractometer is limited by the condition that the refractive index of the prism P must be greater than that of the liquid. The Abbe moreover requires a larger volume of liquid to produce a satisfactory film than the Leitz–Jelley.

It is appropriate at this stage to say something about immersion liquids suitable for refractive index determination. Criteria for the selection of liquids are that they should be miscible in pairs, relatively involatile, low in dispersion and colourless, but this last has to be sacrificed at high refractive index. Liquids that can conveniently be used for making up a standard set are medicinal paraffin (the proprietary Nujol is

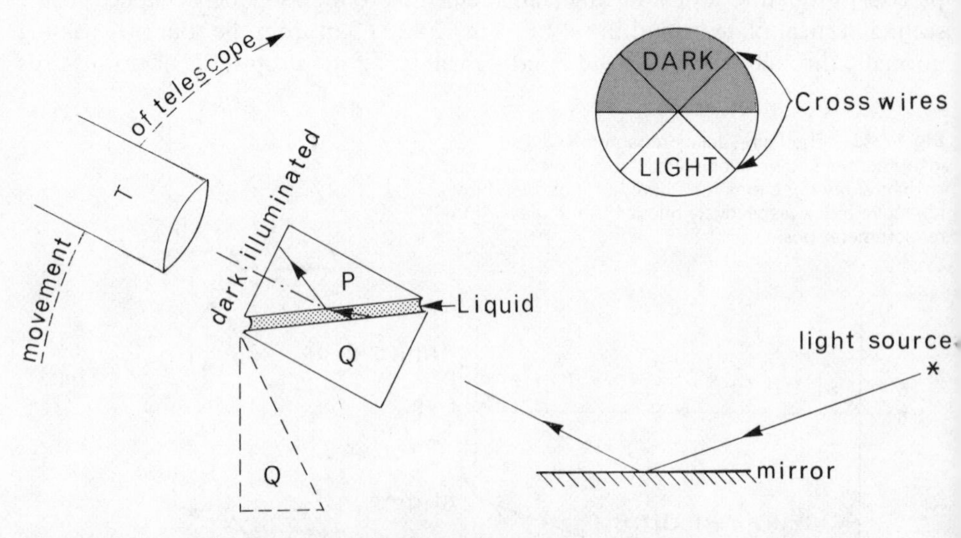

Fig 12.44 The Abbe refractometer. The figure shows schematically the operation of the refractometer. The hinged prism Q is shown in the measuring position and hinged back (broken outline) to admit the liquid. The upper right-hand diagram illustrates the way in which measurements are made by bringing the boundary between light and dark fields into coincidence with the cross-wires and then reading the angular scale.

commonly used) 1·470, α-monobromnapthalene 1·659, and methylene iodide 1·740. Mixtures of Nujol and α-monobromnapthalene and of α-monobromnapthalene and methylene iodide can simply be prepared at refractive index intervals of 0·005 by running appropriate volumes of the pure liquids from burettes into small green glass dropper bottles, mixing thoroughly, checking the refractive index of the mixture on the Leitz–Jelley refractometer, and adjusting the refractive index to the desired value by running in a few more drops of either the lower or higher refractive index component. Mixtures containing methylene iodide should be kept in bottles made of dark brown glass or painted black and stored away from direct sunlight to minimize the photochemical release of iodine. Such a set is suitable for general use in the mineralogical laboratory.

Immersion liquids below the refractive index of Nujol can conveniently be prepared from glycerine (1·47) and water (1·333) mixtures. These should not be kept for more than a few weeks because glycerine takes up water from the atmosphere quite rapidly.

Immersion liquids of refractive index greater than 1·74 (methylene iodide) present something of a problem. All the liquids available are either strongly coloured or hydrolyse rapidly in a moist atmosphere or both. Solutions of sulphur in methylene iodide can be used to extend the range up to 1·78 and they are succeeded by mixtures of methylene iodide and phenyl-diiodoarsine (1·84). Solutions of sulphur and phosphorus in methylene iodide extend to refractive indices as high as 2·06. Various other solutions and low melting point salts have been used as high refractive index immersion media.

It is occasionally possible, when euhedral crystals of a uniaxial or orthorhombic substance are concerned, to measure the principal refractive indices without intervention of a liquid by making use of the *minimum deviation* property of prisms. The minimum deviation experiment is fully described for isotropic substances in elementary textbooks of optics; we shall concern ourselves here only with its extension to anisotropic prisms. If a prismatic crystal of quartz (trigonal) with well-developed faces $\{10\bar{1}0\}$ is set up on a horizontal circle goniometer with its triad axis parallel to the axis of rotation of the stage, two minimum deviation rays will be produced by a parallel beam of monochromatic light incident on the $(1\bar{1}00)$ face and leaving through the $(01\bar{1}0)$ face (Fig 12.45). In this case the vibration direction of one ray will be parallel to the optic axis and that of the other will be parallel to (0001) so that by accurate

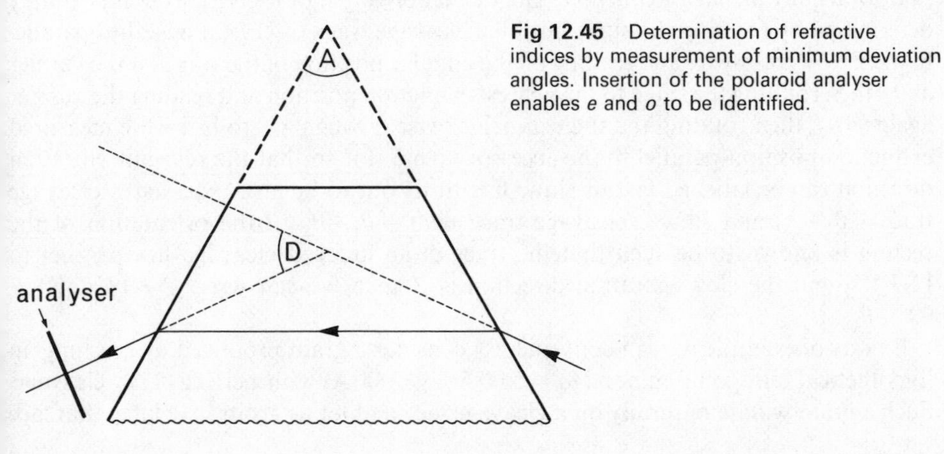

Fig 12.45 Determination of refractive indices by measurement of minimum deviation angles. Insertion of the polaroid analyser enables *e* and *o* to be identified.

measurement of the angles of minimum deviation the refractive indices e and o can be determined. Insertion of a piece of polaroid of known transmission direction in the path of the emergent ray enables its vibration direction to be related to the morphology of the crystal so that e and o can be identified. The angle of minimum deviation D is related to refractive index by the easily derived expression

$$\mu = \frac{\sin \frac{1}{2}(A+D)}{\sin (\frac{1}{2}A)},$$

where A is the prism angle (in this case 60°). It is immediately obvious that the size of prism angle which can yield a minimum deviation position is limited by the condition $\mu \sin (\frac{1}{2}A) < 1$; for refractive indices about 1·5, the largest permissible value of A is about 84°. The method can be extended to determine α, β, and γ for orthorhombic crystals displaying forms such as $\{hk0\}$ and $\{0kl\}$; a pair of faces of the form $\{hk0\}$ would yield either the refractive indices μ_x and μ_z with the minimum deviation ray travelling in the crystal parallel to y or μ_y and μ_z with the ray parallel to x.

Extinction angles

The angular relationship between the extinction positions and any identifiable crystallographic direction in the field of view may yield useful information about either the indices of the plane on which the crystal grain is lying, or, if the attitude of the crystal grain is known, about the orientation or sign of the indicatrix. Furthermore the measurement of extinction angles on sections of partially known orientation may yield diagnostic information of value to the mineralogist.

Basal sections of uniaxial crystals, that is sections lying on a plane perpendicular to the axis of high symmetry, are of course isotropic and no question of extinction position arises. Prismatic crystals or sections, that is sections lying on $(hk0)$, of uniaxial minerals will extinguish when their length, which is parallel to the optic axis, lies parallel or perpendicular to the polarizer vibration direction; such sections are said to have *straight extinction* with respect to their prism axis. A crystal grain or section of a uniaxial crystal lying on some general plane (hkl) will in general have its vibration directions inclined to the direction of intersection of the (hkl) plane with the boundary planes of the crystal (if euhedral) or with cleavage planes: the angle between the slow (or fast) extinction position and the crystallographic direction (crystal edge or cleavage trace) is known as the *extinction angle.* Sections having non-zero extinction angles are said to display *inclined extinction.* The extinction angle of a given section is simply determined by rotating the stage until the cleavage trace or crystal edge lies parallel to the E—W cross-wire and reading off the angular position of the stage on the vernier as θ_1; then rotating the stage to the nearest extinction position and reading the vernier again as θ_2; then rotating the stage anticlockwise through 45° to bring the measured extinction position parallel to the accessory plate slot so that the relevant vibration direction can be labelled fast or slow. If it turns out to be fast, then fast $\wedge$ cleavage trace $= \theta_1 - \theta_2$ and slow $\wedge$ cleavage trace $= 90° - \theta_1 + \theta_2$. If the orientation of the section is known to be such that the trace of an indexed cleavage lies parallel to $[UVW]$ and the slow vibration direction is e' then we can write $e' \wedge [UVW] = 90° - \theta_1 + \theta_2$.

By way of example we can conveniently consider a grain produced by crushing an hypothetical tetragonal mineral ($a = 5{\cdot}00$ Å, $c = 8{\cdot}00$ Å) with perfect $\{111\}$ cleavage. Such a grain will lie naturally on a cleavage face and let us arbitrarily label that face

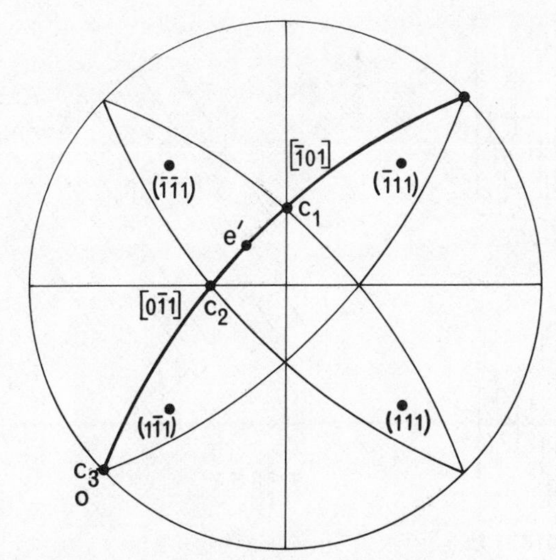

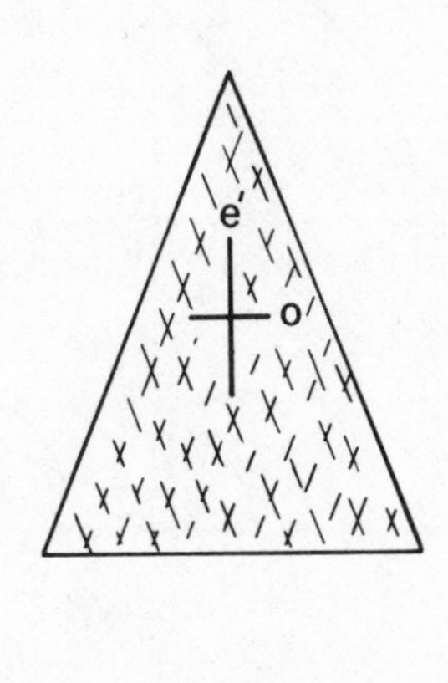

Fig 12.46 Extinction angle in a (111) cleavage flake of a tetragonal mineral. The stereogram shows the location of the cleavage traces c_1, c_2, c_3 at the intersection of the great circles $(1\bar{1}1)$, $(\bar{1}11)$ and $(\bar{1}\bar{1}1)$ with the great circle (111). The vibration directions *e′* and *o* are located by the Biot–Fresnel construction. The drawing on the right shows the appearance of the cleavage flake with *e′* bisecting the *acute* angle between the c_1 $[\bar{1}01]$ and c_2 $[0\bar{1}1]$ cleavage traces.

(111). The extraordinary and ordinary vibration directions will then be the zone axes $e'[\bar{1}\bar{1}2]$ and $o[1\bar{1}0]$ (Fig 12.46). The remaining cleavage planes $(1\bar{1}1)$, $(\bar{1}11)$, and $(\bar{1}\bar{1}1)$ will intersect the plane on which the cleavage fragment lies in the zone axes $C_1[\bar{1}01]$, $C_2[0\bar{1}1]$, and $C_3[1\bar{1}0]$.

Now $\qquad (111) \wedge (1\bar{1}1) = (111) \wedge (\bar{1}11) = 2 \tan^{-1} \sin \tan^{-1} \dfrac{c}{a} = 81°$

and $\qquad (111) \wedge (\bar{1}\bar{1}1) = 2 \tan^{-1} \dfrac{\sqrt{2}c}{a} = 132°.$

It will be recalled that the empirical criterion for cleavage planes to be evident as cleavage traces is that they must be within 20° of being parallel to the microscope axis; in this example the microscope axis is $\perp$ (111) so that for a cleavage plane (*hkl*) to be visible $70° < (111) \wedge (hkl) < 110°$. Therefore only the $(1\bar{1}1)$ and $(\bar{1}11)$ cleavage traces, C_1 and C_2 in the figure, will be visible. The cleavage fragment will display *symmetrical extinction* with equal extinction angles of 22° between the *e′* vibration direction and the traces of the $(1\bar{1}1)$ and $(\bar{1}11)$ cleavages. If the crystal is optically positive, then the extinction angle for the (111) section is described as symmetrical, *slow* $\wedge$ {111} cleavage traces = 22°.

In a thin-section of an aggregate of our exemplary tetragonal mineral most grains will have some general direction parallel to the microscope axis. Extinction in a few grains will be symmetrical or straight with respect to the cleavage traces and there will be no traces of bounding *planes*, only irregular edges, to most grains; in general inclined extinction will be observed and extinction angles will have no practical significance.

Passing now to biaxial crystals and considering first the orthorhombic system, let us take as the first example a substance with the properties $\alpha \parallel y$, $\gamma \parallel x$, $\beta \parallel z$, $2V_\gamma = 60°$,

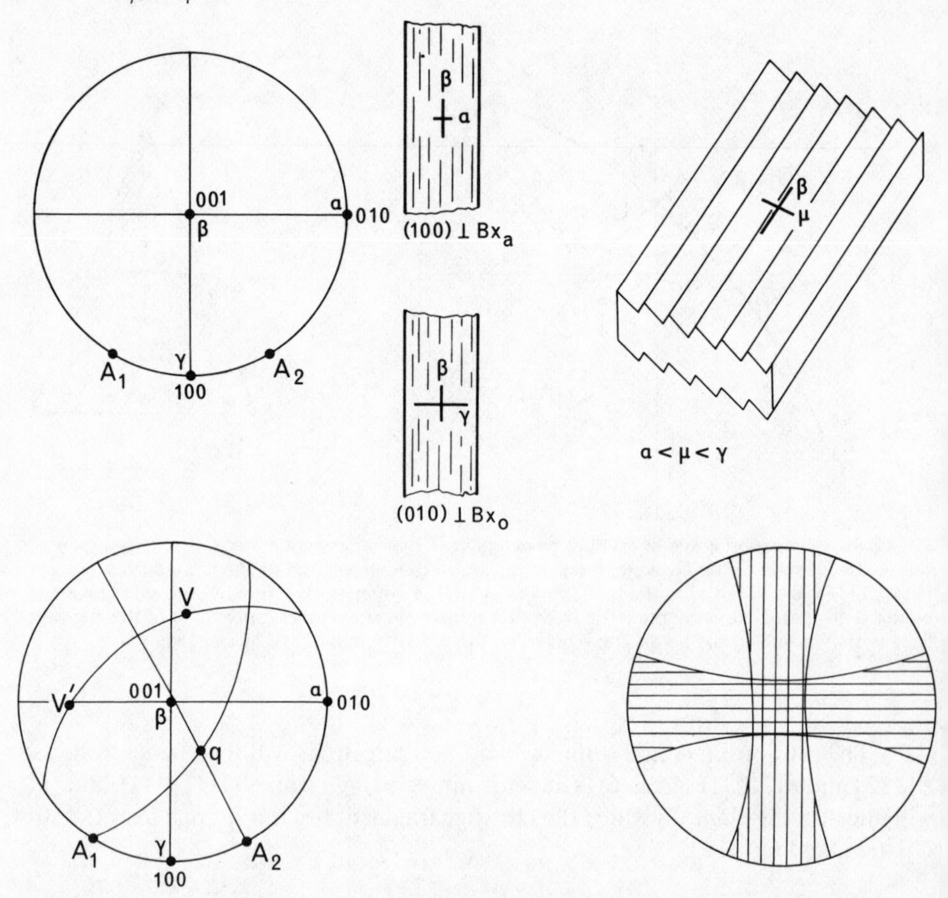

Fig 12.47 Extinction in an orthorhombic mineral with perfect {100} and {001} cleavages. The upper stereogram shows the orientation of the indicatrix. The lower left-hand stereogram shows the vibration directions, *V* and *V′*, for a general direction of propagation *q*. The observation of straight extinction in (100) and (010) cleavage flakes is illustrated; a stepped cleavage flake also displays straight extinction. The lower right-hand stereogram shows the orientation of sections which display two, one, or no cleavage traces; the field of the poles of sections displaying (010) cleavage is shown by vertical hachures and of sections displaying (100) cleavage traces by horizontal hachures.

cleavage perfect parallel to {100} and {010}, crystals prismatic about [001]. A preparation of grains produced by crushing may be expected to contain a majority of grains with straight extinction and some with inclined extinction (Fig 12.47). The crystal grains will show a strong tendency to cleave along the perfect cleavage planes and to lie on those planes. Those grains lying on (100) will have straight extinction relative to the (010) cleavage trace which will be parallel to the length of the grain, the crystallographic *z*-axis; therefore for such grains $slow \wedge z = 0°$. Those grains lying on (010) will likewise have straight extinction, but here relative to the (100) cleavage trace which will lie parallel to the *z*-axis; therefore for such grains $fast \wedge z = 0°$. Some grains will have failed to cleave cleanly and will lie on irregular surfaces composed of (100) and (010) steps; such grains will display straight extinction relative to their probably indistinct cleavage traces and may be length-slow or length-fast depending on the relative development of the (100) and (010) steps. Some grains will lie in completely general orientations.

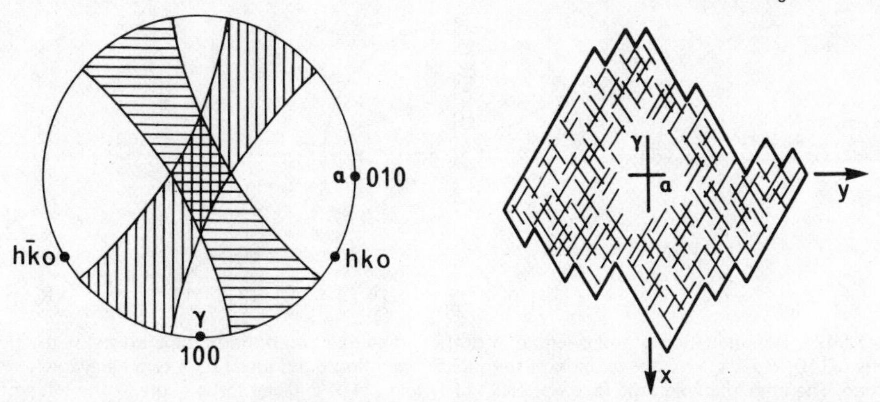

Fig 12.48 Extinction in an orthorhombic mineral with perfect {*hk*0} cleavage. The stereogram shows the orientation of sections displaying two, one, or no cleavage traces; the field of the poles of sections exhibiting (*hk*0) cleavage is indicated by vertical hachures and that of sections exhibiting ($h\bar{k}0$) cleavage by horizontal hachures. The sketch illustrates symmetrical extinction in an (001) section.

A thin-section of a crystal aggregate of such a substance may be expected to contain a majority of grains in general orientations. A stereogram displaying the Biot–Fresnel construction for such a general direction is shown in Fig 12.47. Vibration directions will be V (slow) and V' (fast). Only the (010) cleavage will be apparent as a cleavage trace on the section shown. The extinction angle will be $\theta = \text{slow} \wedge (010)$ cleavage trace. It is apparent from the figure that many general sections will display no cleavage traces.

An orthorhombic substance with prismatic cleavage {*hk*0} will show two cleavage traces on sections within about 20° of (001) and extinction will be approximately symmetrical with respect to them (Fig 12.48). If the form of the cleavage and the unit-cell dimensions are known, observation of whether the vibration direction in the acute angle between the cleavage traces is slow or fast relative to that bisecting the obtuse angle coupled with identification, from the interference figure on a two cleavage section, of the [001] direction as Bx_a, Bx_o, or β enables the orientation of the indicatrix to be completely determined. For example, consider an orthorhombic substance with {110} cleavage such that $(110) \wedge (1\bar{1}0) = 60°$ and suppose that sections displaying the cleavage traces have the slow vibration direction parallel to the bisector of the acute angle between the cleavage traces, that is parallel to [010]. Therefore, either γ or β is $\parallel$ [010]. Observation of the interference figure produced by sections showing two cleavage traces (some searching of the thin-section may be necessary before a more or less centred figure is found) will then define the orientation of the indicatrix completely (Fig 12.49).

In the monoclinic system symmetry controls on the indicatrix are reduced and consequently straight extinction relative to cleavage traces or crystal edges will be restricted to sections in a limited range of special orientations. Let us consider a monoclinic substance with well developed prismatic cleavage {*hk*0} in various selected orientations of the indicatrix. Suppose that $2V_\gamma = 60°$, that the optic axial plane is (010) and that γ lies in the obtuse β angle (i.e., in the angle between the positive directions of the x and z axes). Cleavage traces will be visible as indicated on Fig 12.50(a). Sections whose normals are within about 20° of [001] may display symmetrical extinction with respect to two cleavage traces. It is practically useful to consider the variation of extinction angle ϕ with orientation of sections in the [001]

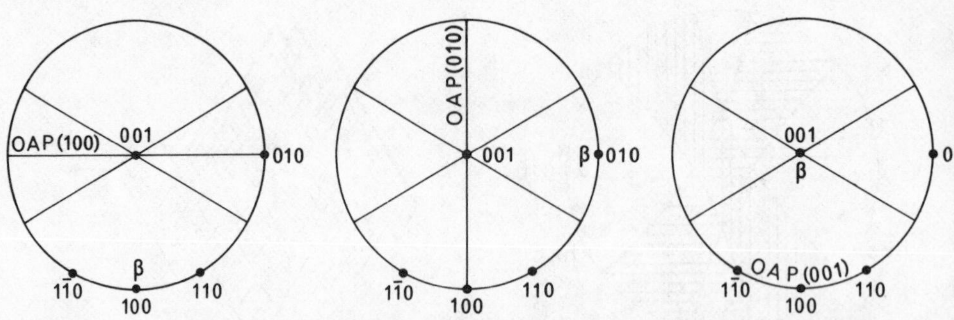

Fig 12.49 Determination of the orientation of the indicatrix of an orthorhombic crystal with perfect {110} cleavage by observation of the interference figure produced by a two cleavage section. The angle between the face normals (110) and (1$\bar{1}$0) is taken to be acute. If the trace of the optic axial plane is found to bisect the acute angle between the cleavage traces, then $\beta \parallel [100]$; if it bisects the obtuse angle, $\beta \parallel [010]$; and if a flash figure is observed $\beta \parallel [001]$. It only remains to determine whether the bisector of the acute angle between the cleavages, the *y* direction, is fast or slow.

trace of OAP	*y*	*x y z*
(100)	fast	$\beta\ \alpha\ \gamma$
	slow	$\beta\ \gamma\ \alpha$
(010)	fast	$\gamma\ \beta\ \alpha$
	slow	$\alpha\ \beta\ \gamma$
(001)	fast	$\gamma\ \alpha\ \beta$
	slow	$\alpha\ \gamma\ \beta$

zone, that is for sections between (100) and (010), only some of which will display a single direction of cleavage trace. The variation of ϕ with orientation of the section is shown in Fig 12.50(b) for two values of $\gamma \wedge c$. The calculated extinction angle ϕ varies from 0° for the (100) section to its maximum value, equal to $\gamma \wedge c$, for the (010) section. The important point is the slowness with which ϕ falls off from its maximum value as θ decreases from 90°. So that the *maximum extinction angle* relative to the cleavage trace observable on examination of a large number (say, 20) of random sections showing a single cleavage trace will be close to $\Phi = \gamma \wedge c$ unless the cleavage (*hk*0) is close to (010).

In the other type of orientation of the indicatrix in the monoclinic system, with the optic axial plane parallel to [010], the extinction angle relative to traces of a prismatic cleavage {*hk*0} in (*hk*0) sections varies rather differently with orientation (Figs 12.50(c), (d)). Here the fall off from the calculated maximum extinction angle, equal to $\gamma \wedge c$ in the example, is more rapid as the plane of the section moves from (010) towards (100) and the maximum extinction angle observable on sections showing one cleavage trace may not provide even an approximate measure of $\gamma \wedge c$. Here again of course sections nearly perpendicular to [001] will display two cleavage traces with symmetrical extinction.

It is appropriate at this stage to make the point that prismatic orthorhombic crystals will display straight extinction relative to the zone axis of the prism in mounts of euhedral crystals and small or zero extinction angles relative to the length of grains in thin-section. Orthorhombic substances with prismatic cleavage will tend to display zero or small extinction angles relative to the cleavage in sections showing a single direction of cleavage trace. On the other hand prismatic monoclinic crystals will in general display inclined extinction relative to the prism axis (unless the prism axis is [010] as is rarely the case) and the maximum extinction angle observable will represent the angle between one of the principal vibration directions and the prism axis. In

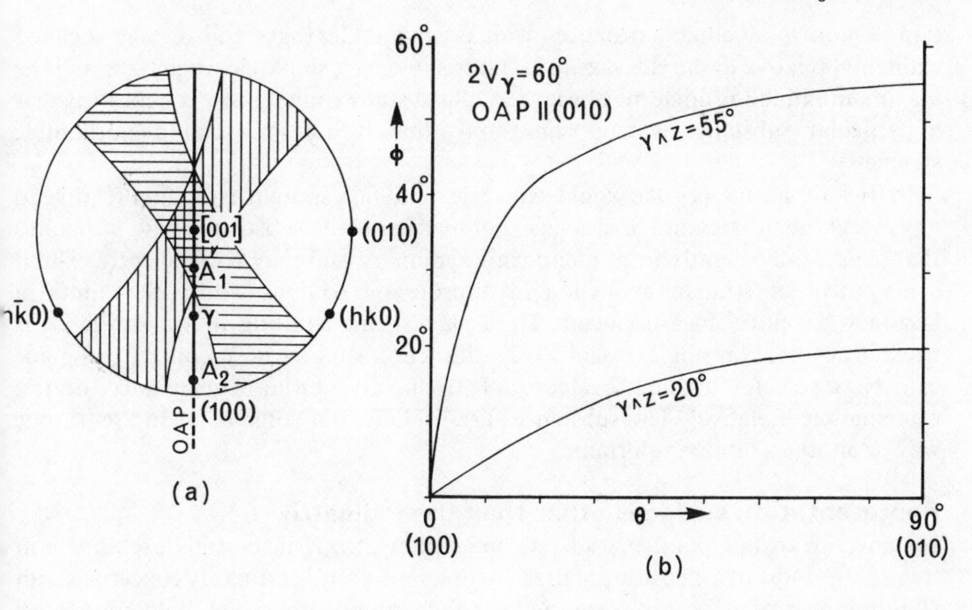

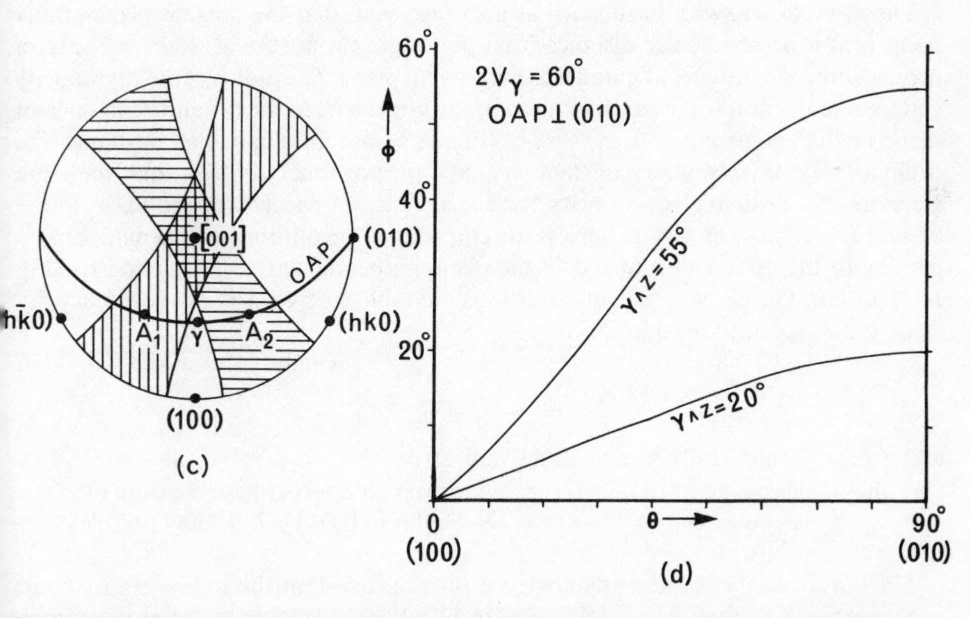

Fig 12.50 Extinction angles in a monoclinic crystal with perfect $\{hk0\}$ cleavage. The stereograms (a) and (c) show the ranges of orientation of the poles of sections displaying two (cross-hatching), chures), or no (blank) cleavages when $(hk0):(100) = 60°$. The stereograms also show the optical orientation for $\gamma \wedge z = 55°$, $2V_\gamma = 60°$ when the optic axial plane is parallel to (010) in (a) and perpendicular to (010) in (c). The graphs show the variation of extinction angle $\phi = \text{slow} \wedge z$ with $\theta = (100):(h'k'0)$ for $(h'k'0)$ sections; (b) illustrates the case where the optic axial plane is parallel to (010) as in (a); and (d) illustrates the case where the optic axial plane is perpendicular to (010) as in (c). On each of the graphs two curves are shown, one for $\gamma \wedge z = 55°$ as in the accompanying stereogram and one for $\gamma \wedge z = 20°$. In the examples illustrated in (a) and (c), where $(hk0):(100) = 60°$, cleavage will only be apparent in $(h'k'0)$ sections with $10° < \theta < 50°$.

thin section monoclinic substances with prismatic cleavages will display inclined extinction relative to the cleavage in sections showing a single cleavage trace and the maximum extinction angle may be large. Optical study cannot show conclusively that a particular substance is monoclinic, only that it has less than orthorhombic symmetry.

In triclinic substances one would expect never to find straight extinction relative to any morphological feature or cleavage, but many triclinic substances have structures that depart only slightly from monoclinic symmetry and may consequently exhibit symmetrical or straight extinction in appropriate sections within the limits of accuracy of optical measurements. The task of demonstrating that a substance is triclinic devolves on single crystal X-ray diffraction studies; no useful general guide can be given for finding evidence of triclinicity during preliminary optical examination. Relatively few substances are triclinic, but some of them are, in one way or another, rather important.

Representation surfaces other than the indicatrix

We have up to this point considered optical anisotropy in crystals exclusively in terms of the indicatrix because polarized light microscopy is primarily concerned with vibration directions. But in other contexts the directions in which light rays travel through the anisotropic medium are of interest and so for the sake of completeness a brief discussion of two other representation surfaces,[8] the ray velocity surfaces and the wave velocity surfaces, is presented here.

The *ray velocity surface* is defined as a surface such that the tangent plane at any point is at a perpendicular distance from the origin proportional to the velocity of propagation of a wave front parallel to the tangent plane. The radius of the ray velocity surface at any point is the *ray direction* and the normal from the origin to the tangent plane at that point is the *wave normal* (Fig 12.51(a)). In a uniaxial substance the ordinary ray travels with constant velocity proportional to o^{-1} and therefore generates an ordinary ray velocity surface that is spherical with radius o^{-1}; the extraordinary ray velocity surface is an ellipsoid of revolution with semi-axes o^{-1} parallel to the optic axis and e^{-1} in the plane perpendicular to the optic axis (Fig 12.51(b), (c)). The ray velocity surface is thus a double surface. The extraordinary ray velocity surface has the equation

$$\frac{x^2}{e^{-2}}+\frac{y^2}{o^{-2}}=1, \quad \text{i.e.} \quad \frac{x^2}{o^2}+\frac{y^2}{e^2}=\frac{1}{e^2o^2},$$

and is thus geometrically similar to the indicatrix

$$\frac{x^2}{o^2}+\frac{y^2}{e^2}=1.$$

The *wave velocity surface* is defined as a surface such that the radius at any point is proportional to the velocity of a wave front whose normal is radial at that point. In Fig 12.51(a) W is a point on the wave velocity surface and R lies on the ray velocity surface. The wave velocity surface is likewise a double surface. The ray and wave velocity surfaces may be seen to be distinct for the simple geometrical reason that the tangent and radius to an ellipse at any point are not in general perpendicular.

The distinction between ray and wave normal directions is made clear by

[8] These are not *representation quadrics* in the sense of chapter 11. In spite of the possibility of confusion the term 'representation surface' is useful.

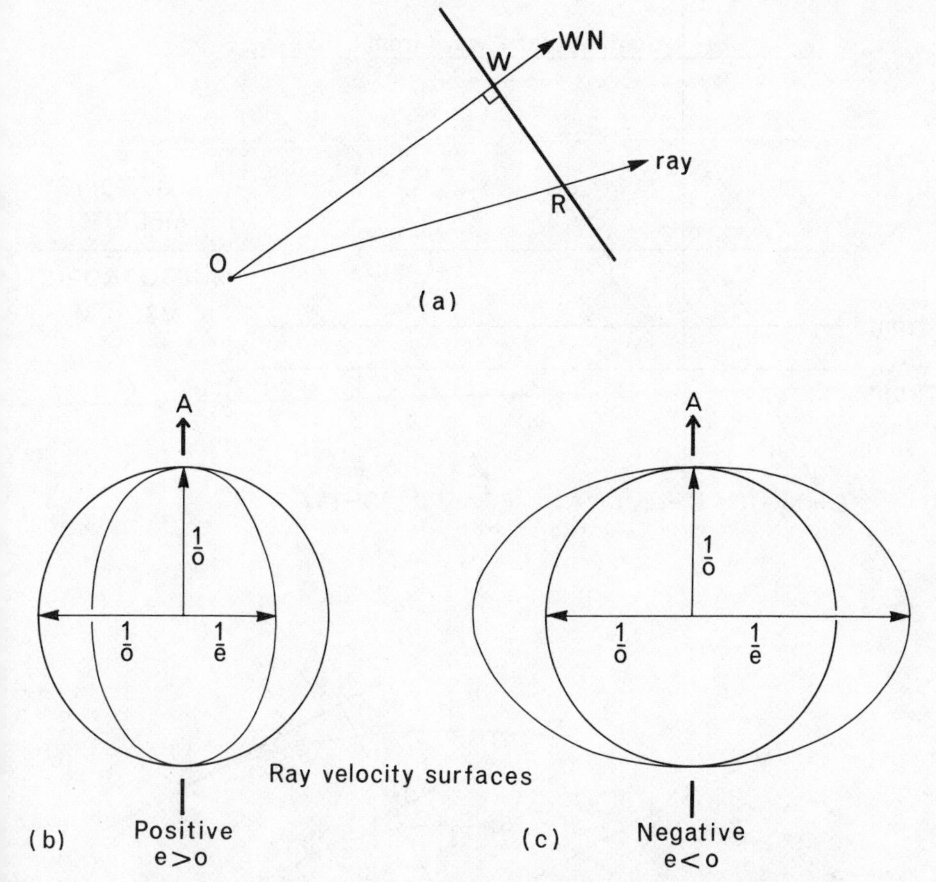

Fig 12.51 Ray velocity surfaces. (a) shows a radius OR of the ray velocity surface and the normal OW from the origin O to the tangent plane at R; the ray velocity surface is constructed so that OW is proportional to the rate of advance of a wave whose wave normal is OW. (b) and (c) show sections of the ray velocity surfaces for positive and negative uniaxial crystals.

considering a plane wave front incident normally on the interface between an isotropic and an anisotropic (uniaxial) medium such that the optic axis of the latter is inclined to the interface. Application of Huyghens' Principle (Fig 12.52(a)) leads to the following conclusions: (i) the ordinary and extraordinary rays diverge, (ii) the ordinary ray passes through the interface undeviated, in accordance with Snell's Law, for the case of normal incidence, (iii) the extraordinary ray is deviated in the case of normal incidence, (iv) the extraordinary ray is not generally perpendicular to its wave front. The passage of the extraordinary wave motion through the anisotropic medium can therefore be described in terms of the rate of advance of the wave front either in the ray direction (ray velocity) or in the wave normal direction (wave velocity). The former gives rise to the ellipsoidal ray velocity surface; the latter to the 'ovoidal' wave velocity surface.

The relationship between the three representation surfaces can be seen by considering a central section of a uniaxial indicatrix, an ellipse with semi-axes o and e (Fig 12.52(b)). If OV represents an extraordinary vibration direction for light propagated in the plane of the section, the corresponding wave front will lie in the plane defined by OV and the normal to the section, so that the wave normal ONN′

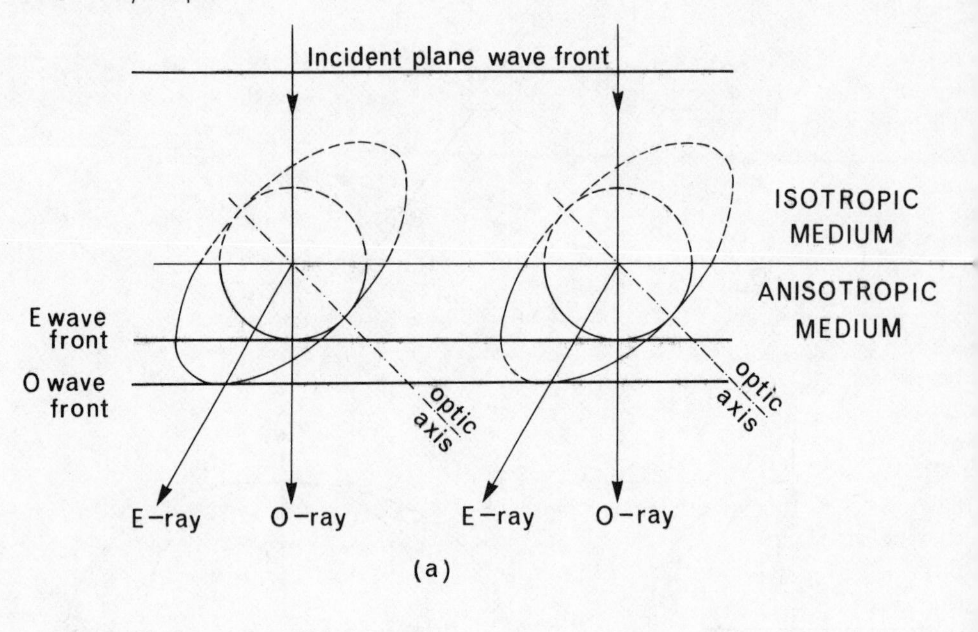

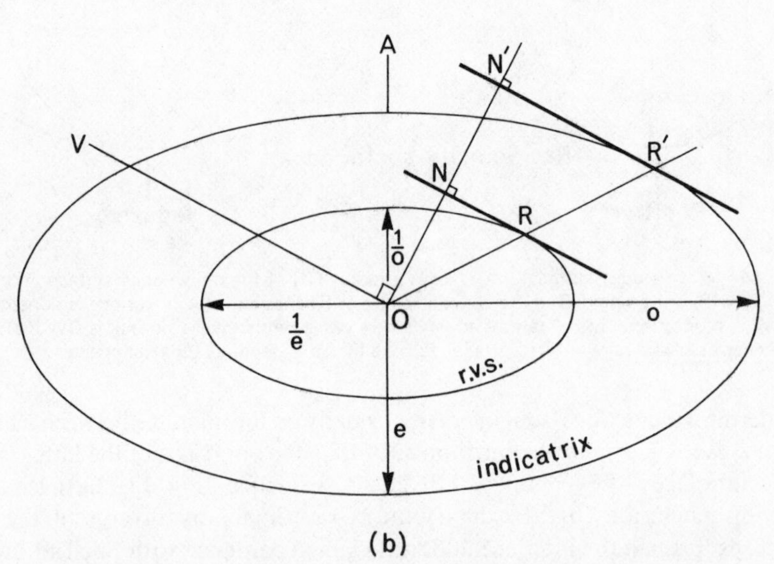

Fig 12.52 Wave velocity surfaces. (a) illustrates the application of Huyghens' principle to a uniaxial crystal. The *e'* vibration direction lies in the plane of the diagram parallel to the E wave front; the *o* vibration direction is perpendicular to the plane of the diagram. (b) illustrates the relationship of the ray velocity surface for the E-ray (the locus of points such as R), the wave velocity surface (the locus of points such as N) and the indicatrix in a grossly exaggerated uniaxial case.

lies in the plane of the section at right-angles to OV. The corresponding ray direction ORR′ will be such that the tangent to the indicatrix at R′, which is parallel to the tangent to the ray velocity surface at R, is perpendicular to ONN′. It follows that in general the extraordinary ray direction is a line joining the origin to the intersection of the appropriate tangent plane to the indicatrix with the plane containing the vibration direction and the wave normal. The ordinary ray direction will lie parallel to the wave normal direction ONN′. Since birefringence is quite a small fraction of the mean refractive index for most crystalline solids, the divergence of ray and wave normal will not be great under the conditions of polarized light microscopy. The indicatrices that appear in the figures of this chapter are all grossly exaggerated for emphasis. The term 'direction of propagation' which has been used throughout refers strictly to the wave normal direction.

It is not worth while here to explore the velocity surfaces for biaxial crystals.

Polarizing devices

The classical polarizing device is the *Nicol prism*, the understanding of which provides an instructive exercise in the application of the ray velocity surface. In practice, the Nicol prism has been almost entirely superseded by *Polaroid film*, which depends on extreme anisotropy of absorption for its polarizing quality. These two devices will be described in turn.

The Nicol prism is constructed from a large cleavage rhombohedron of clear, flawless calcite (the variety *Iceland Spar*) about three times as long as it is broad. The triad axis of symmetry (the optic axis) is symmetrically disposed with respect to the three faces whose angles are all obtuse at the same corner of the cleavage rhombohedron. The two parallel end faces of the rhombohedron are then ground so that the plane angle of one pair of side faces is reduced from its natural value of 70°53′ to 68° (Fig 12.53(a)); this brings the end faces perpendicular to the diagonal plane along which the prism is cut. All three cut surfaces are polished and the two halves are cemented together with a thin film of Canada Balsam.

When unpolarized light, parallel or nearly parallel, to the length of the Nicol prism, is incident on one of the end faces, refraction occurs at the air/calcite interface and

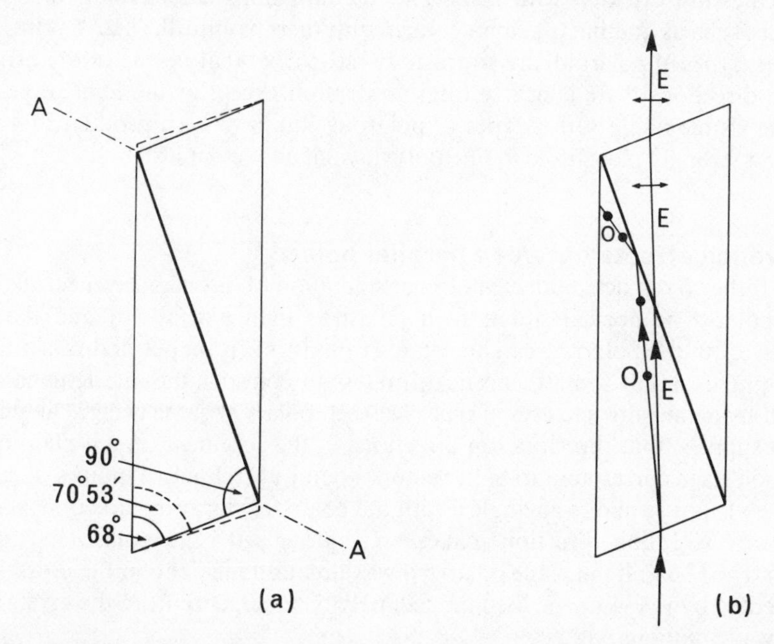

Fig 12.53 The Nicol prism. (a) shows the way in which a cleavage rhombohedron of calcite is cut and its end surfaces ground. (b) shows the paths of the O and E rays through the Nicol prism; the vibration directions for the O and E rays are respectively perpendicular and parallel to the plane of the diagram.

both ordinary and extraordinary rays travel through the leading half of the Nicol prism at angles differently inclined to the incident ray direction. The refractive indices of calcite are $e = 1{\cdot}486$, $o = 1{\cdot}658$ and that of Canada Balsam is 1·530. The *o*-ray is therefore deviated more than the *e*-ray at the air/calcite interface. When it reaches the Balsam film the angular relations are such that the *e*-ray passes through undeviated, but displaced by a minute amount since the film is thin; the *e*-ray travels through the following half of the prism, is refracted at the calcite/air interface, and emerges parallel to the incident ray direction (Fig 12.53(b)). The refractive index for the *o*-ray in calcite is greater than that of Canada Balsam and the *o*-ray impinges on the calcite/balsam interface at a greater angle of incidence than the *e*-ray. The angle of incidence of the *o*-ray exceeds the critical angle for refraction ($\sin^{-1} 1{\cdot}530/1{\cdot}658 = 67°20'$); the *o*-ray is therefore totally reflected at the calcite/balsam interface and passes out through the side of the prism to be absorbed in non-reflecting mounting material. The only light transmitted by the Nicol prism is the *e*-ray which vibrates in the plane containing the incident ray and the optic axis, that is, for practical purposes, in the plane defined by the microscope axis and the bisector of the obtuse angle of the upper end face of the Nicol prism which is visible through the central aperture of the stage.

The disadvantages of the Nicol prism are its expense, its bulk, and that the maximum possible divergence of the emergent beam is only 24°.

All these disadvantages are lacking in *Polaroid film*, which is plastic sheet stretched between rollers almost to breaking point and dyed with an organic dye; the film is mounted between optically worked glass flats. The details of the process are patented but the principles are well known. Cellulose and certain plastics (e.g. Scotch tape or Sellotape) become strongly birefringent on extreme stretching. The dye, which forms a thin film on the surfaces of the plastic, is strongly pleochroic and is absorbed so that its molecules are oriented with respect to the stretching direction in the plastic. Substances such as quinine trisulphate ditriiodide (herapathite, the active principle of the earliest types of polaroid) are so strongly pleochroic that in one of the principal vibration directions there is nearly total absorption except at the long wavelength end of the visible range. Other types of polaroid film have been produced by strain orienting a strongly pleochroic iodine polyvinyl alcohol complex.

Interference effects between parallel polars

We have hitherto restricted ourselves to consideration of interference effects between crossed polars, the normal situation in polarized light microscopy and the most informative, but the polarizer can be rotated in most microscopes and it is a matter of slight practical and some theoretical interest to consider the interference effects produced by an anisotropic crystal plate between polars at some general angle ϕ.

Let us suppose that the vibration directions in the specimen crystal plate on the microscope stage correspond to light vectors y_1 and y_2 (refractive indices μ_1 and μ_2 respectively), that y_1 makes an angle θ with the polarizer vibration direction, and that the analyser vibration direction makes an angle ϕ with the polarizer vibration direction (Fig 12.54). If the plane polarized wave-motion incident on the crystal plate is represented by $y = A \sin \omega t$, then immediately on emergence from the crystal plate the two wave-motions will be

$$y_1 = A \cos \theta \sin \omega t,$$

$$y_2 = A \sin \theta \sin (\omega t + \alpha).$$

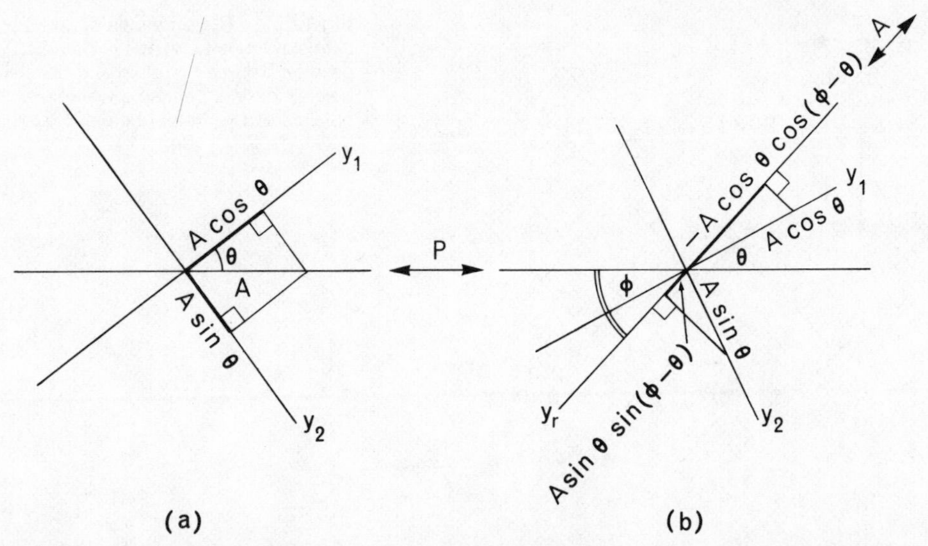

Fig 12.54 Interference between polars inclined at a general angle ϕ. The vibration directions in the crystal are y_1 and y_2 such that y_1 makes an angle θ with the polarizer vibration direction. The analyser vibration direction makes an angle ϕ with the polarizer vibration direction.

The light wave motion transmitted by the analyser will have resultant light vector

$$y_r = A \sin\theta \sin(\phi-\theta)\sin(\omega t+\alpha) - A\cos\theta\cos(\phi-\theta)\sin\omega t.$$

The general equation becomes more tractable in the special case of $\theta = 45°$, then

$$\begin{aligned} y_r &= \frac{A}{\sqrt{2}}\sin(\phi-45°)\sin(\omega t+\alpha) - \frac{A}{\sqrt{2}}\cos(\phi-45°)\sin\omega t \\ &= \frac{A}{2}(\sin\phi-\cos\phi)\sin(\omega t+\alpha) - \frac{A}{2}(\sin\phi+\cos\phi)\sin\omega t \\ &= \frac{A}{2}\sin\phi\,[\sin(\omega t+\alpha)-\sin\omega t] - \frac{A}{2}\cos\phi\,[\sin(\omega t+\alpha)+\sin\omega t] \\ &= A\sin\phi\sin\frac{\alpha}{2}\cos\left(\omega t+\frac{\alpha}{2}\right) - A\cos\phi\cos\frac{\alpha}{2}\sin\left(\omega t+\frac{\alpha}{2}\right). \end{aligned}$$

The two terms in this expression represent respectively the cases of crossed ($\phi = 90°$) and parallel ($\phi = 0°$) polars. For crossed polars, as we found earlier

$$y_r = A\sin\frac{\alpha}{2}\cos\left(\omega t+\frac{\alpha}{2}\right),$$

and for parallel polars,

$$y_r = A\cos\frac{\alpha}{2}\sin\left\{\omega t+\frac{\alpha}{2}\right\}.$$

The condition for total destructive interference between parallel polars at $\theta = 45°$ is then $\cos\frac{1}{2}\alpha = 0$, i.e. $\alpha = (2n+1)\pi$. But we have already shown that $\alpha = (2\pi d/\lambda)(\mu_2-\mu_1)$ so that the condition for destructive interference here becomes $d(\mu_2-\mu_1) = \frac{1}{2}(2n+1)\lambda$. A quartz wedge in the 45° position between parallel polars in monochromatic light will therefore display black bands at distances from its thin end

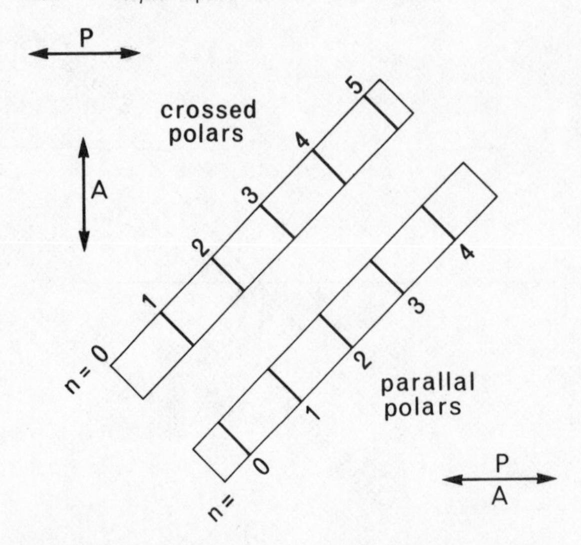

Fig 12.55 The sequence of dark bands in a quartz wedge in the 45° position between crossed and parallel polars: for dark bands between crossed polars $d.\Delta\mu = n\lambda$ and between parallel polars $d.\Delta\mu = \frac{1}{2}(2n+1)\lambda$.

proportional to odd half multiples of λ in contrast to the crossed polars case where the black bands are at distances proportional to λ (Fig 12.55). In white light the quartz wedge between parallel polars will display a sequence of colours similar to that of Newton's scale. In general, the colours displayed at a particular point on the wedge in crossed and parallel polars are complementary, since the condition for destructive interference of a particular wavelength in the one case, $\sin\frac{1}{2}\alpha = 0$, is the condition for maximum transmission of that wavelength in the other, $\cos\frac{1}{2}\alpha = 1$. In particular the sensitive tint bands characteristic of the crossed case become greenish-yellow bands in the parallel case. Rotation of the polarizer from the crossed to the parallel position serves to distinguish between high and low whites: with the polars parallel large values of $d(\mu_2 - \mu_1)$ will still give high whites whereas $d(\mu_2 - \mu_1) \sim 3000$ Å will produce an interference colour similar to first order sensitive tint ($\lambda \sim 6000$ Å cut out for $n = 0$).

Returning briefly in conclusion to the general equation and substituting $\phi = \theta$ we have then

$$y_r = -A \cos\phi \sin\omega t.$$

No interference effects will be observed when the vibration directions in the specimen crystal plate are maintained parallel and perpendicular to the analyser vibration direction irrespective of the angle between the polarizer and analyser vibration directions, but the intensity of light transmitted by the analyser will vary as $\cos^2\phi$. Extinction will of course be observed when the polars are crossed.

Special stages

The most restrictive condition of polarized light microscopy, inability to tilt the specimen crystal in three dimensions with respect to the plane of the stage, can be overcome by the use of either of two types of special stage. These are the *universal stage*, which is specifically designed for work with thin sections of polycrystalline materials, and the *spindle stage*, which is appropriate for single crystal grains.

In essence the universal stage is a device for orienting a crystal plate or polycrystalline specimen in three dimensions by rotation about a number of mutually perpendicular axes. The standard stage has four axes and the microscope stage adds a further vertical axis of rotation. The specimen, cemented between a shortened microscope slide and a cover glass, is sandwiched between glass hemispheres with oil films at the glass/glass interfaces. The inner stage axis A_1 rotates the specimen in its own plane and coincides with the microscope axis when all the other axes are set at zero; A_1 is used initially to set the selected crystal to extinction. The axis A_2 rotates about a horizontal N—S axis and is used to bring one of the mirror planes of the indicatrix vertical and parallel to the N—S cross-wire. The axis A_3 is vertical, parallel to the microscope axis, and is used to bring the second vertical symmetry plane of the indicatrix parallel to the N—S cross-wire. The axis A_4 is horizontal and lies E—W; it is equipped with a graduated drum with a vernier scale and is used to search each vertical symmetry plane in turn for the emergence of optic axes. When the microscope stage with the universal stage attached is rotated about the axis of the former (A_5) so as to bring the crystal into the 45° position a direct determination of $2V$ can be achieved. A general view of the 5-axis universal stage is given in Fig 12.56.

It is not our purpose to describe the techniques associated with the use of the universal stage. Such are adequately dealt with by Emmons (1943) and Muir (1973).

The universal stage was an instrument of central importance in mineralogy and petrology when the only means of estimating the composition of a particular crystal belonging to a solid solution series was by relation of its optical properties to those determinable for analysed samples of uniform composition. The advent of the electron probe (see chapter 15) however about 1960 provided a more certainly reliable and comprehensive means of determining the composition of a single crystal grain. The universal stage remains important for the precise measurement of $2V(\text{to} \pm 2°)$, for determining the orientation of the indicatrix with respect to morphological features, for the study of twinning and for the study of preferred orientation in polycrystalline aggregates.

The descriptions of a great variety of spindle stages are in the literature. The instrument consists essentially of a spindle, the angular rotation of which is measurable, on which is mounted a single crystal, the mounting to which the crystal

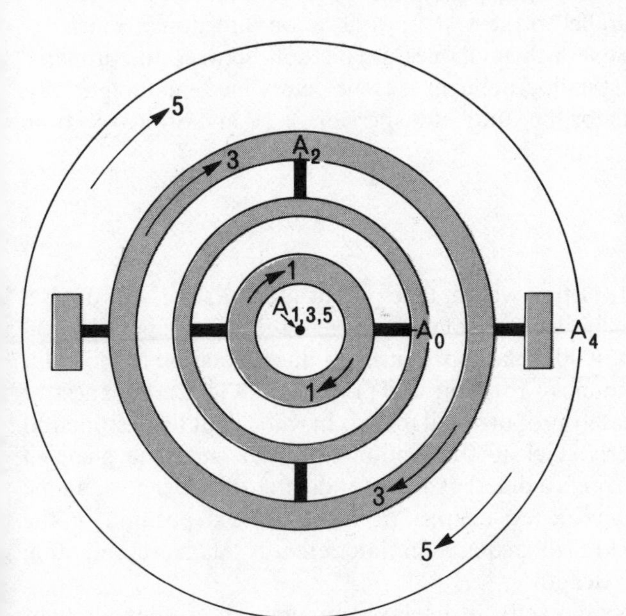

Fig 12.56 The universal stage. The figure shows diagrammatically the relationship of the five axes A_0–A_4 of the stage. The microscope stage provides an additional axis of rotation A_5. When all axes are set at zero A_1 and A_3 are parallel to the microscope axis A_5; A_0 and A_4 are oriented E–W parallel to the microscope stage; and A_2 is oriented N–S. The axis A_0, which is omitted in some models, allows A_1 to be inclined to the microscope axis. Movement about any axis changes the directions of all axes with lower subscripts in the sequence 5, 4, 3, 2, 0, 1.

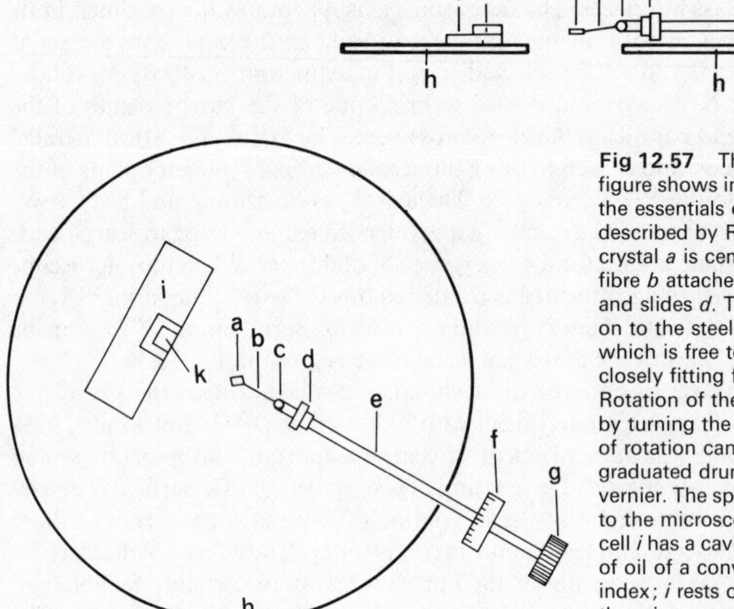

Fig 12.57 The spindle stage. The figure shows in plan and in elevation the essentials of the instrument described by Roy (1965). The crystal *a* is cemented to a fine glass fibre *b* attached to the ball-joint *c* and slides *d*. The slides are screwed on to the steel rod or spindle *e* which is free to rotate inside a closely fitting fixed brass tube. Rotation of the spindle is achieved by turning the knob *g*; the amount of rotation can be read from the graduated drum *f* and its associated vernier. The spindle stage is clamped to the microscope stage *h*. The oil cell *i* has a cavity *k* containing a pool of oil of a convenient refractive index; *i* rests on, but is not fixed to, the microscope stage.

is cemented being angularly adjustable and capable of being centred with respect to the microscope axis; the crystal is in all types immersed in a cell filled with oil of refractive index close to $\frac{1}{2}(\gamma+\alpha)$. One of the best modern spindle stages is that devised by Roy (1965) and illustrated in Fig 12.57.

The spindle stage provides a method of direct estimation of $2V$ and of determination of the orientation of the indicatrix with respect to morphological features. Fibres can be cemented to the crystal parallel to the α, β, or γ vibration directions; transfer of the crystal to an oscillation camera then enables the relation between the principal vibration directions and the crystallographic axes to be determined unambiguously. The spindle stage is invaluable for the study of dispersion of $2V$ and of the variation of absorption with λ.

Optical activity

It was observed by Arago in 1811 that when plane polarized light travels through a parallel-sided plate of quartz in the direction of the optic axis the plane of polarization is rotated. He noticed that some (0001) plates of quartz produced rotation to the right, others to the left; that the amount of rotation was proportional to the thickness of the quartz plate and approximately proportional to λ^{-2}. In white light then extinction will not be observed in a fairly thick (0001) section of quartz since the plane of polarization will be rotated through a different angle for different wavelengths. As the analyser (or polarizer) is rotated a few degrees from the crossed position in the appropriate direction the field will acquire an interference colour representing extinction for a particular wavelength.

In order to explain such optical activity displayed by uniaxial crystals for light travelling parallel to the optic axis it is necessary to make use of the concept of circularly polarized light. Any linear simple harmonic motion such as

$$y = A \sin \omega t$$

can be resolved into two circular motions:

$$x_1 = \frac{A}{2} \cos \omega t, \; y_1 = \frac{A}{2} \sin \omega t, \quad \text{i.e. } x_1^2 + y_1^2 = \left(\frac{A}{2}\right)^2$$

$$x_2 = -\frac{A}{2} \cos \omega t, \; y_2 = \frac{A}{2} \sin \omega t, \quad \text{i.e. } x_2^2 + y_2^2 = \left(\frac{A}{2}\right)^2$$

These are circular motions of equal amplitude but moving in opposite senses. In optically inactive substances they will travel through the crystal parallel to the optic axis with equal velocity and can at any stage of their passage through the crystal or after emergence from the crystal be recombined as a single linear simple harmonic motion.

We shall suppose that in optically active crystals the two circular disturbances of opposite sense travel with different velocities. A phase difference α will then be set up as the two circular motions travel through the crystal plate:

$$x_1 = \frac{A}{2} \cos \omega t, \qquad y_1 = \frac{A}{2} \sin \omega t$$

$$x_2 = -\frac{A}{2} \cos (\omega t + \alpha), \qquad y_2 = \frac{A}{2} \sin (\omega t + \alpha).$$

The resultant will be

$$X = x_1 + x_2 = \frac{A}{2} \{\cos \omega t - \cos (\omega t + \alpha)\}$$

$$= A \sin \frac{\alpha}{2} \sin \left(\omega t + \frac{\alpha}{2}\right)$$

$$Y = y_1 + y_2 = \frac{A}{2} \{\sin \omega t + \sin (\omega t + \alpha)\}$$

$$= A \cos \frac{\alpha}{2} \sin \left(\omega t + \frac{\alpha}{2}\right).$$

These are mutually perpendicular vibrations of the same phase which can be combined (Fig 12.58 (a)) into a single linear simple harmonic motion at an angle ρ to the incident simple harmonic motion such that

$$\tan \rho = \frac{A \sin \frac{1}{2}\alpha}{A \cos \frac{1}{2}\alpha} = \tan \frac{\alpha}{2}$$

i.e.

$$\rho = \frac{\alpha}{2}.$$

The phase difference α can be evaluated in terms of the refractive indices associated with the two circular motions as

$$\alpha = \frac{2\pi d}{\lambda} (\mu_1 - \mu_2)$$

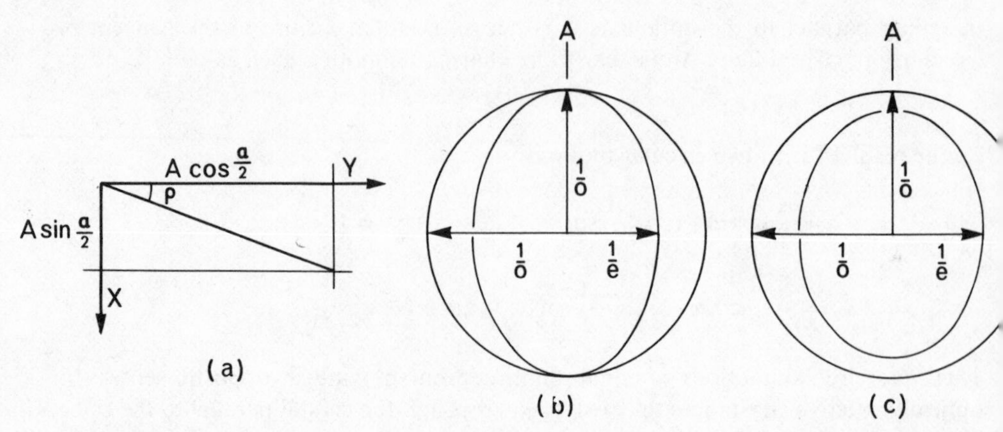

Fig 12.58 Optical activity. (a) shows the combination of the mutually perpendicular vibrations of the same phase produced in an optically active crystal from circular disturbances of opposite sense travelling with different velocities; the angle of rotation of the plane of polarization relative to the incident beam is ρ. (b) and (c) are ray velocity surfaces for an optically inactive and for an optically active uniaxial crystal respectively.

where d is the thickness of the (0001) plate, λ is the wavelength of monochromatic light *in vacuo*, and μ_1, μ_2 are the refractive indices for light travelling parallel to [0001]. The angle of rotation is thus $\rho = (\pi d/\lambda)(\mu_1 - \mu_2)$.

Optical activity can be displayed on a representation surface. The indicatrix is not particularly well suited for the representation of optical activity; the ray velocity surfaces are more appropriate (Fig 12.58). It will be immediately obvious from the figure that the surfaces are parallel but do not touch in the radial direction of the optic axis. We shall not pursue the study of optical activity by consideration of light travelling in directions inclined to the optic axis.

The most striking manifestation of optical activity commonly encountered is in interference figures of thick (0001) sections of quartz. The centre of the axial cross is missing, the centre of the field being filled by a circular area of uniform interference colour.

Optical activity is rarely encountered in polarized light microscopy and is insignificant in the study of transparent solids. Very few substances display marked optical activity because the effect, even when permitted by the symmetry of the structure, is generally too small to be observed. Quartz is the only common strongly active substance, the rotation per mm for Na_D light being 21°43′ corresponding to a refractive index difference of less than 0·0001. Wooster (1949) has discussed the application of observations of optical activity to the determination of crystal class, but the method is not widely applicable.

Reflected light microscopy

Solids in general transmit, absorb, and reflect incident light radiation. We have so far been principally concerned in this chapter with the class of crystalline solids for which transmission is dominant, that is with *transparent* crystalline solids; and we have paid some attention to absorption. A general theory of the interaction of light radiation with crystalline solids has to take the reflexion as well as the transmission and absorptive properties of the crystal into account. We shall not attempt here to provide a comprehensive study of the reflexion of light at a crystal surface, but merely to indicate the general principles of the theory and practice of reflected light microscopy; for a thorough treatment the reader is referred to Galopin and Henry (1972).

Reflected light microscopy is concerned with what are commonly called opaque solids; but it must be borne in mind that opacity is a function of thickness and that many substances which appear opaque in macroscopic specimens become capable of transmitting an appreciable percentage of the intensity of an incident light beam when in the form of a thin film (e.g. gold) or a thin cleavage flake (e.g. MoS_2). We are concerned now with that class of crystalline solids for which the proportion of the intensity of an incident light beam which is transmitted is very small, if not negligible. Such opaque crystals are most usefully prepared for microscopic study in the form of *polished sections*, thick sections of a mineral or a polyphase rock polished mechanically with an abrasive powder to a high degree of flatness over small areas. The basic instrument for the study of polished sections is the *reflected light microscope* in which the light source is reflected by a prism or a half-silvered plate so that it passes through the objective to fall normally on the polished surface of the specimen on the microscope stage. A polarizer is inserted in the optical train between the light source and the specimen so that the incident light beam is polarized in a known direction. The microscope is equipped with an analyser between its objective and ocular. It is useful to have as accessory equipment an instrument for making indentation hardness tests and a photoelectric cell to fit over the ocular for measuring changes in reflected light intensity as the specimen is rotated on the microscope stage relative to the intensity reflected by a calibrated standard.

For an isotropic specimen the *reflectivity* R, usually expressed as a percentage, for plane polarized monochromatic light is given by

$$R = \frac{(n-1)^2+k^2}{(n+1)^2+k^2}$$

where n is its refractive index and k its *absorption coefficient*; the plane polarized incident beam is reflected without change in its plane of polarization. For a uniaxial specimen whose reflecting surface is parallel to the optic axis a plane polarized monochromatic incident light beam is reflected as two mutually perpendicular plane polarized beams with a phase difference; each has a different reflectivity related to a principal refractive index (n_1 or n_2) and a principal absorption coefficient (k_1 or k_2), so that

$$R_1 = \frac{(n_1-1)^2+k_1^2}{(n_1+1)^2+k_1^2}$$

and

$$R_2 = \frac{(n_2-1)^2+k_2^2}{(n_2+1)^2+k_2^2}.$$

Since $|k_2-k_1| > |n_2-n_1|$ for absorbing crystals, the magnitude of the *bireflexion* $|R_2-R_1|$ is greater for opaque than for transparent crystals; for transparent crystals $k_2 \simeq k_1 \simeq 0$. One can attach a *sign* to reflectivity: the convention for a uniaxial crystal to be positive being $R_e > R_o$ by analogy with the convention for optic sign in transmitted light, where for positive crystals $e > o$. In consequence of such anisotropy of reflexion the intensity of the reflected beam will vary systematically as the stage carrying the reflecting specimen is rotated and, if the intensity difference is sufficiently marked to be observable by eye or to be recorded by a photoelectric cell mounted on the ocular, the anisotropy of reflectivity may be used as a diagnostic property.

The diagnostic optical study of opaque crystals is confined to four observations: (i) measurement of reflectivity and its anisotropy, (ii) observation of colour, (iii) measurement of hardness, (iv) observation of morphological characteristics such as shape, cleavage, and twinning. Reflectivity, anisotropy of reflectivity and hardness data have been collected and tabulated for a large number of opaque minerals and form the basis of a diagnostic procedure (Galopin and Henry, 1972). Accurate determination of hardness requires the use of an indentation test, but the relative hardness of two adjacent minerals in a polished specimen can simply be determined by the optical Kalb test which is explained in Fig 12.59.

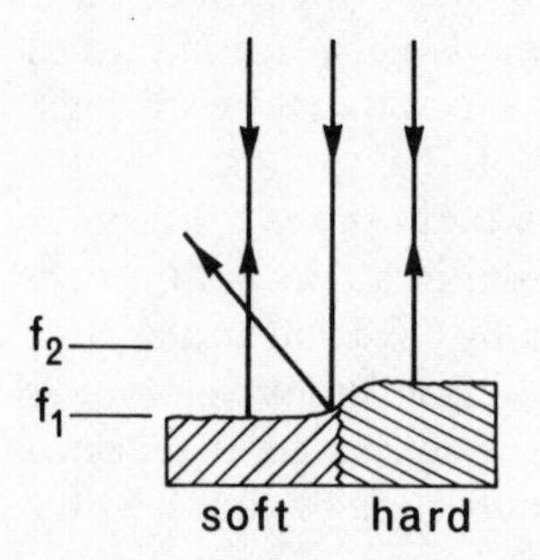

Fig 12.59 The Kalb test. The differing hardness of two adjacent grains results in the departure of the polished surface from perfect flatness, the softer grain being recessed. Reflection of light incident normal to the polished surface gives rise to a bright line at the intergranular boundary. When the microscope stage is lowered away from the objective, the focus of the microscope moves from a position f_1 towards a position f_2 relative to the specimen and in consequence the bright line appears to move *into* the softer grain.

This brief account of reflected light microscopy has been concerned principally with determinative methods and has largely ignored the quantitative study of anisotropy of reflectance as a fundamental technique for the study of the crystalline state. In its present state of development reflected light microscopy can provide little in the way of fundamental information about crystalline solids; it is still essentially a diagnostic technique. In the limit one is always left with some uncertainty about the extent to which the polishing process may have affected the surface layers, which are the reflecting layers, of the specimen. The development of the electron microprobe (chapter 15) has provided a very powerful technique for the chemical characterization of opaque phases, which has superseded in part the diagnostic role of reflected light microscopy; but reflected light microscopy remains the only technique available for the study of the interrelations of opaque mineral phases in an ore-body, whether one mineral has crystallized by reaction between two others or whether one mineral has passed through a polymorphic transformation below its crystallization temperature; and the technique is valuable too for the study of the interrelations of opaque and transparent minerals.

Part II

13
Mineral equilibrium: the thermodynamic basis

Classical thermodynamics is an exact mathematical discipline derivative from three fundamentally independent assumptions known as the *zeroth*, *first*, and *second laws of thermodynamics*. These are peculiar among physical laws in that none of them is susceptible to direct experimental proof; all are empirical assumptions based on observation and generalization of experience. Their justification is the fact that all conclusions from the laws of thermodynamics are without exception in agreement with experimental observations. It is however possible to derive the laws of thermodynamics from the atomic theory and the quantum theory, together with a single very general statistical assumption; this approach leads to the more modern science of *statistical thermodynamics*, which is at once closely dependent on quantum mechanics and on the structural concepts of crystallography.

This chapter will in the main be concerned with the fundamentals of classical thermodynamics and their application to *condensed phases* (that is, to solids and liquids). Only towards the end of the chapter will statistical concepts be introduced and the crystalline nature of solids by implication be recognized. We shall not there present a rigorous exposition of statistical thermodynamics, but merely make use of certain results.

The zeroth law: temperature

Before stating the zeroth law it is necessary to define a thermodynamic *system* as an arbitrarily isolated part of the universe. A system may be homogeneous or heterogeneous; in the latter case it will be composed of a number of homogeneous parts called *phases* each of which is described by specifying its content and a sufficient number of other properties.

The zeroth law may be stated thus: *if two systems are in thermal equilibrium they are said to be at the same temperature*. The zeroth law thus provides a definition of the important property *temperature*. In practical mineralogical terms the zeroth law indicates that if two mineral phases α and β are in thermal equilibrium, there will be no heat flow from one to the other and $T^\alpha = T^\beta$; if on the other hand $T^\alpha \neq T^\beta$ heat will flow from one phase to the other until their temperatures become equal. This is an important principle: it appears trivial only because it is part of our 'commonsense'.

The first law: conservation of energy

Before stating the first law it is necessary to define thermodynamic *state*. The state of a homogeneous system is described by specifying its content and every one of its independent properties. For a heterogeneous system it is necessary to specify the content and independent properties of each phase.

The first law, familiar as the law of conservation of energy, may be stated thus: *when a system passes from an initial state 1 to a final state 2 the change in energy of the system $U(2)-U(1)$ is independent of the form in which energy is supplied to the system, whether it be as heat, electrical energy, mechanical work, or some other form, and independent of the path by which the system passes from state 1 to state 2.* Thus if a system expands by an amount dV against a pressure P, the work done on the system $dw = P\,dV$; and if simultaneously an amount of heat dq is absorbed by the system, the change in energy U of the system will be given by

$$dU = dq + dw$$

$$= dq - P\,dV$$

i.e. $$dq = dU + P\,dV.$$

Integrating between the initial state 1 and the final state 2,

$$\int_1^2 dq = \int_1^2 dU + \int_1^2 P\,dV$$

which may be rewritten as

$$q = U(2) - U(1) + \int_1^2 P\,dV.$$

For a change at constant volume ($dV = 0$) we can therefore write

$$(q)_V = U(2) - U(1).$$

Now it is not experimentally particularly convenient to follow changes in *condensed phases*, that is in solids and liquids, at constant volume. It is experimentally far simpler and generally of greater significance to study changes in condensed phases at constant pressure. It is helpful for this purpose to define a thermodynamic function called *enthalpy* and denoted by the symbol H such that

$$H = U + PV$$

then $$dH = dU + P\,dV + V\,dP$$

and by the first law

$$dU = dq - P\,dV$$

so that

$$dH = dq + V\,dP$$

i.e. $$dq = dH - V\,dP.$$

Integrating between an initial state 1 and a final state 2,

$$\int_1^2 dq = \int_1^2 dH - \int_1^2 V\,dP$$

which may be rewritten as

$$q = H(2) - H(1) - \int_1^2 V\,dP.$$

For a change at constant pressure ($dP = 0$), therefore,

$$(q)_P = H(2) - H(1).$$

The amounts of heat absorbed by a system in changes at constant volume and at constant pressure are thus respectively equal to the differences in energy and in enthalpy between the initial and final states. The role of enthalpy in changes at constant pressure is generally analogous to that of energy in changes at constant volume. Since our prime concern here is the solid state our attention will be concentrated on changes at constant pressure; we shall make much use of enthalpy and correspondingly little of energy.

Heats of reaction: Hess' Law

Consider a chemical reaction represented by

$$\upsilon_{A_1} A_1 + \upsilon_{A_2} A_2 + \cdots \rightarrow \upsilon_{B_1} B_1 + \upsilon_{B_2} B_2 + \cdots$$

where A_1, $A_2, \ldots$ and B_1, $B_2, \ldots$ are the reactants and products respectively and υ_{A_1}, $\upsilon_{A_2}, \ldots \upsilon_{B_1}$, $\upsilon_{B_2}, \ldots$ are the numbers of molecules of each necessary to balance the equation. This formulation can be abbreviated as

$$\sum \upsilon_A A \rightarrow \sum \upsilon_B B.$$

The heat absorbed by such a system $(q)_P$ in the course of unit increase in the extent of reaction at constant pressure is known as the *heat of reaction at constant pressure* or usually just as the *heat of reaction.* Since $(q)_P = H(2) - H(1)$, it is conventional to denote the heat of reaction as ΔH.

For the general reaction $\sum \upsilon_A A \rightarrow \sum \upsilon_B B$ the heat of reaction ΔH is given by

$$\Delta H = \sum \upsilon_B H_B - \sum \upsilon_A H_A$$

where $H_A \ldots, H_B \ldots$ are enthalpies per mole of the reactants and products respectively. In a more specific example

$$C\,(\text{graphite}) + O_2(g) \rightarrow CO_2(g)$$

we can write for the heat of reaction ΔH

$$\Delta H = H(CO_2(g)) - H(C\,(\text{graphite})) - H(O_2(g))$$

and for the reaction

$$CO(g) + \tfrac{1}{2}O_2(g) \rightarrow CO_2(g)$$

the heat of reaction $\Delta H'$ is given by

$$\Delta H' = H(CO_2(g)) - H(CO(g)) - \tfrac{1}{2}H(O_2(g)).$$

Since H is a function of the state of a system, ΔH for successive processes at the same temperature is an additive function. This property of ΔH, known as *Hess' Law*, can be used to evaluate the heat of a reaction that cannot easily be performed quantitatively from other more tractable reactions. For both the reactions mentioned above ΔH is readily measurable. Subtraction of the two equations yields

$$C(\text{graphite}) + \tfrac{1}{2}O_2(g) \rightarrow CO(g)$$

for which the heat of reaction $\Delta H''$ is given by

$$\begin{aligned}\Delta H'' &= H(CO(g)) - H(C(\text{graphite})) - \tfrac{1}{2}H(O_2(g)) \\ &= \Delta H - \Delta H'.\end{aligned}$$

It is important always to observe the convention that ΔH is the excess of the enthalpy of the final state over the enthalpy of the initial state. Regrettably some authors have used the opposite convention.

The heats of other processes are defined analogously. Thus the *heat of melting* of forsterite,

$$Mg_2SiO_4\,(\text{fo}) \rightarrow Mg_2SiO_4\,(\text{l})$$

$$\Delta H = H(\text{l}) - H(\text{fo})$$

and the *heat of transformation* for a polymorphic transformation such as

$$SiO_2\,(\text{tridymite}) \rightarrow SiO_2\,(\text{cristobalite})$$

$$\Delta H = H(\text{crist.}) - H(\text{trid.}).$$

Heats of formation

The definition of enthalpy omits, of necessity, any statement of the zero of enthalpy for a pure substance. Only enthalpy differences can be measured experimentally by calorimetric measurement of heats of reaction and specific heats. It is convenient however to define an arbitrary zero of the enthalpy scale for each pure substance in such a manner as to achieve mutual consistency for all substances. The provision of such an arbitrary reference state then enables heats of reaction to be related as a simple sum and difference to the enthalpies of the pure substances that are the reactants and products in the reaction. Various schemes have been proposed, that most commonly used being based on the assumption that *every chemical element in its stable state at* 298 K *and* 1 *atmosphere has zero enthalpy.* Thus solid copper, liquid mercury, and gaseous oxygen at 298 K and 1 atmosphere are all taken arbitrarily to have zero enthalpy, or in symbolic notation $H^\circ_{298}(\text{Cu, s}) = 0, H^\circ_{298}(\text{Hg, l}) = 0, H^\circ_{298}(O_2, \text{g}) = 0$, where the superscript $^\circ$ indicates unit pressure, the subscript temperature in kelvins, and the last term in parenthesis the physical state of the element. The physical conditions 298 K, 1 atm define the *standard state* on this scheme.

The heat of reaction for

$$Si(s) + O_2(g) \rightarrow SiO_2(qz)$$

is susceptible to direct calorimetric measurement and it has been found that $\Delta H^\circ_{298} = -205{\cdot}4$ kcal. And since

$$\begin{aligned}\Delta H^\circ_{298} &= H^\circ_{298}(\text{qz}) - H^\circ_{298}(\text{Si, s}) - H^\circ_{298}(O_2, \text{g}) \\ &= H^\circ_{298}(\text{qz}) - 0 - 0\end{aligned}$$

the standard heat of formation of quartz at 298 K and 1 atmosphere, $H^\circ_{298}(\text{qz}) = -205{\cdot}4\,\text{kcal mole}^{-1}$. Standard heats of formation are denoted by some authors as ΔH°_f, e.g. $H^\circ_f(\text{qz}) = -205{\cdot}4\,\text{kcal mole}^{-1}$.

The heat of transformation for

$$C(graphite) \rightarrow C(diamond)$$

has been determined as $\Delta H^\circ_{298} = 0{\cdot}453$ kcal. Since graphite is the stable polymorph of carbon in the standard state H°_{298}(graphite) = 0 and the standard heat of formation of diamond H°_{298}(diamond) = $\Delta H^\circ_{298} + H^\circ_{298}$(graphite) = 0·453 kcal mole^{-1}.

The standard heats of formation of silicates can be evaluated by application of Hess' Law to the calorimetrically determined heats of successive reactions. For example,

$$SiO_2(qz)+2MgO(s) \rightarrow Mg_2SiO_4(fo) \qquad \Delta H^\circ_{298} = -15{\cdot}1 \text{ kcal mole}^{-1}$$

$$Si(s)+O_2(g) \rightarrow SiO_2(qz) \qquad \Delta H^\circ_{298} = -205{\cdot}4 \text{ kcal mole}^{-1}$$

$$Mg(s)+\tfrac{1}{2}O_2(g) \rightarrow MgO(s) \qquad \Delta H^\circ_{298} = -143{\cdot}8 \text{ kcal mole}^{-1}$$

Hence $2Mg(s)+Si(s)+2O_2(g) \rightarrow Mg_2SiO_4(fo)$

$$\Delta H^\circ_{298} = -15{\cdot}1-205{\cdot}4+2(-143{\cdot}8) = -508{\cdot}1 \text{ kcal}$$

Therefore the standard heat of formation of forsterite H°_{298}(fo) = −508·1 kcal mole^{-1}.

A comprehensive tabulation of standard heats of formation is given by Rossini *et al* (1961) and there is a more recent compilation of data for minerals by Robie in Clark (1966). A selection of data is listed in Table 13.1.

Standard heats of formation are generally useful for evaluating ΔH for reactions that have not been, or cannot be, performed as well as for attributing particularly large or small heats of reaction to a particular phase.

An alternative standard state for enthalpy is current in the field of silicate thermochemistry. Since silicate compositions can be represented in terms of their constituent oxides some authors have found it convenient to assume that the oxide of every element stable at 298 K and 1 atmosphere has zero enthalpy in that standard state. This convention is used for silicates in Table 13.1.

Temperature dependence of enthalpy

The relation $C_p = (\partial H/\partial T)_P$ defines the *molar specific heat* (or *molar heat capacity*) *at constant pressure.* C_p is determinable by calorimetry and is temperature dependent. For most condensed phases at temperatures greater than about 50 K C_p obeys the empirical equation

$$(C_p)^\circ_T = a+2bT-cT^{-2}$$

in which *a*, *b*, and *c* have to be determined experimentally. Integrating

$$\begin{aligned} H^\circ_T - H^\circ_{298} &= \int_{298}^{T} (C_p)^\circ_T \, dT \\ &= [aT+bT^2+cT^{-1}]^T_{298} \\ &= aT+bT^2+cT^{-1}-B \end{aligned}$$

where $B = 298a+298^2b+c/298$. This expression enables heats of reaction at any temperature to be derived from standard heats of formation provided the requisite specific heat data are available. The evaluation of heats of reaction at any general pressure P requires the application of the second law and is postponed to a later paragraph.

Table 13.1
Thermodynamic Properties of Minerals

Oxides and Hydroxides		$V^{\circ}_{298\cdot15}$ (cm^3 mole^{-1})	$H^{\circ}_{298\cdot15}$ (cal mole^{-1})	$S^{\circ}_{298\cdot15}$ deg^{-1} mole^{-1})
SiO_2	low-quartz	22·690 ± 0·005	−217,650 ± 400	9·88 ± 0·02
SiO_2	low-tridymite	26·53 ± 0·20	−216,900 ± 900	10·50 ± 0·10
SiO_2	low-cristobalite	25·74 ± 0·02	−216,930 ± 800	10·38 ± 0·02
SiO_2	coesite	20·64 ± 0·05	—	9·30 ± 0·50
TiO_2	rutile	18·80 ± 0·02	−225,760 ± 100	12·04 ± 0·04
TiO_2	anatase	20·49 ± 0·03	—	11·93 ± 0·07
Al_2O_3	corundum	25·57 ± 0·01	−400,400 ± 300	12·18 ± 0·03
$AlO(OH)$	boehmite	19·54 ± 0·02	−235,500 ± 3500	11·58 ± 0·05
$AlO(OH)$	diaspore	17·76 ± 0·03	—	8·43 ± 0·04
$Al(OH)_3$	gibbsite	31·96 ± 0·04	−306,380 ± 300	16·75 ± 0·10
$Fe_{.947}O$	wüstite	12·04 ± 0·04	−63,800 ± 400	13·74 ± 0·10
Fe_2O_3	hematite	30·28 ± 0·02	−196,750 ± 1100	20·89 ± 0·05
Fe_3O_4	magnetite	44·53 ± 0·02	−267,400 ± 500	36·03 ± 0·10
MnO	manganosite	13·22 ± 0·01	−92,050 ± 110	14·27 ± 0·10
Mn_2O_3	bixbyite	31·38 ± 0·03	−229,200 ± 2000	26·40 ± 0·50
Mn_3O_4	hausmannite	46·96 ± 0·08	−331,400 ± 400	35·5 ± 1·0
MgO	periclase	11·25 ± 0·01	−143,800 ± 100	6·44 ± 0·04
$Mg(OH)_2$	brucite	24·64 ± 0·03	−221,200 ± 500	15·09 ± 0·05
CaO	lime	16·76 ± 0·01	−151,790 ± 300	9·5 ± 0·2
$Ca(OH)_2$	portlandite	33·06 ± 0·04	−235,610 ± 450	19·93 ± 0·10
H_2O	water (liq.)	18·069 ± 0·003	−68,317 ± 10	16·715 ± 0·03

Spinels, Titanates, etc.				
$MgAl_2O_4$	spinel[N]	39·72 ± 0·03	—	19·26 ± 0·10
$FeAl_2O_4$	hercynite	40·82 ± 0·06	—	25·4 ± 0·2
$MgFe_2O_4$	magnesioferrite[I]	44·25	−341,171 ± 700	29·6 ± 0·6
Fe_2TiO_4	ulvöspinel[I]	45·75	—	40·36 ± 0·60
$FeTiO_3$	ilmenite	31·71 ± 0·05	−295,560 ± 600	25·3 ± 0·3
$MgTiO_3$	geikielite	30·86 ± 0·03	−375,900 ± 400	17·82 ± 0·10
$CaTiO_3$	perovskite	33·72 ± 0·08	−396,900 ± 600	22·4 ± 0·1

Silicates				
Al_2SiO_5	andalusite	51·54 ± 0·01	—	22·28 ± 0·10
Al_2SiO_5	kyanite	44·11 ± 0·02	—	20·02 ± 0·08
Al_2SiO_5	sillimanite	49·91 ± 0·02	—	22·97 ± 0·10
Fe_2SiO_4	fayalite	46·39 ± 0·08	−8282 ± 400*	34·70 ± 0·40
Mn_2SiO_4	tephroite	48·62 ± 0·10	−11,770 ± 600*	39·00 ± 1·00
Mg_2SiO_4	forsterite	43·67 ± 0·08	−15,120 ± 250*	22·75 ± 0·20
$CaMgSiO_4$	monticellite	51·37 ± 0·15	−27,560 ± 600*	—
β–Ca_2SiO_4	larnite	51·60 ± 0·40	−30,190 ± 250*	30·50 ± 0·20
γ–Ca_2SiO_4	calcium olivine	58·63 ± 0·35	−32,743 ± 600*	28·80 ± 0·20
$MnSiO_3$	rhodonite	35·32 ± 0·30	−5920 ± 170*	24·50 ± 0·50
$MgSiO_3$	clino-enstatite	31·47 ± 0·07	−8690 ± 150*	16·22 ± 0·10
$CaMgSi_2O_6$	diopside	66·10 ± 0·10	−36,500 ± 1500*	34·20 ± 0·20
$NaAlSi_2O_6$	jadeite	60·98 ± 0·40	−36,500 ± 1000*	31·90 ± 0·30
$CaSiO_3$	wollastonite	39·94 ± 0·08	−21,250 ± 700*	19·60 ± 0·20
$CaSiO_3$	pseudo-wollastonite	40·08 ± 0·08	—	20·90 ± 0·20
$Ca_2Mg_5Si_8O_{22}(OH)_2$	tremolite	272·95 ± 0·90	−120,840 ± 2500*	131·19 ± 0·30
$NaAlSi_3O_8$	albite	100·21 ± 0·19	−35,900 ± 1500*	50·20 ± 0·40
$KAlSi_3O_8$	orthoclase	108·97	−51,030 ± 1000*	52·47 ± 0·60
$CaAl_2Si_2O_8$	anorthite	100·73 ± 0·15	−21,810 ± 700*	48·45 ± 0·30
$KAlSi_2O_6$	leucite	88·39 ± 0·05	−46,200 ± 1500*	44·05 ± 0·40
$NaAlSiO_4$	nepheline	54·17 ± 0·15	−30,900 ± 1000*	29·72 ± 0·30
$KAlSiO_4$	kaliophilite	59·90 ± 0·08	—	31·85 ± 0·30
$NaAlSi_2O_6.H_2O$	analcite	97·50 ± 0·10	−32,750 ± 700*	56·03 ± 0·60
$CaAl_2Si_2O_7(OH)_2.H_2O$	lawsonite	101·33 ± 0·15	−37,190 ± 600*	56·79 ± 0·50
$Ca_2Al_4Si_8O_{24}.7H_2O$	leonhardite	814·92	−73,740 ± 1500*	220·40 ± 1·60
$Ca_2MgSi_2O_7$	åkermanite	92·82 ± 0·15	−43,830 ± 700*	—
$Mg_3Si_4O_{10}(OH)_2$	talc	134·30 ± 0·80	−44,890 ± 500*	62·34 ± 0·15
$Al_2Si_2O_5(OH)_4$	kaolinite	98·29	−7140 ± 500*	48·53 ± 0·30
$CaTiSiO_5$	sphene	55·70 ± 0·30	−26,850 ± 250*	30·88 ± 0·20

Carbonates				
$FeCO_3$	siderite	29·38 ± 0·02	−178,200 ± 1200	23·9 ± 0·6
$MnCO_3$	rhodocrosite	31·08 ± 0·01	−212,392 ± 800	23·90 ± 0·50
$MgCO_3$	magnesite	28·02 ± 0·01	−266,052 ± 400	15·7 ± 0·2
$CaCO_3$	calcite	36·94 ± 0·02	−288,086 ± 250	22·2 ± 0·2
$CaCO_3$	aragonite	34·16 ± 0·02	−288,134 ± 250	21·2 ± 0·3
$CaMg(CO_3)_2$	dolomite	64·35 ± 0·04	−557,567 ± 800	37·09 ± 0·07
Gases†				
H_2O		24,466·1 ± 1·0	−57,798 ± 10	45·106 ± 0·01
CO_2		24,466·1 ± 1·0	−94,054 ± 30	51·07 ± 0·02

Source: abbreviated from the critical summary of Robie in Clark (1966).

* Heats of formation of silicates are referred to formation from their constituent oxides at 298·15 K and 1 atmosphere.

† The molar volume quoted is that of an ideal gas since for a gas the standard state is the ideal gas at 298·15 K and 1 atmosphere.

N, I normal and inverse spinels respectively.

Experimental determination of heats of reaction: the solution calorimeter

In principle the heat of a reaction is determined by causing the reactants and products separately to undergo simple rapid reactions to produce identical states under such conditions that the heat input or output can be measured precisely. A straightforward example is the polymorphic transformation

$$\mathrm{C(graphite)} \rightarrow \mathrm{C(diamond)}$$

for which $\Delta H = H_D - H_G$. If a known amount of diamond is sparked in pure oxygen at atmospheric pressure in a pressure vessel contained in a calorimeter then, after the appropriate corrections have been made, the heat of combustion of diamond,

$$\mathrm{C(diamond)} + \mathrm{O_2} \rightarrow \mathrm{CO_2(g, 1\ atm)}$$
$$\Delta H' = H_{CO_2} - H_{O_2} - H_D,$$

can be evaluated. By a similar experiment the heat of combustion of graphite, $\Delta H'' = H_{CO_2} - H_{O_2} - H_G$, can be evaluated. The heat of transformation is simply found by subtraction since the products are identical in the two heat of combustion determinations; $\Delta H = \Delta H'' - \Delta H'$. A comprehensive account of combustion calorimetry is to be found in Skinner (1962).

Another way in which the reactants and products of a wide range of reactions can be converted to identical states is by solution in some appropriate solvent. Solution calorimeters employing water, a variety of organic solvents, hydrochloric acid, and hydrofluoric acid have been in use for very many years; the last named of these is particularly appropriate to the study of silicate reactions, because most silicates are rapidly soluble in warm aqueous HF, and will be taken as our example.

The HF solution calorimeter designed by Torgeson and Sahama (1948) is shown in Fig 13.1. The calorimeter itself is a platinum cylinder *a* with two chimneys, one for the platinum stirrer *b* and the other *c* for introduction of the samples, and a cylindrical well *d* for a platinum resistance thermometer and the calibrating heating coils. The

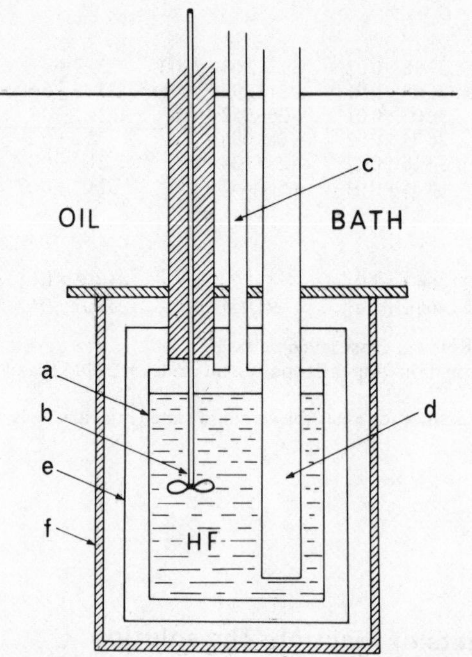

Fig 13.1 The HF solution calorimeter. For description see text.

calorimeter is filled with an aqueous solution containing 20·1 per cent HF and is surrounded by a gold-plated copper heat shield *e*, the whole being enclosed in a heavy gold-plated brass container *f*, which is immersed in an oil bath whose temperature is thermostatically controlled at 73·7 °C. Samples at room temperature contained in gelatin capsules weighted with platinum weights to ensure immediate sinking are dropped into the chimney *c*, the hydrofluoric acid in the calorimeter being at 73·7 °C. Samples of each reactant and product in the chosen reaction are in carefully weighed out stoichiometric proportions and should weigh $\not> 5$ g. For each determination the calorimeter is filled with 856·0 g 20·1 per cent (weight per cent) HF. Equilibration times of up to 30 minutes are found to be convenient. The temperature rise at accurately determined times after the introduction of the sample is measured with the Pt resistance thermometer and converted to the heat evolved in solution by calibration against known energy input from the calibrating heating coils in a separate experiment. Corrections have to be made for the heat of solution of gelatin and for the thermal capacities of the gelatin capsules and the Pt weights between room temperature and 73·7 °C.

An example of the method of calculation is taken from Torgeson and Sahama (1948). It is desired to determine the heat of formation of clinoenstatite ($MgSiO_3$). Heat of solution measurements were performed in the HF calorimeter on $Mg(OH)_2$, SiO_2, and $MgSiO_3$:

$$(1)\ Mg(OH)_2(s, 25^\circ) + 2HF(aq, 73{\cdot}7^\circ) \rightarrow MgF_2(ppt, 73{\cdot}7^\circ) + 2H_2O(aq, 73{\cdot}7^\circ) \qquad \Delta H_1$$

$$(2)\ SiO_2(s, 25^\circ) + 6HF(aq, 73{\cdot}7^\circ) \rightarrow H_2SiF_6(aq, 73{\cdot}7^\circ) + 2H_2O(aq, 73{\cdot}7^\circ) \qquad \Delta H_2$$

$$(3)\ MgSiO_3(s, 25^\circ) + 8HF(aq, 73{\cdot}7^\circ) \rightarrow MgF_2(ppt, 73{\cdot}7^\circ) + H_2SiF_6(aq, 73{\cdot}7^\circ) + 3H_2O(aq, 73{\cdot}7^\circ) \qquad \Delta H_3$$

To obviate the necessity for a heat of mixing correction the $Mg(OH)_2$ sample was dissolved in the calorimeter fluid in which the SiO_2 sample had already been dissolved.

Two more pieces of data are required: the heat of hydration of MgO and the specific heat of H_2O between 25°C and 73·7°C. The former was obtained from earlier heat of solution experiments in (N/1) HCl and the latter by direct calorimetry:

(4) $MgO(s, 25°) + H_2O(l, 25°) \rightarrow Mg(OH)_2(s, 25°)$ ΔH_4

(5) $H_2O(l, 25°) \rightarrow H_2O(l, 75°)$ ΔH_5

The heat of formation of clinoenstatite from its constituent oxides ΔH is then given by

$$\Delta H = \Delta H_1 + \Delta H_2 - \Delta H_3 + \Delta H_4 - \Delta H_5.$$

This expression for ΔH involves, as is usual, differences between large numbers and therefore, although heats of solution may be known very accurately, the resulting heat of formation all too often must be associated with quite a large percentage uncertainty. In this case $\Delta H_1 = -29{,}090 \pm 20$, $\Delta H_2 = -33{,}000 \pm 20$, $\Delta H_3 = -63{,}060 \pm 140$, $\Delta H_4 = -8850 \pm 25$, $\Delta H_5 = 810 \pm 5$ cal, whence $\Delta H = -8690 \pm 150$ cal mole^{-1}.

The second law: direction of change

There is a great diversity of apparently irreconcilable ways in which the second law may be formulated. The statement that will be made here is one of the most mathematical and the most immediately useful. By way of preliminary it is necessary to distinguish between natural, unnatural, and reversible processes. *Natural processes* are such as occur in nature; they proceed in a direction towards equilibrium. An *unnatural process* would proceed in a direction away from equilibrium and does not occur in nature. The limiting case between natural and unnatural processes is a *reversible process* which involves passage in either direction through a continuous sequence of equilibrium states; such processes do not actually occur in nature, but if we make a small change in the conditions we can produce a natural process that differs as little as we choose from a reversible process.

The second law comprises three statements:

(i) The *entropy* S of a system is the sum of the entropies of its parts.

(ii) The increase in entropy dS for a system undergoing a reversible change in which an amount of heat dq is absorbed by the system at a temperature T on the *absolute scale of temperature* is given by

$$dS = \frac{dq}{T}.$$

(iii) The increase in entropy dS for a system undergoing a natural change in which an amount of heat dq is absorbed by the system at a temperature T on the absolute scale of temperature is given by

$$dS > \frac{dq}{T}.$$

The first statement indicates that entropy is, like volume and enthalpy, an *extensive property*.[1] The entropy S of a system comprising one mole each of the phases α and β is given by

$$S = S^{\alpha} + S^{\beta}$$

[1] In contrast the magnitude of an *intensive property* is independent of the quantity of the phase. Examples of intensive properties are temperature, pressure, density.

where S^α and S^β are respectively the molar entropies of phase α and phase β. If the system comprises a moles of phase α and b moles of phase β, then

$$S = aS^\alpha + bS^\beta.$$

In the second and third statements the concept of absolute temperature T is related to entropy change. The intimate relationship between S and T implied by their simultaneous definition in this statement of the second law is central to the science of thermodynamics.

For the development of classical thermodynamics it is not necessary to provide any physical interpretation of the fundamental property entropy. That is one of the prime functions of statistical thermodynamics, in which entropy appears as a measure of 'degree of disorder' and can be given complete quantitative expression in crystalline structures.

Free energy and free enthalpy

Before developing the Second Law it is convenient to introduce two additional thermodynamic functions, F and G defined by

$$F = U - TS$$

$$G = H - TS.$$

F will here be called *free energy*; it is alternatively known as *Helmholtz free energy* or as the *Helmholtz function*. G will here be called *free enthalpy*; it is alternatively known as *Gibbs free energy* or as the *Gibbs function*. Care has to be exercised in reading the literature: not only are a variety of names used for these functions, but some authors denote as F the function that we have represented by G.

Criteria for equilibrium

We take as our starting point the definition of free enthalpy

$$G = H - TS,$$

substitute the defining equation for enthalpy

$$H = U + PV,$$

and differentiate to give

$$dG = dU + PdV + VdP - TdS - SdT.$$

Now by the first law

$$dU + PdV = dq$$

and by the second law

$$dq = TdS$$

for a reversible process, that is for passage through a continuous sequence of equilibrium states. Therefore at equilibrium,

$$dU + PdV = TdS$$

and so $$dG = VdP - SdT.$$

Therefore for a change in a system at equilibrium at constant pressure ($dP = 0$)

and constant temperature ($dT = 0$), $dG = 0$. Moreover for a natural process the second law indicates that

$$dq < T\,dS.$$

Hence $$dU + P dV < T\,dS$$

and so at constant P and T,

$$dG = dU + P dV - T\,dS < 0$$

i.e., in natural processes equilibrium is approached by decreasing G. Therefore at equilibrium at constant pressure and temperature, free enthalpy is at a *minimum.*

In a precisely similar manner it can readily be shown that at equilibrium at constant volume and temperature free energy F is at a minimum.

Dependence of thermodynamic functions on parameters of state

By way of preliminary we introduce four definitions:

(i) specific heat at constant volume $C_v = \left(\frac{\partial U}{\partial V}\right)_V$,

(ii) specific heat at constant pressure $C_p = \left(\frac{\partial H}{\partial T}\right)_P$,

(iii) coefficient of isobaric thermal expansion $\alpha = \frac{1}{V}\left(\frac{\partial V}{\partial T}\right)_P$,

(iv) coefficient of isothermal compressibility $\chi = -\frac{1}{V}\left(\frac{\partial V}{\partial P}\right)_T$.

Definitions (i) and (ii) can easily be seen to be consistent with those familiar in elementary physics. In general $dq = c\,dT$, where c is defined as specific heat under the appropriate conditions. We have already seen that $dq_v = dU$ and that $dq_p = dH$, so that $C_v\,dT = dU$ at constant volume and $C_p\,dT = dH$ at constant pressure.

We have already shown that for a reversible change

$$dG = V\,dP - S\,dT.$$

It follows that

$$\left(\frac{\partial G}{\partial T}\right)_P = -S$$

and $$\left(\frac{\partial G}{\partial P}\right)_T = V.$$

For smooth functions, such as we are concerned with here, the order of successive partial differentiation is immaterial,[2] that is to say there is a *cross-differentiation identity*, which may be written in general terms

$$\left\{\frac{\partial}{\partial x}\left(\frac{\partial z}{\partial y}\right)_x\right\}_y = \left\{\frac{\partial}{\partial y}\left(\frac{\partial z}{\partial x}\right)_y\right\}_x$$

[2] This point is dealt with more fully in Guggenheim (1959), chapter 3.

Therefore $\left\{\frac{\partial}{\partial P}\left(\frac{\partial G}{\partial T}\right)_P\right\}_T = \left\{\frac{\partial}{\partial T}\left(\frac{\partial G}{\partial P}\right)_T\right\}_P$

and hence $-\left(\frac{\partial S}{\partial P}\right)_T = \left(\frac{\partial V}{\partial T}\right)_P$

so that $\left(\frac{\partial S}{\partial P}\right)_T = -V\alpha.$

Returning to the definition of H,

$$H = U + PV$$

differentiating

$$dH = dU + PdV + VdP.$$

and substituting

$$dU + PdV = TdS$$

for a reversible change, we have

$$dH = TdS + VdP.$$

But $dS = \left(\frac{\partial S}{\partial T}\right)_P dT + \left(\frac{\partial S}{\partial P}\right)_T dP,$

therefore
$$dH = T\left(\frac{\partial S}{\partial T}\right)_P dT + \left\{V + T\left(\frac{\partial S}{\partial P}\right)_T\right\} dP$$
$$= T\left(\frac{\partial S}{\partial T}\right)_P dT + V(1 - T\alpha)dP.$$

Comparison with

$$dH = \left(\frac{\partial H}{\partial T}\right)_P dT + \left(\frac{\partial H}{\partial P}\right)_T dP$$

yields $T\left(\frac{\partial S}{\partial T}\right)_P = \left(\frac{\partial H}{\partial T}\right)_P,$

i.e. $\left(\frac{\partial S}{\partial T}\right)_P = \frac{C_p}{T}$

and $\left(\frac{\partial H}{\partial P}\right)_T = V(1 - T\alpha).$

By exactly analogous arguments the remaining six partial differential coefficients

$$\left(\frac{\partial U}{\partial T}\right)_V, \left(\frac{\partial U}{\partial V}\right)_T, \left(\frac{\partial F}{\partial T}\right)_V, \left(\frac{\partial F}{\partial V}\right)_T, \left(\frac{\partial S}{\partial T}\right)_V, \text{ and } \left(\frac{\partial S}{\partial V}\right)_T$$

can be derived. All twelve partial differential coefficients are listed in Table 13.2. Inspection of the table reveals certain points of theoretical and practical significance that are worth noticing.

Entropy appears as the negative temperature coefficient of free enthalpy, the

Table 13.2
Derivatives of the thermodynamic functions

$\left(\dfrac{\partial H}{\partial T}\right)_P = C_p$	$\left(\dfrac{\partial H}{\partial P}\right)_T = V(1-T\alpha)$
$\left(\dfrac{\partial G}{\partial T}\right)_P = -S$	$\left(\dfrac{\partial G}{\partial P}\right)_T = V$
$\left(\dfrac{\partial S}{\partial T}\right)_P = \dfrac{C_p}{T}$	$\left(\dfrac{\partial S}{\partial P}\right)_T = -V\alpha$
$\left(\dfrac{\partial U}{\partial T}\right)_V = C_v$	$\left(\dfrac{\partial U}{\partial V}\right)_T = T\dfrac{\alpha}{\chi}-P$
$\left(\dfrac{\partial F}{\partial T}\right)_V = -S$	$\left(\dfrac{\partial F}{\partial V}\right)_T = -P$
$\left(\dfrac{\partial S}{\partial T}\right)_V = \dfrac{C_v}{T}$	$\left(\dfrac{\partial S}{\partial V}\right)_T = \dfrac{\alpha}{\chi}$

thermodynamic function that is minimized at equilibrium in isothermal isobaric conditions such as obtain in most experimental investigations of systems of condensed phases. The temperature coefficient of entropy is C_p/T and this relationship provides a means of measuring isobaric entropy changes. Integration of the expression for $(\partial S/\partial T)_P$ leads to

$$S_{T_2}-S_{T_1} = \int_{T_1}^{T_2} C_p \, d\ln T.$$

If C_p is measured for a pure phase over small temperature ranges at fairly closely separated temperatures and plotted against log T, the entropy difference between two widely separated temperatures T_1 and T_2 can simply be evaluated from the area beneath the curve of C_p against log T. Alternatively use may be made of the empirical expression

$$C_p = a+2bT-cT^{-2}$$

fitted to the experimental data, whence

$$\begin{aligned} S_{T_2}-S_{T_1} &= \int_{T_1}^{T_2} (aT^{-1}+2b-cT^{-3})dT \\ &= \left[a\ln T+2bT+\frac{c}{2}T^{-2} \right]_{T_1}^{T_2} \\ &= a\ln\frac{T_2}{T_1}+2b(T_2-T_1)+\frac{c}{2}\left(\frac{1}{T_2^2}-\frac{1}{T_1^2}\right). \end{aligned}$$

Thus the entropy S_T of a substance can be evaluated from specific heat measurements over the range zero K to T K except for the undetermined constant S_0. The evaluation of S_0 is the subject of the third law of thermodynamics which we shall not attempt to expound here but merely state that it leads to the conclusion that for all substances $S_0 \geqslant 0$ and for very many crystalline solids $S_0 = 0$. S_T is therefore always positive and consequently the isobaric temperature coefficient of free enthalpy $(\partial G/\partial T)_P = -S$ is invariably negative.

The isothermal pressure coefficient of free enthalpy is V, which represents molar volume if molar free enthalpy is being considered as is usually the case. Molar volumes of crystalline solids can be accurately determined by measurement of unit-cell dimensions. The isobaric temperature coefficient of molar volume $(\partial V/\partial T)_P = V\alpha$ can be measured most conveniently by use of a high-temperature X-ray powder camera. Evaluation of α simultaneously provides a measure of the isothermal pressure coefficient of entropy, $(\partial S/\partial P)_T = -V\alpha$. The isothermal pressure coefficient of molar volume $(\partial V/\partial P)_T = -V\chi$ can be evaluated directly from compression experiments, although not usually conveniently at very high temperatures.

Tables 13.1 and 13.3 list V°_{298}, α, and χ for a selection of crystalline solids. It is immediately obvious that neither thermal expansion nor compressibility makes any very significant contribution to the variation of G with T and P.

Clapeyron's relation

The simplest case of phase equilibrium with which we are concerned is that between two phases of fixed and identical composition, which in practical terms may be a solid and the liquid to which it melts or a pair of polymorphs. Under isobaric conditions G will decrease with rising temperature for each phase, the two curves intersecting at the *transformation temperature* T_t. Since G is minimized at equilibrium the lower curve refers to the stable phase on either side of the transformation temperature, thus in Fig 13.2 phase α is stable at $T \leqslant T_t$ and phase β is stable at $T \geqslant T_t$.

If pressure is allowed to vary the curves α–α and β–β will each be drawn out into a surface and the curve of intersection of the two surfaces in PT space (Fig 13.3) will represent equilibrium between phase α and phase β. The relationship of the slope of the equilibrium curve to the thermodynamic functions of the two phases is given by *Clapeyron's Relation.*

Table 13.3
Coefficients of isobaric thermal expansion (α) and isothermal compressibility (χ) for some minerals

More detailed information on these coefficients is available in the articles by Skinner and Birch in Clark (1966) on which this table is based.

		$10^6\alpha$ (deg^{-1})		$10^6\chi$ (bar^{-1})
		400 °C	800 °C	
SiO_2	quartz	69	−3	2·71
SiO_2	coesite	11	14	—
Al_2SiO_5	andalusite	29	43	—
Al_2SiO_5	kyanite	28	30	—
Al_2SiO_5	sillimanite	18	26	—
Fe_2SiO_4	fayalite	30	31	0·91
Mg_2SiO_4	forsterite	38	44	0·80
$CaMgSiO_4$	monticellite	36	39	—
$MgSiO_3$	clino-enstatite	29	33	1·01
$CaMgSi_2O_6$	diopside	28	32	—
$NaAlSi_2O_6$	jadeite	29	38	0·75
$CaSiO_3$	pseudo-wollastonite	32	36	—
$NaAlSi_3O_8$	albite	27	33	2·02
$CaAl_2Si_2O_6$	anorthite	12	20	—
$KAlSi_3O_8$	microline	17	23	1·92
$Na_3KAl_4Si_4O_{16}$	nepheline	53	72	2·05
$Ca_2MgSi_2O_7$	åkermanite	30	33	—

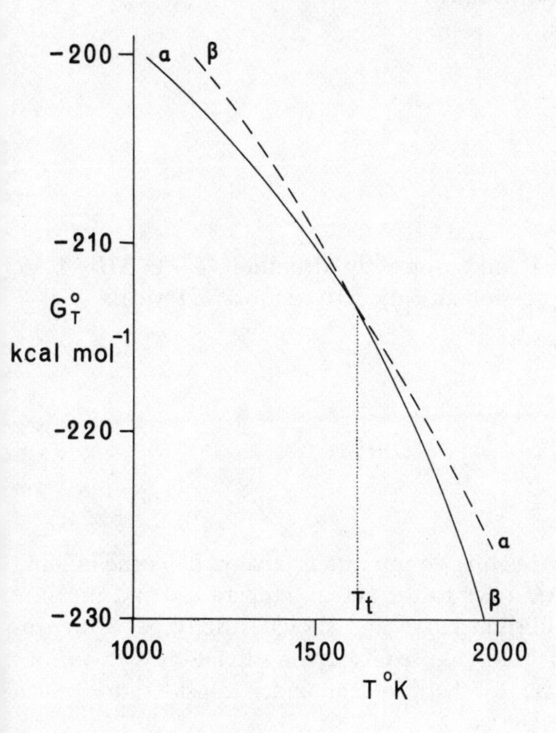

Fig 13.2 Plot of free enthalpy at atmospheric pressure, G°_{T}, against absolute temperature for two phases, α and β. At temperatures below the transformation temperature T_t the phase α is stable and at higher temperatures the phase β is stable.

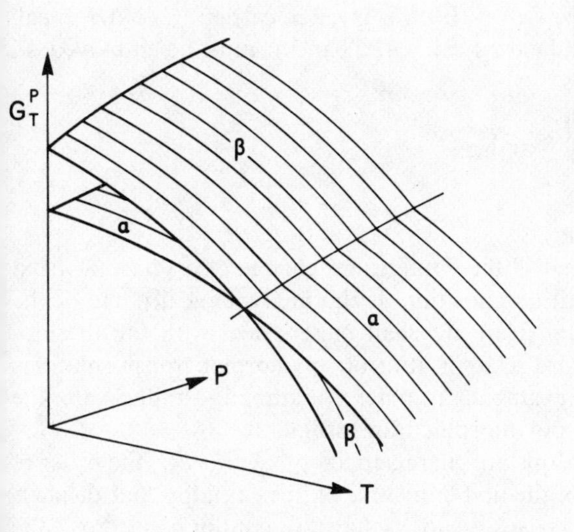

Fig 13.3 Diagrammatic intersection of free enthalpy surfaces for the phases α and β in *GPT* space. The two phases are in equilibrium at the pressures and temperatures given by the curve in which their free enthalpy surfaces intersect.

Let us formulate the reaction as

$$\alpha \rightarrow \beta$$

and define the difference in value of a property J between the two phases as $\Delta J = J^{\beta} - J^{\alpha}$. Now we have already seen that for a single phase

$$dG = -S\,dT + V\,dP$$

and consequently

$$d\Delta G = -\Delta S\,dT + \Delta V\,dP.$$

But at equilibrium

$$d\Delta G = 0$$

therefore $\quad -\Delta S\,dT + \Delta V\,dP = 0.$

i.e. $$\frac{dP}{dT} = \frac{\Delta S}{\Delta V}.$$

Moreover at equilibrium $\Delta G = 0$ and, since by definition $\Delta G = \Delta H - T\Delta S$, $\Delta S = \Delta H/T$. Substitution in the expression already derived for dP/dT yields

$$\frac{dP}{dT} = \frac{\Delta H}{T\,\Delta V}.$$

The two equations

$$\frac{dP}{dT} = \frac{\Delta S}{\Delta V} = \frac{\Delta H}{T\,\Delta V}$$

are alternative forms of *Clapeyron's Relation.* In use it must always be borne in mind that the values of ΔV and ΔS or ΔH refer to the actual temperature and pressure coordinates of points on the equilibrium curve and should strictly be written as ΔV_T^P, ΔS_T^P, and ΔH_T^P. In practice it is usual to use Clapeyron's Relation with various simplifying assumptions appropriate to the problem under consideration; such assumptions will be discussed as they arise.

It is usual to require dP/dT in bar deg^{-1}. If ΔS is given in cal deg^{-1}, or ΔH in cal, and ΔV in cm^3, then the conversion factor 1 cal = 41·8 bar cm^3 must be employed, so that Clapeyron's Relation becomes

$$\frac{dP}{dT} = \frac{41{\cdot}8\,\Delta S}{\Delta V} = \frac{41{\cdot}8\,\Delta H}{T\Delta V}\ \text{bar deg}^{-1}.$$

Polymorphic transformations

A polymorphic transformation, that is the equilibrium change of a phase of fixed composition to another of identical composition, is thermodynamically one of the simplest types of phase equilibrium. Here we shall merely deal with the thermodynamic, as distinct from the structural, classification of polymorphic transformations and consider in detail a single example, graphite ⇄ diamond, to illustrate the application of thermodynamics to polymorphic transformations.

Thermodynamically transformations are characterized by their *order*, the order of a transformation being the order of the first derivative of free enthalpy that displays a discontinuity. Thus *first order transformations* have G continuous across the transformation, whereas the first derivatives of G, $S = -(\partial G/\partial T)_P$ and $V = (\partial G/\partial P)_T$, exhibit discontinuities at the transformation as do higher derivatives. Since there is a discontinuity in S while G is continuous, there is a discontinuity in H, which is known as the latent heat of the transformation; $\Delta G = \Delta H - T_t\Delta S$, but $\Delta G = 0$, therefore $\Delta H = T_t\Delta S$, where T_t is the transformation temperature (Fig 13.4). Included among first order transformations are changes of state, such as melting and vaporization, and certain polymorphic transformations, e.g. graphite ⇌ diamond, aragonite ⇌ calcite, high quartz ⇌ high tridymite.

Second order transformations have G and its first derivatives, S and V, continuous across the transformation; but the second derivatives of G,

$$\frac{\partial^2 G}{\partial T^2} = -\frac{C_p}{T}, \qquad \frac{\partial^2 G}{\partial P^2} = -\chi T \qquad \text{and} \qquad \frac{\partial^2 G}{\partial P . \partial T}$$

exhibit discontinuity at the transformation. Second order transformations have zero latent heat of transformation since ΔG and ΔS are both zero. Examples of the second order are the superconducting transformation in tin (Fig 13.4) and certain other metals at low temperature and, at least ideally, order–disorder transformations.

Third order transformations have C_p continuous, but $\partial C_p/\partial T$ discontinuous; they are exemplified by the Curie point of a variety of ferromagnetic substances. As the order increases, the discontinuity in properties becomes decreasingly significant and it is no longer appropriate to think in terms of a phase change. Transformations of order greater than three have not been recognized with certainty.

It is well known that the interconversion of graphite and diamond is experimentally difficult. The graphitization of diamond can only be achieved, in the absence of a

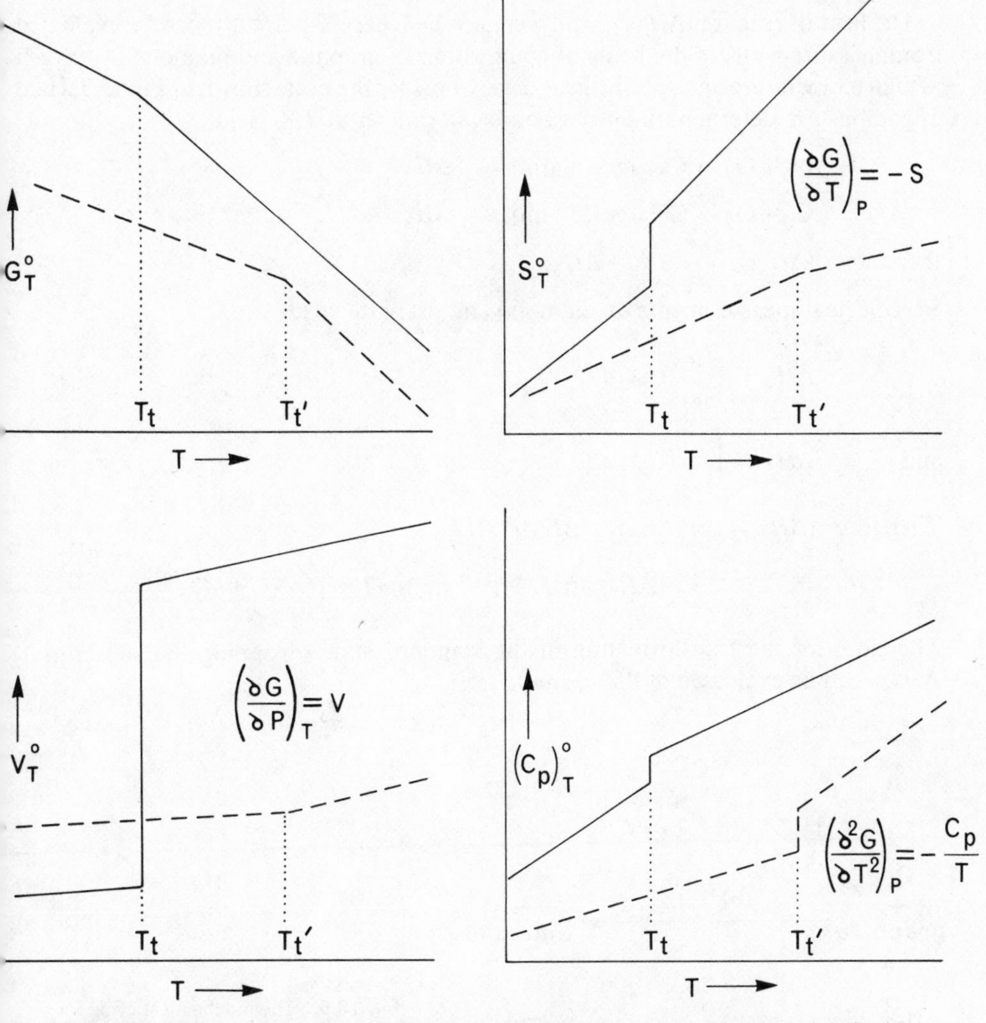

Fig 13.4 Variation with temperature of G, its first derivatives S and V, and its second derivative C_p for a first order transformation at T_t (solid line) and a second order transformation at $T_{t'}$ (broken line).

catalyst, at very high pressures and temperatures, e.g. 33 kb and 1200 °C, 61 kb and 1700 °C. The formation of diamond from graphite and other forms of carbon takes place only at high pressures and temperatures in the presence of a catalyst. But the course of the equilibrium curve in PT space can be estimated without actual realization of interconversion.

The unit-cell dimensions of graphite and of diamond can be determined under atmospheric conditions by measurement of single crystal or powder diffraction patterns: the data for graphite (hexagonal) are $a = 2{\cdot}4612$ Å, $c = 6{\cdot}7079$ Å, and for diamond (cubic) $a = 3{\cdot}5668$ Å. The unit-cell of graphite contains 4 carbon atoms and that of diamond 8 carbon atoms. Therefore

$$\Delta V^\circ_{298} = (\tfrac{1}{8}a_D^3 - \tfrac{1}{4}a_G^2 c_G \sin 60°)\, N = -1{\cdot}91 \text{ cm}^3 \text{ mole}^{-1}$$

where N is Avogadro's number, the subscripts D and G refer to diamond and graphite respectively, and ΔV refers to one mole of carbon involved in the transformation graphite → diamond.

The heat of reaction ΔH°_{298}, which cannot be measured directly, can be evaluated from measurements of the heats of combustion of graphite and diamond in oxygen at high temperature and specific heat data. Consider the cycle shown in Fig 13.5. Heat of combustion determinations in a bomb calorimeter at T K yield

$$C_D + O_2 \rightarrow CO_2(\text{gas, 1 atm}); \qquad \Delta H_2$$

$$C_D + O_2 \rightarrow CO_2(\text{gas, 1 atm}); \qquad \Delta H_3$$

therefore $(\Delta H_1)^\circ_T = -\Delta H_2 + \Delta H_3$.

Specific heat measurements on diamond and graphite yield

$$\Delta H_4 = \int_{298}^{T} (C_p)_D \, dT$$

and

$$\Delta H_5 = \int_{298}^{T} (C_p)_G \, dT.$$

Therefore

$$(\Delta H_1)^\circ_{298} = (\Delta H_1)^\circ_T - \Delta H_4 + \Delta H_5$$

$$= -\Delta H_2 + \Delta H_3 - \int_{298}^{T} (C_p)_D \, dT + \int_{298}^{T} (C_p)_G \, dT.$$

Therefore for the transformation in the standard state (dropping the subscript 1) ΔH°_{298} can be evaluated as 450 cal mole^{-1}.

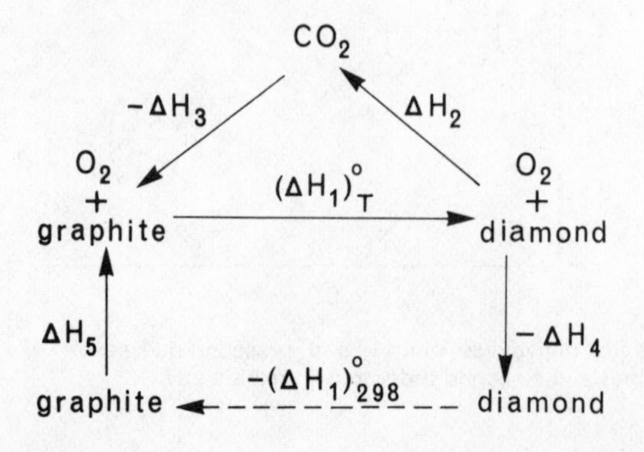

Fig 13.5 Thermochemical cycle for determination of the heat of reaction of graphite → diamond, $(\Delta H_1)^\circ_T$ at the high temperature T K and $(\Delta H_1)^\circ_{298}$ at 298 K.

Specific heat measurements from low temperatures up to 278 °K enable the entropies of diamond and graphite to be evaluated as $(S_D)^\circ_{278} = 0{\cdot}58$, $(S_G)^\circ_{278} = 1{\cdot}36$ cal deg^{-1} mole^{-1}. Therefore $\Delta S^\circ_{278} = -0{\cdot}78$ cal deg^{-1} mole^{-1}.

If, as a first approximation, it is assumed that ΔV and ΔS are independent of temperature and pressure the Clapeyron Relation can be integrated. It is convenient to start with

$$d\Delta G = -\Delta S\,\mathrm{d}T + \Delta V\,dP,$$

which, on putting $\Delta S = \Delta S^\circ_{298}$ and $\Delta V = \Delta V^\circ_{298}$, becomes

$$d\Delta G = -\Delta S^\circ_{298}\,dT + \Delta V^\circ_{298}\,dP.$$

Integration between the standard state and some point PT on the equilibrium curve yields

$$\Delta G - \Delta G^\circ_{298} = -\Delta S^\circ_{298}(T-298) + \Delta V^\circ_{298}(P-1).$$

But $\Delta G = 0$ at equilibrium and we are concerned with equilibrium at high pressures so that $P-1 \sim P$, and therefore

$$-\Delta G^\circ_{298} = -\Delta S^\circ_{298}(T-298) + \Delta V^\circ_{298}\,P.$$

It follows from the definition of free enthalpy that

$$\Delta G^\circ_{298} = \Delta H^\circ_{298} - 298\Delta S^\circ_{298}.$$

Therefore $\quad -\Delta H^\circ_{298} = -\Delta S^\circ_{298}\,T + \Delta V^\circ_{298}\,P$

i.e. $$P = -\frac{\Delta H^\circ_{298}}{\Delta V^\circ_{298}} + \frac{\Delta S^\circ_{298}}{\Delta V^\circ_{298}}\,T.$$

For the graphite → diamond transformation, using the data given above, the first approximation to the equilibrium curve is

$$P = +\frac{453{\cdot}2}{1{\cdot}8823} + \frac{0{\cdot}78}{1{\cdot}8823}\,T \text{ cal cm}^{-3}$$

i.e. $$P = 41{\cdot}843\left(\frac{453{\cdot}2}{1{\cdot}8823} + \frac{0{\cdot}78}{1{\cdot}8823}\,T\right) \text{ bars}$$

i.e. $$P = 10{\cdot}075 + 0{\cdot}01734T \text{ kb.}$$

Since T is of necessity always positive an immediate conclusion from this equation is that the equilibrium curve lies at high pressure at all temperatures.

The accuracy of the calculated equilibrium curve can be improved by taking into account the temperature dependence of ΔC_p, which is possible in this case because high temperature specific heats have been determined for both diamond and graphite. We have already seen that

$$\left(\frac{\partial \Delta H}{\partial T}\right)_P = \Delta C_p \quad \text{and} \quad \left(\frac{\partial \Delta S}{\partial T}\right)_P = \frac{\Delta C_p}{T}$$

hence $$\Delta H^\circ_T = \Delta H^\circ_{298} + \int_{298}^{T} \Delta C_p\,dT \quad \text{and} \quad \Delta S^\circ_T = \Delta S^\circ_{298} + \int_{298}^{T} \frac{\Delta C_p}{T}\,dT.$$

By definition $\Delta G_T^\circ = \Delta H_T^\circ - T\Delta S_T^\circ$

$$\Delta G_T^\circ = \Delta H_{298}^\circ + \int_{298}^{T} \Delta C_p\, dT - T\Delta S_{298}^\circ - T\int_{298}^{T} \frac{\Delta C_p}{T}\, dT$$

And since $\left(\frac{\partial \Delta G}{\partial P}\right)_T = \Delta V$

$$\Delta G_T^P = \Delta G_T^\circ + \int_1^P \Delta V\, dP.$$

If we assume at this level of approximation that ΔV is independent of temperature and pressure, we can put $\Delta V = \Delta V_{298}^\circ$ for all P, T; and if we further assume that P is large, $P-1 \sim P$: the expression for ΔG_T^P then becomes

$$\Delta G_T^P = \Delta G_T^\circ + P\Delta V_{298}^\circ.$$

At equilibrium

$$\Delta G_T^P = 0$$

and therefore $\Delta G_T^\circ + P\Delta V_{298}^\circ = 0$

i.e.
$$\Delta H_{298}^\circ + \int_{298}^{T} \Delta C_p\, dT - T\Delta S_{298}^\circ - T\int_{298}^{T} \frac{\Delta C_p}{T}\, dT + P\Delta V_{298}^\circ = 0$$

i.e.
$$P = -\frac{\Delta H_{298}^\circ}{\Delta V_{298}^\circ} + \frac{\Delta S_{298}^\circ}{\Delta V_{298}^\circ} T + \frac{1}{\Delta V_{298}^\circ}\left\{T\int_{298}^{T} \frac{\Delta C_p}{T}\, dT - \int_{298}^{T} \Delta C_p\, dT\right\}.$$

High-temperature specific heat data for diamond and graphite are given in Table 13.4 in terms of the empirical equation $C_p = a + 2bT - cT^{-2}$. Therefore $\Delta C_p = \Delta a + 2\Delta b\, T - \Delta c\, T^{-2}$,

and
$$\frac{\Delta C_p}{T} = a\, T^{-1} + 2\Delta b - \Delta c\, T^{-3}.$$

Integration yields

$$\int_{298}^{T} \Delta C_p dT = [\Delta a\, T + \Delta b\, T^2 + \Delta c\, T^{-1}]_{298}^{T}$$
$$= \Delta a(T-298) + \Delta b(T^2 - 298^2) + \Delta c(T^{-1} - 298^{-1}),$$

and
$$\int_{298}^{T} \frac{\Delta C_p}{T} dT = \left[\Delta a \ln T + 2\Delta b\, T + \frac{\Delta c}{2} T^{-2}\right]_{298}^{T}$$
$$= \Delta a(\ln T - \ln 298) + 2\Delta b(T-298) + \frac{\Delta c}{2}(T^{-2} - 298^{-2}).$$

Hence
$$T\int_{298}^{T} \frac{\Delta C_p}{T} dT - \int_{298}^{T} \Delta C_p\, dT$$
$$= \Delta a\, T \ln T - \left\{(1 + \ln 298)\Delta a + 596\Delta b + 298^{-2}\frac{\Delta c}{2}\right\} T$$
$$+ \Delta b\, T^2 - \frac{\Delta c}{2} T^{-1} + \{298\Delta a + 298^2\Delta b + 298^{-1}\Delta c\}$$

Table 13.4
Thermochemical data for the transformation graphite → diamond

Diamond
$H^\circ_{298} = 453{\cdot}2$ cal mole^{-1} $S^\circ_{298} = 0{\cdot}585 \pm 0{\cdot}005$ cal deg^{-1} mole^{-1}
$C_p = 2{\cdot}27 + 3{\cdot}06 \times 10^{-3}T - 1{\cdot}54 \times 10^5 T^{-2}$ cal deg^{-1} mole^{-1}
$a = 3{\cdot}5668$ Å $Z = 8$ $V^\circ_{298} = 3{\cdot}4161$ cm^3 mole^{-1}
$\alpha = 4 \times 10^{-6}$ deg^{-1} $\chi = 2 \times 10^{-7}$ bar^{-1}

Graphite
$H^\circ_{298} = 0$ $S^\circ_{298} = 1{\cdot}36 \pm 0{\cdot}02$ cal deg^{-1} mole^{-1}
$C_p = 4{\cdot}03 + 1{\cdot}14 \times 10^{-3}T - 2{\cdot}04 \times 10^5 T^{-2}$ cal deg^{-1} mole^{-1}
$a = 2{\cdot}4612$ Å, $c = 6{\cdot}7079$ Å, $Z = 4$ $V^\circ_{298} = 5{\cdot}2984$ cm^3 mole^{-1}
$\alpha = 25 \times 10^{-6}$ deg^{-1} $\chi = 30 \times 10^{-7}$ bar^{-1}

Graphite → Diamond
$\Delta H^\circ_{298} = 453{\cdot}2$ cal $\Delta S^\circ_{298} = -0{\cdot}78$ cal deg^{-1}
$\Delta C_p = -1{\cdot}76 + 1{\cdot}92 \times 10^{-3}T + 0{\cdot}50 \times 10^5 T^{-2}$ cal deg^{-1}
$\Delta G^\circ_T = 4{\cdot}05T \log T - 10{\cdot}71T - 0{\cdot}96 \times 10^{-3}T^2 - 0{\cdot}25 \times 10^5 T^{-1} + 1060$ cal
$\Delta V^\circ_{298} = -1{\cdot}8823$ cm^3

First approximation		Second approximation		Third approximation*		
T °K	P(kb)	G°_T(cal)	P(kb)	$-\gamma_1$	$-\gamma_2 \times 10^6$	P(kb)
298	15·24	685	15·23	1·8823	7·6060	16·30
500	18·75	880	19·56	1·9063	7·6459	21·10
700	22·21	1123	24·96	1·9301	7·6853	27·32
900	25·68	1383	30·74	1·9538	7·7248	34·26
1100	29·15	1643	36·52	1·9776	7·7643	41·54
1300	32·62	1891	42·04	2·0013	7·8037	48·84

* assuming the equation to be rewritten as $-\Delta G^\circ_T = \gamma_1 P - \gamma_2 P^2$ where

$\gamma_1 = \Delta V^\circ_{298} + (V_D\alpha_D - V_G\alpha_G)(T-298)$

$\gamma_2 = \frac{1}{2}(V_D\chi_D - V_G\chi_G) + \frac{1}{2}(V_D\alpha_D\chi_D - V_G\alpha_G\chi_G)(T-298)$

and therefore

$$\Delta G^\circ_T = -\Delta a\, T \ln T + \left\{(1+\ln 298)\Delta a + 596\Delta b + 298^{-2}\frac{\Delta c}{2} - \Delta S^\circ_{298}\right\} T$$
$$-\Delta b\, T^2 + \frac{\Delta c}{2} T^{-1} + \{\Delta H^\circ_{298} - 298\Delta a - 298^2\Delta b - 298^{-1}\Delta c\}.$$

Values of ΔG°_T at various temperatures and of $P = -\Delta G^\circ_T/\Delta V^\circ_{298}$ at this level of approximation are tabulated in Table 13.4.

The next level of approximation takes into account the dependence of ΔV on temperature and pressure. It follows from the definition of the coefficient of isothermal compressibility $\chi = -V^{-1}(\partial V/\partial P)_T$ that

$$V^P_T = V^\circ_T - \int_1^P \chi V^\circ_T \, dP$$
$$= V^\circ_T(1-\chi P) \quad \text{for } P \geqslant 1.$$

From the definition of isobaric thermal (volume) expansion, $\alpha = V^{-1}(\partial V/\partial T)_P$,

$$V^\circ_T = V^\circ_{298} + \int_{298}^T \alpha V^\circ_{298} \, dT$$
$$= V^\circ_{298}[1+\alpha(T-298)]$$

and hence $V^P_T = V^\circ_{298}[1+\alpha(T-298)](1-\chi P)$.

Therefore for the transformation

$$\Delta V_T^P = (V_{298}^\circ)_D[1+\alpha_D(T-298)](1-\chi_D P)$$
$$-(V_{298}^\circ)_G[1+\alpha_G(T-298)](1-\chi_G P).$$

Now
$$\Delta G_T^P = \Delta G_T^\circ + \int_1^P \Delta V_T^P \, dP$$
$$= \Delta G_T + (V_{298}^\circ)_D[1+\alpha_D(T-298)]\left(P - \frac{\chi_D}{2}P^2\right)$$
$$-(V_{298}^\circ)_G[1+\alpha_G(T-298)]\left(P - \frac{\chi_G}{2}P^2\right).$$

Applying the equilibrium condition

$$\Delta G_T^P = 0$$

we have the equation to the equilibrium curve in PT space,

$$\Delta G_T^\circ + \{\Delta V_{298}^\circ + [(V_{298}^\circ)_D\alpha_D - (V_{298}^\circ)_G\alpha_G](T-298)\}\,P$$
$$-\{(V_{298}^\circ)_D[1+\alpha_D(T-298)]\chi_D - (V_{298}^\circ)_G[1+\alpha_G(T-298)]\chi_G\}\frac{P^2}{2} = 0.$$

Solutions to this quadratic equation are tabulated in Table 13.4.

The third approximation, curve c in Fig 13.6, can be improved by taking into account the temperature and pressure coefficients of α and χ at high temperature and pressure. The resulting improvement would be expected to be slight and will not be explored here.

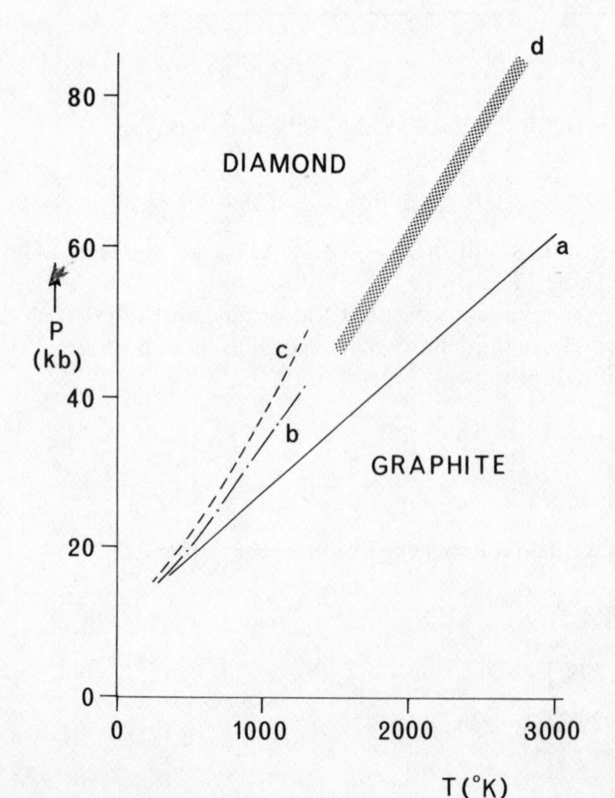

Fig 13.6 The polymorphic equilibrium graphite ⇄ diamond. The solid curve a represents the first approximation with $\Delta S/\Delta V = (\Delta S)_{298}^\circ/(\Delta V)_{298}^\circ$. The dot-dash curve b represents a second approximation which takes into account the temperature dependence of ΔC_p and the broken curve c represents a third approximation which takes into account also the temperature and pressure dependence of ΔV. The shaded curve d summarizes experimental data on the synthesis of diamond from graphite.

The equilibrium curves for graphite → diamond at each level of approximation are plotted on Fig 13.6. It is noticeable that each improvement in the calculation takes the equilibrium curve to successively higher pressures at any given high temperature. These differences are much greater than they will be for the great majority of polymorphic transformations because this is a *transformation of bond type*, the most extreme kind of change possible. In diamond there is strong covalent sp^3 bonding between carbon atoms, whereas in graphite there is comparably strong covalent sp^2 bonding within the layers and very much weaker van der Waals bonding between the structural layers. The effect of this extreme difference in structure on specific heat is not intuitively obvious, but it is generally true to say that the stronger the bonding the smaller the specific heat and consequently the lower the entropy at any temperature. That strongly bonded structures will have markedly low coefficients of thermal expansion and of compressibility is self-evident.

The calculation of a substantially correct *PT* curve for the graphite ⇌ diamond transformation could not lead to an immediate solution of the problem of the synthesis of diamond. Apparatus capable of providing *PT* conditions within the stability field of diamond had been available for some years before the successful synthesis of diamond was performed (Bundy *et al*, 1961). The remaining stumbling block turned out to be kinetic rather than thermodynamic. In this context thermodynamics is concerned only with the relative stability of polymorphs and can provide no information about whether the transformation can be achieved practically by subjecting the reactant to *PT* conditions within the stability field of the product polymorph. Highly ordered covalent structures, that is structures of extremely low entropy as is the case in diamond, do not in general crystallize easily. The difficulty is not in growth but in the formation of stable nuclei of the highly ordered structure: once such stable nuclei are formed growth proceeds at effective rates at high temperatures, the temperature coefficient of crystal growth being exponentially related to temperature. Clearly then the transformation will be facilitated by the introduction of ready-made nuclei of the diamond structure. Finely powdered diamond is not effective, probably because the surface structure of the grains becomes distorted in the crushing process, but the introduction of certain Group VIII metals that form carbides with structures similar to that of diamond is effective. Metals such as Fe, Ni, Rh, Pd, and Pt form carbides of the type R_xC (with $x > 4$) which are cubic with the same atomic coordinates as in diamond and unit-cell edges close to that of diamond. The nucleation and growth of crystals of diamond in the presence of such metal catalysts only takes place if the temperature and pressure exceed those of the eutectic point for the appropriate metal-carbon system and lie within the diamond stability field, that is to say diamond will crystallize in the laboratory only in a molten metallic medium. Synthetic diamonds invariably contain heavy traces of the nucleating metal, but natural diamonds are characteristically almost wholly devoid of metallic impurities; there must be some alternative means of facilitating the nucleation of diamond available in the earth, but as yet undiscovered. Natural diamonds are known to have formed in high temperature environments and it is clear from Fig 13.6 that high temperature implies high pressure for diamond to be stable.

Phases of variable composition

So far we have dealt only with phases of fixed and invariable composition. The arguments of classical thermodynamics are extendable to phases of variable composition without the necessity for *ad hoc* assumptions; and functions descriptive of the compositional variation, such as activity coefficients, can be determined experimentally. However the introduction of certain concepts from statistical thermodynamics makes the description of compositional variation in many cases easier to understand in physical terms. We begin with a purely classical approach and introduce statistical concepts to give physical meaning to the definition of the *ideal solution.*

Partial molar quantities

Let J represent any thermodynamic function and $n_1, n_2 \ldots n_i, n_j, \ldots$ the number of moles of the chemical species $1, 2, \ldots i, j, \ldots$ present in the system. The partial molar quantity $\bar{J}_i$ is then defined as

$$\bar{J}_i = \left(\frac{\partial J}{\partial n_i}\right)_{T,P,n_j}$$

In general the variation of J as a function of temperature, pressure, and composition of the system is expressible as

$$dJ = \left(\frac{\partial J}{\partial T}\right)_{P,n_i,n_j} dT + \left(\frac{\partial J}{\partial P}\right)_{T,n_i,n_j} dP + \sum \left(\frac{\partial J}{\partial n_i}\right)_{T,P,n_j} dn_i$$

i.e.
$$dJ = \left(\frac{\partial J}{\partial T}\right)_{P,n_i,n_j} dT + \left(\frac{\partial J}{\partial P}\right)_{T,n_i,n_j} dP + \sum_i \bar{J}_i \, dn_i$$

In the case of volume, for example, this becomes

$$dV = \alpha V \, dT - \chi V \, dP + \sum_i \bar{V}_i \, dn_i.$$

It should be noticed that for a system containing only a single chemical species $J = n_i \bar{J}_i$ and $\bar{J}_i$ is the magnitude of J per mole.

Chemical potential

Chemical potential was defined by Gibbs as μ_i such that

$$\mu_i = \left(\frac{\partial U}{\partial n_i}\right)_{S,V,n_j}$$

We shall now show that chemical potential is identical with partial molar free enthalpy

$$\bar{G}_i = \left(\frac{\partial G}{\partial n_i}\right)_{T,P,n_j}$$

It follows from the definitions of G and H that

$$G = U + PV - TS$$

hence
$$dG = dU + PdV + VdP - TdS - SdT.$$

But
$$dG = -SdT + VdP + \sum_i \bar{G}_i \, dn_i$$

therefore
$$dU = -PdV + TdS + \sum_i \bar{G}_i \, dn_i.$$

But
$$dU = \left(\frac{\partial U}{\partial V}\right)_{S,n_i,n_j} dV + \left(\frac{\partial U}{\partial S}\right)_{T,n_i,n_j} dS + \sum_i \mu_i \, dn_i$$

Comparison of these two expressions for dU yields the identity

$$\mu_i = \bar{G}_i = \left(\frac{\partial G}{\partial n_i}\right)_{T,P,n_j}$$

and we can therefore write the practically useful equation

$$dG = -S\,dT + V\,dP + \sum_i \mu_i\,dn_i.$$

This expression can be integrated at constant temperature and pressure by the device of allowing each n_i to change by an amount proportional to itself. Suppose that the quantity of the phase is increased in the proportion $(1+d\xi):1$ at constant relative composition. Since G is extensive,

$$dG = G d\xi$$

and since relative composition is maintained constant,

$$dn_i = n_i d\xi \quad \text{for all } i.$$

But for $dT = 0$, $dP = 0$, we have

$$dG = \sum_i \mu_i dn_i$$

therefore $\quad G d\xi = \sum_i \mu_i n_i d\xi.$

Division by $d\xi \neq 0$ yields the integrated expression

$$G = \sum_i \mu_i n_i.$$

It follows that for phases composed of a single chemical species $G = \mu_i n_i$, that is chemical potential corresponds to molar free enthalpy.

That chemical potential is an intensive property can simply be shown by considering two portions of the same phase of masses m_1 and $m_2 = \lambda m_1$, containing n_i and λn_i moles of species i respectively. If the free enthalpies of the two portions are denoted G_1 and G_2, $G_2 = \lambda G_1$ since free enthalpy is extensive, and the chemical potential of i in the two portions, μ_i^1 and μ_i^2 will be given by

$$\mu_i^1 = \left(\frac{\partial G_1}{\partial n_i}\right)_{T,P,n_j} \quad \text{and} \quad \mu_i^2 = \left(\frac{\partial \lambda G_1}{\partial \lambda n_i}\right)_{T,P,n_j}$$

therefore $\quad \mu_i^1 = \mu_i^2.$

The condition for equilibrium, $dT = 0$, $dP = 0$, $dG = 0$ will be modified if the composition of phases in equilibrium is allowed to vary. Let us consider a system composed of several phases, denoted $\alpha, \beta, \ldots$; then for the phase α

$$dG^\alpha = -S^\alpha\,dT + V^\alpha\,dP + \sum_i \mu_i^\alpha\,dn_i^\alpha$$

and similarly for every other phase. For the whole system

$$dG = \sum_\alpha dG^\alpha.$$

At constant temperature and pressure,

$$dG^\alpha = \sum_i \mu_i^\alpha\,dn_i^\alpha.$$

Now suppose that the only change in the system at constant T and P is the transfer of a quantity dn_i of chemical species i from phase α to phase β,

i.e. $\quad dn_i^\alpha + dn_i^\beta = 0 \quad$ and $\quad dn_j^\alpha = dn_j^\beta = \cdots = 0$ for all $j \neq i$

hence $\quad dG^\alpha = \mu_i^\alpha(-dn_i),\ dG^\beta = \mu_i^\beta(+dn_i),$ and $dG^\gamma = dG^\delta = \cdots = 0.$

Therefore for the whole system

$$dG = (\mu_i^\beta - \mu_i^\alpha)dn_i.$$

But at equilibrium at constant temperature and pressure

$$dG = 0$$

therefore $\quad \mu_i^\alpha = \mu_i^\beta,$

that is, the chemical potential of each chemical species i has the same value for that component throughout the system at equilibrium.

Now for a natural process we have shown that

$$dT = 0,\ dP = 0,\ dG < 0$$

and so $\quad (\mu_i^\beta - \mu_i^\alpha)dn_i < 0.$

Since we have specified that dn_i is positive,

$$\mu_i^\beta < \mu_i^\alpha$$

and therefore species i passes from the phase in which its chemical potential is higher into that in which it is lower to produce uniform μ_i throughout the system at equilibrium.

That changes in the chemical potentials of different chemical species are not independent at equilibrium is shown by the following argument. For a single phase system we have shown that

$$G = \sum_i \mu_i n_i.$$

Differentiating,

$$dG = \sum_i \mu_i\, dn_i + \sum_i n_i\, d\mu_i$$

but $\quad dG = -S\,dT + V\,dP + \sum_i \mu_i\, dn_i$

therefore $\quad S\,dT - V\,dP + \sum_i n_i\, d\mu_i = 0.$

This relationship, known as the *Gibbs–Duhem Relation*, indicates that the intensive quantities T, P, μ_i are not independently variable. At equilibrium at constant temperature and pressure it follows that

$$dT = 0,\ dP = 0,\ \sum_i n_i\, d\mu_i = 0,$$

so that in a phase composed of two chemical species labelled 1 and 2,

$$n_1\, d\mu_1 + n_2\, d\mu_2 = 0,$$

i.e. as the chemical potential of one species increases that of the other decreases.

So far we have discussed chemical potential only at constant temperature and pressure. The temperature dependence of μ_i follows from the cross-differentiation identity

$$\left[\frac{\partial}{\partial T}\left(\frac{\partial G}{\partial n_i}\right)_{P,T,n_j}\right]_{P,n_i,n_j} = \left[\frac{\partial}{\partial n_i}\left(\frac{\partial G}{\partial T}\right)_{P,n_i,n_j}\right]_{P,T,n_j}$$

i.e.
$$\left(\frac{\partial \mu_i}{\partial T}\right)_{P,n_i,n_j} = -\left(\frac{\partial S}{\partial n_i}\right)_{P,T,n_j}$$

therefore
$$\left(\frac{\partial \mu_i}{\partial T}\right)_{P,n_i,n_j} = -\bar{S}_i$$

where $\bar{S}_i$ is the partial molar entropy of species i.

A more useful partial differential is

$$\left(\frac{\partial(\mu_i/T)}{\partial T}\right)_{P,n_i,n_j} = \frac{1}{T}\left(\frac{\partial \mu_i}{\partial T}\right) - \frac{\mu_i}{T^2}$$

$$= \frac{-T\bar{S}_i - \mu_i}{T^2}$$

$$= -\frac{\bar{H}_i}{T^2}$$

The pressure dependence of chemical potential follows from

$$\left[\frac{\partial}{\partial P}\left(\frac{\partial G}{\partial n_i}\right)_{P,T,n_j}\right]_{T,n_i,n_j} = \left[\frac{\partial}{\partial n_i}\left(\frac{\partial G}{\partial P}\right)_{T,n_i,n_j}\right]_{P,T,n_j}$$

i.e.
$$\left(\frac{\partial \mu_i}{\partial P}\right)_{T,n_i,n_j} = \left(\frac{\partial V}{\partial n_i}\right)_{P,T,n_j} = \bar{V}_i.$$

The dependence of chemical potential on composition cannot usefully be explored in general terms. Simplifying assumptions based on statistical thermodynamics must be made if expressions that are directly applicable to experimental data are to be developed. We deal under the next heading with one such simplification especially appropriate to condensed phases of variable composition.

Ideal solutions

In classical thermodynamics an ideal solution is defined as a solution in which every component obeys the equation

$$\mu_i = \mu_i^* + RT \ln n_i$$

where the *mol fraction* of component i is defined as $x_i = n_i/\sum n_i$, n_i being the molar quantity of component i in the solution and the summation being taken over all components, and μ_i^* is a constant at constant temperature and pressure. It is usual to distinguish *perfect solutions* (analogous to perfect gas mixtures) as solutions in which every component obeys the ideal solution equation over the whole composition range, $0 \leqslant x_i \leqslant 1$ for all i; μ_i^* is then the molar free enthalpy G_i of pure component i at the given temperature and pressure,

i.e.
$$\lim_{x_i \to 1} \mu_i = \mu_i^*.$$

All other solutions tend towards ideality at extreme dilution: the ideal solution equation for the solvent, usually indicated by the subscript 0, is obeyed only as $x_0 \to 1$

so that $\mu_0^* = G_0$ as before, but the solute components only obey the ideal solution equation as $x_i \to 0$ so that μ_i^* cannot be identified with G_i. We shall here be concerned in the main with ideal solutions that are perfect and we shall describe such simply as ideal solutions. Where we are concerned with solutions that become ideal only when very dilute we shall draw attention to the restriction.

We can alternatively, and profitably, define an ideal solution in terms of a simple statistical model. Suppose that two substances AX and BX have crystal structures that differ only in dimensions, that is to say the atomic coordinates of the X atoms are identical in both unit-cells and the coordinates of the A atoms in AX are the same as those of the B atoms in BX. Let us further suppose that AX and BX form a solid solution and consider an amount of this solid solution containing N_A atoms of A, N_B atoms of B, and $N_A + N_B$ atoms of X. The first coordination shell about a site occupied by an A or a B atom will be occupied by X atoms invariably, but the second coordination shell will be occupied by A and B atoms. The occupational probabilities of a site 1 and of a site 2 in the second coordination shell of 1 will then be

$$\text{probability of A on 1 and A on 2} = \left(\frac{N_A}{N_A + N_B}\right)^2,$$

$$\text{probability of A on 1 and B on 2} = \frac{N_A N_B}{(N_A + N_B)^2},$$

$$\text{probability of B on 1 and A on 2} = \frac{N_A N_B}{(N_A + N_B)^2},$$

$$\text{probability of B on 1 and B on 2} = \left(\frac{N_B}{N_A + N_B}\right)^2.$$

Since we can make no distinction between an A—B and a B—A pair the occupational probabilities of such a pair of sites by pairs of atoms will be:

$$\text{A—A} \quad \frac{N_A^2}{(N_A + N_B)^2}, \quad \text{B—B} \quad \frac{N_B^2}{(N_A + N_B)^2}, \quad \text{A—B} \quad \frac{2N_A N_B}{(N_A + N_B)^2}$$

If we suppose that the coordination number of site 1 with respect to its second coordination shell is z, there will be a total of $\frac{1}{2}z(N_A + N_B)$ pairs of sites such as 1 and 2 in the whole sample of solid solution and the total number of adjacent pairs of A, B atoms will b

$$\text{A—A} \quad \tfrac{1}{2}z\frac{N_A^2}{N_A + N_B}$$

$$\text{B—B} \quad \tfrac{1}{2}z\frac{N_B^2}{N_A + N_B}$$

$$\text{A—B} \quad z.\frac{N_A N_B}{N_A + N_B}$$

There will in general be an interaction energy between the A, B atoms occupying such adjacent sites. If AX and BX are to form a solid solution of any kind, A and B must be chemically similar to the extent of both being cationic (or both anionic) so that the interaction forces will always be repulsive. Let w_{AA} be the increase in potential energy when a pair of A atoms are brought from infinity to adjacent sites in the solid solution and define w_{BB} and w_{AB} similarly. Therefore that part of the potential energy of the sample of solid solution due to the interaction energy of A and B atoms on adjacent sites will be given by

$$\frac{zw_{AA}}{2}\cdot\frac{N_A^2}{N_A+N_B}+\frac{zw_{BB}}{2}\cdot\frac{N_B^2}{N_A+N_B}+zw_{AB}\frac{N_A N_B}{N_A+N_B},$$

whereas for an amount of pure AX containing N_A atoms of A and an amount of pure BX containing N_B atoms of B the contribution of such nearest neighbour interaction to the potential energy will be

$$\frac{zw_{AA}}{2}\cdot N_A+\frac{zw_{BB}}{2}\cdot N_B.$$

Therefore the increase in potential energy when N_A molecules of AX and N_B molecules of BX enter into solid solution is

$$\frac{zw_{AA}}{2}\left(\frac{N_A^2}{N_A+N_B}-N_A\right)+\frac{zw_{BB}}{2}\left(\frac{N_B^2}{N_A+N_B}-N_B\right)+zw_{AB}\frac{N_A N_B}{N_A+N_B}$$
$$=\frac{z\,N_A N_B}{N_A+N_B}\left(w_{AB}-\frac{w_{AA}+w_{BB}}{2}\right).$$

We have considered an amount of solid solution containing N_A+N_B atoms of X and if we define one mole of solid solution as the amount containing N (= Avogadro's number) atoms of X, the *molar excess energy of mixing* is given by

$$\Delta U_m=\frac{Nz\,N_A N_B}{(N_A+N_B)^2}\left(w_{AB}-\frac{w_{AA}+w_{BB}}{2}\right)$$

i.e. $$\Delta U_m=Nz\,x(1-x)\left(w_{AB}-\frac{w_{AA}+w_{BB}}{2}\right)$$

In order to discuss quantitatively the entropy of a solution we have to have recourse to one of the most important results of statistical thermodynamics, a result that we simply assume here without proof:[3]

$$S=k\ln\Omega$$

where k is Boltzmann's constant equal to R/N and Ω is the number of ways in which the atoms can arrange themselves. This equation gives quantitative expression to the qualitative statement that entropy is a measure of degree of disorder.

In the solid solution $A_xB_{1-x}X$ there are N_A+N_B sites that may be occupied by a number N_A of A atoms and a number N_B of B atoms. The number of possible arrangements of the A and B atoms on the N_A+N_B sites is then Ω given by

$$\Omega=\frac{(N_A+N_B)!}{N_A!N_B!}.$$

Therefore $$S=k\ln\frac{(N_A+N_B)!}{N_A!N_B!}.$$

[3] Textbooks of statistical thermodynamics to which the reader is referred for the arguments leading to the equation are Fowler and Guggenheim (1956), Rushbrooke (1960), and chapter 11 of Denbigh (1971).

Since N_A and N_B are very large (of the order of Avogadro's number, $\sim 10^{24}$) the approximation $\ln n! = n \ln n$, known as *Stirling's theorem*, is applicable.

Therefore
$$S = k\{(N_A+N_B)\ln(N_A+N_B) - N_A \ln N_A - N_B \ln N_B\}$$
$$= -k\left\{N_A \ln\frac{N_A}{N_A+N_B} + N_B \ln\frac{N_B}{N_A+N_B}\right\}$$
$$= -k(N_A+N_B)\{x\ln x + (1-x)\ln(1-x)\}$$
$$= -R\{x\ln x + (1-x)\ln(1-x)\} \quad \text{per mole of solid solution.}$$

The entropy evaluated here is not the whole molar entropy of the solid solution at temperature T, but the *configurational entropy* due to the disorder of the arrangement of different atoms on like structural sites. The configurational entropy is the molar entropy that a crystal of the solid solution would have if its arrangement of atoms at temperature T were frozen unchanged at absolute zero. There will in addition be a contribution $\int_0^T C_p . d\ln T$ to the total entropy due to thermal vibrational randomness of all the atoms in the structure. Since there is no randomness in the arrangement of atoms in pure AX and pure BX the configurational entropy for each of these end members of the solid solution will be zero. If it is assumed—and this is a reasonable assumption—that $\int_0^T (C_p)_{SS} d\ln T$ for the solid solution is equal to $x\int_0^T (C_p)_{AX} d\ln T + (1-x)\int_0^T (C_p)_{BX} d\ln T$, then the excess molar entropy of mixing for the solution is given by

$$\Delta S_m = -R[x\ln x + (1-x)\ln(1-x)].$$

Since $0 < x < 1$, $\Delta S_m > 0$.

We are now ready to define an ideal solution in terms of the statistical model as a solution for which $w_{AB} = \frac{1}{2}(w_{AA} + w_{BB})$ and for which the excess molar volume of mixing $\Delta V_m = 0$. The restriction $w_{AB} = \frac{1}{2}(w_{AA} + w_{BB})$ implies that in an ideal solution there is no energetic discrimination between like and unlike pairs of atoms occupying adjacent structurally equivalent positions. It follows from the first of these conditions that for an ideal solution $\Delta U_m = 0$ and, since $\Delta V_m = 0$, ΔH_m is likewise zero. The free enthalpy of mixing $\Delta G_m = \Delta H_m - T\Delta S_m$ will therefore be given for an ideal solution by

$$\Delta G_m = RT\{x\ln x + (1-x)\ln(1-x)\}.$$

The graph of ΔG_m as a function of x is shown in Fig 13.7.

In terms of the classical thermodynamic definition of an ideal solution,

$$\mu_i = \mu_i^* + RT\ln x_i$$

for all i, we have for the solution $A_xB_{1-x}X$,

$$\mu_{AX} = \mu_{AX}^* + RT\ln x$$

and
$$\mu_{BX} = \mu_{BX}^* + RT\ln(1-x)$$

so that for the solution the molar free enthalpy is given by

$$G = x\mu_{AX}^* + (1-x)\mu_{BX}^* + RT\{x\ln x + (1-x)\ln(1-x)\}.$$

But the free enthalpy of a mechanical mixture of x moles of AX and $1-x$ moles of BX is $x\mu_{AX}^* + (1-x)\mu_{BX}^*$ so that the excess free enthalpy of mixing is

$$\Delta G_m = RT\{x\ln x + (1-x)\ln(1-x)\}.$$

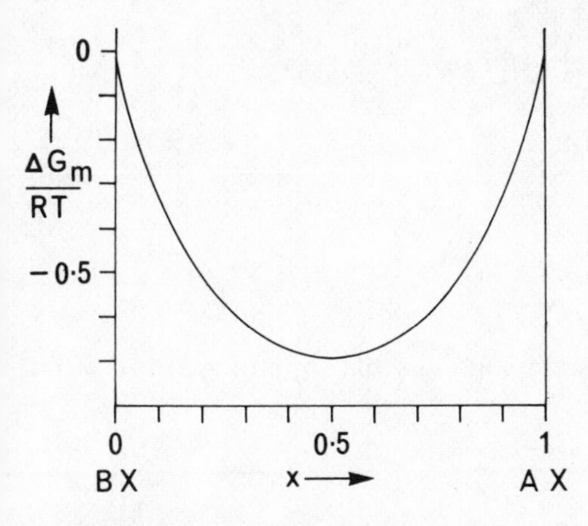

Fig 13.7 Plot of $\Delta G_m/RT = x \ln x + (1-x)\ln(1-x)$ against x for an ideal solution $A_xB_{1-x}X$.

The statistical model is therefore consistent with the classical definition of ideal solution.

In discussing the statistical model we have specifically considered a *solid solution* $A_xB_{1-x}X$, but the definition of an *ideal solution*, although especially appropriate to solid solutions, is equally applicable to liquid solutions.

In the next chapter we shall deal with several examples of ideal or quasi-ideal solid solution. We shall therefore not discuss any specific examples at this point.

Non-ideal solutions: strictly regular solutions

In general non-ideal solutions can be described by equations of the form

$$\mu_i = \mu_i^* + RT \ln a_i,$$

where the *activity* of component i, a_i, is related to the mol-fraction of i by $a_i = \gamma_i x_i$, the *activity coefficient* γ_i being a function of x_i. Such solutions tend to ideality as $\gamma_i \to 1$ and in general $\gamma_i \to 1$ for solutes as $x_i \to 0$.

There are kinds of non-ideal solutions that fit simple statistical models and of these the most important are *strictly regular solutions*,[4] for which $w_{AB} - \frac{1}{2}(w_{AA} + w_{BB}) = w \neq 0$ and $\Delta V_m = 0$. Therefore for a regular solution

$$\Delta H_m = \Delta U_m + \Delta V_m = \Delta U_m = zw \frac{N_A N_B}{N_A + N_B}$$

and

$$\Delta G_m = \Delta H_m - T\Delta S_m$$

$$= zw \frac{N_A N_B}{N_A + N_B} + kT\left(N_A \ln \frac{N_A}{N_A + N_B} + N_B \ln \frac{N_B}{N_A + N_B}\right).$$

If the number of moles of A and B atoms present are respectively n_A and n_B, such that $N_A = Nn_A$ and $N_B = Nn_B$, N being Avogadro's number, then

$$\Delta G_m = zwN \frac{n_A n_B}{n_A + n_B} + RT\left(n_A \ln \frac{n_A}{n_A + n_B} + n_B \ln \frac{n_B}{n_A + n_B}\right)$$

[4] When $w \neq 0$ there must be preferential interaction between A–B pairs relative to A–A and B–B pairs and therefore there cannot be complete randomness of arrangement. However the departure from randomness will be slight if zw is small relative to kT. We follow Guggenheim (1952) in denoting such solutions as *strictly regular* solutions and refer the reader to Fowler and Guggenheim (1956) for discussion of the more general case.

and $$\left(\frac{\partial \Delta G_{\mathrm{m}}}{\partial n_{\mathrm{A}}}\right)_{P,T,n_{\mathrm{B}}} = zwN\left(\frac{n_{\mathrm{B}}}{n_{\mathrm{A}}+n_{\mathrm{B}}}\right)^2 + RT\ln\frac{n_{\mathrm{A}}}{n_{\mathrm{A}}+n_{\mathrm{B}}}$$

therefore $\mu_{\mathrm{AX}} - \mu^*_{\mathrm{AX}} = zwN(1-x)^2 + RT\ln x$

i.e. $\mu_{\mathrm{AX}} = \mu^*_{\mathrm{AX}} + RT\ln x + zwN(1-x)^2$

and similarly

$$\mu_{\mathrm{BX}} = \mu^*_{\mathrm{BX}} + RT\ln(1-x) + zwNx^2.$$

We shall discuss specific examples of solid solutions that approximate to strict regularity in the next chapter.

Order–disorder

We make this digression at this point because the elementary thermodynamic treatment of order–disorder (OD) systems is a simple extension of our discussion of ideal solutions. We have already discussed order–disorder in the alloy β-brass in its crystallographic aspects in chapter 10 and there made use of some of the results that will now be derived. We begin by considering the alloy β-brass at the composition CuZn, pass on to consideration of the alloy $AuCu_3$, and in passing make some comments on the limitations of the elementary treatment of the thermodynamics of OD-systems.

Consider one mole of the alloy AB containing N atoms of element A and N atoms of element B. Suppose that the structure contains equal numbers, N, of two kinds of site, designated α and β, and that in the fully ordered structure all the A atoms lie on α-sites and all the B atoms on β-sites. In the completely disordered structure each type of site will be randomly occupied by $\frac{1}{2}N$ atoms of A and $\frac{1}{2}N$ atoms of B. Let the number of A atoms lying on α-sites in a certain state of disorder be $N_{\mathrm{A}\alpha}$, then in corresponding nomenclature the occupation of N sites of each kind by N atoms of each kind will be

	α-sites	β-sites
A atoms	$N_{\mathrm{A}\alpha}$	$N_{\mathrm{A}\beta} = N - N_{\mathrm{A}\alpha}$
B atoms	$N_{\mathrm{B}\alpha} = N - N_{\mathrm{A}\alpha}$	$N_{\mathrm{B}\beta} = N_{\mathrm{A}\alpha}$

We define a *degree of order* s equal to the difference in the proportion of atoms correctly and incorrectly placed relative to the fully ordered structure,

i.e. $$s = \frac{N_{\mathrm{A}\alpha}+N_{\mathrm{B}\beta}}{2N} - \frac{N_{\mathrm{A}\beta}+N_{\mathrm{B}\alpha}}{2N}$$

$$= \frac{N_{\mathrm{A}\alpha}}{N} - \frac{N-N_{\mathrm{A}\alpha}}{N}$$

$$= \frac{2N_{\mathrm{A}\alpha}}{N} - 1$$

therefore $N_{\mathrm{A}\alpha} = \dfrac{N}{2}(1+s)$.

This definition of degree of order leads to values of s equal to zero and unity for the completely disordered and the fully ordered structures respectively:

	$N_{A\alpha}$	$N_{A\beta}$	$N_{B\alpha}$	$N_{B\beta}$	
$0 < s < 1$	$\frac{1}{2}N(1+s)$	$\frac{1}{2}N(1-s)$	$\frac{1}{2}N(1-s)$	$\frac{1}{2}N(1+s)$	
$s = 0$	$\frac{1}{2}N$	$\frac{1}{2}N$	$\frac{1}{2}N$	$\frac{1}{2}N$	completely disordered
$s = 1$	N	0	0	N	fully ordered

In discussing ideal solutions we were concerned with the arrangement of two kinds of atom on one type of structural site; here we are concerned with the arrangement of two kinds of atom on *two* types of structural site. Just as for ideal solutions the number of ways of arranging A and B atoms on α-sites, irrespective of their arrangement on β-sites, is given by

$$\frac{N!}{\{\frac{1}{2}N(1+s)\}!\{\frac{1}{2}N(1-s)\}!}$$

and the number of possible arrangements of A and B atoms on β-sites is given by an identical expression. The number of possible arrangements of A and B atoms on α- and β-sites is thus

$$\Omega = \left[\frac{N!}{\{\frac{1}{2}N(1+s)\}!\{\frac{1}{2}N(1-s)\}!}\right]^2$$

The excess molar configurational entropy of the system relative to one mole of the fully ordered structure is therefore

$$S_c = k\ln\left[\frac{N!}{\{\frac{1}{2}N(1+s)\}!\{\frac{1}{2}N(1-s)\}!}\right]^2$$

which simplifies on application of Stirling's theorem, N being large (N = Avogadro's number), to

$$S_c = R\{2\ln 2-(1+s)\ln(1+s)-(1-s)\ln(1-s)\}.$$

The excess molar configurational entropy of the completely disordered structure ($s = 0$) over the fully ordered structure ($s = 1$) is thus

$$(S_c)_{max} = 2R\ln 2.$$

We turn now to consider the *configurational energy* of the system and again our argument follows the same pattern as for ideal solutions. Suppose that atoms on α-sites are bonded only to atoms on β-sites and that the coordination number for both types of site is z. The A atoms on α-sites will give rise to $zN_{A\alpha}$ bonds of which the proportion $N_{A\beta}/N$ will be A—A bonds and the remainder A—B bonds. Likewise the B atoms on α-sites will give rise to $zN_{B\alpha}$ bonds of which the proportion $N_{B\beta}/N$ will be B—B bonds and the remainder B—A (indistinguishable from A—B) bonds. The number of each kind of bond per mole will thus be:

$$n_{AA} = zN_{A\alpha}\frac{N_{A\beta}}{N} = zN\frac{1-s^2}{4}$$

$$n_{BB} = zN_{B\alpha}\frac{N_{B\beta}}{N} = zN\frac{1-s^2}{4}$$

$$n_{AB} = z(N_{A\alpha}+N_{B\alpha})-n_{AA}-n_{BB}$$

$$= zN-zN\frac{1-s^2}{2} = zN\frac{1+s^2}{2}.$$

Suppose that the energy U^* of the assemblage of N atoms of A and N atoms of B is reduced by the amounts w_{AA}, w_{BB}, w_{AB} by the formation of each A—A, B—B, A—B bond respectively; the quantities w_{AA} etc are bond energies and are positive. The energy of one mole of the AB crystal is then

$$U = U^* - \{n_{AA} w_{AA} + n_{BB} w_{BB} + n_{AB} w_{AB}\}$$
$$= U^* - \frac{zN}{2}\{\tfrac{1}{2}(w_{AA}+w_{BB})(1-s^2)+w_{AB}(1+s^2)\}$$

The molar energy of the fully ordered state ($s = 1$) is consequently $U^* - zNw_{AB}$ so that the excess molar configurational energy of the system relative to the fully ordered state is

$$U_c = -\frac{zN}{2}\{\tfrac{1}{2}(w_{AA}+w_{BB})(1-s^2)+w_{AB}(1+s^2)\} + zNw_{AB}$$
$$= -\frac{zN}{2}\{\tfrac{1}{2}(w_{AA}+w_{BB})-w_{AB}\}(1-s^2)$$
$$= \frac{zNw}{2}(1-s^2),$$

where $w = w_{AB} - \frac{1}{2}(w_{AA}+w_{BB})$. The excess molar configurational energy of the completely disordered structure ($s = 0$) relative to the fully ordered structure ($s = 1$) is thus $\frac{1}{2}zNw$. The energy difference between the completely disordered and fully ordered states is thus proportional to w, which depends on the pair of elements concerned, and to z, which is of course structure dependent.

Since we are concerned here with a solid system, configurational energy and configurational enthalpy will be effectively equal $U_c \simeq H_c$. Therefore we can write for the excess molar *configurational free enthalpy* of the system relative to the fully ordered structure

$$G_c = H_c - TS_c \simeq U_c - TS_c$$

i.e. $$G_c = \frac{zNw}{2}(1-s^2) - RT\{2\ln 2 - (1+s)\ln(1+s) - (1-s)\ln(1-s)\}.$$

At equilibrium $\partial G_c/\partial s$ will be zero,

i.e. $$\frac{\partial G_c}{\partial s} = -zNws + RT\{\ln(1+s) - \ln(1-s)\} = 0$$

i.e. $$\ln\frac{1+s}{1-s} = \frac{zws}{kT}.$$

This equation can conveniently be solved for selected values of zw/kT by a graphical method. The reader familiar with hyperbolic functions will notice that

$$\ln\frac{1+s}{1-s} = 2\tanh^{-1} s$$

and will be aware that tables of $\tanh^{-1}$ are available. Rewriting the equilibrium condition as

$$\tanh^{-1} s = \frac{zws}{2kT} = x,$$

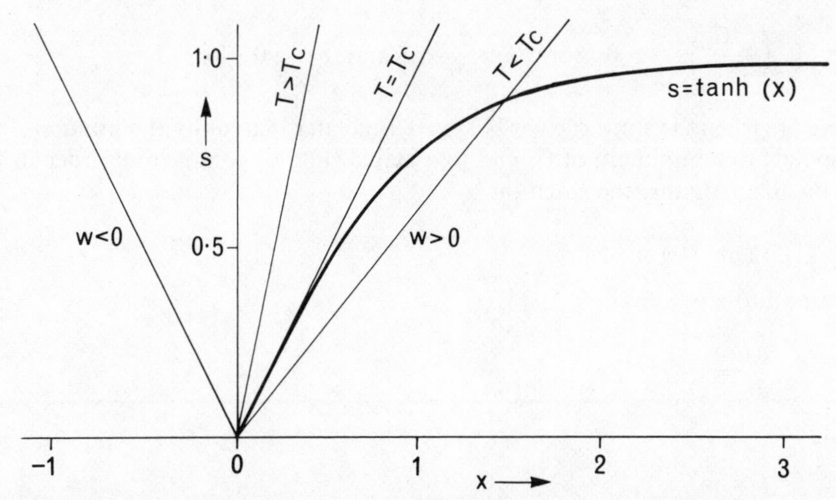

Fig 13.8 Graphical solution of $\ln[(1+s)/(1-s)] = zws/kT$. The figure shows the curve of $s = \tanh(x)$ and a set of lines $s = (2kT/zw)x$ for various values of T. For w positive each line makes two intersections with the curve if $T < T_c$ and only one solution, $s = 0$, for higher temperatures. For w negative there is a single intersection, at $s = 0$, and the alloy is unstable.

we plot a graph (Fig 13.8) of $\tanh^{-1} s$ against x and a set of straight lines through the origin to represent

$$s = \frac{2kT}{zw}x$$

for various slopes $2kT/zw$. The intersection of the curve $s = \tanh(x)$ with any of the straight lines $s = (2kT/zw)x$ represents a solution of the equilibrium equation for a particular value of $2kT/zw$ as illustrated in Fig 10.38.

When the slope of $s = (2kT/zw)x$ is negative there is a single solution at $s = 0$. Since k, T, and z are necessarily positive the negative slope must imply that w is negative, that is $w_{AB} < \frac{1}{2}(w_{AA} + w_{BB})$, which means that the strength of an A—B bond is less than the mean strength of A—A and B—B bonds. When w is negative then the alloy AB is unstable, the stable state consisting of a mixture of crystals of the pure element A and the pure element B. Such *exsolution* is not of immediate concern to us here, but will be discussed in another context in chapter 14.

When the slope of $s = (2kT/zw)x$ is positive, w must be positive and the alloy AB, whether ordered or disordered, is stable. For all positive slopes of the line there is evidently a solution of the equilibrium equation at $s = 0$ and, when the slope has decreased below a certain limiting magnitude, a second solution at $0 < s \leqslant 1$. The stability of the solution at $s = 0$ can simply be investigated by consideration of the sign of the second differential G_c with respect to s. We have already shown that

$$\frac{\partial G_c}{\partial s} = -zNws + RT\{\ln(1+s) - \ln(1-s)\}$$

therefore
$$\frac{\partial^2 G_c}{\partial s^2} = -zNw + \frac{2RT}{1-s^2}$$

hence
$$\left(\frac{\partial^2 G_c}{\partial s^2}\right)_{s=0} = -zNw + 2RT,$$

and so $\left(\frac{\partial^2 G_c}{\partial s^2}\right)_{s=0} > 0$ for $T > \frac{zw}{2k}$ and < 0 for $T < \frac{zw}{2k}$.

Therefore when the slope of $s = (2kT/zw)x$ is greater than unity the solution $s = 0$ corresponds to a minimum of G_c and to a stable state of complete disorder in the alloy. By differentiating the equation

$$\tanh^{-1}(s) = x$$

with respect to x we obtain

$$\frac{ds}{dx} = 1 - s^2$$

so that

$$\left(\frac{ds}{dx}\right)_{s=0} = 1$$

and for $s > 0$

$$\frac{ds}{dx} < 1.$$

The line $s = (2kT/zw)x$ is therefore tangential to the curve $\tanh^{-1}(s) = x$ at the origin when its slope is unity and this situation corresponds to the change in sign of $\partial^2 G_c/\partial s^2$ for the root $s = 0$. Since the slope of $\tanh^{-1}(s) = x$ decreases smoothly from unity as s and x increase from zero this same limit $2kT/zw = 1$ represents the boundary between the case where the equilibrium equation has a single (stable) solution at $s = 0$ and the case where it has two roots, $s = 0$ (unstable) and $0 < s \leqslant 1$. That this second root represents a minimum of G_c can be seen by writing down the condition for $\partial^2 G_c/\partial s^2$ to be positive for $0 < s \leqslant 1$,

$$-zNw + \frac{2RT}{1-s^2} > 0$$

i.e. $$T > \frac{zw}{2k}(1-s^2).$$

There is thus a stable solution of the equilibrium equation for all temperatures between zero K and $T < zw/2k$ for some value of the degree of order s such that $0 < s \leqslant 1$. For each temperature in this range there will be an equilibrium value for the degree of order s and, as we have already seen, for temperatures greater than $zw/2k$, s will be uniformly zero.

The temperature at which, for given values of z and w, the sign of $\partial^2 G_c/\partial s^2$ changes for $s = 0$ is known as the critical temperature $T_c = zw/2k$ for the order–disorder transformation.

In order to explore the temperature dependence of degree of ordering, it is convenient to rewrite the equilibrium equation

$$\tanh^{-1}(s) = \frac{zws}{2kT}$$

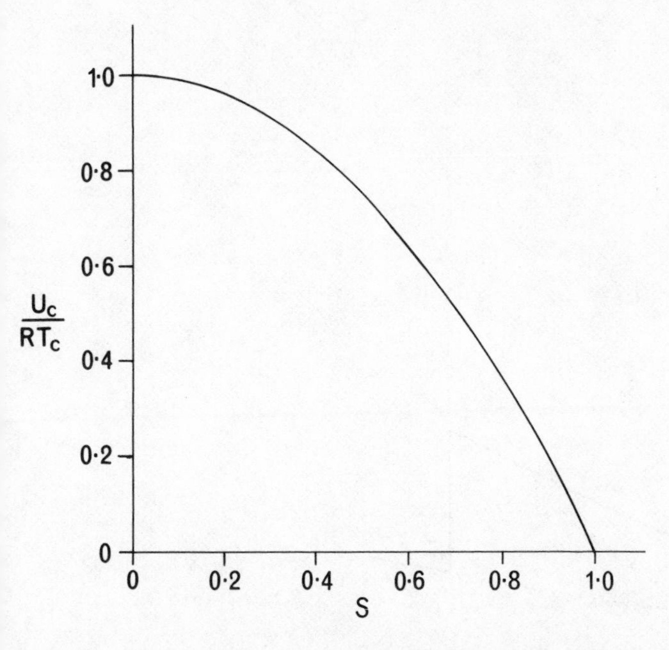

Fig 13.9 Plot of $U_c/RT_c = 1 - s^2$ against s for the Bragg–Williams model of an AB alloy. For a given change δs in the degree of order s, the magnitude of the configurational energy change δU_c decreases sharply as $s \to 0$, a characteristic of a cooperative process.

in terms of the critical temperature T_c in the form

$$\frac{T}{T_c} = \frac{s}{\tanh^{-1}(s)}$$

and to plot s against T/T_c as in Fig 10.38. It is immediately apparent that s changes slowly at first from unity at absolute zero and then increasingly rapidly as the temperature rises, becoming catastrophic as the critical temperature is approached. This increase in the rate of change of s with temperature is accompanied by a progressive decrease in the amount of energy required to produce the same amount of change in the degree of order (Fig 13.9). The expression for configurational energy

$$U_c = \frac{zNw}{2}(1 - s^2)$$

can be rewritten in terms of critical temperature as

$$U_c = RT_c(1 - s^2)$$

and differentiated with respect to s to give

$$\frac{\partial U_c}{\partial s} = -2RT_c s.$$

In the limit as $s \to 0$, that is as $T \to T_c$, $\partial U_c/\partial s \to 0$; this is the essential property of a cooperative process, δU_c for a given change δs decreases sharply as $s \to 0$.

Characteristic of order–disorder transformations is the form of the variation of

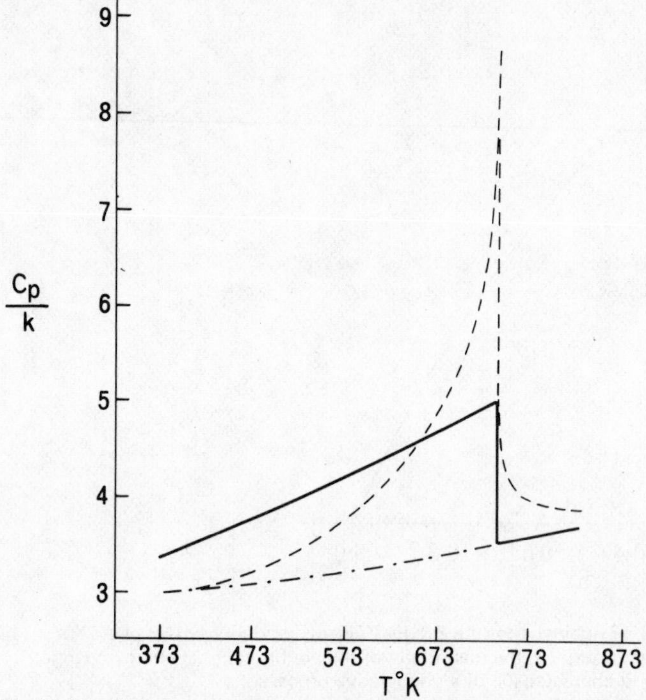

Fig 13.10 Variation of atomic heat capacity with temperature for CuZn. The broken curve represents experimental measurements on the alloy CuZn. The dot-dash curve represents the mean of experimental data for the pure metals Cu and Zn. The solid curve represents the excess atomic heat capacity calculated on the Bragg–Williams model over the mean atomic heat capacities of Cu and Zn.

specific heat with temperature. The configurational contribution to specific heat is given by

$$C_p = \frac{\partial H_c}{\partial T} \simeq \frac{\partial U_c}{\partial T} = -2RT_c s \frac{\partial s}{\partial T}.$$

Configurational specific heat rises from zero at zero K to a peak as T_c is approached and at T_c drops to zero (Fig 13.10). The characteristic shape of the specific heat curve in the neighbourhood of the critical temperature, resembling the Greek letter λ, provides the commonly used name 'lambda point transformation' for this sort of order–disorder transformation; the critical temperature is sometimes referred to as the 'lambda point'.

This elementary thermodynamic treatment of order–disorder in an AB alloy, known as the Bragg–Williams treatment or the 'zeroth approximation', provides a useful preliminary interpretation of experimental data for the alloy CuZn. We have already, in chapter 10, indicated some of the inadequacies of this elementary approach. The more sophisticated treatment, known as the quasi-chemical method or the 'first approximation' is very much more powerful and provides a closer correspondence with experimental evidence. Moreover disordering in the alloy AuCu, (Fig 10.34) which has a normal phase transition rather than a lambda point transformation, cannot be interpreted in Bragg–Williams terms but is amenable to the quasi-chemical approach. The thermodynamics of the quasi-chemical method is however outside the scope of this book; the interested reader is referred to the excellent account in Guggenheim (1952).

We turn now to consider disordering in the alloy $AuCu_3$ using an approximate treatment analogous to the Bragg–Williams treatment for CuZn. The cubic unit-cell of the alloy $AuCu_3$ (Fig 10.34) has two types of site; one, which we designate α, lies at the corners of the unit-cell, 0, 0, 0, and the other, designated β, is at the centre of each face, $0, \frac{1}{2}, \frac{1}{2}; \frac{1}{2}, 0, \frac{1}{2}; \frac{1}{2}, \frac{1}{2}, 0$. When the alloy is fully ordered the gold atom lies on the single α-site and the three copper atoms lie on the β-sites; the lattice type is cubic-P. In the completely disordered alloy the α and β structural sites are all occupied statistically by $Au_{0{\cdot}25}Cu_{0{\cdot}75}$; the lattice type is cubic-F.

Consider one mole of the alloy AB_3 consisting of N atoms of A and $3N$ atoms of B distributed, according to the degree of order in the structure, over N α-sites and $3N$ β-sites. Using the same nomenclature as before, site occupation will be:

	α-sites	β-sites
A atoms	$N_{A\alpha}$	$N_{A\beta} = N - N_{A\alpha}$
B atoms	$N_{B\alpha} = N - N_{A\alpha}$	$N_{B\beta} = 2N + N_{A\alpha}$

In the fully ordered structure $N_{A\alpha} = N$ and in the completely disordered structure $N_{A\alpha} = \frac{1}{4}N$. The degree of order, s, may be defined as a linear function of $N_{A\alpha}$ such that $s = 1$ for the fully ordered and $s = 0$ for the completely disordered structure, that is $N_{A\alpha} = \frac{1}{4}N(1+3s)$. Degree of order as so defined is not equal to the difference between the proportions of correctly and incorrectly placed atoms as was the case for the AB alloy; it is more convenient to define s so that its limits are zero and unity. Site occupation expressed in terms of s is then

	α-sites	β-sites
A atoms	$\frac{1}{4}N(1+3s)$	$\frac{3}{4}N(1-s)$
B atoms	$\frac{3}{4}N(1-s)$	$\frac{3}{4}N(3+s)$

The number of ways of arranging A and B atoms on the N α-sites is simply

$$\frac{N!}{\{\frac{1}{4}N(1+3s)\}!\{\frac{3}{4}N(1-s)\}!}$$

and on the $3N$ β-sites

$$\frac{(3N)!}{\{\frac{3}{4}N(1-s)\}!\{\frac{3}{4}N(3+s)\}!}.$$

The configurational entropy of the system is thus

$$S_c = k \ln \left[\frac{N!}{\{\frac{1}{4}N(1+3s)\}!\{\frac{3}{4}N(1-s)\}!}\right]\cdot\left[\frac{(3N)!}{\{\frac{3}{4}N(1-s)\}!\{\frac{3}{4}N(3+s)\}!}\right]$$

which simplifies on application of Stirling's theorem to

$$S_c = R\{4\ln 4 - \tfrac{3}{4}(1-s)\ln 3 - \tfrac{1}{4}(1+3s)\ln(1+3s) - \tfrac{3}{2}(1-s)\ln(1-s) \\ - \tfrac{3}{4}(3+s)\ln(3+s)\}.$$

The excess molar configurational entropy of the completely disordered state ($s = 0$) over the fully ordered state ($s = 1$) is therefore

$$(S_c)_{max} = R\{4\ln 4 - 3\ln 3\}.$$

The calculation of configurational energy is slightly more complicated than for the AB alloy. In $AuCu_3$ each α-site has 12 nearest neighbours which are all β-sites while each β-site has 4 nearest neighbour α-sites and 8 nearest neighbour β-sites (Fig 10.34). With this structure in mind we take the coordination of an atom on an α-site as z and the coordination of an atom on a β-site as $\frac{1}{3}z$ by atoms on α-sites plus $\frac{2}{3}z$ by atoms on other β-sites; there is no advantage in using the numerical value of z yet. From the N atoms on α-sites there will be zN bonds to nearest neighbour atoms on β-sites, of which

$$zN_{A\alpha}\cdot\frac{N_{A\beta}}{3N} \quad \text{will be A—A bonds,}$$

$$zN_{B\alpha}\cdot\frac{N_{B\beta}}{3N} \quad \text{will be B—B bonds,}$$

and

$$zN_{A\alpha}\cdot\frac{N_{B\beta}}{3N}+zN_{B\alpha}\cdot\frac{N_{A\beta}}{3N} \quad \text{will be A—B bonds.}$$

In order to evaluate the number of bonds between atoms on β-sites it is necessary to subdivide the equivalent β-sites into $\beta_1, \beta_2, \beta_3$ with N atoms on each sub-type of site. (In $AuCu_3$ the β_1, β_2, β_3 sites may be taken to be at the centres of the faces perpendicular to the x, y, z axes respectively.) From the N atoms on β_1-sites there will be $\frac{1}{3}zN$ bonds to atoms on β_2 sites, comprising

$$\frac{z}{3}N_{A\beta_1}\frac{N_{A\beta_2}}{N}=\frac{zN_{A\beta}^2}{27N} \quad \text{A—A bonds,}$$

$$\frac{z}{3}N_{B\beta_1}\frac{N_{B\beta_2}}{N}=\frac{zN_{B\beta}^2}{27N} \quad \text{B—B bonds,}$$

and

$$\frac{z}{3}N_{A\beta_1}\frac{N_{B\beta_2}}{N}+\frac{z}{3}N_{B\beta_1}\frac{N_{A\beta_2}}{N}$$

$$=\frac{2zN_{A\beta}N_{B\beta}}{27N} \quad \text{A—B bonds.}$$

There will in addition be an equal contribution to the number of bonds of each kind from pairs of atoms on β_1 and β_3 sites and on β_2 and β_3 sites so that the total number of bonds of each kind between atoms on β-sites will be $zN_{A\beta}^2/9N$ A—A bonds, $zN_{B\beta}^2/9N$ B—B bonds and $2zN_{A\beta}N_{B\beta}/9N$ A—B bonds. The total number of bonds of each kind per mole will therefore be

$$n_{AA}=\frac{zN_{A\alpha}N_{A\beta}}{3N}+\frac{zN_{A\beta}^2}{9N}=\frac{zN}{8}(1-s^2)$$

$$n_{BB}=\frac{zN_{B\alpha}N_{B\beta}}{3N}+\frac{zN_{B\beta}^2}{9N}=\frac{zN}{8}(9-s^2)$$

$$n_{AB}=\frac{z(N_{A\alpha}N_{B\beta}+N_{B\alpha}N_{A\beta})}{3N}+\frac{2zN_{A\beta}N_{B\beta}}{9N}$$

$$=\frac{zN}{4}(3+s^2).$$

Again suppose that the energy U^* of the assemblage of N atoms of A and $3N$ atoms of B is reduced by the amounts w_{AA}, w_{BB}, w_{AB} on the formation of each A—A, B—B, A—B bond. The energy of one mole of AB_3 will then be

$$U = U^* - (n_{AA}w_{AA} + n_{BB}w_{BB} + n_{AB}w_{AB})$$

which will have the value of $U^* - zN(w_{BB} + w_{AB})$ for the fully ordered state ($s = 1$). The excess molar configurational energy of the structure at degree of order s relative to the fully ordered state is then

$$U_c = -\frac{zN}{8}(1-s^2)w_{AA} - \frac{zN}{8}(9-s^2)w_{BB} - \frac{zN}{4}(3+s^2)w_{AB} + zNw_{BB} + zNw_{AB}$$
$$= \frac{zNw}{4}(1-s^2),$$

where, as for the AB alloys, $w = w_{AB} - \frac{1}{2}(w_{AA} + w_{BB})$. The configurational energy of the completely disordered state ($s = 0$) is thus $\frac{1}{4}zNw$.

We are now in a position to express the excess molar configurational free enthalpy of the system in terms of its degree of order as

$$G_c = H_c - TS_c \simeq U_c - TS_c$$
$$= \frac{zNw}{4}(1-s^2) - RT\{4\ln 4 - \tfrac{3}{4}(1-s)\ln 3 - \tfrac{1}{4}(1+3s)\ln(1+3s)$$
$$- \tfrac{3}{2}(1-s)\ln(1-s) - \tfrac{3}{4}(3+s)\ln(3+s)\}.$$

We obtain the equilibrium condition by differentiating G_c with respect to s,

$$\frac{\partial G_c}{\partial s} = -\tfrac{1}{2}zNws - \tfrac{3}{4}RT\{\ln 3 - \ln(1+3s) + 2\ln(1-s) - \ln(3+s)\},$$

and setting $\partial G_c/\partial s = 0$, whence

$$\ln\frac{(1+3s)(3+s)}{3(1-s)^2} = \frac{2zws}{3kT}.$$

This equation has a root at $s = 0$ for all values of $2zw/3kT$. It also has non-zero roots at values of $2zw/3kT$ in excess of a certain magnitude corresponding to a critical temperature. In order to evaluate these non-zero roots and to determine the critical temperature it is necessary to proceed in a rather different way from the way in which the corresponding problem was solved for the AB alloy. It is clear from Fig 13.11 that $G_c(s) - G_c(0)$ has a single minimum at $s = 0$ at high temperatures. As temperature decreases this simple solution is succeeded by one in which $G_c(s) - G_c(0)$ has two minima, one at $s = 0$ and the other at s^* where $0 < s^* < 1$, separated by maximum. Initially $G_c(s^*) > G_c(0)$ so that the root $s = 0$ still represents the stable state. As temperature decreases (Fig 13.11) a situation arises where $G_c(s^*) = G_c(0)$; we take this as the condition for the *critical temperature*, T_c, of the order–disorder transformation. At temperatures just below T_c, $s = 0$ corresponds to a metastable minimum, $G_c(0) > G_c(s^*)$, and at even lower temperatures $s = 0$ is a maximum of the curve of $G_c(s) - G_c(0)$ with respect to s. In short at all temperatures below T_c the stable minimum lies at s^*, where $0 < s^* < 1$, approaching $s^* = 1$ as T approaches zero K.

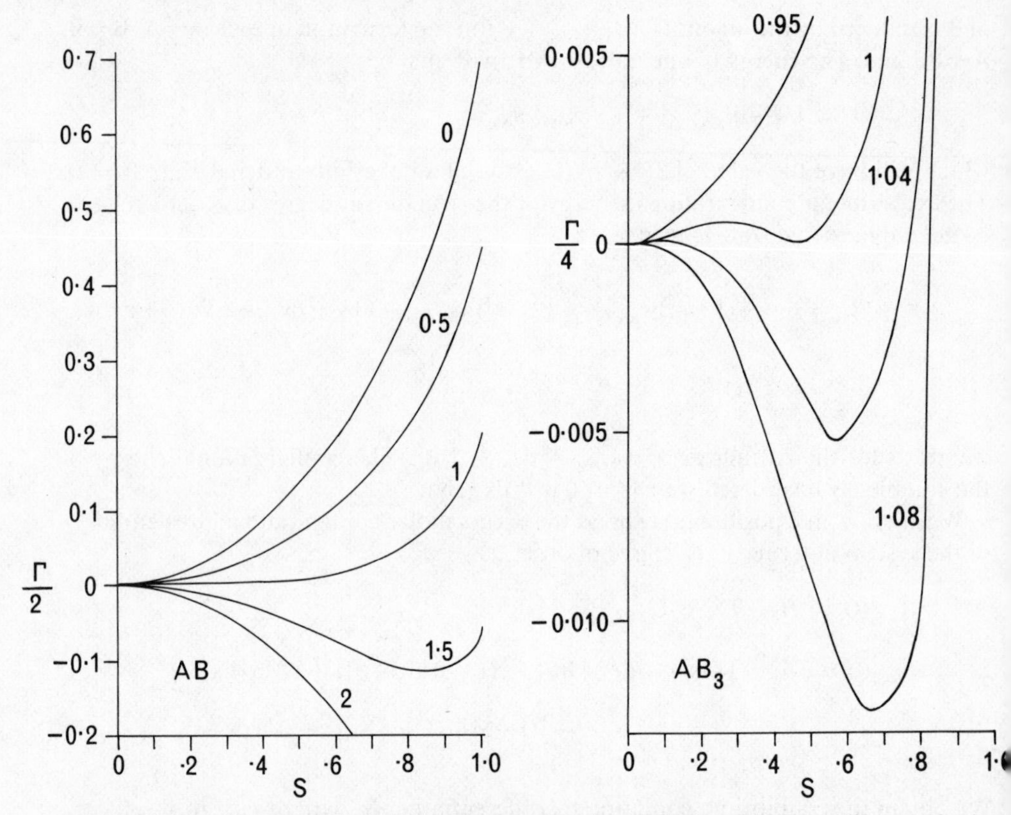

Fig 13.11 The difference in free enthalpy between an alloy of degree of order s and a completely disordered alloy ($s = 0$) plotted against degree of order for the zeroth (Bragg–Williams) approximation for AB and AB_3 alloys. The function plotted is $\Gamma = [G_c(s) - G_c(0)]/RT$; Γ is halved for the AB alloy which contains $2N$ atoms per mole and divided by four for the AB_3 alloy which contains $4N$ atoms per mole. The figure placed against each curve is the value of T_c/T. The scale of the AB_3 diagram is greatly expanded relative to that for AB to display the detail of the curve for $T_c/T = 1$.

The limiting degree of order s^* at a temperature immediately below the critical temperature and the critical temperature T_c are obtainable by putting

$$G_c(s^*) = G_c(0)$$

i.e.
$$\tfrac{1}{4}zNw(1-s^{*2}) - RT\{4\ln 4 - \tfrac{3}{4}(1-s^*)\ln 3 - \tfrac{1}{4}(1+3s^*)\ln(1+3s^*) - \tfrac{3}{2}(1-s^*)\ln(1-s^*) - \tfrac{3}{4}(3+s^*)\ln(3+s^*)\} = \tfrac{1}{4}zNw - RT\{4\ln 4 - 3\ln 3\}$$

i.e.
$$\frac{2zw}{3kT}s^* = \frac{1}{s^*}\{\tfrac{2}{3}(1+3s^*)\ln(1+3s^*) + 4(1-s^*)\ln(1-s^*) + 2(3+s^*)\ln(3+s^*) - 2(3+s^*)\ln 3\}$$

and by use of the equilibrium condition $\partial G_c(s^*)/\partial s = 0$

i.e.
$$\frac{2zw}{3kT}s^* = \ln\frac{(1+3s^*)(3+s^*)}{3(1-s^*)^2}.$$

This pair of simultaneous equations can be solved by successive approximation to give $s^* = 0{\cdot}463$ and $T_c = 0{\cdot}137\, zw/k$.

We are now in a position to compare the essential difference in the nature of the variation of degree of order for AB and AB_3 alloys. In AB alloys s varies smoothly from unity at zero K to zero at the critical temperature and remains zero at all $T > T_c$. In AB_3 alloys there is likewise a critical temperature but s varies smoothly from unity at zero K to $s^* = 0{\cdot}463$ at the critical temperature, where it drops abruptly to zero, its value at all higher temperatures. In AB_3 alloys there are discontinuities at the critical temperature in configurational entropy,

$$S_c(0) - S_c(s^*) = R\{\tfrac{3}{4}(1-s^*)\ln 3 + \tfrac{1}{4}(1+3s^*)\ln(1+3s^*) + \tfrac{3}{2}(1-s^*)\ln(1-s^*) + \tfrac{3}{4}(3+s^*)\ln(3+s^*) - 3\ln 3\}$$

$$= 0{\cdot}78 \text{ cal deg}^{-1}\text{ mole}^{-1}$$

and in configurational energy,

$$U_c(0) - U_c(s^*) = \tfrac{1}{4}zNws^{*2} = 0{\cdot}0536\, zNw.$$

The effect of applying the more sophisticated quasi-chemical method to disordering of an AB_3 alloy is to exaggerate the discontinuity at the critical temperature so that $s^* = 0{\cdot}956$. Experimental data indicate a value of s^* for $AuCu_3$ between 0·8 and 0·7. Although neither the zeroth approximation nor the more powerful first approximation yields a value of s^* in close agreement with experiment, both predict a small decrease in s between zero K and the critical temperature followed by a sudden decrease in s from s^* to zero at the critical temperature; that there is semi-quantitative agreement between theory and experiment does indicate that the, admittedly approximate, theoretical analysis is valid as far as it goes. For a critical examination of the discrepancy between theory and experiment the interested reader is referred to Guggenheim (1952).

The third law: residual entropy

The third law of thermodynamics, which like the other three laws is a generalization of experience and cannot be directly verified by experiment, may be stated in the following form: *For any isothermal process involving only phases in internal equilibrium, or, alternatively, if any phase is in frozen metastable equilibrium, provided the process does not disturb this frozen equilibrium,*

$$\lim_{T\to 0} \Delta S = 0$$

This statement is carefully worded so as not to imply that the entropies of substances tend to zero as the absolute zero of temperature is approached.

We confine this brief account of the third law to one sort of isothermal process, chemical reaction. For many chemical reactions it is found that the calorimetrically determined entropy of reaction is in good agreement with the entropy difference between product and reactant phases calculated from specific heat measurements on each phase extending down to very low temperatures; for example the calorimetrically determined entropy of the reaction $Zn + \tfrac{1}{2}O_2 = ZnO$ at 298·15K is $-24{\cdot}24 \pm 0{\cdot}05$ cal deg^{-1} mole^{-1}, while the entropy difference $S(ZnO) - \tfrac{1}{2}S(O_2) - S(Zn)$ at the same temperature calculated from low-temperature specific heat measurements is $-24{\cdot}07 \pm 0{\cdot}25$ cal deg^{-1} mole^{-1}. But for some kinds of reaction there is always a systematic discrepancy; for example the directly determined and calculated entropies of reaction for $Mg(OH)_2 = MgO + H_2O$ at 298·16 °C are respectively $36{\cdot}67 \pm 0{\cdot}10$ and $35{\cdot}85 \pm 0{\cdot}08$ cal deg^{-1} mole^{-1}, a discrepancy of 0·82 cal deg^{-1} mole^{-1}. Discrepancies of very similar magnitude are found for all reactions involving H_2O and are attributable to the persistence of the randomness of the hydrogen atoms in the ice structure extending down to the lowest attainable temperatures. Such randomness contributes to the entropy of ice an amount known as the *residual entropy*, which must be added to the calorimetrically determined entropy to give the value of S_T to be used in the description of processes involving H_2O at TK. We proceed now to calculate the residual entropy of ice.

The structure of ice was discussed in chapter 10 and illustrated in Fig 10.42. In the structure two possible sites are available for occupation by each hydrogen atom and there are $2N$ hydrogen atoms per mole of H_2O. Therefore the number of possible configurations per mole is 2^{2N}. Consider the configurations about a single oxygen atom,[5] that is in H_4O_2: there are $2^4 = 16$ possible arrangements, of which only 6 have two hydrogen atoms close to the selected oxygen atom. Thus $\frac{6}{16} - \frac{3}{8}$ of the possible arrangements retain the H_2O molecule about one oxygen atom, i.e. $\frac{3}{8}.2^{2N}$ per mole. Further, $\frac{3}{8}$ of these retain the H_2O configuration about two oxygen atoms, i.e. $(\frac{3}{8})^2.2^{2N}$, and so on. So that in one mole, which contains N oxygen atoms, the number of configurations in which every oxygen atom has two hydrogen atoms close to it is $(\frac{3}{8})^N.2^{2N} = (\frac{3}{2})^N$. Therefore the residual molar entropy of ice is $S_0 = k \ln (\frac{3}{2})^N = R \ln \frac{3}{2} =$ 0·806 cal deg^{-1} mole^{-1}. This figure is in satisfactory agreement with the observed discrepancies between the directly determined and calculated entropies of reactions involving ice, 0·82 cal deg^{-1} mole^{-1}, and heavy ice, 0·77 cal deg^{-1} mole^{-1}.

In considering the entropy of any reaction which involves a phase capable of retaining configurational disorder down to low temperatures configurational entropy has to be taken into account.

Equilibrium in polycomponent systems: the Gibbs Phase rule

Before proceeding to the Gibbs Phase Rule it is important that certain terms, which we have already used, should be unambiguously defined. A *phase* is defined as any homogeneous part of a system. Those essentially inhomogeneous parts of the system that form the boundaries between pairs of homogeneous phases are usually of small extent and can safely be neglected here. The *number of phases* in the system will be denoted by $\mathscr{P}$. A *component* is simply a chemical species, but by the *number of components* in the system, denoted by $\mathscr{C}$, we imply the minimum number of substances necessary to specify completely the composition of every phase in the system. Thus in the system $Mg_2SiO_4 - Fe_2SiO_4$, the only phases that can be present are a solid solution $Mg_{2x}Fe_{2(1-x)}SiO_4$ and a liquid solution $Mg_{2y}Fe_{2(1-y)}SiO_4$ so that although there are four chemical elements present, only the amounts of the compounds Mg_2SiO_4 and Fe_2SiO_4 need be specified; therefore $\mathscr{C} = 2$. The *number of degrees of freedom* of a system, denoted by $\mathscr{F}$, is defined as the number of variables of the system that may be freely chosen and must be so chosen if the system is to be in a determinate state.

[5] It is necessary to consider the oxygen atom itself, the four associated hydrogen atoms, and one quarter of each of the four nearest oxygen neighbours.

We have already shown that, at equilibrium, temperature, pressure, and the chemical potential of each component must be uniform throughout the system. Therefore for a system of $\mathscr{P}$ phases, $\alpha, \beta, \ldots$, and $\mathscr{C}$ components, $1, 2, \ldots i \ldots$ we have the following restrictive equations:

$$T^\alpha = T^\beta = T^\gamma = \ldots \quad (\mathscr{P}-1) \text{ equations}$$

$$P^\alpha = P^\beta = P^\gamma = \ldots \quad (\mathscr{P}-1) \text{ equations}$$

$$\left.\begin{array}{l} \mu_1^\alpha = \mu_1^\beta = \mu_1^\gamma = \ldots \\ \vdots \\ \mu_i^\alpha = \mu_i^\beta = \mu_i^\gamma = \ldots \end{array}\right\} (\mathscr{P}-1)\mathscr{C} \text{ equations}$$

$$\vdots$$

There are thus $(\mathscr{C}+2)(\mathscr{P}-1)$ such restrictions. And in addition there is a Gibbs–Duhem Relation applicable to each phase

$$\left.\begin{array}{l} S^\alpha dT - V^\alpha dP + \sum_i n_i^\alpha d\mu_i^\alpha = 0 \\ S^\beta dT - V^\beta dP + \sum_i n_i^\beta d\mu_i^\beta = 0 \\ \vdots \end{array}\right\} \mathscr{P} \text{ equations}$$

Therefore the total number of restrictions is $(\mathscr{C}+2)(\mathscr{P}-1)+\mathscr{P}$.

The number of variables sufficient and necessary to specify the state of each phase is $\mathscr{C}+2$, that is $\mathscr{C}$ compositional variables, pressure, and temperature. Therefore the total number of variables for the system is $\mathscr{P}(\mathscr{C}+2)$.

The number of degrees of freedom $\mathscr{F}$ of the system is the excess of the number of variables over the number of restrictions. Therefore

$$\begin{aligned} \mathscr{F} &= \mathscr{P}(\mathscr{C}+2) - [(\mathscr{C}+2)(\mathscr{P}-1)+\mathscr{P}] \\ &= \mathscr{C}+2-\mathscr{P} \end{aligned}$$

i.e.

$$\mathscr{P}+\mathscr{F} = \mathscr{C}+2.$$

This equation is known as the *Gibbs Phase Rule.*

In the next chapter we shall exemplify the Phase Rule in the context of experimental studies of mineral systems. Many of these examples will refer to experiments performed at constant (atmospheric) pressure. For a system at constant pressure the number of degrees of freedom is reduced by one so that in such conditions $\mathscr{P}+\mathscr{F} = \mathscr{C}+1$.

In the case of metamorphism of rocks within the earth it would appear that mineral assemblages remain stable under rather variable conditions of temperature and pressure. V. M. Goldschmidt therefore suggested that for such systems $\mathscr{F} \geqslant 2$ always; if $\mathscr{F} \geqslant 2$, it follows that $\mathscr{P} \leqslant \mathscr{C}$. The statement $\mathscr{P} \leqslant \mathscr{C}$ is known as *Goldschmidt's Mineralogical Phase Rule*; we shall make little use of this modified phase rule in subsequent chapters.

14
Phase equilibrium: the interpretation of phase diagrams

It is not our purpose here to give an exhaustive phenomenological treatment of phase diagrams. We shall take a selection of illustrative types as the basis for discussion of paths of crystallization and interpretation in terms of the thermodynamics of perfect or strictly regular solutions; the departure of real solutions, liquid or solid, from such simple models is outside our present scope. The reader whose interest is restricted to the qualitative interpretation of phase diagrams can omit the more mathematical sections of this chapter without loss of continuity of reading.

A *phase diagram* is no more than the graphical representation in two dimensions of the application of the Phase Rule, $\mathscr{P}+\mathscr{F}=\mathscr{C}+2$, to a particular system. Since only two dimensions are available some selection of variables must be made, except in the case of a one-component system (for which $\mathscr{F} \leqslant 2$), and it is the nature of the selection that gives rise to the various kinds of diagrams in common use, each appropriate to a particular sort of problem.

Intensive diagrams are planar and have any two of the intensive variables P, T, $\mu_1, \ldots u_i$ as rectangular coordinates, each of the other intensive variables being assigned a conveniently selected constant value. The most commonly used intensive diagram is the PT diagram, which we have already made use of to illustrate the polymorphism of carbon (Fig 13.6), where it provides a complete description of phase relations in the system: $\mathscr{C}=1$ and therefore $\mathscr{P}+\mathscr{F}=3$, so that an area on the diagram, being divariant ($\mathscr{F}=2$), represents the stability field of a single phase and a line or curve, being univariant ($\mathscr{F}=1$), represents the stable coexistence of two phases. Such diagrams are applicable also to systems of more than one component; in general for a system with $\mathscr{C}$ components, an invariant point (P and T fixed, $\mathscr{F}=0$) represents the stable coexistence of $\mathscr{C}+2$ phases, a univariant line ($P=f(T)$, $\mathscr{F}=1$) represents the stable coexistence of $\mathscr{C}+1$ phases, and a divariant area or field (P, T independently variable within the field, $\mathscr{F}=2$) represents the stable coexistence of $\mathscr{C}$ phases. Figure 14.1 shows a PT diagram for a ternary (i.e. three-component) system for which it is necessary to specify the constancy of the chemical potentials of the two components that enter into the fluid phase of variable composition.

We shall make little use of intensive diagrams that have μ_1 and μ_2 plotted as their axes at constant P, T, and μ_i ($i \neq 1, 2$). For an account of such diagrams the reader is referred to Kern and Weisbrod (1967, pp. 248–257) and for a more detailed account to Garrels and Christ (1965).

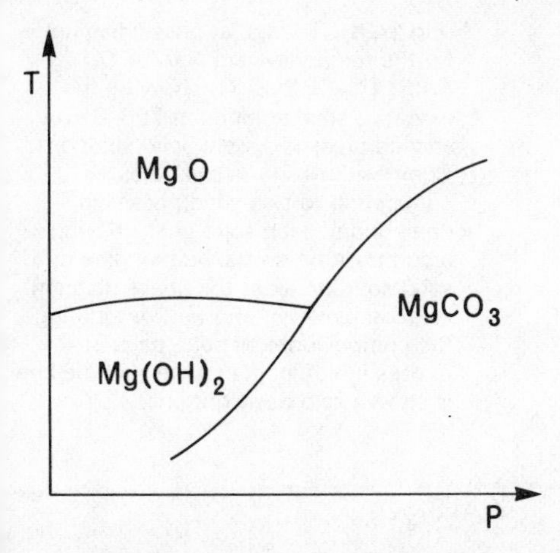

Fig 14.1 Temperature-pressure diagram for the three-component system MgO–H_2O–CO_2 for particular values of the chemical potentials μ_{H_2O} and μ_{CO_2} in the fluid phase.

Extensive or compositional diagrams are plotted at selected constant values of each intensive parameter with molar percentages of components, $n_1, n_2, \ldots n_i, \ldots$ such that $\sum n_i = 100$, as variables. A system with $\mathscr{C}$ components thus requires $\mathscr{C}-1$ dimensions for its representation. A binary system is represented by a linear diagram such as Fig 14.2 which shows the extent of solid solution in the system $NaAlSi_3O_8$–$KAlSi_3O_8$ at given P and T. A ternary system is conventionally represented by an equilateral triangular diagram with each component plotted at a vertex of the triangle; the system

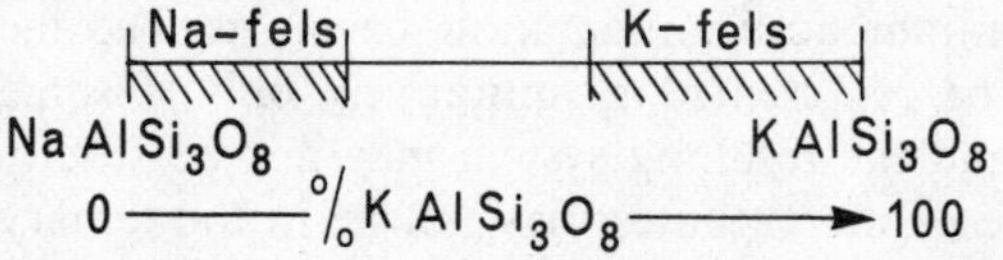

Na-fels K-fels

$NaAlSi_3O_8$ $KAlSi_3O_8$

0 —— $\%KAlSi_3O_8$ —— 100

Fig 14.2 Linear phase diagram for the binary system $NaAlSi_3O_8$–$KAlSi_3O_8$ at a particular temperature and pressure. The extent of solid solution is indicated by the limits of the hachured areas.

of coordinates is explained on Fig 14.3 and the use of such diagrams is illustrated by Fig 14.4 which shows the extent of solid solution in the system $NaAlSi_3O_8$–$KAlSi_3O_8$–$CaAl_2Si_2O_8$ at given P and T. For a quaternary system a perspective drawing of a regular tetrahedron may be used for quantitative representation or alternatively a selection of compositional restrictions may be imposed so that the system can be displayed by means of a number of plane sections.

The most commonly used type of diagram by far is the *composite diagram* in which certain intensive and extensive parameters are plotted as variables, all other

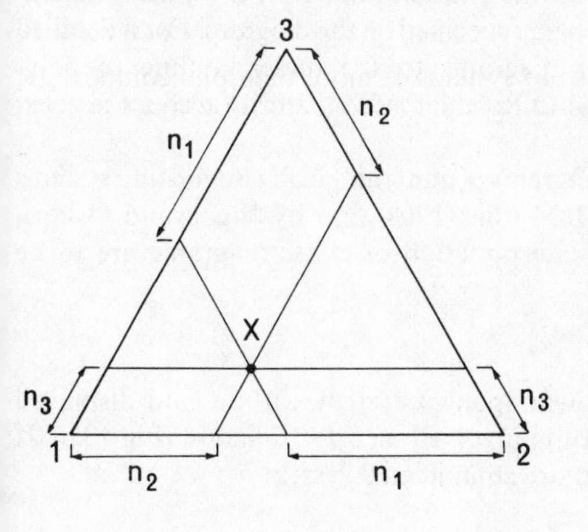

Fig 14.3 Coordinate system employed for triangular phase diagrams of ternary systems. The three components, 1, 2, and 3, are represented by the corners of the equilateral triangle. Compositions represented by points on the base 1–2 of the triangle are devoid of component 3; the corner 3 represents pure component 3; a point such as X within the triangle represents a composition containing a percentage of component 3 proportional to the height of X above the 1–2 base; and likewise for the other two components. The lengths n_1, n_2, n_3 shown on the diagram represent the percentages of components 1, 2, 3 in the composition represented by X if each side of the equilateral triangle is taken to be 100 in the same units.

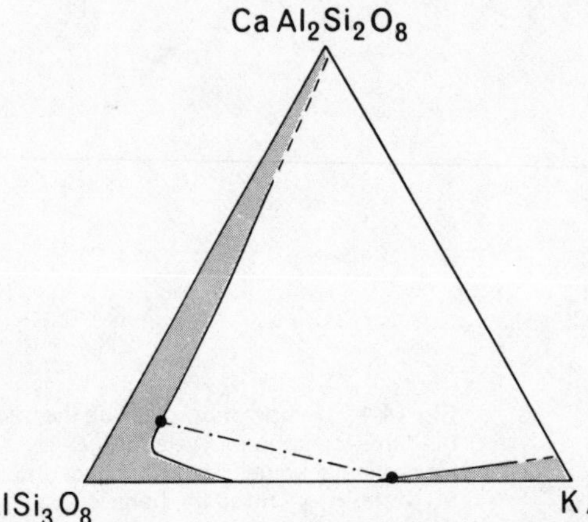

Fig 14.4 Triangular phase diagram for the ternary system $NaAlSi_3O_8$–$KAlSi_3O_8$–$CaAl_2Si_2O_8$ showing the extent of solid solution at 700 °C. The shaded areas represent solid solutions. Compositions in the clear area correspond to two solid phases in equilibrium, each solid phase having a composition on the limiting line of a solid solution area; the phase diagram is incomplete without *tie-lines* joining the compositions of such pairs of phases in equilibrium. One such tie line is shown as a dash-dot line.

parameters being kept constant. Melting, exsolution, and reaction in binary systems are commonly displayed on rectangular coordinates with the single compositional parameter plotted horizontally and T plotted vertically; a series of such diagrams at selected constant pressures gives a complete description of the system. For ternary systems triangular compositional diagrams with temperature plotted as isothermal contours at constant given pressure (Fig 14.22(a)) are commonly used to display melting or exsolution; a series of such diagrams at selected constant pressures gives a complete description of the system.

At this degree of complexity diagrammatic representation begins to become inadequate: quaternary systems can be represented quantitatively only if some compositional restriction is imposed and the resulting system may not be strictly ternary (that is the composition of one or more phases may not lie in the ternary system), and quinary systems will require two compositional restrictions. Of course as $\mathscr{C}$ increases the difficulties of experimentation increase correspondingly and very few systems with $\mathscr{C} > 4$ have been the subject of comprehensive study. It is usual to make a preliminary study of univariant equilibrium in such a polycomponent system and such may be represented on simple PT diagrams; the next stage of experimental complication would be the study of divariant equilibrium and so on.

The *Schairer diagram* overcomes the problem of representing polycomponent systems in two dimensions but in doing so sacrifices quantitative representation. The diagram consists of a network of straight lines each representing a univariant equilibrium ($\mathscr{P} = \mathscr{C} + 1$). The lines intersect in invariant points ($\mathscr{P} = \mathscr{C}$), the temperature and pressure of each invariant point being specified on the diagram. For a detailed account of the application of Schairer diagrams to the understanding of polycomponent systems the reader is referred to Roedder (1959); a brief account is given at the end of this chapter.

Comprehensive collections of phase diagrams of mineralogically interesting systems are provided by Levin, Robbins, and McMurdie (1964) and by Muan and Osborn (1965) where excellent accounts of the interpretation of phase diagrams are to be found also.

One-component systems

We have already discussed a simple one-component system, carbon, and displayed the single univariant solid-solid equilibrium involved on a PT diagram (Fig 13.6). It is convenient here to consider a rather more complicated system.

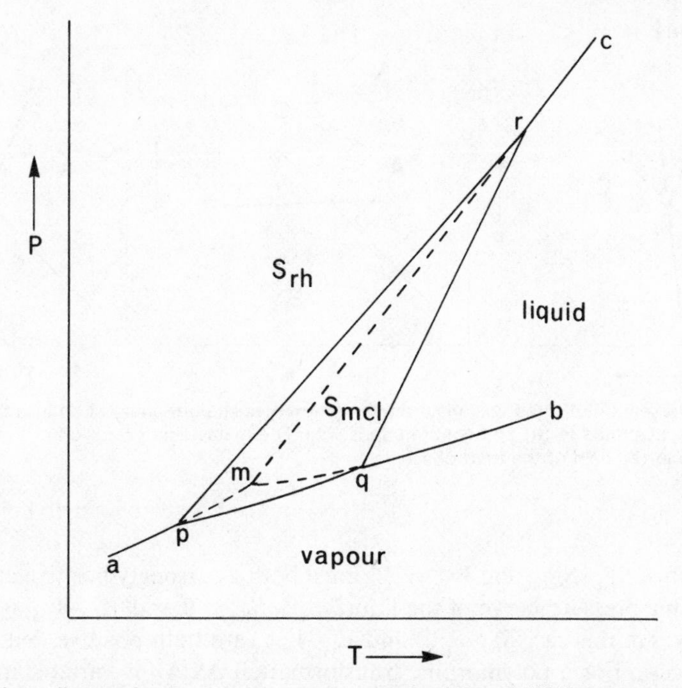

Fig 14.5 Phase diagram for the one-component system sulphur. Solid lines represent equilibria between stable phases; broken lines represent metastable equilibria in the stability field of monoclinic sulphur, i.e. mr is the metastable extension of rc, pm of ap, mq of qb, which correspond respectively to $S_{rh} \rightleftharpoons$ liquid, $S_{rh} \rightleftharpoons$ vapour, liquid $\rightleftharpoons$ vapour (S_{rh} = orthorhombic sulphur; S_{mcl} = monoclinic sulphur).

A phase diagram for sulphur with P and T plotted on rectangular axes is shown in Fig 14.5.[1] Four phases are involved: an orthorhombic solid, a monoclinic solid, a liquid, and a vapour. Each line on the diagram represents the equilibrium coexistence of 2 phases (i.e. $\mathscr{P} = 2$, $\mathscr{C} = 1$, therefore $\mathscr{F} = 1$). Within each area bounded by such univariant lines a single phase exists stably (i.e. $\mathscr{P} = 1$, $\mathscr{C} = 1$, therefore $\mathscr{F} = 2$). At each of the invariant points (alternatively known as triple points) p, q, and r three phases coexist stably (i.e. $\mathscr{P} = 3$, $\mathscr{C} = 1$, therefore $\mathscr{F} = 0$) at a definite pressure and temperature. The line ap is the sublimation curve of orthorhombic sulphur, $S_{rh} \rightleftharpoons S_{vap}$, and expresses the familiar fact that the vapour pressure of a given solid at a given temperature has a fixed value. Similarly the line pq is the sublimation curve of monoclinic sulphur and qb the vapour pressure curve of liquid sulphur. The line pr represents the polymorphic transformation $S_{mcl} \rightleftharpoons S_{rh}$ and the lines qr and rc the melting curves of monoclinic and orthorhombic sulphur respectively. The positive slopes, $dP/dT = \Delta S/\Delta V$, of all the univariant curves indicate that for each phase transformation ΔS and ΔV have the same sign: for the sublimation curves this is not

[1] Coordinates are not given on the pressure and temperature axes because this phase diagram cannot conveniently be drawn quantitatively: the three invariant points p, q, m lie rather close together on the P-axis at low pressures while the invariant point r lies at very much higher pressure. The phase diagram of Fig 14.5 is simply schematic. Experimental data are: p $5{\cdot}0 \times 10^{-6}$ bar, 102°C; q $2{\cdot}4 \times 10^{-5}$ bar, 114°C; m $1{\cdot}7 \times 10^{-5}$ bar, 110·2°C; r $2{\cdot}5 \times 10^{3}$ bar, 151°C; dP/dT for pr is 3°C per kb. Recent work indicates that the temperature of the triple point r, at which S_{rh}, S_{mcl} and liquid are in equilibrium, is as low as 107°C so that the univariant lines mr and qr have negative slopes; this amendment to the phase diagram of sulphur has not been made in Fig 14.5, which follows the traditional form. For recent data the reader is referred to Bell, England, and Kullerud (1966).

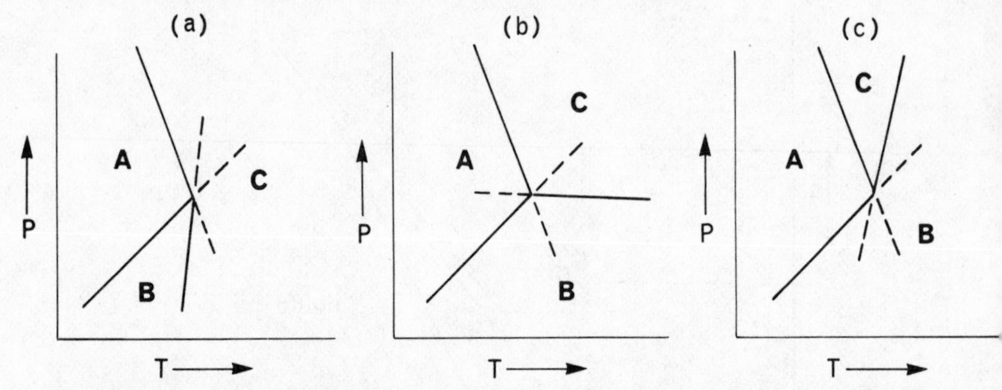

Fig 14.6 Intersection of three univariant curves (shown as straight lines) at an invariant point. The situations illustrated in (a) and (c) are impossible. The metastable extension of a univariant curve must lie in the field of the third phase.

surprising since $S_{vap}-S_{solid}$ and $V_{vap}-V_{solid}$ must both be strongly positive and likewise for the vapour pressure curve of the liquid $S_{vap}-S_{liq} \gg 0$ and $V_{vap}-V_{liq} \gg 0$. For the melting curves in this case $S_{liq}-S_{solid}$ and $V_{liq}-V_{solid}$ are both positive, but that is not so in every case. For a polymorphic transformation $\Delta S/\Delta V$ is variable in sign from substance to substance.

The manner in which univariant curves intersect at an invariant point can simply be demonstrated in the case of a one-component system. The three ways in which three univariant curves, $A \rightleftharpoons B$, $B \rightleftharpoons C$, and $C \rightleftharpoons A$, can intersect are shown in Fig 14.6. In Fig 14.6(a) any point on the metastable extension of the curve $B \rightleftharpoons C$ lies in the field of C; but on the curve $B \rightleftharpoons C$, whether in the stable or metastable part, $G_B = G_C$ while in the field of C, $G_C < G_B$ and $G_C < G_A$: this type of intersection is therefore impossible. In Fig 14.6(b) the metastable extension of $B \rightleftharpoons C$ lies in the field of A so that at any point on the metastable extension $G_B = G_C$, $G_A < G_B$, and $G_A < G_C$: these are reconcilable statements. In Fig 14.6(c) the metastable extension of $B \rightleftharpoons C$ lies in the field of B: the consequences of this arrangement, $G_B = G_C$ and $G_B < G_C$ are irreconcilable. Therefore the metastable extension of a univariant curve must lie in the field of the phase not involved in the equilibrium represented by the curve.

Two-component systems

Since mineralogy is concerned primarily with condensed phases we shall restrict our discussion of systems with two (or more) components at this stage to phase changes that do not involve a vapour phase. Attention will be focused on melting and its converse, crystallization, on polymorphism, and on solid solution and its converse, exsolution.

(1) Melting and crystallization in a simple eutectic system

The simplest type of binary system displays neither polymorphism, nor solid solution; the only condensed phases are a liquid approximating to an ideal solution and the solid crystalline components. The temperature-composition diagram at atmospheric pressure for such a system is shown in Fig 14.7 where the weight percentage of one component is plotted along the compositional axis.

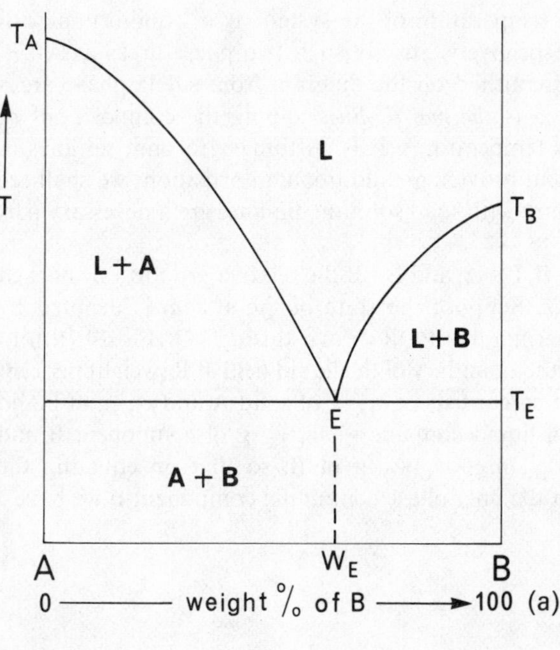

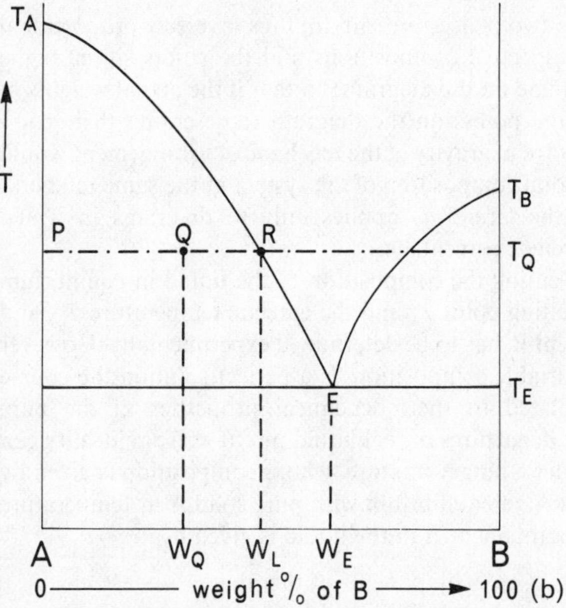

Fig 14.7 Two component (or binary) systems: the simple eutectic system at constant pressure. (a) shows the phase diagram with temperature plotted vertically and composition horizontally; single and two-phase areas are labelled with the phases present at equilibrium at all temperatures and compositions within the area; the eutectic has coordinates w_E, T_E and represents the three-phase equilibrium A+B+L. (b) illustrates the use of the *lever rule* to obtain the proportions of solid phase A and liquid at equilibrium at temperature T_Q and bulk composition w_Q: $m_L . QR = m_A . PQ$.

Since in this diagram we have imposed the condition of constant pressure the phase rule is modified to $\mathscr{P}+\mathscr{F}' = \mathscr{C}+1$. Three phases are involved: liquid, pure solid A, and pure solid B. All three phases will be in equilibrium at a point, the *eutectic*, of fixed temperature T_E and fixed composition w_E. That a single phase is stable within an area of such an isobaric diagram is consistent with the modified phase rule, for $\mathscr{P} = 1$ and $\mathscr{C} = 2$, $\mathscr{F}' = 2$; such an area is that labelled L on Fig 14.7. A mechanical mixture of the two pure solid phases is stable throughout the area below T_E, the relative proportions of the two phases being determined by the bulk composition of the system. In the two-phase areas labelled L+A and L+B a liquid of variable composition, determined by the temperature of the system, is in equilibrium with pure solid A and pure solid B respectively. Strictly such two-phase areas as A+B, L+A, and L+B should be distinguished on the diagram from single phase areas such as L by arbitrarily separated *isothermal tie-lines* joining the compositions of phases in equilibrium at selected temperatures. It is customary to omit tie lines in cases such as this where they would provide no additional information; we shall see subsequently that in ternary systems with solid solution, tie-lines are a necessary part of the phase diagram.

In such two-phase areas as A+B, L+A, and L+B the relative proportions of each phase are given by the *lever rule*. Suppose the state of the system is denoted by $Q(T_Q, w_Q)$ and consider an isothermal line PQR drawn through Q (Fig 14.7(b)) to intersect the pure A axis at P and the boundary of the liquid field at R (weight per cent of B $= w_L$). Suppose that the system consists of m_A g of solid A and m_L g of liquid of composition w_L, then m_L g of liquid contains $\frac{1}{100}m_L w_L$ g of component B and (m_A+m_L) g of system contains $\frac{1}{100}(m_A+m_L)w_Q$ g of B, so that on equating the weights of B in the system and in the only phase containing component B we have

$$m_L w_L = (m_A+m_L)w_Q$$

therefore $$m_L(w_L-w_Q) = m_A w_Q$$

i.e. $$m_L \,.\, \mathrm{QR} = m_A \,.\, \mathrm{PQ}.$$

The proportions by weight of the two phases present are thus inversely proportional to the distances between their respective compositions and the composition of the total system along an isothermal line on the diagram; so that if the actual weights of the two phases were placed at the points on the diagram representing their composition and temperature, the centre of gravity of the mechanical arrangement would lie at the point representing the total composition of the system at the same temperature. It must be stressed that the *lever rule* applies only to diagrams in which composition is represented by weight percentage.

The course of the curve representing the composition of the liquid in equilibrium with pure solid A between its melting point T_A and the eutectic temperature T_E, and the analogous curve for component B, has to be determined experimentally. However if the liquid, the only phase of variable composition, is a perfect solution the course of the curve can simply be related to thermochemical properties of the pure component; this implies that the departures of real liquid mixtures from ideality can be studied comparatively. Consider a binary mixture, whose composition is given by the mol fraction x of component A, in equilibrium with pure solid A at temperature T K. The chemical potential of component A in the liquid is given by

$$\mu_A^L = \mu_A^{L*} + RT\ln x$$

and therefore at equilibrium

$$\mu_A^{S*} = \mu_A^{L*} + RT\ln x$$

i.e. $$\ln x = \frac{\mu_A^{S*} - \mu_A^{L*}}{RT}.$$

Therefore $$\left(\frac{\partial \ln x}{\partial T}\right)_{Pn_A n_B} = \frac{-H_A^{S*} + H_A^{L*}}{RT^2},$$

$$= \frac{(\Delta H_A)_T}{RT^2},$$

where $(\Delta H)_T = H_A^{L*} - H_A^{S*}$ is the heat of melting of pure component A at temperature T. Integration between the melting point of pure A and the conditions under investigation yields

$$[\ln x]_1^x = \int_{T_A}^{T} \frac{(\Delta H_A)_T}{RT^2} dT.$$

But

$$(\Delta H_A)_T = (\Delta H_A)_{T_A} + \int_{T_A}^{T} \Delta C_p \, dT$$
$$= (\Delta H_A)_{T_A} + \Delta C_p (T - T_A)$$

if it is assumed that ΔC_p, the difference in specific heat between liquid and solid A, is constant over the interval T_A to T.

Therefore

$$[\ln x]_1^x = \int_{T_A}^{T} \left(\frac{(\Delta H_A)_{T_A} - \Delta C_p T_A}{RT^2} + \frac{\Delta C_p}{RT} \right) dT$$
$$= \left[\frac{-(\Delta H_A)_{T_A} + \Delta C_p T_A}{RT} + \frac{\Delta C_p}{R} \ln T \right]_{T_A}^{T}$$

therefore

$$\ln x = \frac{-(\Delta H_A)_{T_A} + \Delta C_p T_A}{R} \left(\frac{1}{T} - \frac{1}{T_A} \right) + \frac{\Delta C_p}{R} \ln \frac{T}{T_A}.$$

A better approximation can be achieved by using the empirical expression for ΔC_p as a function of T provided adequate high temperature specific heat data are available for both liquid and solid A, the former being extrapolated below the melting point T_A. On the other hand it is often adequate to make the crude assumption $\Delta C_p = 0$ so that to a first approximation

$$\ln x = -\frac{(\Delta H_A)_{T_A}}{R} \left(\frac{1}{T} - \frac{1}{T_A} \right).$$

Such first approximation equations will be used generally in subsequent paragraphs since our purpose is merely to indicate lines of argument from the perfect solution model rather than to provide expressions for computation.

By analogy the equation to the curve of liquid compositions in equilibrium with pure solid B is

$$\ln (1 - x) = -\frac{(\Delta H_B)_{T_B}}{R} \left(\frac{1}{T} - \frac{1}{T_B} \right).$$

where $(\Delta H_B)_{T_B}$ is the heat of melting and T_B the melting point of pure B. The two curves intersect at the eutectic point (T_E, x_E) so that

$$\ln x_E = -\frac{(\Delta H_A)_{T_A}}{R} \left(\frac{1}{T_E} - \frac{1}{T_A} \right)$$

and

$$\ln (1 - x_E) = -\frac{(\Delta H_B)_{T_B}}{R} \left(\frac{1}{T_E} - \frac{1}{T_B} \right).$$

We now turn to a brief consideration of courses of crystallization and fusion. Consider a liquid whose temperature and composition is given by the point X on Fig 14.8. On cooling the liquid persists until the temperature has fallen to T_X when pure solid A begins to separate. On further cooling under equilibrium conditions the amount of the crystalline solid A increases and the diminishing amount of liquid

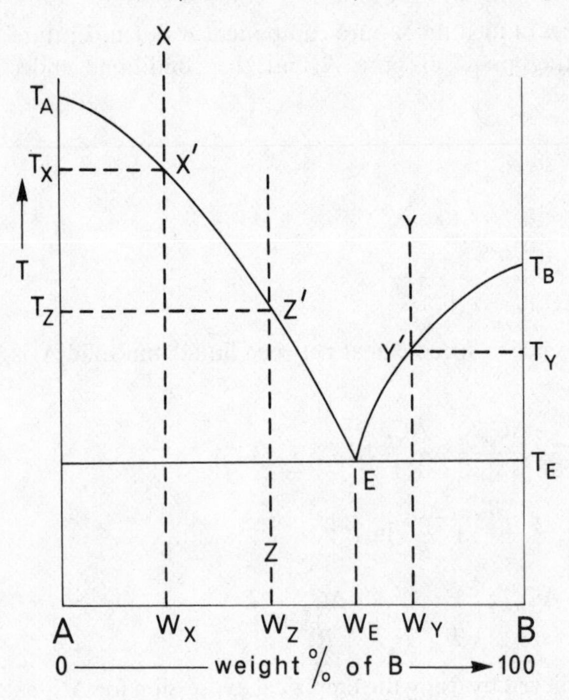

Fig 14.8 Crystallization and melting in a simple eutectic system. The broken lines through X, Y, and Z are lines of constant composition while those through T_X, T_Y, and T_Z are isotherms.

present has a composition given by the intersection of the appropriate isotherm with the curve T_AE, the proportions of solid and liquid being obtainable at any temperature by application of the lever rule. When cooling has proceeded as far as the eutectic temperature the system goes solid as the result of crystallization of an intergrowth of A and B in the proportions given by the eutectic point E. The temperature of the system remains constant at T_E until crystallization is complete. Further cooling produces no further change in the system. Although the solid product can be seen microscopically to consist of crystal grains of pure A and grains of the characteristic eutectic intergrowth of crystalline A and crystalline B, thermodynamically it is to be regarded as a mechanical mixture of the solid phases A and B.

Exactly analogously the liquid Y can be cooled as a liquid as far as T_Y, where solid B begins to crystallize out. The liquid composition follows the path Y' → E on further cooling under equilibrium conditions, more solid Y crystallizing out in the process. At the eutectic temperature T_E the system goes solid by crystallization of an intergrowth of solid A and B. Further cooling produces no further phase change.

On heating a mixture of solid A and solid B in the proportions corresponding to the point Z in Fig 14.8 a trace of liquid will first appear at the eutectic temperature T_E. The temperature of the system will remain constant at T_E until all the grains of B and sufficient of the grains of A have melted to yield a liquid of eutectic composition w_E. The temperature is then free to rise as more of the solid phase A melts to produce a liquid of composition corresponding to the intersection of the appropriate isotherm with the curve T_AE until at the temperature T_Z the last trace of solid disappears, the resultant liquid having the same composition as the total system.

The two curves running from the melting points of the pure components T_A and T_B down to the eutectic E are known jointly as the *liquidus*; the liquidus curves constitute the lower temperature boundary of the liquid field. The isotherm through the eutectic point is known as the *solidus* and represents the upper temperature limit of the solid phases in the absence of liquid.

Realistic departures from equilibrium in such a system have no dramatic effect on the course of crystallization or fusion. Some small degree of undercooling or superheating may be possible and will give rise to a liquid of composition differing slightly from that indicated by the equilibrium curve. But for liquids such as X and Y there will still be a crystallization interval approximately equal to $T_X - T_E$ and $T_Y - T_E$, which may amount to several hundred degrees in metallic, oxide, or silicate systems. And on heating such compositions will still have the large fusion interval that distinguishes them from the fixed temperature melting behaviour of pure components.

(2) Eutectic system with a polymorphous component

The effect of polymorphism in one of the components is merely to produce a break in the liquidus of that component (Fig 14.9) at the transformation temperature T_t. The manner in which the liquidus for the high temperature phase B intersects that for the low temperature phase B′ is shown correctly on the figure. That the alternative mode of intersection, shown inset, is incorrect can readily be seen by considering a point on the stable part of the liquidus of the high temperature phase B, where

$$\mu_B^{S*} = \mu_B^{L*} + RT \ln x$$

and for the metastable low temperature phase B′

$$\mu_{B'}^{S*} = \mu_B^{L*} + RT \ln x'$$

but since B is the stable solid

$$\mu_B^{S*} < \mu_{B'}^{S*}$$

Therefore $\quad x' > x$

so that above the transformation temperature the metastable liquidus of B′ lies at higher mol fraction of B than the stable liquidus of B.

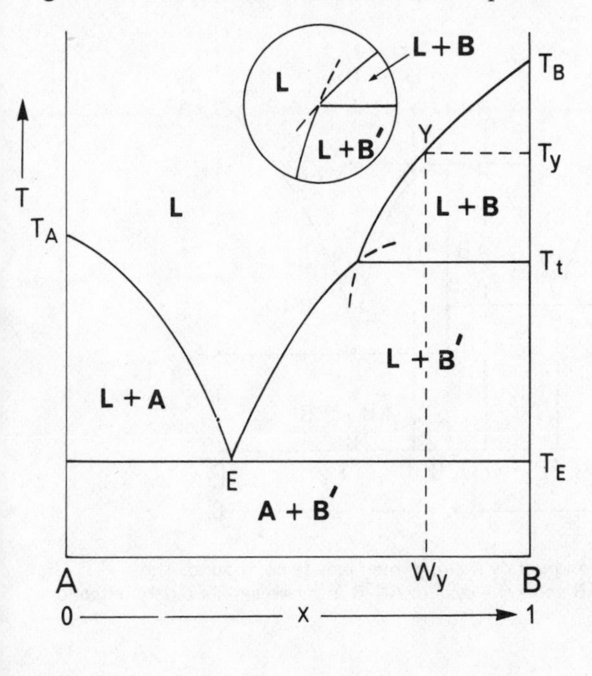

Fig 14.9 A eutectic system in which the component B is polymorphous in the crystalline state. Temperature is plotted vertically and mol fraction (x) of B horizontally. The intersection of liquidus curves at the temperature of the polymorphic transformation is shown correctly on the phase diagram; a thermodynamically impossible intersection is shown inset.

On cooling a composition such as Y solid B will first appear in infinitesimal quantity at the temperature T_Y. On further cooling more of phase B will crystallize until the temperature T_t is reached where the temperature will remain constant under equilibrium conditions until the whole of the high form B has been converted to the low form B′. B′ will then crystallize as the liquid composition moves along the liquidus of B′ towards the eutectic E, where the system will go solid by crystallization of the eutectic intergrowth of A and B′. The special effect of any departure from equilibrium will merely be to permit the persistence of some of the high temperature phase B; subsequent crystallization will be unaffected by this.

(3) Eutectic system with a congruently melting intermediate compound

A binary compound of the components A and B, denoted, $\overline{AB}$, is said to melt *congruently* if it melts at a definite temperature $T_{\overline{AB}}$ to a liquid of its own composition. The interpolation of such a compound in a binary system has the effect of splitting the system A–B into the simple eutectic systems A–$\overline{AB}$ and $\overline{AB}$–B (Fig 14.10). If the compound $\overline{AB}$ has a composition such that its formation from the pure components A and B can be written

$$x_{\overline{AB}}A+(1-x_{\overline{AB}})B \rightarrow \overline{AB}$$

its stability as a solid phase requires

$$\mu_{\overline{AB}}^{S*}-x_{\overline{AB}}\mu_A^{S*}-(1-x_{\overline{AB}})\mu_B^{S*}<0$$

at all temperatures below $T_{\overline{AB}}$. If the liquid phase is a perfect solution the equation

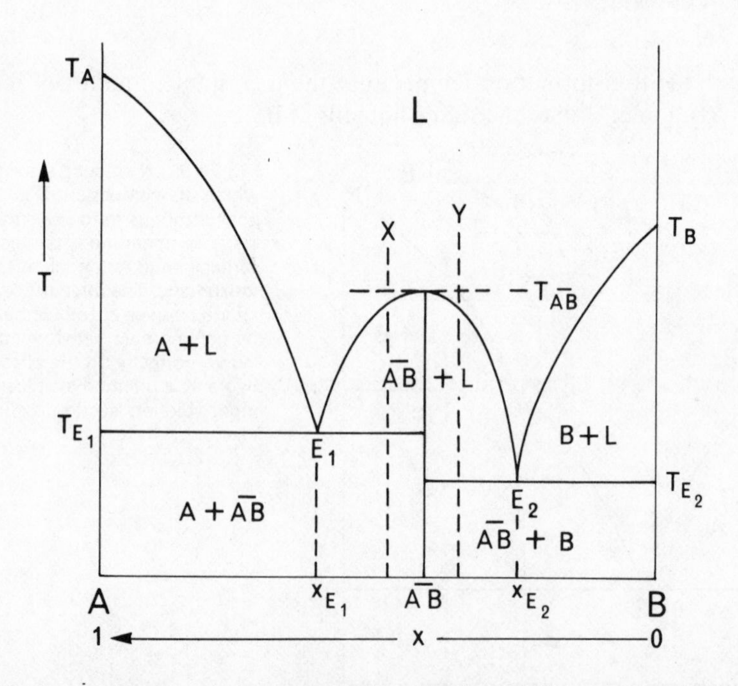

Fig 14.10 A eutectic system with a congruently melting intermediate compound. The interposition of the binary compound $\overline{AB}$ splits the system A—B into two simple binary eutectic systems A—$\overline{AB}$ and $\overline{AB}$—B.

to the liquidus of $\overline{\text{AB}}$ will be given by

$$\begin{aligned}\mu_{\overline{\text{AB}}}^{\text{S}*} &= \mu_{\overline{\text{AB}}}^{\text{L}} \\ &= x_{\overline{\text{AB}}}\mu_{\text{A}}^{\text{L}} + (1-x_{\overline{\text{AB}}})\mu_{\text{B}}^{\text{L}} \\ &= x_{\overline{\text{AB}}}\mu_{\text{A}}^{\text{L}*} + (1-x_{\overline{\text{AB}}})\mu_{\text{B}}^{\text{L}*} + RT\{x_{\overline{\text{AB}}}\ln x + (1-x_{\overline{\text{AB}}})\ln(1-x)\}.\end{aligned}$$

Rearranging and differentiating,

$$x_{\overline{\text{AB}}}\frac{\partial}{\partial T}\ln x + (1-x_{\overline{\text{AB}}})\frac{\partial}{\partial T}\ln(1-x) = \frac{\Delta H_{\overline{\text{AB}}}}{RT^2}$$

where $\Delta H_{\overline{\text{AB}}} = H_{\overline{\text{AB}}}^{\text{L}*} - H_{\overline{\text{AB}}}^{\text{S}*} = x_{\overline{\text{AB}}}H_{\text{A}}^{\text{L}*} + (1-x_{\overline{\text{AB}}})H_{\text{B}}^{\text{L}*} - H_{\overline{\text{AB}}}^{\text{S}*}$

is the heat of fusion of the compound $\overline{\text{AB}}$. Integration assuming $\Delta C_{p_{\overline{\text{AB}}}} = 0$ yields

$$x_{\overline{\text{AB}}}[\ln x]_{x_{\overline{\text{AB}}}}^{x} + (1-x_{\overline{\text{AB}}})[\ln(1-x)]_{x_{\overline{\text{AB}}}}^{x} = -\frac{\Delta H_{\overline{\text{AB}}}}{R}\left[\frac{1}{T}\right]_{T_{\overline{\text{AB}}}}^{T}$$

therefore $x_{\overline{\text{AB}}}\ln\dfrac{x}{x_{\overline{\text{AB}}}} + (1-x_{\overline{\text{AB}}})\ln\dfrac{1-x}{1-x_{\overline{\text{AB}}}} = -\dfrac{\Delta H_{\overline{\text{AB}}}}{R}\left(\dfrac{1}{T} - \dfrac{1}{T_{\overline{\text{AB}}}}\right).$

Differentiating with respect to x yields

$$\frac{x_{\overline{\text{AB}}}}{x} - \frac{1-x_{\overline{\text{AB}}}}{1-x} = \frac{\Delta H_{\overline{\text{AB}}}}{RT^2}\left(\frac{\partial T}{\partial x}\right).$$

Therefore $\partial T/\partial x = 0$ when $x = x_{\overline{\text{AB}}}$ and that this is a maximum is clear from the sign of the second differential

$$\begin{aligned}\frac{\partial^2 T}{\partial x^2} &= \frac{\partial}{\partial x}\left\{\frac{RT^2}{\Delta H_{\overline{\text{AB}}}}\left(\frac{x_{\overline{\text{AB}}}}{x} - \frac{1-x_{\overline{\text{AB}}}}{1-x}\right)\right\} \\ &= \frac{R}{\Delta H_{\overline{\text{AB}}}}\left\{2T\left(\frac{\partial T}{\partial x}\right)\left(\frac{x_{\overline{\text{AB}}}}{x} - \frac{1-x_{\overline{\text{AB}}}}{1-x}\right) - T^2\left(\frac{x_{\overline{\text{AB}}}}{x^2} + \frac{1-x_{\overline{\text{AB}}}}{(1-x)^2}\right)\right\}.\end{aligned}$$

Therefore $$\left(\frac{\partial^2 T}{\partial x^2}\right)_{x=x_{\overline{\text{AB}}}} = -\frac{RT^2}{\Delta H_{\overline{\text{AB}}}}\left(\frac{1}{x_{\overline{\text{AB}}}} + \frac{1}{1-x_{\overline{\text{AB}}}}\right) < 0 \quad \text{since} \quad 0 < x_{\overline{\text{AB}}} < 1.$$

The liquidus of the intermediate compound thus passes through a maximum at the composition of the compound.

Since for practical purposes a binary system with a congruently melting intermediate compound can be split into two simple eutectic systems there is no need for further consideration of courses of crystallization. But it is worth pointing out that very small departures of the composition of the system from the composition of the compound $\overline{\text{AB}}$ will lead to very different temperatures of first appearance of liquid on fusion depending on whether the system lies on the A or the B side of $\overline{\text{AB}}$, and conversely on cooling. Thus on cooling the compositions X and Y in Fig 14.10 will retain a liquid phase down to T_{E_1} and T_{E_2} respectively; in practice $|T_{\text{E}_1} - T_{\text{E}_2}|$ may be several hundred degrees.

(4) Eutectic system with an incongruently melting intermediate compound: the reaction point

A binary compound $\overline{\text{AB}}$ is said to melt *incongruently* if at a certain temperature it dissociates into a liquid and a solid phase of different composition. The existence of

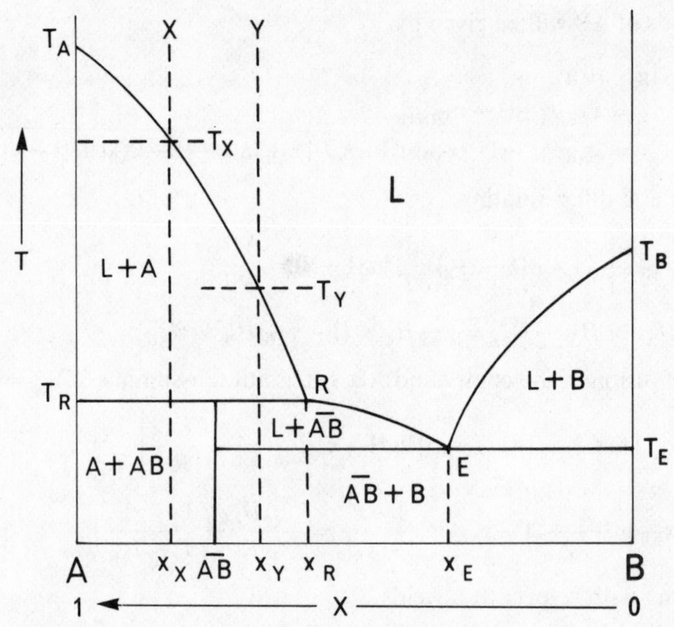

Fig 14.11 A eutectic system with an incongruently melting intermediate compound. The nomenclature corresponds to that of earlier figures. The *reaction point* R has compositional and temperature coordinates x_R and T_R.

such a compound in a binary system has a profound effect on the paths of crystallization of a wide range of liquid compositions in the system. That the phase diagram for such a system is generally of the type illustrated in Fig 14.11 follows immediately from the definition of incongruent melting. The diagram expresses graphically the experimental observation that the compound $\overline{AB}$ decomposes at the temperature T_R into pure solid A and a liquid of composition x_R. The point R (coordinates x_R, T_R) is known as the *reaction point.* Mixtures of the component A and the compound $\overline{AB}$ remains solid on heating until the temperature reaches T_R where the reaction $\overline{AB}(s) \rightarrow A(s)+R(l)$ takes place; as the temperature rises above T_R the composition of the liquid phase moves along the liquidus curve RT_A and the proportion of liquid increases until no solid A remains. A mixture of the component B and the compound $\overline{AB}$ will on the other hand develop a liquid phase on heating to the eutectic temperature T_E: $\overline{AB}$ and B are in a eutectic relationship in just the same way as the intermediate compound and the components in the congruent case considered previously and it is only in its relationship to the other component, A, that the intermediate compound in this case differs. If the bulk composition of the $\overline{AB}$+B mixture lies between the composition of $\overline{AB}$ and the reaction point R the whole of phase B melts as does an adequate amount of the phase $\overline{AB}$ to produce the eutectic liquid E when the temperature has risen to T_E; on further heating more of the phase $\overline{AB}$ melts, the liquid composition moving along the liquidus curve ER; at the temperature T_R the reaction $\overline{AB}(s) \rightarrow A(s)+R(l)$ takes place and the phase $\overline{AB}$ disappears from the system; continued heating enables the temperature to rise as more of the phase A melts, the liquid composition moving along the liquidus curve RT_A, until the system is completely liquid. If the bulk composition lies between R and E the first liquid again appears at T_E but the remaining solid phase $\overline{AB}$ disappears at some temperature less than T_R. If the bulk composition lies between E and the pure component B the first liquid yet again appears when the temperature has risen to T_E; the temperature of the system remains constant at T_E until all of phase $\overline{AB}$ and an adequate amount of phase B have melted to give the eutectic liquid E; further heating produces a rise in temperature as an increasing amount of the solid phase B melts, the composition of the liquid moving along the liquidus curve ET_B until all the solid phase has melted at some temperature less than T_B.

The differing end products of crystallization under equilibrium and non-equilibrium conditions in systems with incongruently melting compounds is of considerable importance both in petrology and in metallurgy. We shall consider the crystallization of two liquids X and Y (Fig 14.11) under equilibrium and non-equilibrium conditions. On cooling the liquid X, solid phase A will first appear at T_X; solid phase A will crystallize and the liquid will move along the A-liquidus to R on further cooling; the temperature will remain constant at T_R until the reaction $A(s)+R(l) \rightarrow \overline{AB}(s)$ has gone to completion and, since $x_X > x_{\overline{AB}}$, the liquid will be wholly exhausted and the system will go solid as $A+\overline{AB}$ at T_R. On cooling the liquid Y solid phase A will first appear at T_Y; solid phase A will crystallize and the liquid will move along the A-liquidus to R on further cooling; the temperature will remain constant at T_R until the reaction $A(s)+R(l) \rightarrow \overline{AB}(s)$ has gone to completion and, since $x_Y < x_{\overline{AB}}$, some liquid will remain in equilibrium with solid $\overline{AB}$; as the temperature falls below T_R more $\overline{AB}$ will crystallize and the liquid composition will move along the $\overline{AB}$-liquidus towards E; the temperature will remain constant at T_E until the system has gone solid by crystallization of the eutectic intergrowth of $\overline{AB}$ and B. The reaction $A(s)+R(l) \rightarrow \overline{AB}(s)$ requires the solution of one solid phase (A) and the deposition of another ($\overline{AB}$); this is obviously a process during which it is practically difficult to maintain equilibrium. The new phase $\overline{AB}$ tends to crystallize on the surface of the dissolving phase A and so to inhibit further reaction. If further reaction becomes impossible after only a small amount of phase A has reacted at the reaction point, the remaining protected crystal grains of phase A cease to belong effectively to the system. Thus under non-equilibrium conditions the liquids X and Y will crystallize phase A at temperatures above T_R, the reaction will then be quickly inhibited and the effective bulk composition of the system will in each case move to R; further cooling will take place as phase $\overline{AB}$ crystallizes and the liquid composition moves along the $\overline{AB}$ liquidus towards E; the system remains at constant temperature T_E until the crystallization of the eutectic intergrowth is complete. The great contrast in the phase assemblages obtained for these two liquids under equilibrium conditions and under conditions where the operation of the reaction point is wholly inhibited becomes very clear if the system is considered quantitatively using rather extreme compositions:

Compositions in weight per cent of component B:

intermediate compound $\overline{AB}$	30%
reaction point R	50%
eutectic point E	60%
liquid X	20%
liquid Y	40%

Final products of crystallization:

	(i) at equilibrium	(ii) at extreme disequilibrium
X	33% $A+67\%\ \overline{AB}$	60% $A^{(1,3)}+28\tfrac{1}{2}\%\ \overline{AB}^{(3)}+11\tfrac{1}{2}\%\ B^{(3)}$
Y	86% $\overline{AB}^{(2)}+14\%\ B^{(2)}$	20% $A^{(1,4)}+57\%\ \overline{AB}^{(4)}+23\%\ B^{(4)}$

Notes:

(1) Early crystallized grains of A enveloped in $\overline{AB}$. Such grains are known in petrography as *armoured relics.*

(2) Microscopic appearance: 67% $\overline{AB}$+33% eutectic intergrowth.

(3) Microscopic appearance: 60% A+13% $\overline{AB}$+27% eutectic intergrowth.

(4) Microscopic appearance: 20% A+27% $\overline{AB}$+53% eutectic intergrowth.

If the assumption is made that the liquid phase is a perfect solution it becomes possible, as in the systems we have considered previously, to interpret the phase diagram for this type of system in terms of measurable thermochemical properties of the substances A, $\overline{AB}$, and B. The equation to the A-liquidus can be derived, as in previous cases, from the equilibrium condition $\mu_A^{S*} = \mu_A^L$ as

$$\ln x = -\frac{\Delta H_A}{R}\left(\frac{1}{T}-\frac{1}{T_A}\right)$$

and similarly for the B-liquidus

$$\ln(1-x) = -\frac{\Delta H_B}{R}\left(\frac{1}{T}-\frac{1}{T_B}\right)$$

Now there is no point in this case in deriving an equation to the $\overline{AB}$-liquidus in the usual way in terms of the heat of melting of the intermediate compound because the melting of $\overline{AB}$ cannot be realized experimentally; instead we require an equation in terms of the heat of reaction either for $\overline{AB}(s) \to A(s)+R(l)$ or for $A(s)+B(s) \to \overline{AB}(s)$.

The equilibrium condition for the $\overline{AB}$-liquidus is

$$\begin{aligned}\mu_{\overline{AB}}^{S*} &= \mu_{\overline{AB}}^{L}\\ &= x_{\overline{AB}}\mu_A^{L*}+(1-x_{\overline{AB}})\mu_B^{L*}+RT\{x_{\overline{AB}}\ln x+(1-x_{\overline{AB}})\ln(1-x)\}\end{aligned}$$

Rearrangement and differentiation with respect to temperature yields

$$x_{\overline{AB}}\frac{\partial}{\partial T}\ln x+(1-x_{\overline{AB}})\frac{\partial}{\partial T}\ln(1-x) = \frac{1}{RT^2}\{x_{\overline{AB}}H_A^{L*}+(1-x_{\overline{AB}})H_B^{L*}-H_{\overline{AB}}^{S*}\}$$

which becomes on integration and substitution of $\Delta H' = x_{\overline{AB}}H_A^{L*}+(1-x_{\overline{AB}})H_B^{L*}-H_{\overline{AB}}^{S*}$ for the sake of convenience

$$x_{\overline{AB}}[\ln x]_{x_R}^{x}+(1-x_{\overline{AB}})[\ln(1-x)]_{x_R}^{x} = -\frac{\Delta H'}{R}\left[\frac{1}{T}\right]_{T_R}^{T}.$$

Therefore $$x_{\overline{AB}}\ln\frac{x}{x_R}+(1-x_{\overline{AB}})\ln\frac{1-x}{1-x_R} = -\frac{\Delta H'}{R}\left(\frac{1}{T}-\frac{1}{T_R}\right).$$

In order to relate $\Delta H'$ to the heat ΔH_R of the reaction that takes place at the reaction point it is first necessary to write the stoichiometric equation for the reaction, which will be

$$\overline{AB}(s) \to \frac{x_{\overline{AB}}-x_R}{1-x_R}\cdot A(s)+\frac{1-x_{\overline{AB}}}{1-x_R}\cdot R(l)$$

hence $$\Delta H_R = \frac{x_{\overline{AB}}-x_R}{1-x_R}H_A^{S*}+\frac{1-x_{\overline{AB}}}{1-x_R}\{x_R H_A^{L*}+(1-x_R)H_B^{L*}\}-H_{\overline{AB}}^{S*}$$

whence $$\Delta H' = x_{\overline{AB}} H_A^{L*} + (1-x_{\overline{AB}}) H_B^{L*} + \Delta H_R - \frac{x_{\overline{AB}} - x_R}{1-x_R} H_A^{S*} - \frac{1-x_{\overline{AB}}}{1-x_R} \times \{x_R H_A^{L*} + (1-x_R) H_B^{L*}\}$$

$$= \frac{x_{\overline{AB}} - x_R}{1-x_R}(H_A^{L*} - H_A^{S*}) + \Delta H_R$$

$$= \left(\frac{x_{\overline{AB}} - x_R}{1-x_R}\right) \Delta H_A + \Delta H_R$$

Or alternatively $\Delta H'$ may be related to the heat of combination ΔH_C of the pure components A and B to form the intermediate compound $\overline{AB}$,

$$x_{\overline{AB}} A(s) + (1-x_{\overline{AB}}) B(s) \rightarrow \overline{AB}(s)$$

$$\Delta H_C = H_{\overline{AB}}^{S*} - x_{\overline{AB}} H_A^{S*} - (1-x_{\overline{AB}}) H_B^{S*} \quad \text{at} \quad T_R.$$

Then $$\Delta H' = x_{\overline{AB}} H_A^{L*} + (1-x_{\overline{AB}}) H_B^{L*} - \Delta H_C - x_{\overline{AB}} H_A^{S*} - (1-x_{\overline{AB}}) H_B^{S*}$$

$$= x_{\overline{AB}} \Delta H_A + (1-x_{\overline{AB}}) \Delta H_B - \Delta H_C.$$

Using either expression, ΔH_R or ΔH_C could be obtained from experimentally determined T, x points on the $\overline{AB}$ liquidus by plotting $x_{\overline{AB}} \ln x + (1-x_{\overline{AB}}) \ln(1-x)$ against T^{-1} to yield (if the liquid is a perfect solution and if the assumption $\partial \Delta H / \partial T = 0$, i.e. $\Delta C_p = 0$ is valid for each enthalpy difference involved) a straight line of slope $-\Delta H'/R$, ΔH_A, and ΔH_B being obtained from the A liquidus and the B liquidus respectively.

(5) System displaying complete miscibility in the solid state: zoning or coreing

This is the type of such mineralogically well-known systems as olivine and the plagioclase feldspars. The isobaric phase diagram consists of a smoothly curved convex liquidus and a smooth concave solidus separated by a lenticular two-phase area (Fig 14.12). The liquidus and solidus intersect at the margins of the diagram: the

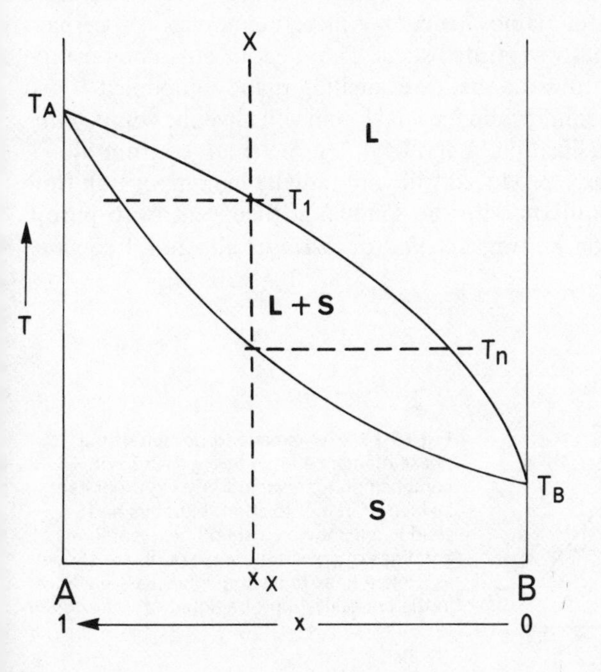

Fig 14.12 A binary system with complete miscibility in the solid state.

pure components A and B each melt at a fixed temperature, T_A and T_B respectively. That intermediate compositions crystallize and melt over a temperature interval is expressed graphically by the separation of liquidus and solidus for $0 < x < 1$. In the previous section we discussed what may be described as a case of discontinuous reaction at a definite reaction temperature; here we are concerned with continuous reaction over a temperature interval that may be as much as 100 °C. That there will be a significant difference in the end products of crystallization under equilibrium and extreme disequilibrium conditions is to be expected.

We consider first the crystallization of a liquid X (Fig 14.12) under equilibrium conditions. On cooling the system remains liquid until a temperature T_1 is reached where an infinitesimal amount of solid, whose composition is given by the intersection of the T_1 isotherm with the solidus, appears. The diagram states that on further cooling the already crystallized solid reacts with liquid to produce an increased amount of solid relatively enriched in the lower melting-point component B and a diminished amount of liquid likewise enriched in component B (Fig 14.13) so that when the temperature has fallen by a finite amount from the temperature T_1 of first appearance of solid to the temperature T_2

$$\frac{\text{wt \% of solid}}{\text{wt \% of liquid}} = \frac{\text{DE}}{\text{CD}}.$$

This process of reaction at infinitesimally separated temperatures continues until the temperature T_n is reached where the amount of liquid remaining is infinitesimally small; the final liquid is strongly enriched, as the first solid was strongly impoverished, in the lower melting-point component B. Under equilibrium conditions then uniform crystalline grains of composition X are produced at T_n, all earlier crystallized grains having reacted with the liquid as the temperature has fallen from T_1 to T_n.

That equilibrium is difficult to maintain in such a system is intuitively obvious. Time must be allowed for reaction to be completed before the temperature of the system can fall infinitesimally and this condition applies over the whole range from T_1 to T_n. If equilibrium is not maintained reaction will be incomplete, or perhaps even vestigial, at each infinitesimally separated stage. The effective bulk composition of the system will move steadily towards the lower melting-point component B and under conditions of complete disequilibrium the final liquid will have the composition of pure component B. The crystallization interval will thus be enlarged from $T_1 - T_n$ to $T_1 - T_B$ and the crystalline grains produced will vary radially in composition from w_1^s (the intersection of the T_1 isotherm with the solidus) at their centres to pure B marginally. Such smooth variation, known as *zoning* (or *coreing* in alloys), is frequently

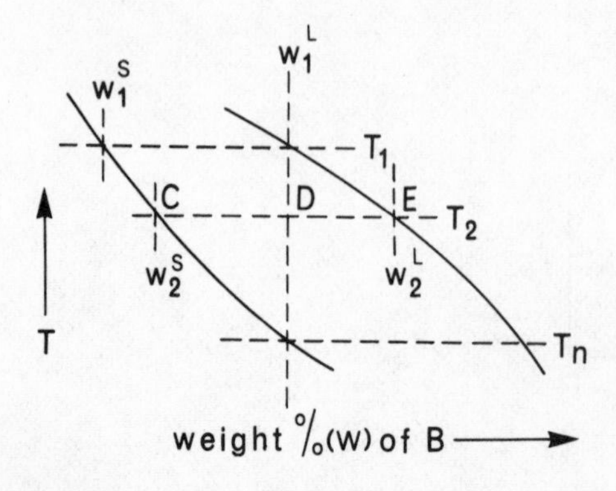

Fig 14.13 An expanded portion of the phase diagram shown in Fig 14.12 with composition represented in weight per cent instead of in mol fraction. Liquidus and solidus are shown as solid lines; isotherms and lines of constant composition are shown as broken lines to illustrate successive stages in the crystallization of a liquid of composition W_1^L.

observed in such commonly encountered solid solutions as the olivines, plagioclase feldspars, pyroxenes, etc of rocks that have crystallized from natural melts as well as in the products of laboratory crystallization as diverse as Cu–Ni alloys, ClC_6H_5–BrC_6H_5 'mixed crystals', and Na_2CO_3–Na_2SO_4 'mixed crystals'.

The simplest model for such a system is to assume that both liquid and solid phases are perfect solutions. Then, if x^S and x^L are respectively the mol fraction of component A in the solid and liquid phase respectively (Fig 14.12), the equilibrium condition at some temperature T in the melting or crystallization interval is $\mu_A^S = \mu_A^L$ and simultaneously $\mu_B^S = \mu_B^L$,

i.e. $$\mu_A^{S*} + RT \ln x^S = \mu_A^{L*} + RT \ln x^L$$

whence $$\ln \frac{x^L}{x^S} = -\mu_A^{L*} + \mu_A^{S*}$$

therefore by differentiation with respect to T and integration between limits

$$\ln \frac{x^L}{x^S} = -\frac{\Delta H_A}{R}\left(\frac{1}{T} - \frac{1}{T_A}\right)$$

if it is assumed that $\Delta C_{p_A} = C_{p_A}^{L*} - C_{p_A}^{S*} = 0$. And by analogy

$$\ln \frac{1-x^L}{1-x^S} = -\frac{\Delta H_B}{R}\left(\frac{1}{T} - \frac{1}{T_B}\right).$$

Thus if the course of the liquidus and solidus curves has been determined experimentally the heats of melting ΔH_A and ΔH_B can be determined from the slope of plots of $\ln (x^L/x^S)$ and $\ln (1-x^L)/(1-x^S)$ respectively against T^{-1}. Contrariwise if ΔH_A and ΔH_B are known the liquidus and solidus curves can be constructed by simultaneous solution of the two equations at conveniently chosen temperature intervals between T_A and T_B.

(6) System displaying complete miscibility in the solid state with a minimum in the liquidus

Such a system is familiar in the silicate field in the alkali feldspars, where it is complicated by the incongruent melting of K-feldspar. We take, for the sake of simplicity, a hypothetical system with components A and B (Fig 14.14). The phase diagram states the experimental observation that if the liquidus displays a minimum so does the solidus and the minima are coincident. All compositions other than the pure components and the minimum melting composition exhibit a melting (and crystallization) interval. As in the immediately preceding example the maintenance of equilibrium is practically difficult and consequently zoning is commonly observed in such systems, but the limits are different: here the final liquid composition in disequilibrium crystallization and the marginal composition of a crystal grain will, in the limit, be the composition of the minimum and not that of the lower melting-point component. On heating a solid solution belonging to such a system a liquid phase will first appear at some temperature $\geqslant T_{min}$ and it is evident from Fig 14.14 that for a wide range of compositions this will be a temperature below that of the melting point of either component. The existence of such minimum-melting systems among the common rock-forming minerals plays a critical role in the interpretation of the generation of silicate liquids from rocks deeply buried in the earth.

The perfect solution model, which we have consistently used hitherto, cannot give rise to a liquidus with a minimum. The next simplest model for real solutions, the strictly regular solution, which we have already considered in chapter 13, does

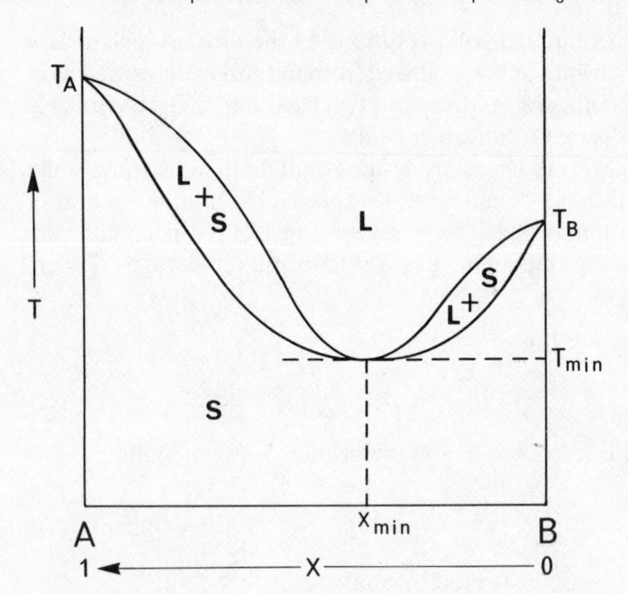

Fig 14.14 A binary system with complete miscibility in the solid state and a minimum in the liquidus. The solidus necessarily has a minimum coincident with that of the liquidus; the coordinates of the azeotropic point are labelled x_{min}, T_{min}.

however provide an adequate model for the interpretation of the phase diagram of Fig 14.14. If we assume that the solid solution is a strictly regular solution, then at any temperature T

$$\mu_A^S = \mu_A^{S*} + RT \ln x^S + \alpha^S(1-x^S)^2$$

and
$$\mu_B^S = \mu_B^{S*} + RT \ln(1-x^S) + \alpha^S(x^S)^2$$

where x^S is the mol fraction of component A in the solid solution and $\alpha^S = Nzw^S$. If we further assume that the liquid is a strictly regular solution then at any temperature T

$$\mu_A^L = \mu_A^{L*} + RT \ln x^L + \alpha^L(1-x^L)^2$$

and
$$\mu_B^L = \mu_B^{L*} + RT \ln(1-x^L) + \alpha^L(x^L)^2$$

where x^L is the mol fraction of component A in the liquid and $\alpha^L = Nz^L w^L$.

The condition for the liquid and solid solutions to be in equilibrium is

$$\mu_A^L = \mu_A^S \qquad \text{and} \qquad \mu_B^L = \mu_B^S.$$

The first of these equations on substitution of the expressions for chemical potential in the two phases yields

$$\ln \frac{x^L}{x^S} + \frac{\alpha^L(1-x^L)^2 - \alpha^S(1-x^S)^2}{RT} = -\frac{\mu_A^{L*} - \mu_A^{S*}}{RT}$$

which becomes on differentiation with respect to temperature

$$\frac{\partial}{\partial T}\left(\ln \frac{x^L}{x^S} + \frac{\alpha^L(1-x^L)^2 - \alpha^S(1-x^S)^2}{RT}\right) = \frac{\Delta H_A}{RT^2},$$

where $\Delta H_A = H_A^{L*} - H_A^{S*}$ is the heat of melting of pure A.

Integrating between pure component A and x, T,

$$\left[\ln \frac{x^L}{x^S} + \frac{\alpha^L(1-x^L)^2 - \alpha^S(1-x^S)^2}{RT}\right]_{x=1,\,T=T_A}^{x,\,T} = -\frac{\Delta H_A}{R}\left[\frac{1}{T}\right]_{T_A}^{T}$$

i.e.
$$\ln\frac{x^{\mathrm{L}}}{x^{\mathrm{S}}}+\frac{\alpha^{\mathrm{L}}(1-x^{\mathrm{L}})^2-\alpha^{\mathrm{S}}(1-x^{\mathrm{S}})^2}{RT}=-\frac{\Delta H_{\mathrm{A}}}{R}\left(\frac{1}{T}-\frac{1}{T_{\mathrm{A}}}\right)$$

and likewise
$$\ln\frac{1-x^{\mathrm{L}}}{1-x^{\mathrm{S}}}+\frac{\alpha^{\mathrm{L}}(x^{\mathrm{L}})^2-\alpha^{\mathrm{S}}(x^{\mathrm{S}})^2}{RT}=-\frac{\Delta H_{\mathrm{B}}}{R}\left(\frac{1}{T}-\frac{1}{T_{\mathrm{B}}}\right).$$

These two equations represent simultaneously the liquidus and solidus curves on the assumption of $\Delta C_p = 0$. Differentiation of both equations with respect to T yields simultaneous equations from which $\partial x^{\mathrm{S}}/\partial T$ and $\partial x^{\mathrm{L}}/\partial T$ are separable and can be inverted to give:

$$\frac{\partial T}{\partial x^{\mathrm{L}}}=\frac{T(x^{\mathrm{S}}-x^{\mathrm{L}})\{RT-2\alpha^{\mathrm{L}}x^{\mathrm{L}}(1-x^{\mathrm{L}})\}}{x^{\mathrm{L}}(1-x^{\mathrm{L}})[\alpha^{\mathrm{L}}\{x^{\mathrm{S}}-2x^{\mathrm{S}}x^{\mathrm{L}}+(x^{\mathrm{L}})^2\}-\alpha^{\mathrm{S}}x^{\mathrm{S}}(1-x^{\mathrm{S}})+\Delta H_{\mathrm{A}}x^{\mathrm{S}}+\Delta H_{\mathrm{B}}(1-x^{\mathrm{S}})]}$$

$$\frac{\partial T}{\partial x^{\mathrm{S}}}=\frac{T(x^{\mathrm{S}}-x^{\mathrm{L}})\{RT-2\alpha^{\mathrm{S}}x^{\mathrm{S}}(1-x^{\mathrm{S}})\}}{x^{\mathrm{S}}(1-x^{\mathrm{S}})[\alpha^{\mathrm{L}}x^{\mathrm{L}}(1-x^{\mathrm{L}})-\alpha^{\mathrm{S}}\{x^{\mathrm{L}}-2x^{\mathrm{S}}x^{\mathrm{L}}+(x^{\mathrm{S}})^2\}+\Delta H_{\mathrm{A}}x^{\mathrm{L}}+\Delta H_{\mathrm{B}}(1-x^{\mathrm{L}})]}$$

It is evident that $\partial T/\partial x^{\mathrm{L}} = \partial T/\partial x^{\mathrm{S}} = 0$ for $x^{\mathrm{L}} = x^{\mathrm{S}}$, that is the liquidus and solidus curves touch at their common maximum or minimum. Whether the common point is a maximum or a minimum can be found from the sign of the second differential coefficient which may be obtained, by differentiating again, substituting

$$\frac{\partial x^{\mathrm{S}}}{\partial x^{\mathrm{L}}}=\frac{\partial T}{\partial x^{\mathrm{L}}}\bigg/\frac{\partial T}{\partial x^{\mathrm{S}}}$$

and putting
$$\frac{\partial T}{\partial x^{\mathrm{L}}}=\frac{\partial T}{\partial x^{\mathrm{S}}}=0$$

and $x^{\mathrm{S}} = x^{\mathrm{L}}$, as
$$\frac{\partial^2 T}{\partial (x^{\mathrm{L}})^2}=\frac{2RT(\alpha^{\mathrm{S}}-\alpha^{\mathrm{L}})\{RT-2\alpha^{\mathrm{L}}x(1-x)\}}{\{RT-2\alpha^{\mathrm{S}}x(1-x)\}\{(\alpha^{\mathrm{L}}-\alpha^{\mathrm{S}})x(1-x)+\Delta H_{\mathrm{A}}x+\Delta H_{\mathrm{B}}(1-x)\}}$$

and a similar expression applies for $\partial^2 T/\partial(x^{\mathrm{S}})^2$. Since α^{L} and α^{S} are small relative to RT and ΔH_{A} and ΔH_{B} are positive for melting, $\partial^2 T/\partial x^2$ in each case has the sign of $\alpha^{\mathrm{S}}-\alpha^{\mathrm{L}} = N(z^{\mathrm{S}}w^{\mathrm{S}}-z^{\mathrm{L}}w^{\mathrm{L}})$. On general structural grounds a liquid is to be expected to be less discriminating about the occupation of neighbouring sites so that in general $w^{\mathrm{S}} > w^{\mathrm{L}}$ and $z^{\mathrm{S}} \simeq z^{\mathrm{L}}$. Therefore for liquid-solid equilibrium $\alpha^{\mathrm{S}}-\alpha^{\mathrm{L}} > 0$ and consequently $\partial^2 T/\partial(x^{\mathrm{L}})^2$ and $\partial^2 T/\partial(x^{\mathrm{S}})^2$ will be positive so that both liquidus and solidus are at minima at their point of contact, which is known as the *azeotropic point* $(x_{\mathrm{az}}, T_{\mathrm{az}})$.

The temperature and composition of the azeotropic point can be evaluated simply by inserting the condition $x^{\mathrm{S}} = x^{\mathrm{L}} = x_{\mathrm{az}}$ in the simultaneous equations to the solidus and liquidus. Thus

$$(1-x)^2=-\frac{\Delta H_{\mathrm{A}}(T_{\mathrm{A}}-T_{\mathrm{az}})}{(\alpha^{\mathrm{L}}-\alpha^{\mathrm{S}})T_{\mathrm{A}}}$$

and
$$x^2=-\frac{\Delta H_{\mathrm{B}}(T_{\mathrm{B}}-T_{\mathrm{az}})}{(\alpha^{\mathrm{L}}-\alpha^{\mathrm{S}})T_{\mathrm{B}}}$$

therefore
$$\left(\frac{1-x}{x}\right)^2=\frac{\Delta S_{\mathrm{A}}(T_{\mathrm{A}}-T_{\mathrm{az}})}{\Delta S_{\mathrm{B}}(T_{\mathrm{B}}-T_{\mathrm{az}})},$$

since at the melting point $\Delta G_A = \Delta H_A - T_A \Delta S_A = 0$. Experimental determination of the azeotropic point thus makes possible the evaluation of $\Delta H_A/\Delta H_B$, if only the melting points of the pure components are known, and of $\alpha^L - \alpha^S$, if one heat of melting has been separately determined.

(7) Exsolution in solid solutions

Hitherto we have dealt only with solid solutions in which the components are miscible in all proportions. Such complete miscibility is of common occurrence as is *partial miscibility* with which we are concerned here. It will be shown that the extent of miscibility in general increases with increased temperature. The petrologist, who is concerned with the extremely slow cooling of solid phases possible only under natural conditions, thus commonly comes across partial miscibility in the context of a phase that has crystallized at high temperature and had to adjust to equilibrium or near-equilibrium conditions at lower temperatures by *exsolution* of a second phase.

We have already shown that for a *perfect binary solution* the molar free enthalpy of mixing[2] is given by

$$\Delta G_m = RT\{x \ln x + (1-x)\ln(1-x)\}.$$

Since T is necessarily positive and x is a positive fraction ΔG_m is negative for all values of T and x. The solution is therefore stable relative to a mixture of the pure components over the whole compositional range. Moreover ΔG_m varies smoothly from zero at $x = 0$, through a minimum at $x = \frac{1}{2}$ (since $(\partial \Delta G_m/\partial x)_T = RT \ln(x/1-x)$ and $(\partial^2 \Delta G_m/\partial x^2)_T = RT/x(1-x)$) to zero again at $x = 1$; there are no points of inflexion in the curve (i.e. there are no solutions of $RT/x(1-x) = 0$ for $T > 0$, $0 < x < 1$) and therefore the solution is also stable relative to any mixture of solutions of particular composition over the whole compositional range. Perfect solutions can thus never exhibit partial miscibility.

For a *strictly regular solution* however the molar free enthalpy of mixing is a less simple function of mol fraction. We have already shown that

$$\Delta G_m = RT\{x \ln x + (1-x)\ln(1-x)\} + \alpha x(1-x)$$

per mole, where $\alpha = zwN$.

Therefore $$\left(\frac{\partial \Delta G_m}{\partial x}\right)_T = RT\left\{\ln\frac{x}{1-x} + \frac{zw}{kT}(1-2x)\right\}.$$

The condition for ΔG_m to be at a maximum or minimum is then

$$\ln\frac{x}{1-x} = \frac{zw}{kT}(2x-1).$$

This expression has one solution $x = \frac{1}{2}$ which is independent of the magnitude of zw/kT.

$$\left(\frac{\partial^2 \Delta G_m}{\partial x^2}\right)_T = RT\left(\frac{1}{x(1-x)} - \frac{2zw}{kT}\right).$$

[2] In discussion of exsolution equilibria we are concerned with minimization of free enthalpy of mixing ΔG_m and not of the total free enthalpy of the system ΔG since the solution only differs from a mechanical mixture configurationally in the two models we consider. Whether $(G_{AX})_T^0$ and $(G_{BX})_T^0$ are equal or not is as irrelevant to the argument as it is inaccessible to experiment.

so that the condition for a point of inflexion (known in this context as a spinode) is

$$x(1-x) = \frac{kT}{2zw}$$

i.e.
$$x = \tfrac{1}{2} \pm \tfrac{1}{2}\sqrt{\left(1 - \frac{2kT}{zw}\right)}.$$

There will therefore be no points of inflexion if $(2kT/zw) > 1$ and two points of inflexion symmetrically disposed about $x = \frac{1}{2}$ if $(2kT/zw) < 1$. This change of behaviour defines a *critical temperature* $T_c = zw/2k$ characteristic of the system. Rewriting the expression for the second differential in terms of T_c and substituting $x = \frac{1}{2}$

$$\left[\left(\frac{\partial^2 \Delta G_m}{\partial x^2}\right)_T\right]_{x=\frac{1}{2}} = 4RT\left(1 - \frac{T_c}{T}\right)$$

makes it clear that for $T > T_c$ there is a minimum at $x = \frac{1}{2}$ and for $T < T_c$ a maximum at $x = \frac{1}{2}$. The change in the form of ΔG_m as a function of x above and below T_c is illustrated in Fig 14.15.

Thus in a strictly regular solution at temperatures greater than T_c, ΔG_m as a function of x passes through a minimum at $x = \frac{1}{2}$ and has no real points of inflexion; partial miscibility is therefore impossible. However at temperatures below T_c there is a maximum at $x = \frac{1}{2}$ and symmetrically disposed points of inflexion at $x = \frac{1}{2} \pm \frac{1}{2}\sqrt{\{1-(T/T_c)\}}$. Therefore there must be symmetrical minima at $x = \frac{1}{2} \pm \delta$, where $\frac{1}{2}\sqrt{\{1-(T/T_c)\}} < \delta < \frac{1}{2}$. Values of x corresponding to the minima of ΔG_m are given by rewriting the condition $(\partial \Delta G_m/\partial x)_T = 0$ in terms of T_c as

$$\frac{2(2x-1)}{\ln\{x/(1-x)\}} = \frac{T}{T_c}.$$

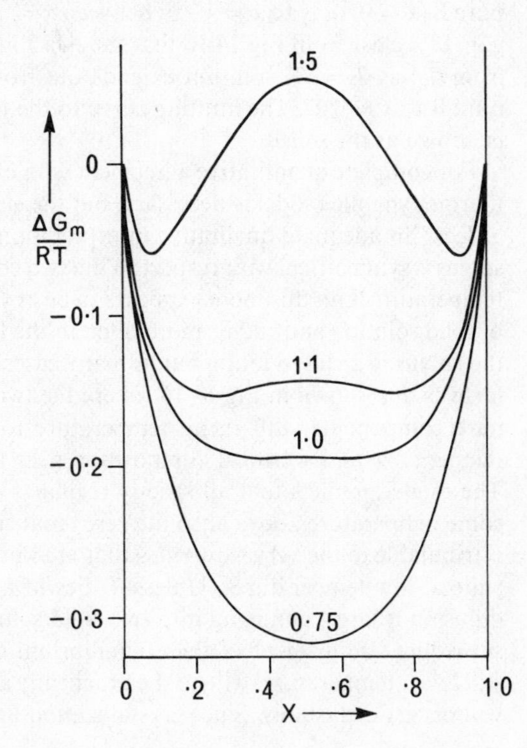

Fig 14.15 The variation of molar free enthalpy of mixing, plotted as $\Delta G_m/RT$, with mol fraction for a strictly regular solution at various temperatures. Against each curve is indicated the value of $T_c/T = zw/2kT$.

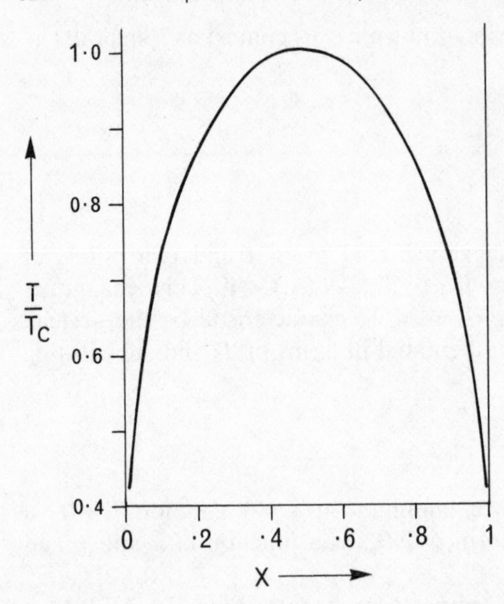

Fig 14.16 The solvus. The plot of T/T_c against mol fraction for a strictly regular solution represents the limiting curve of solid solution; within the solvus two phases are in equilibrium.

From the plot of T/T_c as a function of x shown in Fig 14.16 the compositions of the minima at any temperature below T_c can be read off provided the magnitude of T_c is known.

It is evident from Fig 14.15 that a mixture of the two solutions of minimum ΔG_m will have a lower free enthalpy than a solution in the compositional range $\frac{1}{2}-\delta < x < \frac{1}{2}+\delta$ at any temperature $T < T_c$. Therefore in this temperature region solid solution will extend from pure A ($x = 1$) only as far as $x = \frac{1}{2}+\delta$ and from pure B ($x = 0$) only to $x = \frac{1}{2}-\delta$; between $x = \frac{1}{2}-\delta$ and $x = \frac{1}{2}+\delta$ there is a miscibility gap. It is clear from Fig 14.16 that the miscibility gap widens fairly rapidly as T falls from T_c; at $T = \frac{1}{2}T_c$ solution extends out from pure A only to $x = 0.98$ and from pure B to $x = 0{\cdot}02$. The limiting curve to the two phase region of the miscibility gap is known as the *solvus*.

For complete quantitative agreement with experimental data a more sophisticated thermodynamic model is necessary, but the strictly regular solution model provides at least an adequate qualitative interpretation of exsolution. In predicting that the solvus is symmetrical with respect to the two components and closed only at its upper temperature limit this model appears to be at variance with experiment in some cases of solid solution and rather more often in the liquid state (e.g. nicotine–water, where the solvus is a closed loop), but in many cases a solvus is found of the same general form as that shown in Fig 14.16, where the two branches move out from close to the pure components with rising temperature to meet, with zero slope, at a critical temperature and a critical composition near the centre of the compositional range. The model predicts that all strictly regular solutions will have a miscibility gap at some temperature above absolute zero: that such is not found to be so may well be attributable to the very slow rates that are known to obtain for diffusion in the solid state at low temperatures. Unless T_c lies in a temperature range where the rate of diffusion is large, unmixing into two solid solutions of equilibrium composition and subsequent adjustment of these equilibrium compositions to those appropriate to still lower temperatures will not be practically attainable; a single solid solution phase will persist metastably. Since crystallization under natural conditions in the earth is

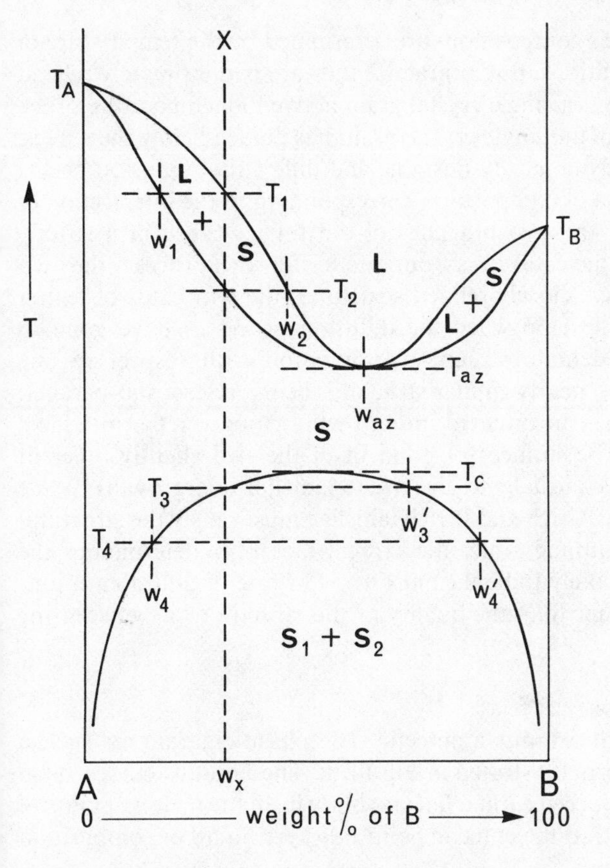

Fig 14.17 A binary system in which both solid and liquid phases are strictly regular solutions. The liquidus, solidus, and solvus are shown as solid lines; the broken lines of constant temperature or composition serve to illustrate the crystallization and subsequent exsolution of the liquid X. The azeotropic and critical temperatures are shown by the broken isothermal lines T_{az} and T_c; the composition of the azeotropic point is labelled w_{az}.

characteristically extremely slow, petrology is much concerned with the phenomenon of exsolution.

We conclude this section with a brief account of the crystallization and subsequent exsolution of a liquid in a binary system where both liquid and solid phases are strictly regular solutions. We consider first the crystallization under equilibrium conditions of the liquid X in Fig 14.17. Crystallization will begin when the system has cooled to T_1 with the appearance of an infinitesimal amount of solid of composition w_1. On further cooling the amount of the solid phase will increase steadily and its composition will move along the solvus becoming richer in component B until the last liquid (composition w_2) disappears at the temperature T_2. The resulting single solid phase persists until the system has cooled to T_3 where an infinitesimal amount of solid solution of composition w'_3 is in equilibrium with solid solution of composition w_x. When the temperature has fallen to T_4 the compositions of the two solid solutions will have moved along the solvus to w_4 and w'_4 while their proportions, given by the lever rule, will have become

$$\frac{\text{wt of A-rich solution}}{\text{wt of B-rich solution}} = \frac{w'_4 - w_x}{w_x - w_4}$$

On further cooling the compositions of the two solid solutions will become more extreme, their proportions approaching $(100 - w_x)/w_x$ in the limit, until eventually the rate of exsolution becomes too slow for further change to occur.

In thermodynamic terms the two exsolved phases are simply two phases and therefore, by the phase rule, their compositions are determined by the temperature of the system under isobaric conditions. But in practice they are two intimately related phases occurring within what was a single crystal grain above the temperature of the solvus. When the temperature of the single crystal grain has passed below the solvus, the structure becomes thermodynamically unstable and differential diffusion begins to produce regions of the crystal of composition corresponding to the intersection of the appropriate isotherm with the two branches of the solvus. Since in a strictly regular solution, such as we have taken as our model for the process, the two components are required to have closely related structures, the two kinds of region of uniform composition that develop when the diffusion processes have gone to completion will be in some definite structural orientation with respect to one another, like elements of the two nearly similar structures being at least sub-parallel: the single crystal grain becomes an oriented intergrowth. Moreover the interfaces between the two phases must be surfaces of good fit of the two slightly different structures. Exsolution is observed usually to give rise to lamellar intergrowths. While the phase interfaces between the A-rich and B-rich lameilae must satisfy the structural criterion of good fit on the atomic scale, the critical factor in determining the interfacial orientation is more likely to be the anisotropy of rate of diffusion of ions through the host structure rather than the quality of the structural fit between the host and exsolved lamellae.

(8) Systems in which the solvus intersects the solidus

We first consider a system that exhibits a eutectic. The phase diagram for such a system will be of the general form illustrated in Fig 14.18. The liquidus curves, as in the simple eutectic system considered earlier, fall smoothly from the melting points of the two components, T_A and T_B, to the eutectic point E, where liquid of composition

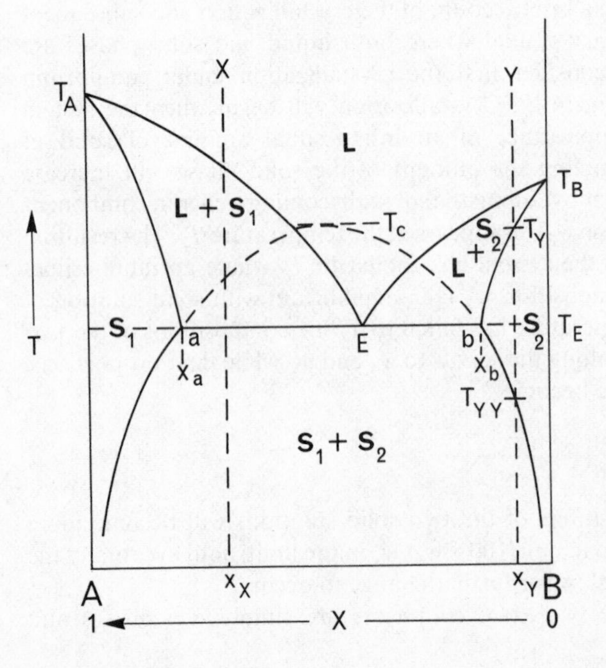

Fig 14.18 A binary eutectic system with partial miscibility in the solid state. The metastable extension of the solvus above the solidus (i.e. $T_c > T_E$) is shown by the broken curve.

x_E is in equilibrium at T_E not with the pure components but with solid solutions of combustion x_a and x_b. Below the eutectic temperature the two branches of the solvus, defining the limits of the miscibility gap, fall off from a and b towards the pure components. The metastable part of the solvus above T_E is shown by a broken line in the figure. The areas T_A—a—A and T_B—b—B represent the T—x stability fields of the solid solutions.

A liquid such as X begins to crystallize solid solution S_1 at the temperature of intersection of its isocompositional line x_X with the liquidus T_A—E. On further cooling under equilibrium conditions liquid and solid move along the liquidus and solidus until they reach respectively E and a at the temperature T_E. The temperature of the system remains constant at T_E until all the liquid has disappeared and only crystalline S_1 of composition x_a and S_2 of composition x_b are present. Further cooling results in adjustment along the solvus of the compositions of the solid solutions by exsolution of S_2 from S_1 on the branch through a and of S_1 from S_2 on the branch through b.

The liquid Y behaves differently by crystallizing the solid solution S_2 above T_E, the composition of liquid and solid solution moving along the liquidus and solidus respectively until at T_Y the liquid is exhausted. The resulting single solid phase S_2 cools without change of composition (x_Y) until the solvus is reached at T_{YY}, where S_1 begins to exsolve. Further cooling produces changes only in the proportions and compositions of the host S_2 and the exsolved S_1 lamellae.

That there is a miscibility gap in the solid solution implies non-ideality and we shall assume for simplicity a strictly regular solid solution. The liquid may or may not be ideal: we shall assume a strictly regular solution, which reduces to a perfect solution when $\alpha^L = 0$. In these terms this case differs from that of the previous section in being associated with a larger positive value of α^S. The simultaneous equations to the S_1 liquidus and solidus are then

$$\ln\frac{x_1^L}{x_1^S}+\frac{\alpha^L(1-x_1^L)^2-\alpha^S(1-x_1^S)^2}{RT} = -\frac{\Delta H_A}{R}\left(\frac{1}{T}-\frac{1}{T_A}\right)$$

and $$\ln\frac{1-x_1^L}{1-x_1^S}+\frac{\alpha^L(x_1^L)^2-\alpha^S(x_1^S)^2}{RT} = -\frac{\Delta H_B}{R}\left(\frac{1}{T}-\frac{1}{T_B}\right)$$

where x_1^L and x_1^S are the mol fractions of components A on the liquidus and solidus respectively at T K. And likewise for the S_2 liquidus and solidus

$$\ln\frac{x_2^L}{x_2^S}+\frac{\alpha^L(1-x_2^L)^2-\alpha^S(1-x_2^S)^2}{RT} = -\frac{\Delta H_A}{R}\left(\frac{1}{T}-\frac{1}{T_A}\right)$$

and $$\ln\frac{1-x_2^L}{1-x_2^S}+\frac{\alpha^L(x_2^L)^2-\alpha^S(x_2^S)^2}{RT} = -\frac{\Delta H_B}{R}\left(\frac{1}{T}-\frac{1}{T_B}\right).$$

At the eutectic the two branches of the liquidus intersect at $T = T_E$, $x_1^L = x_2^L = x_E$. Substitution and elimination of ΔH_A yields

$$\ln x_1^S - \ln x_2^S = \frac{\alpha^S}{RT_E}(x_1^S - x_2^S)(2 - x_1^S - x_2^S)$$

and elimination of ΔH_B yields

$$\ln(1-x_1^S) - \ln(1-x_2^S) = \frac{\alpha^S}{RT_E}(x_2^S - x_1^S)(x_2^S + x_1^S).$$

Therefore $\ln x_1^S - \ln x_2^S - \ln(1-x_1^S) + \ln(1-x_2^S)$

$$= \frac{\alpha^S}{RT_E}(x_1^S - x_2^S)\{(2 - x_1^S - x_2^S) + (x_2^S + x_1^S)\}$$

$$= \frac{2\alpha^S}{RT_E}(x_1^S - x_2^S).$$

Now the condition that x_1^S lies on the solvus is, as we have shown earlier,

$$2(2x_1^S - 1)\ln\frac{1-x_1^S}{x_1^S} = \frac{T}{T_c}$$

but $$T_c = \frac{zw}{2k} = \frac{\alpha^S}{2R}$$

so that on substitution and rearrangement we can write

$$\ln x_1^S - \ln(1-x_1^S) = \frac{\alpha^S}{RT}(2x_1^S - 1)$$

and similarly

$$\ln x_2^S - \ln(1-x_2^S) = \frac{\alpha^S}{RT}(2x_2^S - 1)$$

therefore $$\ln x_1^S - \ln x_2^S - \ln(1-x_1^S) + \ln(1-x_2^S) = \frac{2\alpha^S}{RT}(x_1^S - x_2^S).$$

Therefore each branch of the solidus intersects the corresponding branch of the solvus at the eutectic temperature as shown in Fig 14.18.

The other type of system that we have to discuss under this head has a reaction point (known in this context as a *peritectic*) and is of the general form illustrated in Fig 14.19. We begin by considering the crystallization of four liquids. A liquid whose composition lies to the left of a crystallizes initially a solid solution S_1, the composition of liquid and solid moving along the S_1 liquidus and solidus respectively until the liquid is exhausted at some temperature above T_P; on further cooling to some temperature below T_P exsolution of S_2 from S_1 begins. A liquid of composition between a and b crystallizes solid solution S_1 initially, the compositions of liquid and solid phases adjusting to temperature until at the peritectic temperature T_P, S_1 has the composition a and the liquid has the peritectic composition P; the temperature of the system remains constant at T_P until the reaction $S_1(a) + L(P) \rightarrow S_2(b)$ has gone to completion with complete exhaustion of the liquid; the resulting phase assemblage, solid solution S_1 of composition a and solid solution S_2 of composition b in proportions given by the lever rule, adjusts to further lowering of temperature by exsolution of lamellae of S_2 from S_1 and lamellae of S_1 from S_2 in accordance with the course of the solvus shown. A liquid of composition between b and P likewise reaches T_P where the peritectic reaction exhausts the early crystallized solid solution S_1 of composition a; just below T_P the system consists of a liquid phase lying on the S_2 liquidus in equilibrium with a solid solution lying on the S_2 solidus; the liquid is shortly thereafter exhausted and the system then consists of a single solid solution phase S_2, which on further cooling starts to exsolve S_1. A liquid of composition to the right of P crystallizes S_2 and becomes exhausted at some temperature above T_B where its isocompositional line intersects the S_2 solidus; if equilibrium can be

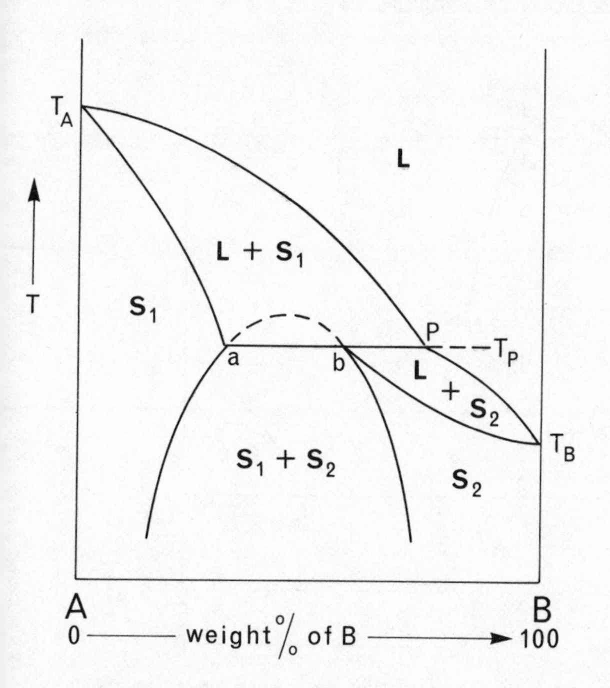

Fig 14.19 A binary system with a reaction point and partial miscibility in the solid state. In such a system the reaction point is known as a peritectic point.

maintained at a sufficiently low temperature exsolution of S_1 may be recorded.

The phase diagram shown in Fig 14.18 obviously differs from that of Fig 14.19 in the relationship of the solvus to the liquidus and solidus: in the former the critical temperature of the solvus lies above the liquidus, but in the latter between the liquidus and the solidus. We have seen that the eutectic diagram can be interpreted approximately in terms of strictly regular liquid and solid solutions with $\alpha^S \gg \alpha^L$, or in more general terms the departure of the solid solution from ideality is markedly greater than that of the liquid. We merely state, without quantitative explanation, that the peritectic diagram can be interpreted in terms of a liquid that is only slightly non-ideal and a solid solution that is markedly non-ideal; the slight departure of the liquidus from the smooth curve of Fig 14.12 is obvious.

Disequilibrium crystallization of a liquid in the eutectic system cannot enable the liquid to persist at temperatures below T_E. Zoning of the solid solution crystallized above T_E and failure to exsolve below T_E are the only possible significant consequences of disequilibrium. In the peritectic diagram however zoning above T_P will be followed by failure of the reaction $L + S_1 \rightarrow S_2$ to go to completion at T_P (as in the case of the reaction point) with the consequence that all liquids will cease crystallization only when the temperature has fallen to T_B; zoning of S_2 and failure to exsolve will also be features of disequilibrium crystallization in such a system.

(9) Liquid immiscibility

A liquid solution may exhibit a miscibility gap in just the same way as we have found in solid solutions. The strictly regular solution model is adequate to account for the existence of liquid immiscibility but, in the silicate and oxide field especially, it fails to explain the common asymmetry of the solvus; recourse must be had to more general, and less simple, models for real solutions. Liquid immiscibility is not nowadays considered to have frequently exerted any significant influence on the crystallization of minerals from melts in nature, but its common occurrence in

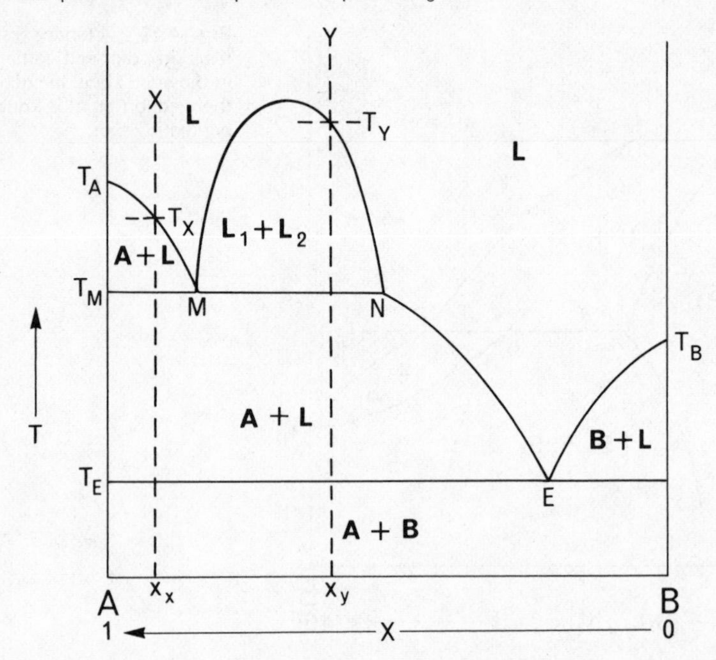

Fig 14.20 Liquid immiscibility in a binary eutectic system. Within the curve which stretches from the monotectic point M to the point N two liquid phases are in equilibrium.

silica-rich liquids at atmospheric pressure is a source of inconvenience to the experimentalist. Evidence of liquid immiscibility is well known in the study of meteorites and in the field of metallurgy in general.

Figure 14.20 illustrates the simplest kind of system that displays liquid immiscibility, a liquid solvus on the A-rich side of the diagram and a simple eutectic between A and B. Liquid immiscibility only affects the crystallization of liquids with $x > x_N$. A liquid X whose composition lies between pure A and the point M, where two liquids and a solid are in equilibrium (*the monotectic point*), crystallizes A as the temperature and composition of the liquid move along the liquidus towards M; the temperature of the system remains constant at T_M until the reaction M(l) → N(l) + A(s) has gone to completion with considerable diminution (given by the lever rule) in the amount of liquid present; as the temperature of the system falls from T_M the composition of the liquid moves along the liquidus from N to E; the small amount of liquid, of composition E, remaining at T_E then crystallizes as a eutectic intergrowth. A liquid Y whose composition lies between M and N begins to exsolve into two liquid phases at the temperature at which its isocompositional line intersects the liquid solvus; further cooling produces greater disparity in the compositions of the two liquid phases until T_M is reached where L_1 has the composition M and L_2 the composition N; temperature remains constant at T_M until the reaction M(l) → A(s) + N(l) has gone to completion; crystallization of A from liquids along N—E and eutectic crystallization follow. It should be noted that liquids with $1 > x > x_M$ begin to crystallize A at temperatures above T_M while for liquids with $x_M \geqslant x \geqslant x_N$ the first crystalline phase A appears at T_M; of course for liquids with $x_N > x > x_E$ the first solid appears at $T < T_M$.

(10) Thermochemical consequences of ionization in liquid and solid solutions

The quantitative thermodynamic interpretation of binary systems that has been developed in this chapter has been based on the implicit assumption that the pure components persist as definite unionized molecular entities in solution, whether liquid or solid. This is manifestly inappropriate in the field of ionic inorganic salts where it is known from structural crystallographic studies that in a solid solution of AX and BY the cations A and B are distributed over similar structural sites and the anions X and Y likewise. Physical, especially spectrographic, studies of the corresponding liquids indicate that ionization can usually be treated as complete there too. In solid solutions of complex oxy-salts, such as silicates, the pattern of ionization is usually determinable on a *prima facie* basis from the known structures of the components, but in the liquid phase the question of the magnitude of the anionic groupings has all too frequently not been resolved by spectrographic study: that pyroxene melts contain long $(SiO_3)_n^{2n-}$ chains and feldspar melts extensive $(AlSi_3O_8)_n^{n-}$ rafts, where n is in each case very large, has not been demonstrated but is to be expected on general grounds. We shall not deal with specific examples here but merely discuss the principles of the quite straightforward correction for ionization in terms of simple systems following the treatment of Bradley (1962, 1964) rather than that of Temkin (1945).

We consider first the simple eutectic system with no solid solution and suppose that component 1 is the simple salt $R_{a_1}^{b_1+}X_{b_1}^{a_1-}$. Then

$$\mu_1^L = a_1\mu_{1+}^L + b_1\mu_{1-}^L$$

where μ_{1+} and μ_{1-} are the chemical potentials of the cations R and the anions X in the liquid phase. Assuming that the liquid is a perfect solution

$$\mu_1^L = a_1\mu_{1+}^{L*} + b_1\mu_{1-}^{L*} + RT\ln x_{1+}^{a_1}x_{1-}^{b_1}$$
$$= \mu_1^{L*} + RT\ln x_{1+}^{a_1}x_{1-}^{b_1}.$$

Hence at equilibrium

$$\mu_1^{S*} = \mu_1^{L*} + RT\ln x_{1+}^{a_1}x_{1-}^{b_1}$$

Therefore $$[\ln x_{1+}^{a_1}x_{1-}^{b_1}]_{T_1}^{T} = -\frac{\Delta H_1}{R}\left[\frac{1}{T}\right]_{T_1}^{T}$$

where T_1 is the melting point of pure component 1, i.e., pure $R_{a_1}^{b_1+}X_{b_1}^{a_1-}$. Thus at T_1

$$x_{1+} = \frac{a_1}{a_1+b_1} \quad \text{and} \quad x_{1-} = \frac{b_1}{a_1+b_1},$$

therefore $$\ln x_{1+}^{a_1}x_{1-}^{b_1} - \ln\frac{a_1^{a_1}b_1^{b_1}}{(a_1+b_1)^{a_1+b_1}} = -\frac{\Delta H_1}{R}\left(\frac{1}{T}-\frac{1}{T_1}\right)$$

i.e. $$\ln Q_1 = -\frac{\Delta H_1}{R}\left(\frac{1}{T}-\frac{1}{T_1}\right)$$

where $$Q_1 = \frac{x_{1+}^{a_1}x_{1-}^{b_1}(a_1+b_1)^{a_1+b_1}}{a_1^{a_1}b_1^{b_1}}$$

This is an equation of the same form as in the unionized case considered previously where $Q_1 = x_1$.

The evaluation of Q_1 may be illustrated by a perfect liquid solution between KCl (mol fraction x_1) and Na_2SO_4 (mol fraction x_2). The total number of ions per mole of liquid phase will be $2x_1+3x_2$.

Therefore $x_{1+} = x_{1-} = \dfrac{x_1}{2x_1+3x_2}$, $x_{2+} = \dfrac{2x_2}{2x_1+3x_2}$, and $x_{2-} = \dfrac{x_2}{2x_1+3x_2}$;

and $\mathrm{a}_1 = \mathrm{b}_1 = 1$, $\mathrm{a}_2 = 2$, and $\mathrm{b}_2 = 1$.

Hence $$Q_1 = \frac{4x_1^2}{(2x_1+3x_2)^2} = \left(\frac{1-x_2}{1+\frac{1}{2}x_2}\right)^2,$$

since $x_1+x_2 = 1$,

and $$Q_2 = \left(\frac{1-x_1}{1-\frac{1}{3}x_1}\right)^3.$$

If the two components have a common ion, as will often be the case, ionization has no effect on the equations to the two branches of the liquidus. Consider a perfect liquid solution of NaCl (mol fraction x_1) and NaF (mol fraction x_2). The total number of ions per mole of liquid phase will be $2(x_1+x_2) = 2$.

Hence $x_{1+} = x_{2+} = \dfrac{x_1+x_2}{2} = \frac{1}{2}$, $x_{1-} = \dfrac{x_1}{2}$, and $x_{2-} = \dfrac{x_2}{2}$.

Since $\mathrm{a}_1 = \mathrm{b}_1 = \mathrm{a}_2 = \mathrm{b}_2 = 1$,

$$Q_1 = \tfrac{1}{2}\frac{x_1}{2}2^2 = x_1$$

and $$Q_2 = \tfrac{1}{2}\frac{x_2}{2}2^2 = x_2$$

as would be so if NaCl and NaF were unionized in the melt.

If the anions of the components, whose stoichiometric compositions are represented by $R_{\mathrm{a}_1}^{\mathrm{b}_1+}X_{\mathrm{b}_1}^{\mathrm{a}_1-}$ and $S_{\mathrm{a}_2}^{\mathrm{b}_2+}Y_{\mathrm{b}_2}^{\mathrm{a}_2-}$, are polymeric, $(X)_{n_1}^{\mathrm{a}_1 n_1-}$ and $(Y)_{n_2}^{\mathrm{a}_2 n_2-}$ where n_1 and n_2 are very large, a liquid phase which is a perfect solution of x_1 moles of RX and x_2 moles of SY will contain $\{\mathrm{a}_1+(\mathrm{b}_1/n_1)\}x_1+\{\mathrm{a}_2+(\mathrm{b}_2/n_2)\}x_2 \simeq \mathrm{a}_1x_1+\mathrm{a}_2x_2$ ions per mole.

Hence $x_{1+} = \dfrac{\mathrm{a}_1x_1}{\mathrm{a}_1x_1+\mathrm{a}_2x_2}$, $x_{2+} = \dfrac{\mathrm{a}_2x_2}{\mathrm{a}_1x_1+\mathrm{a}_2x_2}$.

At equilibrium

$$\begin{aligned}\mu_1^{S*} &= \mathrm{a}_1\mu_{1+}^{L*} + \frac{\mathrm{b}_1}{n_1}\mu_{1-}^{L*} + RT\ln x_{1+}^{\mathrm{a}_1}x_{1-}^{\mathrm{b}_1/n_1} \\ &= \mu_1^{L*} + RT\ln x_{1+}^{\mathrm{a}_1}x_{1-}^{\mathrm{b}_1/n_1} \\ &= \mu_1^{L*} + RT\ln x_{1+}^{\mathrm{a}_1}\end{aligned}$$

since for n_1 very large $\mathrm{b}_1/n_1 \to 0$.

Therefore $$[\ln x_{1+}^{\mathrm{a}_1}]_{T_1}^{T} = -\frac{\Delta H_1}{R}\left[\frac{1}{T}\right]_{T_1}^{T}.$$

Now at T_1 $$x_{1+} = \frac{\mathrm{a}_1}{\mathrm{a}_1+(\mathrm{b}_1/n_1)} \simeq 1,$$

hence $$\ln x_{1+}^{a_1} = -\frac{\Delta H_1}{R}\left(\frac{1}{T}-\frac{1}{T_1}\right)$$

so that here

$$Q_1 = x_{1+}^{a_1} = \left(\frac{1-x_2}{1-\{(a_1-a_2)/a_1\}x_2}\right)^{a_1}$$

and similarly

$$Q_2 = x_{2+}^{a_2} = \left(\frac{1-x_1}{1+\{(a_1-a_2)/a_2\}x_1}\right)^{a_2}.$$

By way of further exemplification we shall consider ionization in one other type of system only, the case of equilibrium between perfect liquid and solid solutions (Fig 14.12). Since the components in such a system very commonly have a common ion and similar structures we shall assume that the components can be represented as $R_a^{b+}\ X_b^{a-}$ and $S_a^{b+}\ X_b^{a-}$. We can write for the chemical potential of component 1 in the liquid (μ_{1L}) and solid (μ_{1S}) phases,

$$\mu_{1L} = \mu_{1L}^* + RT \ln x_{1L+}^{a} x_{1L-}^{b}$$

$$\mu_{1S} = \mu_{1S}^* + RT \ln x_{1S+}^{a} x_{1S-}^{b}$$

Therefore at equilibrium

$$\left[\ln\left\{\frac{x_{1L+}^{a} x_{1L-}^{b}}{x_{1S+}^{a} x_{1S-}^{b}}\right\}\right]_{x_1=1}^{x_1} = -\frac{\Delta H_1}{R}\left[\frac{1}{T}\right]_{T_1}^{T}.$$

Now a solution, whether solid or liquid, containing x_1 moles of $R_a^{b+}\ X_b^{a-}$ and x_2 moles of $S_a^{b+}\ X_b^{a-}$ gives rise to $(a+b)(x_1+x_2) = a+b$ ions per mole and therefore

$$x_{1+} = \frac{a}{a+b}x_1, \quad x_{2+} = \frac{a}{a+b}x_2, \quad \text{and} \quad x_{1-} = x_{2-} = \frac{b}{a+b}.$$

Therefore $$\left[\ln\left\{\frac{a^a b^b x_{1L}^a}{a^a b^b x_{1S}^a}\right\}\right]_{x_1=1}^{x_1} = -\frac{\Delta H_1}{R}\left[\frac{1}{T}\right]_{T_1}^{T}$$

and so $$a \ln\frac{x_{1L}}{x_{1S}} = -\frac{\Delta H_1}{R}\left(\frac{1}{T}-\frac{1}{T_1}\right)$$

i.e. $$Q_1 = \left(\frac{x_{1L}}{x_{1S}}\right)^a \quad \text{and similarly} \quad Q_2 = \left(\frac{x_{2L}}{x_{2S}}\right)^a$$

in contrast to the unionized case where

$$Q_1 = \frac{x_{1L}}{x_{1S}} \quad \text{and} \quad Q_2 = \frac{x_{2L}}{x_{2S}}.$$

The correction for ionization is equally relevant to the non-ideal equations necessary for the description of other types of binary system.

(11) *PTX* diagrams for binary systems

We have so far restricted our treatment of binary systems to isobaric conditions. For their complete representation in *PTX* space binary systems require either a three-dimensional diagram or a series of *TX* diagrams for different pressures, the latter providing a more easily decipherable record of quantitative experimental data.

Since for most compositions the molar volume of the liquid is greater than that of the solid phase (H_2O is an exception with V°_{273} (ice) = 19·6, V°_{273} (water) = 18·0 cm^3 mole^{-1}) ΔV for melting is usually positive. Therefore $dP/dT = \Delta S/\Delta V$ is usually positive since the entropy of a liquid is in general greater than that of the corresponding solid. Liquidus curves thus tend to lie at higher temperatures at higher pressures and may in addition exhibit changes of form such as a change from incongruent to congruent melting in an intermediate compound. The solvus in contrast is usually associated with a negligible ΔV and exsolution is consequently not pressure-sensitive.

In experimental mineralogy (see chapter 16) it is often convenient to apply pressure through the medium of a chemically reactive fluid phase. The effect on a melting system of increasing the pressure of water vapour will be different from that of a wholly inert gas because in general water is more soluble in silicate liquids than in solid silicates (i.e. pG (anhydrous silicate) + $qG(H_2O) > rG$ (silicate with dissolved H_2O) + sG (H_2O vapour with dissolved silicate), where p, q, r, s are stoichiometric coefficients) and consequently all temperatures on the liquidus will in general be depressed at high p_{H_2O}; such depression may give rise at sufficiently high p_{H_2O} to the significant change in behaviour of a system that results from the intersection of solidus and solvus.

Three-component systems

We shall discuss three-component, or *ternary*, systems in rather less detail than we have provided for binary systems. We shall confine ourselves to a selection of types of ternary systems illustrative of significant points of interpretation, types of invariant point and so on. Figs 14.22–29 show several hypothetical types of ternary isobaric melting diagrams; in each case the liquidus is represented by means of isothermal contours and the nature of the solid phase in equilibrium with liquid in each field of the liquidus surface is indicated.

When three phases are in equilibrium in a ternary system their relative proportions are given by application of the *centre of gravity principle*, which is a logical extension of the lever rule. Suppose the composition P to be composed of the three phases a, b, and c, all of which are in the ternary system A–B–C as shown in Fig 14.21. If the weight of the system is m_P and the weight of each phase present is respectively m_a, m_b, and m_c then the amount of component A in phase a is $m_a x_a/100$, in phase b $m_b x_b/100$, in phase c $m_c x_c/100$, and in the whole system $m_P x_P/100$. Therefore $m_P x_P = m_a x_a + m_b x_b + m_c x_c$ and similarly for component B $m_P y_P = m_a y_a + m_b y_b + m_c y_c$. These two equations are identical with those necessary to express the statement that P is the centre of gravity of a system of masses m_a, m_b, and m_c geometrically disposed at the apices of the plane triangle abc. It follows that if the line aP is produced to meet the side bc at u, then the weight percentage of phase a is $100Pu/au$, of phase b $100Pv/bv$, and of phase c $100Pw/cw$.

(1) Ternary system without compounds or solid solution

The phase diagram for such a system is shown in Fig 14.22, where the hypothetical components are labelled A, B, and C. Each bounding binary system is a simple eutectic system, the binary eutectics being labelled E_1, E_2, and E_3. Since no temperature axis is available the fact that the liquidus falls from each pure component towards a binary eutectic is indicated by an arrow directed towards falling

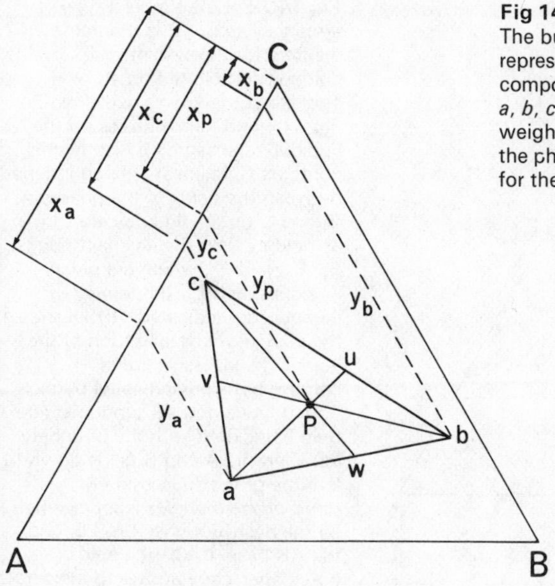

Fig 14.21 The centre of gravity principle. The bulk composition of the system is represented by the point P and the compositions of three phases by the points *a*, *b*, *c*. The triangular diagram is plotted in weight percentage. The weight percentage of the phase *a* is given by 100 *Pu*/*au* and likewise for the other phases.

temperature. If the mol fractions of the components A, B, and C in the liquid phase are denoted by x_A, x_B, and x_C respectively, the equation to the *field* of A on the liquidus, i.e. that part of the liquidus surface where solid A is in equilibrium with liquid, will be

$$\mu_A^{S*} = \mu_A^{L*} + RT \ln x_A$$

therefore $$\ln x_A = -\frac{\Delta H_A}{R}\left(\frac{1}{T} - \frac{1}{T_A}\right)$$

and similarly for the fields of B and C,

$$\ln x_B = -\frac{\Delta H_B}{R}\left(\frac{1}{T} - \frac{1}{T_B}\right)$$

$$\ln x_C = -\frac{\Delta H_C}{R}\left(\frac{1}{T} - \frac{1}{T_C}\right)$$

Differentiating at constant pressure,

$$\frac{dx_A}{dT} = \frac{\Delta H_A x_A}{RT^2}$$

therefore $$\frac{dx_A}{dT} > 0$$

if $\Delta H_A > 0$ as is generally so for heats of melting. Therefore the A liquidus falls away to lower temperatures as the composition of the liquid moves away from the A corner by increase of x_B and x_C with concomitant decrease of x_A. The B and C liquidus surfaces behave likewise. Temperature contours on the three fields of the liquidus

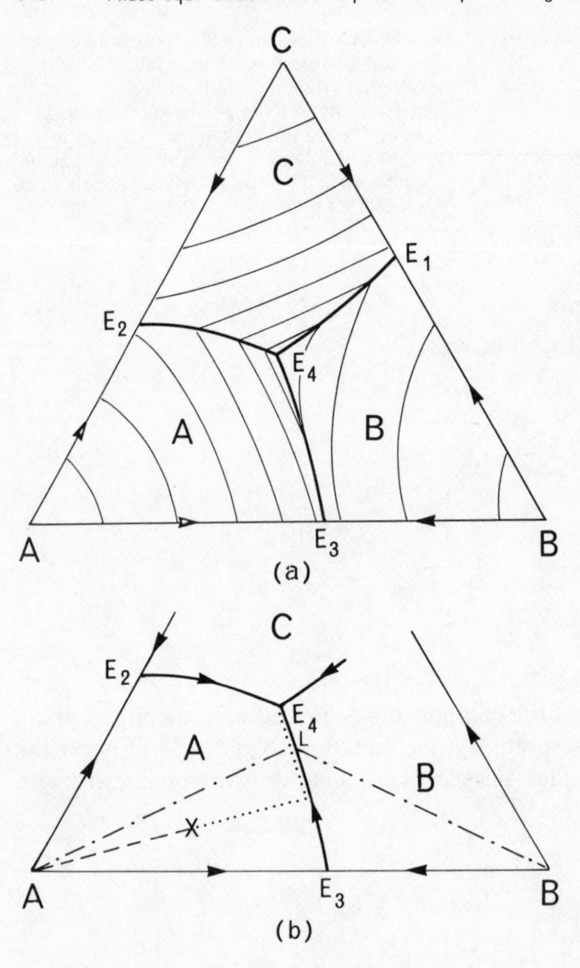

Fig 14.22 The simplest type of ternary system. (a) is the phase diagram for the system with composition plotted on an equilateral triangle as explained in Fig 14.3 and temperatures on the liquidus represented by isothermal contours (thin lines); the boundaries between the *fields* of the phases A, B, and C on the liquidus are shown as bold curves; the invariant points, E_1, E_2, and E_3, which are binary eutectics, and E_4, the ternary eutectic, are indicated. (b) illustrates the course of crystallization of the liquid X; successive liquid compositions are indicated by the dotted line along AX produced and then coincident with the boundary between the A and B fields down to E_4; one of the succession of *three-phase triangles* is represented by the dash-dot lines LA, LB, and the AB base. In (b) the field boundaries carry arrows to indicate the direction of falling temperature and such arrows are shown on all subsequent ternary phase diagrams.

surface are shown by broken lines on the figure for a hypothetical case. The field boundary between the A and B fields is given by the simultaneous solution of

$$\ln x_A = -\frac{\Delta H_A}{R}\left(\frac{1}{T}-\frac{1}{T_A}\right)$$

and $$\ln x_B = -\frac{\Delta H_B}{R}\left(\frac{1}{T}-\frac{1}{T_B}\right),$$

whence $$\frac{dx_A}{dT} = \frac{\Delta H_A x_A}{RT^2} \quad \text{and} \quad \frac{dx_B}{dT} = \frac{\Delta H_B x_B}{RT^2}$$

but $$x_A + x_B + x_C = 1$$

and therefore

$$\begin{aligned}\frac{dx_C}{dT} &= -\frac{dx_A}{dT} - \frac{dx_B}{dT}\\ &= -\frac{1}{RT^2}(\Delta H_A x_A + \Delta H_B x_B)\\ &< 0.\end{aligned}$$

Therefore the field boundary falls in temperature as it moves away from E_3 towards compositions with $x_C > 0$. And likewise the other two field boundaries fall in temperature from E_1 and E_2 respectively. Arrows on the figure point towards lower temperatures.

That the field boundaries are non-linear can simply be seen by solving the equations to the A and the B liquidus surfaces to give the equation to the A/B boundary by elimination of T as

$$\frac{\ln x_A}{\Delta H_A} - \frac{\ln x_B}{\Delta H_B} = \frac{1}{R}\left(\frac{1}{T_A} - \frac{1}{T_B}\right)$$

Therefore $$\frac{dx_B}{dx_A} = \frac{\Delta H_B}{\Delta H_A} \cdot \frac{x_B}{x_A}$$

but for a straight line on a triangular diagram dx_B/dx_A = constant.

The Phase Rule, $\mathscr{P} + \mathscr{F} = \mathscr{C} + 2$, requires that the three solid phases should be in equilibrium with the liquid phase ($\mathscr{P} = 4$) under isobaric conditions at a fixed temperature in a ternary system ($\mathscr{C} = 3$), i.e. $\mathscr{F} = 3+2-4 = 1$, but $\mathscr{P}$ is fixed so that $\mathscr{F}' = 0$. The mutual intersections of the field boundaries are therefore coincident, their point of common intersection being designated a *ternary eutectic point* (E_4 in Fig 14.22) and being characterized as the point of lowest temperature on the liquidus surface; in particular, and we shall see that this is distinctive, all three field boundaries approach the ternary eutectic by decrease in temperature. At the ternary eutectic, $x_A = (x_A)_E$, $x_B = (x_B)_E$, $x_C = (x_C)_E$, and $T = T_E$, where

$$\ln (x_A)_E = -\frac{\Delta H_A}{R}\left(\frac{1}{T_E} - \frac{1}{T_A}\right)$$

$$\ln (x_B)_E = -\frac{\Delta H_B}{R}\left(\frac{1}{T_E} - \frac{1}{T_B}\right)$$

$$\ln (x_C)_E = -\frac{\Delta H_C}{R}\left(\frac{1}{T_E} - \frac{1}{T_C}\right)$$

and $$(x_A)_E + (x_B)_E + (x_C)_E = 1.$$

A typical liquid, such as X in Fig 14.22(b), begins to crystallize, in this case solid A, when its temperature falls to the liquidus temperature appropriate to its composition. As A crystallizes the liquid composition moves along the extension of the line AX away from A, the proportions of solid A and residual liquid being given by the lever rule. When the liquid composition reaches the A/B field boundary, the solid phase B begins to crystallize. As the liquid composition moves down the thermal valley that is the field boundary, A and B crystallize simultaneously, the proportions of the three phases in equilibrium being given by application of the centre of gravity principle to the *three phase triangle* l, A, B when the composition of the liquid has reached l. Simultaneous crystallization of A and B continues until the temperature has fallen to that of the ternary eutectic T_E where all three solid phases crystallize together and the temperature cannot fall below T_E until all the liquid has completely disappeared. In such a system the lower limit of the crystallization interval (or melting interval if temperature is rising) for all compositions other than the pure components is the temperature of the ternary eutectic. Disequilibrium conditions produce no significant variation in the sequence of events during crystallization in this type of system.

(2) Ternary system with a congruently melting binary compound and no solid solution: van Alkemade's Theorem

The phase diagram for a hypothetical system is shown in Fig 14.23. The bounding binary systems A–C and B–C are simple eutectic systems with eutectics at E_2 and E_1. The system A–B is a binary eutectic system with a congruently melting intermediate compound (binary type (3)) and has two eutectics E_3 and E_4 with a maximum of the liquidus curve at the composition of the compound $\overline{AB}$. That a field boundary runs from each eutectic point with falling temperature into the ternary system follows from the argument developed in discussion of Fig 14.22. What remains to be determined is the nature of the temperature variation along the $\overline{AB}$/C field boundary, which is unrelated to any binary eutectic, and thence the nature of the invariant points at which $A+\overline{AB}+C+l$ and $B+\overline{AB}+C+l$ are in equilibrium under isobaric conditions.

If the composition of the compound $\overline{AB}$ is given by $x_A = x_{\overline{AB}}$, $x_B = 1-x_{\overline{AB}}$, $x_C = 0$, its congruent melting point by $T_{\overline{AB}}$, and its heat of melting by $\Delta H_{\overline{AB}}$, the equation to the $\overline{AB}$ liquidus surface

$$\begin{aligned}\mu^{S*}_{\overline{AB}} &= \mu^{L}_{\overline{AB}}\\ &= x_{\overline{AB}}\mu^{L}_{A}+(1-x_{\overline{AB}})\mu^{L}_{B}\\ &= \mu^{L*}_{\overline{AB}}+RT\{x_{\overline{AB}}\ln x_A+(1-x_{\overline{AB}})\ln x_B\}\end{aligned}$$

which integrates to

$$x_{\overline{AB}}\ln\frac{x_A}{x_{\overline{AB}}}+(1-x_{\overline{AB}})\ln\frac{x_B}{1-x_{\overline{AB}}} = -\frac{\Delta H_{\overline{AB}}}{R}\left(\frac{1}{T}-\frac{1}{T_{\overline{AB}}}\right).$$

Differentiation with respect to x_A yields

$$\frac{x_{\overline{AB}}}{x_A}+\frac{1-x_{\overline{AB}}}{x_B}\cdot\frac{dx_B}{dx_A} = \frac{\Delta H_{\overline{AB}}}{RT^2}\cdot\frac{dT}{dx_A}.$$

Therefore $$\frac{dT}{dx_A} = \frac{RT^2}{\Delta H_{\overline{AB}}}\left(\frac{x_{\overline{AB}}}{x_A}+\frac{1-x_{\overline{AB}}}{x_B}\cdot\frac{dx_B}{dx_A}\right).$$

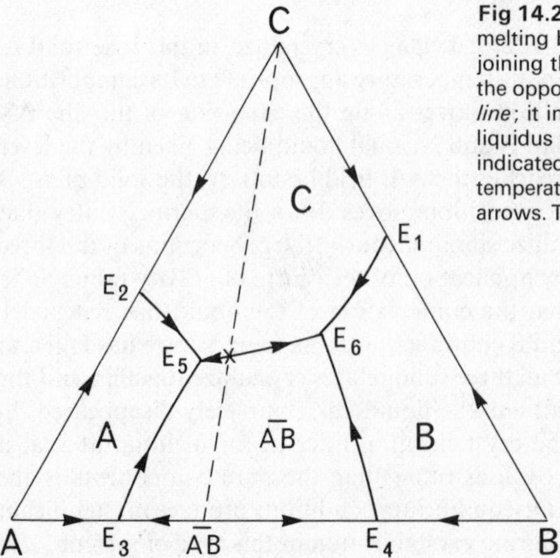

Fig 14.23 A ternary system with a congruently melting binary compound $\overline{AB}$. The broken line joining the composition of the compound $\overline{AB}$ to the opposite corner C is known as a *van Alkemade line*; it intersects the boundary between the liquidus fields of $\overline{AB}$ and C in a *saddle point* indicated by a cross. Directions of falling temperature on field boundaries are indicated by arrows. The system has ternary eutectics E_5 and E_6.

Imposing the restriction x_C = constant, that is considering plane sections of the phase diagram parallel to the T axis and the A–B join,

$$\left(\frac{\partial x_B}{\partial x_A}\right)_{x_C} = -1 \qquad \text{since} \quad x_A + x_B + x_C = 1$$

therefore $$\left(\frac{\partial T}{\partial x_A}\right)_{x_C} = \frac{RT^2}{\Delta H_{\overline{AB}}} \cdot \left(\frac{x_{\overline{AB}}}{x_A} - \frac{1-x_{\overline{AB}}}{x_B}\right)$$

and hence for

$$\left(\frac{\partial T}{\partial x_A}\right)_{x_C} = 0,$$

$$\frac{x_A}{x_B} = \frac{x_{\overline{AB}}}{1-x_{\overline{AB}}}.$$

That this condition represents a maximum is evident from the sign of the second differential when $(\partial T/\partial x_A)_{x_C} = 0$:

$$\left(\frac{\partial^2 T}{\partial x_A^2}\right)_{x_C} = -\frac{RT^2}{\Delta H_{\overline{AB}}} \left(\frac{x_{\overline{AB}}}{x_A^2} + \frac{1-x_{\overline{AB}}}{x_B^2}\right) < 0.$$

There is thus a ridge running across the $\overline{AB}$ liquidus surface along the line $x_A/x_B = x_{\overline{AB}}/(1-x_{\overline{AB}})$, that is along the join $\overline{AB}$–C. This is a general result which applies equally to the case, which will be considered subsequently, where the $\overline{AB}$ liquidus along the $\overline{AB}$–C join is metastable. It may be expressed in words as *van Alkemade's Theorem*: the boundary between the liquidus fields of any two phases passes through a temperature maximum where it intersects the line joining the compositions of the two phases on a triangular diagram even though the boundary at this point may be metastable.

In the system displayed on Fig 14.23 therefore the boundary between the $\overline{AB}$ and the C fields passes through a temperature maximum where it intersects the $\overline{AB}$–C join and the intersection is taken to lie on the stable part of the field boundary. Such maxima are known topographically as *saddle-points.* The invariant points E_5 and E_6 are therefore both eutectics. Each ternary eutectic lies, as it must, within the triangle defined by the compositions of the three solid phases that are in equilibrium at the eutectic point (i.e. E_5 in triangle A–$\overline{AB}$–C and E_6 in triangle $\overline{AB}$–B–C; consequently the system A–B–C is divisible into two ternary sub-systems A–$\overline{AB}$–C and $\overline{AB}$–B–C, the course of crystallization of any liquid being describable throughout in terms of compositions wholly within one sub-system. Each *compatibility triangle* such as A–$\overline{AB}$–C or $\overline{AB}$–B–C thus has a ternary invariant point associated with it.

Since no new principles are introduced we shall not describe the course of crystallization of any liquid in this system, but merely state that compositions in the triangle A–$\overline{AB}$–C contain a liquid phase at $T \geqslant T_{E_5}$, the composition of the last liquid to crystallize being given by E_5, while compositions in the triangle $\overline{AB}$–B–C go to E_6 at T_{E_6}.

(3) Systems with a ternary reaction point

If the system shown in Fig 14.23 is modified so that the stable part of the A/$\overline{AB}$ field

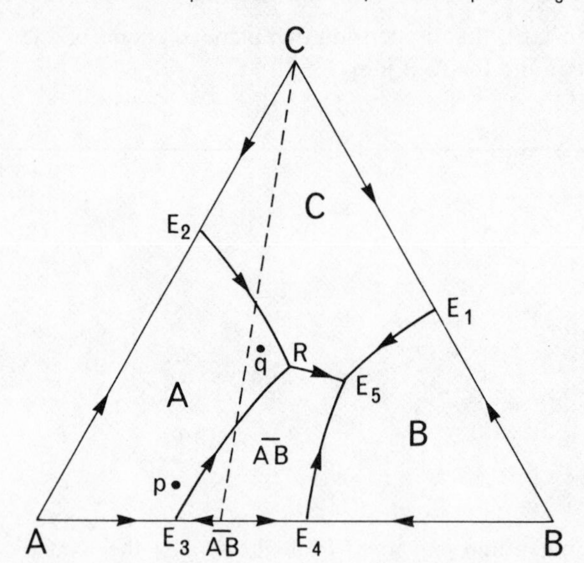

Fig 14.24 A ternary system with a congruently melting binary compound $\overline{AB}$ and a ternary reaction point. In this system the boundary between the liquidus fields of $\overline{AB}$ and A crosses the van Alkemade line which joins the compositions $\overline{AB}$ and C. The system has a ternary reaction point R and a ternary eutectic E_5. The crystallization of the liquids p and q is discussed in the text.

boundary intersects the $\overline{AB}$–C joins some significant consequences result. Such a system is shown in Fig 14.24. The bounding binary systems of course remain simple eutectic systems. The field boundary curves from E_1, E_2, E_3, and E_4 fall in temperature as they move into the ternary system in conformity with van Alkemade's rule. The maximum on the $\overline{AB}$/C field boundary lies to the left of the stable portion and therefore temperature falls smoothly from the point labelled R to E_5. The isobaric invariant point R at which the solid phases A, $\overline{AB}$, and C are in equilibrium with liquid lies outside the corresponding compatibility triangle A–$\overline{AB}$–C; R is known as a *reaction point* and is characterized by having two field boundaries falling towards it and one falling away from it on the liquidus surface. Compositions within the compatibility triangle A–$\overline{AB}$–C terminate their crystallization at the reaction point; those in the compatibility triangle $\overline{AB}$–B–C at the ternary eutectic E_5.

To exemplify the operation of the reaction point we consider the crystallization of the two liquids p and q shown on Fig 14.24. Liquid p lies above the field of A in the compatibility triangle A–$\overline{AB}$–C. The sequence of events in its crystallization is: (i) A↓ as liquid moves along A–p produced, (ii) A↓+$\overline{AB}$↓ as liquid moves along the A/$\overline{AB}$ field boundary, (iii) at R the four phases A(s)+$\overline{AB}$(s)+C(s)+R(l) are in equilibrium and temperature cannot fall below T_R until one phase has vanished; that the vanishing phase must be liquid follows from the argument that the composition p can be expressed in terms of A(s)+$\overline{AB}$(s)+C(s), but not in terms of $\overline{AB}$(s)+C(s)+R(l) because R and p lie in different compatibility triangles. The liquid q lies above the field of A in the other compatibility triangle $\overline{AB}$–B–C. The sequence of events in its crystallization is: (i) A↓ as liquid moves along A–q produced, (ii) A↓+C↓ as liquid moves along the A/C field boundary, (iii) at R the four phases A(s)+C(s)+$\overline{AB}$(s)+R(l) are in equilibrium and now it is the solid phase A that must disappear by resorption because both q and R lie in the compatibility triangle that excludes component A, (iv) when A has been completely resorbed temperature falls below T_R and $\overline{AB}$↓+C↓ as the liquid moves along the $\overline{AB}$/C field boundary, (v) at E_5 the system goes solid as $\overline{AB}$(s)+C(s)+B(s).

We have considered crystallization under equilibrium conditions. If equilibrium were not maintained the course of crystallization of liquid p would not be substantially

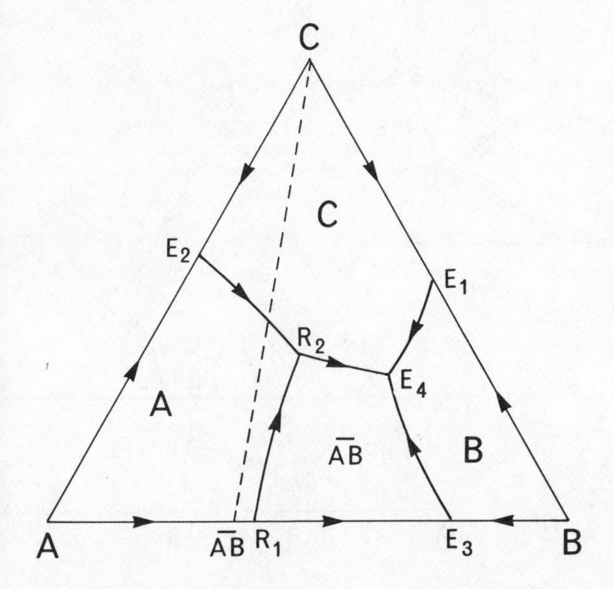

Fig 14.25 A ternary system with an incongruently melting binary compound $\overline{AB}$.

altered since resorption of a solid phase is not required at the reaction point; and in the case of liquid q failure of A to resorb at the reaction point would affect the nature and proportions of solid phases in the product but the final liquid would, as under equilibrium conditions, have the composition of the ternary eutectic E_5.

A ternary reaction point arises also if the binary compound $\overline{AB}$ has incongruent melting relations as illustrated in Fig 14.25. This is no more than an extension of the previous case, the $\overline{AB}$ field here lying wholly on the B side of the $\overline{AB}$–C join. Since no new principle is involved we shall not consider the course of crystallization of any particular liquid compositions, but merely state that here too compositions in the compatibility triangle A–$\overline{AB}$–C cease crystallization at the reaction point R_2 while those in the triangle $\overline{AB}$–B–C go to the ternary eutectic E_4.

(4) Systems with a ternary distribution point

If the field of the binary compound $\overline{AB}$ on the liquidus surface lies wholly within the ternary system and does not reach the A–B join, as illustrated in Fig 14.26, a third type of isobaric invariant point arises. The bounding binary systems are simple eutectic systems and the field boundaries fall in temperature, as usual, as they move into the ternary system. Van Alkemade's rule indicates that in both the examples illustrated the temperature of the A/$\overline{AB}$ field boundary and of the B/$\overline{AB}$ field boundary falls as the composition of the liquid moves away from the A–B join. The point labelled D therefore is characterized by a single field boundary falling towards it and two falling away; such an isobaric invariant point is known as a *distribution point* and completes the range of possible types of ternary isobaric invariant point:

$\equiv$ E, $\equiv$ R, $\equiv$ D.

At the distribution point four phases are in equilibrium: A(s)+B(s)+$\overline{AB}$(s)+D(l). Temperature can only fall from T_D if one phase vanishes. There are two possibilities: solid phase A or solid phase B may be resorbed. If phase A is resorbed, three phases remain in equilibrium B(s)+$\overline{AB}$(s)+D(l) and this is only possible if the overall composition lies in the triangle $\overline{AB}$–B–D. If phase B is resorbed, the three phases

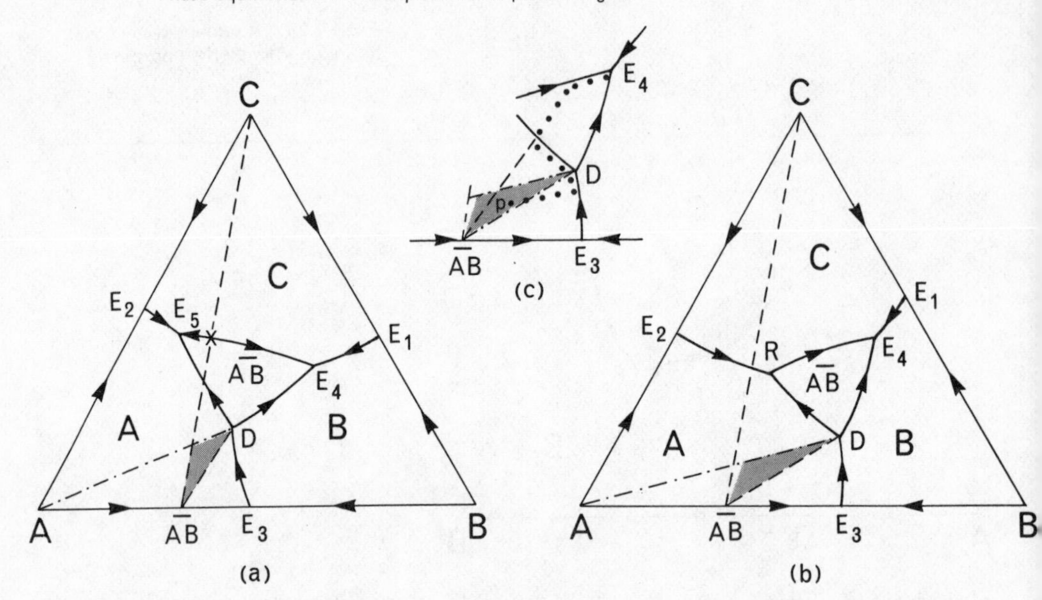

Fig 14.26 Ternary systems with a binary compound $\overline{AB}$ which has no liquidus field in the binary system A–B. Each system has three invariant points: that shown in (a) has a ternary *distribution point* D and two ternary eutectics E_4 and E_5 while that shown in (b) has a ternary distribution point D, a ternary reaction point R and one ternary eutectic E_4. In (a) and (b) the stippled area covers the range of compositions which, under equilibrium conditions, cannot follow the full length of the A–$\overline{AB}$ field boundary; (c) shows the course of crystallization of such a liquid p, successive liquid compositions being represented by dots.

remaining are A(s) + $\overline{AB}$(s) + D(l) and this is only possible for overall compositions in the triangle A–$\overline{AB}$–D. Thus for compositions in the compatibility triangle A–$\overline{AB}$–C the liquid moves from the distribution point along the A/$\overline{AB}$ field boundary and for compositions in $\overline{AB}$–B–C along the $\overline{AB}$/B field boundary. In both the systems illustrated the isobaric invariant point B(s) + $\overline{AB}$(s) + C(s) + E_4(l) is a ternary eutectic; but in (a) there is a saddle-point on the $\overline{AB}$/C field boundary which gives rise to a second ternary eutectic E_5 in place of the reaction point R shown in (b).

One feature of these two phase diagrams to which attention can conveniently be drawn here is that for compositions in the stippled area the liquid cannot proceed from D all the way along the A/$\overline{AB}$ field boundary to E_5 or R and thence to E_4 because such a composition will lie outside the three-phase triangle A–$\overline{AB}$-liquid on the A/$\overline{AB}$ boundary before E_5 or R is reached. When the $\overline{AB}$-liquid join passes through the point representing the composition of the system only two phases can be present, $\overline{AB}$(s) and liquid, that is to say A(s) has been gradually resorbed as the liquid has moved along the A/$\overline{AB}$ field boundary from D (i.e. the field boundary is in a reaction relationship to solid phase A). If only two phases are present, the Phase Rule no longer requires the liquid composition to be constrained to an isobaric univariant line. The liquid therefore moves across the $\overline{AB}$ field along the extension of the line joining $\overline{AB}$ to the overall composition of the system until the $\overline{AB}$/C boundary is reached, when simultaneous crystallization of $\overline{AB}$ and C begins as the liquid moves to the ternary eutectic E_4. The course of crystallization of such a liquid is illustrated in the enlarged portion of the phase diagram Fig 14.26(c).

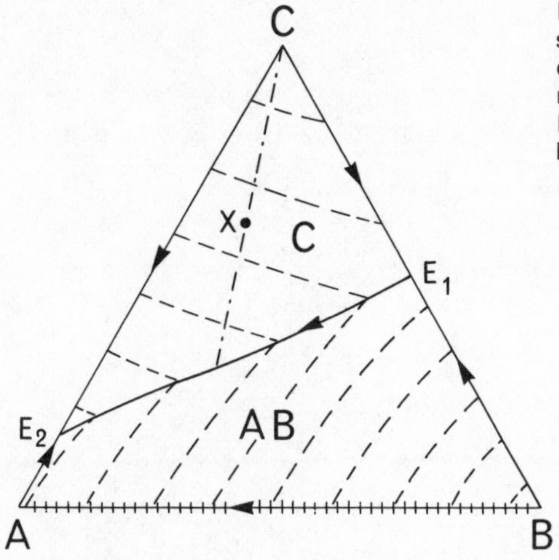

Fig 14.27 A ternary system with a binary solid solution. Complete miscibility between the components A and B in the solid state is represented by cross-hatching of the A–B join. Isotherms on the liquidus surface are represented by broken curves.

(5) Systems with a ternary compound

Since no new principle arises we shall not deal here with such systems.

(6) Ternary system with a binary solid solution

We shall deal first with the case in which one of the bounding binary systems is of the type illustrated in Fig 14.12, that is solid and liquid phases are both perfect solutions, and the other two bounding binary systems are simple eutectic systems. We shall suppose that the ternary liquid is a perfect solution and that solid solution does not extend into the ternary system. The phase diagram for such a system is of the type shown in Fig 14.27. The C liquidus surface slopes down from the corner representing the component C towards the two binary eutectics E_1 and E_2. The liquidus surface that represents equilibrium between liquid and solid solution AB slopes from the A–B join towards the binary eutectics E_1 and E_2. The AB/C field boundary and the liquidus for the binary system are shown arbitrarily with temperature falling from right to left on the diagram. A liquid such as X, whose composition lies in the C-field of the liquidus, follows a normal course, C crystallizing as the liquid composition moves with falling temperature along C–X projected to the C/AB field boundary; the liquid then moves a certain distance along the field boundary towards E_2. The diagram as it stands provides no information about the composition of the binary solid solution in equilibrium with any ternary liquid. For this reason also the diagram cannot account for the course of crystallization of a liquid whose composition lies in the AB field except to give the temperature of beginning of crystallization. We now turn to considering ways of displaying this additional information on the phase diagram.

We shall show in chapter 16 how *tie lines* joining the compositions of liquid and solid solutions in equilibrium can be determined experimentally. Tie lines such as ab can then be plotted (Fig 14.28(a)) on the triangular diagram: b represents the composition of the AB solid solution in equilibrium with the liquid of composition a and since a lies on the AB liquidus surface the temperature corresponding to the isothermal tie line is T_a. Now clearly it would not be feasible to plot tie lines for more than a small selection of points such as a on the AB liquidus; the resulting network of tie lines would soon become too complex to be decipherable. Direct plotting of tie

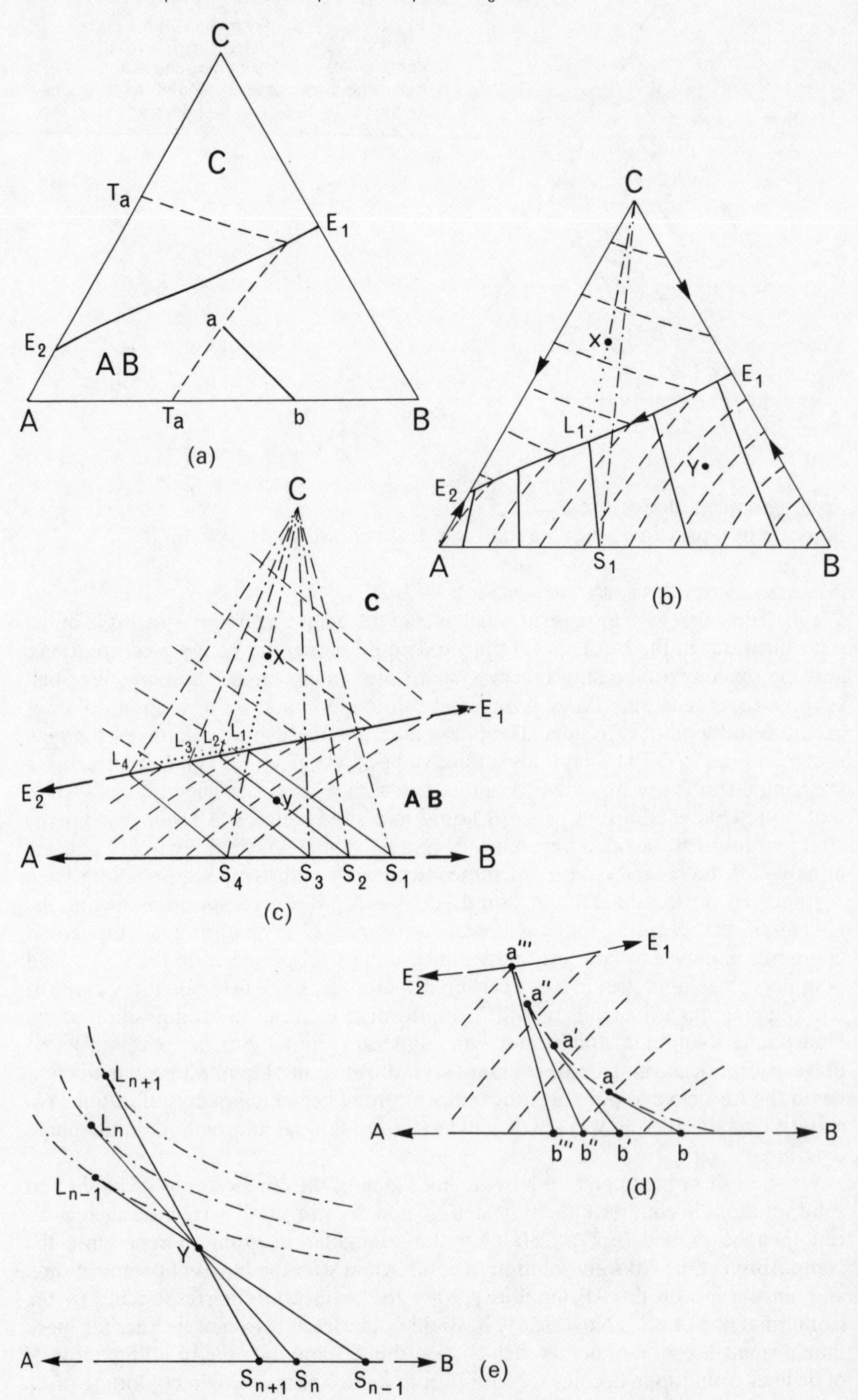
C
C
T_a
E_1
a
E_2
AB
A
T_a
b
B
(a)
C
X
E_1
L_1
Y
E_2
A
S_1
B
(b)
C
C
X
E_1
L_4
L_3
L_2
L_1
E_2
y
AB
A
B
S_4
S_3
S_2
S_1
(c)
a'''
E_1
E_2
a''
a'
a
A
B
b'''
b''
b'
b
(d)
L_{n+1}
L_n
L_{n-1}
Y
A
B
S_{n+1}
S_n
S_{n-1}
(e)

lines is useful if a limited range of liquid compositions only, such as those on the AB/C field boundary, is important. If such a set of tie lines is determined and plotted at selected intervals a phase diagram such as Fig 14.28(b) results and from this diagram the description of the course of crystallization of the liquid X can be completed. The liquid reaches the field boundary at the point labelled L_1 where it is in equilibrium with solid C and solid solution AB of composition S_1, the compositions of the three phases in equilibrium at this temperature being represented by the three-phase-triangle L_1S_1C. As the system cools successive equilibria are represented by a succession of such three-phase triangles, shown exaggerated on Fig 14.28(c), the temperature being given by interpolation between the liquidus isotherms shown. The three-phase triangles pivot about the point representing the phase of fixed composition C until X lies on the side joining C to the composition of the solid solution, here S_4. The composition of the final liquid is given by L_4 and the temperature of completion of crystallization by the isotherm on the liquidus surface through L_4, it being assumed throughout that equilibrium is maintained so that there is complete adjustment of the composition of the solid solution by reaction with the liquid as temperature falls. Just as in the binary case discussed previously equilibrium cannot easily be maintained in practice and zoning is commonly observed.

The same set of tie lines is adequate for a partial description of the course of crystallization of a liquid such as Y (Fig 14.28(b)) whose composition lies on the AB field of the liquidus. The temperature of beginning of crystallization can be read from the diagram by interpolation between the isotherms drawn on the AB–liquidus surface but the composition of the first solid remains indeterminate. When the liquid reaches the field boundary one degree of freedom is lost by the appearance of a third phase C and the equilibrium is represented by a three-phase triangle such as L_2S_2C in Fig 14.28(c). The composition of the solid solution as C begins to crystallize is thus represented by the tie line (interpolated if necessary) from the field boundary that passes through Y. Successive three-phase triangles describe successive equilibria until a temperature is reached where the join between C and the solid solution passes through Y; at this temperature crystallization ceases, the final liquid having a composition and temperature given by interpolation between L_3 and L_4.

If the earlier stages of crystallization of a liquid whose composition lies in the AB field are of interest it becomes necessary to determine experimentally tie lines for liquids on the AB liquidus away from the field boundary and to represent on the phase diagram the resulting large number of tie lines as *fractionation curves* defined

Fig 14.28 Crystallization of liquids in the system shown in Fig 14.27. In (a) an isothermal *tie-line* ab joins a liquid of composition a to an A–B solid solution of composition b in equilibrium at temperature T_a; the T_a isotherm on the AB and C liquidus surfaces is shown as a broken line. In (b) isotherms on the liquidus surfaces together with a few tie-lines between liquid compositions on the C/AB field boundary and the solid solution line A–B are shown; the three-phase triangle L_1S_1C for the beginning of crystallization of the AB solid solution from the liquid X is outlined. In (c), which is distorted for clarity, a succession of three-phase triangles, CL_nS_n, illustrating successive stages in the simultaneous crystallization of C and the solid solution AB are outlined; CL_n and CS_n joins are shown as dot-dash lines and the L_nS_n tie-lines as thin solid lines; the compositional path of the liquid phase from X is indicated by dotted lines. In (d) the dot-dash curve a–a′–a″–a‴ is a *fractionation curve*, its tangents ab, a′b′, a″b″, a‴b‴, shown as solid lines, are tie lines between liquid and solid solution, and the broken curves are isotherms on the AB liquidus surface. The crystallization of the liquid Y of diagram (b) is illustrated in (e) where fractionation curves are shown as dot-dash curves and the solid lines represent successive tie-lines pivoting through the bulk composition Y of the system to join successive liquid and solid solution compositions L_nS_n, etc.

by the statement: the tangent to a fractionation curve represents the tie line for a liquid composition represented by the point of contact of the tangent (Fig 14.28(d)). Interpolation between adjacent fractionation curves provides a complete account of the equilibrium compositions of all liquid and solid solutions. Thus a liquid such as Y in Fig 14.28(b) begins to crystallize at T_y an AB solid solution whose composition is given by the intersection of the A–B join with the tangent at Y to the fractionation curve passing through Y. Since we are concerned with two-phase (liquid + solid solution) equilibrium until the liquid reaches the AB/C field boundary, successive equilibria will be represented by the pencil of tie lines passing through Y, each tie line being terminated by its intersection S_n with the A–B join and by the point L_n at which it is tangential to the fractionation curve through L_n (Fig 14.28(e)).

Fractionation curves, as their name implies, enable the course of disequilibrium crystallization to be read immediately from the diagram. If the infinitesimal amount of solid solution first crystallized is wholly inactive, the effective composition of the system will move an infinitesimal distance along the tie line from Y, that is along the fractionation curve and so under conditions of complete disequilibrium the com-

Fig 14.29 The ternary system $MgO-FeO-SiO_2$. In the phase diagram (a) isotherms on the liquidus are shown as thin solid lines where they have been precisely determined and as thin broken lines where their position is conjectural; liquidus field boundaries are shown as solid lines, except for that between the olivine and magnesio-wüstite fields which is shown broken because its course has not been precisely determined; to avoid complication of the diagram the binary solid solutions $MgSiO_3-FeSiO_3$ (pyroxene), $Mg_2SiO_4-Fe_2SiO_4$ (olivine), and MgO–FeO (magnesio-wüstite) are not indicated by the conventional cross-hatched lines. Diagrams (b), (c), and (d) are isothermal sections of the system at 1550, 1450, and 1350 °C respectively; in each tie-lines between the three phases of variable composition (liquid, $RSiO_3$ solid solutions and R_2SiO_4 solid solutions) are shown as thin solid lines.

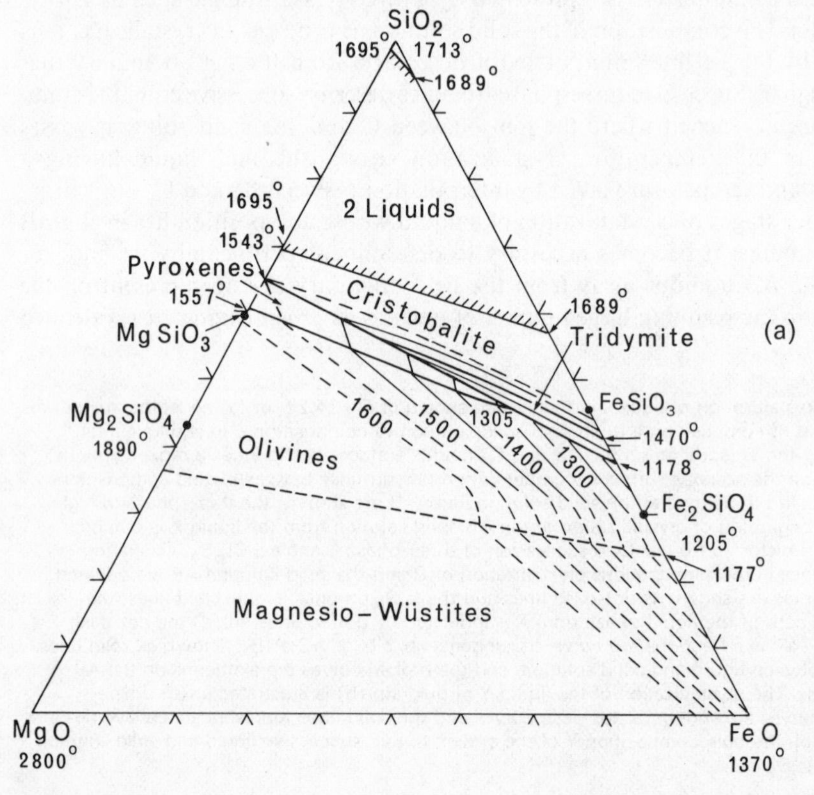

position of the liquid will follow the fractionation curve through Y to the AB/C field boundary and the crystals of AB will be zoned.

An alternative method of diagrammatic representation of equilibria in a ternary system with solid solution is by means of a series of isothermal sections showing the intersection of the liquidus surface and the course of isothermal tie lines. This method is particularly useful when not enough tie lines have been determined experimentally to justify the drawing of fractionation curves on the liquidus field of the solid solution. Figure 14.29(a) shows field boundaries and isothermal contours on the liquidus surface, but no tie lines, for the system MgO—FeO—SiO_2. Figure 14.29(b), (c), (d) are isothermal sections of the same system showing the range of liquid compositions and tie lines between the three phases of variable composition, liquid, solid solution R_2SiO_4, and solid solution $RSiO_3$ at three selected temperatures; isothermal sections at many more temperatures are necessary for a full description of the system (see Bowen and Schairer, 1935).

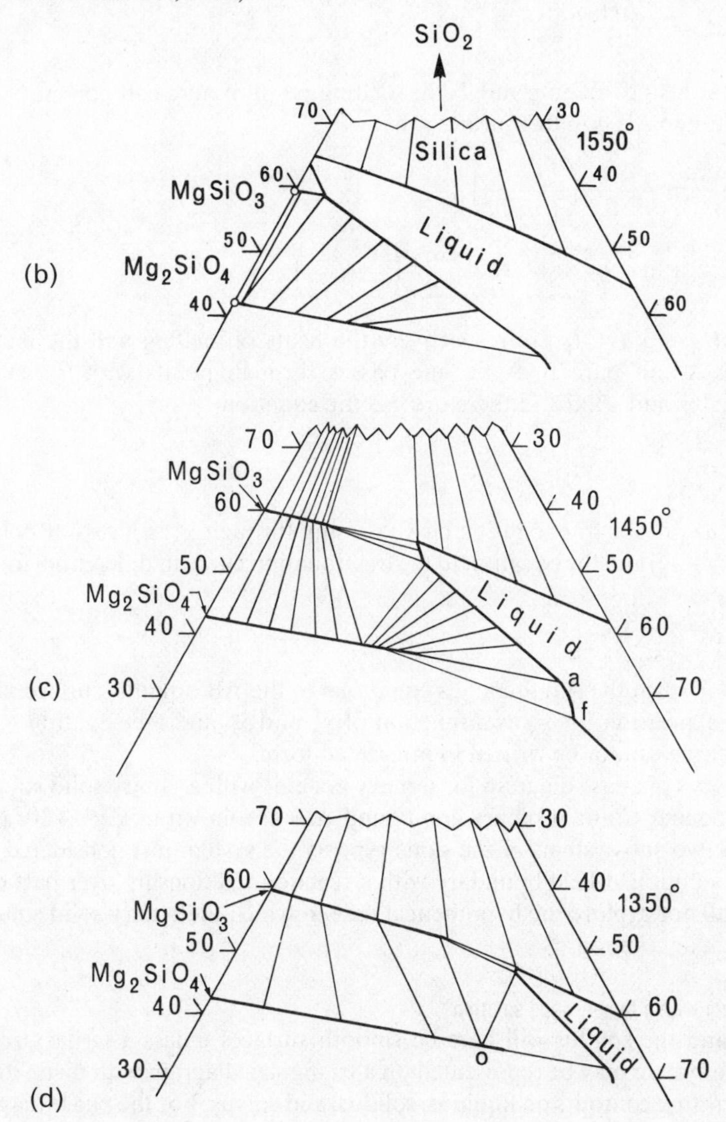

The assumption of perfect ternary liquid solution leads to

$$\mu_A^L = \mu_A^{L*} + RT \ln x_A^L$$
$$\mu_B^L = \mu_B^{L*} + RT \ln x_B^L$$
$$\mu_C^L = \mu_C^{L*} + RT \ln x_C^L$$
$$(x_A^L + x_B^L + x_C^L = 1)$$

and of perfect binary solid solution to

$$\mu_A^S = \mu_A^{S*} + RT \ln x_A^S$$
$$\mu_B^S = \mu_B^{S*} + RT \ln x_B^S$$
$$(x_A^S + x_B^S = 1).$$

The equation to the C liquidus surface is therefore

$$\ln x_C^L = -\frac{\Delta H_C}{R}\left(\frac{1}{T} - \frac{1}{T_C}\right)$$

where ΔH_C is the heat of melting and T_C the melting point of pure component C; and the equations to the AB liquidus surface are

$$\ln \frac{x_A^L}{x_A^S} = -\frac{\Delta H_A}{R}\left(\frac{1}{T} - \frac{1}{T_A}\right)$$

and $$\ln \frac{x_B^L}{x_B^S} = \ln \frac{1 - x_A^L - x_C^L}{1 - x_A^S} = -\frac{\Delta H_B}{R}\left(\frac{1}{T} - \frac{1}{T_B}\right)$$

where ΔH_A, ΔH_B and T_A, T_B are respectively the heats of melting and the melting points of pure A and pure B. A tie line passes through points with x_A and x_C coordinates x_A^L, x_C^L and x_A^S, 0 and therefore has the equation

$$x_C = \frac{x_C^L}{x_A^L - x_A^S} \cdot x_A - \frac{x_A^S x_C^L}{x_A^L - x_A^S}.$$

The slope dx_C/dx_A of the fractionation curve passing through x_A^L, x_C^L at that point is therefore $x_C^L/(x_A^L - x_A^S)$ and the equation to the fractionation curve in differential form is

$$\frac{dx_C^L}{dx_A^L} = \frac{x_C^L}{x_A^L - x_A^S}.$$

Elimination of T from the simultaneous equations to the AB liquidus surface yields an intractable expression for x_A^S as a function of x_A^L and x_C^L and the equation to the fractionation curve cannot be written in integrated form.

Two other types of phase diagram for ternary systems with a binary solid solution (between component C and a binary compound $\overline{AB}$) are shown in Fig 14.30: (a) is separable into two sub-systems of the same type as the system just considered and (b) has a solid-solution/A field boundary with a reaction relationship over part of its course. We shall not explore the hypothetical case in which the binary solid solution is a strictly regular solution.

(7) Ternary system with a ternary solid solution

The liquidus and the solidus will here be smooth surfaces unless a ternary solvus intervenes. The system may be represented on a triangular diagram with distinctively marked temperature contours on liquidus, solidus, and solvus. For the phase diagram

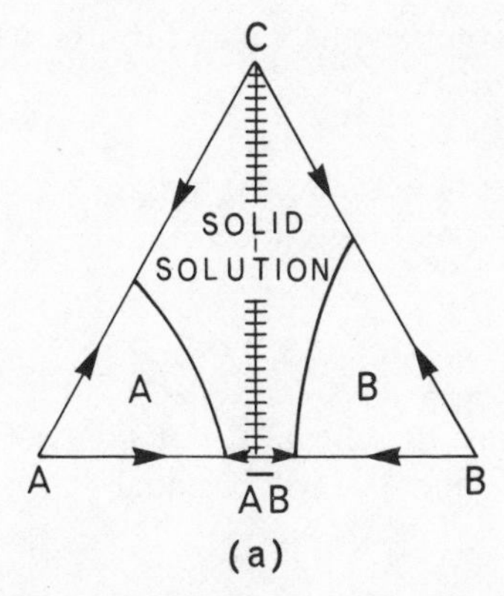

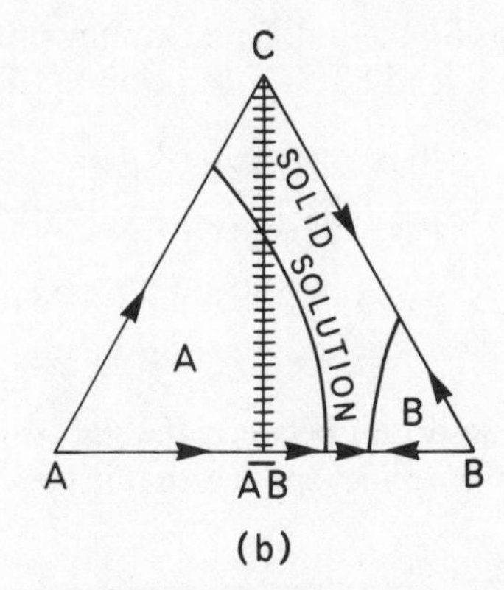

Fig 14.30 Two types of ternary system with a binary solid solution between one component and a binary compound $\overline{AB}$. In (a) the binary compound is congruently melting and so is the whole solid solution range $\overline{AB}$–C. In (b) the binary compound is incongruently melting and the solid solution changes from incongruent to congruent melting between its extreme compositions $\overline{AB}$ and C.

to be fully informative fractionation curves on the liquidus and on the solvus must be shown. Such diagrams can only be determined with immense experimental labour and none has yet been published for a mineralogical system.

(8) Exsolution from a ternary solid solution

As in the binary case exsolution implies that the solid solution is not perfect. Exsolution may however be observable in only one or two of the bounding binary systems because the critical temperature for exsolution is below the temperature at which the rate of exsolution is adequate to achieve any result even in conditions of such extremely slow cooling as are presented geologically. No more will be said here because no new principle is involved.

(9) Ternary system with two binary solid solutions

In a system such as that shown in Fig 14.29 two binary solid solutions have fields on the liquidus and if, as is so here, the fields are adjacent then three phases of variable composition are in equilibrium along the field boundary. The special point of interest about such a system is the disposition in TX^2 space of the divariant surface over which the two solid solutions are in equilibrium with one another; this surface extends to lower temperatures in the subsolidus space from the univariant field boundary in which it intersects the liquidus.

Consider a solid solution S1 whose components 1 and 2 have a common anion and similar structures so that they can be represented as $R_a^{b+} X_b^{a-}$ and $S_a^{b+} X_b^{a-}$. If the chemical potential of component 1 in the solid solution S1 is denoted by μ_{1S1}, then

$$\mu_{1S1} = a\,\mu_{1S1+} + b\,\mu_{1S1-}$$

and $$\mu_{1S1+} = \mu^*_{1S1+} + RT \ln x_{1S1+}$$

And likewise

$$\mu_{2S1} = a\,\mu_{2S1+} + b\,\mu_{2S1-}$$

$$\mu_{2S1+} = \mu^*_{2S1+} + RT \ln x_{2S1+}$$

If the other solid solution S2 has structurally similar components with the same cations as S1 but a different common anion, its components 1 and 2 can be written as $R_c^{b+}Y_b^{c-}$ and $S_c^{b+}Y_b^{c-}$ respectively. Then

$$\mu_{1S2} = c\mu_{1S2+} + b\mu_{1S2-}$$

$$\mu_{2S2} = c\mu_{2S2+} + b\mu_{2S2-}$$

and
$$\mu_{1S2+} = \mu^*_{1S2+} + RT\ln x_{1S2+}$$

$$\mu_{2S2+} = \mu^*_{2S2+} + RT\ln x_{2S2+}$$

At equilibrium between the two solid solutions, the chemical potentials of their common chemical species, that is the cations R^{b+} and S^{b+}, will be equal

i.e.
$$\mu_{1S1+} = \mu_{1S2+}$$

and
$$\mu_{2S1+} = \mu_{2S2+}$$

Hence
$$\ln\frac{x_{1S1+}}{x_{1S2+}} = \frac{\mu^*_{1S2+} - \mu^*_{1S1+}}{RT}$$

and
$$\ln\frac{x_{2S1+}}{x_{2S2+}} = \frac{\mu^*_{2S2+} - \mu^*_{2S1+}}{RT}$$

Subtraction of these two equations yields an integrable expression:

$$\ln\left(\frac{x_{1S1+}}{x_{2S1+}}\cdot\frac{x_{2S2+}}{x_{1S2+}}\right) = \frac{1}{RT}(\mu^*_{2S1+} - \mu^*_{1S1+} + \mu^*_{2S2+} - \mu^*_{1S2+}).$$

Now
$$\mu^*_{2S1+} - \mu^*_{1S1+} = \frac{1}{a}(\mu^*_{2S1} - \mu^*_{1S1}) - \frac{b}{a}(\mu^*_{2S1-} - \mu^*_{1S1-})$$

but the anions 1S1– and 2S1– are both X^{a-} and therefore $\mu^*_{1S1-} = \mu^*_{2S1-} = \mu^*_{X^{a-}}$,

whence
$$\mu^*_{2S1+} - \mu^*_{1S1+} = \frac{1}{a}(\mu^*_{2S1} - \mu^*_{1S1})$$

and similarly

$$\mu^*_{2S2+} - \mu^*_{1S2+} = \frac{1}{c}(\mu^*_{2S2} - \mu^*_{1S2}).$$

Therefore
$$\ln\left(\frac{x_{1S1+}}{x_{2S1+}}\cdot\frac{x_{2S2+}}{x_{1S2+}}\right) = \frac{1}{RT}\left\{\frac{1}{a}(\mu^*_{2S1} - \mu^*_{1S1}) - \frac{1}{c}(\mu^*_{2S2} - \mu^*_{1S2})\right\}.$$

Now, for the reaction between components

$$\underset{(1S1)}{\frac{1}{a}R_aX_b} + \underset{(2S2)}{\frac{1}{c}S_cY_b} = \underset{(2S1)}{\frac{1}{a}S_aX_b} + \underset{(1S2)}{\frac{1}{c}R_cY_b}$$

the free enthalpy of reaction at T K and P bars is given by

$$\Delta G_T^P = \frac{1}{a}(\mu^*_{2S1} - \mu^*_{1S1}) - \frac{1}{c}(\mu^*_{2S2} - \mu^*_{1S2}).$$

Hence
$$\ln\left(\frac{x_{1S1+}}{x_{2S1+}}\cdot\frac{x_{2S2+}}{x_{1S2+}}\right) = \frac{\Delta G_T^P}{RT}.$$

But $$x_{1S1+} = \frac{a}{a+b}x_{1S1}, \qquad x_{2S1+} = \frac{a}{a+b}x_{2S1},$$

$$x_{1S2+} = \frac{c}{c+b}x_{1S2}, \qquad \text{and} \quad x_{2S2+} = \frac{c}{c+b}x_{2S2};$$

and putting $x_1 = x_{1S1}$ and $x_2 = x_{1S2}$ so that $1-x_1 = x_{1S2}$ and $1-x_2 = x_{2S2}$,

$$\ln\left(\frac{x_1}{1-x_1}\cdot\frac{1-x_2}{x_2}\right) = \frac{\Delta G_T^P}{RT}.$$

Now ΔG_T^P is not in general independent of temperature and pressure, but we are concerned here with a rather special kind of reaction. R_aX_b and S_aX_b have similar structures and may be expected to differ little in entropy, S^*_{1S1} and S^*_{2S1}, at any temperature; and likewise for R_cY_b and S_cY_b with entropies S^*_{1S2} and S^*_{2S2}. Therefore $\Delta S = (1/a)(S^*_{2S1} - S^*_{1S1}) - (1/c)(S^*_{2S2} - S^*_{1S2})$ is small and may be taken to be effectively independent of pressure and temperature, i.e. $\Delta S_T^P \simeq \Delta S^\circ_{298}$. Moreover $\Delta V = (1/a)(V^*_{2S1} - V^*_{1S1}) - (1/c)(V^*_{2S2} - V^*_{1S2})$ represents the difference in the change of molar volume when the same cationic substitution, S for R, occurs in different structures so that ΔV is small and may be taken to be effectively independent of pressure and temperature, i.e. $\Delta V_T^P \simeq \Delta V^\circ_{298}$. Now

$$d\Delta G = \left(\frac{\partial \Delta G}{\partial T}\right)_P dT + \left(\frac{\partial \Delta G}{\partial P}\right)_T dP$$
$$= -\Delta S\, dT + \Delta V\, dP,$$

which in this case becomes

$$d\Delta G = -\Delta S^\circ_{298}\, dT + \Delta V^\circ_{298}\, dP.$$

Therefore $\Delta G_T^P = \Delta G^\circ_{298} - \Delta S^\circ_{298}(T-298) + \Delta V^\circ_{298} P$

if $P \gg 1$ bar.

Hence $$\ln\left(\frac{x_1}{1-x_1}\cdot\frac{1-x_2}{x_2}\right) = \frac{\Delta G^\circ_{298} + 298\Delta S^\circ_{298} + \Delta V^\circ_{298} P}{RT} - \frac{\Delta S^\circ_{298}}{R}$$
$$= \frac{\Delta H^\circ_{298} + \Delta V^\circ_{298} P}{RT} - \frac{\Delta S^\circ_{298}}{R}.$$

Thus at constant temperature and pressure the factor $x_1(1-x_2)/(1-x_1)x_2$ is a constant K_T^P. If ΔH°_{298}, ΔV°_{298}, and ΔS°_{298} are known (see chapter 13) the variation of K_T^P with pressure and temperature can be evaluated.

Quaternary systems

We shall restrict our discussion of quarternary systems to an examination of two complementary methods of representing a particular system, $CaSiO_3$–Ca_2SiO_4–$Ca_2Al_2SiO_7$–FeO. For convenience of labelling the diagrams the abbreviations in common use in cement chemistry are used: C ≡ CaO, F ≡ FeO, A ≡ Al_2O_3, S ≡ SiO_2. In abbreviated form the components are then CS–C_2S–C_2AS–F. There is one binary compound in the system, C_3S_2, and two binary solid solutions, melilite (abbreviated *mel*) with components $Ca_2Al_2SiO_7$ and $Ca_2FeSi_2O_7$, i.e. C_2AS and C_2FS_2 (on the CS–F join), and olivine[3] (abbreviated *ol*) with components Ca_2SiO_4 and $CaFeSiO_4$,

[3] Used here in the structural sense (see chapter 10).

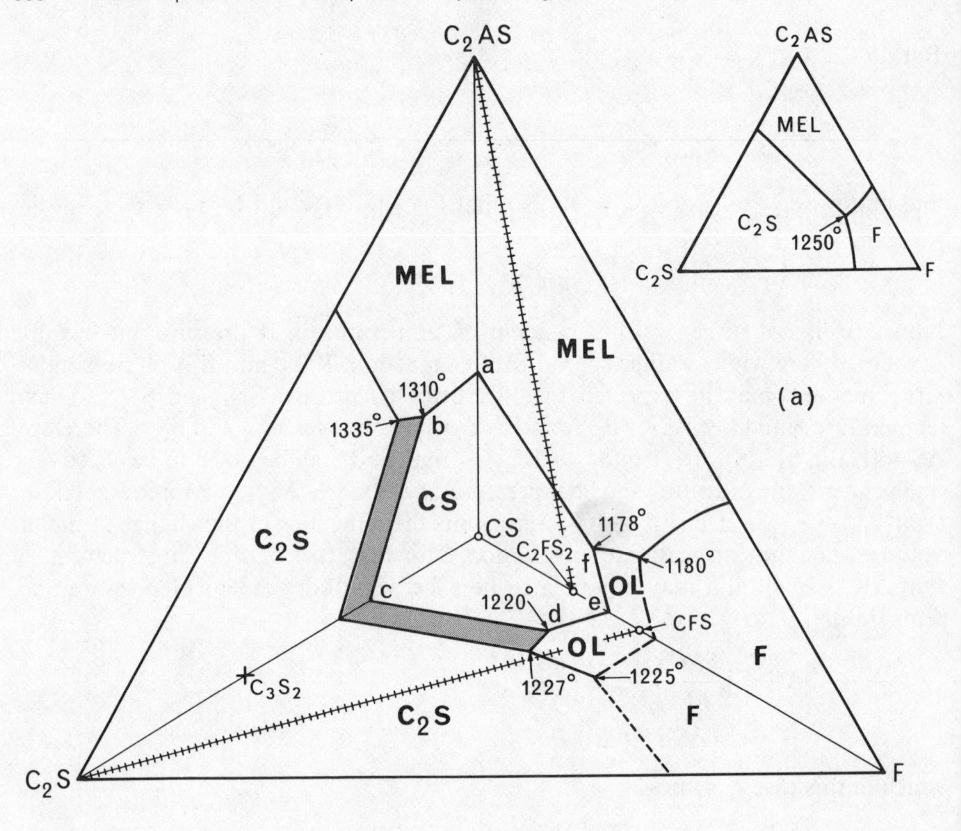

Fig 14.31 The quarternary system $CaSiO_3$–Ca_2SiO_4–$Ca_2Al_2SiO_7$–FeO. In (a) the four components are situated at the apices of a regular tetrahedron; the phase relations are displayed by removing the Ca_2SiO_4–$Ca_2Al_2SiO_7$–FeO(C_2S–C_2AS–F) face of the tetrahedron and displaying phase relations in that ternary system in the separate diagram shown, reduced, on the right. The shaded area represents the liquidus field of C_3S_2. In (b) the same system is displayed as a *Schairer diagram,* which is a formal grid of univariant lines intersecting in invariant points. Throughout this figure 'cement nomenclature' is employed: C = CaO, F = FeO, A = Al_2O_3, S = SiO_2.

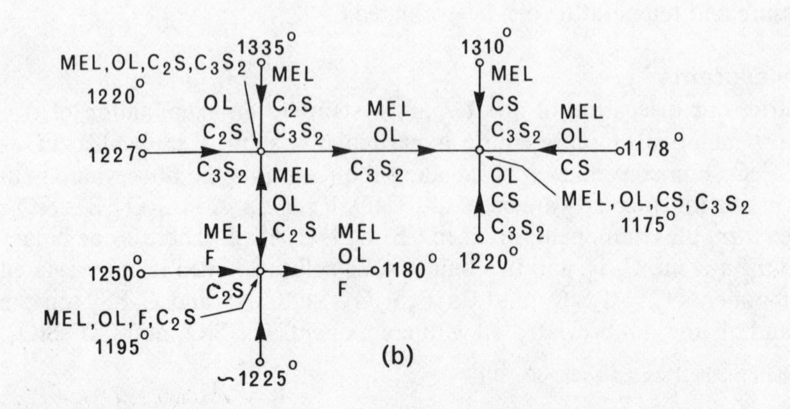

i.e. C_2S and CFS (on the CS–F join). The component Ca_2SiO_4 is polymorphic: we are concerned here with only two of its several polymorphs, α-Ca_2SiO_4 (indicated as CS) and γ-Ca_2SiO_4 which is the end-member of the olivine solid solution series.

Figure 14.31(a) shows the four components at the apices of a regular tetrahedron with its front face removed and displayed separately. For a quaternary system under isobaric conditions $\mathscr{F}' = 5 - \mathscr{P}$ and therefore two-phase equilibrium is represented by a volume in TX^3 space; thus all liquid compositions in the volume between the CS apex and the surface *abcdef* begin their course of crystallization by crystallizing CS and every point in this volume is associated with a definite liquidus temperature. Such a surface as *abcdef* represents three-phase equilibrium between liquid and two solid phases and is shown on the diagram only by its intersection with the faces of the tetrahedron. Two such surfaces intersect in univariant lines of four-phase equilibrium which are indicated only by their points of intersection with the faces of the tetrahedron, points such as *b*, *d*, *f*. Such univariant lines intersect within the tetrahedron in invariant points (not shown) at which five phases are in equilibrium. The compositional ranges of the two solid solution series are indicated + + + + + and the composition of the compound C_3S_2 by +. The three-dimensional liquidus fields are not labelled but it is obvious that C_3S_2, the iron-rich end of the melilite series, and all but the iron-rich end of the olivine series melt incongruently.

Phase relations within the tetrahedron are made clearer by the Schairer diagram shown in Fig 14.31(b) where univariant curves of four-phase equilibrium between liquid and three indicated solid phases are shown conventionally as straight lines on a rectangular grid. The extremities of each line are points on the faces of the tetrahedron, whose temperatures are stated. The univariant lines intersect at invariant points of stated temperature but unrecorded composition within the tetrahedron. Directions of falling temperature along the univariant lines are indicated by arrows from which it can be seen immediately that $1 + \text{mel} + \text{ol} + C_2S + C_3S_2$ and $1 + \text{mel} + \text{ol} + C_2S + F$ are in equilibrium at quaternary reaction points while $1 + \text{mel} + \text{ol} + C_3S_2 + CS$ are in equilibrium at a quaternary eutectic at 1175 °C, the lowest temperature on the liquidus.

Two such diagrams, the one emphasizing composition and the other emphasizing temperature, read in conjunction can yield an overall view of melting relations in a quaternary system, but it is evident that much less detail of temperature-composition relationships can be displayed diagrammatically than for systems of fewer components. Some attempt at displaying additional experimental data may be made by plotting liquidus isotherms on ternary sections such as C_2S–CFS–C_2AS, C_3S_2–C_2FS_2–C_2AS, etc. In general such sections will not be ternary systems since liquids within such a compositional plane will be in equilibrium with solid phases whose compositions may lie outside the plane. If a univariant curve of four-phase equilibrium represents equilibrium between a liquid (of necessity in the plane of the section) and three solid phases all of which have compositions in the plane of the section, then the intersection of the univariant curve with the plane of the section is a ternary invariant point; but if the composition of one or more of the phases lies outside the plane, the intersection is known as a *piercing point*.

15
Compositional analysis

The mineralogist, and the inorganic chemist, frequently needs to be able to determine the composition of a substance in which he is interested. Much of what we have said in earlier chapters implies the ability to determine the chemical composition of a substance, whether a stoichiometric compound of fixed composition or a solid solution. In chapters 7 and 12 we have indicated ways in which certain physical methods may be used to determine, more or less approximately, the compositions of specimens of rather simple solid solutions; all such physical methods depend ultimately on more comprehensive and precise methods of compositional analysis. In this chapter we provide a brief account of the principles and field of application of the principal methods of compositional analysis. For experimental details the reader is referred to analytical manuals and recent review articles; references are given for each method discussed.

Separation techniques

Most methods of mineral analysis require that the mineral whose composition is to be determined must first be separated from the other minerals that coexist with it in the host rock. Very occasionally a rock is truly monomineralic; separation is then a negligible problem as is often so for the products of inorganic synthesis. More often rocks and ore bodies contain, on the hand-specimen scale, three or more principal phases and perhaps as many accessory minerals.

The first stage of separation is to crush the rock to reduce it to single mineral grains. The rock specimen is first broken into walnut-sized fragments either by hammering or better by use of a hydraulic rock-splitter. The fragments are then ground until the whole sample passes a 90-mesh sieve (i.e. $<10^{-3}$ cm^2 minimum cross-section) by a manually operated percussion mortar, a roll mill, or a swing mill. Since very small particles interfere with separation procedures it is customary to subject the resulting powder to ultrasonic scrubbing in an aqueous medium, decanting the suspended fines and repeating the process until virtually nothing further goes into suspension.

The two properties on which the most generally useful separation techniques depend are magnetic susceptibility and specific gravity; these are complementary and are usually applied in this order. The prepared rock powder is passed through an isodynamic magnetic separator (such as the type designed and marketed by S. G. Frantz). This instrument consists in essence of a d.c. electromagnet with long pole

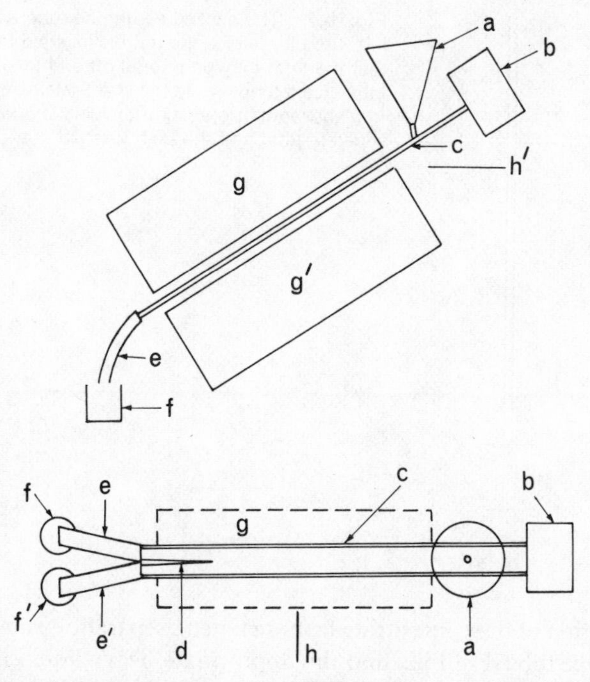

Fig 15.1 The isodynamic magnetic separator; the upper diagram is a side elevation and the lower diagram a plan. Key: *a* is the funnel through which the crushed sample is introduced; *b* is a vibrator rigidly attached to the chute *c*; the solid fence *d* directs the streams of more and less strongly magnetic grains into the conduits *ee'* which lead to separate collecting vessels *ff'*; *gg'* are the upper and lower poles of the variable electromagnet between which the chute *c* runs; *h* and *h'* are axes about which the chute may be inclined and tilted.

pieces between which the mineral grains are passed on a vibrating chute (Fig 15.1). The electromagnet and chute assembly are mounted so that the angle of tilt about axes parallel and perpendicular to the length of the chute can be varied. The grains of greater susceptibility (unless they are diamagnetic) will be pulled gradually towards the higher side of the chute in the course of their passage through the magnetic field while the grains of lower susceptibility will travel along the lower side of the chute; the two streams of grains are directed into separate collecting vessels. The current through the electromagnet is varied to achieve a significant separation in a particular case. Before putting the prepared rock powder through the Frantz separator it is necessary to remove strongly magnetic grains with a powerful permanent magnet; if this is not done they will stick to the upper pole piece and inhibit the flow of grains along the chute. One important feature of the Frantz separator is the speed with which a large amount of material can be reduced to a relatively small fraction in which the mineral to be analysed is concentrated. Further concentration may be achieved by recirculating the selected fraction through the separator and more complete extraction by recirculating the rejected fraction.

Magnetic separation rarely produces better than a 90 per cent concentrate and recourse must be had to other methods to obtain a pure preparation from the magnetic concentrate. The most generally useful of these is density separation in a heavy liquid such as methylene iodide (CH_2I_2) diluted with carbon tetrachloride (giving densities up to 3·32 g cm^{-3}) or Clerici Solution (aqueous thallous formate malonate) diluted with water (giving densities up to 4·28 g cm^{-3} at room temperature and 4·65 g cm^{-3}

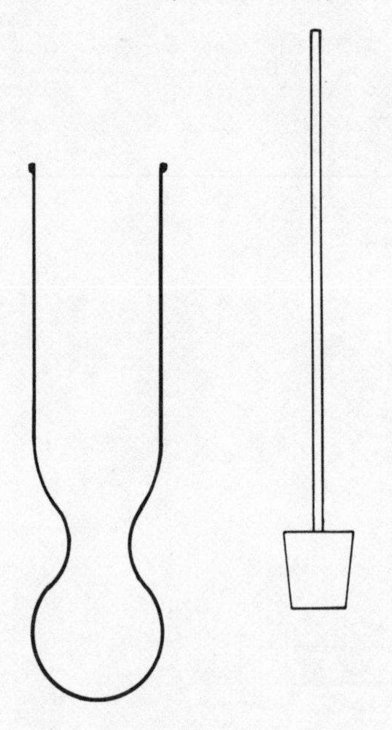

Fig 15.2 The necked centrifuge tube is shown on the left; the plunger on the right is inserted into the neck so that the float may be poured off and the upper part of the tube cleaned out while the sink remains trapped in the bulb. In a subsequent operation the sink is poured into a separate filter and the bulb washed out.

at 50 °C). A portion of the concentrate from magnetic separation is introduced into a necked centrifuge tube (Fig 15.2) and the appropriate heavy liquid is added, having been previously diluted to a density just greater than that of the required mineral. The tube is then centrifuged at about 4000 rev/min for a few minutes. If a clear separation into 'float' and 'sink' has been achieved, the stalked stopper is inserted gently and pressed home to seal off the bulb containing the high density fraction; the float, containing the required mineral, is poured off and filtered; the sides of the centrifuge tube are washed with diluent (*hot* water in the case of Clerici), the stalked stopper is removed, and the contents of the bulb are separately filtered. The float is reintroduced into the centrifuge tube and a liquid of density slightly less than that of the required mineral is added. After centrifuging the required mineral should be in the 'sink'. After filtering and washing the preparation of the required mineral is examined under the polarizing microscope and, if impurities are detected, it is successively recentrifuged in liquids closer to its own density. If the required mineral occurs in very small grains it may be necessary at this stage to break down composite grains by further grinding; centrifugal separation of the resulting fine powder may be facilitated by introduction of an ultrasonic probe into the tube before centrifuging in order to separate clusters of mineral grains.

A variety of other techniques have been applied successfully to particular separation problems and some are in general use in some laboratories. These include the superpanner, the shaking table, the electrostatic separator, and the flotation cell (especially for the separation of sulphides).

Thorough general accounts of mineral separation techniques with extensive bibliographies are provided by Muller (1967) and Wager and Brown (1960).

Analysis by classical chemical methods, colorimetry, flame photometry and related techniques

In classical chemical analysis each element is separated by means of its characteristic chemical reactions from all others present in a solution of the mineral sample and then determined either gravimetrically or volumetrically. The methods of classical analysis of silicates are described in detail by Groves (1951) and Kolthoff and Sandell (1950); we restrict ourselves here to a description in the very broadest terms of the procedure for analysis on a modified 'classical' scheme of a mineral containing only the major elements of the crust, a silicate such as an amphibole. It is emphasized that the scheme outlined here would be substantially modified and elaborated by a skilled analyst seeking to produce determinations of high accuracy, but it serves to indicate the nature of the technique.

A weighed amount of the mineral is fused with Na_2CO_3, the fusion product is evaporated twice with HCl, and the insoluble residue is ignited, weighted, evaporated with HF (to volatilize SiF_4), and weighed again; SiO_2 is given by the loss in weight. The combined filtrates are precipitated twice with NH_4OH and the insoluble residue, the so-called R_2O_3 precipitate, is ignited, weighed, and fused with $K_2S_2O_7$; the fusion product is dissolved in dilute H_2SO_4 and made up to a definite volume, say 250 ml, from which aliquots are treated separately, (i) for Mn a 50 ml portion is acidified and heated with KIO_4, made up to a fixed volume, and the intensity of the colour due to the permanganate ion is determined colorimetrically, (ii) for Ti another 50 ml portion is treated with hydrogen peroxide in acid solution, made up to a known volume, and the intensity of the yellow colour due to pertitanic acid determined colorimetrically, (iii) the remaining 150 ml is used to determine total Fe by passing through a *silver reductor* to reduce Fe^{3+} to Fe^{2+} and titrating the resulting ferrous solution against a standard solution of the oxidizing agent ceric sulphate. The combined filtrates from ammonia precipitation are precipitated twice with ammonium oxalate and the insoluble residue of CaC_2O_4 is heated at 500 °C and weighed as $CaCO_3$. The combined filtrates from oxalate precipitation are precipitated twice with $(NH_4)_2HPO_4$ and the insoluble residue is ignited and weighed as $Mg_2P_2O_7$. Phosphorus is determined on a separate portion of the specimen which is taken into solution in HF and $HClO_4$, evaporated, dissolved in aqueous $HClO_4$ treated with ammonium molybdate and ammonium vanadate, and made up to a known volume; the intensity of the resulting yellow molybdivanadophosphoric acid complex is determined colorimetrically. Aluminium is determined by subtraction of the weights of total iron as Fe_2O_3, total Mn as Mn_3O_4, TiO_2, and P_2O_5 (appropriately scaled to the weight of sample used for main fusion) from the weight of the R_2O_3 precipitate.

The hydrogen content of the specimen is determined on a separate portion by measurement of the weight of water evolved on very strong heating either alone or in the presence of an oxidizing flux.

The remaining two major elements of rock-forming minerals, the alkalis Na and K, are determined by *flame photometry*. In essence flame photometry is an elaboration of the familiar flame test; alternatively it may be regarded as a simplified sort of atomic emission spectrometry (q.v.). Electrons, excited by the thermal energy of the flame, in falling back to the ground state emit light of wavelength characteristic of the element concerned. A photocell measures the intensity of the monochromatic light emitted. Relatively few elements are excited at the temperature of the flame in contrast to the almost complete coverage of the periodic table, except for elements of very low atomic number, provided by the higher temperatures of the carbon arc (or spark discharge) used in atomic emission spectroscopy proper. For the routine determination of Na and K a relatively cool flame, such as that provided by a coal-gas/air mixture, is desirable because few other elements will be excited. The sample in solution is sucked into the flame at a constant rate by the inflowing air and coal-gas at constant pressure. The light emitted by the flame is intensified by mirrors and passed through a filter

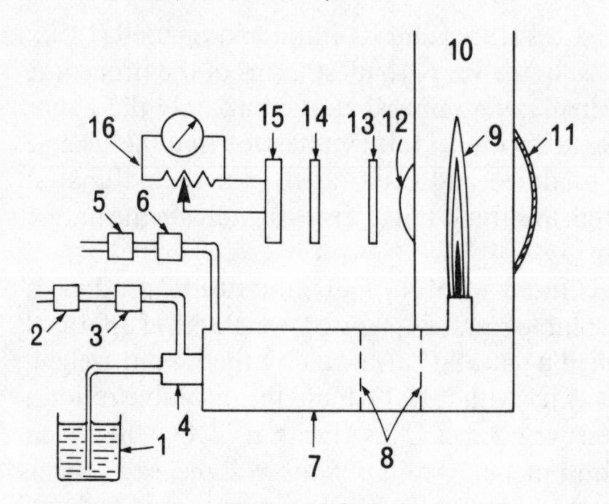

Fig 15.3 Diagrammatic representation of a simple type of flame photometer. Key: 1—beaker containing the solution to be analysed; 2—air flow control valve; 3—air pressure gauge; 4—atomizer; 5—gas flow control valve; 6—gas pressure stabilizer; 7—mixing chamber; 8—baffle plates; 9—broad flat flame; 10—chimney; 11—concave reflector; 12—lens; 13—glass heat absorber; 14—optical filter; 15—photocell; 16—calibrated potentiometer and galvanometer.

which transmits a band containing the wavelength to be measured in the simpler type of instrument and through a prism or diffraction grating in more elaborate instruments. The intensity of the light transmitted by the filter or deviated through a particular angle by the prism or grating is measured by a barrier-layer photocell. Figure 15.3 shows the essential components of a simple type of flame photometer. The instrument is calibrated with respect to solutions of accurately known composition. For determination of Na and K the sample is evaporated with H_2SO_4–HF, taken up in warm water, treated with ammoniacal ammonium carbonate to precipitate Fe, Al, Ti, Mg, Ca, etc, made up to a known (large) volume, and filtered before being injected into the flame.

Colorimetric analysis, which is used for Mn, Ti, and P in the modified 'classical' scheme outlined above, can be applied directly to the determination of many other major and trace elements; indeed a complete scheme for the determination of the common crustal elements largely by colorimetric methods has been provided by Shapiro and Brannock (1956). The most comprehensive accounts of colorimetric analysis are those of Sandell (1950) and Boltz (1958). We restrict ourselves here to a brief account of the essentials of the method.

When monochromatic light passes through a thickness l of a coloured solution, in which the concentration of the coloured substance is c, the intensity I_0 of the incident beam is reduced to I where

$$\log_{10}\frac{I}{I_0} = -kcl,$$

k being a constant known as the *extinction coefficient* (the Lambert–Beer Law). Light from a tungsten-filament projection lamp is passed through a monochromator, which may be a coloured filter or better a prism or diffraction grating, and then through the specimen solution contained in a parallel-sided glass cell. The light transmitted falls on a photocell, the amplified response from which is compensated by a potentiometer graduated directly in units of $\log_{10} I/I_0$. It is always necessary to run a reagent blank in an identical glass cell. Calibration against standard solutions has to be repeated fairly frequently because of fluctuations in the colour temperature of the lamp and ageing of the photocell. In the *Spekker* type of spectrophotometer the output of one photocell receiving light that has passed through the coloured solution is balanced in a bridge circuit against the output of another photocell receiving light from the same source diminished in intensity by an iris diaphragm; a calibrated cam-shaped shutter placed immediately before the specimen cell and adjusted to give compensation gives a direct measurement of log I/I_0. This type of instrument requires less frequent calibration against standard solutions.

Colorimetric, like gravimetric and volumetric analysis usually requires prior separation to remove interfering elements, but for many elements the presence of a considerable variety of other elements in the solution can be tolerated and in a few cases specific colour-forming reactions are available. Satisfactory procedures have been established for most elements in colorimetric as in gravimetric/volumetric analysis. A critical account of these analytical methods is provided by Vincent (1960). We defer discussion of detection limits and accuracy to the end of this chapter.

Optical emission spectrometry

The basis of the technique is that chemical compounds are vaporized and decomposed into their constituent atoms at the high temperature of a carbon arc. In the arc a small proportion (usually < 1 per cent) of the atoms of every element present are in excited electronic states. Transitions from excited states to the ground state are accompanied by emission of radiation, which is usually in the visible or ultraviolet wavelength range. The wavelength of the radiation emitted is characteristic of the element involved and its intensity is dependent on the concentration of the element in the arc. The spectrometer is simply a device for dispersing the total emission from the arc and recording the intensity of each spectral line.

A sample of 10–50 mg in weight, usually mixed with carbon and an internal standard, the whole being finely ground, is introduced into a small cup drilled into one of the carbon poles between which the arc is to be struck; this pole is usually the anode. The cathode, another carbon rod, is situated a few mm above the anode and a d.c. arc (3–20 amp, 100–250 volt) is struck between them. Light passes from the arc through a slit and falls on a dispersing device, which may be either a prism or a diffraction grating (Fig 15.4). In most of the more sophisticated instruments the light

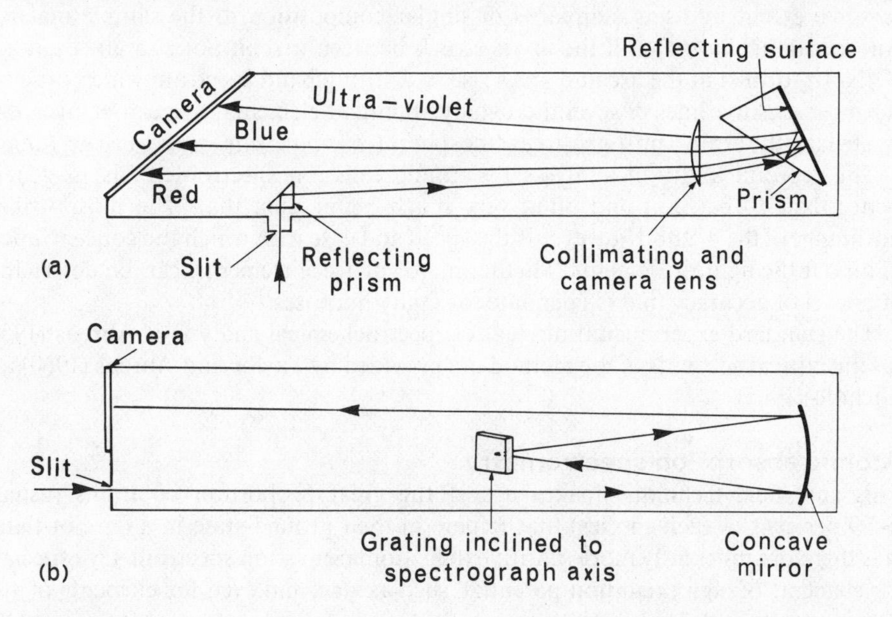

Fig 15.4 The atomic emission spectrograph: (a) shows a prism instrument and (b) a grating instrument.

beam is then reflected back to pass again through the dispersing device before it falls on the recorder, which is most commonly a photographic plate, but may be a slit and phototube assembly.

For a quantitative analysis a *step-sector* is rotated in front of the slit of the spectrograph so that each spectral line shows a sequence of steps of graded density, each step corresponding to twice the exposure time and therefore twice the intensity of the preceding inner step. The density of the photographic record of spectral lines is measured with a *photodensitometer*, an instrument containing a light source and a photocell in circuit with a galvanometer. The differential response of the galvanometer to light passing through a spectral line on the plate relative to that for a clear area of the plate (for which the galvanometer deflection is kept constant) is a measure of the intensity of the spectral line, but the relationship is non-linear and it is necessary to calibrate the instrument by making use of the known relative intensity of adjacent steps on spectral lines determined by the geometry of the step-sector. In this way calibration curves for successive steps are set up to relate galvanometer deflection to spectral intensity.

Standardization is achieved by measuring the intensities of a selected line in the spectrum of the chosen element (for instance the Cr line at 4254 Å) in a series of samples of known composition and by measuring the intensities of a particular line in the spectrum of the internal standard (for instance the Pd line at 4473 Å) and then plotting, in this case (log intensity of Cr 4254)/(log intensity of Pd 4473) against log Cr content, to establish a *working curve*. The concentration of Cr in a set of unknowns can thence be determined from the working curve once $\log I_{Cr4254}/\log I_{Pd4473}$ has been determined for each sample.

In the use of atomic emission spectrometry as an analytical tool one of the difficulties is interference between elements with neighbouring spectral lines and another is the formation of compounds, with their own characteristic emission spectra, in the arc if the arc-temperature is ill-chosen. Both these difficulties can be overcome to some extent by using standards of similar composition to the sample material. Another difficulty is that if the arc is struck between carbon poles in air, cyanogen (C_2N_2) is formed in the arc and gives rise to a strong band spectrum which obscures the most sensitive lines of several crustally abundant elements; this can be overcome by striking the arc in a nitrogen-free atmosphere (*controlled atmosphere arc excitation*).

The reproducibility of analyses by atomic emission spectrometry is in general better than 10 per cent and often very much better than that. The most striking advantage of the method however is the speed and ease with which the concentrations of all but the lightest elements, whether major or trace elements, can be determined at a level of accuracy that is adequate for many purposes.

The standard experimental manual of spectrochemical analysis is Ahrens (1950). Useful critical accounts of the method are provided by Taylor and Ahrens (1960) and Nicholls (1967).

Atomic absorption spectrometry

This analytical technique makes use of the great proportion of atoms (usually > 99 per cent of each species) that remain in their ground-state in an arc or flame. It is therefore inherently more sensitive than atomic emission spectrometry especially for elements of high excitation potential, such as zinc, and even for elements of very low excitation potential, such as caesium. Atoms in their ground state are able to absorb light energy only at particular wavelengths (*resonance lines*) corresponding to

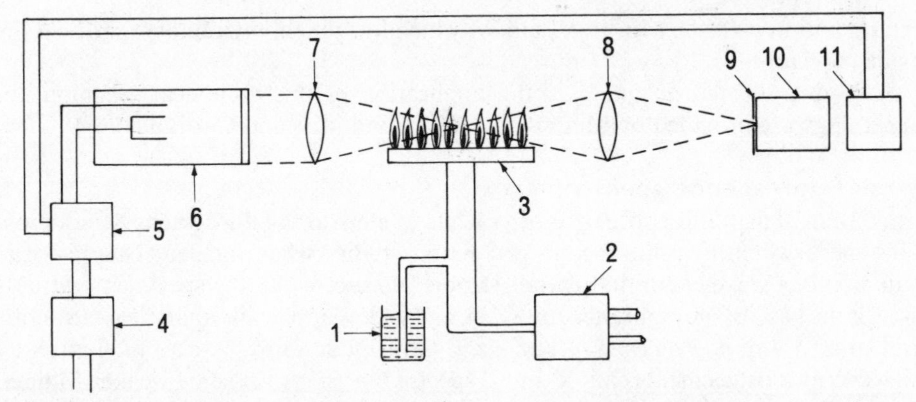

Fig 15.5 Diagrammatic representation of an atomic absorption spectrometer. Key: 1—beaker containing the solution to be analysed; 2—gas and air pressure controls and gauges; 3—gas burner; 4—A.C. voltage controller; 5—electronic chopper; 6—hollow cathode lamp with end-window of fused quartz; 7, 8—fused quartz lenses; 9—slit; 10—monochromator; 11—radiation detector tuned to chopper.

the energies of transitions from the ground state to excited states. Since the spectral width of such resonance lines is very small ($\sim 10^{-3}$ Å) it is necessary to use a monochromatic light source of appropriate wavelength for the analytical determination of each element. A practically convenient source is a high intensity *hollow-cathode lamp*, which has a tungsten rod anode and a hollow cylindrical cathode of the element under investigation enclosed in a glass envelope with a fused silica end-window and a neon or argon atmosphere at low pressure.

Monochromatic radiation from the hollow-cathode lamp (Fig 15.5) is passed through an oxy-acetylene or oxy-hydrogen flame (temperature $\sim 3000\,°C$) into which a solution of the sample is aspirated (as in flame-photometry). The incident light beam is focused on the centre of the flame and the emergent beam is focused on a slit, whence it passes through a monochromator adjusted to filter off background radiation and then impinges on a photomultiplier device which provides a measurement of intensity. To eliminate the contribution of the flame continuum to the emergent beam over the range of the selected resonance line the a.c. light source is modulated by mechanical or electronic chopping and a differential reading of the photomultiplier output is taken as a measure of intensity of absorption.

For quantitative analysis a known weight of the mineral, or other, sample is dissolved to form a known volume of solution. A convenient and widely applicable method for minerals is fusion with LiF and boric acid, followed by digestion in sulphuric acid, but for the more volatile elements sample preparation has to be more elaborate. Calibration may be made either against pure solutions of the selected element or by a *spiking* technique (that is, observations of the effect of additions of known amounts of the element to be determined on the photomultiplier output). The concentration in the flame of the element to be determined may be enhanced by liquid–liquid extraction techniques.

The method works best at low concentrations because the absorption–concentration curve flattens off at high concentration. It is restricted to elements with resonance lines at wavelengths > 2000 Å, because at shorter wavelengths absorption in the air-path is excessive. Some 40 elements however have resonance lines in the range 2000 to 10,000 Å. The accuracy of the method is affected by the formation of compounds in the flame at $\sim 3000\,°C$; this is a practical obstacle to the determination of Si, Al, Mo, Ti, V, and W, which form highly refractory oxides in the oxidizing flame.

Fully documented accounts of the application of this analytical technique in mineralogy are provided by McLaughlin (1967) and May and Cuttitta (1967).

X-ray fluorescence spectrometry

Irradiation of matter by primary X-rays leads to emission of fluorescent X-radiation. This effect, which is a nuisance in diffraction studies when incident characteristic radiation of a wavelength that excites strong fluorescence in the specimen crystal is used, is the basis of the highly accurate, fast, and widely applicable analytical technique known as *X-ray fluorescence spectrometry.* On irradiation by X-rays of sufficiently high energy most elements emit K and L spectra together with some weaker M lines. The method is applicable to all elements with atomic number greater than 10 (Ne), K spectra being utilized for the lower elements and L spectra for elements with $Z \geqslant 63$ (Eu); for the lightest elements to which the technique is applicable, Na (11) to Ca (20), it is necessary to evacuate the whole path of the fluorescent radiation.

The source of primary X-rays is a sealed-off tube with its target close to its filament and operated at a higher voltage than for diffraction work. The target is constructed of an element of high atomic number and high thermal conductivity. As the atomic number of the target is increased the proportion of energy in the continuous X-ray spectrum is increased relative to that in the superimposed characteristic spectrum and it is the continuum that is used to induce fluorescence in the sample (these are of course the opposite of the requirements for an X-ray source to be used for diffraction studies). Targets of Cr (24), Mo (42), or Au (79) are in most common use. The target element cannot of course be determined in the fluorescent emission from the sample.

The primary X-ray beam impinges on the sample to be analysed. The preparation of the sample is a matter of some importance if reproducible and accurate analyses are to be made. The sample, which is a cylindrical disc about 30 mm in diameter, may be either a pressed powder (with its back and edge coated with boric acid) or a solution in lithium borate glass prepared under standard conditions by fusion at 980 °C.

The fluorescent X-radiation emitted by the sample (Fig 15.6) passes through a collimator (*the primary collimator*) to fall on a crystal monochromator or *analysing crystal.* This is either a cleavage flake or a portion of a tabular crystal, for which the *d*-spacing of lattice planes parallel to the large face is known. Convenient analysing

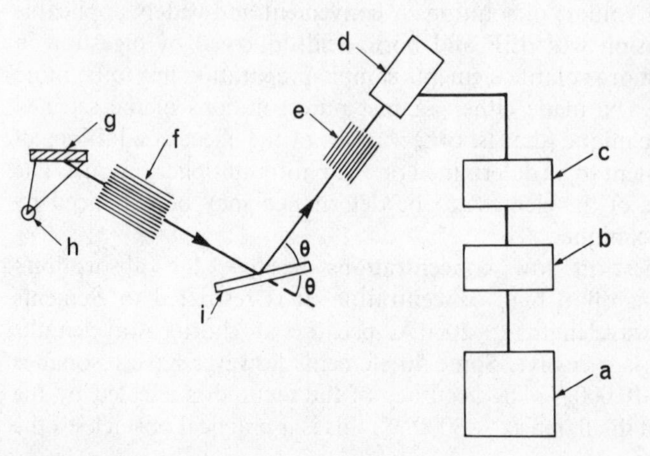

Fig 15.6 Diagrammatic representation of an X-ray fluorescence spectrometer. Key: a—counter and recorder; b—pulse height selector; c—amplifier; d—detector; e—secondary collimator; f—primary collimator; g—sample; h—high intensity X-ray source; i—analysing crystal.

crystals are LiF(200), LiF(220), Ge(111), pentaerythritol(002), $(NH_4)H_2PO_4$(110), and potassium acid phthalate(001). The analysing crystal will only diffract when the wavelength and angle of incidence of the fluorescent radiation from the sample are such as to satisfy the Bragg Equation, $\lambda = 2d \sin \theta$. The X-radiation diffracted by the analysing crystal passes through a collimator (*the secondary collimator*), mounted so that its axis is inclined at the same angle, θ, as the axis of the primary collimator to the plane of the analysing crystal but on the other side of the normal to that plane. A *detector* is aligned with the secondary collimator (Fig 15.6). The analysing crystal and the secondary collimator-detector assembly are mounted on a goniometer so that they can be rotated in such a manner that they always make angles θ and 2θ respectively with the axis of the primary collimator. A practically important refinement is the use of an analysing crystal bent so that its diffracting planes are cylindrical surfaces (this implies the use of an adequately flexible substance) of appropriate radius to permit the slightly divergent beam emitted from the primary collimator to be focused on the detector, so conserving intensity.

The function of the *detector*, which may be either a scintillation counter or a gas–flow–proportional counter, is to convert the energy of individual X-ray quanta into electrical energy. The electrical output of the detector is amplified before being passed through a *pulse height selector*, which rejects harmonics of the Bragg Equation as well as spurious pulses due to cosmic radiation, and then into a counter which counts and prints out the number of pulses in a fixed time.

By using a motor to rotate the goniometer, in which the secondary collimator and the detector are geared to travel at twice the angular rate of the analysing crystal, and coupling the output of the counter to a pen recorder, intensity of fluorescent emission from the sample is plotted against Bragg angle, which can be converted to wavelength since the appropriate *d*-spacing for the analysing crystal is known. Such a spectrogram provides a semi-quantitative analysis of the sample material by comparison with the known wavelengths and relative intensities of the K, L, and (for the heavier elements) M emission lines of all the elements.

Before the counter output for identified spectral lines can be converted to elementary concentrations corrections have to be made for absorption and for the presence of interfering elements; and the counting error has to be evaluated statistically. Such minutiae, practically important though they are, lie outside our scope and the reader is referred to Norrish and Chappell (1967) and to May and Cuttitta (1967) for fully-referenced critical accounts of this analytical method.

Electron probe microanalysis

This is the most versatile of all the methods currently available for the determination of the elementary composition of solids. It is especially appropriate in the fields of mineralogy and metallurgy in that prior separation of pure phases is not required, the sample being submitted in the form of a polished section of any thickness greater than about 60 μm. Under optimum conditions a volume as small as 1 μm^3 (1 μm^2 area and 1 μm thick) can be analysed; exsolution phenomena and zoning (coring) in minerals (and metals) become susceptible to quantitative compositional study by this technique, which is applicable to all elements of atomic number $\geqslant 9$ (fluorine).

In the simplest terms this analytical technique can be described by saying that a small area of the specimen is made the target of a specialized kind of X-ray tube, the emission from which is analysed by a single crystal spectrometer (as in X-ray fluorescence analysis) and fed into a counting device, the output from which, after the appropriate corrections have been made, can be interpreted as an elementary concentration by reference to an analysis similarly performed on a standard of known composition.

An electron beam from a source about 50 μm across is passed through a sequence of magnetic lenses (a condenser and an objective lens) which focus it to an area which may be as small as 1 μm^2 on the surface of the specimen. The material to be analysed may be a substantial fragment with a polished surface, as used for reflected light studies, or a thin section (about twice the standard thickness used for transmitted light studies, i.e. about 60 μm, is convenient for polishing); but in either case the essential requirement is that the surface of the specimen should be flat, free of scratches and highly polished. The polished surface of the specimen is coated by vacuum evaporation with a conducting film (usually carbon $\sim$ 200 Å thick) which maintains the specimen at near earth potential during bombardment by electrons. The specimen and all relevant standards are clamped in a mechanical stage equipped with micrometer drives (calibrated to 1 μm) along mutually perpendicular axes in the plane of the polished surface. The stage assembly has associated with it a light–optical system (Fig 15.7) which enables the specimen to be viewed in transmitted light between crossed polars or in reflected light so that by operation of the micrometers a particular area of a phase in a polyphase specimen can be selected for analysis.

The X-radiation emitted by the irradiated area of the specimen passes to a crystal spectrometer, similar to that described in the section on X-ray fluorescence analysis (but here always of the curved crystal type), and thence to a proportional counter with a bank of ancillary electronics. The whole path of the electron beam and of the X-radiation selected for analysis is *in vacuo.*

A very useful feature of the electron probe microanalyser is the provision of a *scanning system.* Magnetic deflection coils or electrostatic deflection plates introduced into the electron–optical system are programmed to cause the electron beam to traverse an area of the specimen of 500 μm $\times$ 500 μm in a raster pattern (i.e. backwards and forwards along parallel lines which are successively displaced as in the building up of the image on a television screen) once every 2–3 seconds. The spot on a cathode-ray tube is scanned synchronously over a larger area, $\sim$ 10 $\times$ 10 cm, and an image of the selected area of the specimen is formed in which brightness is approximately proportional to concentration of the selected element. By using a long-persistence phosphor in the screen the image can be inspected visually or it can be photographed with an exposure of several minutes to provide a permanent record of the distribution of the selected element in the specimen. Alternatively the scanning system can be made to cause the electron beam to make a linear traverse across the specimen and the ratemeter can be coupled to the vertical deflection control of the cathode-ray tube to give on the screen a concentration profile of the selected element along the chosen line. This technique is particularly applicable to the study of zoning and frozen diffusion equilibria in minerals and metals.

The output of the electronic backing of the proportional counter simply yields a measurement of intensity (as counts per unit time) for a particular wavelength band. By measuring the intensity of the continuum adjacent to the selected spectral peak and subtracting this *background* intensity from that measured for the peak a reduced intensity is arrived at. The ratio of the reduced intensities measured for the selected element in specimen and standard is, to a first approximation only, equal to the ratio of the atomic concentrations of that element in specimen and sample. If a standard is used whose composition is close to that of the sample the approximation becomes

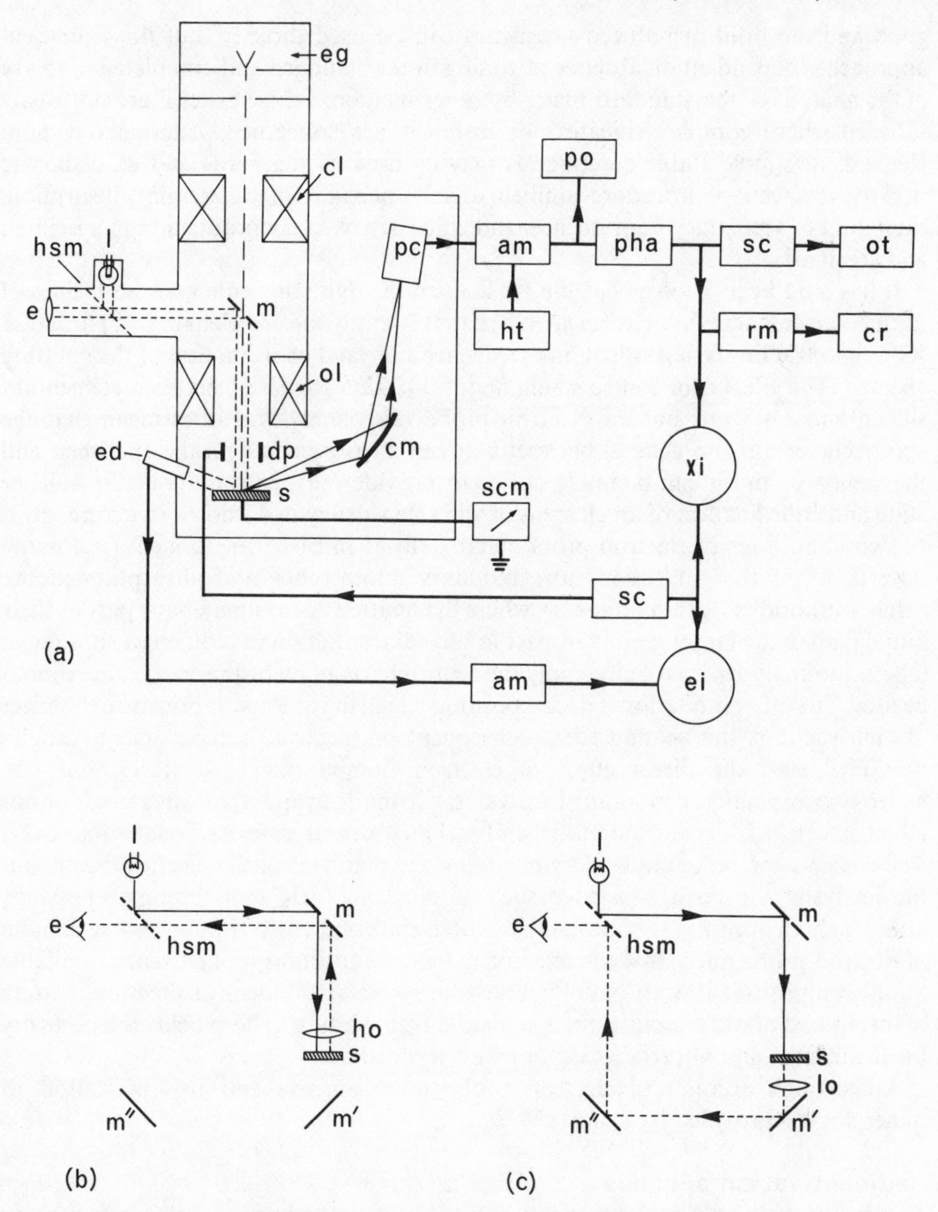

Fig 15.7 Schematic diagram of the components of an electron probe microanalyser. *Electron-optical system*: eg electron gun; cl condenser lens; ol objective lens; s specimen; scm specimen current meter. *X-ray spectrometer*: cm crystal monochromator; pc proportional counter. *Electronics*: ht counter HT supply; am pre-amplifier and amplifier; po pulse oscilloscope; pha pulse height analyser; sc scaler; ot output typewriter; rm ratemeter; cr chart recorder. *Scanning system*: dp deflection plates; sc scanning circuitry; Xi X-ray image display; ed electron detector; am amplifier; ei electron image display. The *light microscope system* is indicated in very general terms in (a) and two of several modes of operation are shown schematically in the lower diagrams: l lamp; hsm half-silvered mirror; m mirror with central aperture to permit passage of electron beam; ho high-power objective; lo low-power objective; m′, m″ mirrors; e eyepiece. The arrangement for observation of the specimen in reflected light at high magnification is shown in (b) and in transmitted light at low magnification in (c).

good and the ratio of reduced intensities can be used directly, but this empirical approach is dependent on absence of zoning in the standard and complete reliability of the analysis of the standard made by other methods; these criteria are not easily satisfied where complex silicates, for instance, are concerned. Alternatively pure elements or simple stable compounds may be used as standards and an elaborate iterative correction procedure applied to take account of differential absorption, excitation of secondary fluorescence, and efficiency of X-ray production in specimen and standard.

It has long been known that the K_α lines of the light elements and the L lines of the heavy elements exhibit a *chemical shift*, that is to say the wavelength of a particular X-ray spectral line is dependent on the valence state and coordination of the emitting element. The effect is of course small, about 0·0022 Å for Si K_α between elementary silicon and a silicate, but for electron probe microanalysis it does mean that the spectrometer must in general be reset between measurements on the specimen and the standard, and it can be made use of to provide information about the valence state and coordination of an element of variable valency in a known structure.

Two limitations of electron probe microanalysis must be mentioned. One is the uncertainty of the corrections for secondary fluorescence and absorption in the neighbourhood of a grain boundary where the emitted X-rays may have part of their initial path in the target grain and part in the adjacent grain of a different substance: if accurate analyses are required only the central parts of inclusion-free grains should be used. The other limitation is decomposition which may be due variously to the effect of high vacuum, the heating effect consequent on electron bombardment (usually $\not> 300\,^\circ C$), and the direct effect of electron bombardment; hydrates, but not hydroxy-compounds, commonly lose water, Na and K evaporate from some, but not all, silicates (e.g. feldspars and micas) and carbonates, and many carbonates lose CO_2. These effects can be minimized by enlarging the diameter of the electron beam and the analysis of even a substance such as $Na_2Ca(CO_3)_2$ then becomes possible, although high accuracy is unattainable. But in spite of these limitations the technique of electron probe microanalysis remains the most important tool currently available for the compositional study of polyphase solid systems; it is ideally suited to the study of metals and of wide application in mineralogy, obviating in both fields the necessity for the tedious and uncertain task of phase separation.

A thorough account of electron probe microanalysis and its application to mineralogy is provided by Long (1967).

Radioactivation analysis

This is a technique of important but limited application based on the radioactivity acquired by certain elements as a consequence of neutron bombardment. Its main use in mineralogy is as a very sensitive method of trace element analysis.

The first step in radioactivation analysis is to irradiate weighed amounts of specimen and standard separately in the reactor in a slow neutron flux. Under such conditions certain isotopes undergo a nuclear reaction, the simplest and analytically most important being of the (n, γ) type. After irradiation specimen and standard are separately taken wholly into solution in an appropriate solvent and quite large weighed amounts of the natural inactive form of the derivative element are introduced into each solution to act as a *carrier*. The necessary chemical separations are then performed to isolate some compound of the element from both specimen and standard solution; the use of the carrier makes it unnecessary for these precipitations to be complete. Each precipitate is weighed and then the intensity of its γ (or β) activity is measured. The content of the element X in the sample is then given by

$$\text{content of X in standard} \times \frac{\text{counts per unit weight of sample precipitate}}{\text{counts per unit weight of standard precipitate}}.$$

The γ-ray spectrograph can in many cases be used to simplify the analytical procedure by obviating the need for chemical separation of the element under investigation. This instrument is used in conjunction with published tables of wavelength for γ-ray spectra which enable a strong line attributable to the element under investigation to be identified and counted in specimen and sample; the identification can, and should, be confirmed by counting over a sufficient time to establish a value for the half-life of the nuclide concerned.

By way of example it may be mentioned that radioactivation analysis has been used to determine traces of Sr in sea-water by counting the γ-emission for the short-lived isotope ^{87}Sr ($t_{\frac{1}{2}} = 2{\cdot}9$ hours) and, a week later, the β-emission for the longer lived isotope ^{89}Sr ($t_{\frac{1}{2}} = 54$ days).

A critical and fully referenced account of the method is given by Mapper (1960).

Mössbauer spectrometry

The valence state and structural environment of iron in its solid compounds is amenable to study by a type of nuclear resonance spectroscopy dependent on the *Mössbauer effect.* The isotope ^{57}Fe, which makes up 2.14 per cent of natural iron, has a very sharp resonance absorption peak for γ-radiation of the appropriate wavelength. However the absorption spectrum is dependent in detail on the environment of the Fe nuclei in the solid specimen. Decay of the unstable isotope ^{57}Co dispersed by diffusion in an iron matrix provides a suitable source of γ-radiation; only if the environment of the iron nuclei in source and specimen are identical will resonance absorption occur, otherwise radiation will be transmitted and can be measured. By imparting a velocity to the source relative to the specimen the wavelength of the γ-radiation is changed by the Doppler effect and by employing a sequence of positive and negative relative velocities intensity of γ-radiation transmitted by the specimen can be plotted against wavelength. That this is possible depends on the extremely narrow spectral width of the resonance absorption peaks and on the availability of a device for moving the source at speeds between 1 and 10^{-3} cm s^{-1} accurate to $< 10^{-3}$ cm s^{-1}.

The Mössbauer spectrum of most iron minerals is complex but it is susceptible to analysis by standard curve-fitting procedures. The absorption maxima that emerge enable distinctions to be made on the basis of their wavelengths between Fe^{2+} and Fe^{3+} and between various structural environments for each oxidation state, thus Fe^{3+} in regular tetrahedral, distorted tetrahedral, regular octahedral, and distorted octahedral coordination to oxygen can be distinguished and likewise for Fe^{2+}.

Mössbauer spectrometry is a valuable ancillary analytical technique for providing information about the oxidation state of iron in solids, providing simultaneously useful structural information. Although about thirty isotopes exhibit the effect its practical application is largely confined to ^{57}Fe. Useful reviews of Mössbauer spectrometry are given by Greenwood (1967) and Wheatley (1970).

Indirect methods of analysis

In special circumstances the chemical composition of a mineral may be determinable without actual chemical or physical determination of the concentration of any of its constituent elements. In the case of a binary solid solution series, such as the common olivines whose compositions can be represented to a first approximation in terms of the end-members Mg_2SiO_4 and Fe_2SiO_4 only, the composition of a particular olivine can be estimated by measurement of $2V$ on the Universal-stage, or better by measurement of the Bragg angle of a sensitive peak on the X-ray powder diffractometer record. In the garnet (cubic) solid solution series, which has five commonly significant end-members, composition may be estimated, but only very approximately, by measurement of refractive index, density and unit-cell edge coupled with the quite simple chemical determination of Mn.

Such indirect methods for the determination of composition have considerable currency in mineralogy, but it must always be borne in mind that they are dependent in the first place on correct identification of the mineral and on the assumption that it has what may be called a 'normal' composition. If one is dealing with a well-known and frequently analysed type of mineral from a well-known environment such methods are safe enough, but for an uncommon mineral or a little studied environment they are inadequate.

Estimated accuracy of analytical methods

For long chemical methods have been regarded as the standard against which the various physical methods of analysis are compared, but in recent years some physical methods have been developed to a level where they are superior in accuracy. Chemical analyses made by skilled and experienced analysts are reproducible within close limits, but cooperative studies of analyses of the same complex material performed by different analysts using a variety of chemical procedures have revealed notable discrepancies. The changes in accepted analytical procedure that have resulted have eliminated some systematic errors and some of the spread, but the level of accuracy, as distinct from reproducibility, remains barely satisfactory for some elements. Since chemical methods are relatively time consuming, analyses are not usually done in duplicate by different procedures and no estimate can be made of the accuracy of a particular analysis. In contrast most of the physical methods are susceptible to statistical evaluation of counting errors and are moreover much faster so that repeat analyses are the rule rather than the exception.

In general it can be said that for chemical methods of analysis the standard deviation expressed as a percentage of the weight percentage of the element found is < 2 per cent for elements present in concentrations in excess of 5 per cent and increases sharply at lower concentrations. A study by Fairbairn and Schairer (1952) of twelve analyses by different analysts of a very carefully prepared homogeneous glass of known composition revealed that the mean for SiO_2 (at the 70 per cent level) was 0·4 per cent low, while the mean for Al_2O_3 (at the 15 per cent level) was high by an exactly compensating amount; for the remaining elements determined, Ca, Mg, Na, and K, the discrepancy between the analytical mean and the true value was within or close to the standard deviation of the twelve analyses.

For emission spectrographic determination of major constituents Ingamells and Suhr (1963) found variation within ± 5 per cent of the amount present and accuracy such that the mean is within 5 per cent of the accepted value for most elements. For trace elements, where emission spectrometry has in the past been pre-eminent, accuracy falls with decreasing content of the element concerned to an order of magnitude for most elements at levels of concentration of a few ppm. Atomic absorption spectrometry is in general capable of significantly greater accuracy and lower limits of detection.

X-ray fluorescence spectrometry has an accuracy of ~ 1 per cent of the amount of the element for elements whose concentration is greater than ~ 2 per cent by weight (Norrish and Hutton, 1969); accuracy falls off gradually with decreasing concentration and is comparable with, or slightly better than that of emission spectrometry at the trace element level. The comparative simplicity of X-ray spectra however makes the identification and estimation of small amounts of such elements as the rare earths easier and more certain by this method.

Electron probe microanalysis has comparable accuracy, about 1 per cent of the amount present if > 2 per cent, falling off gradually at lower concentrations. The method is applicable over the whole concentration range down to very low limits of detection for elements of high atomic number (Sweatman and Long, 1968). The method is of course capable of detecting very small differences in concentration along a scanning line, differences an order of magnitude smaller than the accuracy of the absolute determination.

16
Mineral equilibrium and synthesis: experimental methods

In this chapter we deal in general terms with the techniques of mineral synthesis, with experimental criteria for the attainment of equilibrium, with the characterization of the products of synthetic experiments, and in conclusion with the determination of phase diagrams for simple systems.

Reactants

Natural crystalline minerals are not often suitable as starting materials for phase equilibrium studies because, even when very finely powdered, they tend to react rather slowly with one another so that in the time available for laboratory experiments equilibrium cannot be achieved close to the equilibrium temperature and in consequence a falsely high equilibrium temperature may be determined. A second disadvantage of using natural minerals as reactants is that many of the more interesting minerals are solid solutions and so their use immediately increases the number of components in the experimental system to more than the minimum number necessary to define the problem under investigation in its simplest form. These two considerations generally rule out the use of natural minerals as reactants except as the final stage of a synthetic study, a stage which is often omitted. It is usual to make use of reactants which are simple in composition and in a highly reactive state; homogeneous glasses or gels of precisely known composition are particularly useful.

Glasses are prepared by successive fusion, quenching, and grinding of precisely weighed-out mixtures of synthetic oxides of high purity (Schairer, 1959). At least four fusions are usually necessary before homogeneity of a silicate glass is achieved. It is quite simple to test the resulting glass for homogeneity by placing crushed fragments in a liquid whose refractive index is near that of the glass and applying the Becke line test under the most favourable conditions for detecting small differences in refractive index. Al_2O_3 is particularly slow to dissolve in silicate melts so that aluminous glasses may have to be re-fused as many as six times before homogeneity is achieved.

Gels are prepared by mixing in appropriate volumes standardized metal nitrate solutions with either silica sol or tetraethyl orthosilicate. The resultant solution is made alkaline with aqueous ammonia and left until it forms a thick gel, which will take a time ranging from a few hours to several days in different cases. The gel is slowly dried at 50–75 °C for 1–2 days and then at progressively higher temperatures

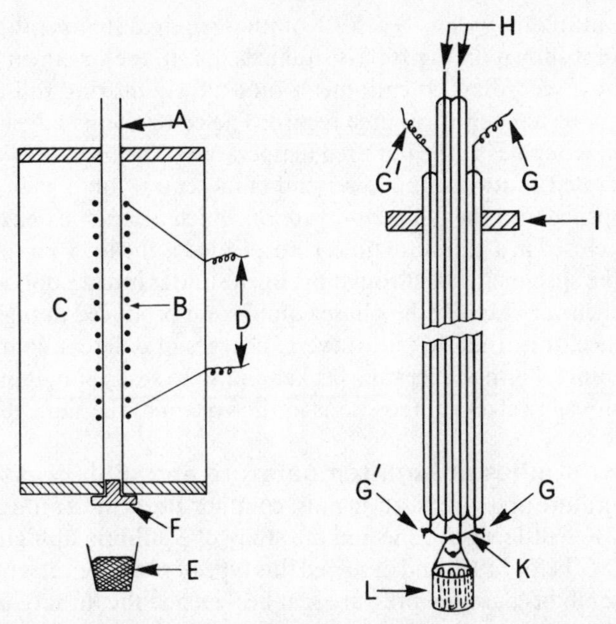

Fig 16.1 A vertical tube furnace for high-temperature studies at atmospheric pressure. The diagram on the left shows the furnace. A is the ceramic tube, around which is wound spirally the platinum resistance wire B; the furnace is packed tightly with magnesia C which acts as a thermal insulator; D represents the connection of B to a low-voltage, high-amperage electricity supply; for quenching, the plug F is removed to allow the charge to fall into a pool of mercury in the beaker E. The diagram on the right shows, on a larger scale, the apparatus suspended in the vertical tube A. The ceramic bucket L holds a number of charges sealed in noble-metal capsules; the thermocouple leads H are carried down through ceramic or silica glass tubes to form a couple close to the charges; the leads GG′, also in ceramic or silica glass tubes, carry the current used to melt the wire K and so allow the bucket L to fall into E at the end of the run; the four tubes carrying GG′HH are cemented together and to the disc I which rests on the top of the tube A, I being positioned so that L is at the 'hot spot' of the furnace.

until the water is totally expelled and the nitrate decomposed at 900 °C. In some cases it is convenient to crystallize the gel by heating to an even higher temperature. A detailed account of the preparation of silicate gels is given by Luth and Ingamells (1965); various laboratories use variants of this procedure, but all produce in the end essentially similar gels.

For decomposition studies, such as dehydration or decarbonation, the natural mineral hydroxide or carbonate may be a satisfactory starting material if it can be cleanly separated from the other minerals with which it occurs.

Apparatus for high temperature studies at atmospheric pressure

Many silicate and oxide systems can conveniently be studied in a platinum-wound vertical tube-furnace of the type described by Shairer (1959). The platinum (or for temperatures below about 1000 °C nichrome) spiral heating element (Fig 16.1) is wound on a ceramic cylinder and surrounded by tightly packed magnesia, which serves as a thermal insulator. Temperature is measured by a $Pt{-}Pt_{90}Rh_{10}$ thermocouple hanging in the central tube adjacent to the specimen. Thermocouple and specimen are carefully positioned so as to be in the 'hot spot' of the furnace, that is in the central portion of the length of the tube where the axial temperature gradient is small. The thermocouple controls the energy supply to the heating circuit through an electronic controller, operating mercury relays, which maintains the furnace temperature to within ±0·5 °C of the required temperature. A second thermocouple (not shown on Fig 16.1) is situated close to the specimen and its output is printed out by a recording potentiometer on a slowly moving roll of chart paper to give a continuous temperature–time record. The controller is adjusted so that the recording thermocouple is at the required temperature. The furnace is equipped with a manually operated shutter at the lower end of its central tube; at the end of a run the shutter is opened and the palladium wire on which a ceramic beaker, containing the specimen enclosed in a platinum tube, is suspended is melted by a separate electric circuit so that the specimen falls through the open shutter into a pool of cold mercury or water to quench very rapidly the phase equilibrium produced at high temperature within the furnace. It is usual to place several charges of different composition, each in its own platinum tube, in the ceramic beaker and so to study simultaneously several (perhaps as many as twelve) compositions in the system at the same temperature.

Apparatus for studies at high temperature and high pressure

The high-temperature pressure vessel in most common use is the *Tuttle cold seal bomb*, named after O. F. Tuttle who pioneered the study of equilibria at high temperatures and high pressures before 1950 and designed this type of pressure vessel; it is described as a cold seal bomb because the pressure seal lies outside the furnace and remains at a much lower temperature than the portion of the bomb which contains the charge under study. Such a bomb is constructed from a rod of 'stellite' or 'Nimonic' alloy or 'René metal', about 1 inch in diameter, by drilling an axial hole of $\frac{1}{8}$ inch to $\frac{1}{4}$ inch diameter to within $\frac{3}{4}$ inch of the far end of the rod; the open end is finished in a conical opening and tapped with a screw thread to take the conical connexion to the pressure line so that the pressure line is sealed on by a cone-in-cone joint; the closed end is drilled to a depth of about $\frac{1}{2}$ inch to accommodate a thermocouple. The charge, in a sealed platinum, gold, or silver capsule (cut from an extruded tube to a length of about $\frac{3}{4}$ inch and welded at both ends), is dropped to the bottom of the axial hole and the remaining space in the axial hole is filled by a spacing rod of the same alloy as the bomb is made from. The function of the spacing rod is to decrease the volume of gas in the bomb and so reduce the explosion hazard as well as to hold the specimen capsule in place at the far end of the axial hole. The bomb is inserted in a vertical or horizontal tube furnace so that the specimen is centrally placed in the 'hot spot' of the furnace. Pressure is applied to the charge by collapse of its enclosing capsule which is enveloped in the 'pressure medium', usually water vapour pumped into the bomb through capillary pressure tubing from a reservoir maintained at the required pressure by a manually operated jack or a pneumatic pump. A Bourdon pressure gauge is incorporated into the pressure line. The experimental arrangement is shown in Fig 16.2. Quenching of the charge at the end of the run is usually achieved by removing the furnace from the pressure vessel, or vice versa, and directing a high pressure blast of cold air on the bomb while maintaining pressure by opening the valve between the pressure line and the pressure reservoir. In special cases it may be desirable to drop pressure while maintaining high temperature or to drop pressure and temperature together; special techniques have been developed for such cases. After quenching, the capsule is removed from the bomb and cut open; the charge is then examined optically and by X-ray powder diffractometry or occasionally by single crystal X-ray methods. The limiting conditions of operation of the cold seal bomb are about 1000 °C at 2 kb or 700 °C at 5 kb, the limiting criterion being the hot-strength of the alloy from which the bomb is made. By constructing the bomb of larger diameter alloy rod, about

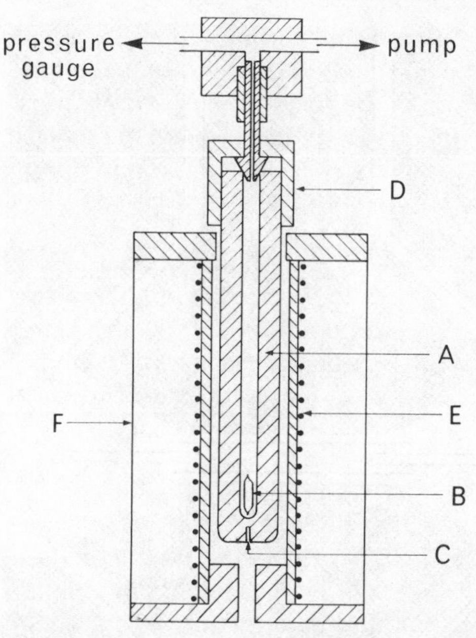

Fig 16.2 The Tuttle cold seal bomb. The bomb A is drilled with an axial hole which contains the charge sealed into the noble-metal capsule B and with an axial hole C to accommodate the recording thermocouple; D is the cone-in-cone pressure seal which lies outside the furnace F; the spiral furnace-winding E is packed with magnesia powder which acts as a thermal insulator in the furnace F; pressure is applied through a gas or vapour raised to the required high pressure by a jack or pump and recorded by a Bourdon gauge inserted in the cold pressure line.

2 inch, and improving the cold seal and using argon as the pressure medium the cold seal bomb can be taken up to much higher pressures, e.g. 10 kb at 700 °C (Luth and Tuttle, 1963).

An important modification of the Tuttle cold seal bomb is the *internally heated cold seal bomb* in which a small furnace is placed within a pressure vessel whose outer surface is cooled by circulating water. The charge is situated in the axial aperture of the furnace. Such devices are now in common use for work up to 10 kb at 1200 °C.

For higher pressures recourse must be had to some type of uniaxial device, such as the *simple squeezer* as modified to operate at high temperatures by Griggs and Kennedy (1956) and illustrated in Fig 16.3. Pressure is applied through pistons made of materials of very high compressive strength. Such materials are in general brittle and to avoid brittle fracture of the pistons they are conically shaped so that they are always supported by a confining pressure, the conical surface being lubricated by thin copper foil. In such a device equilibrium appears to be attained more rapidly than in a hydrostatic pressure vessel such as the cold seal bomb. Attainment of equilibrium is facilitated by oscillation of one of the pistons during the run (Dachille and Roy, 1960). The difficulty with this type of apparatus is that pressure cannot be measured directly and it is difficult to estimate the relative magnitudes of the hydrostatic and shearing stresses. The pressure applied to the charge is calculated from the pressure applied to the pistons, a correction being made for friction; the pressure gradients set up in the charge between the pistons lead to considerable uncertainty. Temperature, applied by an external split furnace, can be estimated quite accurately. Quenching is achieved by opening the split furnace and playing an air blast on the inter-piston block at the level of the charge. The charge is removed from the squeezer as a hard lenticular disc, which is then crushed for optical and X-ray diffractometric examination. The PT range of the simple squeezer extends to 80 kb at 500 °C or 20 kb at 1000 °C.

The versatility of the simple squeezer has been improved by incorporation of an internal graphite furnace (Boyd and England, 1960), which enables the PT range for equilibrium studies to be extended to 120 kb at 500 °C or 60 kb at 2000 °C.

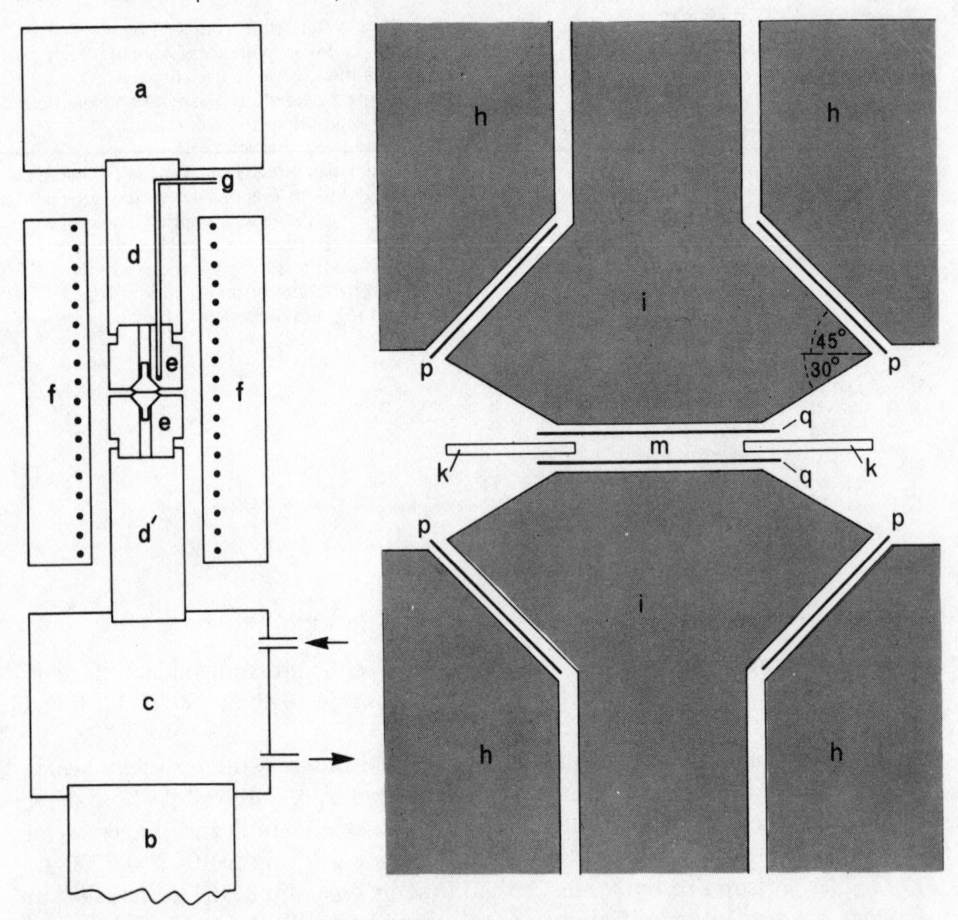

Fig 16.3 The simple squeezer. On the left a schematic diagram of the apparatus and on the right an exploded view of the piston assembly. Key: *a* the fixed plate and *b* the ram of a hydraulic press; *c* the water cooling chamber; *dd′* steel pushers, *e* piston and holder assembly, *f* split cylindrical furnace, *g* thermocouple passing through upper pusher and holder, *h* stellite holders, *i* pistons constructed either of stellite (45° cone as shown) or of cemented carbide (20° cone), *k* steel ring, *m* sample, *p* copper foil to act as lubricant at high pressure, *q* $Pt_{10}Rh_{90}$ foil to prevent reaction with piston faces and to facilitate sample removal.

The next stage of complication is the two-stage pressure vessel (Boyd, 1962) in which pressure is applied to the charge through a piston whose outer shank is supported by the confining pressure of a sleeve of KBr; KBr undergoes a 10 per cent reduction of volume on transformation at 19 kb.

The final stage of complication so far attained (Fig 16.4) is the tetrahedral anvil press (Hall, 1958) in which a pyrophyllite tetrahedron, enclosing a charge of up to 2 ml in volume and a graphite furnace, is compressed on each of its triangular faces by four anvils (i.e. pistons tapered to fit together closely). One piston is compressed by a hydraulic ram while the other three are forced into a steel cone so designed that pressure is evenly distributed on the faces of the tetrahedron. The pistons are lubricated by pyrophyllite gaskets, through one of which a thermocouple is introduced to measure the temperature of the charge. For work at ultra-high pressures the number of anvils can be increased, for instance to six in an octahedral configuration

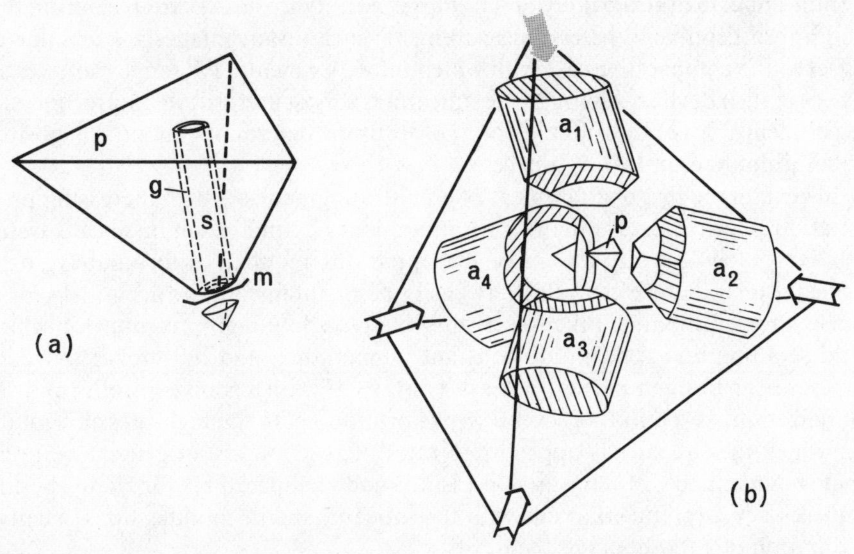

Fig 16.4 The tetrahedral anvil press. (a) shows the pyrophyllite tetrahedron *p* in which a hole passes from one vertex to the centre of the opposite face; the hole is lined with a graphite sleeve *g* which acts as the furnace and the sample *s* is placed within the graphite sleeve; the metal strip (Ta or Ni) *m* serves to make electrical contact between the graphite sleeve and the three lower anvils. (b) shows the tetrahedral disposition of the four anvils a_1–a_4 about the pyrophyllite tetrahedron; the three lower anvils a_2–a_4 are confined by a steel cone and a force is applied by a hydraulic ram to the upper anvil a_1, the force applied by the ram being shown by a solid bold arrow and the wedge reaction of the confining cone on the other three anvils by open arrows. The three lower anvils form one terminal of the graphite furnace and the upper anvil a_1 the other terminal. The pyrophyllite tetrahedron flows at high pressure to form a pressure seal between the faces of the four anvils. A thermocouple passes between adjacent anvils into the specimen cylinder.

or twenty-four in an icositetrahedral configuration. A useful survey of high pressure apparatus is provided by Bradley (1969).

Experimental methods

The purpose of any synthetic experiment is to bring a charge of known composition to equilibrium under a particular set of physical conditions (*P*, *T*, etc) in such a manner that the composition of the charge remains unchanged and the physical conditions remain constant and uniform throughout the volume of the charge. The experiment may last for only a few minutes or for as long as several months. At the conclusion of the experiment it is necessary to freeze the phase equilibrium by rapid quenching or by sudden release of pressure. The charge may be maintained at constant composition by sealing it, by welding, into a capsule which is impervious to the external atmosphere. Constancy of temperature is maintained by using the output of a thermocouple to control the supply of electric current to the furnace; the controller is adjusted early in the course of a run so that a second thermocouple inserted in a hole drilled in the bomb reads the required temperature. In the cold seal bomb and its modifications, where pressure is applied through a gaseous medium, maintenance of constant temperature implies maintenance of constant pressure; in such circumstances pressure is essentially uniform and hydrostatic throughout the body of the charge and is susceptible to measurement by a gas-pressure gauge in the external pressure line. In the squeezer and anvil types of apparatus however pressure is apt to be non-uniform over the body of the charge, non-hydrostatic, variable during the run, and not susceptible to precise measurement; such disadvantages are to some extent offset by the comparative ease with which the experimental PT range can be extended by use of such devices. Although the squeezer is a most useful device for reconnaissance experiments, it is not suitable for the accurate determination of the position of equilibrium lines in P–T–X^n space.

In general the rate of attainment of equilibrium increases with increasing pressure. At atmospheric pressure the rate of attainment of equilibrium in silicate systems is relatively slow, but the nature of the apparatus required is fortunately such that constant temperature can relatively easily be maintained quite accurately for many weeks or even months. Likewise at the relatively low high-pressures for which the cold seal bomb is appropriate constant temperature, and by implication constant pressure, can be maintained to at least $\pm 5\,°C$ ($\pm 1°$ with a good controller and voltage stabilization) for periods of several weeks or months. In the high-pressure conditions for which the squeezer is appropriate rates of reaction are in general quite fast so that maintenance of constant pressure and temperature for long periods is unnecessary; experimental runs with the squeezer and its modifications usually have a duration of a few hours at the most.

We have already described the quenching of the charge at the end of a run in an atmospheric pressure apparatus and a cold seal bomb. Quenching is achieved in squeezer and anvil devices by swinging open the split furnace and playing an air-blast on the pistons or, if the charge will not react with water, by plunging the piston assembly and charge into a large bath of cold water.

After quenching, a portion of the charge is mounted in an oil of suitable refractive index for examination with the polarizing microscope. The phases present can often be identified by refractive index, other optical properties, and habit. Another portion of the charge is subjected routinely to X-ray diffractometry. It is usually considered advisable to use these two techniques in conjunction: while diffractometry will in general provide a rapid and reliable means of identification of the crystalline phases present in quantity, optical examination is necessary for recognition of the presence of glass (indicative of a liquid phase in the equilibrium assemblage) as well as for the identification of phases present in only small amounts. Moreover optical examination may, if a characteristic habit or twinning is involved, provide a distinction between phases whose diffraction patterns are difficult to distinguish with certainty in the diffractometer record of the polyphase quenched charge.

For a complete description of the equilibrium phase assemblage in the quenched charge it is of course necessary to know the composition of any glass or solid solution that may be present. The only composition-dependent property of a glass that can conveniently be determined is the refractive index. In a binary system measurement of refractive index is adequate to determine the composition of the glass provided that the refractive indices of an adequate number of glasses (produced by complete melting and quenching of known compositions) in the system have been determined. For glasses in ternary and higher systems composition can only be determined for a glass in a polyphase assemblage by taking its refractive index in conjunction with the determined compositions of all the solid phases and the bulk composition of the charge; even so it is not always possible to estimate the composition of the glass. This point will be taken up in the next section.

The composition of a binary solid solution can most conveniently be determined by measurement on the diffractometer trace of the d-spacing of a reflexion which is particularly sensitive to compositional change. The chosen reflexion must be clear of any reflexion due to any other phase that may be present. For accurate measurement of the *d*-spacing of the chosen reflexion it is usually necessary to introduce an internal standard into the diffractometer mount and for this purpose a separate portion of the quenched charge may be reserved. For ternary solid solutions the measurement of two or more peaks on the diffractometer record may enable the two compositional parameters to be determined, but often the accuracy of the resulting determinations is inadequate. Measurement of refractive indices may be used to supplement the *d*-spacings derived from the diffractometer record; but all too often no improvement in the accuracy of the determination of composition can be achieved. If the composition of a solid solution, or glass, cannot be determined satisfactorily by such methods, recourse may be had to the electron probe microanalyser; but there is a snag in that the crystal grains produced in routine synthetic experiments are near the minimum size that can be analysed accurately with the electron probe.

Criteria for equilibrium

When a homogeneous glass is subjected to a temperature and pressure at which a solid phase A is stable, it may first crystallize a metastable phase B. The crystallization of a metastable phase in such circumstances is governed by the empirical *Ostwald's step rule*: a liquid or solution will tend preferentially to crystallize a phase or phases of higher entropy than the stable assemblage if that is possible. The reaction by which the stable assemblage is produced will then not be glass → stable assemblage, but metastable assemblage → stable assemblage and the latter, involving solid and possibly very stable structures, may well be a markedly slow process. If the run is too short in duration, only the metastable assemblage will be recorded in the quenched charge. A rather longer run may be expected to produce an excessive number of phases ($\mathscr{P} > \mathscr{C}+2-\mathscr{F}$) belonging to both the stable and metastable assemblages. Runs of longer duration may yield the stable assemblage if they are long enough; in some systems under dry conditions at atmospheric pressure runs of many weeks may be needed to achieve equilibrium. As long as any change with time is detectable in the proportions or nature of the phases in the quenched charge it is certain that equilibrium has not yet been attained. In some systems certain particularly unreactive phases become recognizable as metastable phases and, provided they occur only in small quantity, it may nevertheless be possible to work out a phase diagram of diminished accuracy for the system.

If the starting material is a gel it will crystallize on its path to the selected pressure and temperature of the run and the phases that crystallize will in general be determined by ease of nucleation rather than by considerations of thermodynamic stability. Such metastable phases can have only a transient existence but they may nevertheless persist for times of the same order of magnitude as the duration of short synthetic runs. Recognition of metastable phases as such is not easy: they will disappear with time, but the time required for their disappearance may be longer than that of a normal experimental run. Careful examination of the phase assemblage produced at successively increasing temperature at the chosen pressure and composition will usually reveal which phases are metastable. Confirmation may be obtained by comparing the phase assemblage produced by raising pressure to the chosen value at atmospheric temperature and then increasing temperature to the chosen value with the assemblage produced by increasing temperature to the chosen value at atmospheric pressure and then raising the pressure to the chosen value.

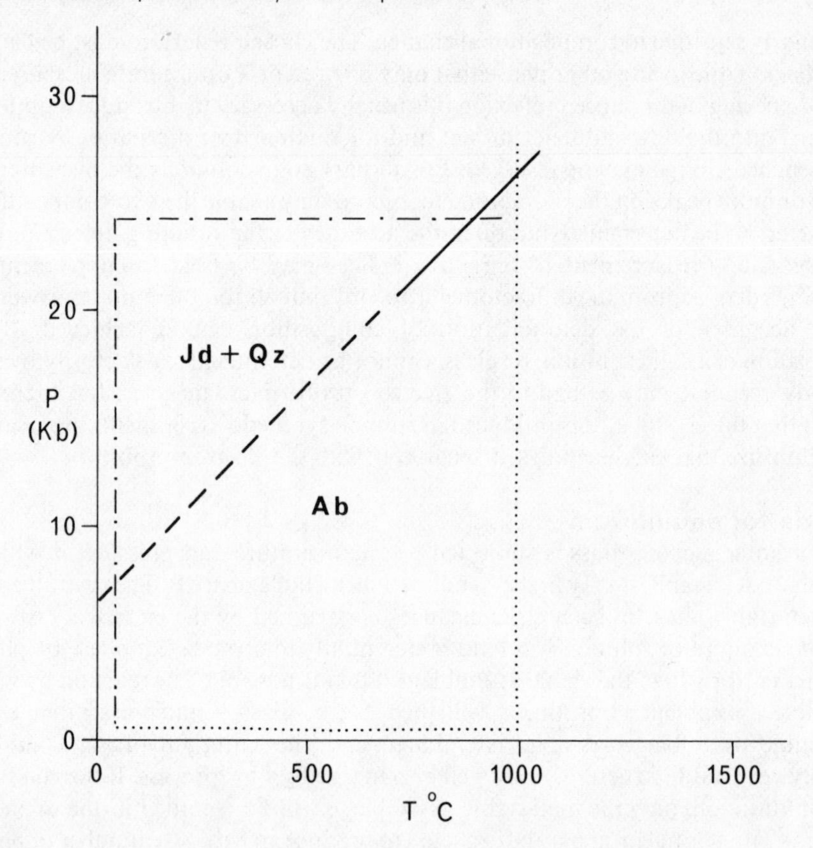

Fig 16.5 P–T diagram for the jadeite+quartz ⇄ albite equilibrium. The dotted lines indicate the course for an albite charge to demonstrate the reaction Ab→Jd+Qz. The dash-dot lines indicate the course for a jadeite+quartz charge to demonstrate the reverse reaction Jd+Qz→Ab. The equilibrium line, shown as a bold solid line at high *PT* (and extended to lower *PT* as a broken line where it has not been proved experimentally), has been determined by a series of such experiments.

Ideally a univariant *PT* curve or a divariant *PTX* surface, and so on, should be based on reversibility of the relevant reaction. But this is often not a practically realizable criterion in silicate systems where crystallization of a glass or gel to the stable assemblage is very much more rapid than the conversion of one assemblage of crystalline solid phases to another. It follows that the path along which the system passes to the chosen pressure and temperature may be critical where reaction between crystalline solid phases is slow. The equilibrium $A \rightleftharpoons B$ is thus usually studied by observations on the reactions $C \rightarrow A$ and $C \rightarrow B$; it is essential to demonstrate in each case that the change of reaction behaviour of C corresponds to the equilibrium $A \rightleftharpoons B$.

Sometimes it is possible to use natural crystalline minerals as starting materials and then it is possible to define a *PT* curve or surface with greater reliability. Birch and LeComte (1960) studied the equilibrium albite⇌jadeite+quartz ($NaAlSi_3O_8 \rightleftharpoons NaAlSi_2O_6 + SiO_2$) by taking an albite charge through the albite field and across the field boundary; that is to say they raised the temperature of the charge to a selected value and then increased the pressure to put it into the jadeite+quartz field (Fig 16.5). They also raised the pressure of charges composed of intimately mixed equimolecular

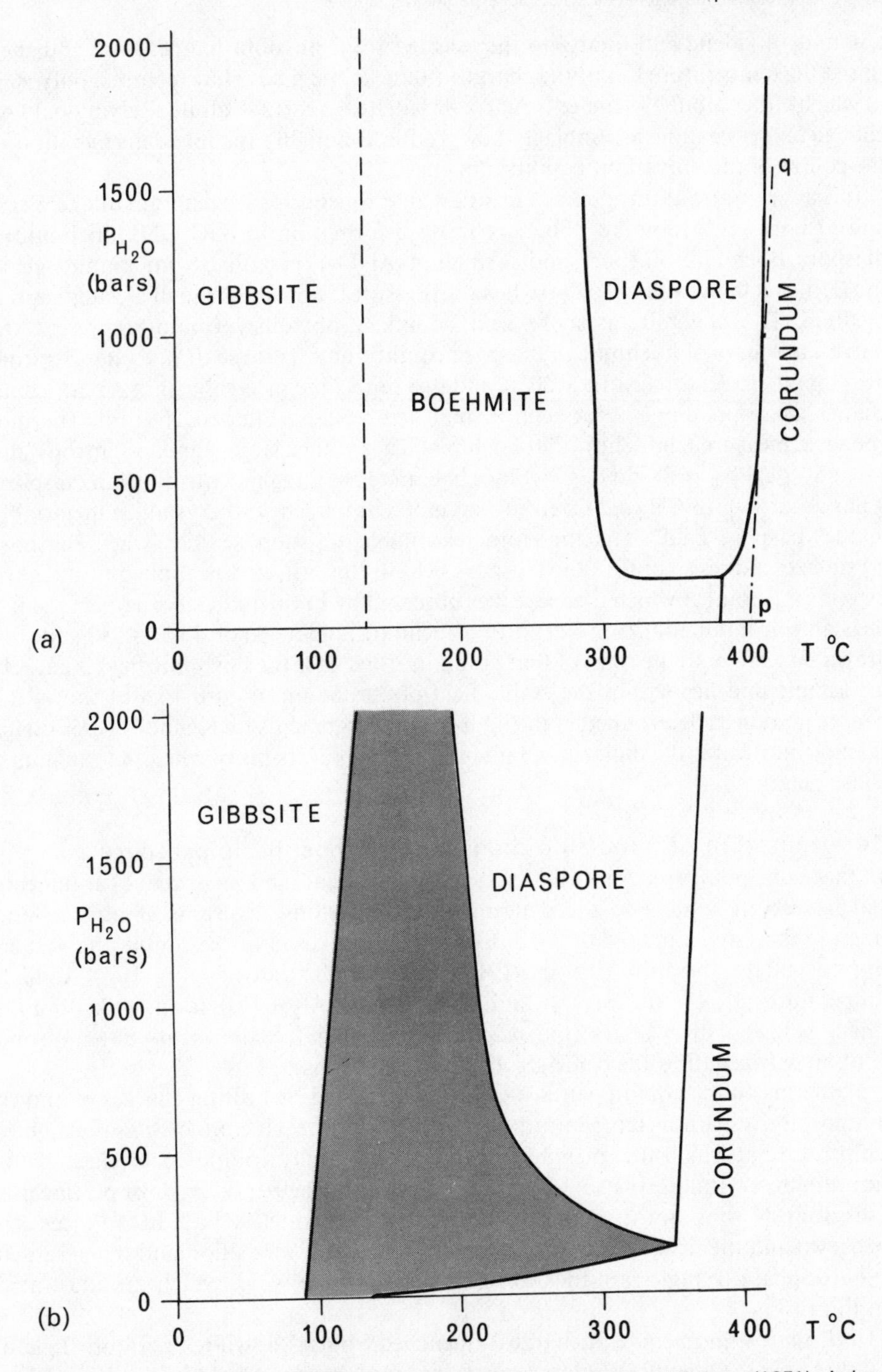

Fig 16.6 (a) *P–T* plot for the system Al_2O_3–H_2O according to Ervin and Osborn (1951). At low temperatures gibbsite $Al(OH)_3$ is stable and at high temperatures corundum Al_2O_3 is stable. At intermediate temperatures the two polymorphs of AlO.OH, boehmite or disapore, crystallize in synthetic experiments. (b) *P–T* plot for the system Al_2O_3–H_2O according to Kennedy (1959). Gibbsite, diaspore, and corundum are the only stable solid phases. The shaded area represents that part of the diaspore stability field in which boehmite crystallizes more readily, as a metastable phase, than the stable phase diaspore.

amounts of jadeite and quartz to the selected value at room temperature and then raised the temperature to put the charge into the albite field. They recorded only runs in which either albite → jadeite + quartz or jadeite + quartz → albite. Relying only on change in a crystalline assemblage they produced a highly reliable determination of the course of the univariant equilibrium curve.

It is appropriate at this point to mention an erroneous phase diagram due to Ervin and Osborn (1951) for the polymorphic transformation in AlO(OH), boehmite ⇌ diaspore. Boehmite, diaspore, and corundum (Al_2O_3) crystallize from alumina gel in the *PT* fields shown in Fig 16.6(a) when the pressure is applied through a water vapour medium. The boehmite/diaspore field boundary obviously cannot represent the univariant curve of boehmite ⇌ diaspore equilibrium: its slope dP/dT changes from near infinity to zero in less than 50 °C and it is simply inconceivable that $\Delta S/\Delta V$ could change correspondingly over such a small temperature interval, if at all. Thermochemical measurements show that boehmite has a rather large standard entropy and that $\Delta S^{\circ}_{298}/\Delta V^{\circ}_{298} = 88$ bar deg^{-1}. Since boehmite has a higher entropy than diaspore, it has according to *Ostwald's step rule* an inherent tendency to crystallize metastably in the diaspore field. It is therefore reasonable to suppose that when diaspore crystallizes it does so only in the diaspore field. If this criterion is applied to the most extreme *PT* point at which diaspore was observed by Ervin and Osborn to crystallize, the stable field boundary for diaspore ⇌ boehmite is a line pq of slope +88 bar deg^{-1} drawn through or to the right of that point. In either case the boehmite field is entirely metastable and lies within the stable field of corundum. Figure 16.6(b) shows the phase diagram as determined in the light of these arguments by Kennedy (1959). This example illustrates the dangers of failing to test for reversibility when determining a phase diagram.

Determination of a melting diagram at atmospheric pressure

In this concluding section we discuss some special and instructive arguments that have been applied to the determination of melting diagrams at atmospheric pressure in systems involving solid solution. We take as our example the system diopside–albite–anorthite (Bowen, 1915) which is illustrated in Fig 16.7. At high temperature albite ($NaAlSi_3O_8$) and anorthite ($CaAl_2Si_2O_8$) form a continuous binary solid solution series (the *plagioclase* feldspars); there is no ternary solid solution between diopside ($CaMgSi_2O_6$) and plagioclase.

Isotherms on the liquidus surface can be constructed by heating charges of known composition at various temperatures and recording for each composition the highest temperature at which the quenched charge is not wholly composed of glass. If the nature of the crystalline phase is determined optically, whether diopside or plagioclase, a diagram showing isotherms on the diopside and plagioclase liquidus surfaces can be drawn and the field boundary outlined in terms of composition and temperature.

Three-phase-triangles are then determinable in three ways, which are illustrated on Fig 16.7:

(1) *the composition method.* A charge is quenched from a known temperature T_A and, if a three-phase assemblage, glass + plagioclase + diopside, should be produced, then the composition of the glass will be given by the known temperature from which the charge was quenched and that of the plagioclase may be determined by X-ray diffractometry. As indicated in Fig 16.7 the tie line between liquid A and solid solution of composition B can be drawn provided the compositional and thermal course of the field boundary on the liquidus is known.

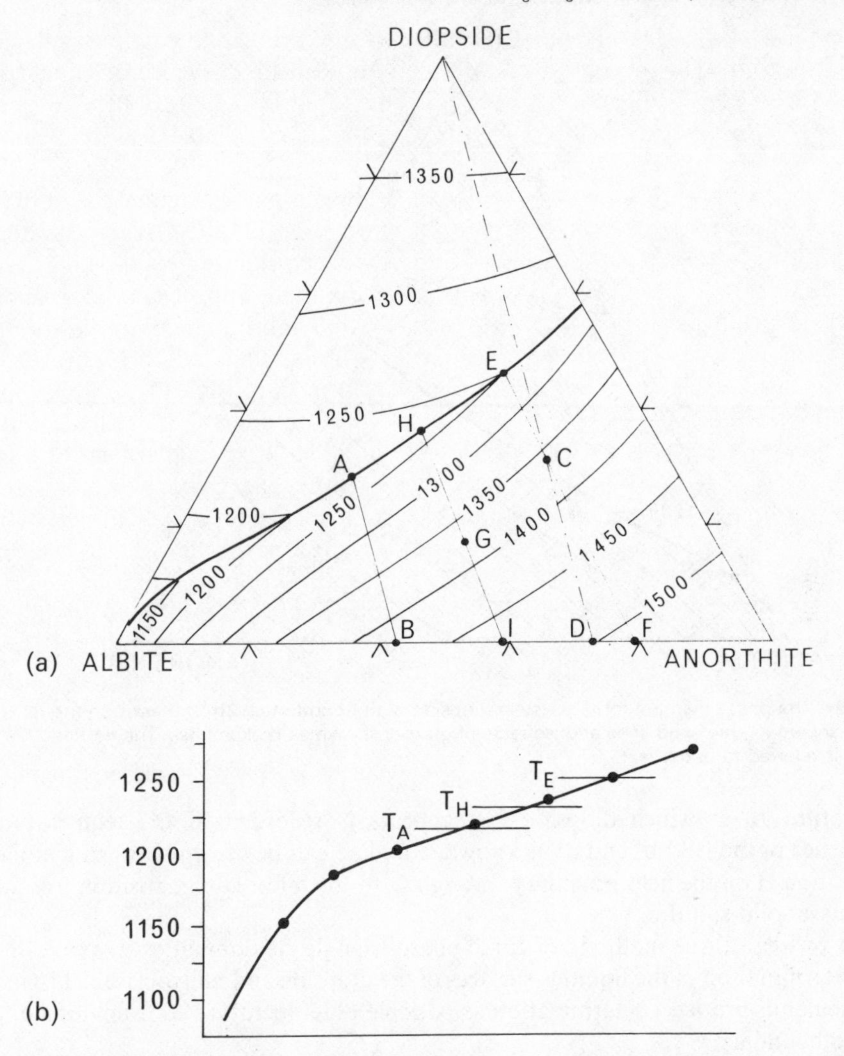

Fig 16.7 (a) shows the phase diagram for the system diopside ($CaMgSi_2O_6$)–albite ($NaAlSi_3O_8$)–anorthite ($CaAl_2Si_2O_8$) with isotherms on the liquidus indicated. The tie-lines A–B, H–G–I, E–C–F, and (diopside)–C–D are referred to in the text. (b) shows the temperature-composition curve for the diopside-plagioclase field boundary with the temperatures of the liquids, A, H, E indicated; the compositional axis runs from near albite along the field boundary to the anorthite-diopside join.

(2) *temperature of beginning of melting.* Charges composed of a mixture C of known proportions of synthetic diopside and synthetic plagioclase of known composition D are heated to various temperatures and the lowest temperature T_E at which glass is observed in the quenched charge is recorded. Again if the temperature dependence of the field boundary is known the tie line EF between a liquid on the field boundary and the binary solid solution can be drawn.
(3) *temperature of beginning of crystallization.* Mixtures of known composition G in the plagioclase field are quenched from successively lower temperatures and the

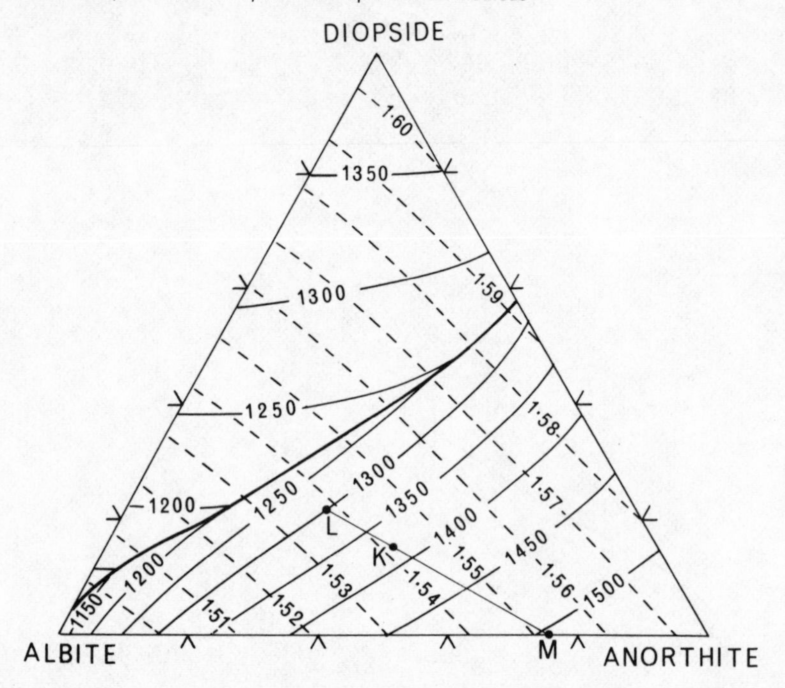

Fig 16.8 The phase diagram for the system diopside-albite-anorthite with isotherms on the liquidus shown as thin solid lines and isofracts of glasses shown as broken lines. The tie-line L–K–M is referred to in the text.

temperature T_H at which diopside first appears is recorded. If the temperature dependence of the field boundary is known, a tie line can be drawn between a liquid composition H on the field boundary through G to the inferred composition I of the plagioclase solid solution.

Each of these three methods of three-phase-triangle determination requires the prior determination of the liquidus surfaces of the diopside and plagioclase fields and in particular as precise a determination as possible of temperature coordinates along the field boundary.

Determination of tie lines between liquid compositions within the plagioclase field and plagioclase solid solutions depends on determinations of isotherms on the plagioclase liquidus surface and isofracts (lines of equal refractive index) of homogeneous glasses in the same compositional range. If a mixture K (Fig 16.8) of known proportions of plagioclase of known composition and diopside yields a glass L of determined refractive index in equilibrium with plagioclase (of undetermined composition) at a known temperature, a tie line between liquid L and solid solution M can be drawn from the intersection of isofracts and isotherms on the plagioclase liquidus surface. Alternatively the composition M of the plagioclase may be determined by diffractometry instead. From a great many of such observations fractionation curves can be drawn.

Appendices

Appendix A: Constructions in the Stereographic Projection

I: To project a pole P at ρ° from N along a given diameter QOR (Fig A.1)

The pole P lies in the plane PNS, which intersects the plane of projection in the diameter QOR of the primitive circle. Imagine the sphere of projection to be rotated through 90° about QOR (Fig A.1(a)): S will move to S′, N to N′, and P to P′. S′ lies on the primitive circle 90° away from Q and R.

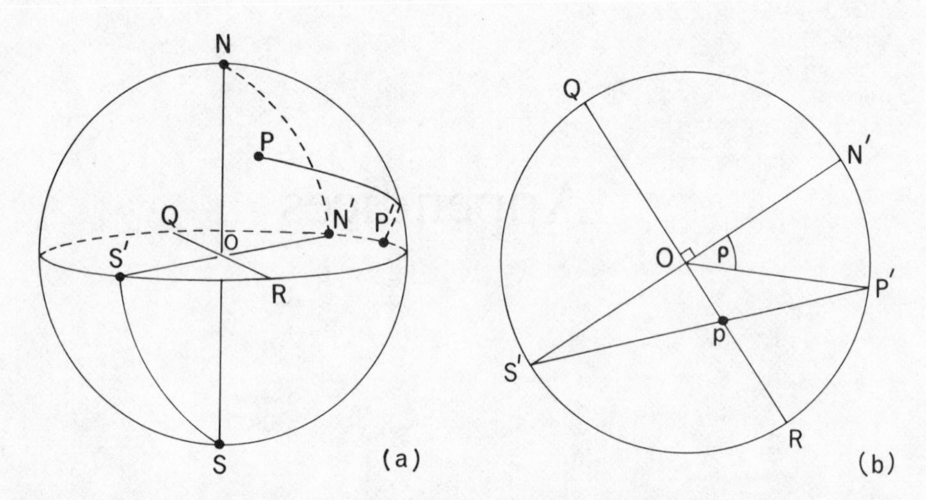

Fig A.1 To project a pole P at ρ° from N along a given diameter QOR. In the spherical projection (a) P is located in the plane NRSQ; on rotation through 90° about QOR this plane becomes N′RS′Q and P moves to P′. The plane diagram (b) shows the location of p at the intersection of S′P′ with QOR. Rotation back to the original orientation does not move p; p is the stereographic projection of P.

The projection of P is then made (Fig A.1(b)) in the following manner:
1. Draw the diameter N′OS′ perpendicular to QOR,
2. Draw OP′ such that $\widehat{N'OP'} = \rho^\circ$,
3. Join S′P′ to intersect QOR in p.

Imagine now that the sphere of projection is rotated back through 90° about the diameter QOR so that S′ moves to S, N′ to N, and P′ to P; p remains unchanged in position and is therefore the projection of the pole P.

Of course we could more simply have placed p at a distance $r \tan \frac{1}{2}\rho$ from O along OR, but we have described this construction to emphasize its principle, which is made use of in subsequent more complicated constructions. There too the reader will find apparently unnecessary elaborations, which are introduced primarily to establish principles.

II: To find the projection of the opposite P_0 of the pole P (Fig A.2)

The opposite of a pole P is the pole situated at the opposite extremity of the diameter of the sphere of projection through P. P_0 thus represents the direction at 180° to the direction represented by P. If P represents the normal to a face (*hkl*), then P_0 represents the normal to the face $(\bar{h}\bar{k}\bar{l})$.

It is evident from the section of the sphere of projection containing P_0, S, and P drawn in Fig A.2(a) that if P is projected using the south pole and P_0 using the north pole as projection point, then $Op = Op'_0$. It is therefore a simple matter to plot p'_0 by drawing the diameter through p and marking on it a point at an equal distance from O on the other side of O from p.

If the south pole is used as the projection point for both P and P_0 then P_0 projects outside the primitive circle and its position can be found by making use of the fact that PP_0 is a diameter and so the angle $PSP_0 = 90^\circ$. If the primitive circle is imagined to be the section of the sphere of

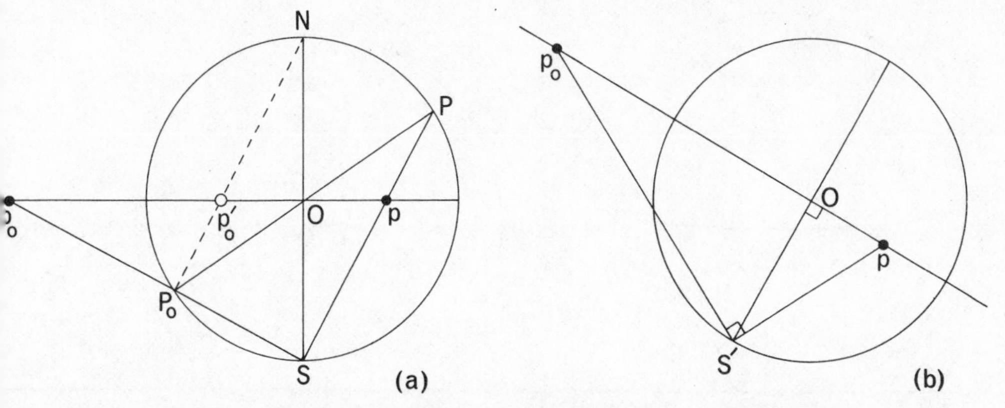

Fig A.2 To find the projection of the opposite P_0 of the pole P. (a) is the section of the sphere of projection containing POP_0 and perpendicular to the plane of projection; p and p_0 are respectively the stereographic projections of P and P_0 from the south pole; p_0' is the stereographic projection of P_0 from the north pole. (b) shows the sphere of projection rotated about the diameter through p to bring the south pole to S'; the intersection of Op with the perpendicular to S'p through S' locates p_0.

projection containing P, S, and N, the projection of the opposite of the pole P can be constructed in the following manner (Fig A.2(b)):

1. Draw OS' perpendicular to Op,
2. Draw $S'p_0$ perpendicular to pS',
3. p_0 lies at the intersection of $S'p_0$ with pO produced.

III: To construct a great circle through two poles (Fig A.3)

We shall first discuss the general case and then pass on to a special case.

a To construct a great circle through two general poles

A great circle is defined by specifying two non-opposite poles, P and Q (Fig A.3(a)), which lie on the great circle; the great circle also passes by definition through the centre of the sphere of projection. However, to draw a circle—and in general the projection of a great circle is an arc of a circle—it is necessary to locate three points lying on the circle. Now since the plane of a great circle passes through the centre, the opposite of any pole lying on the great circle also lies on the great circle. The construction is made in the following manner (Fig A.3(b)):

1. Construct the opposite p_0 of the pole p by construction II,
2. Construct the perpendicular bisectors gt and gt_0 of qp and qp_0,
3. The geometrical centre of the projected great circle is then g.

b To construct a great circle through two poles, one of which lies on the primitive

If the pole P through which the great circle is to pass lies on the primitive, its opposite P_0 lies on the primitive at the opposite end of the diameter of the primitive through P. The geometrical centre of the projected great circle therefore lies on the perpendicular bisector of this diameter POP_0, that is on the diameter perpendicular to the diameter POP_0. The great circle is constructed in the following manner (Fig A.3(c)):

1. Draw the diameter through P,
2. Draw the diameter ROR_0 perpendicular to POP_0,
3. Construct the perpendicular bisector gt of Pq,
4. The geometrical centre of the required great circle lies at the intersection g of gt and ROR_0.

Some part of the great circle will project outside the primitive unless the point of projection is changed from the south to the north pole for this part. That part of the great circle which lies within the primitive when projected from the north pole will contain the opposite of every pole on the other part of the great circle which lies within the primitive circle when projected from the south pole; the opposite of every pole lying on the arc PmP_0 lies on the arc P_0m_0P, POP_0 being

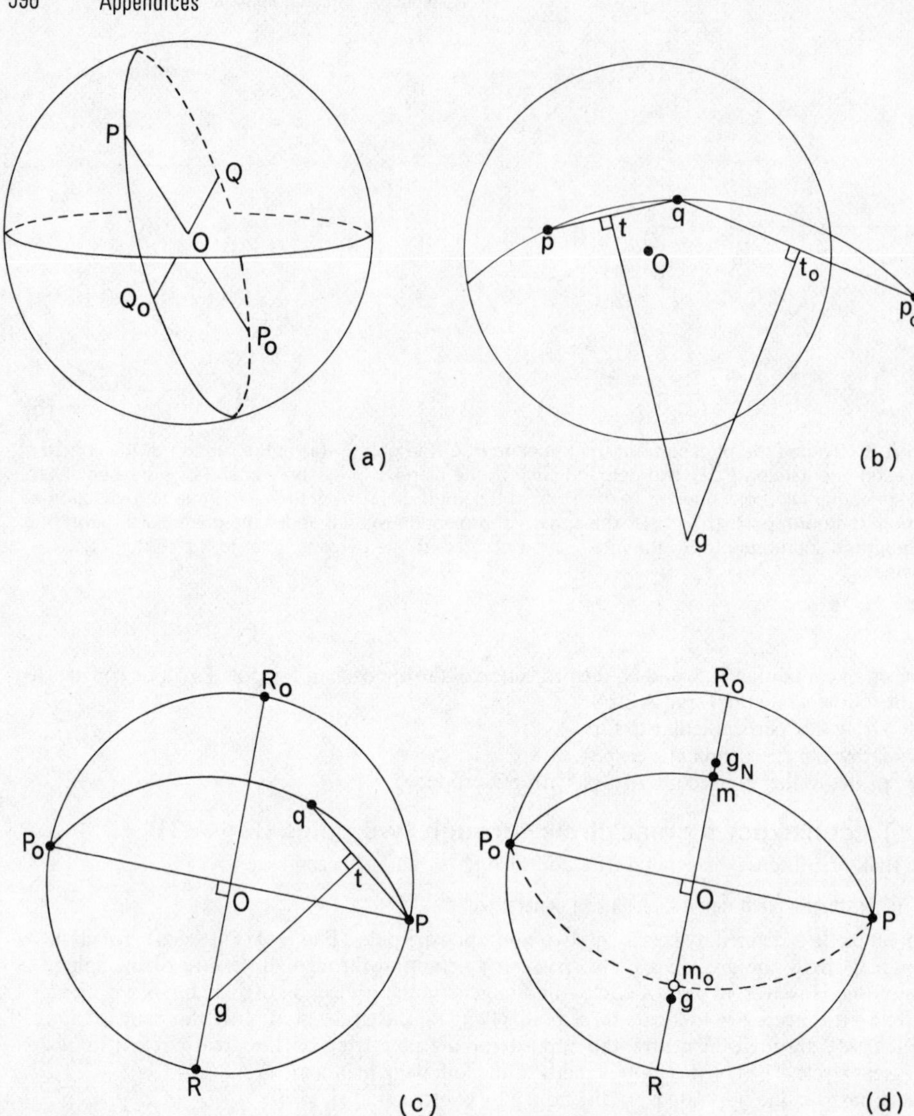

Fig A.3 To construct a great circle through two poles. (a) shows the sphere of projection with the plane containing P, Q, and their opposites P_0, Q_0 outlined. (b) illustrates the construction for the general case where p and q lie within the primitive; p_0 is the opposite of p; gt and gt_0 are the perpendicular bisectors of pq and qp_0 respectively; g is the geometrical centre of the required great circle. (c) illustrates the construction for the special case where one pole, P, lies on the primitive; ROR_0 is the diameter of the primitive perpendicular to OP; gt is the perpendicular bisector of qP. (d) illustrates the construction of that part of the great circle which has to be projected from the north pole in order to be brought within the primitive; the geometrical centre of this part of the great circle lies on the diameter ROR_0 of the primitive perpendicular to OP at g_N such that $Og = Og_N$. A great circle projected from the north pole is shown conventionally by a broken arc.

the diameter of the primitive in which the great circle intersects the equatorial plane in Fig A.3(d). In particular if m lies on the diameter perpendicular to POP_0 then m_0 also lies on that diameter on the opposite side of O so that $Om = Om_0$. It is apparent from the figure that the geometrical centres g and g_N of the great circle projected from the south and north poles respectively lie on the diameter ROR_0 on either side of O so that $Og = Og_N$.

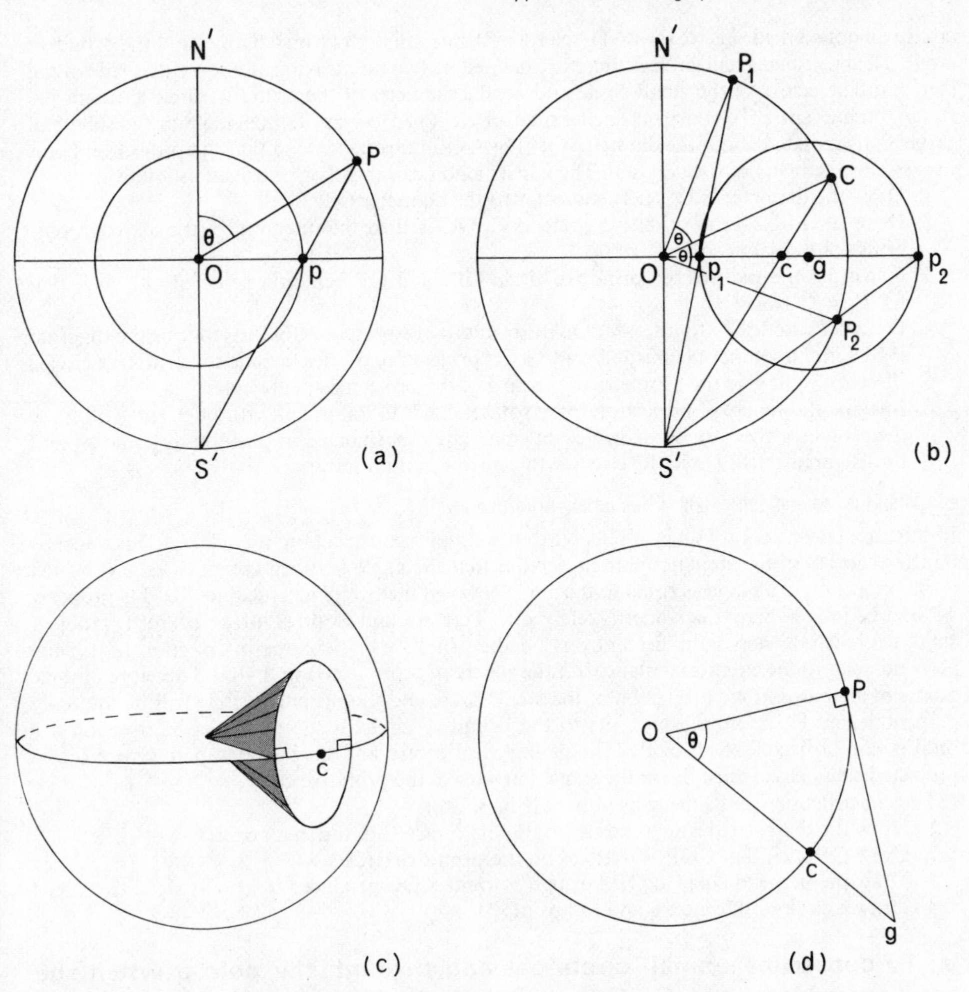

Fig A.4 To construct a small circle of radius $\theta°$ about a given pole. (a) illustrates the special case where the given pole is the north pole; the required small circle has its geometrical centre at O and passes through p where $\widehat{OS'p} = \frac{1}{2}\theta$. (b) illustrates the general case where the stereographic centre of the small circle is c; rotation through 90° about Oc brings the south pole to S′; S′c produced meets the primitive in C; $\widehat{COP_1} = \widehat{COP_2} = \theta°$; p_1 and p_2 are the stereographic projections from S′ of P_1 and P_2; the geometrical centre of the small circle lies on OC at g such that $p_1g = gp_2$. When the stereographic centre of the small circle lies on the primitive (c) a simple construction, shown in (d), becomes available: $\widehat{COP} = \theta$ and g lies at the intersection of the normal to OP through P with OC produced.

IV: To construct a small circle of radius $\theta°$ about a pole (Fig A.4)

a When the pole is the north pole

In this case the plane of the small circle is parallel to the equatorial plane and both the stereographic and geometrical centre of the projected small circle will coincide at N. All that is required to determine the radius of the projected circle is a pole plotted at $\theta°$ from N. The construction is performed in the following manner (Fig A.4(a)):

1. Plot the projection p of a pole P at $\theta°$ from N along any diameter by construction I,
2. Draw a circle with centre O and radius Op.

b When the stereographic centre lies within the primitive circle

In this case the geometrical centre g and the projection c of the stereographic centre of the small circle will not coincide (as we showed in chapter 2), but will both lie on a diameter of the primitive circle. The projected small circle is therefore defined by two points lying at θ° on either side of the stereographic centre of the small circle and on the diameter of the primitive circle through the stereographic centre. To construct such a small circle it is necessary to imagine that the sphere of projection is rotated about the diameter Oc (Fig A.4(b)) through 90° so that the projection point moves into the plane of the diagram. The construction can then be performed as follows:

1. Draw the diameter N′OS′ perpendicular to the diameter Oc.
2. Draw S′c to meet the primitive circle at C; OC is then the direction of the stereographic centre of the small circle.
3. Construct the radii of the primitive circle OP_1 and OP_2 on either side of OC such that $\widehat{COP_1} = \widehat{COP_2} = \theta$.
4. Draw $S'P_1$ and $S'P_2$ to intersect Oc in p_1 and p_2 respectively. (Notice that in the diagram P_2 lies in the southern hemisphere and so its projection, p_2, lies outside the primitive circle.)
5. Bisect p_1p_2 to give the geometrical centre g of the projected small circle.
6. Imagine the sphere of projection to be rotated back to its original attitude. This will leave the geometrical centre of the projected small circle unchanged in position at g and p_1 and p_2 also unaffected. Draw the circle with centre g and radius gp_1.

c When the stereographic centre lies on the primitive circle

In this case construction IVb is applicable, but a simpler construction is available. This depends on the property of the stereographic projection that the angle between two arcs, each of which is the projection of a plane, is equal to the angle between the two planes (chapter 2). This property will not be proved here; the reader is referred to Terpstra and Codd (1961, p. 12) for a proof. A small circle whose stereographic centre is on the primitive circle is the intersection of a plane perpendicular to the equatorial plane with the sphere of projection (Fig A.4(c)). Therefore at their points of intersection the projection of the small circle and the primitive circle will be mutually perpendicular. Therefore the tangents to the primitive circle at the points of intersection are radii of the projected small circle. The geometrical centre and the stereographic centre of the projected small circle must lie on the same diameter of the primitive circle.

The construction can be made as follows (Fig A.4(d)):

1. Draw the diameter through the stereographic centre C of the small circle.
2. Draw OP such that $\widehat{COP} = \theta$. P lies on the primitive circle.
3. Draw the perpendicular to OP through P to meet OC produced in g.
4. Draw a circle with centre g and radius gP.

V: To construct a small circle passing through the pole p when the stereographic centre C of the small circle lies on the primitive circle (Fig A.5)

Since the plane of the small circle is perpendicular to the equatorial plane the small circle will be symmetrical across the equatorial plane. It follows that if the small circle is projected using both north and south poles so that its projection lies wholly within the primitive circle, then that part projected from the north pole will be coincident with that projected from the south pole. The pole whose projection is coincident with p when the north pole is used as the projection point will, like p, lie on the small circle; to draw the projection of the small circle this pole has to be re-projected using the south pole as the projection point.

Since the geometrical centre of the projected small circle lies on OC the construction can be effected by imagining the sphere of projection to be rotated through 90° about its diameter Op and proceeding as follows (Fig A.5):

1. Draw the diameter through Op.
2. Draw the diameter N′OS′ perpendicular to Op.
3. Draw S′p to meet the primitive circle in P.
4. Draw N′P to meet Op produced in p_1 and imagine the sphere to be rotated back to its original attitude.
5. Draw the perpendicular bisector of pp_1 to meet OC produced in g.
6. Draw the circle with centre g and radius gp.

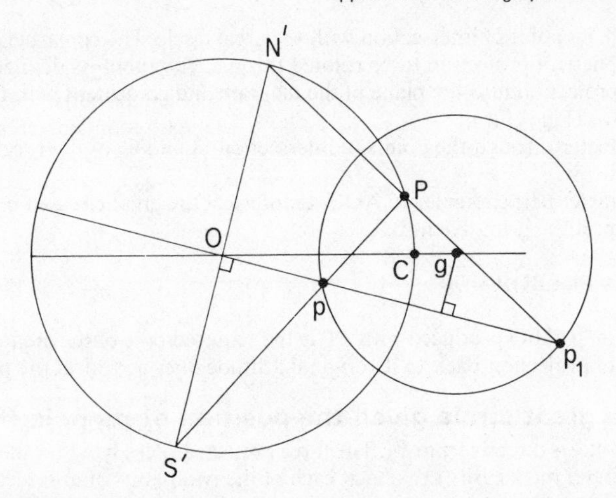

Fig A.5 To construct a small circle passing through the pole p when its stereographic centre C lies on the primitive circle. The sphere of projection is imagined to be rotated about Op to bring the north and south poles to N′ and S′. S′p meets the primitive in P and N′P meets Op produced in p_1. The perpendicular bisector of pp_1 meets OC produced in g, which is the geometrical centre of the required small circle.

VI: To find the pole of a great circle (Fig A.6)

By definition the pole of a great circle lies 90° from every point on the great circle and so represents the normal to the plane of the great circle. In Fig A.6(a) a great circle meets the primitive circle at the opposite ends of the diameter AC. Any direction perpendicular to the directions represented by the poles A and C must lie in a plane perpendicular to the line AC. This plane, which is perpendicular to the equatorial plane, will pass through the south pole and will therefore project as the diameter bOp of the primitive circle perpendicular to the diameter AOC. The pole of the great circle must also be perpendicular to the pole B, which lies at the intersection of the great circles ABC and BP.

The projection of the pole of a great circle can therefore be found by drawing the diameter of the primitive circle which is perpendicular to the diameter through the points of intersection of the great circle with the primitive circle and then measuring a distance equivalent to 90° along

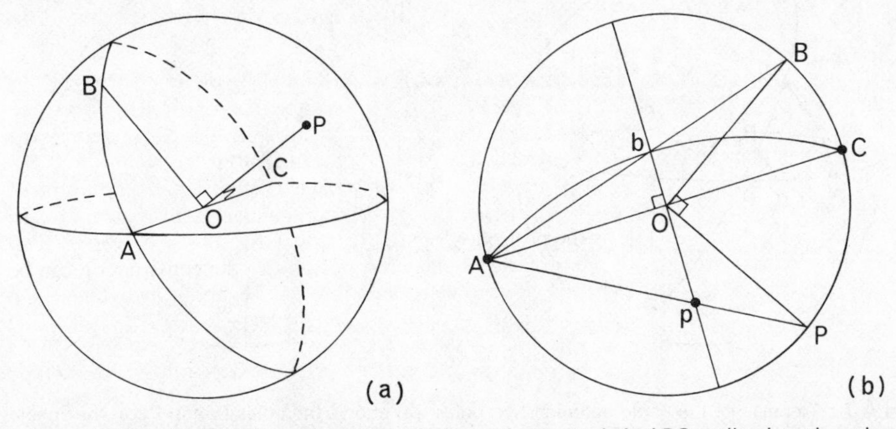

Fig A.6 To find the pole of a great circle. (a) shows the great circle ABC outlined on the sphere of projection; OP is the pole of this great circle. The stereographic projection (b) shows the great circle AC which intersects the diameter of the primitive perpendicular to AC in b; rotation of the sphere of projection through 90° about bO brings the south pole to A so that b is the stereographic projection of B; P lies on the primitive and $\widehat{BOP} = 90°$; AP intersects bO produced in p, which is the pole of the great circle. This figure serves also to illustrate the construction of a great circle whose pole is p.

the diameter from b, its point of intersection with the great circle. The construction is performed by imagining the sphere of projection to be rotated through 90° about its diameter bOp so as to bring the point of projection into the plane of the diagram and coincident with the point A; and proceeding as follows (Fig A.6(b)):

1. Draw the diameter through the points of intersection, A and C, of the great circle with the primitive circle.
2. Draw the diameter perpendicular to AOC, to intersect the great circle in b.
3. Draw Ab to meet the primitive in B.
4. Join OB.
5. Draw OP such that $\widehat{BOP} = 90°$.
6. Draw AP.
7. The intersection p of bO produced with AP is the required pole of the great circle. Rotation of the sphere of projection back to its original attitude does not affect the position of p.

VII: To draw a great circle given the position of its pole (Fig A.6)

This construction follows directly from VI. The three projected poles lying on the projected great circle that can be found most easily are one at each of the two points of intersection of the great circle with the primitive circle (these are at the ends of the diameter perpendicular to the diameter through the projection of the pole) and one on the diameter through the projection of the pole P. If the sphere of projection is imagined to be rotated through 90° about Op the construction can be performed as follows:

1. Draw the diameter Op.
2. Draw the diameter AC perpendicular to Op.
3. Draw the line Ap to meet the trace of the sphere of projection in P.
4. Draw OB such that $\widehat{BOP} = 90°$.
5. Draw AB.
6. The point of intersection b of AB with pO produced is the projection of a pole lying on the great circle.
7. Imagine the sphere of projection to be rotated back to its original attitude and draw the great circle through A, b, and C by construction IIIb.

VIII: To measure the angle between two poles (Fig A.7)

The angle between the directions represented by the poles P_1 and P_2 is $P_1\widehat{O}P_2$ (Fig A.7(a)). On the stereogram this angle is represented by the arc p_1p_2 of the great circle through p_1 and p_2; this

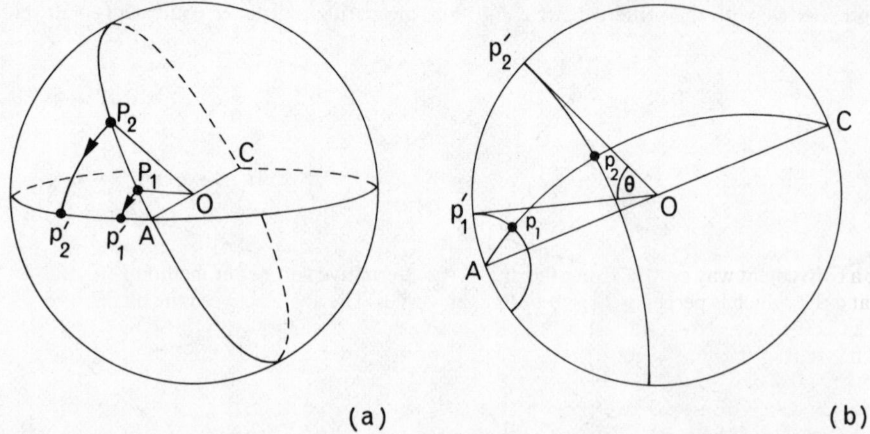

Fig A.7 To measure the angle between two poles. (a) shows the poles P_1 and P_2 on the sphere of projection; the great circle P_1P_2 intersects the plane of projection in the diameter AOC of the primitive; rotation about AOC brings P_1 and P_2 on to the primitive at p_1' and p_2'; $p_1'\widehat{O}p_2' = P_1\widehat{O}P_2$. (b) illustrates the construction: p_1 and p_2 are the stereographic projections of the given poles P_1 and P_2, and the great circle on which they lie intersects the primitive in AOC; a small circle with stereographic centre A is drawn through p_1 to intersect the primitive in p_1' and another small circle with the same stereographic centre through p_2 intersects the primitive in p_2'; the required angle is $p_1'Op_2'$.

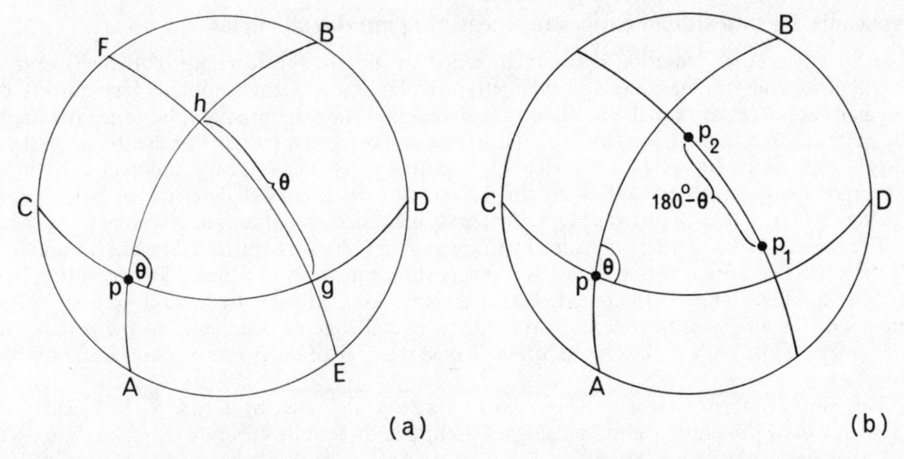

Fig A.8 To find the angle between two great circles, ApB and CpD as shown in (a). The great circle whose pole is p intersects the two great circles in h and g respectively. Measurement of the angle between h and g by the method illustrated in Fig A.7 gives the required angle. (b) illustrates an alternative construction: the poles p_1 and p_2 of the great circles ApB and CpD are located by the method illustrated in Fig A.6 (b); the angle between p_1 and p_2, found by the method illustrated in Fig A.7, is then the supplement of the required angle.

arc cannot be measured to give the angle directly. One method of measuring the angle is to rotate the poles on the sphere of projection about the diameter AC, keeping the equatorial plane and the projection points fixed, until the great circle containing p_1 and p_2 becomes the primitive circle; the poles P_1 and P_2 will then coincide with their projections and direct measurement of the angle P_1OP_2 becomes possible. During rotation the path of each pole will be a circle centred on the diameter that is the axis of rotation. Or in other words each pole will describe an arc of a small circle whose plane is perpendicular to the axis of rotation and whose stereographic centre is the axis of rotation. Thus to measure the angle between the projected poles p_1 and p_2 the steps in the construction are (Fig A.7(b)):

1. Construct the great circle Ap_1p_2C by construction IIIa.
2. Draw small circles, each with stereographic centre A to pass through p_1 and p_2 respectively by construction V.
3. Measure the angle $p'_1Op'_2$ between the points of intersection of the two small circles with the primitive. This is equal to the required angle P_1OP_2.

IX: To find the angle between two great circles (Fig A.8)

The angle between two great circles can be measured by drawing the tangents to the projections of the great circles at their point of mutual intersection (chapter 2), but this is neither an accurate nor a convenient way of measuring the angle. An alternative and better method is to draw in the great circle which is perpendicular to both great circles. The angle between the planes of the great circles is then given by the angle between their lines of intersection with this plane.

The steps in the construction needed to measure the angle between the two great circles ApB and CpD (Fig A.8(a)) are:

1. Construct the great circle EghF whose pole p is the intersection of the two great circles by construction VII.
2. Measure the angle between g and h, which are the intersections of the great circle EghF with the great circles AphB and CpgD respectively by construction VIII.

Yet another method is to find the poles of the two great circles and then to measure the angle between them. The angle between two planes is equal to the supplement of the angle between their normals or poles. The steps of the construction are (Fig A.8(b)):

1. Find the poles p_1 and p_2 of the two great circles, ApB and CpD, by construction VI.
2. Measure the angle between p_1 and p_2 by construction VIII.

Appendix B: Two simple devices for measuring interfacial angles

The two devices we describe here are intended as aids in the teaching of the elements of morphological crystallography and symmetry, enabling the student to obtain very quickly the angular data necessary for the plotting of a stereogram. In careful hands the measured angles are quite accurate enough for plotting on a stereogram of $2\frac{1}{2}$ inch radius. For greater accuracy of angular measurement it is necessary to use more refined optical goniometers, with a correspondingly greater expenditure of time; such instruments are well described by Terpstra and Codd (1961). The two-circle optical goniometer still has occasional uses in research.

The *contact goniometer* was invented by Carangeot in 1780. It is illustrated in Fig B.1 and fully described in the caption of that figure. It is a very simple instrument suitable for measuring large crystals and the wooden crystal models used in elementary teaching. Its accuracy (about $\pm 2°$) is limited by the ability of the operator to hold the crystal edge perpendicular to the plane of the goniometer while keeping the crystal firmly against the edge of the protractor and the adjustable arm.

The simple *optical goniometer* shown in Fig B.2 was designed by Dr. J. V. P. Long of the Department of Mineralogy and Petrology, Cambridge. It is quite accurate enough (about $\pm 1°$) for elementary teaching and for this purpose has two advantages over more elaborate instruments: its simplicity enables the student to set the crystal and obtain angular measurements very quickly, and its construction is such that it can be made very cheaply. The instrument, which is fully described in the caption to Fig B.2, is so designed that an image of the lamp filament is visible through the viewing tube when a reflecting surface lies normal to the axis of the viewing tube. Thus if a crystal is set with a zone axis parallel to the horizontal axis of the instrument, successive faces in the zone will be brought into the reflecting position as the shaft is rotated. By noting the angular readings on the dial r at which reflexions coincide with the cross-wires interfacial angles in this zone may be measured.

The procedure for setting a crystal on the simple optical goniometer may be described with reference to Fig B.2(c). (1) Set the instrument so that the axis A of the universal joint t is vertical. (2) Mount the crystal with plasticine so that face 1 in the selected zone is approximately horizontal and so that an edge between two adjacent faces in this zone is parallel to the axis of the horizontal shaft p. (3) Locate the image of the filament on the cross-wires by turning the horizontal shaft and adjusting the tilt of the crystal with the axis B of the universal joint. (4) Turn the horizontal shaft so as to bring the next face 2 into the reflecting position and locate the image of the filament on the cross-wires by adjusting the axis A of the universal joint. (5) Slide the magnet v on the steel block q so that the crystal remains in the centre of the field of view throughout 360° rotation of the horizontal shaft. When the crystal is set, measurements of interfacial angles

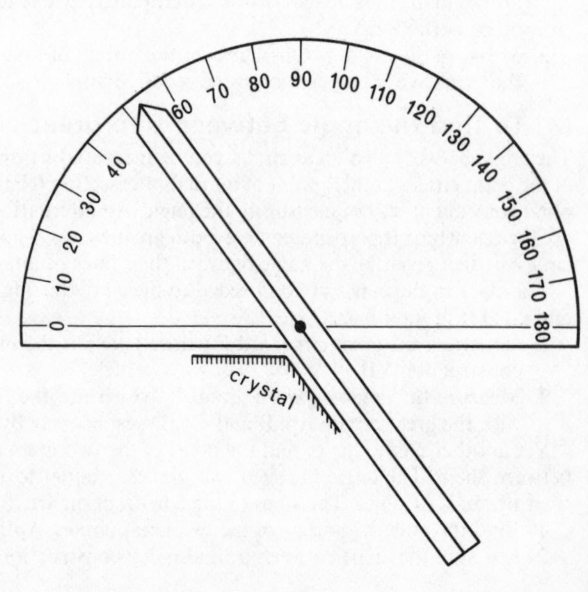

Figure B.1 The contact goniometer is constructed from a perspex protractor to which a perspex arm is attached so as to pivot about the centre of the protractor. The crystal is placed between the edge of the protractor on the 0° side and the arm so that the crystal edge is perpendicular to the plane of the protractor and the angle between face normals is read off. The goniometer is calibrated in degrees or half degrees (only 10° divisions are shown on the figure for simplicity). The interfacial angle of the crystal shown in exploded relationship to the goniometer is thus measured as $52\frac{1}{2}°$.

in the selected zone can be made. The crystal may then be remounted for measurement of another zone. The crystal should have dimensions in the range 5–10 mm and bright faces for convenient use of this instrument; suitable examples are octahedra of magnetite and cleavage rhombohedra of calcite.

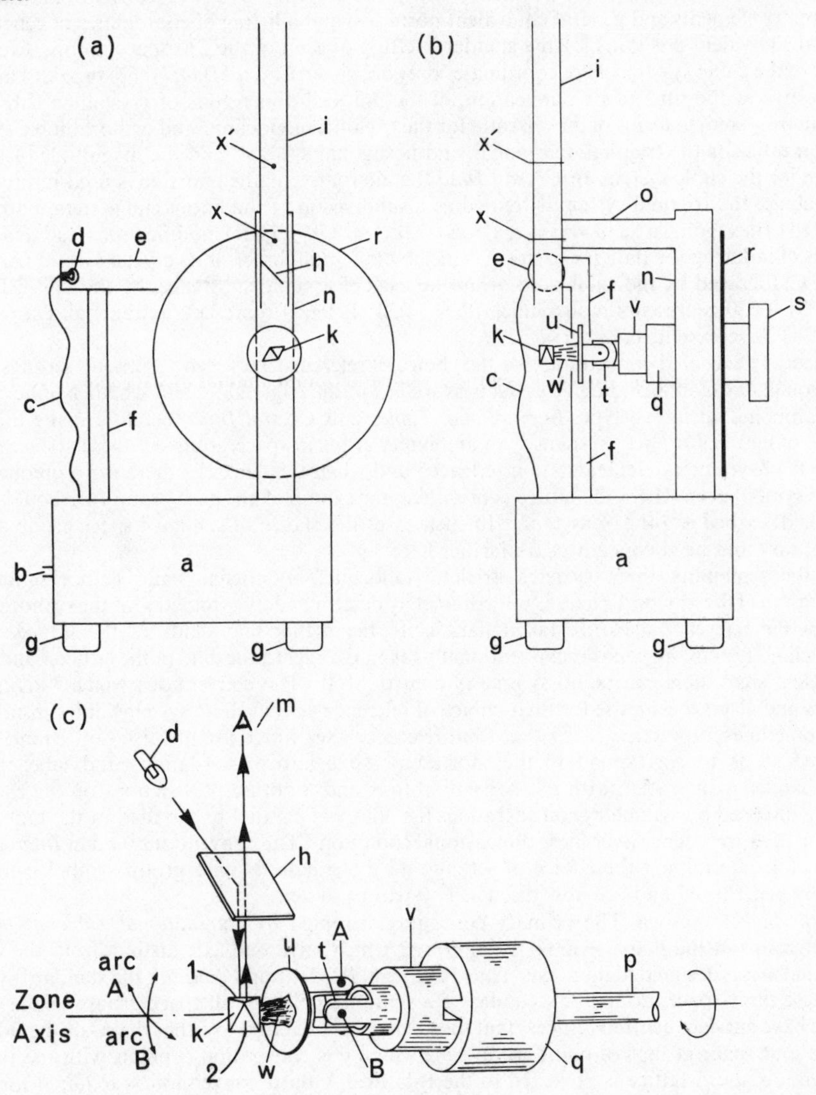

Fig B.2 The simple optical goniometer. (a) and (b) are respectively front and side elevations and (c) shows the detail of the crystal adjustment and optical paths. The transformer, contained in an alloy box *a*, is fed from the mains at *b* and its output passes through the flex *c* to the torch bulb *d*, which is set in the cylindrical tube *e* mounted adjustably on the rod *f*; the transformer box sits on rubber feet *g*. The incident light beam is reflected by the half-silvered cover-slip *h*, mounted in the square-section tube *i*, to impinge on the crystal *k*. When a face of the crystal is set horizontal the light beam is reflected up the viewing tube *i*, passing undeviated through the half-silvered plate to be observed at *m*. The viewing tube *i* is rigidly attached by the arm *o* to the pillar of square-section tube *n*. A horizontal shaft *p*, carrying at one of its ends the steel cylinder *q* and at the other the graduated dial *r* (a 360° protractor) and attached knob *s*, rotates on bearings set in the pillar *n*. The orientation of the crystal *k* can be adjusted by manual manipulation of the universal joint *t* which is cemented with epoxy resin to the steel disc *u* and to the magnet *v*. The crystal is centred by manual adjustment of the position of the magnet *v* on the steel cylinder *q*. The crystal is attached to the steel disc *u* by the plasticine *w*. The viewing tube carries cross-wires *x*, *x*′.

Appendix C: Rules for selecting standard settings of space groups in *International Tables for X-ray Crystallography* and in *Crystal Data*

International Tables for X-ray Crystallography, vol. 1 (1969) gives two conventional diagrams (symmetry elements and general equivalent positions) and a listing of coordinates of general and special equivalent positions for the standard setting of each of the 230 space groups, except for those of the cubic system where coordinates only are given. *Crystal Data* (1963) provides a listing, exhaustive at the time of its publication, of the unit-cell dimensions of crystalline substances, system by system in terms of the a/b ratio for the triclinic, monoclinic, and orthorhombic systems, the c/a ratio for the trigonal, tetragonal, and hexagonal systems, and the magnitude of the cell edge a for the cubic system. In *Crystal Data* the alternative name anorthic is used in preference to triclinic; the trigonal system is treated as a subdivision of the hexagonal system, both being referred to hexagonal axes ($a = b \neq c, \alpha = \beta = 90°, \gamma = 120°$). These modifications lead to a simple means of indexing the data for each substance by the initial letter of its crystal system (A, M, O, T, H, C) followed by the axial ratio or, in the case of a cubic substance, by the unit-cell edge. Thus M–1·020 represents monoclinic, $a/b = 1{\cdot}020$; T–0·690 represents tetragonal, $c/a = 0{\cdot}690$; and C–8·29 represents cubic, $a = 8{\cdot}29$ Å.

Generally accepted conventions for the choice of reference axes lead to unique settings in the tetragonal, hexagonal, and cubic systems as well as in the trigonal system whether of hexagonal or rhombohedral lattice type. *International Tables* and *Crystal Data* therefore make the same choice of unit-cell in these systems.[1] An ambiguity arises in space groups of these systems where two sets of symmetry elements are interleaved and when this happens there is no unique space group symbol even where the lattice is primitive; for example the space group $I\bar{4}c2$ could just as well be described as $I\bar{4}b2$ or as $I\bar{4}a2_1$. In such cases the choice of standard space group symbol is arbitrary and need concern us no further here.

In the remaining three systems, triclinic (anorthic), monoclinic, and orthorhombic, the orientation of the x, y, and z axes is not completely determined by symmetry: in the orthorhombic system the reference axes are taken parallel to the orthogonal diads of the lattice, in the monoclinic system the y-axis is conventionally taken parallel to the diad of the lattice,[2] and in the triclinic system there can be no symmetry control at all. The conventions which *International Tables* and *Crystal Data* use for their choice of reference axes in these systems differ in principle and sometimes in practice. In *Crystal Data* reference axes which are not fixed by symmetry are selected so as to correspond to the shortest possible lattice translations and labelled as a right-handed axial system with $c < a < b$, α obtuse, and β obtuse. Restrictions on the choice of setting imposed by symmetry considerations (i.e. that y is parallel to the diad in the monoclinic system) take precedence over these dimensional conditions. The conventions used in *International Tables* for determining the choice of setting and the standard space group symbol cannot be usefully generalized and are now discussed system by system.

Orthorhombic System. The primary convention adopted by *International Tables* is that for space groups in the point groups 222 and $mm2$ which have one axis distinct from the others; then that axis is designated the z-axis. Thus $P222_1$ and $P2_12_12$ and $Pmm2$ are the standard symbols for these three space groups. A secondary convention applies to all space groups of the system which have one-face centred lattices: that the C-lattice is preferred to either the A- or the B-lattice. In the four space groups of point group $mm2$ where this convention conflicts with the primary convention, the A-lattice is preferred to the B-lattice. A third convention is required for space groups which have non-primitive lattices and two sets of symmetry elements interleaved: the standard symbol is chosen so as to show that symmetry element which appears first in the following sequence, $m, a, b, c, n, d, 2, 2_1$. Thus the symbol $Cmcm$ is preferred to $Cbnn$. Apart from the application of these three conventions *International Tables* are not consistent in their choice of standard setting, the arbitrary standard settings of the old International Tables (1935) being preferred to the formulation of elaborate new rules.

The dimensional convention $c < a < b$ is rigidly adopted by *Crystal Data* for the orthorhombic system. In consequence A-, B-, and C-lattices are all permissible and the distinct axis in space

[1] For rhombohedral space groups in the trigonal system *International Tables* give space group symbols, conventional diagrams, and coordinates of general and special equivalent positions referred to hexagonal axes and *in addition* coordinates of general and special equivalent positions referred to rhombohedral axes.

[2] *International Tables* provide a description of each monoclinic space group also in the alternative setting with the z-axis parallel to the diad of the lattice.

groups of the point groups 222 and *mm* may be *x*, or *y*, or *z*. For all three point groups of the system the space group symbols produced by application of this convention may be in non-standard form.

Monoclinic System. The convention that the non-primitive lattice is a C-lattice is consistently adopted in *International Tables.* Since space groups with a C-lattice have two sets of symmetry elements interleaved the same convention is adopted as for the analogous situation in the orthorhombic system. A further convention is applied to space groups with a primitive lattice and a glide plane: the glide plane is taken to be a *c*-glide.

The convention $c < a$ and β obtuse is rigidly adopted by *Crystal Data.* In consequence the non-primitive lattice type may be C, A, or I and so some space group symbols will be non-standard.

Triclinic (*Anorthic*) *System.* No conventions are required for the choice of reference axes for the two space groups $P1$ and $P\bar{1}$ in *International Tables.* However in *Crystal Data,* where a/b is tabulated for triclinic substances, conventions are required to determine which lattice repeats are to be labelled a and b. The conventions adopted are that the three shortest non-coplanar lattice repeats are taken as the directions of the reference axes so as to form a right-handed axial system with $c < a < b$, α obtuse, β obtuse. For a discussion of conventions for choosing a unit-cell in the triclinic system the reader is referred to Kelsey and McKie (1964).

Appendix D: Spherical trigonometry: the equations for a general triangle

Let a sphere, centre O, intersect three of its radii in A, B, and C, the angles between the radii being $\widehat{BOC} = a$, $\widehat{COA} = b$, $\widehat{AOB} = c$ (Fig D.1). The sphere intersects the planes BOC, COA, AOB in the great circle arcs BC, CA, AB which form the sides of the *spherical triangle* ABC.

Select a point D on OB produced such that AD $\perp$ OA and a point E on OC produced such that AE $\perp$ OA. From the plane triangle ODE,

$$DE^2 = OE^2 + OD^2 - 2OE.OD\cos\widehat{DOE}$$

and from the plane triangle ADE,

$$DE^2 = AE^2 + AD^2 - 2AE.AD\cos\widehat{DAE}.$$

Thus $$OE^2 + OD^2 - 2OE.OD\cos\widehat{DOE} = AE^2 + AD^2 - 2AE.AD\cos\widehat{DAE}.$$

But AD $\perp$ OA and AE $\perp$ OA, therefore

$$OD^2 - AD^2 = OA^2 = OE^2 - AE^2,$$

whence $$OA^2 = OE.OD.\cos\widehat{DOE} - AE.AD.\cos\widehat{DAE}$$

i.e. $$\cos\widehat{DOE} = \frac{OA}{OE}\cdot\frac{OA}{OD} + \frac{AE}{OE}\cdot\frac{AD}{OD}\cdot\cos\widehat{DAE}$$

i.e. $$\cos\widehat{DOE} = \cos\widehat{AOE}.\cos\widehat{AOD} + \sin\widehat{AOE}.\sin\widehat{AOD}.\cos\widehat{DAE}.$$

Thus $$\cos a = \cos b.\cos c + \sin b.\sin c.\cos A.$$

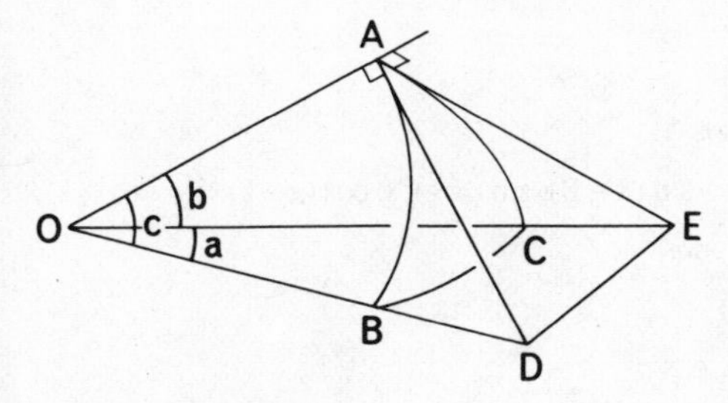

Fig D.1 A, B, C are points on the surface of a sphere of centre O, and define the spherical triangle ABC. The plane ADE is perpendicular to OA. The inter-radial angles a, b, c correspond to the angular lengths of the sides BC, CA, AB of the spherical triangle.

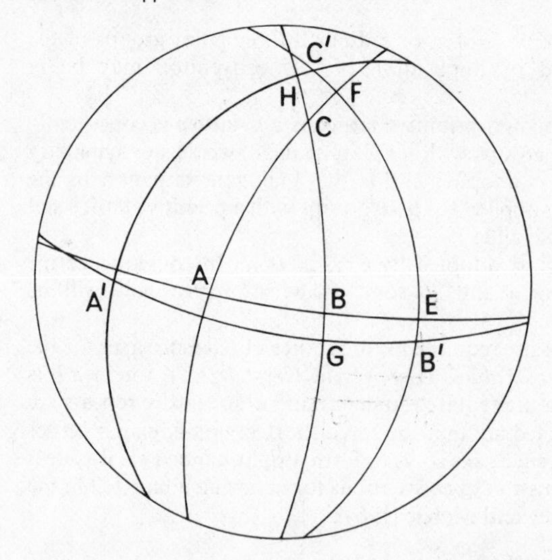

Fig D.2 Stereographic projection showing the bounding great circles of the spherical triangle ABC. The great circles whose poles are A, B, C are respectively B'C', C'A', A'B'; their intersections define the polar triangle A'B'C' of the triangle ABC. ABC is itself the polar triangle of A'B'C'. The two great circles through A intersect the great circle whose pole is A in E and F. The two great circles through A' intersect the great circle whose pole is A' in G and H.

By exactly similar argument we can derive the expressions

$$\cos b = \cos c . \cos a + \sin c . \sin a . \cos B$$

and

$$\cos c = \cos a . \cos b + \sin a . \sin b . \cos C.$$

In order to derive the remaining equations for the general spherical triangle it is necessary to construct the *polar triangle* A'B'C' of the triangle ABC, which is defined as the triangle formed by the intersection of the great circles whose poles are A, B, and C (Fig D.2). Since A is the pole of the great circle B'C' and C the pole of the great circle A'B',

$$A{:}B' = C{:}B' = \tfrac{1}{2}\pi.$$

Therefore B' is the pole of the great circle AC. The triangle ABC is therefore the polar triangle of its own polar triangle A'B'C'.

Let the great circles AB and AC intersect the great circle B'C' in E and F respectively. Then E:F = A and

$$C'{:}E = F{:}B' = \tfrac{1}{2}\pi,$$

whence $a' = B'{:}C' = B'{:}F + F{:}C' = B'{:}F + E{:}C' - E{:}F.$

Thus $a' = \pi - A.$

Similarly if the great circles A'B' and A'C' intersect the great circle BC in G and H respectively.

$$G{:}C = B{:}H = \tfrac{1}{2}\pi$$

and $A' = G{:}H = G{:}C + C{:}H = G{:}C + B{:}H - B{:}C.$

Thus $A' = \pi - a.$

By analogous arguments it can be shown that

$$b' = \pi - B \quad \text{and} \quad c' = \pi - C,$$

$$B' = \pi - b \quad \text{and} \quad C' = \pi - c.$$

Now for the general spherical triangle A'B'C',

$$\cos a' = \cos b' . \cos c' + \sin b' . \sin c' . \cos A'.$$

Therefore $\cos(\pi - A) = \cos(\pi - B) . \cos(\pi - C) + \sin(\pi - B) . \sin(\pi - C) . \cos(\pi - a)$

i.e. $\cos A = -\cos B . \cos C + \sin B . \sin C . \cos a$

and similarly

$$\cos B = -\cos C . \cos A + \sin C . \sin A . \cos b$$

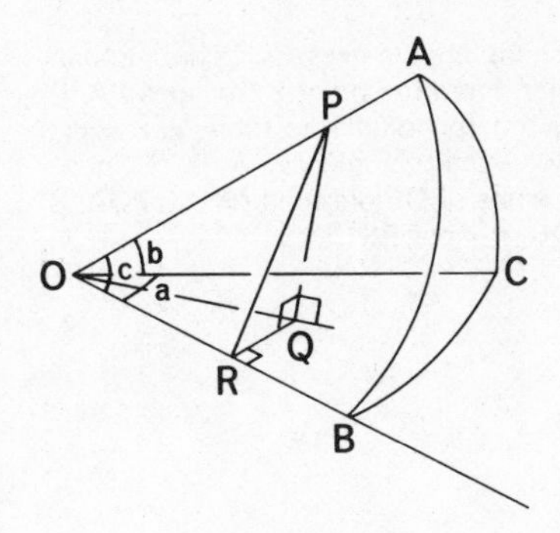

Fig D.3 A, B, C are points on the surface of a sphere of centre O. PQ is perpendicular to the plane BOC and QR is perpendicular to OB. The angle $\widehat{PRQ}$ is thus equal to the angle at the corner B of the spherical triangle ABC.

and $\quad \cos C = -\cos A . \cos B + \sin A . \sin B . \cos c.$

To prove the remaining equations for the general triangle it is convenient to use a different construction. Fig D.3 shows three radii from the centre of the sphere O intersecting the surface of the sphere in A, B, and C. As before the angles $\widehat{BOC}$, $\widehat{COA}$, $\widehat{AOB}$ are a, b, c respectively. From a point P on OA a perpendicular is drawn to the plane BOC to meet the plane in Q and QR is drawn perpendicular to OB to intersect OB in R. Since $\widehat{OQP}$, $\widehat{ORQ}$, and $\widehat{PQR}$ are right angles

$$OP^2 = OQ^2 + PQ^2,$$
$$OQ^2 = OR^2 + RQ^2,$$
$$PR^2 = RQ^2 + PQ^2.$$

Thus $\quad OP^2 = OR^2 + PR^2.$

Therefore $\quad \widehat{ORP} = \tfrac{1}{2}\pi.$

Now $\quad \widehat{PRQ} = B$

therefore $\quad PQ = PR \sin B$

$$= OP \sin c . \sin B.$$

In a precisely similar manner it can be shown that

$$PQ = OP \sin b . \sin C.$$

Thus $\quad \dfrac{\sin B}{\sin b} = \dfrac{\sin C}{\sin c}.$

By an analogous argument it can be shown that

$$\frac{\sin A}{\sin a} = \frac{\sin B}{\sin b}.$$

Appendix E: Three-dimensional analytical geometry

I: Derivation of the expression $l^2 + m^2 + n^2 = 1$, where l, m, n, are the direction cosines of a line referred to orthogonal axes

The *direction cosines* of a line are defined as the cosines of the angles that the line makes with the positive direction of each of the three reference axes, x, y, and z.

We consider a line defined by its direction cosines l, m, n and draw a line OP parallel to it and passing through the origin O (Fig E.1). From the point P a perpendicular is dropped to each axis to intersect the x-axis in A, the y-axis in B, and the z-axis in C. The direction cosines of OP are then by definition,

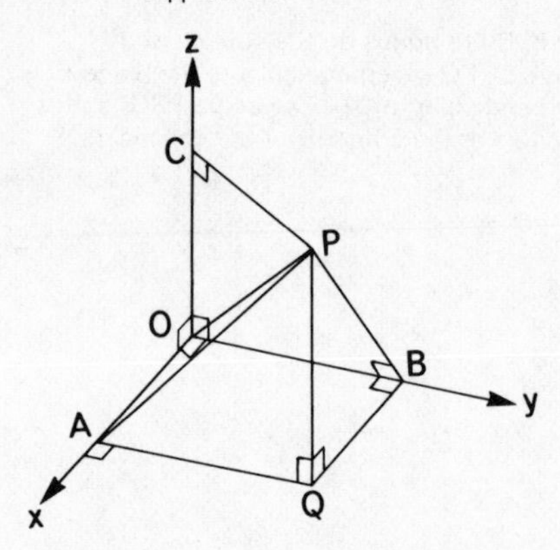

Fig E.1 The reference axes x, y, z are mutually perpendicular. From the point P the lines PA, PB, PC are drawn perpendicular to the x, y, z axes respectively, which they meet in A, B, C. The direction cosines of OP are then $l = \cos \widehat{POA}$, $m = \cos \widehat{POB}$, $n = \cos \widehat{POC}$.

$$l = \cos \widehat{POA} = \frac{OA}{OP}, \qquad m = \cos \widehat{POB} = \frac{OB}{OP}, \qquad n = \cos \widehat{POC} = \frac{OC}{OP}.$$

But for orthogonal axes by Pythagoras' Theorem (Fig E.1),

$$OA^2 + OB^2 + OC^2 = OP^2.$$

Therefore $$\left(\frac{OA}{OP}\right)^2 + \left(\frac{OB}{OP}\right)^2 + \left(\frac{OC}{OP}\right)^2 = 1$$

and so $$l^2 + m^2 + n^2 = 1.$$

II: Derivation of the expression for the angle between two lines given their direction cosines referred to orthogonal axes

We consider two lines OP_1 and OP_2 defined by their direction cosines $l_1 m_1 n_1$ and $l_2 m_2 n_2$ respectively, P_1 and P_2 being each at a distance r from the origin O (Fig E.2). The coordinates of P_1 are then $x_1 = rl_1$, $y_1 = rm_1$, $z_1 = rn_1$ and of P_2 are $x_2 = rl_2$, $y_2 = rm_2$, $z_2 = rn_2$. If the origin is moved from O to P_1, the coordinates of P_2 referred to P_1 as origin will be $x_2' = x_2 - x_1$, $y_2' = y_2 - y_1$, $z_2' = z_2 - z_1$. Therefore

$$\begin{aligned}(P_1P_2)^2 &= (x_2 - x_1)^2 + (y_2 - y_1)^2 + (z_2 - z_1)^2 \\ &= r^2\{(l_2 - l_1)^2 + (m_2 - m_1)^2 + (n_2 - n_1)^2\} \\ &= r^2\{2 - 2(l_1 l_2 + m_1 m_2 + n_1 n_2)\}.\end{aligned}$$

Now $$\cos\theta = 1 - 2\sin^2 \tfrac{1}{2}\theta$$

i.e. $$\cos\theta = 1 - 2\left(\frac{P_1P_2}{2r}\right)^2.$$

Thus $$\cos\theta = l_1 l_2 + m_1 m_2 + n_1 n_2.$$

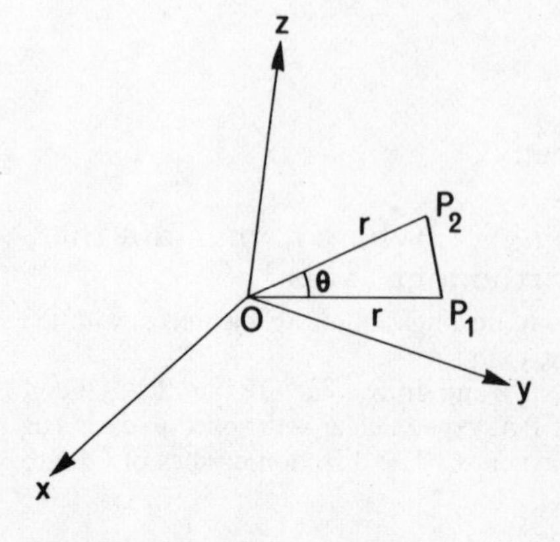

Fig E.2 The reference axes x, y, z are mutually perpendicular. P_1 and P_2 are placed at the same distance r from the origin O so as to define two lines OP_1 and OP_2, which make an angle θ with one another.

Appendix F: Crystal setting

Any account of X-ray diffraction by single crystals, such as that provided in chapter 8, would be incomplete without some description of the techniques used for mounting and setting a crystal in a particular orientation. No attempt is made here to give a comprehensive survey of all the techniques available for mounting and setting single crystals; we restrict ourselves to those techniques which we ordinarily use and which we find both easy to teach to others and suitable for setting a crystal very quickly in a chosen orientation with respect of the camera geometry. In this appendix we assume an understanding of the reciprocal lattice and the reflecting sphere, without which it would be pointless to engage in practical single crystal X-ray diffraction studies.

The first task is always to select a single crystal and to identify in it some prominent zone axis. For this purpose optical goniometry, in the case of well shaped (euhedral) crystals, may be useful and polarized light microscopy, in the case of transparent crystals whether euhedral or not, is particularly useful. By the use of either of these techniques it is usually possible, for instance, to identify the principal axis of a uniaxial crystal; we shall assume initially that it is desired to mount and set such a crystal about such an axis. Transparent crystals of lower symmetry and opaque shapeless crystals of any system present a more difficult problem, the solution of which is achieved by the use of just the same methods but takes more time.

Crystal mounting

We start with a crystal lying in a pool of refractive index oil on a microscope slide on the stage of a binocular microscope. The magnification should be in the range $\times 10$ to $\times 40$. The crystal is eased to the edge of the pool of oil and beyond with the aid of a fine needle point mounted in a wooden handle (a fine artists' brush from which the camel hairs have been removed makes a very convenient handle). A drop of amyl acetate—or some other volatile solvent in which the crystal is insoluble—is placed close to the crystal and the crystal is pushed with the needle point through the pool of amyl acetate and out towards the edge of the slide. The crystal is moved with the needle point until the selected zone axis is approximately perpendicular to the edge of the microscope slide.[1] When the last trace of amyl acetate has evaporated and the crystal is quite dry, a small drop of a slow setting adhesive (Durofix thinned with amyl acetate is eminently suitable) is placed on the slide close to the crystal. A glass fibre, about 15 mm long, $< 0{\cdot}05$ mm diameter, and with one of its ends pushed into a pea-sized blob of plasticine, is then dipped in the drop of adhesive and quickly brought into contact with the crystal so that the glass fibre is approximately parallel to the selected zone axis of the crystal. If this rather delicate operation has been correctly performed the crystal will be stuck firmly on the tip of the glass fibre within rather less than one minute of making contact (if the operation is performed too slowly, the crystal will merely be held by surface tension to the surface skin of the adhesive droplet and will fall off if tapped against the slide; it is then necessary to start again). The crystal and its fibre are then mounted on a set of crystallographic arcs (Fig 8.1) by pressing the blob of plasticine against the column at the top of the arc assembly.

Crystal setting with a zone axis parallel to the spindle axis of the arcs

For both oscillation and Weissenberg photography it is necessary to align a zone axis in the crystal parallel to the camera axis. Initially we shall consider situations in which the selected zone axis diverges by no more than a few degrees of arc from the camera axis; later on we shall discuss gross mis-setting.

The crystal is first centred at the intersection of the camera axis with the incident X-ray beam. The spindle carrying the arcs is then rotated manually until the plane of one arc is approximately parallel to the incident X-ray beam, the plane of the other arc being then perpendicular to the incident X-ray beam; the spindle is locked in this position and a Laue photograph is taken. If the selected zone axis has been set precisely parallel to the camera axis, the reciprocal lattice net corresponding to the normals to planes lying in this zone will be horizontal and so the reflexions

[1] Instead of a hand-held needle point a 'micro-manipulator' may be used. Micro-manipulators, which are available commercially, are in essence three-dimensional pantographs; they are useful for dealing with exceptionally small crystals but are an unnecessary luxury for persons with steady hands dealing with crystals of the sort of size usually used for X-ray diffraction.

(a)

(b)

(c)

(d)

(e)

mushroom

collimator

film

X

O Z

∥ axis

2ψ

⊥a

corresponding to the intersection of this equatorial net with the reflecting sphere will lie in a horizontal plane; thus when the cylindrical Laue photograph is laid flat after development this prominent 'zone' of reflexions will be seen to lie on a straight (horizontal) line perpendicular to the (vertical) camera axis (Fig F.1 (a)). If, however, the selected zone axis is inclined to the camera axis, the reciprocal lattice nets normal to it will not be horizontal. The reflections in the equatorial layer will then lie on the surface of a shallow cone, whose axis is the zone axis and whose apex is at the centre of the reflecting sphere. The incident X-ray beam will lie on the surface of this cone. The cone will thus intersect the cylindrical film in a curve which passes through the point of intersection of the forward direction of the incident beam with the film. The shape of such a curve is, in general, complex and it is convenient to discuss it with respect to two simple situations: where the zone axis is in the plane perpendicular to the incident X-ray beam containing the camera axis (Fig F.1 (b)) and where the zone axis is in the plane containing the incident X-ray beam and the camera axis (Fig F.1 (c)).

In Figs F.1 (a)–(d) the film is shown in each case with the observer imagined to be looking *towards* the X-ray tube and the edge of the film which was uppermost in the camera uppermost in the figure. When removing the film from the cassette for development it is necessary to indicate its orientation in the camera; this can conveniently be done by holding the cassette as it was in the camera with the exit-hole towards oneself and scratching a distinctive mark (such as X or √) on the top right-hand corner of the film. Some means of indicating the intersection of the film with the plane perpendicular to the camera axis containing the incident X-ray beam (the 'horizontal' plane) is required and this is achieved by slipping a brass 'mushroom' (Fig F.1 (e)) over the collimator when setting photographs are being taken; the 'mushroom' casts an arcuate shadow on each side of the film, each shadow being symmetrical about the equatorial line (Figs F.1 (a)–(d)).

If the arc whose plane is perpendicular to the incident X-ray beam (i.e., its axis is parallel to the incident beam) is observed looking towards the X-ray tube, then the mis-setting illustrated in Fig F.1 (b) will require for its correction an anticlockwise adjustment of that arc only. The mis-setting illustrated in Fig F.1 (c) is such that the equatorial net of the reciprocal lattice (which of course passes through the origin O) slopes upwards towards the X-ray tube; in this case the equatorial reflexions lie on a curve which is symmetrical about the vertical through the exit-hole of the film. Correction of this mis-setting is achieved by adjustment of the arc whose plane is parallel to that of the incident X-ray beam and the camera axis (i.e., its axis is perpendicular to the incident beam) in a clockwise sense when the observer moves to the right through 90° from his previous position.

We now consider the general case where adjustments on both arcs are required (Fig F.1 (d)). We shall show how the amount of adjustment on each arc can be deduced from measurements on the film in this general case and in the special cases considered qualitatively in the preceding paragraph. The problem is complicated by the dependence of the inclination of the upper arc (Fig 8.1) to the camera axis on the inclination of the lower arc to the camera axis; only when the scale reading of the lower arc is zero is the axis of the upper arc perpendicular to the camera axis.

Fig F.1 Setting a crystal on an oscillation camera so that a selected zone axis Z is parallel to the spindle axis of the camera. (b)–(d) show diagrammatically the appearance of the film for various mis-settings; each film carries a tick in the top right-hand corner to indicate its orientation in the camera during exposure; the shadows cast by the 'mushroom', illustrated in (e), are indicated by arcs at the sides of each film, and serve to define the equatorial line; no reflexions are shown on the diagrams, only a continuous curve representing the zero layer reflexions. In (a) the crystal is perfectly set. (b) shows the curves of the zero layer when the zone axis Z is inclined at an angle $\rho = 10°$ (or 20°) to the spindle axis O in the plane normal to the incident X-ray beam; the stereogram on the right shows the great circle corresponding to the zero layer reciprocal lattice net. (c) shows the curves of the zero layer when the zone axis Z is inclined at an angle $\psi = 10°$ (and 20°) to the spindle axis in the plane of the incident beam and the spindle axis; the stereogram on the right shows the zero-layer reciprocal lattice net projected as a small circle of radius $90° - \psi$ passing through the opposite of X. (d) shows a general case of mis-setting and the orientation of the corresponding zero-layer reciprocal lattice net. In (b)–(d) the displacement of the zero-layer curve at points corresponding to $\theta = \pm 45°$ are indicated as Δ_1 on the left and Δ_2 on the right-hand side of the film: in (b) $\Delta_2 = -\Delta_1$ and $\rho = 2\Delta_1$, in (c) $\Delta_2 = \Delta_1$ and $\psi = 2\Delta_1$, and in (d) $\rho = \Delta_1 - \Delta_2$ and $\psi = \Delta_1 + \Delta_2$ where Δ_1, Δ_2 are measured in mm (positive above the equatorial line) and ρ, ψ are in degrees for a camera of radius 28·65 mm.

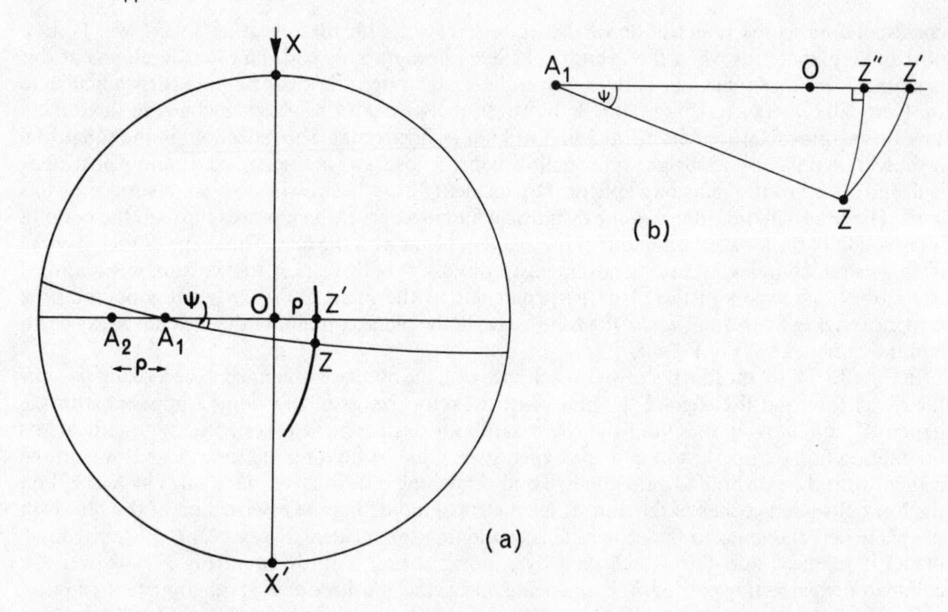

Fig F.2 In the stereogram (a) the forward direction of the incident beam is XX′, the spindle axis of the camera is O, the selected zone axis is Z, and the initial attitude of the axis of the upper arc is A_1. Movement of the crystal through the angle ψ on the upper arc causes Z to move on the small circle (shown bold) through Z, whose stereographic centre is A_1, until it reaches Z′ on the great circle normal to XX′. After bringing the zone axis to Z′ it is moved into coincidence with O by a movement ρ on the lower arc, which will move the axis of the upper arc from A_1 to A_2. (b) exaggerates the distinction between the small circle ZZ′ and the great circle XZ″, which intersects the great circle normal to XX′ in Z″. In calculating the corrections to be made to the arcs it is assumed that Z′ and Z″ are coincident.

In Fig F.2(a), the zone axis is represented by the pole Z, the camera or oscillation axis by the pole O at the centre of the stereogram, and the incident X-ray beam by the poles XX′; the lower arc is taken to be set with its axis parallel to the incident beam and the axis of the upper arc is represented by the pole A_1. To bring the zone axis into coincidence with the camera axis it is necessary to move the crystal through the angle ψ in a clockwise sense about the axis of the upper arc, where ψ is the angle between the great circles A_1Z and A_1O. The effect of this adjustment is to transfer the pole of the zone axis to Z′, which lies in the plane normal to the axis XX′ of the lower arc; Z′ thus lies at the intersection of the small circle through Z whose stereographic centre is A_1 with the great circle whose pole is X. The crystal is then rotated through the angle ρ in an anticlockwise sense about the axis of the lower arc to transfer the zone axis from Z′ to O; this adjustment of the lower arc moves the axis of the upper arc from A_1 to A_2 through the angle ρ in the plane normal to X. The great circle XZ is normal to the plane A_1O which it intersects in Z″ (Fig F.2(b)). We shall assume for simplicity that Z′ and Z″ are coincident, which is tantamount to assuming that the zone axis Z is normal to the axis A_1 of the upper arc; if this were precisely so, then $ZZ'' = \psi$ and $OZ'' = \rho$. This simplifying assumption leads to underestimation of the adjustments, ρ and ψ, of the arcs but greatly simplifies the calculation of the magnitudes, necessarily approximate, of ρ and ψ from measurements of the setting photograph.

We now consider the general case of a reflexion R produced by a lattice plane in the zone whose axis is Z (Fig F.3(a)). Since the direction R of the reflected beam and the forward direction X′ of the incident X-ray beam lie on the surface of a cone whose axis is the zone axis Z, $ZX' = ZR = \eta$, where η is the semi-angle of the cone, and $RX' = 2\theta$, where θ is the Bragg angle of the reflexion. Therefore for the general spherical triangle RX′Z,

$$\cos ZR = \cos RX' . \cos ZX' + \sin RX' . \sin ZX' . \cos \widehat{RX'Z}$$

i.e. $$\cos \eta = \cos 2\theta . \cos \eta + \sin 2\theta . \sin \eta . \cos \varepsilon$$

where $\widehat{RX'Z} = \varepsilon$;

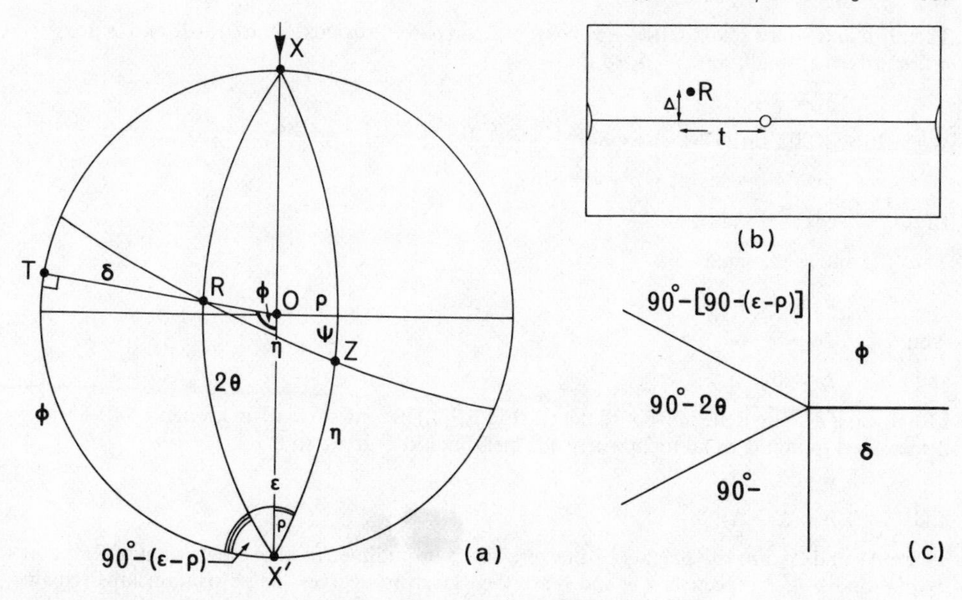

Fig F.3 The stereogram (a) illustrates the general case of a reflexion R produced by a lattice plane in the zone whose zone axis is Z; the spindle axis of the camera is O; and the forward direction of the incident beam is XX′. The angular distance of the intersection of the great circle X′Z with the great circle normal to XX′ from Z is ψ and from O is ρ. The Naperian diagram for the right-angled spherical triangle TX′R ($\widehat{RTX'} = 90°$) is shown in (c). The definition of the coordinates of the reflexion R on the film are shown in (b): $\Delta = r \tan \delta$ and $t = \phi r$, where r is the camera radius. Angular relations in (a) are grossly exaggerated for clarity.

whence $\quad 2 \sin^2 \theta \cos \eta = 2 \sin \theta \cos \theta . \sin \eta . \cos \varepsilon$

so that $\quad \cos \varepsilon = \cot \eta . \tan \theta.$

Adoption of the convention that ψ is positive when $\eta < 90°$ (i.e. the correction of magnitude ψ is made in the clockwise sense when the observer moves to the right through 90° from the position facing the X-ray tube) leads to the equation $\eta = 90° - \psi$, so that

$$\cos \varepsilon = \tan \psi . \tan \theta.$$

The reflected beam R intersects the cylindrical film at a point with coordinates t, Δ (Fig F.3(b)), where t is the horizontal distance of the reflexion from the centre of the exit hole of the main beam and Δ is the vertical distance of the reflexion above the equatorial line. Now the coordinate $t = \phi r$, where r is the camera radius and $\phi = \widehat{X'OR} = XT$ (Fig F.3(a)); and $\Delta = r \tan \delta$, where $\delta = TR$. From the Naperian triangle TX′R (Fig F.3(c))

$$\sin \delta = \sin 2\theta . \cos (\varepsilon - \rho)$$

and $\quad \sin (\varepsilon - \rho) = \tan \phi . \cot 2\theta$

i.e. $\quad \tan \phi = \tan 2\theta . \sin (\varepsilon - \rho).$

Since we already have the relationship $\cos \varepsilon = \tan \psi . \tan \theta$, we can calculate δ and ϕ for any scattering angle 2θ if ψ and ρ are known for the zone and so the shape of the zero layer curve of a mis-set crystal can be calculated. But the converse procedure of calculating ρ and ψ from the observed shape of the zero layer curve, which is the problem with which we are concerned here, is laborious; it may be simplified by considering merely the coordinates of the zero layer curve for $\theta = \pm 45°$. When $2\theta = 90°$, $\cos \varepsilon = \tan \psi$ and

$$\begin{aligned} \sin \delta_1 &= \cos (\varepsilon - \rho) \\ &= \cos \varepsilon . \cos \rho + \sin \varepsilon . \sin \rho \\ &= \tan \psi . \cos \rho + \sin (\cos^{-1} \tan \psi) . \sin \rho. \end{aligned}$$

Then if ρ and ψ are small, $\tan\psi \to \psi$, $\cos\rho \to 1$, $\sin\rho \to \rho$, $\sin(\cos^{-1}\tan\psi) \to 1$, and, since δ is necessarily also small, $\sin\delta_1 \to \delta_1$ so that

$$\delta_1 = \psi + \rho$$

When $2\theta = -90°$, $\sin 2\theta = -1$ so that

$$\sin\delta_2 = -\cos\varepsilon.\cos\rho - \sin\varepsilon.\sin\rho$$

and $$\cos\varepsilon = -\tan\psi.$$

Then if ρ and ψ are small,

$$\delta_2 = \psi - \rho.$$

Thus $$2\psi = \delta_1 + \delta_2$$

and $$2\rho = \delta_1 - \delta_2.$$

Moreover, if δ is small the film coordinate Δ (Fig F.3(b)) becomes $\Delta = r\delta$ and, when $r = 28{\cdot}65$ mm, Δ measured in mm is equal to 2δ where δ is measured in degrees, so that

$$\psi = \Delta_1 + \Delta_2$$

and $$\rho = \Delta_1 - \Delta_2,$$

where Δ_1 and Δ_2 are the heights of the zero layer in millimetres above the equatorial line at $\theta = 45°$ and $-45°$ respectively; ψ and ρ are then given in degrees. These relationships remain good approximations for cylindrical cameras of 30 mm radius, always assuming that ρ and ψ are small angles.

It is, in practice, often convenient to be able to derive the corrections to be made to the two arcs directly from the observed shape of the zero layer curve. The displacement of the curve from the equatorial line at $2\theta = 180°$ is dependent only on the arc whose plane is parallel to the incident X-ray beam. Thus if the zero layer curve lies above the equatorial line at the edges of the film (the exit-hole or the shadow of the back-stop being in the centre of the film), then the zone axis is tilted away from the collimator (as illustrated in Fig F.1(c)). The curve close to the intersection of the forward direction of the incident beam with the film (i.e. the exit-hole or the shadow of the back-stop) is approximately a straight line; the arc whose plane is normal to the incident beam has to be rotated to achieve the required correction in the sense which will cause this line to become parallel to the equatorial line. Thus in the cases illustrated in Fig F.1(b) an anticlockwise rotation of $\rho°$ is required to correct the mis-setting of the crystal.

The zero layer curve intersects the equatorial line in general in two points, at $\theta = 0$ and at a point whose θ value depends on the relative magnitudes of the mis-settings on the two arcs. If $|\psi| = |\rho|$ the second point of intersection is at $2\theta = +90°$ or at $2\theta = -90°$. If the second point of intersection lies at $|2\theta| > 90°$, the greater error is in the arc whose plane is perpendicular to the incident beam (as illustrated in Fig F.1(d)); but if the second intersection is at $|2\theta| < 90°$ the greater error is in the arc whose plane is parallel to the incident beam. It is thus only necessary to measure the displacements, Δ_1 and Δ_2, of the zero layer curve from the equatorial line at $2\theta = 90°$ and at $2\theta = -90°$; to take their sum and difference $\Delta_1 + \Delta_2$ and $\Delta_1 - \Delta_2$; and so to deduce the necessary corrections, ρ and ψ, for the mis-setting of the crystal.

The equations derived earlier, $\psi = \Delta_1 + \Delta_2$ and $\rho = \Delta_1 - \Delta_2$, apply only for small mis-settings and are only good approximations when the angle between the axis of the upper arc and the camera axis approaches 90°. They do however always lead to corrections in the right sense so that by use of these simple equations successive correction will achieve the desired result of bringing the zone-axis into coincidence with the camera axis. Moreover the range of possible angles between the axis of the upper arc and the camera axis, $90° \pm 30°$, has little effect on the accuracy of the applied corrections, provided the corrections are small. In practice the initial correction of a grossly mis-set crystal can only be very approximate, but thereafter the simple equations become quite accurate so that the crystal can be accurately set in a small number of operations.

For precision setting of a crystal, as is necessary before moving film photographs are taken, the precise positioning of the equatorial line on the film is essential. This can readily be achieved when the crystal is very nearly set by use of the *double Laue photograph.* A Laue photograph is taken with the arcs as nearly as possible parallel and perpendicular to the incident beam, then the crystal is rotated anti-clockwise about the camera axis through 179° and a second Laue

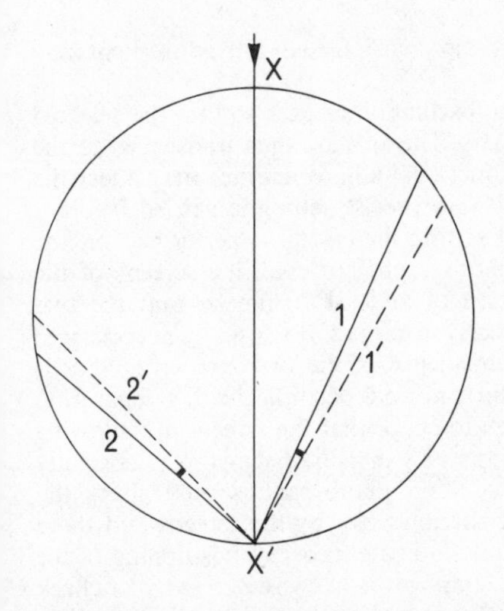

Fig F.4 The double Laue method. The figure shows the reflecting circle for two positions of the crystal. The directions labelled 1 and 2 represent reciprocal lattice rows in the zero layer net for the initial orientation of the crystal; those labelled 1′ and 2′ are the corresponding directions after the crystal has been rotated about the spindle axis (normal to the plane of the figure) in the anticlockwise sense through 179° (the angles between 1 and 1′, 2 and 2′ are grossly exaggerated for clarity). The directions of the reflected beams are given by the corresponding radii of the reflecting circle.

photograph of about one half of the exposure time of the first is taken on the same film. The reflexions of the first exposure will lie outside the weaker reflexions of the second exposure on the right-hand side of the film and relatively inside on the left-hand side (Fig F.4). The displacement of the reflexions coupled with their intensity difference enables the zero layer curves corresponding to the two exposures of the film to be readily recognized. Had the rotation between the first and second exposures of the film been precisely 180° the two curves would have been mirrored in the equatorial line; but making the rotation just less than 180° serves to separate the reflexions and so make the interpretation of the double Laue photograph easier. The relative displacement is however, small, so that the equatorial line can be precisely located and, in particular, measurement of the separation of the two curves at $2\theta = \pm 90°$ yields precise values of $2\Delta_1$ and $2\Delta_2$, which are both necessarily small, so that the final corrections ψ and ρ can be made accurately. Successive use of double Laue photographs until the two curves become linear and coincident with the equatorial line enables the selected zone axis to be set precisely parallel to the camera axis. We emphasize the point that the double Laue technique is only profitably applicable when the rough setting methods described earlier have brought the selected zone axis almost into coincidence with the camera axis. Rough setting is usually adequate for the taking of oscillation photographs; precision setting by the double Laue method is a necessary preliminary to the taking of Weissenberg photographs.

In the initial stages of the setting of a grossly mis-set crystal it is often advantageous to use oscillation rather than Laue photographs even though fewer reflexions are recorded. Layer lines are discernible on an oscillation photograph of even a grossly mis-set crystal; and from even the very rough estimate that can be made of the layer line spacing in such a case it is usually possible to see whether these layer lines refer to the zone axis one is trying to set parallel to the camera axis. Against this advantage must be balanced the information—or rather at this stage of setting, the suggestions—about the symmetry of the crystal that can be gleaned from a Laue photograph. In practice it is usually a question of deciding whether it will be easier to set the crystal by recognition of a characteristic, usually large, layer line spacing on an oscillation photograph, or by recognition of some characteristic symmetry, usually a mirror plane, on a Laue photograph. In making the decision one bears in mind what one already knows from morphology and optics about the probable orientation of the crystal. If one approach is not rapidly fruitful, one tries the other.

When the selected zone axis has been set parallel to the camera axis, it may then be necessary to locate a particular symmetry direction within the zero layer. This is most simply done by inspecting a setting Laue photograph for symmetry, adjusting the orientation of the arcs by rotation of the camera spindle until the selected symmetry axis is parallel to the incident beam, and taking another Laue photograph to confirm. If the selected symmetry axis lies at a distance s millimetres from the centre of the exit-hole on the zero layer line of the setting Laue, it will be necessary to adjust the camera spindle in the appropriate sense through $s(90/\pi r)$ degrees, where r is the camera radius in millimetres, i.e., $0{\cdot}955s$ for a camera of 30 mm radius and s for a camera of 28·65 mm radius. The Laue photograph taken after the adjustment has been made will display the Laue symmetry of the crystal along this direction. At this stage a further fine adjustment may be necessary.

For *Weissenberg* photography the crystal is set on an oscillation camera so that the selected zone axis is parallel—within about 1°—to the camera axis. The arcs are then transferred to the Weissenberg camera. Transference of arcs from one camera spindle to another may affect the alignment of the crystal by as much as 1° and anyway very precise setting is needed before a Weissenberg photograph is taken so the final stage of setting the crystal is performed on the Weissenberg camera. For this purpose double Laue photographs are used, the screens of the Weissenberg camera being moved back to a separation of at least 20 mm so that the two zero layer curves of the double Laue photograph are clearly displayed. It is usually necessary to take at least two double Laue photographs before coincidence of the two zero layer lines is achieved. It is then necessary to return the screens to their normal positions, with a separation of 2–4 mm, before taking a ±20° oscillation photograph to check that the screens are allowing free passage of the diffracted beams of the zero layer line and that the back-stop is correctly placed to trap the undeviated X-ray beam. This oscillation photograph should show the zero layer reflexions symmetrically placed between the shadows cast by the screens and there should be no excessive blackening of the film, which would indicate incorrect positioning of the back-stop. Before taking upper layer Weissenberg photographs it is likewise necessary to check with a ±20° oscillation photograph that the screens have been correctly positioned so that the reflexions of the upper layer line are symmetrically disposed between the shadows cast by the screens and that the back-stop has been correctly positioned.

For *precession* photography it is necessary to set a zone axis parallel to the incident beam so that the spindle axis has to be parallel to some direction in the corresponding zone. It is moreover convenient to have a reciprocal lattice row of the selected zone set parallel to the spindle axis. Where the selected reciprocal lattice row is parallel to a zone axis (e.g. the reciprocal lattice row *h*00 is parallel to [100] in an orthorhombic crystal) it is often convenient to set the crystal with that zone axis parallel to the spindle axis on an oscillation camera. The arcs are then transferred to a precession camera so that the selected reciprocal lattice row is approximately parallel to the spindle axis of the precession camera. In the course of setting the crystal on the oscillation camera it may become clear which direction in the zero layer will have to be set parallel to the incident beam on the precession camera; this will usually be so when the direction is a symmetry axis.

If the crystal is not accurately aligned with a reciprocal lattice row parallel to the spindle axis and a zone axis parallel to the incident X-ray beam, it can be brought into alignment by taking

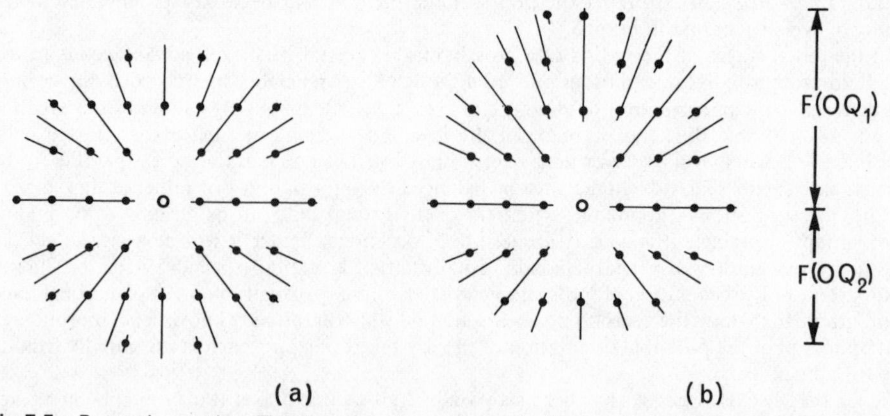

Fig F.5 Precession setting. The diagrams show schematically the appearance of 10° precession photographs taken with unfiltered radiation and without a screen. In (a) the crystal is precisely aligned with a reciprocal lattice row parallel to the spindle axis and a zone axis parallel to the incident beam; the distant ends of the white radiation streaks lie on a circle. In (b) the crystal is mis-set so that the distant ends of the white radiation streaks lie on a non-circular figure; the long and short axes of this figure measured on the film perpendicular to the line parallel to the spindle axis are $F(OQ_1)$ and $F(OQ_2)$; the quantity required for correcting the mis-setting is $F\Delta = F(OQ_1) - F(OQ_2)$. To avoid confusion in these diagrams non-zero layer streaks have been omitted.

a precession photograph with $\bar{\mu} = 10°$, white radiation, and no screen. If the crystal is precisely aligned the zero layer reflexions will lie within a circle of radius $2F \sin \bar{\mu}$, which will be clearly defined by the disposition of the associated white radiation streaks (Fig F.5(a)). If the crystal is slightly mis-set, the circle is distorted into a loop (Fig F.5(b)), the asymmetry of which can be utilized to correct the mis-setting. Consider the section (Fig F.6(a)) through the reflecting sphere containing the incident beam XIO and the direction IN_1, which is inclined at $\bar{\mu}$ to IO. Then OP_1 represents the line of intersection of the zero layer reciprocal lattice net with the plane of the figure when the crystal is perfectly set; OP_1' represents the line of intersection of the zero layer reciprocal lattice net with the plane of the figure for the mis-set crystal. Let IN_1' be the normal from I to OP_1' and ε the magnitude of the angle N_1IN_1'. Then since $\widehat{OIN_1} = \widehat{N_1IP_1} = \bar{\mu}$ and

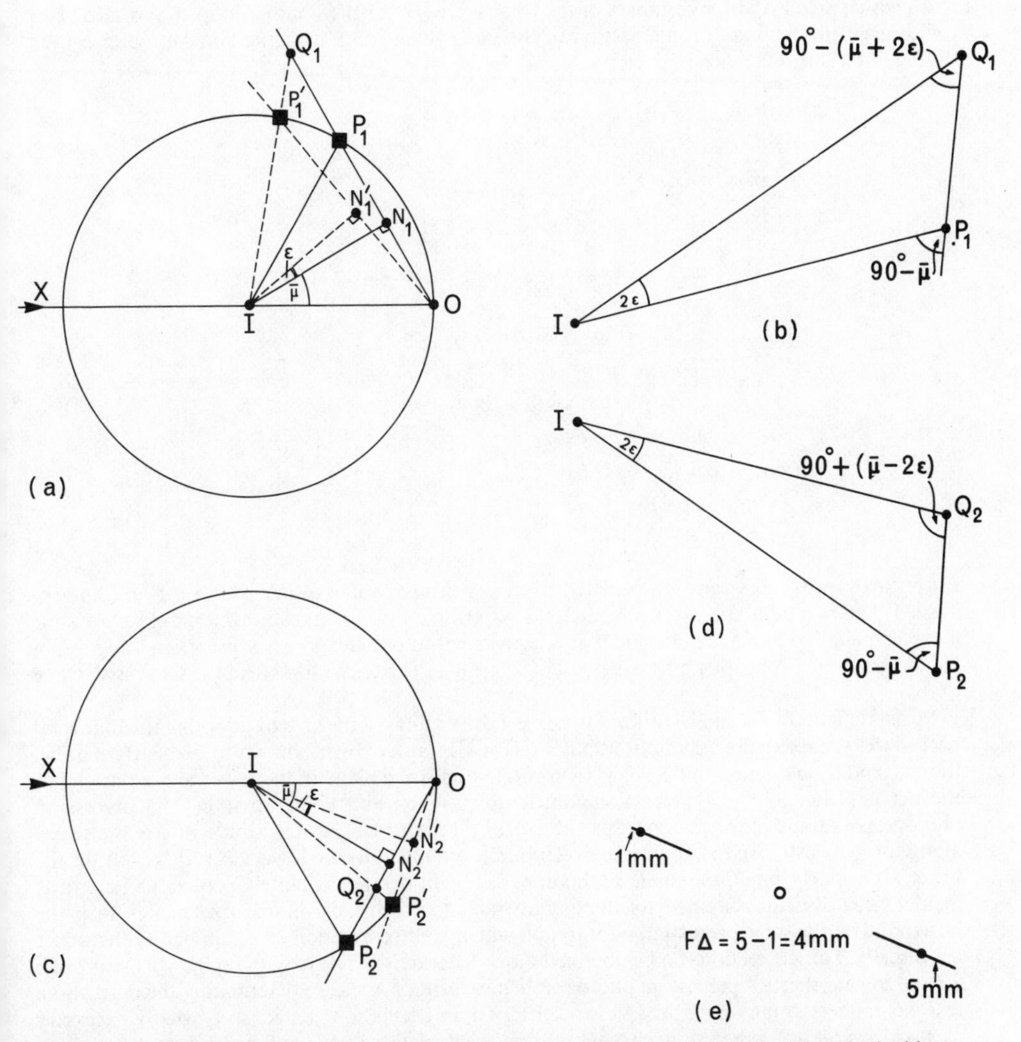

Fig F.6 Precession setting. (a) is a section through the reflecting sphere containing the incident beam XIO and the normal IN_1 to the reciprocal lattice row OP_1 for the perfectly set crystal; when the crystal is mis-set (exaggerated on the figure), P_1 becomes P_1' and N_1 becomes N_1'; Q_1 is the intersection of IP_1' produced with OP_1 produced. The plane triangle IP_1Q_1 shown in (b) is an enlargement of the same triangle in (a). (c) and (d) are the corresponding diagrams when P_2 lies below XIO. (e) illustrates a convenient means of evaluating $F\Delta$ by measurement of the white radiation streaks beyond a pair of centrosymmetrically related $K\alpha$ reflexions lying in an appropriate direction; the centre of the film is indicated by the open circle.

$\widehat{OIN'_1} = \widehat{N'_1IP'_1} = \bar{\mu}+\varepsilon$, $\widehat{P_1IP'_1} = 2\varepsilon$ and $\widehat{IP_1N_1} = 90°-\bar{\mu}$. Let Q_1 be the intersection of IP'_1 produced with OP_1 produced. Then in the triangle IP_1Q_1 (Fig F.6(b)) $\widehat{IQ_1P_1} = 90°-\bar{\mu}-2\varepsilon$ and by the sine rule

$$\frac{P_1Q_1}{IP_1} = \frac{\sin 2\varepsilon}{\cos(\bar{\mu}+2\varepsilon)}.$$

Similarly it can be shown that when the normal to the tangent plane at O for $\bar{\mu} = 0$ lies below XIO (Fig F.6(c), (d)), then

$$\frac{P_2Q_2}{IP_2} = \frac{\sin 2\varepsilon}{\cos(\bar{\mu}-2\varepsilon)}.$$

The difference $F\Delta$ in the distances of the opposite edges of the limiting loop of the zero layer reflexions from the point of intersection of the undeviated beam with the film can be measured on the film. This difference is given by

$$\begin{aligned} F\Delta &= F(OQ_1-OQ_2) \\ &= F(OP_1+P_1Q_1-OP_2+P_2Q_2) \\ &= F(P_1Q_1+P_2Q_2), \end{aligned}$$

since $\quad OP_1 = OP_2 = 2\sin\bar{\mu}$.

And since $IP_1 = IP_2 = 1$,

$$\begin{aligned} F\Delta &= F\sin 2\varepsilon\left(\frac{1}{\cos(\bar{\mu}+2\varepsilon)}+\frac{1}{\cos(\bar{\mu}-2\varepsilon)}\right) \\ &= F\sin 2\varepsilon\left(\frac{\cos(\bar{\mu}-2\varepsilon)+\cos(\bar{\mu}+2\varepsilon)}{\cos(\bar{\mu}+2\varepsilon).\cos(\bar{\mu}-2\varepsilon)}\right) \\ &= \frac{2F\sin 2\varepsilon.\cos\bar{\mu}.\cos 2\varepsilon}{\cos^2\bar{\mu}.\cos^2 2\varepsilon-\sin^2\bar{\mu}.\sin^2 2\varepsilon} \\ &= \frac{F\sin 4\varepsilon.\cos\bar{\mu}}{\cos^2 2\varepsilon-\sin^2\bar{\mu}}. \end{aligned}$$

From this expression ε can be calculated from a measured value of $F\Delta$, F and $\bar{\mu}$ being known. Table F.1 gives a tabulation of $F\Delta$ and ε for $F = 60$ mm, $\bar{\mu} = 10°$. As the variation of ε with $\bar{\mu}$ for a particular value of $F\Delta$ is quite small, it is convenient in practice to use a precession angle $\bar{\mu}$ of 10°. A graph of $F\Delta$ against ε provides a quick means of determining ε from a measured value of $F\Delta$.

Measurements of $F\Delta$ can be utilized to correct the settings of the arcs by adopting the following procedure. We make the assumption that a reciprocal lattice row is known to be nearly parallel to the spindle axis. The spindle axis is then rotated until the plane of the lower arc is parallel to the incident beam and a 10° precession photograph is taken with white radiation. This precession photograph should show at least this reciprocal lattice row. The magnitude of $F\Delta$ measured along this row will give the angular correction ε to be made to the lower arc.[2] If the tilt on the lower arc is fairly small, the plane of the upper arc will be approximately perpendicular to the incident beam so measurement of the angle which the reciprocal lattice row makes with the horizontal on the same precession photograph will give the correction to be applied to the upper arc directly. This correction can be obtained more accurately by rotating the spindle axis through 90°, taking another 10° precession photograph, measuring $F\Delta$ and so determining the correction, ε, to be applied to the upper arc. As for setting on an oscillation camera it is usually necessary to take several 10° precession photographs in each of these orientations and so by making successive corrections bring the selected reciprocal lattice row into precise alignment with the spindle axis of the camera; in particular when the reading on the lower arc departs far from zero, the calculated correction ε to the upper arc is not accurate and so the upper arc can only be corrected by successive approximation.

[2] It is often convenient to evaluate $F\Delta$ by selecting two centrosymmetrically related $K\alpha$ reflexions and then measuring the difference in the lengths of the white radiation streaks extending outwards from these $K\alpha$ spots (Fig F.6(e)).

When a reciprocal lattice row has been set accurately parallel to the spindle axis, any desired reciprocal lattice net containing this row can be recorded by setting the spindle dial at the appropriate angle. Errors in this setting may be determined by measuring $F\Delta$ in the 'vertical' direction of the film to determine the angle ε through which the setting of the dial must be rotated in order to bring a prominent reciprocal lattice net parallel to the plane of the film.

If the orientation of the crystal with respect to the reciprocal lattice is unknown, then a series of 10° precession photographs has to be taken with white radiation at 5° intervals of the dial setting until a major reciprocal lattice net is located. The same procedure may be used to locate reciprocal lattice directions in a crystal fragment of unknown orientation. The crystal is mounted on arcs, both arcs being set at zero, and 10° precession photographs are taken at 5° intervals of dial reading until a reciprocal lattice net with a high density of points is located. The crystal is then remounted so that this net is parallel to the plane of one of the arcs and the normal setting procedure is begun.

Table F.1. Angular setting error in precession photographs. The table gives values of ε corresponding to measured values of $F\Delta$ for $\bar{\mu} = 10°$ and $F = 60$ mm. An extended table is given in *International Tables*, vol 2 (1959), p. 200.

F (mm)		*F* (mm)	
0	0	10	2°20′
1	0°14′	11	2°34′
2	0°28′	12	2°47′
3	0°42′	13	3°01′
4	0°56′	14	3°15′
5	1°10′	15	3°29′
6	1°24′	16	3°43′
7	1°38′	17	3°56′
8	1°52′	18	4°10′
9	2°06′	19	4°24′

Appendix G: Units and constants

In this book we have not attempted to make use of SI units systematically for two reasons. Firstly, the Ångstrom is in crystallography, and especially in crystal chemistry, a more useful unit than the nanometre (1 nm = 10 Å). Secondly, almost all the literature of thermochemistry is written in terms of the calorie rather than the joule (1 J = 0·239 cal). For these reasons of convenience and readability of the literature we have preferred a system of mixed units in the text. In this appendix we relate these units to SI units and give the values of some useful constants.

Summary of SI units

length	metre	m (nanometre = 10^{-9} m)
mass	kilogramme	kg
time	second	s
electric current	ampere	A
thermodynamic temperature	kelvin	K
energy	joule	$J = kg\,m^2\,s^{-2}$
force	newton	$N = kg\,m\,s^{-2} = J\,m^{-1}$
electric charge	coulomb	$C = A\,s$
electric potential difference	volt	$V = kg\,m^2\,s^{-3}\,A^{-1} = J\,A^{-1}\,s^{-1}$
customary temperature, t	degree Celsius	°C t°C = TK − 273·15
pressure	newton per square metre	$N\,m^{-2}$

Conversion factors

length:	Ångstrom unit	1 Å = 10^{-8} cm = 10^{-1} nm
	micron	1 μ = 10^{-4} cm = 1 μm
density:	grammes per cc	1 $g\,cm^{-3}$ = $10^3\,kg\,m^{-3}$
pressure:	atmosphere	1 atm = 1·0133 bar = 101·33 $kN\,m^{-2}$
	bar	1 bar = $10^5\,N\,m^{-2}$
energy:	erg	1 erg = 10^{-7} J
	calorie	1 cal = 4·184 J = 41·84 bar cm^3
	electron volt	1 eV = 1·6021.10^{-19} J = 23·061 kcal $mole^{-1}$
free energy:	1 kcal deg^{-1} $mole^{-1}$ = 4·184.10^3 $J\,K^{-1}\,mol^{-1}$	
entropy:	entropy unit	1 e.u. = 1 cal deg^{-1} $mole^{-1}$ = 4·184 $J\,K^{-1}\,mol^{-1}$

Constants

π = 3·14159	π^{-1} = 0·31831
1° = 0·01745 radian	1 radian = 57·296°
e = 2·71828	
ln 10 = 2·30259	$\log_{10}$ e = 0·43429
Planck's constant	h = 6·6256.10^{-27} erg s = 6·6256.10^{-34} J s
Avogadro's number	N = 6·02252.10^{23} $mole^{-1}$
Boltzmann's constant	k = 1·3805.10^{-16} erg deg^{-1} = 1·3805.10^{-23} $J\,K^{-1}$
gas constant	$R = kN$ = 1·9872 cal deg^{-1} $mole^{-1}$ = 8·314 $J\,K^{-1}\,mol^{-1}$
electron rest mass	m = 0·911.10^{-27} g
unit of atomic mass	1 a.m.u. = 1·66042.10^{-24} g
charge on the proton	e = 4·8030.10^{-10} e.s.u. = 1·602.10^{-19} C
velocity of light *in vacuo*	c = 2·9979.10^{10} cm s^{-1} = 2·9979.10^8 m s^{-1}

Bibliography

Chapter 1

Gillispie, C. C. (1972). *Dictionary of Scientific Biography*, Scribners, New York.

Phillips, F. C. (1971). *An Introduction to Crystallography*, 4th edn, Oliver and Boyd, Edinburgh.

Chapter 3

Bloss, F. D. (1971). *Crystallography and Crystal Chemistry*. Holt, Rinehart and Winston, New York.

Chapter 4

Buerger, M. J. (1956). *Elementary Crystallography*, Wiley, New York.

Donnay, J. D. H., G. Donnay, E. G. Cox, O. Kennard, and M. V. King (1963). *Crystal Data*, 2nd edn, Amer. Cryst. Assoc.

Hilton, H. (1963). *Mathematical Crystallography*, Dover, New York.

International Tables for X-ray Crystallography, vol. 1, 3rd edn (1969).

Chapter 5

Buerger, M. J. (1942). *X-ray Crystallography*, Wiley, New York.

Phillips, F. C. (1971). *An Introduction to Crystallography*, 4th edn, Oliver and Boyd, Edinburgh.

Chapter 6

International Tables for X-ray Crystallography, vol. 1, 3rd edn (1969).

International Tables for X-ray Crystallography, vol. 3 (1962); 2nd edn (1968).

James, R. W. (1967). *The Optical Principles of the Diffraction of X-rays*, Bell, London.

Jenkins, F. A., and H. E. White (1957). *Fundamentals of Optics*, McGraw-Hill, New York.

Lipson, H., and C. A. Taylor (1958). *Fourier Transforms and X-ray Diffraction*, Bell, London.

Longhurst, R. S. (1957). *Geometrical and Physical Optics*, Longmans, London.

Chapter 7

Christophe-Michel-Levy, M., and A. Sandrea (1953). La hogbomite de Frain (Tchécoslovaquie). *Bull. Soc. franc. Minér. Crist.*, **76,** 430–433.

Fang, J. H., and F. D. Bloss (1966). *X-ray Diffraction Tables*, Southern Illinois University Press, Carbondale and Edwardsville, Ill.

Klug, H. P., and L. E. Alexander (1954). *X-ray Diffraction Procedures*, Wiley, New York.

Lipson, H., and H. Steeple (1970). *Interpretation of X-ray Powder Diffraction Patterns*, Macmillan, London.

McKie, D. (1963). The högbomite polytypes. *Min. Mag.*, **33,** 563–580.

Yoder, H. S., and Th. G. Sahama (1957). Olivine X-ray determinative curve. *Amer. Min.* **42,** 475–491.

Chapter 8

Alcock, N. W., and G. M. Sheldrick (1967). The determination of accurate unit-cell dimensions from inclined Weissenberg photographs. *Acta Cryst.*, **23,** 35–38.

Buerger, M. J. (1942). *X-ray Crystallography*, Wiley, New York.

Buerger, M. J. (1964). *The Precession Method in X-ray Crystallography*, Wiley, New York.

Henry, N. F. M., H. Lipson, and W. A. Wooster (1960). *The Interpretation of X-ray Diffraction Photographs*, 2nd edn, Macmillan, London.

Jeffery, J. W. (1971). *Methods in X-ray Crystallography*, Academic Press, London and New York.

Main, P., and M. M. Woolfson (1963). Accurate lattice parameters from Weissenberg photographs. *Acta Cryst.*, **16,** 731–733.

Nuffield, E. W. (1966). *X-ray Diffraction Methods*, Wiley, New York.

Woolfson, M. M. (1970). *An Introduction to X-ray Crystallography*, Cambridge University Press, London.

Chapter 9

Bacon, G. E. (1962). *Neutron Diffraction.* Oxford University Press, London.

Hirsch, P. B., A. Howie, R. B. Nicholson, D. W. Pashley, and M. J. Whelan (1965). *Electron Microscopy of Thin Crystals*, Butterworths, London.

Woolfson, M. M. (1970). *An Introduction to X-ray Crystallography*, Cambridge University Press, London.

Zussman, J. (1967). *Physical Methods in Determinative Mineralogy*, Academic Press, London and New York.

Chapter 10

Ahrens, L. H. (1952). The use of ionization potentials, part 1: ionic radii of the elements. *Geochim. et Cosmochim. Acta*, **2,** 155–169.

Allred, A. L., and E. G. Rochow (1958). A scale of electronegativity based on electrostatic force. *Journ. Inorg. Nucl. Chem.*, **5,** 264–268.

Birle, J. D., G. V. Gibbs, P. B. Moore, and J. V. Smith (1968). Crystal structures of natural olivines. *Amer. Min.*, **53,** 807–824.

Cotton, F. A., and G. Wilkinson (1966). *Advanced Inorganic Chemistry*, Interscience, New York.

Coulson, C. A. (1963). *Valence*, 2nd edn, Oxford University Press, London.

Goldschmidt, V. M. (1926). Geochemische Verteilungsgesetze der Elemente. *Skr. Norske vid.-akad. Oslo, I, Math.-nat. kl.*, no. 2.

Greenwood, N. N. (1968). *Ionic Crystals, Lattice Defects and Nonstoichiometry*, Butterworths, London.

Hume-Rothery, W., R. E. Smallman, and C. W. Haworth (1969). *The Structure of Metals and Alloys*, Institute of Metals, London.

Kittel, C. (1971). *Introduction to Solid State Physics*, 4th edn, Wiley, New York.

Mandelcorn, L. (1964). *Non-stoichiometric Compounds*, Academic Press, New York.

Murrell, J. N., S. F. A. Kettle, and J. M. Tedder (1971). *Valence Theory*, Wiley, New York.

Pauling, L. (1960). *The Nature of the Chemical Bond*, 3rd edn, Cornell University Press, New York.

Phillips, C. S. G., and R. J. P. Williams (1965–6). *Inorganic Chemistry*, 2 vols, Clarendon Press, Oxford.

Tosi, M. P., and F. G. Fumi (1964). Ionic sizes and Born repulsive parameters in the NaCl type alkali halides. *Journ. Phys. Chem. Solids*, **25,** 31–43, 45–52.

Wyckoff, R. W. G. (1963). *Crystal Structures*, vol. 1, 2nd edn, Interscience, New York.

Chapter 11

Buerger, M. J. (1956). *Elementary Crystallography.* Wiley, New York.

Clark, S. P. (1966). *Handbook of Physical Constants*, Geol. Soc. Amer., Memoir 97.

Megaw, H. D. (1971). Crystal structures and thermal expansion, *Materials Res. Bull.*, **6,** 1007–1018.

Nye, J. F. (1960). *Physical Properties of Crystals*, Oxford University Press, London.

Shewmon, P. G. (1963). *Diffusion in Solids*, McGraw-Hill, New York.

Smith, G. S., and L. E. Alexander (1963). Refinement of the atomic parameters of α-quartz. *Acta Cryst.*, **16,** 462–471.

Wooster, W. A. (1949). *A Text-book on Crystal Physics*, Cambridge University Press, London.

Wooster, W. A., and A. Breton (1970). *Experimental Crystal Physics*, Clarendon Press, Oxford.

Chapter 12

Bloss, F. D. (1961). *An Introduction to the Methods of Optical Crystallography*, Holt, Rinehart, and Winston, New York.

Born, M., and E. Wolf (1964). *Principles of Optics*, Pergamon, Oxford.

Ditchburn, R. W. (1963). *Light*, Blackie, London.

Emmons, R. C. (1943). *The Universal Stage.* Geol. Soc. Amer. Memoir 8.

Galopin, R., and N. F. M. Henry (1972). *Microscopic Study of Opaque Minerals*, Heffer, Cambridge.

Hartshorne, N. H., and A. Stuart (1970). *Crystals and the Polarizing Microscope*, 4th edn, Arnold, London.

Jenkins, F. A., and H. E. White (1957). *Fundamentals of Optics*, McGraw-Hill, New York.

Lipson, S. G., and H. Lipson (1969). *Optical Physics*, Cambridge University Press, London.
Longhurst, R. S. (1957). *Geometrical and Physical Optics*, Longmans, London.
Muir, I. D. (1973). *The Universal Stage*, Longmans, London.
Nye, J. F. (1960). *Physical Properties of Crystals*, Oxford University Press, London.
Preston, T. (1928). *The Theory of Light*, Macmillan, London.
Roy, N. N. (1965). A modified spindle stage permitting the direct determination of $2V$. *Amer. Min.*, **50,** 1441–1449.
Wooster, W. A. (1949). *A Text-book on Crystal Physics*, Cambridge University Press, London.

Chapter 13

Bundy, F. P., H. P. Bovenkerk, H. M. Strong, and R. H. Wentorf, Jr., (1961). Diamond–graphite equilibrium line from growth and graphitization of diamond. *Journ. Chem. Phys.*, **35,** 383–391.
Clark, S. P. (1966). *Handbook of Physical Constants*, Geol. Soc. Amer., Memoir 97.
Denbigh, K. G. (1971). *The Principles of Chemical Equilibrium*, Cambridge University Press, London.
Fowler, R. H., and E. A. Guggenheim (1956). *Statistical Thermodynamics*, Cambridge University Press, London.
Guggenheim, E. A. (1952). *Mixtures*, Oxford University Press, London.
Guggenheim, E. A. (1959). *Thermodynamics*, 4th edn, North-Holland, Amsterdam.
Rossini, F. D., D. D. Wagman, W. H. Evans, S. Levine, and I. Jaffe (1961). *Selected Values of Thermodynamic Properties*, National Bureau of Standards, circ. 500, Washington, D.C.
Rushbrooke, G. S. (1960). *Introduction to Statistical Mechanics*, Oxford University Press, London.
Skinner, H. A. (editor), (1962). *Experimental Thermochemistry*, vol. 2. Wiley–Interscience, New York.
Torgeson, D. R., and G. Th. Sahama (1948). A hydrofluoric acid solution calorimeter and the determination of the heats of formation of Mg_2SiO_4, $MgSiO_3$, $CaSiO_3$. *Journ. Amer. Chem. Soc.*, **70,** 2156–2160.

Chapter 14

Bell, P. M., J. L. England, and G. Kullernd (1966). High pressure differential thermal analysis. *Carnegie Institution Year Book*, **65,** 354–356.
Bowen, N. L., and J. F. Schairer (1935). The system $MgO–FeO–SiO_2$. *Amer. Journ. Sci.*, **29,** 151–217.
Bradley, R. S. (1962, 1964). Thermodynamic calculations on phase equilibria involving fused salts. *Amer. Journ. Sci.*, **260,** 374–382, 550–554; **262,** 541–544.
Garrels, R. M., and C. L. Christ (1965). *Solutions, Minerals, and Equilibria*. Harper and Row, New York.
Kern, R., and A. Weisbrod (1967). *Thermodynamics for Geologists*. Translated by D. McKie. Freeman-Cooper, San Francisco.
Levin, E. M., C. R. Robbins, and H. F. McMurdie (1964). *Phase Diagrams for Ceramists*. Amer. Ceram. Soc. (Supplement published 1969).
Muan, A., and E. F. Osborn (1965). *Phase Equilibria among Oxides in Steelmaking*, Addison-Wesley, Reading, Mass.
Roedder, E. (1959). Silicate melt systems. *Physics and Chemistry of the Earth*, **3,** 224–297.
Temkin, M. (1945). Mixtures of fused salts as ionic solutions. *Acta Physicochimica U.R.S.S.*, **20,** 411–420.

Chapter 15

Ahrens, L. H. (1950). *Spectrochemical Analysis*, Addison-Wesley, Cambridge, Mass.
Boltz, D. F. (1958). *Colorimetric Determination of Nonmetals*, Interscience, New York.
Fairbairn, H. W., and J. F. Schairer (1952). A test of the accuracy of chemical analysis of silicate rocks. *Amer. Min.*, **37,** 744–757.
Greenwood, N. N. (1967). The Mössbauer spectra of chemical compounds. *Chemistry in Britain*, **3,** 56–72.
Groves, A. W. (1951). *Silicate Analysis*, 2nd edn, Allen and Unwin, London.
Ingamells, C. O., and N. H. Suhr (1963). Chemical and spectrochemical analysis of standard silicate samples. *Geochim. et Cosmochim. Acta*, **27,** 897–910.

Kolthoff, I. M., and E. B. Sandell (1950). *Textbook of Quantitative Inorganic Analysis*, Macmillan, London.

Long, J. V. P. (1967). Electron probe microanalysis. In J. Zussman (ed.), *Physical Methods in Determinative Mineralogy*, Academic Press, London and New York.

McLaughlin, R. J. W. (1967). Atomic absorption spectroscopy. In J. Zussman (ed.), *Physical Methods in Determinative Mineralogy*. Academic Press, London and New York.

Mapper, D. (1960). Radioactivation analysis. In A. A. Smales and L. R. Wager (eds.), *Methods in Geochemistry*, Interscience, New York and London.

May, I., and F. Cuttitta (1967). New instrumental techniques in geochemical analysis. In P. H. Abelson (ed.), *Researches in Geochemistry*, vol. 2, Wiley, New York.

Muller, L. D. (1967). Laboratory methods of mineral separation. In J. Zussman (ed.), *Physical Methods in Determinative Mineralogy*, Academic Press, London and New York.

Nicholls, G. D. (1967). Emission spectrography. In J. Zussman (ed.), *Physical Methods in Determination Mineralogy*, Academic Press, London and New York.

Norrish, K., and B. W. Chappell (1967). X-ray fluorescence spectrography. In J. Zussman (ed.), *Physical Methods in Determinative mineralogy*, Academic Press, London and New York.

Norrish, K., and J. T. Hutton (1969). An accurate X-ray spectrographic method for the analysis of a wide range of geological samples. *Geochim. et Cosmochim. Acta*, **33**, 431–453.

Sandell, E. B. (1950). *Colorimetric Determination of Traces of Metals*, 2nd edn, Interscience, New York.

Shapiro, L., and W. W. Brannock (1956). Rapid analysis of silicate rocks. *U.S. Geol. Surv. Bull.*, 1036–C.

Sweatman, T. R., and J. V. P. Long (1968). Quantitative electron-probe microanalysis of rock-forming minerals. *Journ. Petrol.*, **10**, 332–379.

Taylor, S. R., and L. H. Ahrens (1960). Spectrochemical analysis. In A. A. Smales and L. R. Wager (eds), *Methods in Geochemistry*, Interscience, New York and London.

Vincent, E. A. (1960). Analysis by gravimetric and volumetric methods, flame photometry, colorimetry and related techniques. In A. A. Smales and L. R. Wager (eds), *Methods in Geochemistry*, Interscience, New York and London.

Wager, L. R., and G. M. Brown (1960). Collection and preparation of material for analysis. In A. A. Smales and L. R. Wager (eds), *Methods in Geochemistry*, Interscience, New York and London.

Wheatley, P. J. (1970). *The Chemical Consequences of Nuclear Spin*, North-Holland, Amsterdam.

Chapter 16

Birch, F., and P. Le Comte (1960). Temperature-pressure plane for albite composition. *Amer. Journ. Sci.*, **258**, 209–217.

Bowen, N. L. (1915). The crystallization of haplobasaltic, haplodioritic and related magmas. *Amer. Journ. Sci.*, **40**, 161–185.

Boyd, F. R. (1962). Phase equilibria in silicate systems at high pressures and temperatures: apparatus and selected results. In R. H. Wentorf (ed.), *Modern Very High Pressure Techniques*, Butterworths, London.

Boyd, F. R., and J. L. England (1960). Apparatus for phase-equilibrium measurements at pressures up to 50 kilobars and temperatures up to 1750 °C. *Journ. Geophys. Research*, **65**, 741–748.

Bradley, C. C. (1969). *High Pressure Methods in Solid State Research*, Butterworths, London.

Dachille, F., and R. Roy (1960). Influence of 'displacive-shearing' stresses on the kinetics of reconstructive transformations effected by pressure in the range 0–100,000 bars. In J. H. de Boer (ed.), *Reactivity of Solids*, Elsevier, Amsterdam.

Ervin, G., and E. F. Osborn (1951). The system Al_2O_3–H_2O. *Journ. Geol.*, **59**, 381–394.

Griggs, D. T., and G. C. Kennedy (1956). A simple apparatus for high pressures and temperatures. *Amer. Journ. Sci.*, **254**, 722–735.

Hall, H. T. (1958). Some high-pressure, high-temperature apparatus design considerations: equipment for use at 100,000 atmospheres and 3000 °C. *Rev. Sci. Inst.*, **29**, 267–275.

Kennedy, G. C. (1959). Phase relations in the system Al_2O_3–H_2O at high temperatures and pressures. *Amer. Journ. Sci.*, **257**, 563–573.

Luth, W. C., and C. O. Ingamells (1965). Gel preparation of starting materials for hydrothermal experimentation. *Amer. Min.*, **50**, 255–258.

Luth, W. C., and O. F. Tuttle (1963). Externally heated cold-seal pressure vessels for use up to 10,000 bars and 750 °C. *Amer. Min.*, **48,** 1401–1403.
Schairer, J. F. (1959). Phase equilibria with particular reference to silicate systems. In J. O'M. Bockris (ed.), *Physicochemical Measurements at High Temperatures*, Butterworths, London.

Appendices

Donnay, J. D. H., G. Donnay, E. G. Cox, O. Kennard, and M. V. King (1963). *Crystal Data.* 2nd edn. American Crystallographic Association. [3rd edn 2 vols (1973)]
International Tables for X-ray Crystallography, vol. 1, 3rd edn (1969).
International Tables for X-ray Crystallography, vol. 2 (1959).
Kelsey, C. H., and D. McKie (1964). The unit-cell of aenigmatite. *Min. Mag.*, **33,** 986–1001.
Terpstra, P., and L. W. Codd (1961). *Crystallometry.* Academic Press, New York.

Index